Electrodynamics

Masud Chaichian · Ioan Merches
Daniel Radu · Anca Tureanu

Electrodynamics

An Intensive Course

 Springer

Masud Chaichian
Department of Physics
University of Helsinki
Helsinki
Finland

Ioan Merches
Faculty of Physics
Alexandru Ioan Cuza University
Iasi
Romania

Daniel Radu
Faculty of Physics
Alexandru Ioan Cuza University
Iasi
Romania

Anca Tureanu
Department of Physics
University of Helsinki
Helsinki
Finland

ISBN 978-3-662-56853-8 ISBN 978-3-642-17381-3 (eBook)
DOI 10.1007/978-3-642-17381-3

© Springer-Verlag Berlin Heidelberg 2016
Softcover reprint of the hardcover 1st edition 2016
This work is subject to copyright. All rights are reserved by the Publisher, whether the whole or part of the material is concerned, specifically the rights of translation, reprinting, reuse of illustrations, recitation, broadcasting, reproduction on microfilms or in any other physical way, and transmission or information storage and retrieval, electronic adaptation, computer software, or by similar or dissimilar methodology now known or hereafter developed.
The use of general descriptive names, registered names, trademarks, service marks, etc. in this publication does not imply, even in the absence of a specific statement, that such names are exempt from the relevant protective laws and regulations and therefore free for general use.
The publisher, the authors and the editors are safe to assume that the advice and information in this book are believed to be true and accurate at the date of publication. Neither the publisher nor the authors or the editors give a warranty, express or implied, with respect to the material contained herein or for any errors or omissions that may have been made.

Printed on acid-free paper

This Springer imprint is published by Springer Nature
The registered company is Springer-Verlag GmbH Berlin Heidelberg

The grand aim of all science is to cover the greatest number of empirical facts by logical deduction from the smallest number of hypotheses or axioms.

Albert Einstein

Preface

Electrodynamics is the unified description of electric and magnetic phenomena, in terms of the electromagnetic field, conceived as carrier of energy. This study can be performed in two ways: either phenomenologically, for media at rest or moving slowly as compared to the speed of light, or in a relativistically-covariant form, for any possible speed. As a matter of fact, electrodynamics is intrinsically linked to the theory of relativity. Terms like "non-relativistic" or "relativistic" electrodynamics are meaningless, since the two approaches differ only by the mathematical formalism and not in a deep conceptual way. As the reader will further learn, the theory of relativity appeared and developed *within* the study of the electrodynamics of moving media, allowing us to state that electrodynamics *is* relativistic.

The above two possible ways of describing the electromagnetic phenomena represent the basis of the structure of this book. The pedagogical principle of moving from simple to complex, in strong connection with the chronological order of the elaboration of theories, has also been respected.

The first part of the book contains five chapters devoted to the basic principles, fundamental notions and the laws of electrostatics, the theory of stationary currents, the equations of the electromagnetic field, the generation and propagation of the electromagnetic waves, as well as to a border discipline, the magnetofluid dynamics.

The second part consists of three chapters. As soon as the experimental basis and the principles of the theory of relativity are given, the reader gets familiar with the fundamental consequences of the Lorentz transformations. Next, the analytical and tensor formalisms are used in the relativistically-covariant formulation of the fundamental phenomena of mechanics and electrodynamics.

The third part concludes with an introduction to the general theory of relativity, with application to the general-relativistic study of the electromagnetic field.

The book also contains six appendices, whose purpose is to provide the reader with the fundamentals of the necessary mathematics used in the book: tensor calculus, Dirac's *delta* function, the Green function method, differential operators in various coordinate systems, etc.

This book is intended for both undergraduate and graduate students who have physics as major subject. Since any general course in physics should contain a chapter regarding the theory of the electromagnetic field, it can also be used by those studying engineering, mathematics, chemistry, astrophysics, and related subjects. There are only few disciplines whose applicability can be compared to that of electrodynamics. Practically, all electric and electronic devices used around the world work on the basis of the laws of electromagnetism.

The electromagnetic field theory has special harmony and beauty, fully confirming the statement *"Great truths are simple"*. Indeed, it was Maxwell, one of the greatest physicists of all times, who succeeded in creating the first unified description of electric and magnetic phenomena by his famous system of equations. Maxwell's theory contained intrinsically also the relativistic invariance of the special relativity, a fact which was discovered only a few decades later. Nowadays, the attempts and successes in unifying all the known forces (except the gravitational one) in Nature into a single theory, the so-called grand unified theory (GUT), or further developments which include also the gravitational force, such as string theory, in essence have the same aim. It is worthwhile to mention that, chronologically speaking, the electromagnetic field was also the first one to be quantized, by Paul Dirac in 1927. This achievement initiated the development of one of the most efficient theoretical tools in all physical disciplines, the *quantum theory of fields*.

This book is the outcome of the authors' lectures and teaching experiences over many years in different countries and for students of diverse fields of physics, engineering, and applied mathematics. The authors believe that the presentation and the distribution of the topics, the various applications presented in different areas and the set of solved and proposed problems, make this book a comprehensive tool for students and researchers.

During the preparation of this book, the authors have benefited from discussing various topics with many of their colleagues and students. It is a pleasure to express our deep gratitude to all of them and to acknowledge the stimulating discussions and their useful advice.

October 2016

Masud Chaichian
Ioan Merches
Daniel Radu
Anca Tureanu

Contents

Short History

More than two millennia passed between the incipient, qualitative observations on electric and magnetic phenomena, due to *Thales of Miletus* (c. 620–c. 546 BCE), *Democritus* (c. 460–c. 370 BCE), *Aristotle* (384–322 BCE), and up to the first quantitative result, Coulomb's law, written towards the end of the eighteenth century.

During antiquity were known the compass, the magnetic properties of the solid bodies, as well as the electricity produced by friction.

In the Middle Ages appeared some investigations on terrestrial magnetism. *Petrus Peregrinus* (fl. 1261–1269), in "Epistola de magnete" (1269) defined the terms *north pole* and *south pole* of a magnet and described how to identify them. The English physician *William Gilbert* (1544–1603) in his book "De magnete" (1600) recognized that the Earth was itself a magnet and defined the terrestrial magnetic poles. He described many experiments on electric and magnetic static phenomena and invented the electroscope, which was the first electrical measuring instrument.

Gilbert's research considerably enlarged the experimental study of electrostatics and magnetostatics. Later on, in the famous work of *Otto von Guericke* (1602–1686) "Experimenta Nova Magdeburgica de Vacuo Spatio" (1672), for the first time substances were divided into conductors and insulators. Another milestone in experimental research took place in 1731, when *Stephen Gray* (1666–1736) succeeded in sending electricity through metal wires. *Charles François du Fay* (1698–1739) discovered that charged bits of metal foil can attract or repel each other, and concluded that there were two kinds of charges, which were then called "fluids". It took almost one century to settle the controversy whether there existed one or two types of electric charges.

During the eighteenth century the Leyden/electric jar was invented in 1765 by *Pieter van Musschenbroeck* (1692–1761). Many discoveries in electricity and magnetism are due to *Benjamin Franklin* (1706–1790). He proved that lightening is an electric discharge and invented the lightening rod. Inspired by the Leyden jar, he invented the plane capacitor. He also introduced the concepts of "positive" and "negative" electricity and discovered the law of conservation of charge. The investigations carried out in this period are based upon the hypothesis of the existence of an imponderable medium called *æther*, as a medium for the electric, magnetic, and light phenomena.

A new era in the study of electric and magnetic phenomena began in 1785, when *Charles Augustin de Coulomb* (1736–1806) postulated his famous law. An impressive amount of research on the subject is connected with the names of *Carl Friedrich Gauss* (1777–1855), *Pierre-Simon de Laplace* (1749–1827), *Siméon Denis Poisson* (1781–1840), *Alessandro Volta* (1745–1827), *Humphry Davy* (1778–1829), *Georg Simon Ohm* (1789–1854), *James Prescott Joule* (1818–1889), etc.

In 1820 the Danish physicist *Hans Christian Ørsted* (1777–1851) discovered the magnetic effect of the electric current. Thus, for the first time, a connection between electric and magnetic phenomena was established. Ørsted's research was successfully continued by *André-Marie Ampère* (1775–1836), who discovered in 1823 his circuital law connecting the circulation of the magnetic field around a closed loop and the electric current passing through the loop, found the connection between a circular current and a magnetic foil, and established the formula for the force of interaction between two currents.

One of the most influential figures in the development of modern electrodynamics was *Michael Faraday's* (1791–1867). In 1831 he discovered the phenomenon of *electromagnetic induction*, which was a crucial step in the unification of electric and magnetic processes. In 1833 he established the laws of electrolysis, and in 1836 – the theory of electric and magnetic field lines. He also introduced the notion of a *field*, as a continuous material medium, defined at each point by its *intensity*. Unlike the mechanical interpretation of his predecessors, Faraday believed that the electric interactions do not propagate "instantly", but in a finite time interval, step by step, by *contiguity* or *adjacency*, through the medium of the field. He defined the magnetic permeability and discovered the dia- and para-magnetism. In 1832, Faraday submitted to the Royal Society of London a sealed envelope which was opened after more than one hundred years, in 1937. The content of the letter showed that he prefigured already then the existence of the electromagnetic waves. An excerpt of that letter reads as follows: *"I am inclined to compare the diffusion of magnetic forces from a magnetic pole, to the vibrations upon the surface of disturbed water, or those of air in the phenomena of sound, i.e., I am inclined to think the vibratory theory will apply to these phenomena, as it does to sound, and most probably to light. By analogy I think it may possibly apply to the phenomena of induction of electricity of tension also."* In 1846, Faraday published his paper "Thoughts on Ray Vibrations" in which he expounded on the conception of electromagnetic pulses or waves, and which Maxwell considered to be identical in substance with his own theory of electromagnetism.

The revolutionary ideas of Faraday were brilliantly developed by *James Clerk Maxwell* (1831–1879). His fundamental work was "A Treatise on Electricity and Magnetism", published in 1873. There he wrote down his famous system of equations (whose modern form was given by Oliver Heaviside (1850–1925) in 1881), and used them to elaborate the electromagnetic theory of light and to postulate the existence of electromagnetic waves. Maxwell's theory denied the Newtonian concept of instantaneous action at a distance, but still conceived the

field as being a state of elastic tension of the *æther*. By generalizing the fundamental laws of stationary currents in electromagnetism, Maxwell defined the notion
of *displacement current* and thus was led to acknowledge the fact that the electromagnetic phenomena can also take place in vacuum. The year 1888 is the year
of the triumph of Maxwell's theory, when *Heinrich Rudolf Hertz* (1857–1894)
produced electromagnetic waves, proving their reflection, refraction, diffraction,
and interference.

The discovery of the electron in 1897, by *Joseph John Thomson* (1856–1940),
led to the elaboration of the *microscopic theory* of electromagnetic phenomena. The
most prominent contributions in this respect are due to *Hendrik Antoon Lorentz*
(1853–1928), *Henri Poincaré* (1854–1912), and *Paul Langevin* (1872–1946). The
electronic theory made possible the explanation of some phenomena, such as dia-,
para-, and ferromagnetism, the polarization of dia-, para-, and ferroelectric substances, light dispersion, etc., which cannot be explained in the framework of
Maxwell's macroscopic theory.

The elaboration of the electrodynamics of moving media, by *Hertz* and *Lorentz*
at the end of the nineteenth century, led to contradictions concerning the hypothetical absolute, quiescent, ubiquitous cosmic æther. These contradictions were
solved in 1905, by the special theory of relativity of *Albert Einstein* (1879–1955).
The unification of the notions of space and time is a great conceptual leap in
theoretical physics. The principles of this theory, its formalism, and some of its
applications are discussed in the second part of the book.

The special theory of relativity was born within the study of electrodynamics,
and the next aim of Einstein was to incorporate gravity in this relativistic framework. In the first paper on this subject in 1907, Einstein introduced the *equivalence
principle*, which is the cornerstone of the general theory of relativity. Until 1915, he
developed the mathematical structure of the theory, based on Riemannian geometry
and tensor calculus, and finally came to the famous equations which encapsulate the
interplay between gravitational field and matter. As John A. Wheeler succinctly
summarized the core of the theory, "spacetime tells matter how to move; matter tells
spacetime how to curve." General relativity is the first modern theory of gravity and
the basis of the fast developing field of cosmology. In the third part of the book we
introduce the fundamentals of general relativity, as the natural continuation and
development of the theories and ideas elaborated on in the previous chapters.

The quantization of the electromagnetic field (*quantum electrodynamics*), the
study of the interaction between conducting fluids and electromagnetic field
(*magnetofluid dynamics*), microwave propagation, cosmology, etc., are subjects and
disciplines developed during the twentieth century and even nowadays.

Part I
Electrodynamics: Phenomenological Approach

Chapter 1
Electrostatic Field

1.1 Electrostatic Field in Vacuum

1.1.1 Coulomb's Law

In 1773, Henry Cavendish (1731–1810) established, by analogy to Newton's universal law of gravity, the formula expressing the interaction force between two point charges. Cavendish's research remained unknown for more than one century, until 1879, when it was published by James Clerk Maxwell (1831–1879).

Experimentally, the electrostatic interaction law was discovered by the French engineer Charles Augustin de Coulomb (1736–1806), by means of a torsion balance and an electric pendulum. His paper was published in 1785, much before the publication of Cavendish's discovery, and so the formula is known as *Coulomb's law*.

Coulomb's law represents the inverse-square law of variation of electric force with distance, i.e. two point charges q_1 and q_2, placed in vacuum, interact with each other with an electrostatic force

$$\mathbf{F}_{12} = k_e \frac{q_1 q_2}{|\mathbf{r}_1 - \mathbf{r}_2|^3}(\mathbf{r}_1 - \mathbf{r}_2) = k_e \frac{q_1 q_2}{r_{12}^3}\mathbf{r}_{12}, \tag{1.1}$$

where the vectors $\mathbf{r}_1, \mathbf{r}_2, \mathbf{r}_{12}$ are shown in Fig. 1.1.

The unit for electric charge in the International System of Units (abbreviated SI) is the *coulomb*, with the symbol C, representing the quantity of electricity carried in 1 second by a current of 1 ampere. The ampere is the only electromagnetic fundamental unit of SI and it will be defined in Chap. 2. The constant k_e depends on the system of units. For example, in SI it is $k_e = \frac{1}{4\pi\epsilon_0}$, where $\epsilon_0 = 8.854 \times 10^{-12}\,\mathrm{F} \cdot \mathrm{m}^{-1}$ is the *vacuum permittivity*. (By the term "vacuum" is understood "free space".) In the above value of the permittivity, the symbol F stands for *farad*, which is the SI unit for the electric capacity, defined as the capacitance of a capacitor between the plates of

© Springer-Verlag Berlin Heidelberg 2016
M. Chaichian et al., *Electrodynamics*, DOI 10.1007/978-3-642-17381-3_1

Fig. 1.1 Coulomb force of interaction between two point charges.

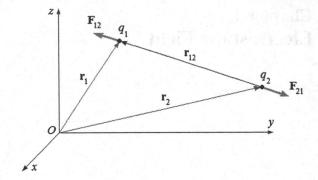

which there appears a potential difference of 1 volt when it is charged by a quantity of electricity of 1 coulomb.

Since $\mathbf{r}_{12} = -\mathbf{r}_{21}$, this means that $\mathbf{F}_{12} = -\mathbf{F}_{21}$, consequently the electrostatic interactions satisfy the action and reaction principle. But, unlike the universal gravity force (Newton's law), the electrostatic force can be either attractive, or repulsive.

For an arbitrary dielectric medium, ϵ_0 is replaced in the formula (1.1) by the absolute permittivity ϵ, with the expression $\epsilon = \epsilon_0 \epsilon_r$, where ϵ_r is the relative permittivity of the medium.

If n point charges $q_1, q_2, \ldots, q_n$, characterised by the position vectors $\mathbf{r}_i$, $i = 1, \ldots, n$, interact with the point charge q, then the resulting force that acts on q is

$$\mathbf{F}(\mathbf{r}) = \sum_{i=1}^{n} k_e \frac{qq_i}{|\mathbf{r} - \mathbf{r}_i|^3} (\mathbf{r} - \mathbf{r}_i), \tag{1.2}$$

where $\mathbf{r}$ determines the position of q.

1.1.2 Charge Density

If we consider a continuous distribution of electric charge, we can characterize this distribution by the notion of *charge density*. Since the charge distribution over a body depends on its geometric form, we may distinguish three cases:

(a) Linear distribution:

$$\lambda = \lim_{\Delta l \to 0} \frac{\Delta q}{\Delta l} = \frac{dq}{dl}, \qquad q = \int_C \lambda \, dl; \tag{1.3}$$

(b) Superficial distribution:

$$\sigma = \lim_{\Delta S \to 0} \frac{\Delta q}{\Delta S} = \frac{dq}{dS}, \qquad q = \int_S \sigma \, dS; \tag{1.4}$$

(c) Volume (space) distribution:

$$\rho = \lim_{\Delta\tau \to 0} \frac{\Delta q}{\Delta\tau} = \frac{dq}{d\tau}, \qquad q = \int_V \rho \, d\tau, \tag{1.5}$$

where $d\tau \equiv d\mathbf{r} = dx \, dy \, dz$ is the volume element.

The concept of charge density can be defined also for point charges, using the *Dirac δ function* (see Appendix E). By means of the *delta* distribution, the density of the point charge q, situated at the point $P_0(\mathbf{r}_0)$, is written as

$$\rho(\mathbf{r}) = q\delta(\mathbf{r} - \mathbf{r}_0). \tag{1.6}$$

By integrating over the infinite three-dimensional volume V, which includes the point P_0, we obtain

$$\lim_{V \to \infty} \int_V \rho(\mathbf{r}) d\tau = q \int \delta(\mathbf{r} - \mathbf{r}_0) \, d\mathbf{r} = q.$$

The density of a discrete system of n point charges is

$$\rho(\mathbf{r}) = \sum_{i=1}^{n} q_i \delta(\mathbf{r} - \mathbf{r}_i). \tag{1.7}$$

The Coulomb force of interaction between a point charge q and a body filling up a volume V', assuming that the charge of density ρ is continuously distributed inside the body, is obtained by dividing the body in infinitely small domains of volume $d\tau' = d\mathbf{r}'$ and charge dq', and then integrating the elementary force between q and dq', with the result

$$\mathbf{F} = \frac{q}{4\pi\epsilon_0} \int_{V'} \frac{\mathbf{r} - \mathbf{r}'}{|\mathbf{r} - \mathbf{r}'|^3} \rho(\mathbf{r}') d\mathbf{r}', \tag{1.8}$$

where $\mathbf{r}$ is the radius-vector of the charge q.

1.1.3   Electrostatic Field Strength

If an electric force acts everywhere in a spatial domain, we say that in this domain there exists an *electric field*. If the electric field does not vary in time, it is called *stationary* or *electrostatic*.

The electrostatic field created by a point charge Q is characterized by its *intensity* $\mathbf{E}$, which is defined as the ratio between the force $\mathbf{F}$ acting on a positive test charge q, placed in the field, and this test charge, i.e.

$$\mathbf{E}(\mathbf{r}) = \frac{\mathbf{F}(\mathbf{r})}{q} = \frac{Q}{4\pi\epsilon_0} \frac{\mathbf{r} - \mathbf{R}}{|\mathbf{r} - \mathbf{R}|^3}, \tag{1.9}$$

Fig. 1.2 The intensity $\mathbf{E}(\mathbf{r})$ of the electrostatic field generated by a positive charge Q.

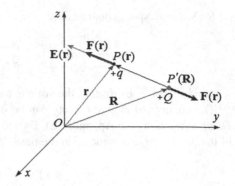

where the charge Q, generating the field, is located at the point $P'(\mathbf{R})$, as in Fig. 1.2. If the charge Q is placed at the origin of coordinates, then

$$\mathbf{E}(\mathbf{r}) = \frac{Q}{4\pi\epsilon_0}\frac{\mathbf{r}}{r^3}. \tag{1.10}$$

The field intensity $\mathbf{E}(\mathbf{r})$ of n point charges q_i, $i = 1, 2, \ldots, n$, produced at a point $P(\mathbf{r})$, is

$$\mathbf{E}(\mathbf{r}) = \frac{1}{4\pi\epsilon_0}\sum_{i=1}^{n} q_i \frac{\mathbf{r} - \mathbf{r}_i}{|\mathbf{r} - \mathbf{r}_i|^3}. \tag{1.11}$$

If the charge is continuously distributed on both the volume and the surface of a body, the field it produces at a point $P(\mathbf{r})$ will be

$$\mathbf{E}(\mathbf{r}) = \frac{1}{4\pi\epsilon_0}\left[\int_{V'} \rho(\mathbf{r}')\frac{\mathbf{r} - \mathbf{r}'}{|\mathbf{r} - \mathbf{r}'|^3}\, d\tau' + \int_{S'} \sigma(\mathbf{r}')\frac{\mathbf{r} - \mathbf{r}'}{|\mathbf{r} - \mathbf{r}'|^3}\, dS'\right]. \tag{1.12}$$

The vector quantity $\mathbf{E}$ is called also *electric field strength* or, simply, *electric field*.

1.1.4   Field Lines

Let us consider the curve C given by its parametric equations $x_i = x_i(s)$, $i = 1, 2, 3$. If at any point of C the field $\mathbf{E}$ is tangent to the curve, then the curve is a *line of the field* $\mathbf{E}$. (This definition is valid for *any* vector field.) The electrostatic field lines are also called *lines of force*.

Denoting by $d\mathbf{s}$ an oriented element of the field line, then by definition we may write $d\mathbf{s} \times \mathbf{E} = 0$, or, in projection on axes of an orthogonal Cartesian reference system,

$$\frac{dx}{E_x} = \frac{dy}{E_y} = \frac{dz}{E_z}, \tag{1.13}$$

which are the *differential equations* of the electrostatic field lines, written in Cartesian coordinates. In spherical coordinates, the differential equations of the field lines are

$$\frac{dr}{E_r} = \frac{rd\theta}{E_\theta} = \frac{r\sin\theta\,d\varphi}{E_\varphi}, \tag{1.14}$$

while in cylindrical coordinates they become

$$\frac{d\rho}{E_\rho} = \frac{\rho d\varphi}{E_\varphi} = \frac{dz}{E_z}, \tag{1.15}$$

etc.

If the solution of the system of equations (1.13) is unique, then through each point of the considered domain passes only one line of force. By definition, the sense of the field lines is given by the sense of the field.

1.1.5 Flux of the Electrostatic Field

The *flux* of the field **E** through the surface S is defined as

$$\Phi_e = \int_S \mathbf{E} \cdot d\mathbf{S} = \int_S \mathbf{E} \cdot \mathbf{n}\,dS = \int_S E_n dS, \tag{1.16}$$

where **n** is the unit vector of $d\mathbf{S}$ and by E_n is denoted the projection of **E** on the direction of **n**. In case of a point charge, using (1.10) we can write

$$\Phi_e = \frac{q}{4\pi\epsilon_0} \int_S \frac{1}{r^3}\mathbf{r} \cdot \mathbf{n}\,dS = \frac{q}{4\pi\epsilon_0} \int_S \frac{dS\cos\alpha}{r^2} = \frac{q}{4\pi\epsilon_0} \int_\Omega d\Omega, \tag{1.17}$$

where α is the plane angle between the vectors **r** and **n**, and $d\Omega$ is the elementary solid angle through which dS is "seen" from the point charge q (Fig. 1.3). If the surface S is closed, then $\oint d\Omega$ equals 4π when the charge is inside the surface, and zero, when the charge is outside. Then, from (1.17) we have

$$\Phi_e = \oint_S \mathbf{E} \cdot d\mathbf{S} = \begin{cases} \frac{q}{\epsilon_0}, & \text{if } q \text{ is inside } S; \\ 0, & \text{if } q \text{ is outside } S, \end{cases} \tag{1.18}$$

which is the *integral form of Gauss's law*. The law was formulated by Carl Friedrich Gauss (1777–1835). Remarkably, from this theorem follows immediately the Coulomb law itself, which we postulated in Sect. 1.1.1 as a purely empirical law. Indeed, by placing one of the charges in Fig. 1.1 inside a closed spherical surface, one immediately finds Eq. (1.1).

Fig. 1.3 The elementary
solid angle $d\Omega$ through
which dS is seen from the
point charge q.

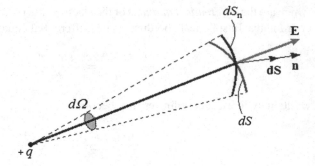

If the distribution of n point charges inside the closed surface S is discrete, Gauss's
law gives

$$\Phi_e = \frac{1}{\epsilon_0} \sum_{i=1}^{n} q_i, \tag{1.19}$$

while for a continuous distribution inside a domain of volume V,

$$\Phi_e = \frac{1}{\epsilon_0} \int_V \rho \, d\tau. \tag{1.20}$$

Applying the divergence theorem (A.32), we find that

$$\Phi_e = \oint_S \mathbf{E} \cdot \mathbf{n} \, dS = \int_V \nabla \cdot \mathbf{E} \, d\tau,$$

which, by comparison with (1.20), leads to the *differential form of Gauss's law*:

$$\nabla \cdot \mathbf{E} = \frac{1}{\epsilon_0} \rho. \tag{1.21}$$

If we denote $\mathbf{D} = \epsilon_0 \mathbf{E}$, Gauss's law can be re-written as

$$\text{div} \, \mathbf{D} = \rho. \tag{1.22}$$

It results from (1.21) that $\text{div} \, \mathbf{E} \neq 0$ at the points where $\rho \neq 0$. These points
are called generically *sources* of the electric field. They can be positive ($\rho > 0$)
or negative ($\rho < 0$). Consequently, the electrostatic field is a *source field*, electric
charges being the sources of the field. By convention, the field lines enter into a
negative source and come out from a positive one. (According to a more precise
terminology, a positive charge is termed *source*, while a negative charge is called
sink.) Besides, for any configuration of positive and/or negative electric charges,
there is at least one field line which goes to or comes from infinity, i.e. there is
always one "open" field line at least.

1.1.6 Electrostatic Field Potential

Let us consider a vector field $\mathbf{a}(\mathbf{r}, t)$. If there exists a scalar function $\varphi(\mathbf{r}, t)$, such that

$$\mathbf{a}(\mathbf{r}, t) = -\nabla\varphi(\mathbf{r}, t),$$

then we say that $\mathbf{a}$ is a *potential* field, while φ is the *potential* of this field. The field is called *conservative* if φ does not depend explicitly on time.

Such a field is, for example, the electrostatic field of a point charge. If the charge is placed at the origin, one can write

$$\mathbf{E}(\mathbf{r}) = \frac{q}{4\pi\epsilon_0}\frac{\mathbf{r}}{r^3} = -\frac{q}{4\pi\epsilon_0}\nabla\left(\frac{1}{r}\right) = -\nabla\left(\frac{1}{4\pi\epsilon_0}\frac{q}{r}\right).$$

With the notation

$$V(\mathbf{r}) = \frac{1}{4\pi\epsilon_0}\frac{q}{r}, \tag{1.23}$$

we have

$$\mathbf{E} = -\nabla V. \tag{1.24}$$

Formula (1.23) defines the potential of the field produced in vacuum by the point charge q, at a distance r from q, while (1.24) expresses the connection between the field $\mathbf{E}$ and its potential V.

If the charge is at a point defined by the radius-vector $\mathbf{R}$, then the potential at the point $P(\mathbf{r})$ will be

$$V(\mathbf{r}) = \frac{1}{4\pi\epsilon_0}\frac{q}{|\mathbf{r} - \mathbf{R}|}. \tag{1.25}$$

The potential produced at $P(\mathbf{r})$ by n point charges $q_1, q_2, \ldots, q_n$ is

$$V(\mathbf{r}) = \frac{1}{4\pi\epsilon_0}\sum_{i=1}^{n}\frac{q_i}{|\mathbf{r} - \mathbf{r}_i|}, \tag{1.26}$$

while the potential created by a continuous charge distribution (a charged body) is

$$V(\mathbf{r}) = \frac{1}{4\pi\epsilon_0}\left[\int_{V'}\frac{\rho(\mathbf{r}')\,d\tau'}{|\mathbf{r} - \mathbf{r}'|} + \oint_{S'}\frac{\sigma(\mathbf{r}')\,dS'}{|\mathbf{r} - \mathbf{r}'|}\right], \tag{1.27}$$

i.e. the electrostatic field satisfies the *principle of superposition*. Having in view Eq. (1.24) and using the fact that ∇ operates only on quantities dependent on $\mathbf{r}$, it can be easily seen that $-\nabla$ applied to (1.27) leads to $\mathbf{E}$ given by (1.12).

Let us now calculate the circulation of the vector $\mathbf{E}$ along the closed curve C. Using the properties of total differentials, we can write

$$\oint_C \mathbf{E} \cdot d\mathbf{l} = - \oint_C \nabla V \cdot d\mathbf{l} = - \oint_C dV = 0. \tag{1.28}$$

Applying the Stokes theorem (A.34), we obtain

$$\oint_C \mathbf{E} \cdot d\mathbf{l} = \int_S (\nabla \times \mathbf{E}) \cdot d\mathbf{S} = 0,$$

meaning that the curl of the vector $\mathbf{E}$ vanishes,

$$\nabla \times \mathbf{E} = 0, \tag{1.29}$$

at each point of the electrostatic field. This is another way to express the conservative property of the electrostatic field. Any field satisfying a condition similar to (1.29) is called *irrotational* or *vorticity-free*.

Let us write again the circulation of the field $\mathbf{E}$, but this time taken between two points A and B of an arbitrary contour:

$$\int_A^B \mathbf{E} \cdot d\mathbf{l} = - \int_A^B dV = V_A - V_B. \tag{1.30}$$

Since the integral on the l.h.s. of (1.30) is numerically equal to the mechanical work done by the electric forces to displace a unit charge from A to B,

$$L = \int_A^B \mathbf{F} \cdot d\mathbf{l} = q \int_A^B \mathbf{E} \cdot d\mathbf{l}, \tag{1.31}$$

from Eq. (1.30) we find

$$V_A = V_B + \int_A^B \mathbf{E} \cdot d\mathbf{l}. \tag{1.32}$$

The potential difference $V_A - V_B$ is uniquely determined (the electric field $\mathbf{E}$ is determined by the charge distribution, and the circulation does not depend on the integration path), but the potential at a point (say, A) does not have this property. To determine uniquely the potential at the point A, we displace the point B to infinity and *choose* $V(\infty) = 0$. Then we have

$$V_A = \int_A^\infty \mathbf{E} \cdot d\mathbf{l}. \tag{1.33}$$

This relation defines the potential at a point of the electrostatic field. The interpretation of Eq. (1.33) is that the potential at a point A is numerically equal to the mechanical work done by electric forces to displace a unit point charge from that point to infinity.

Let us write the potential of the field produced by a point charge, placed at the origin of coordinates, as

$$V(\mathbf{r}) = \frac{q}{4\pi\epsilon_0} \int_r^\infty \frac{\mathbf{r} \cdot d\mathbf{l}}{r^3}.$$

If α is the plane angle between $\mathbf{r}$ and $d\mathbf{l}$, then $\mathbf{r} \cdot d\mathbf{l} = r dl \cos \alpha = r dr$, leading to

$$V(\mathbf{r}) = \frac{q}{4\pi\epsilon_0} \int_r^\infty \frac{dr}{r^2} = \frac{1}{4\pi\epsilon_0} \frac{q}{r}. \tag{1.34}$$

If the charge is located at the point $P'(\mathbf{r}')$, then

$$V(\mathbf{r}) = \frac{1}{4\pi\epsilon_0} \frac{q}{|\mathbf{r} - \mathbf{r}'|}. \tag{1.35}$$

Remark that the form $V \sim 1/r$ of the Coulomb potential is valid only in the case of three spatial dimensions. Generally, for n space dimensions ($n = 1, 3, 4, \ldots$), the corresponding Coulomb potential V behaves as $V \sim 1/r^{(n-2)}$, while for $n = 2$, the potential goes as $V \sim \log r$.

The unit for potential and potential difference in SI is the *volt*, with the symbol V, thus named in honour of Alessandro Volta (1745–1827). The volt is the potential difference between two points of a conducting wire carrying a constant current of 1 ampere, when the power dissipated between these points is equal to 1 watt. Equivalently, it is the potential difference between two parallel, infinite planes spaced 1 metre apart that create an electric field of 1 newton per coulomb, or the potential difference between two points that will impart one joule of energy per coulomb of charge that passes through it.

1.1.7 Equipotential Surfaces

Let us consider a stationary surface S, whose equation is given by the constancy of the potential on it, i.e.

$$V(x_1, x_2, x_3) = V_0 = \text{const.},$$

where $x_1 = x$, $x_2 = y$, and $x_3 = z$. By differentiation, one obtains

$$dV = \sum_{i=1}^3 \frac{\partial V}{\partial x_i} dx_i = \nabla V \cdot d\mathbf{r} = -\mathbf{E} \cdot d\mathbf{r} = 0. \tag{1.36}$$

Since $d\mathbf{r}$ is in the plane tangent to the surface S, (1.36) tells us that at each point of the surface, the vector $\mathbf{E}$ is oriented along the normal to the surface. By giving different values to V_0 we obtain a family of surfaces, called *equipotential surfaces*. In other words, by *equipotential surface* we mean the locus of all points having the same potential.

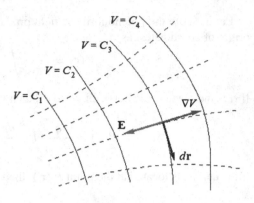

Fig. 1.4 Equipotential
surfaces and field lines.

Formula (1.36) also says that the electrostatic field lines are oriented along the normal to the equipotential surface at each point of the field. Consequently, the family of field lines is orthogonal to the family of equipotential surfaces (Fig. 1.4).

The shape of the equipotential surfaces depends on the geometry of the charge distribution that generates the field. For example, Eq. (1.23) shows that the equipotential surfaces of a point charge are concentric spheres, $r = $ const., with the source charge at the centre.

1.1.8 *Equations of the Electrostatic Potential*

1.1.8.1 Differential Form

From (1.21), (1.24), and (A.50), we obtain

$$\Delta V(\mathbf{r}) = -\frac{1}{\epsilon_0}\rho(\mathbf{r}), \tag{1.37}$$

which is a second order non-homogeneous elliptic partial differential equation, called *Poisson's equation*, thus named after the French mathematician and physicist Siméon Denis Poisson (1781–1840), who published it in 1813. At the points where no charges are present, Eq. (1.37) reduces to the homogeneous *Laplace's equation*, named after Pierre-Simon de Laplace (1749–1827),

$$\Delta V(\mathbf{r}) = 0. \tag{1.38}$$

The solutions of the Laplace equation are called *harmonic functions*. It follows from (1.38) that the function $V(\mathbf{r})$, together with its first and second order derivatives, must be continuous within a certain three-dimensional domain D, including its boundary. By solving the Poisson (or Laplace) equation, we determine $V(\mathbf{r})$ and then $\mathbf{E}(\mathbf{r})$ by the formula $\mathbf{E} = -\nabla V$.

In view of (1.6), Poisson's equation for a single point charge is

$$\Delta V(\mathbf{r}) = -\frac{1}{\epsilon_0} q \delta(\mathbf{r} - \mathbf{r}_0), \tag{1.39}$$

while for n point charges it becomes

$$\Delta V = -\frac{1}{\epsilon_0} \sum_{i=1}^{n} q_i \delta(\mathbf{r} - \mathbf{r}_i). \tag{1.40}$$

The scalar field V in the domain D is uniquely determined by the Poisson equation, together with the boundary conditions on the closed surface S which borders the domain D.

1.1.8.2   Integral Representation

By means of the second Green identity (A.41), one can find the *integral representations* of Poisson's and Laplace's equations. Introducing (1.35) into (1.39) and dividing by $q \neq 0$, we have

$$\Delta \left(\frac{1}{|\mathbf{r} - \mathbf{r}'|} \right) = -4\pi \delta(\mathbf{r} - \mathbf{r}'). \tag{1.41}$$

Let us now apply Δ to the expression (1.27). Using (1.41) and (E.21), we obtain

$$\begin{aligned}
\Delta V &= \frac{1}{4\pi\epsilon_0} \int_{V'} \rho(\mathbf{r}') \Delta \left(\frac{1}{|\mathbf{r} - \mathbf{r}'|} \right) d\tau' \\
&= -\frac{1}{\epsilon_0} \int_{V'} \rho(\mathbf{r}') \delta(\mathbf{r} - \mathbf{r}') d\tau' = -\frac{1}{\epsilon_0} \rho(\mathbf{r}),
\end{aligned}$$

i.e. Poisson's equation.

Assume, again, that the charge is continuously distributed inside the three-dimensional domain D' (including its boundary) and consider two continuous, differentiable, and singularity-free functions $\varphi(\mathbf{r})$ and $\psi(\mathbf{r})$. Then we have, using (A.41) and in the notation of Fig. 1.5,

$$\int_{V'} \left(\varphi \Delta' \psi - \psi \Delta' \varphi \right) d\tau' = \oint_{S'} \left(\varphi \frac{\partial \psi}{\partial n'} - \psi \frac{\partial \varphi}{\partial n'} \right) dS', \tag{1.42}$$

where $d\tau'$ and dS' are a volume element of D' and a surface element of S', respectively, and Δ' implies derivatives with respect to the components of $\mathbf{r}'$.

Fig. 1.5 A continuously
distributed electric charge.

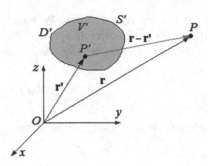

If we choose

$$\varphi(\mathbf{r}') \equiv \frac{1}{4\pi} \frac{1}{|\mathbf{r} - \mathbf{r}'|} \quad \text{and} \quad \psi(\mathbf{r}') \equiv V(\mathbf{r}'), \tag{1.43}$$

then (1.42) leads to

$$\int_{V'} \left[\frac{1}{4\pi} \frac{1}{|\mathbf{r} - \mathbf{r}'|} \Delta' V(\mathbf{r}') - V(\mathbf{r}') \Delta' \left(\frac{1}{4\pi} \frac{1}{|\mathbf{r} - \mathbf{r}'|} \right) \right] d\tau'$$
$$= \frac{1}{4\pi} \oint_{S'} \left[\frac{1}{|\mathbf{r} - \mathbf{r}'|} \frac{\partial V}{\partial n'} - V(\mathbf{r}') \frac{\partial}{\partial n'} \left(\frac{1}{|\mathbf{r} - \mathbf{r}'|} \right) \right] dS'.$$

Observing that

$$\nabla \left(\frac{1}{|\mathbf{r} - \mathbf{r}'|} \right) = -\nabla' \left(\frac{1}{|\mathbf{r} - \mathbf{r}'|} \right), \tag{1.44}$$

$$\nabla \cdot \nabla \left(\frac{1}{|\mathbf{r} - \mathbf{r}'|} \right) = \Delta \left(\frac{1}{|\mathbf{r} - \mathbf{r}'|} \right) = -\nabla \cdot \nabla' \left(\frac{1}{|\mathbf{r} - \mathbf{r}'|} \right) = -\nabla' \cdot \nabla \left(\frac{1}{|\mathbf{r} - \mathbf{r}'|} \right)$$

$$= \nabla' \cdot \nabla' \left(\frac{1}{|\mathbf{r} - \mathbf{r}'|} \right) = \Delta' \left(\frac{1}{|\mathbf{r} - \mathbf{r}'|} \right), \tag{1.45}$$

and using (1.37) and (1.41), we find

$$\int_{V'} \left[-\frac{1}{4\pi\epsilon_0} \frac{1}{|\mathbf{r} - \mathbf{r}'|} \rho(\mathbf{r}') + V(\mathbf{r}') \delta(\mathbf{r} - \mathbf{r}') \right] d\tau'$$
$$= \frac{1}{4\pi} \oint_{S'} \left[\frac{1}{|\mathbf{r} - \mathbf{r}'|} \frac{\partial V}{\partial n'} - V(\mathbf{r}') \frac{\partial}{\partial n'} \left(\frac{1}{|\mathbf{r} - \mathbf{r}'|} \right) \right] dS'.$$

Integrating the second term in the l.h.s according to (E.21), we arrive at

$$V(\mathbf{r}) = \frac{1}{4\pi\epsilon_0} \int_{V'} \frac{\rho(\mathbf{r}')}{|\mathbf{r} - \mathbf{r}'|} d\tau'$$

$$+ \frac{1}{4\pi} \oint_{S'} \left[\frac{1}{|\mathbf{r} - \mathbf{r}'|} \frac{\partial V}{\partial n'} - V(\mathbf{r}') \frac{\partial}{\partial n'} \left(\frac{1}{|\mathbf{r} - \mathbf{r}'|} \right) \right] dS', \qquad (1.46)$$

which is the *integral representation of Poisson's equation*.

If there is no charge distribution in D', i.e. $\rho = 0$, Eq. (1.46) leads to the *integral representation of Laplace's equation*:

$$V(\mathbf{r}) = \frac{1}{4\pi} \oint_{S'} \left[\frac{1}{|\mathbf{r} - \mathbf{r}'|} \frac{\partial V}{\partial n'} - V(\mathbf{r}') \frac{\partial}{\partial n'} \left(\frac{1}{|\mathbf{r} - \mathbf{r}'|} \right) \right] dS'. \qquad (1.47)$$

If the charges are located in a finite region, but we extend the integration domain to the whole three-dimensional space and use the condition $V(\infty) = 0$, we are left with

$$V(\mathbf{r}) = \frac{1}{4\pi\epsilon_0} \int \frac{\rho(\mathbf{r}')}{|\mathbf{r} - \mathbf{r}'|} d\tau',$$

as expected. From now on, as in the above formula, whenever we do not indicate the domain of integration, it will mean integration over the whole space.

We can rewrite (1.46) and (1.47) by means of the *Green function* $G(\mathbf{r}, \mathbf{r}')$, and the Dirichlet or Neumann conditions on the boundary surface S' (see Appendix F). By definition, the Green function of our problem is the solution of the equation

$$\Delta G(\mathbf{r}, \mathbf{r}') = -\delta(\mathbf{r} - \mathbf{r}'). \qquad (1.48)$$

Comparing (1.48) and (1.41), we find that

$$G(\mathbf{r}, \mathbf{r}') = \frac{1}{4\pi} \frac{1}{|\mathbf{r} - \mathbf{r}'|} + \eta(\mathbf{r}, \mathbf{r}'), \qquad (1.49)$$

where $\eta(\mathbf{r}, \mathbf{r}')$ is a solution of the Laplace equation inside the domain D', i.e. $\Delta\eta(\mathbf{r}, \mathbf{r}') = 0$. Introducing now the expression (1.49) into (1.46) and (1.47), we obtain, respectively

$$V(\mathbf{r}) = \frac{1}{\epsilon_0} \int_{V'} \rho(\mathbf{r}')G(\mathbf{r}, \mathbf{r}') \, d\tau' \qquad (1.50)$$

$$+ \oint_{S'} \left[G(\mathbf{r}, \mathbf{r}') \frac{\partial V(\mathbf{r}')}{\partial n'} - V(\mathbf{r}') \frac{\partial G(\mathbf{r}, \mathbf{r}')}{\partial n'} \right] dS', \quad \text{for } \rho \neq 0,$$

and

$$V(\mathbf{r}) = \oint_{S'} \left[G(\mathbf{r}, \mathbf{r}') \frac{\partial V(\mathbf{r}')}{\partial n'} - V(\mathbf{r}') \frac{\partial G(\mathbf{r}, \mathbf{r}')}{\partial n'} \right] dS', \quad \text{for } \rho = 0. \qquad (1.51)$$

From the above expression, it would seem that for a complete determination of $V(\mathbf{r})$, one must know both the functions V and $\partial V/\partial n$ on the boundary surface S'. Actually, imposing conditions on both the potential and the electric field on the boundary S' would mean an overspecification of the problem: the existence and uniqueness theorem requires knowledge of only one of these functions. Therefore, we shall use the natural Dirichlet or Neumann boundary conditions, i.e.

$$G(\mathbf{r}, \mathbf{r}')\Big|_{S'} = 0 \quad \text{for Dirichlet's problem;} \tag{1.52}$$

$$\frac{\partial G(\mathbf{r}, \mathbf{r}')}{\partial n'}\bigg|_{S'} = C(\text{const.}) \quad \text{for Neumann's problem.} \tag{1.53}$$

The Dirichlet condition is equivalent to specifying the potential on the boundary, while the Neumann condition is equivalent to specifying the electric field.

We do not take the Neumann condition (1.53) with zero on the r.h.s, since integrating (1.48) over a three-dimensional domain and using the properties of the δ distribution, we find

$$\int_{V'} \nabla' \cdot (\nabla' G) d\tau' = \oint_{S'} \nabla' G \cdot d\mathbf{S}' = \oint_{S'} \frac{\partial G}{\partial n'} dS' = -1. \tag{1.54}$$

Thus, (1.53) is the only choice compatible with (1.54), leading to $C = -1/S'$, where S' is the boundary surface of integration. Noting that

$$\langle V \rangle = \frac{1}{S'} \oint_{S'} V(\mathbf{r}') dS' \tag{1.55}$$

is the mean value of the potential on S', the integral representations of Poisson's equation, with Dirichlet and Neumann boundary conditions, are respectively

$$V(\mathbf{r}) = \frac{1}{\epsilon_0} \int_{V'} \rho(\mathbf{r}') G_D(\mathbf{r}, \mathbf{r}') d\tau' - \oint_{S'} V(\mathbf{r}') \frac{\partial G_D(\mathbf{r}, \mathbf{r}')}{\partial n'} dS', \tag{1.56}$$

$$V(\mathbf{r}) = \frac{1}{\epsilon_0} \int_{V'} \rho(\mathbf{r}') G_N(\mathbf{r}, \mathbf{r}') d\tau' + \oint_{S'} G_N(\mathbf{r}, \mathbf{r}') \frac{\partial V(\mathbf{r}')}{\partial n'} dS' + \langle V \rangle. \tag{1.57}$$

The integral representations of the Laplace equation, with Dirichlet and Neumann boundary conditions, are similarly obtained.

In general, finding the Green function of an operator is a difficult task. There are, however, some special methods of finding the solution of the Laplace and Poisson equations, that do not involve the explicit integration of these equations, such as: the method of electric images, the expansion of the potential in series of orthogonal functions, the method of conjugate functions, the method of inversion, the method of conformal mapping, the wedge problem, etc. We shall consider some of these applications in Sect. 1.3.

1.1.9 Electrostatic Field Energy

The mechanical work done to bring a point charge $+q$ from infinity to a point A, where the potential is $V(A)$, is performed against the electrostatic force, and, consequently, it transforms into potential energy accumulated by the charge:

$$W_e^{pot}(A) = qV(A). \tag{1.58}$$

Let us calculate, by means of (1.58), the potential energy of interaction between two point charges q_1 and q_2, situated in vacuum, at a distance r from each other. If q_2 is absent, no work is performed if one brings q_1 from infinity to a point A. If q_2 is brought to a point B, at a distance r from A, with q_1 at A, one performs the mechanical work $W_e^{(1)} = q_2 V_2$, where V_2 is the potential of the field created by q_1 at B. Similarly, the work done to bring q_1 at A, with q_2 being already at B, is $W_e^{(2)} = q_1 V_1$, where V_1 is the potential of the field created by q_2 at A. Since

$$V_1 = \frac{1}{4\pi\epsilon_0}\frac{q_2}{r} \quad \text{and} \quad V_2 = \frac{1}{4\pi\epsilon_0}\frac{q_1}{r},$$

it follows that

$$W_e = W_e^{(1)} = W_e^{(2)} = \frac{1}{4\pi\epsilon_0}\frac{q_1 q_2}{r}, \tag{1.59}$$

or, equivalently,

$$W_e = \frac{1}{2}\left(W_e^{(1)} + W_e^{(2)}\right) = \frac{1}{2}\left(q_1 V_1 + q_2 V_2\right). \tag{1.60}$$

This relation can be generalized for n charges $q_1, q_2, \ldots, q_n$, as

$$W_e = \frac{1}{2}\sum_{i=1}^{n} q_i V_i, \tag{1.61}$$

where V_i is the potential of the charges q_k $(k \neq i)$ at the location of the point charge q_i:

$$V_i = \frac{1}{4\pi\epsilon_0}\sum_{k\neq i}\frac{q_k}{r_{ik}}, \tag{1.62}$$

r_{ik} being the distance between q_i and q_k. Thus, we may write (1.61) in the form

$$W_e = \frac{1}{8\pi\epsilon_0}\sum_{i,k=1}^{n}\frac{q_i q_k}{r_{ik}} \quad (i \neq k). \tag{1.63}$$

If the charges are continuously distributed in a three-dimensional domain, or on a surface, the total interaction energy of the system is

$$W_e = \frac{1}{2} \int_V \rho(\mathbf{r}) V(\mathbf{r}) \, d\tau \quad \text{or} \quad W_e = \frac{1}{2} \int_S \sigma(\mathbf{r}) V(\mathbf{r}) dS, \tag{1.64}$$

respectively.

The electrostatic field energy can also be expressed in terms of the field intensity $\mathbf{E}$. If charges are continuously distributed in the finite three-dimensional domain D, of volume V, including the boundary S, then Gauss's law (1.21) together with (A.32) and (A.43), give

$$W_e = \frac{1}{2} \int_V \rho V \, d\tau = \frac{\epsilon_0}{2} \int_V V \left(\nabla \cdot \mathbf{E} \right) d\tau = \frac{\epsilon_0}{2} \int_V E^2 d\tau + \frac{\epsilon_0}{2} \oint_S V \mathbf{E} \cdot d\mathbf{S}.$$

If we extend the integration over the whole space (in other words, the integration is performed on a sphere with radius $R \to \infty$) and one observes that $V E_n \to 0$ as $1/R^3$, while $dS \sim R^2$, then we are left with

$$W_e = \frac{\epsilon_0}{2} \int E^2 d\tau = \frac{1}{2} \int \mathbf{E} \cdot \mathbf{D} \, d\tau. \tag{1.65}$$

Here we used the notation $\mathbf{D} = \epsilon_0 \mathbf{E}$ for the *electric displacement in vacuum*. Unlike relations (1.64), which show that the electrostatic field energy differs from zero only at the points where there are charges, formula (1.65) tells us that $W_e \neq 0$ even at points where there is no charge, but there is a field $\mathbf{E} \neq 0$. This means that the electrostatic field is an *energy carrier*.

There is one more essential distinction between (1.64) and (1.65). According to (1.64), the energy may be either positive, or negative, depending on the sign of the charge, while (1.65) says that the energy cannot take negative values ($E^2 > 0$). Furthermore, by (1.63) the energy of a point charge is zero, while by (1.65) this energy is infinite. The explanation consists in the fact that (1.63) does not take into account the interaction between charges and their own field, while (1.65) expresses namely the *self-energy* of the charge.

Let us calculate the total energy of the electrostatic field produced by two point charges, q_1 and q_2. If $\mathbf{E}_1$ and $\mathbf{E}_2$ are the fields created by the charges, taken separately, then the total field is $\mathbf{E} = \mathbf{E}_1 + \mathbf{E}_2$, such that

$$W_e = \frac{\epsilon_0}{2} \int E^2 d\tau = \frac{\epsilon_0}{2} \int E_1^2 d\tau + \frac{\epsilon_0}{2} \int E_2^2 d\tau + \epsilon_0 \int \mathbf{E}_1 \cdot \mathbf{E}_2 d\tau.$$

The first two terms give the self-energy of the charges, while the last term expresses the *mutual energy of interaction*.

Since $(\mathbf{E}_1 - \mathbf{E}_2)^2 \geq 0$, we have $\mathbf{E}_1^2 + \mathbf{E}_2^2 \geq 2(\mathbf{E}_1 \cdot \mathbf{E}_2)$, meaning that the self-energy of the charges is always greater than (in particular, equal to) their mutual energy.

1.1.10 Electrostatic Dipole

By *dipole* we mean a system of two point charges, equal in magnitude and having opposite signs, their mutual distance being negligible as compared to the distance from the dipole to the point where the action of the system is determined. Using (1.25) and the notation in Fig. 1.6, the potential of the dipole at the point $P(\mathbf{r})$ is

$$V(\mathbf{r}) = \frac{q}{4\pi\epsilon_0} \left(\frac{1}{|\mathbf{r} - \mathbf{r}' - \mathbf{l}|} - \frac{1}{|\mathbf{r} - \mathbf{r}'|} \right).$$

Since, by definition $|\mathbf{l}| \ll |\mathbf{r}|$, we may expand in series the first term and obtain

$$|\mathbf{r} - \mathbf{r}' - \mathbf{l}|^{-1} = \left[(\mathbf{r} - \mathbf{r}')^2 - 2\mathbf{l} \cdot (\mathbf{r} - \mathbf{r}') + l^2 \right]^{-1/2}$$

$$= \frac{1}{|\mathbf{r} - \mathbf{r}'|} \left[1 - \frac{2\mathbf{l} \cdot (\mathbf{r} - \mathbf{r}')}{|\mathbf{r} - \mathbf{r}'|^2} + \frac{l^2}{|\mathbf{r} - \mathbf{r}'|^2} \right]^{-1/2}$$

$$= \frac{1}{|\mathbf{r} - \mathbf{r}'|} \left[1 + \frac{\mathbf{l} \cdot (\mathbf{r} - \mathbf{r}')}{|\mathbf{r} - \mathbf{r}'|^2} + \ldots \right] \simeq \frac{1}{|\mathbf{r} - \mathbf{r}'|} - \mathbf{l} \cdot \nabla \left(\frac{1}{|\mathbf{r} - \mathbf{r}'|} \right),$$

where the terms in l^2, l^3, etc. have been neglected. If we define $\mathbf{p} = q\mathbf{l}$ as the *electric dipole moment*, one can write the dipole potential as

$$V(\mathbf{r}) = -\frac{1}{4\pi\epsilon_0} \mathbf{p} \cdot \nabla \left(\frac{1}{|\mathbf{r} - \mathbf{r}'|} \right) = \frac{1}{4\pi\epsilon_0} \frac{\mathbf{p} \cdot (\mathbf{r} - \mathbf{r}')}{|\mathbf{r} - \mathbf{r}'|^3}. \tag{1.66}$$

For the necessities of the infinitesimal calculus one defines the *point dipole*. The moment of such a system is

Fig. 1.6 Geometrical representation of an electric dipole.

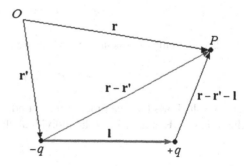

Fig. 1.7 Forces acting on
the charges of an electric
dipole placed in a
non-uniform electric field.

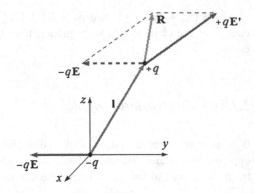

$$p = \lim_{\substack{l \to 0 \\ q \to \infty}} (q\mathbf{l}). \tag{1.67}$$

Let us calculate the electric field of a dipole. Using (1.24) and (A.54), we have

$$\mathbf{E} = -\nabla V = -\frac{1}{4\pi\epsilon_0}\nabla \left[\frac{\mathbf{p}\cdot(\mathbf{r}-\mathbf{r}')}{|\mathbf{r}-\mathbf{r}'|^3}\right]$$

$$= \frac{1}{4\pi\epsilon_0}\left\{\frac{3[\mathbf{p}\cdot(\mathbf{r}-\mathbf{r}')](\mathbf{r}-\mathbf{r}')}{|\mathbf{r}-\mathbf{r}'|^5} - \frac{\mathbf{p}}{|\mathbf{r}-\mathbf{r}'|^3}\right\},$$

or, if the origin of the coordinate system is chosen at the location of the point dipole,

$$\mathbf{E}(\mathbf{r}) = \frac{1}{4\pi\epsilon_0}\left[\frac{3(\mathbf{p}\cdot\mathbf{r})\mathbf{r}}{r^5} - \frac{\mathbf{p}}{r^3}\right]. \tag{1.68}$$

Suppose now that the dipole is situated in a non-uniform electric field and let us calculate the resultant $\mathbf{R}$ of the forces acting on the dipole. Figure 1.7 shows that

$$\mathbf{R} = q(\mathbf{E}' - \mathbf{E}). \tag{1.69}$$

Expanding $\mathbf{E}'$ in MacLaurin series, we have

$$\mathbf{E}' = \mathbf{E}(\mathbf{r} + \mathbf{l}) = \mathbf{E}(\mathbf{r}) + (\mathbf{l}\cdot\nabla)\mathbf{E} + \dots$$

and, keeping only the linear term and using it in (1.69), we find

$$\mathbf{R} = q(\mathbf{l}\cdot\nabla)\mathbf{E}. \tag{1.70}$$

This result shows that the dipole rotates under the action of a torque determined by the forces $-q\mathbf{E}$ and $+q\mathbf{E}$ on the individual charges and moves along its axis. If the field $\mathbf{E}$ is homogeneous, then $\mathbf{R} = 0$, which means that the action of the field reduces to the torque

$$\mathcal{M} = \mathbf{l} \times (q\mathbf{E}) = \mathbf{p} \times \mathbf{E}. \tag{1.71}$$

The energy of the dipole placed in the external field $\mathbf{E}$ is

$$W_e = qV(\mathbf{r} + \mathbf{l}) - qV(\mathbf{r}) \simeq q\mathbf{l} \cdot \nabla V = \mathbf{p} \cdot \nabla V = -\mathbf{p} \cdot \mathbf{E}. \tag{1.72}$$

Electrical Double Layer

A large number of dipoles with their momenta parallel and oriented in the same direction form a *double layer* (see Fig. 1.8). The potential of the field of such a system, at a certain point P, is

$$V(P) = \frac{1}{4\pi\epsilon_0} \left[\int_{\Sigma} \frac{\sigma dS}{r} + \int_{\Sigma'} \frac{\sigma' dS'}{r'} \right].$$

Since the distance between the surfaces Σ and Σ' is very small, we may integrate over a median surface S. Noting that

$$\frac{1}{r'} \simeq \frac{1}{r} + \mathbf{l} \cdot \nabla \left(\frac{1}{r} \right), \tag{1.73}$$

the potential at P reads

$$V(P) = \frac{1}{4\pi\epsilon_0} \int_S \sigma \left(\frac{1}{r} - \frac{1}{r'} \right) dS = -\frac{1}{4\pi\epsilon_0} \int_S \sigma l \mathbf{n} \cdot \nabla \left(\frac{1}{r} \right) dS.$$

But

$$-\mathbf{n} \cdot \nabla \left(\frac{1}{r} \right) dS = \frac{\mathbf{n} \cdot \mathbf{r}}{r^3} dS = \frac{dS \cos \alpha}{r^2} = d\Omega,$$

and, if we call

$$\tau_e = \sigma l$$

the *moment* or *power of the double layer*, we finally have

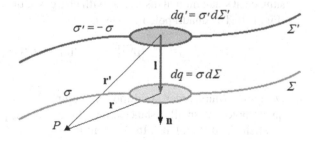

Fig. 1.8 Geometrical representation of an electrical *double layer*.

$$V(P) = \frac{1}{4\pi\epsilon_0} \int_\Omega \tau_e d\Omega. \tag{1.74}$$

If the double layer is homogeneous, $\tau_e = $ const. and (1.74) leads to

$$V(P) = \frac{1}{4\pi\epsilon_0}\tau_e\Omega, \tag{1.75}$$

where Ω is the solid angle under which the whole surface S is seen from P. It follows from (1.75) that inside a closed double layer with $\tau_e = $ const., the potential is τ_e/ϵ_0 for both the positive and the negative interior surfaces while outside the double layer it is zero. This means that the electrical double layers present surfaces of discontinuity for the potential function.

1.1.11 Electrostatic Multipoles

A set of electric charges of both signs, situated at small mutual distances as compared to the point where the effect of the system is considered, distributed discretely or continuously, give rise to an *electrostatic multipole*. By definition, an *n-order multipole* is a system of 2^n poles, formed by two $n-1$ order multipoles and having charges situated at the corners of a spatial geometric system with the sides $l_1, l_2, \ldots, l_n$. The moment of such a system is

$$\left|\mathbf{p}^{(n)}\right| = n! \lim_{\substack{l_1\ldots l_n \to 0 \\ q\to\infty}} (ql_1l_2 \ldots l_n). \tag{1.76}$$

Here are a few examples of multipoles:

(a) *Monopole* (zeroth-order multipole), with a single point charge;
(b) *Dipole* (first-order multipole), of moment (see Sect. 1.1.10):

$$\mathbf{p}^{(1)} = \lim_{\substack{l\to 0 \\ q\to\infty}} (q\mathbf{l});$$

(c) *Quadrupole* (second-order multipole), formed by two opposite, parallel dipoles, situated at small mutual distances, with charges at the corners of a parallelogram (Fig. 1.9). The quadrupole moment is

$$\left|\mathbf{p}^{(2)}\right| = 2! \lim_{\substack{l_1,l_2\to 0 \\ q\to\infty}} (ql_1l_2); \tag{1.77}$$

(d) *Octupole* (third-order multipole), composed of two opposite, parallel quadrupoles, at small distances, the charges being placed at the corners of a parallelepiped (Fig. 1.10). Its moment is

Fig. 1.9 Geometrical representation of an *electric quadrupole*.

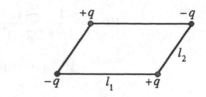

Fig. 1.10 Geometrical representation of an *electric octupole*.

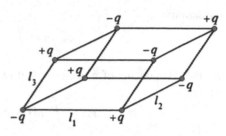

$$\left|\mathbf{p}^{(3)}\right| = 3! \lim_{\substack{l_1,l_2,l_3 \to 0 \\ q \to \infty}} (q l_1 l_2 l_3). \tag{1.78}$$

1.1.11.1 Potentials of a Continuous Distribution of Stationary Electric Charges

Consider a continuous charge distribution inside a three-dimensional domain D', of volume V', bounded by the surface S'. (The case of a discrete distribution is treated in a similar way, integrals being replaced by sums.) As we know, the potential of the field of such a distribution, at a point $P(\mathbf{r})$, with the notations from Fig. 1.5, is

$$V(\mathbf{r}) = \frac{1}{4\pi\epsilon_0} \int_{V'} \frac{\rho(\mathbf{r}')d\tau'}{|\mathbf{r} - \mathbf{r}'|}. \tag{1.79}$$

It is convenient to choose the origin of the coordinate system at an arbitrary point of D' and, assuming $|\mathbf{r}'| \ll |\mathbf{r}|$, we expand in series the ratio $1/|\mathbf{r} - \mathbf{r}'|$. The result is:

$$\frac{1}{|\mathbf{r} - \mathbf{r}'|} = \frac{1}{r} - \frac{x_i'}{1!} \frac{\partial}{\partial x_i'} \left(\frac{1}{r}\right) + \frac{1}{2!} x_i' x_k' \frac{\partial^2}{\partial x_i \partial x_k} \left(\frac{1}{r}\right) - \cdots$$

$$= \sum_{l=0}^{\infty} \frac{(-1)^l}{l!} (\mathbf{r}' \cdot \nabla)^{(l)} \left(\frac{1}{r}\right), \tag{1.80}$$

where we used the summation convention, i.e. the repeated indices i and k are summed over. Introducing the expansion into (1.79), we have

$$V(\mathbf{r}) = \frac{1}{4\pi\epsilon_0} \sum_{l=0}^{\infty} \frac{(-1)^l}{l!} \int_{V'} \rho(\mathbf{r}')(\mathbf{r}' \cdot \nabla)^{(l)} \left(\frac{1}{r}\right) d\tau'$$

$$= V^{(0)} + V^{(1)} + V^{(2)} + \ldots = \sum_{l=0}^{\infty} V^{(l)}(\mathbf{r}). \qquad (1.81)$$

By definition,

$$p_i = \int_{V'} \rho(\mathbf{r}')x_i' d\tau' \qquad (1.82)$$

are the components of a vector called *electric dipole moment* of the distribution, while

$$p_{ik} = \int_{V'} \rho(\mathbf{r}') \left(3x_i' x_k' - r'^2 \delta_{ik}\right) d\tau' \qquad (1.83)$$

are the components of a second-order symmetric tensor called *electric quadrupole moment* of the distribution, and so on. Expression (1.83) is easily obtained from (1.81), taking $l = 2$ and subtracting the null term:

$$r'^2 \delta_{ik} \frac{\partial^2}{\partial x_i \partial x_k} \left(\frac{1}{r}\right) = r'^2 \frac{\partial^2}{\partial x_i \partial x_i} \left(\frac{1}{r}\right)_{r \neq 0} = 0.$$

In general, the 2^l-*order multipole electric moment* is defined by

$$p_{i_1 \ldots i_l} = (l + 1) \int_{V'} \rho(\mathbf{r}')x_{i_1}' \ldots x_{i_l}' \, d\tau'. \qquad (1.84)$$

The potential V of the distribution can be then written as

$$V(\mathbf{r}) = \frac{1}{4\pi\epsilon_0} \sum_{l=0}^{\infty} \frac{(-1)^l}{(l+1)!} p_{i_1 \ldots i_l} \partial_{i_1} \ldots \partial_{i_l} \left(\frac{1}{r}\right).$$

We are now able to interpret the terms $V^{(0)}, V^{(1)}, V^{(2)} \ldots$ in (1.81), as follows:

$$V^{(0)} = \frac{1}{4\pi\epsilon_0} \frac{Q}{r}$$

is the *monopole potential* produced at distance r by the charge $Q = \int_{V'} \rho(\mathbf{r}')d\tau'$, situated at the origin;

$$V^{(1)} = -\frac{1}{4\pi\epsilon_0} p_i \frac{\partial}{\partial x_i} \left(\frac{1}{r}\right) = -\frac{1}{4\pi\epsilon_0} \mathbf{p} \cdot \nabla \left(\frac{1}{r}\right)$$

is the *dipole potential* (see (1.66)), the dipole being at the origin;

$$V^{(2)} = \frac{1}{4\pi\epsilon_0} \frac{1}{6} p_{ik} \frac{\partial^2}{\partial x_i \partial x_k} \left(\frac{1}{r}\right) = \frac{1}{4\pi\epsilon_0} \frac{1}{6} p_{ik} \left(\frac{3x_i x_k}{r^5} - \frac{\delta_{ik}}{r^3}\right)$$

is the *quadrupole potential*, assuming again that the system is at the origin, and so on.

1.1.11.2 Energy of a Continuous Distribution of Stationary Charges in an External Electric Field

Consider a system of discrete electric charges, distributed in the three-dimensional domain D and placed in the electrostatic field $\mathbf{E}$. The field $\mathbf{E}$ is considered as external, or background field, therefore $\nabla \cdot \mathbf{E} = 0$. The potential energy of the charge q_k, at the point where the potential is $V(\mathbf{r}_k)$, is $W_k = q_k V(\mathbf{r}_k)$, which means that the total energy of the system is

$$W_e = \sum_k W_k = \sum_k q_k V(\mathbf{r}_k).$$

For a continuous distribution, the last formula becomes

$$W_e = \int_{V'} \rho(\mathbf{r}')V(\mathbf{r}')d\tau'. \tag{1.85}$$

Expanding now $V(\mathbf{r}')$ in Taylor series around the origin O, chosen at an arbitrary point of D, we obtain

$$V(\mathbf{r}') = V(0) + x_i' \left(\frac{\partial V}{\partial x_i}\right)_0 + \frac{1}{2} x_i' x_k' \left(\frac{\partial^2 V}{\partial x_i \partial x_k}\right)_0 + \ldots$$

$$= V(0) - x_i' E_i(0) - \frac{1}{2} x_i' x_k' \left(\frac{\partial E_k}{\partial x_i}\right)_0 - \ldots$$

$$= V(0) - x_i' E_i(0) - \frac{1}{6}(3x_i' x_k' - r'^2 \delta_{ik}) \left(\frac{\partial E_k}{\partial x_i}\right)_0 + \ldots, \tag{1.86}$$

where we added the null term

$$r'^2 \delta_{ik} \left(\frac{\partial E_k}{\partial x_i}\right)_0 = r'^2 (\nabla \cdot \mathbf{E})_0 = 0.$$

Introducing this expression for $V(\mathbf{r}')$ into (1.85), we find finally that

$$W_e = QV(0) - p_i E_i(0) - \frac{1}{6} p_{ik} \left(\frac{\partial E_k}{\partial x_i} \right)_0 + \dots = \sum_{l=0}^{\infty} W^{(l)}. \qquad (1.87)$$

From the expression (1.87) we can deduce the significance of the terms $W^{(0)}$, $W^{(1)}$, $W^{(2)}$ etc., as follows:

(a) $W^{(0)} = QV(0)$ is the potential energy of the distribution, assuming that the whole charge is concentrated at the origin;

(b) $W^{(1)} = -\mathbf{p} \cdot \mathbf{E}(0)$ is the dipole potential energy of the system (see (1.72)), with the dipole at the origin;

(c) $W^{(2)} = -\frac{1}{6} p_{ik} \left(\frac{\partial E_k}{\partial x_i} \right)_0$ is the quadrupole potential energy of the charges, with the quadrupole at the origin, etc.

The terms $W^{(0)}$, $W^{(1)}$, $W^{(2)}$, ... indicate the specific way of interaction between the field and various multipole formations: charge versus field potential, dipoles versus field intensity, quadrupoles versus the field gradient, etc.

The multipole theory is important not only in electrostatics or magnetostatics, but also in the multipole radiation theory, nuclear physics, solid state physics, etc.

1.2   Electrostatic Field in Polarized Media

1.2.1   Dielectric Polarization

A dielectric is an insulator which becomes *polarized* when introduced in an electric field. By *polarization* we mean the appearance of dipoles in each volume element of the body. If the dipoles exist already in the material, they orient themselves under the action of the field.

The polarization of a medium is studied by means of a vector quantity called *polarization density* or, simply, *polarization*, defined as the electric dipole moment per unit volume:

$$\mathbf{P} = \lim_{\Delta \tau \to 0} \frac{\Delta \mathbf{p}}{\Delta \tau} = \frac{d\mathbf{p}}{d\tau}, \qquad (1.88)$$

where $\mathbf{p}$ is the dipole moment. Using this definition, we write the potential created by an infinitesimal, three-dimensional domain of a continuous distribution of electric dipoles. In view of (1.66), we may write the potential of a continuous dipole distribution as

$$V(\mathbf{r}) = \frac{1}{4\pi\epsilon_0} \int_{V'} \mathbf{P}(\mathbf{r}') \cdot \nabla' \left(\frac{1}{|\mathbf{r} - \mathbf{r}'|} \right) d\tau', \qquad (1.89)$$

or, using (A.43) and the divergence theorem,

$$V(\mathbf{r}) = \frac{1}{4\pi\epsilon_0} \int_{V'} \frac{(-\nabla' \cdot \mathbf{P})}{|\mathbf{r} - \mathbf{r}'|} d\tau' + \frac{1}{4\pi\epsilon_0} \oint_{S'} \frac{P_n(\mathbf{r}')}{|\mathbf{r} - \mathbf{r}'|} dS', \qquad (1.90)$$

where $P_n(\mathbf{r}')$ is the polarization component orthogonal to the surface element dS' around the point defined by $\mathbf{r}'$.

Comparing (1.90) with (1.27) which gives the potential of a continuous charge distribution, one can write by analogy

$$V(\mathbf{r}) = \frac{1}{4\pi\epsilon_0} \left[\int_{V'} \frac{\rho_p(\mathbf{r}')d\tau'}{|\mathbf{r} - \mathbf{r}'|} + \oint_{S'} \frac{\sigma_p(\mathbf{r}')dS'}{|\mathbf{r} - \mathbf{r}'|} \right], \qquad (1.91)$$

where

$$\rho_p(\mathbf{r}') = -\nabla' \cdot \mathbf{P} \quad \text{and} \quad \sigma_p(\mathbf{r}') = P_n \qquad (1.92)$$

are the *spatial* and *superficial charge densities*, respectively, appearing as a result of the polarization. Remark, however, that

$$Q_p = \int_{V'} \rho_p(\mathbf{r}')d\tau' + \oint_{S'} \sigma_p(\mathbf{r}')dS' = -\int_{V'} \nabla' \cdot \mathbf{P} d\tau' + \oint_{S'} \mathbf{P} \cdot d\mathbf{S}' = 0,$$

meaning that the total polarization charge is zero.

The above considerations tell us that a continuous dipole distribution, in a finite domain, from the point of view of exterior electrostatic actions, behaves like a continuous distribution of both spatial and superficial *fictitious* charges. These charges appear only *as a result of* and *during* the polarization. To distinguish polarization charges from free charges, one may call the first *bound charges*.

1.2.2 Gauss's Law for Dielectric Media

Let ρ be the density of free point charges located in a dielectric, and

$$\rho_p = -\nabla \cdot \mathbf{P}$$

the density of polarization charge, due to the field produced by the free charges. Gauss's law in the dielectric then reads

$$\nabla \cdot \mathbf{E} = \frac{1}{\epsilon_0}(\rho + \rho_p) = \frac{1}{\epsilon_0}(\rho - \nabla \cdot \mathbf{P}).$$

Combining the two divergence terms, we obtain

$$\nabla \cdot (\epsilon_0 \mathbf{E} + \mathbf{P}) = \rho,$$

or, in integral form,

$$\oint_S (\epsilon_0 \mathbf{E} + \mathbf{P}) \cdot d\mathbf{S} = Q. \tag{1.93}$$

The vector field defined as

$$\mathbf{D} = \epsilon_0 \mathbf{E} + \mathbf{P} \tag{1.94}$$

is called *electric displacement*.

In vacuum, $\mathbf{P} = 0$ since there is no polarization, such that $\mathbf{D} = \epsilon_0 \mathbf{E}$. Using the definition (1.94), we have

$$\oint_S \mathbf{D} \cdot d\mathbf{S} = Q, \tag{1.95}$$

which is the integral form of Gauss's law for dielectric media.

The *differential form* of Gauss's law reads:

$$\nabla \cdot \mathbf{D} = \rho, \tag{1.96}$$

where we emphasize that ρ is the density of *free charge*.

This law stands as a fundamental element used by Maxwell in elaborating his theory of the electromagnetic field.

1.2.3 Types of Dielectrics

From the point of view of the polarization property, a medium can be *homogeneous*, when polarization does not depend on the point, or *non-homogeneous*, when it does. If the polarization of the medium does not depend on direction, then the medium is *isotropic*; if it depends, the medium is *anisotropic*.

In general, the polarization $\mathbf{P}$ is a function of the field strength, $\mathbf{P} = \mathbf{P}(\mathbf{E})$. If the field is not very strong and the existence of a *spontaneous* polarization $\mathbf{P}_0$ is presumed, we may write

$$\mathbf{P} \simeq \mathbf{P}_0 + \alpha \mathbf{E},$$

where α is a coefficient called *polarizability*. If $\mathbf{P}_0 = 0$ and α does not depend on the applied field, the medium is said to be *linear*. From now on we shall consider only linear media.

The experimental data show that in case of linear, homogeneous, and isotropic media there exists the relation

$$\mathbf{D} = \epsilon \mathbf{E}, \quad \text{with } \epsilon = \epsilon_0 \epsilon_r, \tag{1.97}$$

which represents a *constitutive relation*. Other constitutive equations of electromagnetism we shall encounter in Chap. 2, Sect. 2.2.1, relating the magnetic induction $\mathbf{B}$

and the magnetic field intensity $\mathbf{H}$. The constitutive equations describe the response of bound charges and currents to the applied fields.

Using (1.94), we find that

$$\mathbf{P} = \epsilon_0(\epsilon_r - 1)\mathbf{E} = \epsilon_0 \kappa \mathbf{E} = \alpha \mathbf{E}. \tag{1.98}$$

Here,

$$\kappa = \epsilon_r - 1$$

is the *electric susceptibility* of the medium, which is related to the polarizability by the relation

$$\alpha = \epsilon_0 \kappa.$$

In isotropic, non-homogeneous media, the permittivity depends on the point, $\mathbf{D} = \epsilon(x, y, z)\mathbf{E}$, meaning that the coefficients κ and α are also functions of the point:

$$\mathbf{P} = \epsilon_0 \kappa(x, y, z)\mathbf{E} = \alpha(x, y, z)\mathbf{E}. \tag{1.99}$$

The *ferroelectric media* are characterized by a polarization $\mathbf{P}_0 \neq 0$ in the absence of the external electric field and a nonlinear response of the polarization to the applied electric field:

$$P = P_0 + \alpha_1 E + \alpha_2 E^2 + \alpha_3 E^3 + \ldots, \tag{1.100}$$

where $\alpha_1 \ll \alpha_2 \ll \alpha_3 \ll \ldots$.

In linear anisotropic media, the relation between $\mathbf{P}$ and $\mathbf{E}$ has the form

$$P_i = \epsilon_0 \kappa_{ik} E_k, \qquad i, k = 1, 2, 3, \tag{1.101}$$

where κ_{ik} is the *electric susceptibility tensor* of the medium and Einstein's summation convention has been used. We can then write for the electric displacement

$$D_i = \epsilon_{ik} E_k = \epsilon_0(\epsilon_r)_{ik} E_k = \epsilon_0 \delta_{ik} E_k + P_i,$$

and

$$D_i = \epsilon_0 \left[(\epsilon_r)_{ik} - \delta_{ik} \right] E_k = \epsilon_0 \kappa_{ik} E_k,$$

where

$$\kappa_{ik} = (\epsilon_r)_{ik} - \delta_{ik}. \tag{1.102}$$

The electric polarization can be:

(a) *diaelectric*, as a result of the displacement of positive and negative charge centres in atoms (molecules) under the action of the field $\mathbf{E}$;
(b) *paraelectric*, due to the orientation of dipoles with $\mathbf{p} \neq 0$ when the field is absent, but having $\mathbf{P} = 0$ as a result of thermal motion;
(c) *ferroelectric*, characterized by $\mathbf{P}_0 \neq 0$ in the absence of an external electric field.

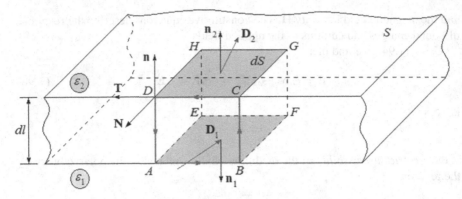

Fig. 1.11 An elementary prism *ABCDEFGH* used to determine the jump relations for the normal and tangent components of the fields **E** and **D**. Remark that, while $\mathbf{n}_1$ and $\mathbf{n}_2$ are the normal versors of $d\mathbf{S}_1$ and $d\mathbf{S}_2$, the unit vectors **T** and **N** are mutually orthogonal, but otherwise arbitrary in the tangent plane.

1.2.4 Jump Conditions for the Components of the Fields **E** and **D**

When passing through interfaces which separate media with different permittivities, the fields **E** and **D** vary rapidly from one point to another, and, as a result, the field lines undergo discontinuous variations (refractions). This shows that the interfaces are *surfaces of discontinuity* for the field components. The relations describing the variation of tangent and normal components of the fields **E**, **D** when passing through such surfaces are called *jump conditions*.

Consider two homogeneous and isotropic neighbouring dielectric media, of permittivities ϵ_1 and ϵ_2, separated by an interface in which it is assumed that there are electric charges.

Let $\mathbf{E}_1$, $\mathbf{D}_1$ and $\mathbf{E}_2$, $\mathbf{D}_2$ be the fields in the two media. Let us also delimit on the two sides of the interface an elementary right prism, of basis dS, thickness dl, and volume $d\tau$. The unit vectors of the two bases are shown in Fig. 1.11.

Each of the fields **E** and **D** can be decomposed into two vector components: one normal and the other tangent to dS. In the following we shall find the jump relations for the normal and tangent components of the fields **E** and **D**.

1.2.4.1 Jump Conditions for the Normal Components

Gauss's law (1.95), applied to the elementary prism ABCDEFGH (see Fig. 1.11), yields

$$\mathbf{D} \cdot d\mathbf{S} = dQ,$$

where

$$\mathbf{D} \cdot d\mathbf{S} = d\Phi_1 + d\Phi_2 + d\Phi_l,$$

$d\Phi_1$ and $d\Phi_2$ being the elementary fluxes through the bases, $d\Phi_l$ – the flux through the lateral sides, and

$$dQ = \rho d\tau = \rho dS dl.$$

If one chooses $\mathbf{n}_2 = -\mathbf{n}_1 = \mathbf{n}$, we have

$$d\Phi_1 = \mathbf{D}_1 \cdot d\mathbf{S}_1 = -\mathbf{D}_1 \cdot \mathbf{n} dS = -D_{1n} \, dS,$$
$$d\Phi_2 = \mathbf{D}_2 \cdot d\mathbf{S}_2 = \mathbf{D}_2 \cdot \mathbf{n} dS = D_{2n} \, dS.$$

In the limit $dl \to 0$, the area of the lateral surfaces is infinitesimal compared to the bases and the lateral flux practically vanishes:

$$\lim_{dl \to 0} d\Phi_l = 0.$$

The charge gets distributed in fact on the surface dS, with the surface density σ. Since the charge must remain finite and equal to dQ, we have

$$dQ = \sigma dS.$$

Introducing all these results into Gauss's law, we obtain

$$(\mathbf{D}_2 - \mathbf{D}_1) \cdot \mathbf{n} = D_{2n} - D_{1n} = \sigma. \tag{1.103}$$

This relation expresses the discontinuity of the normal component of the electric displacement. Since $\mathbf{D}_1 = \epsilon_1 \mathbf{E}_1$ and $\mathbf{D}_2 = \epsilon_2 \mathbf{E}_2$, we may write also

$$\epsilon_2 E_{2n} - \epsilon_1 E_{1n} = \sigma. \tag{1.104}$$

We conclude that the normal components of $\mathbf{E}$ and $\mathbf{D}$ vary *discontinuously* across the interface. If the surface density of the free charge is zero on the interface, $\sigma = 0$, but $\epsilon_1 \neq \epsilon_2$, D_n presents no jump, but E_n still varies discontinuously.

1.2.4.2 Jump Conditions for the Tangent Components

Let us calculate the circulation of the field $\mathbf{E}$ along the elementary closed contour $ABCDA$ in Fig. 1.11. Using (1.28), we can write

$$\oint_{ABCDA} \mathbf{E} \cdot d\mathbf{s} = \int_{AB} \mathbf{E}_1 \cdot d\mathbf{s}_1 + \int_{BC} \mathbf{E} \cdot d\mathbf{s} + \int_{CD} \mathbf{E}_2 \cdot d\mathbf{s}_2 + \int_{DA} \mathbf{E} \cdot d\mathbf{s} = 0.$$

We have $d\mathbf{s}_1 = -\mathbf{T}ds$ and $d\mathbf{s}_2 = \mathbf{T}ds$, and the contribution of the circulation on the sides BC and DA vanishes in the limit $dl \to 0$, leading to

$$(\mathbf{E}_2 - \mathbf{E}_1) \cdot \mathbf{T} = E_{2T} - E_{1T} = 0. \tag{1.105}$$

Observing that $\mathbf{T} = \mathbf{N} \times \mathbf{n}$, we obtain

$$(\mathbf{E}_2 - \mathbf{E}_1) \cdot (\mathbf{N} \times \mathbf{n}) = \mathbf{N} \cdot [\mathbf{n} \times (\mathbf{E}_2 - \mathbf{E}_1)] = 0,$$

or, by eliminating the trivial case of coplanarity of the vectors,

$$\mathbf{n} \times (\mathbf{E}_2 - \mathbf{E}_1) = 0.$$

Since $E_{1T} = D_{1T}/\epsilon_1$ and $E_{2T} = D_{2T}/\epsilon_2$, the relation (1.105) can be written as

$$\frac{D_{2T}}{\epsilon_2} - \frac{D_{1T}}{\epsilon_1} = 0. \tag{1.106}$$

Recall that the direction of $\mathbf{T}$ is arbitrary in the tangent plane, therefore we can conclude that the tangent component of the field $\mathbf{E}$ varies continuously, while the tangent component of $\mathbf{D}$ makes a jump of magnitude ϵ_1/ϵ_2. There is no jump, of course, if $\epsilon_1 = \epsilon_2$.

Observations:

(a) Since the electrostatic field is conservative, the jump relations (1.104) and (1.105) give rise to the following *boundary conditions* for the potential:

$$\left(\frac{\partial V}{\partial T}\right)_1 = \left(\frac{\partial V}{\partial T}\right)_2,$$
$$\epsilon_1 \left(\frac{\partial V}{\partial n}\right)_1 - \epsilon_2 \left(\frac{\partial V}{\partial n}\right)_2 = \sigma. \tag{1.107}$$

Considering two points situated on each side of the interface, we have

$$\int_1^2 \mathbf{E} \cdot d\mathbf{s} = -\int_1^2 dV = V_1 - V_2.$$

If the two points are infinitesimally close to the interface and the electric field is considered to be finite on the surface, we find

$$V_1 = V_2,$$

i.e. the potential is continuous across the interface on which there is a surface distribution of charges.

(b) From the electrostatic point of view, materials are divided into conductors and dielectrics (insulators). All the internal points of a conductor, in view of Ohm's law (see (2.19)), are characterized by $\mathbf{E} = 0$. As a result, on the surface of a conductor,

$$\sigma = D_n = \epsilon E_n,$$

which means

$$V_S = \text{const.}; \qquad \sigma = -\epsilon \left(\frac{\partial V}{\partial n} \right)_S, \tag{1.108}$$

i.e. $\mathbf{E}$ is oriented along the normal to the conducting surface.

Inside a dielectric the electric field is not null. At every internal point the potential obeys the Poisson equation, as well as the jump conditions (1.104) and (1.105) on the boundary.

1.3 Special Methods of Solving Problems in Electrostatics

As mentioned in Sect. 1.1.8.2, the integration of the Poisson equation is not difficult if the associated Green's function is known. But finding Green's function is not an easy task, therefore some special methods of solving electrostatics problems have been elaborated. Here we present some of these methods.

The general procedure consists in identifying the boundary conditions satisfied by the potential V of the electrostatic field, using the following criteria:

(a) The potential is finite everywhere, except for the points where the sources (point charges) are located;
(b) The potential is continuous everywhere, including the surface of a conducting or dielectric body, except for double layers;
(c) On the surface of a conductor one must know either $V = V|_S = \text{const.}$, or (see (1.108))

$$\sigma = -\epsilon \left(\frac{\partial V}{\partial n} \right)_S;$$

(d) At the interface between two dielectric media, if $\sigma = 0$, we have (see (1.107))

$$\epsilon_1 \left(\frac{\partial V}{\partial n} \right)_1 - \epsilon_2 \left(\frac{\partial V}{\partial n} \right)_2 = 0; \tag{1.109}$$

(e) The potential is null at infinity, and sources are at a finite distance.

1.3.1 Method of Electric Images

This method is used to determine some essential electric quantities (potential, charge density, force of interaction, etc.) for one (or more) point charges, in the presence of certain separation surfaces, such as conductors connected to the ground, or maintained at a constant potential.

The method of images is based on the fact that, depending on the geometry of the problem, a small number of point charges, conveniently chosen and located, can simulate the boundary conditions required by the problem. These are called *image charges*.

This method gives the possibility to determine, among other things, the charge density σ on the surface of a simply-shaped conductor (plane, sphere, etc.), in the presence of one or more point charges, *without* really solving Poisson's equation.

The essence of the method of images is the uniqueness of the solution of Poisson's equation with given boundary conditions. A complicated configuration is mimicked by a simple one, with identical boundary conditions. As a result, the solution of the complicated problem is found by solving the simple one. There are a few aspects to be taken into account when using this method:

- The boundary conditions naturally separate the space into a domain of interest, D, where the potential has to be determined, and a complementary domain, D', which is "hidden behind" the boundary. The method does not produce the correct potential for the region D', but this is of no concern, since one is interested in the potential only in D;
- The Poisson equation in the domain of interest has to be the same in the original problem and in the changed configuration, therefore no addition of charge or any other modification can be made in D;
- In the domain D' we can add charges at chosen locations, change permittivities etc., such that the new configuration in D' produces exactly the boundary conditions at the interface between D and D'.

All these requirements insure that the Poisson equation with its boundary conditions does not change in D, while the problem is rendered much simpler by the changes made in the complementary domain D'. Below we shall demonstrate the method with a few illuminating examples.

1.3.1.1 Plane Conductor

Let us consider a metallic plane surface, connected to the ground (i.e. $V = 0$), situated in vacuum, under the influence of the positive point charge q (Fig. 1.12), placed at a distance d from the metalic plate, at the point A. The problem is to find the potential above the plate.

The potential at an arbitrary point P is given by the sum of two potentials: the potential of the charge q and the potential of the charge distribution on the plane, due

Fig. 1.12 Method of *electric images*: a plane conductor.

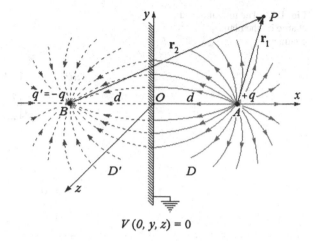

$$V(0, y, z) = 0$$

to the influence of q:

$$V_P = \frac{1}{4\pi\epsilon_0} \frac{q}{|\mathbf{r}_1|} + \frac{1}{4\pi\epsilon_0} \int_S \frac{\sigma(\mathbf{r})}{|\mathbf{r}|} dS. \tag{1.110}$$

It is hard to determine the induced charge distribution $\sigma(\mathbf{r})$. But if we think in general terms, we actually have to solve Poisson's equation with a point charge at the point $(d, 0, 0)$, satisfying the boundary conditions $V = 0$ on the plane $x = 0$ and $V \to 0$ at infinity (i.e. far away from the charge, where $x^2 + y^2 + z^2 \gg d^2$). Thus, the boundary condition separates the space into the domain D which is above the plate, and the complementary domain D' below.

The task is to find a new and simple configuration of charges in D', which would simulate the boundary conditions. We notice that these conditions are fulfilled by a configuration in which the metalic plate is removed and an "image" point charge $q' = -q$ is placed at the point B, symmetric to A with respect to the plane $x = 0$. For this new configuration, the potential at any arbitrary point P of the domain D is

$$V_P = \frac{1}{4\pi\epsilon_0} \left(\frac{q}{r_1} - \frac{q}{r_2} \right) \tag{1.111}$$

$$= \frac{q}{4\pi\epsilon_0} \left\{ \left[(x-d)^2 + y^2 + z^2 \right]^{-1/2} - \left[(x+d)^2 + y^2 + z^2 \right]^{-1/2} \right\}.$$

The uniqueness of the solution of Poisson's equation with given boundary conditions ensures that this potential is the one produced by the original configuration as well. Hence, formula (1.111) can now be used to calculate the charge density on the original metallic plate. According to (1.108), we have

$$\sigma = -\epsilon_0 \left(\frac{\partial V}{\partial n} \right)_S = -\epsilon_0 \left(\frac{\partial V}{\partial x} \right) \bigg|_{x=0} = -\frac{q}{2\pi} \frac{d}{r_0^3}, \tag{1.112}$$

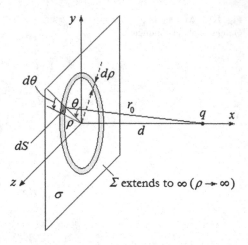

Fig. 1.13 The total induced
charge on the plane
conductor.

where $r_0 = \left(d^2 + y^2 + z^2\right)^{1/2}$. The total charge induced on the plate can be obtained
by integration of (1.112) on the whole plane $x = 0$. Not surprisingly, the result is
$-q$.

Indeed, we have (see Fig. 1.13):

$$\int_{\Sigma} \sigma \, dS = -\frac{qd}{2\pi} \int_{\Sigma} \frac{dS}{r_0^3} = -\frac{qd}{2\pi} \int_0^{\infty} \frac{\rho \, d\rho}{\left(\rho^2 + d^2\right)^{3/2}} \int_0^{2\pi} d\theta = -q.$$

Remark, however, that the method gives the correct result for the potential only
above the metallic plate, i.e. in the region where the original charge was placed.

1.3.1.2   Two Orthogonal Metallic Half-Planes

Let us consider two orthogonal and grounded metallic half-planes, and a point electric
charge q placed inside the "corner", at a distance a from the half-plane A and b from
the half-plane B, as in Fig. 1.14. We have to find the potential at any point inside the
metallic corner.

In this case, the method of images requires us to remove the metallic plates and
to place point charges outside the corner, such that the boundary conditions of the
original problem are exactly matched, i.e. $V = 0$ on the half-planes A and B. The
symmetry of the problem and the experience acquired in the previous case suggest
that we shall have to place three image charges, as illustrated in Fig. 1.14.

The potential at an arbitrary point P inside the metallic corner is

$$V = \frac{q}{4\pi\epsilon_0} \left(\frac{1}{r_1} + \frac{1}{r_2} - \frac{1}{r_3} - \frac{1}{r_4} \right),$$

Fig. 1.14 A point charge q close to a grounded metallic corner.

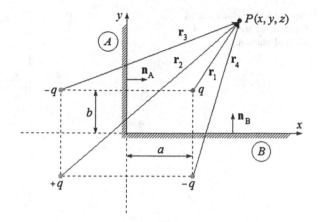

and the electric field at the same point is

$$\mathbf{E} = \frac{q}{4\pi\epsilon_0}\left(\frac{\mathbf{r}_1}{r_1^3} + \frac{\mathbf{r}_2}{r_2^3} - \frac{\mathbf{r}_3}{r_3^3} - \frac{\mathbf{r}_4}{r_4^3}\right).$$

To find the induced charge on the half-plane A, we shall use $\mathbf{E} = \frac{\sigma}{\epsilon_0}\mathbf{n}$. Let us then find the electric field on the plane A, where $r_1 = r_3$ and $r_2 = r_4$. From Fig. 1.15 we observe that $\mathbf{r}_1 - \mathbf{r}_3 = -2a\mathbf{n}_A$ and $\mathbf{r}_2 - \mathbf{r}_4 = 2a\mathbf{n}_A$. Thus, the electric field at a point on the half-plane A will be

$$\mathbf{E}_A = \frac{q}{4\pi\epsilon_0}\left(\frac{\mathbf{r}_1 - \mathbf{r}_3}{r_1^3} + \frac{\mathbf{r}_2 - \mathbf{r}_4}{r_2^3}\right) = -\frac{q}{2\pi\epsilon_0}a\left(\frac{1}{r_1^3} - \frac{1}{r_2^3}\right)\mathbf{n}_A,$$

where

$$r_1 = \sqrt{a^2 + (y-b)^2 + z^2}, \qquad r_2 = \sqrt{a^2 + (y+b)^2 + z^2}.$$

The density of charge is then

$$\sigma_A = -\frac{qa}{2\pi}\left(\frac{1}{r_1^3} - \frac{1}{r_2^3}\right),$$

while the total charge induced on the half-plane A is

$$q_A = -\frac{qa}{2\pi}\int_0^\infty dy \int_{-\infty}^\infty \left\{ \frac{1}{\left[a^2 + (y-b)^2 + z^2\right]^{1/2}} - \frac{1}{\left[a^2 + (y+b)^2 + z^2\right]^{1/2}} \right\} dz$$

$$= -\frac{2}{\pi}q \arctan\frac{b}{a}.$$

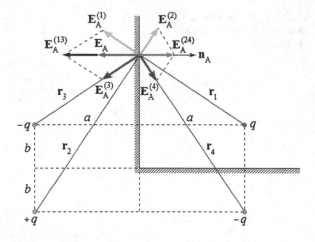

Fig. 1.15 The electric field E_A at an arbitrary point of the half-plane A.

An analogous calculation for the half-plane B gives

$$q_B = -\frac{2}{\pi} q \, \arctan \frac{a}{b}.$$

The total charge induced on the two half-planes is

$$q' = q_A + q_B = -\frac{2}{\pi} q \left(\arctan \frac{b}{a} + \arctan \frac{a}{b} \right) = -q,$$

as it would be expected.

1.3.1.3   Plane Interface Between Two Dielectrics

Let us consider now that the plane $x = 0$ separates two dielectric media of permittivities ϵ_1 for $x > 0$ and ϵ_2 for $x < 0$. The point charge q is situated in the same place as in Fig. 1.12, at the point A, and we have to find the potential at any point in space.

Since there are no free charges on the separation surface, the continuity of the potential and electric field across the boundary implies

$$V_1|_S = V_2|_S, \quad \epsilon_1 \left(\frac{\partial V_1}{\partial n} \right)_S = \epsilon_2 \left(\frac{\partial V_2}{\partial n} \right)_S, \tag{1.113}$$

where $V_1 = V(x > 0)$ and $V_2 = V(x < 0)$.

To find the potential V_1 in the domain D_1 by the method of images, we shall consider the space filled with the dielectric of permittivity ϵ_1 and mimic the effect of the other dielectric upon the region $x > 0$ by an "image charge" q' symmetrically situated at the point B (see Fig. 1.12). The value of the charge q' has to be determined.

The potential in the region D_1 will read

$$V_1 = \frac{1}{4\pi\epsilon_1} \left(\frac{q}{r_1} + \frac{q'}{r_2} \right).$$

The potential V_2, in the region D_2, will be simulated by a configuration in which the whole space is filled with the dielectric of permittivity ϵ_2, and with a charge q'' placed at the original location of the charge q:

$$V_2 = \frac{1}{4\pi\epsilon_2} \frac{q''}{r_2}.$$

Now we have to find the values of the fictitious charges q' and q''. To this end, we shall use the boundary conditions (1.113) and consider the point P on the interface, at the origin of the coordinate system (see Fig. 1.12). We arrive at the system of equations

$$q - q' = q'',$$
$$\epsilon_2(q + q') = \epsilon_1 q'',$$

with the solution

$$q' = \frac{\epsilon_1 - \epsilon_2}{\epsilon_1 + \epsilon_2} q, \qquad q'' = \frac{2\epsilon_2}{\epsilon_1 + \epsilon_2} q.$$

The superficial density of the polarization charge on the plane $x = 0$, in view of (1.92), (1.94), and (1.113), as well as the constitutive relation $\mathbf{D} = \epsilon\mathbf{E}$, is then

$$\sigma_p = P_{1n} - P_{2n} = (\epsilon_1 - \epsilon_0)E_{1n} - (\epsilon_2 - \epsilon_0)E_{2n}$$
$$= (\epsilon_2 - \epsilon_0) \left(\frac{\partial V_2}{\partial x} \right)_{x=0} - (\epsilon_1 - \epsilon_0) \left(\frac{\partial V_1}{\partial x} \right)_{x=0}.$$

Performing the calculations, we obtain

$$\sigma_p = \frac{\epsilon_0}{4\pi} \left[\frac{q(x - d)}{\epsilon_1 \left[(x - d)^2 + y^2 + z^2\right]^{3/2}} + \frac{q'(x + d)}{\epsilon_1 \left[(x + d)^2 + y^2 + z^2\right]^{3/2}} \right.$$
$$\left. - \frac{q''(x - d)}{\epsilon_2 \left[(x - d)^2 + y^2 + z^2\right]^{3/2}} \right]_{x=0} = \frac{\epsilon_0 q d}{4\pi\epsilon_1 r_0^3} \left[-1 + \frac{\epsilon_1 - \epsilon_2}{\epsilon_1 + \epsilon_2} + \frac{2\epsilon_1}{\epsilon_1 + \epsilon_2} \right]$$
$$= \frac{\epsilon_0 q d}{2\pi\epsilon_1 r_0^3} \frac{\epsilon_1 - \epsilon_2}{\epsilon_1 + \epsilon_2}, \tag{1.114}$$

where r_0 has the same significance as in (1.112). If $\epsilon_1 = \epsilon_2$, we have $\sigma = 0$, as expected.

Fig. 1.16 Method of *electric images*: a spherical conductor.

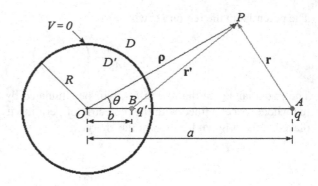

1.3.1.4 Spherical Conductor

Consider a conducting sphere of radius R, grounded and placed in vacuum, with its centre situated at the distance a from a point charge q placed at the point A. We have to find the potential at an arbitrary point P outside the sphere.

Obviously, in this case the boundary condition is $V = 0$ on the spherical shell. The shell separates the space into the domain D which is outside of the sphere and the domain D' which is inside the sphere. We can modify the charge configuration inside the sphere, in order to simulate the null potential on the spherical shell. To this end, we place an "image" charge q' at the point B, located at the distance b ($b < R$) from the centre O of the sphere (Fig. 1.16). The potential at the point P is similar to that given by (1.112), i.e.

$$V_P = \frac{1}{4\pi\epsilon_0}\left(\frac{q}{r}+\frac{q'}{r'}\right). \tag{1.115}$$

Using the condition $V|_S = 0$, we have to find both the value of the fictitious charge q' and the distance b where it is placed. We have

$$\frac{q}{\sqrt{a^2+R^2-2aR\cos\theta}}+\frac{q'}{\sqrt{b^2+R^2-2bR\cos\theta}}=0,$$

or

$$\frac{q}{a}\left(1+\frac{R^2}{a^2}-\frac{2R}{a}\cos\theta\right)^{-1/2}+\frac{q'}{R}\left(1+\frac{b^2}{R^2}-\frac{2b}{R}\cos\theta\right)^{-1/2}=0.$$

This equality becomes an identity if one chooses

$$ab = R^2 \quad \text{and} \quad q' = -\frac{R}{a}q, \tag{1.116}$$

which leads to the formula for the potential at an arbitrary point P outside the sphere:

$$V_P = \frac{1}{4\pi\epsilon_0} \left(\frac{1}{r} - \frac{R}{a}\frac{1}{r'} \right). \tag{1.117}$$

The relation (1.116) shows that q' is located at a point which is the inverse of the point A (where q is) with respect to the sphere.

The charge density on the sphere is then

$$\sigma = -\epsilon_0 \left(\frac{\partial V}{\partial \rho} \right)_{\rho=R} = \frac{q}{4\pi b^2} \frac{\frac{b^2}{R^2} - 1}{\left(\frac{R^2}{b^2} - 2\frac{R}{b}\cos\theta + 1 \right)^{3/2}}. \tag{1.118}$$

This result can be expressed in terms of a only. In view of (1.116), we find

$$\sigma = -\frac{q}{4\pi} \frac{a^2 - R^2}{R r_0^3}, \tag{1.119}$$

where

$$r_0 = r|_{\rho=R} = \sqrt{a^2 + R^2 - 2aR\cos\theta}.$$

The solution of this problem by the method of images led us also to a general result, valid for any two point charges of different signs: the locus of points of null potential for such a system is a sphere with the centre on the line which connects the two charges, but located not in between the charges.

1.3.2   Integration of the Laplace Equation by the Method of Separation of Variables

Using the Laplace equation, it is now our purpose to determine the electrostatic field potential produced by a uniformly charged body. Since many problems of electrostatics present a spherical or cylindrical geometry, we exemplify the method in spherical and cylindrical coordinates.

The method of separation of variables is employed very often in solving partial differential equations. The idea is to seek the solution by writing it as the product of functions, each of them depending only on one coordinate.

1.3.2.1   Laplace Equation in Spherical Coordinates

Let r, θ, φ be the spherical coordinates of the point where the potential is to be determined. Using (D.28), we write the Laplace equation as

$$\frac{1}{r^2} \left\{ \frac{\partial}{\partial r} \left(r^2 \frac{\partial V}{\partial r} \right) + \frac{1}{\sin\theta} \left[\frac{\partial}{\partial \theta} \left(\sin\theta \frac{\partial V}{\partial \theta} \right) + \frac{1}{\sin\theta} \frac{\partial^2 V}{\partial \varphi^2} \right] \right\} = 0,$$

or, in a more convenient form,

$$\frac{1}{r}\frac{\partial^2}{\partial r^2}(rV) + \frac{1}{r^2 \sin\theta}\frac{\partial}{\partial\theta}\left(\sin\theta\frac{\partial V}{\partial\theta}\right) + \frac{1}{r^2\sin^2\theta}\frac{\partial^2 V}{\partial\varphi^2} = 0. \qquad (1.120)$$

We seek a solution of this equation of the form

$$V(r, \theta, \varphi) = R(r)\Theta(\theta)\Phi(\varphi). \qquad (1.121)$$

Introducing (1.121) into (1.120) and using the substitution

$$R(r) = \frac{1}{r}U(r),$$

one obtains

$$\frac{1}{r}\frac{d^2 U}{dr^2}\Theta\Phi + \frac{U}{r^3}\left[\frac{1}{\sin\theta}\frac{d}{d\theta}\left(\sin\theta\frac{d\Theta}{d\theta}\right)\right]\Phi + \frac{U}{r^3}\frac{\Theta}{\sin^2\theta}\frac{d^2\Phi}{d\varphi^2} = 0,$$

or, by amplifying with $r^2\sin^2\theta/U\Phi\Theta$,

$$\frac{r^2\sin\theta}{U}\frac{d^2 U}{dr^2} + \frac{\sin\theta}{\Theta}\frac{d}{d\theta}\left(\sin\theta\frac{d\Theta}{d\theta}\right) + \frac{1}{\Phi}\frac{d^2\Phi}{d\varphi^2} = 0. \qquad (1.122)$$

On the one hand, by definition of Φ, the expression $\frac{1}{\Phi}\frac{d^2\Phi}{d\varphi^2}$ depends on φ only. On the other hand, according to Eq. (1.122), it depends only on r and θ. This is possible only if

$$\frac{r^2\sin\theta}{U}\frac{d^2 U}{dr^2} + \frac{\sin\theta}{\Theta}\frac{d}{d\theta}\left(\sin\theta\frac{d\Theta}{d\theta}\right) = -\frac{1}{\Phi}\frac{d^2\Phi}{d\varphi^2} = \gamma, \qquad (1.123)$$

where γ is a constant. The general solution of the equation

$$\Phi'' + \gamma\Phi = 0$$

is

$$\Phi = C_1 e^{\sqrt{-\gamma}\varphi} + C_2 e^{-\sqrt{-\gamma}\varphi}. \qquad (1.124)$$

In order to have Φ as a periodical function, of period 2π, $\sqrt{\gamma}$ must be a positive integer number, i.e. $\gamma = m^2$, with $m = 0, \pm 1, \pm 2, \ldots$

In electrostatic applications one usually considers only the first term of the solution (1.124), that is

$$\Phi = \text{const.} \times e^{im\varphi}, \qquad (1.125)$$

or some real combinations of $\cos m\varphi$ and $\sin m\varphi$.

Equation (1.123) also gives

$$\frac{r^2}{U}\frac{d^2U}{dr^2} + \frac{1}{\Theta \sin\theta}\frac{d}{d\theta}\left(\sin\theta\frac{d\Theta}{d\theta}\right) - \frac{m^2}{\sin^2\theta} = 0.$$

Repeating the procedure shown above and denoting the separation constant by $l(l+1)$, we arrive at the following two equations in r and θ:

$$\frac{d^2U}{dr^2} = \frac{l(l+1)}{r^2}U, \tag{1.126}$$

$$\frac{1}{\sin\theta}\frac{d}{d\theta}\left(\sin\theta\frac{d\Theta}{d\theta}\right) + l(l+1)\Theta - \frac{m^2}{\sin^2\theta}\Theta = 0. \tag{1.127}$$

If in the θ-equation we make the substitution $x = \cos\theta$, we obtain the generalized Legendre equation, whose solutions are the associated Legendre functions. The finite solutions exist only if l is a positive integer number, such that $-l \le m \le +l$. One can show that the *associated Legendre functions* satisfy the formula

$$P_l^m(x) = \frac{1}{2^l l!}\left(1 - x^2\right)^{\frac{m}{2}}\frac{d^{l+m}}{dx^{l+m}}\left(x^2 - 1\right)^l, \quad \text{where} \quad x = \cos\theta. \tag{1.128}$$

The solutions of (1.126) are sought in the form

$$U(r) = Ar^\alpha. \tag{1.129}$$

After performing the derivatives, one obtains

$$[\alpha(\alpha - 1) - l(l+1)]\, U(r) = 0. \tag{1.130}$$

This relation says that α can be either $l+1$, or l, leading to

$$U(r) = Ar^{l+1} + \frac{B}{r^l},$$

and, consequently,

$$R(r) = \frac{1}{r}U(r) = Ar^l + Br^{-(l+1)}. \tag{1.131}$$

Since the sum of two solutions of the Laplace equation is also a solution, we finally have

$$V(r, \theta, \varphi) = \sum_{l=0}^{\infty}\sum_{m=-l}^{+l}\left(A_{lm}r^l + B_{lm}r^{-(l+1)}\right) Y_{lm}(\theta, \varphi), \tag{1.132}$$

where we have denoted by $Y_{lm}(\theta, \varphi)$ the *spherical harmonics*

$$Y_{lm}(\theta, \varphi) = (-1)^m \sqrt{\frac{2l+1}{4\pi} \frac{(l-m)!}{(l+m)!}} P_l^m(\cos\theta)e^{im\varphi}, \qquad (1.133)$$

while A_{lm} and B_{lm} are coefficients of the expansion, i.e. constants which have to be determined from the boundary conditions.

Special cases:

(a) Inside a conductor the potential must be finite, meaning that for $r = 0$ we have to take $B_{lm} = 0$, and (1.132) yields

$$V(r, \theta, \varphi) = \sum_{l=0}^{\infty} \sum_{m=-l}^{+l} A_{lm} r^l Y_{lm}(\theta, \varphi). \qquad (1.134)$$

(b) Outside a conductor the potential must satisfy two conditions: be finite, and become zero at infinity. It means that for $r \to \infty$ we are obliged to take $A_{lm} = 0$, and (1.132) gives

$$V(r, \theta, \varphi) = \sum_{l=0}^{\infty} \sum_{m=-l}^{+l} B_{lm} r^{-(l+1)} Y_{lm}(\theta, \varphi). \qquad (1.135)$$

(c) If the potential is symmetric with respect to the z-axis (azymuthal symmetry), we have $m = 0$ and (1.132) leads to

$$V(r, \theta, \varphi) = \sum_{l=0}^{\infty} \left(A_{lm} r^l + B_{lm} r^{-l-1}\right) \sqrt{\frac{2l+1}{4\pi}} P_l^0(\cos\theta)$$

$$= \sum_{l=0}^{\infty} \left(C_l r^l + D_l r^{-l-1}\right) P_l^0(\cos\theta), \qquad (1.136)$$

where $P_l^0 = P_l$ are the Legendre polynomials:

$$P_0 = 1, \quad P_1 = \cos\theta, \quad P_2 = \frac{1}{2}\left(3\cos^2\theta - 1\right), \text{ etc.}$$

The Legendre polynomials form a complete orthogonal set, therefore they can be used for the series expansion of quantities with azymuthal symmetry.

As an application, let us expand in series the factor $\frac{1}{|r-r'|}$, appearing in the formula giving the potential of a point charge:

$$\frac{1}{|\mathbf{r} - \mathbf{r}'|} = \frac{1}{\left(r^2 + r'^2 - 2rr' \cos \theta\right)^{1/2}}$$

$$= \frac{1}{r} \left[1 + \frac{r'}{r} \cos \theta + \left(\frac{r'}{r}\right)^2 \frac{3 \cos^2 \theta - 1}{2} + \cdots \right],$$

which can also be written as

$$\frac{1}{|\mathbf{r} - \mathbf{r}'|} = \frac{1}{r} \sum_{l=0}^{\infty} \left(\frac{r'}{r}\right)^l P_l(\cos \theta) = \sum_{l=0}^{\infty} \frac{(r')^l}{r^{l+1}} P_l(\cos \theta). \qquad (1.137)$$

Comparing this result with the analysis presented in Sect. 1.1.11, we realize that the first term of (1.137) (multiplied by $\frac{q}{4\pi\epsilon_0}$) stands for the monopole potential, the second – for dipole potential, the third – for quadrupole potential, etc.

1.3.2.2   Laplace Equation in Cylindrical Coordinates

Using formula (D.34) from Appendix D, we write the Laplace equation in cylindrical coordinates as

$$\frac{1}{\rho} \frac{\partial}{\partial \rho} \left(\rho \frac{\partial V}{\partial \rho} \right) + \frac{1}{\rho^2} \frac{\partial^2 V}{\partial \varphi^2} + \frac{\partial^2 V}{\partial z^2} = 0. \qquad (1.138)$$

We separate the variables and seek the solution in the form

$$V(\rho, \varphi, z) = R(\rho)\Phi(\varphi)Z(z). \qquad (1.139)$$

From similar considerations as in the case of spherical coordinates, the following ordinary differential equations for the functions R, Φ, and Z are obtained:

$$\frac{d^2 Z}{dz^2} + C_1 Z = 0,$$

$$\frac{d^2 \Phi}{d\varphi^2} - C_2 \Phi = 0, \qquad (1.140)$$

$$\rho \frac{d}{d\rho} \left(\rho \frac{dR}{d\rho} \right) - \left(C_1 \rho^2 - C_2 \right) R = 0.$$

The single valuedness of Φ leads again to the requirement $C_2 = -m^2$, while C_1 is arbitrary. For convenience, we put $C_1 = -k^2$ and consider k a real and positive constant. As a result, Eqs. (1.140)$_{1,2}$ have solutions of the form $\sinh kz$, $\cosh kz$, and $\sin m\theta$, $\cos m\theta$, respectively, while (1.140)$_3$ becomes

$$\frac{d^2 R}{d\rho^2} + \frac{1}{\rho} \frac{dR}{d\rho} + \left(k^2 - \frac{m^2}{\rho^2} \right) R = 0, \qquad (1.141)$$

which is known as *Bessel's equation*. This equation can be written in a more convenient form by using the change of variable $x = k\rho$, as

$$x \frac{d}{dx}\left(x \frac{dR}{dx}\right) + \left(x^2 - m^2\right) R = 0. \tag{1.142}$$

The solution is sought for as the power series

$$R = \sum_{n=0}^{\infty} a_n x^{n+p}, \tag{1.143}$$

where the coefficients a_n and the index p are to be determined. Introducing (1.143)
into (1.142), we have

$$\sum_{n=0}^{\infty} \left[(n+p)^2 - m^2 + x^2\right] a_n x^{n+p} = 0.$$

This leads to the following recurrence relations for a_n:

$$\left(p^2 - m^2\right) a_0 = 0,$$
$$\left[(1+p)^2 - m^2\right] a_1 = 0,$$
$$\left[(2+p)^2 - m^2\right] a_2 + a_0 = 0, \tag{1.144}$$
$$\vdots$$
$$\left[(n+p)^2 - m^2\right] a_n + a_{n-2} = 0, \qquad (n \geq 2).$$

If we admit $a_0 \neq 0$ (which implies $a_1 = 0$), it follows that $p = \pm|m|$. Consider the
case $p \geq 0$. In general, we can write

$$a_{2n+1} = 0, \qquad\qquad n = 0, 1, 2, \ldots,$$
$$a_{2n} = -\frac{a_{2n-2}}{2^2 n(n + |m|)}, \qquad n = 1, 2, 3, \ldots.$$

Replacing n by $n - 1, n - 2, \ldots$ successively, and multiplying the resulting relations,
one obtains

$$a_{2n} = \frac{(-1)^n a_0}{2^{2n} n!(|m| + 1)(|m| + 2) \ldots (|m| + n)}, \qquad n = 1, 2, 3, \ldots. \tag{1.145}$$

For the choice

$$a_0 = 1/2^{|m|}|m|!,$$

we obtain a solution of Bessel's equation known as the *Bessel function of the first kind*:

$$J_{|m|}(x) = \sum_{n=0}^{\infty} \frac{(-1)^n x^{2n+|m|}}{2^{2n+|m|} n! (n+|m|)!}. \tag{1.146}$$

For $p = -|m|$, one obtains another solution $J_{-|m|}$ which, for m integer, is linearly dependent on $J_{|m|}$. Since (1.142) is a second-order differential equation, there exists one more linearly independent solution $N_p(\rho)$, called *Bessel's function of the second kind*. The general solution of Bessel's equation then reads

$$R(\rho) = A J_m(k\rho) + B N_m(k\rho). \tag{1.147}$$

The solution of the z-equation is of the form Ae^{kz}, where k can be either positive and negative.

Now we can write the solution of the Laplace equation as

$$V(\rho, \varphi, z) = \sum_{m=-\infty}^{+\infty} \int_{-\infty}^{+\infty} [A_m(k) J_m(k\rho) + B_m(k) N_m(k\rho)] \, e^{kz} e^{im\varphi} \, dk. \tag{1.148}$$

1.3.3 Two-Dimensional Electrostatic Problems and Conformal Mapping

Some problems of electrostatics can be studied in two dimensions. Such situations occur in the systems where charge distributions obey certain symmetry rules. If, for instance, the symmetry is cylindrical, the problem is studied in a plane orthogonal to the symmetry axis, and then the solution is extended to three dimensions. Such problems are called *two-dimensional (plane)* and are elegantly solved with the help of complex functions.

Consider the complex variable $z = x + iy$. If to each value of z is associated a complex number $f(z) = \varphi(x, y) + i\psi(x, y)$, where the functions φ and ψ are real, we call $f(z)$ a function of the complex variable z. Let us calculate

$$\lim_{\Delta z \to 0} \frac{f(z + \Delta z) - f(z)}{\Delta z} = \lim_{\substack{\Delta x \to 0 \\ \Delta y \to 0}} \left[\frac{\varphi(x + \Delta x, y + \Delta y) - \varphi(x, y)}{\Delta x + i\Delta y} \right.$$

$$\left. + i \frac{\psi(x + \Delta x, y + \Delta y) - \psi(x, y)}{\Delta x + i\Delta y} \right] \tag{1.149}$$

$$= \lim_{\substack{\Delta x \to 0 \\ \Delta y \to 0}} \left[\frac{\left(\frac{\partial \varphi}{\partial x} \Delta x + i \frac{\partial \psi}{\partial y} \Delta y \right) + i \left(\frac{\partial \psi}{\partial x} \Delta x - i \frac{\partial \varphi}{\partial y} \Delta y \right)}{\Delta x + i\Delta y} \right].$$

The limit (1.149) does not depend on the order in which the limit is taken, on the real or the imaginary direction, if $\varphi(x, y)$ and $\psi(x, y)$ satisfy the *Cauchy–Riemann conditions*

$$\frac{\partial \varphi}{\partial x} = \frac{\partial \psi}{\partial y} \quad \text{and} \quad \frac{\partial \varphi}{\partial y} = -\frac{\partial \psi}{\partial x}. \tag{1.150}$$

A holomorphic function is a complex function which is differentiable at every point of a domain. A function $f(z) = \varphi(x, y) + i\psi(x, y)$, whose real and imaginary parts are real-differentiable functions, is holomorphic if and only if the Cauchy–Riemann relations (1.150) are satisfied throughout the domain of definition. If $f(z)$ is holomorphic at every point of a vicinity of z, then $f(z)$ is *analytical* at z.

In view of the Cauchy–Riemann conditions (1.150), we have

$$\frac{df(z)}{dz} = \frac{\partial \varphi}{\partial x} + i\frac{\partial \psi}{\partial x} = \frac{\partial \psi}{\partial y} - i\frac{\partial \varphi}{\partial y} = \frac{\partial \varphi}{\partial x} - i\frac{\partial \varphi}{\partial y}, \tag{1.151}$$

as well as

$$\Delta\varphi = 0, \qquad \Delta\psi = 0, \tag{1.152}$$

$$\frac{\partial \varphi}{\partial x}\frac{\partial \psi}{\partial x} + \frac{\partial \varphi}{\partial y}\frac{\partial \psi}{\partial y} = (\nabla\varphi) \cdot (\nabla\psi) = 0, \tag{1.153}$$

meaning that φ and ψ are *harmonic functions*, while the families of curves $\varphi =$ const., $\psi =$ const. are *orthogonal*. The function $f(z)$ is also called a *complex potential*, being denoted by w.

We shall apply these general considerations to solving certain problems of electrostatics by the method of complex functions and conformal transformations.

Let us consider an electrostatic field $\mathbf{E}(x, y)$ defined in the (x, y) plane where the electric charge density is zero. Since in vacuum $\nabla \times \mathbf{E} = 0$ as well as $\nabla \cdot \mathbf{E} = 0$, we can express $\mathbf{E}$ both in terms of a scalar potential V as

$$\Delta V = 0,$$

and in terms of a vector potential $\mathbf{U}$ as

$$\mathbf{E} = \nabla \times \mathbf{U}.$$

We may take $\mathbf{U} = (0, 0, U)$ and, projecting on the coordinate axes the vector relation $\nabla \times \mathbf{U} = -\nabla V$, we obtain

$$\frac{\partial U}{\partial x} = \frac{\partial V}{\partial y}, \qquad \frac{\partial U}{\partial y} = -\frac{\partial V}{\partial x}, \tag{1.154}$$

meaning that $U(x, y)$ and $V(x, y)$ can be regarded as the real and imaginary parts, respectively, of the complex potential

$$w(z) = U(x, y) + iV(x, y), \qquad z = x + iy, \qquad (1.155)$$

satisfying the Cauchy–Riemann conditions (1.150). According to (1.13), the *lines of force* are defined by

$$\frac{dx}{E_x} = \frac{dy}{E_y},$$

which in this case is equivalent to $U = $ const. The family of curves $V = $ const. are, in turn, *equipotential lines*. According to (1.153), the two families of curves are orthogonal.

Remark also that, on the one hand,

$$E^2 = E_x^2 + E_y^2 = \left(\frac{\partial V}{\partial x}\right)^2 + \left(\frac{\partial V}{\partial y}\right)^2,$$

and on the other hand, according to (1.151) and (1.154),

$$\left|\frac{dw}{dz}\right|^2 = \left(\frac{dw}{dz}\right)\left(\frac{dw}{dz}\right)^* = \left(\frac{\partial U}{\partial x} + i\frac{\partial V}{\partial x}\right)\left(\frac{\partial U}{\partial x} - i\frac{\partial V}{\partial x}\right)$$
$$= \left(\frac{\partial U}{\partial x}\right)^2 + \left(\frac{\partial V}{\partial x}\right)^2 = \left(\frac{\partial V}{\partial x}\right)^2 + \left(\frac{\partial V}{\partial y}\right)^2.$$

Comparing the last two relations, we find

$$E = \left|\frac{dw}{dz}\right|. \qquad (1.156)$$

In other words, to know the complex potential associated with a certain electrostatic configuration means to determine the field of that configuration using (1.156). In its turn, the complex potential is determined by the form of the plane cross section of a conductor or, still, by the charge distribution that generates the field.

The relation (1.155) maps the plane (x, y) to the plane (U, V). One also observes that to the net of orthogonal curves $U(x, y) = $ const., $V(x, y) = $ const. in the (x, y)-plane corresponds a net of orthogonal straight lines $U = $ const., $V = $ const. in the (U, V)-plane. A map which preserves the angle between curves, when passing from the plane (x, y) to the plane (U, V), is called *conformal transformation*.

Since $dw = dU + idV = |w'(z)|(dx + idy)$, the relation between the arc elements of the two representations is

$$dL^2 = dU^2 + dV^2 = |w'(z)|^2 (dx^2 + dy^2) = |w'(z)|^2 dl^2.$$

To illustrate the advantages of the complex method, we shall consider further some simple examples.

Fig. 1.17 The equipotential lines and the lines of force corresponding to the choice (1.157) for the complex potential.

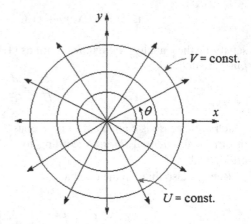

1.3.3.1 Linear, Uniformly Charged Conductor

Let the complex potential be

$$w(z) = Ai \ln z + B, \qquad (1.157)$$

where A is a real constant and B a complex number, $B = C_1 + iC_2$. Writing z in trigonometric form as

$$z = r\, e^{i\theta}$$

and applying (1.155), we have

$$U = -A\theta + C_1 , \qquad V = A \ln r + C_2, \qquad (1.158)$$

showing that the lines of force $U =$ const. ($\theta =$ const.) are radii outgoing from the coordinate origin, while the equipotential lines $V =$ const. ($r =$ const.) are concentric circles with the common centre at the origin of coordinates, as in Fig. 1.17.

Using (1.156), we find the modulus of the field:

$$E^2 = \left| \frac{dw}{dz} \right|^2 = \frac{A^2}{|z|^2} = \frac{A^2}{r^2}. \qquad (1.159)$$

The same result is obtained by expressing the field components in polar coordinates,

$$E_r = -\frac{\partial V}{\partial r} , \qquad E_\theta = -\frac{1}{r}\frac{\partial V}{\partial \theta},$$

then calculating the derivatives and finally using the relation $E^2 = E_r^2 + E_\theta^2$.

The shape of the equipotential and field lines shows that the charge configuration whose potential was analyzed corresponds to a linear conductor (the electrostatic potential is constant on the surface of a conductor), oriented along the z-axis, uniformly charged and long enough to have the same configuration of the lines $U = $ const., $V = $ const. at any cross section $z = $ const.

To determine the constant A one applies the divergence theorem to a cylindrical surface of radius r and length l, with the axis along the conducting wire:

$$\Phi = \int \mathbf{E} \cdot d\mathbf{S} = E_r(2\pi r l) = \frac{q}{\epsilon_0}.$$

Since $E_r = -A/r$, we obtain

$$A = -\frac{\lambda}{2\pi\epsilon_0},$$

where $\lambda = q/l$ is the linear charge density.

1.3.3.2 Two Orthogonal Conducting Plates

Consider the map

$$w(z) = kz^2 = k(x + iy)^2, \tag{1.160}$$

k being a constant. Applying the already known procedure, we find

$$U(x, y) = k\left(x^2 - y^2\right), \qquad V(x, y) = 2kxy. \tag{1.161}$$

The net of curves $x^2 - y^2 = $ const. and $xy = $ const. represent two families of equilateral hyperbolas, as in Fig. 1.18. The components of the electrostatic field are

Fig. 1.18 The equipotential lines and the lines of force corresponding to two orthogonal charged plates, whose complex potential is given by (1.160).

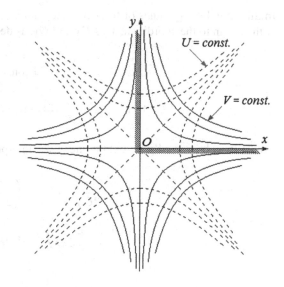

$$E_x = -\frac{\partial V}{\partial x} = -2ky, \quad E_y = -\frac{\partial V}{\partial y} = -2kx,$$

so that,

$$E^2 = E_x^2 + E_y^2 = 4k^2 \left(x^2 + y^2\right) = 4k^2 r^2. \tag{1.162}$$

The same result is obtained if one uses (1.156):

$$E^2 = \left(\frac{dw}{dz}\right)\left(\frac{dw}{dz}\right)^* = 4k^2|z|^2 = 4k^2 r^2.$$

This means that the magnitude of the field intensity is proportional to the magnitude of the radius vector of some point in the (x, y)-plane. Such a field can be found in the vicinity of a system of orthogonal plane conductors, and is used for focusing a flux of electric charges, as in an electron microscope, particle accelerators, etc.

1.3.3.3   Cylindrical Dipole

As a final example, let us take the complex potential of the form

$$w(z) = \frac{A^2}{z}, \tag{1.163}$$

where A is a real constant. One can write

$$U(x, y) = \frac{A^2 x}{x^2 + y^2}, \quad V(x, y) = -\frac{A^2 y}{x^2 + y^2}.$$

In this case, both the lines of force and the equipotential curves are circles, tangent at the origin to the coordinate axes Ox and Oy, as depicted in Fig. 1.19. Indeed,

$$U(x, y) = \frac{A^2 x}{x^2 + y^2} = \text{const.} \equiv C,$$

$$V(x, y) = -\frac{A^2 y}{x^2 + y^2} = \text{const.} \equiv D, \tag{1.164}$$

can be straightforwardly brought to the easier interpretable form

$$\left(x - \frac{A^2}{2C}\right)^2 + y^2 = \left(\frac{A^2}{2C}\right)^2,$$

$$x^2 + \left(y + \frac{A^2}{2D}\right)^2 = \left(\frac{A^2}{2D}\right)^2. \tag{1.165}$$

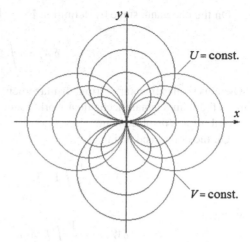

Fig. 1.19 The equipotential lines and the lines of force corresponding to a cylindrical dipole.

The system with such a configuration of the field and equipotential lines is an infinite cylindrical dipole, placed at the origin and oriented along the axis Oz.

The magnitude of the field intensity is calculated by means of (1.156). The result is

$$E = \frac{A^2}{r^2},$$

which is a Coulomb-type electric field.

1.4   Mechanical Action of the Electrostatic Field on Dielectric Media. Electrostriction

Consider a dielectric medium placed in an external electric field. The polarization phenomenon implies mechanical displacements within the molecules (atoms) of the dielectric. If these molecules are initially neutral, in the presence of the field will appear a continuously distributed polarization charge. Let ρ_e and ρ be the charge and mass densities, respectively. Assuming the dielectric to be linear and isotropic but not homogeneous, the permittivity will be a function of position and mass density, which is denoted here by $\rho(\mathbf{r})$:

$$\epsilon = \epsilon\,(\mathbf{r}, \rho(\mathbf{r}))\,. \tag{1.166}$$

The external field acts on the dielectric with a force $\mathbf{F}_e$, called *ponderomotive force*. Let $\mathbf{f}_e$ be its volume density. Our purpose is to calculate $\mathbf{f}_e$. To this end, we shall use two ways of writing the energy variation of the dielectric in the external field.

On the one hand, since by definition $\mathbf{F}_e = \int \mathbf{f}_e d\tau$, we have

$$\delta L = -\delta W_e = \int \mathbf{f}_e \cdot \delta\mathbf{s}\, d\tau, \tag{1.167}$$

where $\delta\mathbf{s}$ is an elementary virtual displacement of some particle of the dielectric, and δL is the infinitesimal mechanical work done by all the particles of the dielectric as a result of this displacement.

On the other hand, using (1.65), i.e.

$$W_e = \frac{1}{2}\int \mathbf{E}\cdot\mathbf{D}d\tau = \frac{1}{2}\int \epsilon E^2 d\tau,$$

we have

$$\delta W_e = -\frac{1}{2}\int E^2\delta\epsilon\, d\tau + \int \mathbf{E}\cdot\delta\mathbf{D}d\tau.$$

But

$$\int \mathbf{E}\cdot\delta\mathbf{D}d\tau = -\int \nabla V\cdot\delta\mathbf{D}d\tau = -\int \nabla\cdot(V\delta\mathbf{D})d\tau + \int V\nabla\cdot(\delta\mathbf{D})d\tau.$$

If we extend the integration domain over the whole space, apply the divergence theorem, and remember that at infinity the potential is zero, the first integral on the r.h.s. vanishes, while the second can be re-written using the Gauss law in the form (1.96), $\nabla\cdot(\delta\mathbf{D}) = \delta\rho_e$, where by ρ_e we denoted the charge density. Then the energy variation is

$$\delta W_e = -\frac{1}{2}\int E^2\delta\epsilon\, d\tau + \int V\,\delta\rho_e\, d\tau. \tag{1.168}$$

To compare (1.167) and (1.168), we must express the *local variations* $\delta\epsilon$ and $\delta\rho_e$ in terms of $\delta\mathbf{a}$. From the mechanics of continuous media it is known that the local variations are connected to the substantial variations $D\epsilon$ and $D\rho_e$ by the relations

$$\frac{D\epsilon}{dt} = \frac{\partial\epsilon}{\partial t} + (\mathbf{u}\cdot\nabla)\epsilon, \qquad \frac{D\rho_e}{dt} = \frac{\partial\rho_e}{\partial t} + (\mathbf{u}\cdot\nabla)\rho_e,$$

where $\mathbf{u} = \delta\mathbf{s}/\delta t$ is the velocity of the virtual displacement of the particle of fluid. We may write also

$$D\epsilon = \delta\epsilon + \delta\mathbf{s}\cdot\nabla\epsilon, \qquad D\rho_e = \delta\rho_e + \delta\mathbf{s}\cdot\nabla\rho_e. \tag{1.169}$$

If we assume that the observer moves together with the fluid particle, the permittivity ϵ will be a function of only the mass density ρ, such that

$$D\epsilon = \frac{\partial\epsilon}{\partial\rho}D\rho,$$

or, by means of (1.169) and the equation of continuity $\frac{\partial \rho}{\partial t} = -\nabla \cdot \left(\rho \frac{\partial s}{\partial t} \right)$,

$$\delta\epsilon + \delta\mathbf{s} \cdot \nabla\epsilon = \frac{\partial\epsilon}{\partial\rho}(\delta\rho + \delta\mathbf{s} \cdot \nabla\rho) = -\rho\frac{\partial\epsilon}{\partial\rho}\nabla \cdot (\delta\mathbf{s}),$$

that is

$$\delta\epsilon = -\delta\mathbf{s} \cdot \nabla\epsilon - \rho\frac{\partial\epsilon}{\partial\rho}\nabla \cdot (\delta\mathbf{s}). \tag{1.170}$$

Using a similar procedure, the equation of continuity for the electric charge leads to

$$\delta\rho_e = -\nabla \cdot (\rho_e\delta\mathbf{s}). \tag{1.171}$$

Plugging (1.170) and (1.171) into relation (1.168), we find

$$\delta W_e = \frac{1}{2} \int E^2 \nabla\epsilon \cdot \delta\mathbf{s}\, d\tau + \frac{1}{2} \int E^2 \rho \frac{\partial\epsilon}{\partial\rho} \nabla \cdot (\delta\mathbf{s}) d\tau - \int V \nabla \cdot (\rho_e\delta\mathbf{s}) d\tau.$$

But

$$\int E^2 \rho \frac{\partial\epsilon}{\partial\rho} \nabla \cdot (\delta\mathbf{s}) d\tau = \int \nabla \cdot \left(E^2 \rho \frac{\partial\epsilon}{\partial\rho} \delta\mathbf{s} \right) d\tau - \int \nabla \left(E^2 \rho \frac{\partial\epsilon}{\partial\rho} \right) \cdot \delta\mathbf{s}\, d\tau,$$

$$\int V \nabla \cdot (\rho_e\delta\mathbf{s}) d\tau = \int \nabla \cdot (V\rho_e\delta\mathbf{s}) d\tau - \int \rho_e\delta\mathbf{s} \cdot \nabla V d\tau,$$

and thus, using the divergence theorem and extending to infinity the integration domain, we obtain

$$\delta W_e = -\int \left[\rho_e\mathbf{E} - \frac{1}{2}E^2 \nabla\epsilon + \frac{1}{2}\nabla \left(\rho \frac{\partial\epsilon}{\partial\rho} E^2 \right) \right] \cdot \delta\mathbf{a}\, d\tau. \tag{1.172}$$

Comparing now (1.167) and (1.172), the *ponderomotive force density* is

$$\mathbf{f}_e = \rho_e\mathbf{E} - \frac{1}{2}E^2 \nabla\epsilon + \frac{1}{2}\nabla \left(\rho \frac{\partial\epsilon}{\partial\rho} E^2 \right). \tag{1.173}$$

If in our discussion we had considered the existence of other forces, for example of gravitational or hydrostatic nature, then we should have added extra terms on the r.h.s. of (1.173).

   Relation (1.173) leads us to an interesting interpretation. In fluid mechanics, the volume density forces $\mathbf{f}$ can be represented in terms of a system of tensions (stresses) applied on the boundary surface of the fluid. These tensions form a second-order tensor T_{ik}, called *stress tensor*. The forces and stresses are related by equations of motion called *Cauchy's equations*, which read

$$\rho a_i = \rho F_i + \frac{\partial T_{ik}}{\partial x_k}, \tag{1.174}$$

where a_i is the component of the acceleration of the particle of fluid and $\mathbf{F}$ is the *specific mass force*, $\rho \mathbf{F} = \mathbf{f}$. At equilibrium ($\mathbf{a} = 0$), Eq. (1.174) yields

$$f_i = -\frac{\partial T_{ik}}{\partial x_k}. \tag{1.175}$$

In the framework of this formalism, let us project (1.173) on the x_i-axis, using also Gauss's formula, $\nabla \cdot \mathbf{D} = \rho_e$:

$$(\mathbf{f})_i = E_i \frac{\partial D_k}{\partial x_k} - \frac{1}{2} E^2 \frac{\partial \epsilon}{\partial x_i} + \frac{1}{2} \frac{\partial}{\partial x_i} \left(\rho \frac{\partial \epsilon}{\partial \rho} E^2 \right).$$

Using the fact that $\mathbf{E}$ is conservative, while the dielectric is linear, non-homogeneous, and isotropic, after some re-arrangements of terms we obtain

$$(\mathbf{f}_e)_i = -\frac{\partial T_{ik}^{(e)}}{\partial x_k},$$

where the stress tensor has the form

$$T_{ik}^{(e)} = -\epsilon E_i E_k + \frac{1}{2} E^2 \left(\epsilon - \rho \frac{\partial \epsilon}{\partial \rho} \right) \delta_{ik}. \tag{1.176}$$

The quantities $T_{ik}^{(e)}$ are the components of a symmetric, second-order tensor, called by Maxwell *electric stress tensor*.

Anticipating, we mention that in the case of the magnetostatic field one obtains Maxwell's *magnetic stress tensor*

$$T_{ik}^{(m)} = -\mu H_i H_k + \frac{1}{2} H^2 \left(\mu - \rho \frac{\partial \mu}{\partial \rho} \right) \delta_{ik}, \tag{1.177}$$

where $\mathbf{H}$ is the magnetic field intensity and μ is the magnetic permeability of the medium.

Application

Consider a dielectric fluid, characterized by $\rho_e = 0$ and $\epsilon = \epsilon(\rho)$. In the case of an isotropic dielectric, the permittivity ϵ is related to the polarizability α by the *Clausius–Mossotti formula*:

$$\frac{\epsilon - \epsilon_0}{\epsilon + 2\epsilon_0} = \frac{C}{3} \rho, \quad C = \frac{\alpha}{m}, \tag{1.178}$$

where α is the molecular polarizability and m is the mass of a molecule. The formula is thus named in honour of Ottaviano-Fabrizio Mossotti (1791–1863) who studied the relation between the dielectric constants of two different media, and of Rudolf Clausius (1822–1888), who gave the formula in 1879. It is straightforward to show that the expression $\rho \frac{\partial \epsilon}{\partial \rho}$ which appears in (1.176) can then be written as

$$\rho \frac{\partial \epsilon}{\partial \rho} = \frac{1}{3\epsilon_0}(\epsilon - \epsilon_0)(\epsilon + 2\epsilon_0). \tag{1.179}$$

We also assume that the dielectric liquid is at equilibrium, under the action of electrostatic, hydrostatic, and gravitational forces:

$$\mathbf{f}_e + \mathbf{f}_{hydro} + \mathbf{f}_{grav} = 0. \tag{1.180}$$

Since ϵ depends only on ρ, we have $\nabla \epsilon = \frac{\partial \epsilon}{\partial \rho} \nabla \rho$, such that

$$\mathbf{f}_e = -\frac{1}{2}E^2 \nabla \epsilon + \frac{1}{2}\nabla \left(\rho \frac{\partial \epsilon}{\partial \rho} E^2 \right) = \frac{1}{2}\rho \nabla \left(E^2 \frac{\partial \epsilon}{\partial \rho} \right).$$

We also have

$$\mathbf{f}_{hydro} = -\nabla p,$$
$$\mathbf{f}_{grav} = \rho g \nabla h,$$

where p is the pressure inside the fluid, g is the gravitational acceleration and h is the height inside the fluid. The equilibrium equation (1.180) then becomes

$$\frac{1}{2}\rho \nabla \left(E^2 \frac{\partial \epsilon}{\partial \rho} \right) - \nabla p + \rho g \nabla h = 0. \tag{1.181}$$

In Fig. 1.20 is shown a liquid in which is partially immersed a plane condenser connected to a static potential difference. Denote by p the pressure at the level where the electric field is $\mathbf{E}$, and by p_0 the pressure at the level where the field is null. Neglecting the action of the gravitational field, we have

$$p = p_0 + \frac{1}{2}\rho \left(E^2 \frac{\partial \epsilon}{\partial \rho} \right),$$

and finally, making use of (1.179),

$$p = p_0 + \frac{1}{6\epsilon_0}E^2(\epsilon - \epsilon_0)(\epsilon + 2\epsilon_0). \tag{1.182}$$

Consequently, at the free surface of the dielectric liquid appears an excess of "hydrostatic" pressure which diminishes its volume, as shown in Fig. 1.20. The effect

Fig. 1.20 Schematic
representation of the
electrostriction phenomenon.

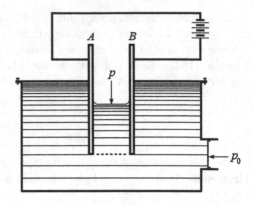

of variation (decrease) of the volume of a dielectric under the action of an electrostatic
field is called *electrostriction*. This phenomenon must not be confounded with the
inverse piezoelectricity, which appears only in the case of anisotropic media.

Electrostriction has many applications, such as the ultrafine adjustment of certain
dielectric layers known as *optical plates*: half-wave plates and quarter-wave plates.

The *magnetostriction* phenomenon, meaning the volume variation of a medium
under the action of a magnetic field, is approached in a similar way.

1.5   Solved Problems

Problem 1. Determine the shape of the equipotential surfaces of the electric field
produced by a uniformly distributed charge $\lambda = $ const. along a straight wire of length
$2c$.

Solution. Let us consider the coordinate system as in Fig. 1.21. The potential at
an arbitrary fix point P in the (x, y)-plane is

$$V_P = k_e \int_{-c}^{+c} \frac{dq}{r} = k_e \lambda \int_{-c}^{+c} \frac{d\xi}{r}, \qquad k_e = \frac{1}{4\pi\epsilon_0}, \tag{1.183}$$

where $d\xi$ is a length element of the wire. The relation

$$r = \left[(x - \xi)^2 + y^2\right]^{1/2}$$

suggests the substitution

$$u = x - \xi + r,$$

which gives

$$\frac{dr}{d\xi} = -\frac{x - \xi}{r} = 1 - \frac{u}{r}. \tag{1.184}$$

Fig. 1.21 A uniformly
distributed charge q along a
straight wire of length $2c$.

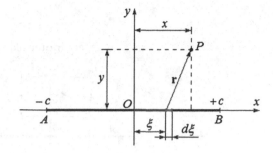

But $dr = du + d\xi$, such that

$$\frac{d\xi}{r} = -\frac{du}{u}.$$ (1.185)

The potential at the point P then becomes

$$V_P = -k_e\lambda \int_{u_1}^{u_2} \frac{du}{u} = k_e\lambda \ln \frac{x+c+r_1}{x-c+r_2}.$$ (1.186)

To find the equipotential surfaces we equate the argument of the logarithm in (1.186) to a constant K. Denoting $x + c = x_1$ and $x - c = x_2$, we have

$$x_1 + r_1 = K(x_2 + r_2),$$ (1.187)

where

$$r_1^2 = x_1^2 + y^2 \quad \text{and} \quad r_2^2 = x_2^2 + y^2,$$

and, in view of (1.187), it follows that

$$r_2 - x_2 = K(r_1 - x_1).$$ (1.188)

From (1.187) and (1.188) we find

$$r_1 + r_2 = 2c\frac{K+1}{K-1} \equiv 2a = \text{const.}$$ (1.189)

Therefore, the locus of the points which satisfy the condition $V_P = \text{const.}$ (K fixed) is an ellipse with foci at the ends of the wire. Rotating the figure around the x-axis and giving to K different values, we obtain *homofocal ellipsoids of revolution* with foci at A and B. The field lines are orthogonal to the equipotential surface at any point of the field (see Fig. 1.22).

It is easy to show that $K = \frac{a+c}{a-c}$, and the final result is

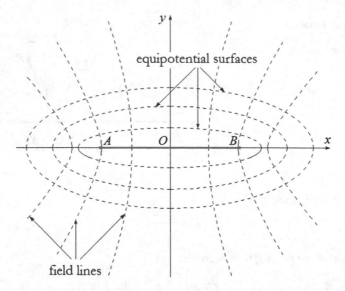

Fig. 1.22 Plane projection of the field lines and equipotential surfaces for a uniformly distributed charge q along a straight wire of length $2c$.

$$V_P = \frac{1}{4\pi\epsilon_0}\lambda\ln\frac{a+c}{a-c}. \tag{1.190}$$

In particular, if the points A and B approach each other until they coincide, the ellipsoids degenerate into concentric spheres, while if they move away to infinity the equipotential surfaces are coaxial cylinders with the wire as common axis.

Problem 2. Show that the quadrupole moment tensor of a homogeneous charge distribution with axial symmetry has one single distinct component, and calculate the potential of the system.

Solution. The geometry of the problem suggests to use cylindrical coordinates ρ, φ, z. Taking Oz as symmetry axis, and observing that the charge density ρ^e is independent of φ, $\rho^e = \rho^e(\rho, z)$, we have by definition (see (1.83)):

$$p_{ik} = \int \rho^e(\rho, z)\left(3x_i x_k - r^2\delta_{ik}\right)d\tau. \tag{1.191}$$

The tensor p_{ik} is symmetric, in general being defined by six components. Its non-diagonal components are

$$p_{12} = p_{xy} = 3\int \rho^e(\rho, z)\rho^3 d\rho dz \int_0^{2\pi}\sin\varphi\cos\varphi d\varphi = 0,$$

$$p_{23} = p_{yz} = 3\int \rho^e(\rho, z)\rho^2 z d\rho dz \int_0^{2\pi}\sin\varphi d\varphi = 0,$$

$$p_{31} = p_{zx} = 3 \int \rho^e(\rho, z)\rho^2 z d\rho dz \int_0^{2\pi} \cos\varphi d\varphi = 0.$$

Since $x = \rho\cos\varphi$, $y = \rho\sin\varphi$, and $r^2 = \rho^2 + z^2$, the three diagonal components of p_{ik} are

$$p_{33} = p_{zz} = 2\pi \int \rho^e(\rho, z)\left(2z^2 - \rho^2\right)\rho d\rho dz,$$

$$p_{11} = p_{xx} = \int \rho^e(\rho, z)\left[3x^2 - (\rho^2 + z^2)\right]\rho d\rho d\varphi dz$$

$$= -\pi \int \rho^e(\rho, z)\left(2z^2 - \rho^2\right)\rho d\rho dz = -\frac{p_{33}}{2},$$

$$p_{22} = p_{yy} = \int \rho^e(\rho, z)\left[3y^2 - (\rho^2 + z^2)\right]\rho d\rho d\varphi dz = p_{xx} = -\frac{p_{33}}{2}.$$

Thus, only p_{33} is independent, the other two nonvanishing components being expressible in terms of p_{33}. Obviously, we have

$$Tr(p_{ik}) = \sum_{i=1}^{3} p_{ii} = 0.$$

The quadrupole potential is (see Sect. 1.1.11):

$$V^{(2)} = \frac{k_e}{6}p_{ik}\left(\frac{3x_i x_k}{r^5} - \frac{\delta_{ik}}{r^3}\right), \quad \text{where} \quad k_e = \frac{1}{4\pi\epsilon_0}. \tag{1.192}$$

Introducing the components of the quadrupole moment tensor and denoting $p_{33} = p$, we have

$$V^{(2)} = k_e\left[p_{11}\left(\frac{3x^2}{r^5} - \frac{1}{r^3}\right) + p_{22}\left(\frac{3y^2}{r^5} - \frac{1}{r^3}\right) + p_{33}\left(\frac{3z^2}{r^5} - \frac{1}{r^3}\right)\right]$$

$$= k_e\frac{p}{6r^3}\left[\frac{3z^2}{r^2} - \frac{3}{2}\left(\frac{x^2 + y^2}{r^2}\right)\right].$$

Since $x^2 + y^2 = r^2 - z^2$ and $z = r\cos\theta$, we finally obtain

$$V^{(2)} = \frac{1}{8\pi\epsilon_0}\frac{p}{r^3}P_2(\cos\theta), \tag{1.193}$$

where

$$P_2(\cos\theta) = \frac{3\cos^2\theta - 1}{2}$$

is the Legendre polynomial of second degree.

Problem 3. Determine the potential V and the electrostatic field intensity $\mathbf{E}(r)$ in two points, one internal and the other external to a sphere of radius R, uniformly electrized with charge Q.

Solution. The potential V results as a solution of Poisson's equation

$$\Delta V = \begin{cases} -\frac{\rho}{\epsilon_0}, & \text{inside the sphere,} \\ 0, & \text{outside the sphere,} \end{cases} \tag{1.194}$$

where $\rho = \frac{3Q}{4\pi R^3}$ is the charge density inside the sphere.

Since the formulation of the problem suggests a spherical geometry, we shall use spherical coordinates to express the Laplacian of the potential:

$$\Delta V = \frac{1}{r^2}\left\{\frac{\partial}{\partial r}\left(r^2\frac{\partial V}{\partial r}\right) + \frac{1}{\sin\theta}\left[\frac{\partial}{\partial\theta}\left(\sin\theta\frac{\partial V}{\partial\theta}\right) + \frac{1}{\sin\theta}\frac{\partial^2 V}{\partial\varphi^2}\right]\right\}. \tag{1.195}$$

Because of the radial symmetry, the potential V depends only on r and (1.195) reduces to

$$\Delta V = \frac{1}{r^2}\left[\frac{\partial}{\partial r}\left(r^2\frac{\partial V}{\partial r}\right)\right]. \tag{1.196}$$

Case I - The potential and the field intensity at an arbitrary point inside the sphere.

Denoting these quantities by $V_i(r)$ and $E_i(r)$, and using (1.194) and (1.196), we find

$$\frac{1}{r^2}\left[\frac{\partial}{\partial r}\left(r^2\frac{\partial V_i}{\partial r}\right)\right] = -\frac{\rho}{\epsilon_0}. \tag{1.197}$$

The first integration of (1.197) gives

$$r^2\frac{\partial V_i}{\partial r} = -\frac{\rho}{\epsilon_0}\frac{r^3}{3} + C_1, \tag{1.198}$$

where C_1 is a constant which is to be determined. The second integration of (1.197) then leads to

$$V_i(r) = -\frac{\rho r^2}{6\epsilon_0} - \frac{C_1}{r} + C_2. \tag{1.199}$$

The field intensity $\mathbf{E}_i = -\nabla V_i$ results immediately from (1.198):

$$E_i(r) = -\left(\frac{\partial V}{\partial r}\right)_i = \frac{\rho r}{3\epsilon_0} - \frac{C_1}{r^2}. \tag{1.200}$$

Case II - The potential and the field intensity at an arbitrary point outside the sphere.

Let $V_e(r)$ and $E_e(r)$ be the quantities we are looking for. At an arbitrary point outside the sphere the potential V_e satisfies Laplace's equation $\Delta V_e = 0$, i.e.

$$\frac{1}{r^2}\left[\frac{\partial}{\partial r}\left(r^2\frac{\partial V_e}{\partial r}\right)\right] = 0. \tag{1.201}$$

Integrating, one obtains the field

$$r^2\frac{\partial V_e}{\partial r} = C_3 \Rightarrow \frac{\partial V_e}{\partial r} = \frac{C_3}{r^2} = -E_e(r), \tag{1.202}$$

while a second integration yields

$$V_e(r) = -\frac{C_3}{r} + C_4. \tag{1.203}$$

The constants of integration $C_1, \ldots, C_4$ are determined by imposing the boundary conditions

$$V_e(r \to \infty) = 0, \qquad V_i(r \to 0) = \text{finite},$$

as well as the continuity conditions

$$V_i(R) = V_e(R), \qquad E_i(R) = E_e(R).$$

Using these conditions, we find

$$C_1 = 0, \quad C_4 = 0, \quad C_2 = \frac{\rho R^2}{2\epsilon_0}, \quad C_3 = -\frac{\rho R^3}{3\epsilon_0}.$$

Now we are able to write the final form of both the potential and the field, inside and outside the sphere:

$$V_i(r) \equiv V(r)|_{r \leq R} = \frac{\rho R^2}{2\epsilon_0}\left(1 - \frac{r^2}{3R^2}\right) = \frac{3Q}{8\pi\epsilon_0 R}\left(1 - \frac{r^2}{3R^2}\right),$$

$$E_i(r) \equiv E(r)|_{r \leq R} = \frac{\rho r}{3\epsilon_0} = \frac{Q}{4\pi\epsilon_0}\frac{r}{R^3}, \tag{1.204}$$

and

$$V_e(r) \equiv V(r)|_{r > R} = \frac{\rho R^3}{2\epsilon_0}\frac{1}{r} = \frac{Q}{4\pi\epsilon_0 r},$$

$$E_e(r) \equiv E(r)|_{r > R} = \frac{\rho R^3}{3\epsilon_0 r^2} = \frac{Q}{4\pi\epsilon_0 r^2}. \tag{1.205}$$

The graphical representations of the potential and the field, for both situations, are given in Figs. 1.23 and 1.24.

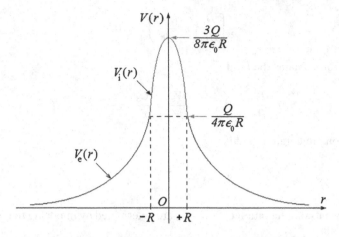

Fig. 1.23 Graphical representation of the potential V as a function of the distance r (Problem 3).

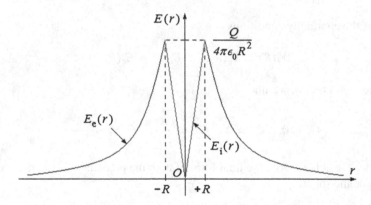

Fig. 1.24 Graphical representation of the field E as a function of the distance r (Problem 3).

Observation:

In agreement with Gauss's law, the potential generated by the uniformly charged sphere at an arbitrary point outside the sphere is the same as the potential of a point charge Q situated at the centre of the sphere.

Problem 4. Determine the field intensity at some point on the axis of a uniformly charged disk. The radius of the disk is R and its superficial charge density is $\sigma = \text{const}$.

Solution. A conveniently chosen coordinate system (see Fig. 1.25) allows us to write the potential at some point of the z-axis:

Fig. 1.25 A uniformly charged disk (Problem 4).

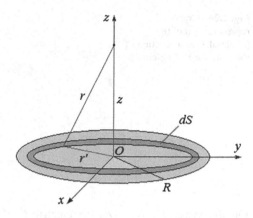

$$V(0, 0, z > 0) = \frac{1}{4\pi\epsilon_0} \int_S \frac{dq}{r} = \frac{1}{4\pi\epsilon_0} \int_S \frac{\sigma \, dS}{r} = \frac{1}{4\pi\epsilon_0} \int_0^R \frac{\sigma 2\pi r' dr'}{r} \qquad (1.206)$$

$$= \frac{\sigma}{2\epsilon_0} \int_0^R \frac{r' dr'}{\sqrt{r'^2 + z^2}} = \frac{\sigma}{2\epsilon_0} \sqrt{r'^2 + z^2} \Big|_0^R = \frac{\sigma}{2\epsilon_0} \left(\sqrt{R^2 + z^2} - z \right).$$

For $z \gg R$, the series expansion

$$\sqrt{R^2 + z^2} - z = z \left[\sqrt{1 + \left(\frac{R}{z}\right)^2} - 1 \right] \simeq z \left(1 + \frac{1}{2} \frac{R^2}{z^2} - 1 \right) = \frac{R^2}{2z}$$

leads to

$$V(z \gg R) = \frac{\sigma}{2\epsilon_0} \frac{R^2}{2z} = \frac{q}{4\pi\epsilon_0 z}. \qquad (1.207)$$

This result was expected, because for $z \gg R$ the disk appears point-like.

By symmetry reasons, the potential must have the same values for $z < 0$:

$$V(0, 0, z < 0) = \frac{\sigma}{2\epsilon_0} \left(\sqrt{R^2 + z^2} + z \right). \qquad (1.208)$$

The potential V has a singular behaviour at the point $z = 0$ (see Fig. 1.26). For $z < 0$ the function $V(z)$ is ascending and the slope of the graph is positive, while for $z > 0$, $V(z)$ is descending and the slope is negative. In other words, the potential $V(z)$ presents a sudden change in the sign of the slope at the origin of the coordinate axes. Since $\mathbf{E} = -\nabla V$, it results that at the point $z = 0$ the field intensity also presents a discontinuity. Let us find this discontinuity.

For $z = 0$ we have

$$V(z)\big|_{z=0} = \frac{\sigma R}{2\epsilon_0}. \qquad (1.209)$$

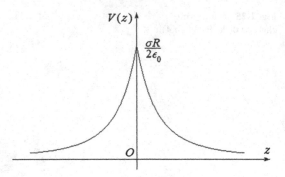

Fig. 1.26 Graphical representation of the potential V as a function of the distance z (Problem 4).

On the symmetry axis Oz, the electric field $\mathbf{E}$ has only the z-component, $\mathbf{E} = (0, 0, E_z)$:

$$E_{z+} = E_z(z > 0) = -\frac{dV(z > 0)}{dz} = \frac{\sigma}{2\epsilon_0}\left(1 - \frac{z}{\sqrt{z^2 + R^2}}\right), \qquad (1.210)$$

and

$$E_{z-} = E_z(z < 0) = -\frac{dV(z < 0)}{dz} = -\frac{\sigma}{2\epsilon_0}\left(1 + \frac{z}{\sqrt{z^2 + R^2}}\right). \qquad (1.211)$$

Calculating the limit of (1.210) and (1.211) at the point $z = 0$ we arrive at

$$\lim_{\substack{z \to 0 \\ z > 0}} E_z = \lim_{z \to 0} E_{z+} = \frac{\sigma}{2\epsilon_0}, \qquad (1.212)$$

and

$$\lim_{\substack{z \to 0 \\ z < 0}} E_z = \lim_{z \to 0} E_{z+} = -\frac{\sigma}{2\epsilon_0}. \qquad (1.213)$$

Relations (1.212) and (1.213) show that the values of the electric field created by a charged disk at the point $z = 0$ are identical to those of a uniformly charged ($\sigma = $ const.) infinite plane, situated in vacuum. This result is as expected, since for $z \to 0$ the disk appears as an infinite plane. In addition,

$$(E_{z+} - E_{z-})\big|_{z=0} = \frac{\sigma}{2\epsilon_0} - \left(-\frac{\sigma}{2\epsilon_0}\right) = \frac{\sigma}{\epsilon_0}, \qquad (1.214)$$

meaning that at the point $z = 0$, while passing from one side of the disk to the other, the electric field jumps, and the value of the jump is $\frac{\sigma}{\epsilon_0}$ (Fig. 1.27).

Fig. 1.27 Jump of the
electric field while passing
from one side of the disk to
the other (Problem 4).

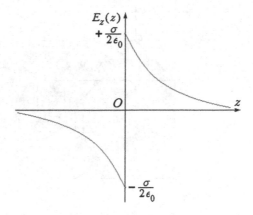

Problem 5. A point charge q is placed at point $P_0(x_0, y_0, z_0)$ inside a grounded
metallic parallelipipedic box ($0 \le x \le a, 0 \le y \le b, 0 \le z \le c$). Determine the electrostatic potential inside the box.

Solution. To avoid expressing the charge density of a single point charge and work
with it from the beginning – which is a difficult task – we start by assuming that the
charge q is uniformly distributed in an elementary parallelepiped whose walls are
parallel to the walls of the box (see Fig. 1.28). Then, in due course, we shall make
this volume tend to zero. Under this assumption, the charge density is

$$\rho(x, y, z) = \frac{q}{8hkl}, \tag{1.215}$$

for $x_0 - h < x < x_0 + h$, $y_0 - k < y < y_0 + k$, $z_0 - l < z < z_0 + l$ and 0 for the
rest of the values of x, y, z.

The potential $V(x, y, z)$ inside the box is obtained by solving the Poisson equation

$$\frac{\partial^2 V}{\partial x^2} + \frac{\partial^2 V}{\partial y^2} + \frac{\partial^2 V}{\partial z^2} = -\frac{1}{\epsilon_0}\rho(x, y, z), \tag{1.216}$$

with the boundary conditions

$$\begin{cases} V(0, y, z) = V(a, y, z) = 0, & \text{for any } y, z \text{ inside the box;} \\ V(x, 0, z) = V(x, b, z) = 0, & \text{for any } x, z \text{ inside the box;} \\ V(z, y, 0) = V(x, y, c) = 0, & \text{for any } x, y \text{ inside the box.} \end{cases} \tag{1.217}$$

The homogeneous equation corresponding to Poisson's equation (1.216) is the
Laplace equation

$$\frac{\partial^2 V}{\partial x^2} + \frac{\partial^2 V}{\partial y^2} + \frac{\partial^2 V}{\partial z^2} = 0, \tag{1.218}$$

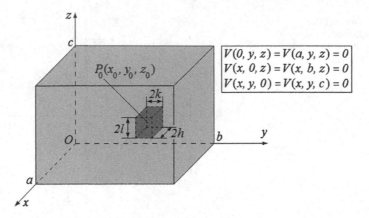

Fig. 1.28 Schematic representation of the parallelepipedic box from Problem 5.

whose solution can be obtained by the Fourier's method (the method of separation of variables). Due to the rectangular symmetry of the problem, the solution of the Eq. (1.218) must be sought of the form

$$V(x, y, z) = X(x)Y(y)Z(z),$$ (1.219)

where X, Y, and Z are functions which will be determined using the boundary conditions. Using the standard procedure, we introduce (1.219) into (1.218) and the resulting equation is divided by $X(x)Y(y)Z(z)$, obtaining

$$\frac{1}{X(x)}\frac{d^2X(x)}{dx^2} + \frac{1}{Y(y)}\frac{d^2Y(y)}{dy^2} + \frac{1}{Z(z)}\frac{d^2Z(z)}{dz^2} = 0.$$ (1.220)

This relation can be satisfied only if each of the three terms in the right-hand side is a constant, such that the algebraic sum of the three constants vanishes:

$$\frac{1}{X(x)}\frac{d^2X(x)}{dx^2} = \chi^2,$$

$$\frac{1}{Y(y)}\frac{d^2Y(y)}{dy^2} = -\lambda^2,$$ (1.221)

$$\frac{1}{Z(z)}\frac{d^2Z(z)}{dz^2} = -\mu^2,$$

with

$$\chi^2 = \lambda^2 + \mu^2.$$ (1.222)

The solutions of the last two equations of (1.221) have the general form

$$Y(y) = A_1 \cos(\lambda y) + B_1 \sin(\lambda y),$$
$$Z(z) = A_2 \cos(\mu z) + B_2 \sin(\mu z),$$

where A_i and B_i, $i = 1, 2$ are arbitrary integration constants, which will be determined using the boundary conditions. Thus, the conditions

$$V(x, 0, z) \left[= X(x)Y(0)Z(z)\right] = 0, \quad \forall x \in [0, a], \ z \in [0, c],$$
$$V(x, y, 0) \left[= X(x)Y(y)Z(0)\right] = 0, \quad \forall x \in [0, a], \ y \in [0, b]$$

lead to $A_i = 0$, $i = 1, 2$, while from the conditions

$$V(x, b, z) \left[= X(x)Y(b)Z(z)\right] = 0, \quad \forall x \in [0, a], \ z \in [0, c],$$
$$V(x, y, c) \left[= X(x)Y(y)Z(c)\right] = 0, \quad \forall x \in [0, a], \ y \in [0, b]$$

it follows that λ and μ have to be of the form

$$\lambda \equiv \lambda_m = \frac{m\pi}{b}, \quad m \in \mathbb{Z},$$
$$\mu \equiv \mu_n = \frac{n\pi}{c}, \quad n \in \mathbb{Z}.$$

We have thus found that the potential $V(x, y, z)$, which is the solution of the Poisson equation (1.216), has to be written as a double series, as follows:

$$V(x, y, z) = \sum_{m=1}^{\infty} \sum_{n=1}^{\infty} u_{mn}(x) \sin\left(\frac{m\pi}{b}y\right) \sin\left(\frac{n\pi}{c}z\right), \qquad (1.223)$$

where $u_{mn}(x)$, $m, n \in \mathbb{Z}$ are unknown functions of x, which have to be determined. Introducing (1.223) into (1.216), we obtain the differential equation for $u_{mn}(x)$:

$$\sum_{m=1}^{\infty} \sum_{n=1}^{\infty} \left[\frac{d^2 u_{mn}(x)}{dx^2} - \alpha_{mn}^2 u_{mn}(x) \right] \sin\left(\frac{m\pi}{b}y\right) \sin\left(\frac{n\pi}{c}z\right) = -\frac{1}{\epsilon_0} \rho(x, y, z),$$

$$(1.224)$$

where

$$\alpha_{mn} = \pi \sqrt{\frac{m^2}{b^2} + \frac{n^2}{c^2}}. \qquad (1.225)$$

Using the orthogonality conditions

$$\int_0^b \sin\left(\frac{n\pi}{b}y\right) \sin\left(\frac{m\pi}{b}y\right) dy = \frac{b}{2}\delta_{mn},$$
$$\int_0^c \sin\left(\frac{n\pi}{c}z\right) \sin\left(\frac{m\pi}{c}z\right) dz = \frac{c}{2}\delta_{mn},$$

it follows from (1.216) that

$$\frac{d^2 u_{mn}(x)}{dx^2} - \alpha_{mn}^2 u_{mn}(x) = -\frac{4}{bc\epsilon_0} \int_0^b \int_0^c \rho(x, \eta, \zeta) \sin\left(\frac{m\pi}{b}\eta\right)$$

$$\times \sin\left(\frac{n\pi}{c}\zeta\right) d\eta d\zeta. \tag{1.226}$$

The general solution of this equation can be obtained by the method of variation of parameters. The solutions of the characteristic equation attached to the homogeneous equation corresponding to (1.226) are $r_{1,2} = \pm\alpha_{mn}$, and the fundamental system of solutions of the homogeneous equation reads

$$\begin{cases} u_{mn}^{(1)}(x) = \sinh(\alpha_{mn}x), \\ u_{mn}^{(2)}(x) = \cosh(\alpha_{mn}x). \end{cases}$$

According to the method of variation of parameters, the general solution of the non-homogeneous equation is

$$u_{mn}(x) = u_{mn}^{(1)}(x) A_{mn}(x) + u_{mn}^{(2)}(x) B_{mn}(x), \tag{1.227}$$

where $A'_{mn}(x)$ and $B'_{mn}(x)$ are the solutions of the system

$$u_{mn}^{(1)}(x) A'_{mn}(x) + u_{mn}^{(2)}(x) B'_{mn}(x) = 0,$$

$$\frac{du_{mn}^{(1)}(x)}{dx} A'_{mn}(x) + \frac{du_{mn}^{(2)}(x)}{dx} B'_{mn}(x)$$

$$= -\frac{4}{bc\epsilon_0} \int_0^b \int_0^c \rho(x, \eta, \zeta) \sin\left(\frac{m\pi}{b}\eta\right) \sin\left(\frac{n\pi}{c}\zeta\right) d\eta d\zeta,$$

which can be re-written as

$$\sinh(\alpha_{mn}x) A'_{mn}(x) + \cosh(\alpha_{mn}x) B'_{mn}(x) = 0,$$

$$\alpha_{mn}\cosh(\alpha_{mn}x) A'_{mn}(x) + \alpha_{mn}\sinh(\alpha_{mn}x) B'_{mn}(x)$$

$$= -\frac{4}{bc\epsilon_0} \int_0^b \int_0^c \rho(x, \eta, \zeta) \sin\left(\frac{m\pi}{b}\eta\right)$$

$$\times \sin\left(\frac{n\pi}{c}\zeta\right) d\eta d\zeta.$$

Extracting $B'_{mn}(x)$ from the first equation and introducing it in the second one, we have

$$\alpha_{mn} A'_{mn}(x) = -\frac{4\cosh(\alpha_{mn}x)}{bc\epsilon_0} \int_0^b \int_0^c \rho(x, \eta, \zeta) \sin\left(\frac{m\pi}{b}\eta\right) \sin\left(\frac{n\pi}{c}\zeta\right) d\eta d\zeta, \tag{1.228}$$

where the well-known relation $\cosh^2 \psi - \sinh^2 \psi = 1$, $\forall \, \psi \in \mathbb{R}$, has been used. Integrating (1.228), one obtains straightforwardly

$$A_{mn}(x) = \int A'_{mn}(x)dx = -\int \left[\int_0^b \int_0^c \rho(x, \eta, \zeta) \sin\left(\frac{m\pi}{b}\eta\right) \sin\left(\frac{n\pi}{c}\zeta\right) d\eta d\zeta \right]$$

$$\times \frac{4}{\alpha_{mn}bc\epsilon_0} \cosh(\alpha_{mn}x)dx + \mathcal{A}_{mn}, \qquad (1.229)$$

where $\mathcal{A}_{mn}$ are arbitrary constants of integration. Then

$$B'_{mn}(x) = -A'_{mn}(x)\tanh(\alpha_{mn}x) = \frac{4\sinh(\alpha_{mn}x)}{\alpha_{mn}bc\epsilon_0}$$

$$\times \int_0^b \int_0^c \rho(x, \eta, \zeta) \sin\left(\frac{m\pi}{b}\eta\right) \sin\left(\frac{n\pi}{c}\zeta\right) d\eta d\zeta, \qquad (1.230)$$

which by integration yields

$$B_{mn}(x) = \int B'_{mn}(x)dx = \int \left[\int_0^b \int_0^c \rho(x, \eta, \zeta) \sin\left(\frac{m\pi}{b}\eta\right) \sin\left(\frac{n\pi}{c}\zeta\right) d\eta d\zeta \right]$$

$$\times \frac{4}{\alpha_{mn}bc\epsilon_0} \sinh(\alpha_{mn}x)dx + \mathcal{B}_{mn}, \qquad (1.231)$$

where $\mathcal{B}_{mn}$ are also arbitrary constants of integration. Both $\mathcal{A}_{mn}$ and $\mathcal{B}_{mn}$ have to be determined from the boundary conditions $u_{mn}(0) = u_{mn}(a) = 0$.

The general solution of (1.224) is then

$$u_{mn}(x) = \sinh(\alpha_{mn}x) \left\{ -\frac{4}{\alpha_{mn}bc\epsilon_0} \int \left[\int_0^b \int_0^c \rho(x, \eta, \zeta) \sin\left(\frac{m\pi}{b}\eta\right) \right. \right.$$

$$\times \sin\left(\frac{n\pi}{c}\zeta\right) d\eta \, d\zeta \right] \cosh(\alpha_{mn}x)dx + \mathcal{A}_{mn} \Bigg\} + \cosh(\alpha_{mn}x) \left\{ \frac{4}{\alpha_{mn}bc\epsilon_0} \right.$$

$$\times \int \left[\int_0^b \int_0^c \rho(x, \eta, \zeta) \sin\left(\frac{m\pi}{b}\eta\right) \sin\left(\frac{n\pi}{c}\zeta\right) d\eta \, d\zeta \right] \sinh(\alpha_{mn}x)dx + \mathcal{B}_{mn} \Bigg\}$$

$$= \mathcal{A}_{mn}\sinh(\alpha_{mn}x) + \mathcal{B}_{mn}\cosh(\alpha_{mn}x) - \frac{4\sinh(\alpha_{mn}x)}{\alpha_{mn}bc\epsilon_0} \int_0^x \int_0^b \int_0^c$$

$$\times \rho(\xi, \eta, \zeta)\sin\left(\frac{m\pi}{b}\eta\right)\sin\left(\frac{n\pi}{c}\zeta\right)\cosh(\alpha_{mn}\xi)d\xi d\eta d\zeta + \frac{4\cosh(\alpha_{mn}x)}{\alpha_{mn}bc\epsilon_0}$$

$$\times \int_0^x \int_0^b \int_0^c \rho(\xi, \eta, \zeta)\sin\left(\frac{m\pi}{b}\eta\right)\sin\left(\frac{n\pi}{c}\zeta\right)\sinh(\alpha_{mn}\xi)d\xi d\eta d\zeta$$

$$= \mathcal{A}_{mn}\sinh(\alpha_{mn}x) + \mathcal{B}_{mn}\cosh(\alpha_{mn}x) - \frac{4}{\alpha_{mn}bc\epsilon_0} \int_0^x \int_0^b \int_0^c$$

$$\times \rho(\xi, \eta, \zeta)\sin\left(\frac{m\pi}{b}\eta\right)\sin\left(\frac{n\pi}{c}\zeta\right)\left[\sinh(\alpha_{mn}x)\cosh(\alpha_{mn}\xi)\right.$$

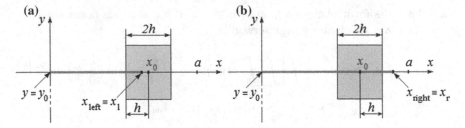

Fig. 1.29 An auxiliary construction used to explain the solution to Problem 5.

$$- \cosh(\alpha_{mn}x)\sinh(\alpha_{mn}\xi)\Big]d\xi d\eta d\zeta \tag{1.232}$$

$$= \mathcal{A}_{mn}\sinh(\alpha_{mn}x) + \mathcal{B}_{mn}\cosh(\alpha_{mn}x)$$

$$- \frac{4}{\alpha_{mn}bc\varepsilon_0}\int_0^x\int_0^b\int_0^c \rho(\xi,\eta,\zeta)\sin\left(\frac{m\pi}{b}\eta\right)\sin\left(\frac{n\pi}{c}\zeta\right)\sinh[\alpha_{mn}(x-\xi)]d\xi d\eta d\zeta.$$

For $x < x_0$, the last term goes to zero when we take the volume of the small charged paralellipiped to zero. To explain this, let us consider the corresponding simpler problem in two dimensions (Fig. 1.29).

For $x < x_0$, i.e. on the left side of x_0, for any value of x, say x_s (which cannot be greater than x_0), when passing to the limit $h \to 0$, x_s will always remain outside the rectangle of side $2h$. But, if $x > x_0$, no matter how long is the integration segment over x, say $x_d < a$, when taking the limit $h \to 0$, the little rectangle becomes more and more narrow until it becomes the segment described by equation $x = x_0$, of length $2k$, and oriented orthogonal to x-axis. Then it will always exist a small rectangle of sides $2h$, $2k$, inside the integration interval, in which $\rho = \frac{\lambda}{4hk} \neq 0$, where λ is a linear charge density. In other words, for any x_d, with $x_0 < x_d < a$, there exists $h < x_d - x_0 \neq 0$. The generalization of this reasoning to three dimensions is straightforward.

Consequently, for $x < x_0$ the last term of (1.232) goes to zero when taking the limit, while for $x > x_0$, in view of (1.215), we may write

$$I(h,k,l) \equiv \int_0^x\int_0^b\int_0^c \rho(\xi,\eta,\zeta)\sin\left(\frac{m\pi}{b}\eta\right)\sin\left(\frac{n\pi}{c}\zeta\right)\sinh[\alpha_{mn}(x-\xi)]d\xi d\eta d\zeta$$

$$= \frac{q}{8hkl}\int_{x_0-h}^{x_0+h}\int_{y_0-k}^{y_0+k}\int_{z_0-l}^{z_0+l}\sin\left(\frac{m\pi}{b}\eta\right)\sin\left(\frac{n\pi}{c}\zeta\right)\sinh[\alpha_{mn}(x-\xi)]d\xi d\eta d\zeta$$

$$= \frac{q}{8hkl}\int_{x_0-h}^{x_0+h}\sinh[\alpha_{mn}(x-\xi)]\,d\xi\int_{y_0-k}^{y_0+k}\sin\left(\frac{m\pi}{b}\eta\right)\,d\eta\int_{z_0-l}^{z_0+l}\sin\left(\frac{n\pi}{c}\zeta\right)$$

$$\times d\zeta = \frac{-q}{8hkl}\frac{bc\pi^{-2}}{mn\alpha_{mn}}\Big\{\cosh[\alpha_{mn}(x-\xi)]\Big\}_{x_0-h}^{x_0+h}\left[\cos\left(\frac{m\pi}{b}\eta\right)\right]_{y_0-k}^{y_0+k}$$

$$\times \left[\cos\left(\frac{n\pi}{c}\zeta\right)\right]_{z_0-l}^{z_0+l} = \frac{qbc\pi^{-2}\alpha_{mn}^{-1}}{8hklmn}\Big\{\cosh[\alpha_{mn}(x-x_0+h)] - \cosh[\alpha_{mn}$$

$$\times (x - x_0 - h)]\} \left\{ \cos \left[\frac{m\pi}{b}(y_0 + k) \right] - \cos \left[\frac{m\pi}{b}(y_0 - k) \right] \right\}$$
$$\times \left\{ \cos \left[\frac{n\pi}{c}(z_0 + l) \right] - \cos \left[\frac{n\pi}{c}(z_0 - l) \right] \right\},$$

and, taking the limit,

$$\lim_{\substack{h \to 0 \\ k \to 0 \\ l \to 0}} I(h, k, l) = \frac{qbc\alpha_{mn}^{-1}}{\pi^2 mn} \sinh[\alpha_{mn}(x - x_0)] \sin\left(\frac{m\pi}{b}y_0\right) \sin\left(\frac{n\pi}{c}z_0\right)$$

$$\times \lim_{h \to 0} \frac{\sinh(\alpha_{mn}h)}{h} \lim_{k \to 0} \frac{\sin\left(\frac{m\pi}{b}k\right)}{k} \lim_{l \to 0} \frac{\sin\left(\frac{n\pi}{c}l\right)}{l} = q \sinh[\alpha_{mn}(x - x_0)]$$

$$\times \sin\left(\frac{m\pi}{b}y_0\right) \sin\left(\frac{n\pi}{c}z_0\right) \lim_{h \to 0} \frac{\sinh(\alpha_{mn}h)}{\alpha_{mn}h} \lim_{k \to 0} \frac{\sin\left(\frac{m\pi}{b}k\right)}{\frac{m\pi}{b}k} \lim_{l \to 0} \frac{\sin\left(\frac{n\pi}{c}l\right)}{\frac{n\pi}{c}l}$$

$$= q \sinh[\alpha_{mn}(x - x_0)] \sin\left(\frac{m\pi}{b}y_0\right) \sin\left(\frac{n\pi}{c}z_0\right).$$

Therefore, we can write the functions $u_{mn}(x)$ as

$$u_{mn}(x) = \mathcal{A}_{mn} \sinh(\alpha_{mn}x) + \mathcal{B}_{mn} \cosh(\alpha_{mn}x) \qquad (1.233)$$
$$+ \begin{cases} 0, & x < x_0 \\ -\frac{4q}{\alpha_{mn}bc\epsilon_0} \sinh[\alpha_{mn}(x - x_0)] \sin\left(\frac{m\pi}{b}y_0\right) \sin\left(\frac{n\pi}{c}z_0\right), & x > x_0. \end{cases}$$

Time has come to determine the constants of integration. The boundary condition $u_{mn}(0) = 0$ gives $\mathcal{B}_{mn} = 0$, while the condition $u_{mn}(a) = 0$ leads to

$$\mathcal{A}_{mn} = \frac{4q}{\alpha_{mn}bc\epsilon_0} \frac{\sinh[\alpha_{mn}(a - x_0)] \sin\left(\frac{m\pi}{b}y_0\right) \sin\left(\frac{n\pi}{c}z_0\right)}{\sinh(\alpha_{mn}a)}.$$

We are now able to write the solution of Poisson's equation in the form

$$V = \frac{4q}{bc\epsilon_0} \sum_{m=1}^{\infty} \sum_{n=1}^{\infty} \frac{\sinh[\alpha_{mn}(a - x_0)] \sin\left(\frac{m\pi}{b}y_0\right) \sin\left(\frac{n\pi}{c}z_0\right)}{\alpha_{mn} \sinh(\alpha_{mn}a)}$$

$$\times \sinh(\alpha_{mn}x) \sin\left(\frac{m\pi}{b}y\right) \sin\left(\frac{n\pi}{c}z\right), \quad \text{for } x < x_0,$$

$$V = \frac{4q}{bc\epsilon_0} \sum_{m=1}^{\infty} \sum_{n=1}^{\infty} \frac{\sinh[\alpha_{mn}(a - x_0)] \sin\left(\frac{m\pi}{b}y_0\right) \sin\left(\frac{n\pi}{c}z_0\right)}{\alpha_{mn} \sinh(\alpha_{mn}a)}$$

$$\times \sinh(\alpha_{mn}x) \sin\left(\frac{m\pi}{b}y\right) \sin\left(\frac{n\pi}{c}z\right) - \frac{4q}{bc\epsilon_0} \sum_{m=1}^{\infty} \sum_{n=1}^{\infty} \frac{\sinh[\alpha_{mn}(x - x_0)]}{\alpha_{mn}}$$

$$\times \sin\left(\frac{m\pi}{b}y_0\right) \sin\left(\frac{n\pi}{c}z_0\right) \sin\left(\frac{m\pi}{b}y\right) \sin\left(\frac{n\pi}{c}z\right), \quad \text{for } x > x_0. \qquad (1.234)$$

We can still write some terms in a compact form by using the identity

$$\sinh(\alpha - \beta)\sinh\gamma + \sinh(\beta - \gamma)\sinh\alpha + \sinh(\gamma - \alpha)\sinh\beta = 0,$$

with $\alpha = \alpha_{mn}a$, $\beta = \alpha_{mn}x_0$, $\gamma = \alpha_{mn}x$. The final result is

$$V(x, y, z)\Big|_{x<x_0} = \frac{4q}{bc\epsilon_0} \sum_{m=1}^{\infty} \sum_{n=1}^{\infty} \frac{\sinh[\alpha_{mn}(a - x_0)]\sin\left(\frac{m\pi}{b}y_0\right)\sin\left(\frac{n\pi}{c}z_0\right)}{\alpha_{mn}\sinh(\alpha_{mn}a)}$$

$$\times \sinh(\alpha_{mn}x)\sin\left(\frac{m\pi}{b}y\right)\sin\left(\frac{n\pi}{c}z\right),$$

$$V(x, y, z)\Big|_{x>x_0} = \frac{4q}{bc\epsilon_0} \sum_{m=1}^{\infty} \sum_{n=1}^{\infty} \frac{\sinh(\alpha_{mn}x_0)\sin\left(\frac{m\pi}{b}y_0\right)\sin\left(\frac{n\pi}{c}z_0\right)}{\alpha_{mn}\sinh(\alpha_{mn}a)}\sinh[\alpha_{mn}(a - x)]$$

$$\times \sin\left(\frac{m\pi}{b}y\right)\sin\left(\frac{n\pi}{c}z\right), \tag{1.235}$$

with α_{mn} given by (1.225).

1.6 Proposed Problems

1. The potential of the electrostatic field created in vacuum by a dipole of moment $\mathbf{p} = const.$ is

$$V = \frac{1}{4\pi\epsilon_0} \frac{\mathbf{p} \cdot \mathbf{r}}{r^3}.$$

   Determine the electric field $\mathbf{E}$ and find the field lines in spherical coordinates.
2. Determine the field produced by a uniformly charged straight circular cylinder, of radius R and infinite length. The superficial charge density is σ.
3. The charge Q is distributed on a sphere of radius R. In the neighbourhood of the sphere there is a point charge $+q$. Find:
   a) The charge density σ on the sphere, in particular at *punctum remotum* and *punctum proximum* with respect to the point charge;
   b) The force of interaction between the point charge and the charged sphere.
4. A point charge q is situated at the distance a with respect to a conducting sphere of radius R $(a > R)$. Using the method of electrical images, determine the charge distribution on the sphere in the following cases:
   a) The sphere is insulated and grounded;
   b) The sphere is insulated and charged with charge Q;
   c) The sphere is grounded and $R \to \infty$;
   d) The sphere is neutral, insulated, and placed in a uniform electric field $\mathbf{E}_0$.
5. Determine the equipotential surfaces and the force lines for the case of a complex potential $w(z) = \sqrt{z}$. What is the contour of the grounded conductor which corresponds to such a potential?

6. A point charge q is placed at the centre of a homogeneous dielectric sphere of radius R and relative permittivity ϵ_r. If the sphere is situated in vacuum, determine the polarization charge density on the surface of the sphere.

7. Determine the Fourier components of the electrostatic potential V and of the electric field $\mathbf{E}$ produced by a point charge q.

8. Determine the quadrupole electric moment of a uniformly charged ellipsoid.

9. The permittivity of an inhomogeneous dielectric sphere of radius R, placed in vacuum, varies according to the law $\epsilon(r) = \epsilon_0 \left(\frac{r}{R} + 2 \right)$. Find the electrostatic field intensity created by the charge Q, uniformly distributed in the volume of the sphere.

10. Suppose that the potential on the surface of a straight circular cylinder of radius R depends on φ only: $V(R, z, \varphi) = f(\varphi)$, where $f(\varphi)$ is a given function. If there are no sources inside the cylinder, determine the potential in this space.

Chapter 2
Fields of Stationary Currents

2.1 Magnetostatic Field in Vacuum

2.1.1 Stationary Electric Current

If at the ends A and B (or at any two points) of a conductor is applied a potential difference or voltage, this will produce in the conductor an oriented displacement of electric charges. In other words, in the conductor will circulate an *electric current*. This phenomenon exists as long as $V_A \neq V_B$. If the potential difference is constant in time, the current is called *continuous* or *stationary*.

The electric current is characterized by its *intensity*. The *instantaneous intensity* I is defined as

$$I = \frac{dq}{dt}, \tag{2.1}$$

being numerically equal to the rate of electric charge passing in unit time the cross section of the conductor.

To describe the local behaviour of the electric current (in other words, the behaviour at a certain point inside the conductor), one introduces the *current density* $\mathbf{j}$ by the relation

$$I = \int_S \mathbf{j} \cdot d\mathbf{S}, \tag{2.2}$$

where S is the area of the conductor's cross section. The current density $\mathbf{j}$ is oriented along the direction of displacement of positive charges. Thus, the modulus of the current density is numerically equal to the intensity of the current passing a surface of unit area, orthogonal to the direction of the displacement of charges.

If $|\mathbf{j}| = $ const. and $\mathbf{j} \parallel d\mathbf{S}$, Eq. (2.2) leads to $j = I/S$. If $S \to 0$, I being finite, it follows that $j \to \infty$. Therefore, a (geometrically) linear current represents a line of singularity for the quantity $\mathbf{j}$.

© Springer-Verlag Berlin Heidelberg 2016
M. Chaichian et al., *Electrodynamics*, DOI 10.1007/978-3-642-17381-3_2

In SI, the unit for electric current is the *ampere*, denoted by A, and defined as that constant current which, if maintained in two straight parallel conductors of infinite length, of negligible circular cross section, and placed 1 metre apart in vacuum, would produce between these conductors a force equal to 2×10^{-7} newton per metre of length. The ampere is one of the seven fundamental units of SI, and the only one specific to electromagnetism.

2.1.2 Fundamental Laws

2.1.2.1 Electric Charge Conservation and Continuity Equation

One of the fundamental principles of physics is the *electric charge conservation law*. This principle is quantitatively expressed by the *continuity equation*.

Let Q be a charge continuously distributed at the time t_0 in the three-dimensional domain D_0, of volume V_0, bounded by the surface S_0, and let $\rho(\mathbf{r}_0, t_0)$ be the corresponding charge density. The form and dimensions of the considered volume, as well as the density, vary in time, so that at the time t the same charge Q of density $\rho(\mathbf{r}, t)$ will occupy the domain D, of volume V, bounded by the surface S (Fig. 2.1). The invariance of charge is represented by the equality $Q(t_0) = Q(t)$, or

$$\int_{V_0} \rho(\mathbf{r}_0, t_0)d\tau_0 = \int_{V} \rho(\mathbf{r}, t)d\tau. \tag{2.3}$$

To find the connection between $\rho(\mathbf{r}, t)$ and $\rho(\mathbf{r}_0, t_0)$, one observes that the position of a particle P, at time t, depends on both the time t and its initial position $\mathbf{r}_0$, that is

$$\mathbf{r} = \mathbf{r}(\mathbf{r}_0, t). \tag{2.4}$$

Assume that to each particle P_0 of D_0 corresponds only one particle P of D. This fact is mathematically expressed by

Fig. 2.1 Time evolution of the domain D, of volume V, bounded by the surface S.

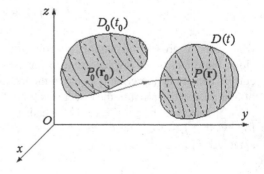

$$J = \frac{\partial(x_1, x_2, x_3)}{\partial(x_1^0, x_2^0, x_3^0)} \neq 0, \tag{2.5}$$

where J is the functional determinant (Jacobian) of the transformation (2.4). The condition (2.3) can be written in a more convenient form by setting the same integration domain on both sides. To this end, we use (2.4) and write

$$d\tau = dx_1 dx_2 dx_3 = \frac{\partial(x_1, x_2, x_3)}{\partial(x_1^0, x_2^0, x_3^0)} dx_1^0 dx_2^0 dx_3^0 = J d\tau_0. \tag{2.6}$$

Denoting $\rho(\mathbf{r}_0, t_0) = \rho_0$, $\rho(\mathbf{r}, t) = \rho$, we obtain

$$\int_{V_0} (\rho_0 - J\rho) d\tau_0 = 0.$$

Since this equality must be valid for any V_0, it follows that the integrand must be identically null, i.e.

$$J\rho = \rho_0. \tag{2.7}$$

This is the *continuity equation* in the formulation of Jean-Baptiste le Rond d'Alembert (1717–1783).

The continuity equation can be written in an alternative form, as a first-order partial differential equation, frequently encountered in the theory of continuous media (fluids). Let us take the total derivative with respect to time of (2.7):

$$\rho \frac{dJ}{dt} + J \frac{d\rho}{dt} = 0. \tag{2.8}$$

The Jacobian can be expressed in the form

$$J = \sum_{j=1}^{3} \frac{\partial x_i}{\partial x_j^0} A_i^j, \tag{2.9}$$

where A_i^j is the algebraic complement of the element $\frac{\partial x_i}{\partial x_j^0}$ in the determinant J, when the determinant J is expanded with respect to the elements of the ith row. (Note that there is no summation over the index i in (2.9)). If in J the elements of the row i are replaced by the elements of another row, say k, the determinant vanishes. Then (2.9) can be re-written as

$$\sum_{j=1}^{3} \frac{\partial x_k}{\partial x_j^0} A_i^j = J \delta_{ik}. \tag{2.10}$$

Using the derivative rule for determinants, we find

$$\frac{dJ}{dt} = \sum_i \sum_j \frac{d}{dt}\left(\frac{\partial x_i}{\partial x_j^0}\right) A_i^j$$

$$= \sum_i \sum_j \frac{\partial}{\partial x_j^0}\left(\frac{dx_i}{dt}\right) A_i^j = \sum_i \sum_j \frac{\partial v_i}{\partial x_j^0} A_i^j,$$

where v_i with $i = 1, 2, 3$ are the components of the instantaneous velocity of a charged particle whose motion we are following. In view of (2.10), we obtain

$$\frac{dJ}{dt} = \sum_i \sum_j \sum_k \frac{\partial v_i}{\partial x_k} \frac{\partial x_k}{\partial x_j^0} A_i^j = J \sum_i \sum_k \frac{\partial v_i}{\partial x_k} \delta_{ik}$$

$$= J \sum_i \frac{\partial v_i}{\partial x_i} = J \, \nabla \cdot \mathbf{v}. \tag{2.11}$$

This formula was established by Leonhard Euler (1707–1783). Substituting (2.11) into (2.8) and simplifying by $J (\neq 0)$, we get

$$\frac{d\rho}{dt} + \rho \nabla \cdot \mathbf{v} = 0. \tag{2.12}$$

But $\rho = \rho(\mathbf{r}, t)$, hence

$$\frac{d\rho}{dt} = \frac{\partial \rho}{\partial t} + \frac{\partial \rho}{\partial x_i} \frac{dx_i}{dt} = \frac{\partial \rho}{\partial t} + \mathbf{v} \cdot \nabla \rho.$$

By means of (A.43), we can then write Eq. (2.12) as

$$\frac{\partial \rho}{\partial t} + \nabla \cdot (\rho \mathbf{v}) = 0. \tag{2.13}$$

This is the *differential form* of the continuity equation and (2.13) connects the velocity field $\mathbf{v}(\mathbf{r}, t)$ with the scalar field of charge density $\rho(\mathbf{r}, t)$.

The physical interpretation of the continuity equation is revealed by integrating (2.13) over a three-dimensional domain D, fixed with respect to the observer, of volume V and bounded by the surface S:

$$\int_V \frac{\partial \rho}{\partial t} d\tau = - \int_V \nabla \cdot (\rho \mathbf{v}) d\tau.$$

As the integration domain is fixed we can apply the divergence theorem (A.32), with the result:

$$\frac{d}{dt} \int_V \rho \, d\tau = - \oint_S \rho \mathbf{v} \cdot d\mathbf{S}. \tag{2.14}$$

Fig. 2.2 Intuitive
representation of the
conservation of charge.

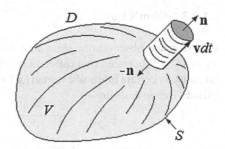

This is the integral form of the law of conservation of electric charge: *the rate of change of the electric charge in the domain D equals the difference between the outgoing and ingoing charges, per unit time, through the boundary surface S.* In other words, *the charge leaves the domain D, in unit time, as a flux of the vector field $\rho\mathbf{v}$ through the boundary surface S.*

Obviously, the charges (positive and/or negative) can enter and leave the volume V through the boundary surface S. Then, the l.h.s. of (2.14) represents the rate of variation of the charge within the volume V. To emphasize the physical significance of the r.h.s. of (2.14), let $\mathbf{v}$ be the velocity of the charge[1] passing through S during the time interval dt. The charge dQ is proportional to the volume of an infinitesimal cylinder, of base $d\mathbf{S}$ and generatrix $\mathbf{v}dt$ (Fig. 2.2):

$$dQ = \rho\mathbf{v} \cdot d\mathbf{S}dt.$$

The quantity $\rho\mathbf{v} \cdot d\mathbf{S}$ is then the charge which flows in unit time through the surface element $d\mathbf{S}$, and the quantity $\oint_S \rho\mathbf{v} \cdot d\mathbf{S}$ represents the total charge which leaves or enters the volume V per unit time. Choosing the outward normal, the quantity $\rho\mathbf{v} \cdot d\mathbf{S}$ is positive if the charges leave the volume V ($\mathbf{v} \cdot \mathbf{n} > 0$), and negative if the charges enter the volume ($\mathbf{v} \cdot \mathbf{n} < 0$). The physical significance of the vector field $\rho\mathbf{v}$ becomes clear by observing that its direction coincides to that of the velocity, while its magnitude represents the positive charge passing, in unit time, a unit surface orthogonal to the direction of velocity, i.e. the density of current:

$$[\rho v] = \left[\frac{Q}{V}\frac{l}{t}\right] = \left[\frac{Q/t}{V/l}\right] = \left[\frac{I}{S}\right] = [j].$$

Therefore, we may write the continuity equation in the form:

$$\frac{\partial \rho}{\partial t} + \nabla \cdot \mathbf{j} = 0. \tag{2.15}$$

---

[1]Here, the notion of *particle* must be understood in the framework of the theory of continuous media. In this context, by a *charged particle* we do not mean a "customary" charged particle (like an electron, or a proton, etc.), because in the theory of continuous deformable media, a particle is an *infinitely small entity* (see Sect. 3.2).

2.1.2.2   Ohm's Law

Experimental observations show that the ratio of the potential difference (voltage) $V_1 - V_2 \neq 0$ between two different points of a conductor and the resulting current intensity, in certain limits of external factors (pressure, temperature, etc.), is a constant called *electric resistance*:

$$\frac{V_1 - V_2}{I} = R = \text{const.} \tag{2.16}$$

This is the integral form of *Ohm's law*, written for a fragment of a homogeneous circuit. The law is named after the German physicist Georg Simon Ohm (1789–1854), who proposed a first form of it in 1827 based on his experimental observations.

The electric resistance of a conductor, assumed to be homogeneous and having a constant cross section, depends on the nature, form, and dimensions of the conductor by the relation

$$R = \rho \frac{l}{S}, \tag{2.17}$$

where l is the length of the conductor, S – the area of the cross section, and ρ – the *resistivity* of the material. The inverse of ρ is denoted by λ and it is known as *electric conductivity*, $\lambda = 1/\rho$. The SI unit for electric resistance is the *ohm*, with the symbol Ω. The ohm is the electric resistance between two points of a conductor when a constant potential difference of 1 volt, applied to these points, produces in the conductor a current of 1 ampere, the conductor not being the seat of any electromotive force.

The integral form of Ohm's law can also be written as

$$V_1 - V_2 = \int_1^2 \mathbf{E} \cdot d\mathbf{l} = RI. \tag{2.18}$$

To establish the *differential (local)* form of Ohm's law, valid for a fragment of circuit, we shall use (2.18):

$$\int_1^2 \mathbf{E} \cdot d\mathbf{l} = \int_1^2 E_l dl = I \frac{l}{\lambda S} = j \frac{l}{\lambda}.$$

If the studied sample is a homogeneous and isotropic conducting cylinder, then $E_l = \text{const.}$, while the vectors $\mathbf{j}$ and $\mathbf{E}$ have the same orientation, and the last formula yields

$$\mathbf{j} = \lambda \mathbf{E}, \tag{2.19}$$

which is the *differential (local) form* of Ohm's law. It shows that, if there is a field $\mathbf{E} \neq 0$ at some point of a conductor, then at this point will also exist a current of density $\mathbf{j}$. Ohm's law is also valid in case of variable fields and of non-homogeneous isotropic conductors. In case of anisotropic bodies, the conductivity is a symmetric tensor λ_{ik}, and Ohm's law reads

$$j_i = \lambda_{ik} E_k, \quad i, k = 1, 2, 3. \tag{2.20}$$

If the only acting forces are of electrostatic type, in a very short time the potential becomes equal everywhere, meaning that there is no electric current. To maintain the current in a conductor, it is necessary to apply a force field of non-electrostatic nature (mechanical, chemical, thermal, nuclear, etc.), denoted by $\mathbf{E}^{ext}$. Ohm's law (2.19) then becomes

$$\mathbf{j} = \lambda(\mathbf{E} + \mathbf{E}^{ext}), \tag{2.21}$$

which stands for the *local generalized* Ohm's law. The *integral* form of this law is

$$\int_1^2 \mathbf{E} \cdot d\mathbf{l} + \int_1^2 \mathbf{E}^{ext} \cdot d\mathbf{l} = R\,I. \tag{2.22}$$

The integral $\int_1^2 \mathbf{E}^{ext} \cdot d\mathbf{l}$ represents the *electromotive force* between the points 1 and 2.

2.1.2.3 Joule–Lenz Law

This law refers to the caloric effect of the electric current. The integral form of this law is

$$W = (V_1 - V_2)It = RI^2 t, \tag{2.23}$$

or

$$P = RI^2,$$

where W is the caloric energy (the heat) released in the conductor by the electric current and P is the power of the current.

The differential form of this law is written in terms of the *power density p* of the electric current. The power density is defined by the expression $P = \int_V p\,d\tau$, where $d\tau$ is a volume element of the conductor in which the heat is released. If the conductor is a homogeneous cylinder, of length l and cross section S, we have

$$p = \frac{RI^2}{Sl} = \frac{1}{\lambda}\,j^2, \tag{2.24}$$

or

$$P = \int_V \frac{j^2}{\lambda}\,d\tau.$$

The law was discovered in studies of resistive heating by the English physicist James Prescott Joule (1818–1889) in 1841 and, independently, by the Russian physicist of German Baltic origin Heinrich Lenz (1804–1865) in 1842.

2.1.3 Magnetic Field of a Stationary Electric Current

In 1820, Hans Christian Ørsted (1777–1851) discovered the magnetic effect of a
stationary current. In his experiments he showed that the interactions between currents (or between currents and permanent magnets) are of the same nature as the
interaction between magnetized bodies. Jean-Baptiste Biot (1774–1862) and Félix
Savart (1791–1841), in 1820, and André-Marie Ampère (1775–1836), between 1820
and 1826, discovered the basic experimental laws relating the magnetic flux density
(magnetic induction) **B** to the currents and also established the law of force between
one current and another.

The magnetic field of a stationary current is a *magnetostatic field*. It can vary
from one point to another, but remains constant in time at any point. Such a field
can be described by the same procedure as the electrostatic field, with the essential
difference that, unlike the electric charges, no free magnetic charges (or magnetic
monopoles) have ever been observed.

It is proven experimentally that, at large distances, the magnetic field produced
by a circular current has the same properties as the field produced by a permanent
magnet (magnetic dipole).

There are two important laws, which we shall present in the following, describing
the magnetic action of a stationary electric current.

The **Biot–Savart–Laplace law** expresses the *magnetic flux density* (*magnetic
induction*) **B** of the magnetic field produced by the stationary electric current I, in
vacuum at some point P (Fig. 2.3):

$$\mathbf{B} = k \int_C \frac{I d\mathbf{l} \times \mathbf{r}}{r^3}. \tag{2.25}$$

Here, $d\mathbf{l}$ is an oriented line element of the wire conducting the current, $I\,d\mathbf{l}$ is a *current
element*, $\mathbf{r}$ is the radius vector of the point P with respect to the current element, C is
the current (wire) contour (AB in Fig. 2.3a), and the magnitude and dimension of the

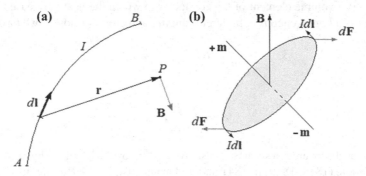

Fig. 2.3 (a) A *current element* $I\,d\mathbf{l}$ which produces at the point P the magnetic field **B**; (b) Forces
acting on an elementary electric circuit, placed in the magnetic field **B**.

constant k depends on the system of units used. For instance, in Gaussian units the constant is empirically found to be $k = 1/c$, where c is the speed of light in vacuum, while in SI units, $k = \mu_0/4\pi$, where $\mu_0 = 4\pi \times 10^{-7} \text{H} \cdot \text{m}^{-1} = 4\pi \times 10^{-7} \text{N} \cdot \text{A}^{-2}$ is the *magnetic permeability* of the vacuum. Remark that (2.25) is an inverse square law, just like Coulomb's law in electrostatics. However, the vector character is very different: the intensity of the electrostatic field $\mathbf{E}$ is a genuine vector (a polar vector), whereas the magnetic induction $\mathbf{B}$ is a pseudovector (an axial vector, as it is obtained by the cross product of two polar vectors).

Formula (2.25) can be written by means of the current density $\mathbf{j}$. Note that

$$I d\mathbf{l} = j S d\mathbf{l} = \mathbf{j} S dl = \mathbf{j} d\tau,$$

leading to

$$\mathbf{B} = \frac{\mu_0}{4\pi} \int_V \frac{\mathbf{j} \times \mathbf{r}}{r^3} d\tau, \tag{2.26}$$

where V is the volume occupied by the current. The ratio $\mathbf{B}/\mu_0$ is usually denoted by $\mathbf{H}$,

$$\mathbf{H} = \mathbf{B}/\mu_0, \tag{2.27}$$

and it is called *intensity of the magnetic field* in vacuum.

The force law describing the interaction between an electric circuit C traveled by the current I and a magnetic field of induction $\mathbf{B}$ (Fig. 2.3b) was discovered by Ampère. The field acts on the "dipole" (i.e. circuit) tending to orient it along the field. On each current element $I d\mathbf{l}$ acts the elementary *Laplace force*

$$d\mathbf{F} = I d\mathbf{l} \times \mathbf{B},$$

such that the force acting on the whole circuit is

$$\mathbf{F} = \oint_C I d\mathbf{l} \times \mathbf{B}. \tag{2.28}$$

Ampère studied the interaction between loop circuits and found that the force between two loops traveled by the currents I_1 and I_2 (see Fig. 2.4) is

$$\mathbf{F}_{12} = \frac{\mu_0}{4\pi} I_1 I_2 \oint_{C_1} \oint_{C_2} \frac{d\mathbf{l}_1 \times (d\mathbf{l}_2 \times \mathbf{r}_{12})}{r_{12}^3}. \tag{2.29}$$

After some vector manipulations, the formula can be put in a form symmetric in $d\mathbf{l}_1$ and $d\mathbf{l}_2$, emphasizing the law of action and reaction:

$$\mathbf{F}_{12} = -\frac{\mu_0}{4\pi} I_1 I_2 \oint_{C_1} \oint_{C_2} \frac{(d\mathbf{l}_1 \cdot d\mathbf{l}_2) \mathbf{r}_{12}}{r_{12}^3} = -\mathbf{F}_{21}. \tag{2.30}$$

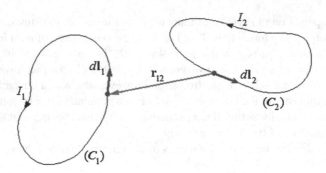

Fig. 2.4 Interaction between two current-carrying loops.

2.1.4 Magnetic Dipole

An elementary circuit behaves like a magnetic dipole of moment $\mathbf{m} = I\,\mathbf{S}$, where I is the current intensity and $\mathbf{S}$ is the oriented surface of the circuit (loop), the direction being given by the right-hand rule. A finite circuit may be covered by a net of imaginary, elementary circuits, each of them behaving like a magnetic dipole. As a matter of fact, such a system is a magnetic double layer, called *magnetic sheet*.

As electric and magnetic dipoles are almost perfectly analogous, we shall review some fundamental definitions and formulas.

The *magnetic dipole moment* is defined by

$$\mathbf{m} = \mathcal{M}\mathbf{l} \quad (d\mathbf{m} = \mathbf{l}\,d\mathcal{M})\,, \tag{2.31}$$

where $\mathcal{M}$ is the *fictitious magnetic charge* (this quantity is analogous to the electric charge q in electrostatics).

The *magnetic scalar potential*, at some point of the field produced by the magnetic dipole, is

$$V_m = \frac{\mu_0}{4\pi}\frac{\mathbf{m}\cdot\mathbf{r}}{r^3} = -\frac{\mu_0}{4\pi}\mathbf{m}\cdot\nabla\left(\frac{1}{r}\right)\,, \tag{2.32}$$

allowing to calculate the induction $\mathbf{B}$ of the magnetic field produced by the dipole by taking the gradient of the scalar potential:

$$\mathbf{B} = -\nabla V_m = \frac{\mu_0}{4\pi}\left[\frac{3(\mathbf{m}\cdot\mathbf{r})\mathbf{r}}{r^5} - \frac{\mathbf{m}}{r^3}\right]\,. \tag{2.33}$$

The resultant of forces acting on a magnetic dipole is (see Eq. (1.70)) $\mathbf{R}_m = \mathcal{M}(\mathbf{l}\cdot\nabla)\mathbf{B}$ and, for a uniform field,

$$\mathbf{R}_m = \mathbf{m}\times\mathbf{B}. \tag{2.34}$$

The *energy* of the magnetic dipole in the external magnetic field of induction **B** is

$$W_m = -\mathbf{m} \cdot \mathbf{B}. \tag{2.35}$$

Magnetic Double Layer

Let us find the scalar potential of a magnetic double layer, using the almost perfect analogy between the electric and magnetic dipoles. Since an electric or a magnetic double layer is nothing else but a large number of dipoles with parallel moments, the principle of superposition ensures the extension of the analogy to double layers as well, noting that

$$\mathbf{p} = q\mathbf{l} \quad \leftrightarrow \quad \mathbf{m} = \mathcal{M}\mathbf{l}$$

$$V_e^{dipole} = \frac{1}{4\pi\epsilon_0} \frac{\mathbf{p} \cdot \mathbf{r}}{r^3} \quad \leftrightarrow \quad V_m^{dipole} = \frac{\mu_0}{4\pi} \frac{\mathbf{m} \cdot \mathbf{r}}{r^3} \tag{2.36}$$

$$V_e^{double\ layer} = \frac{1}{4\pi\epsilon_0} \int_S \frac{\mathbf{n} \cdot \mathbf{r}}{r^3} \tau_e\, dS \quad \leftrightarrow \quad V_m^{double\ layer} = \frac{\mu_0}{4\pi} \int_S \frac{\mathbf{n} \cdot \mathbf{r}}{r^3} \tau_m\, dS$$

$$\mathbf{E}^{dipole} = \frac{1}{4\pi\epsilon_0} \left[\frac{3(\mathbf{p} \cdot \mathbf{r})\mathbf{r}}{r^5} - \frac{\mathbf{p}}{r^3} \right] \quad \leftrightarrow \quad \mathbf{B}^{dipole} = \frac{\mu_0}{4\pi} \left[\frac{3(\mathbf{m} \cdot \mathbf{r})\mathbf{r}}{r^5} - \frac{\mathbf{m}}{r^3} \right]$$

$$\mathbf{E}^{double\ layer} = -\nabla V_e^{double\ layer} \quad \leftrightarrow \quad \mathbf{B}^{double\ layer} = -\nabla V_m^{double\ layer},$$

where S is the median surface between the surfaces Σ si Σ' of the double layer (see Fig. 2.5, where σ_m si $\sigma'_m = -\sigma_m$ represent the density of fictitious magnetic charges on the two surfaces of the double layer).

The scalar potential of the magnetic field created by the double layer at the point P (see Fig. 2.5) is

Fig. 2.5 A magnetic double layer (*magnetic sheet*).

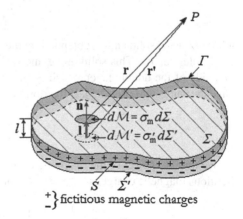

$$dM = \sigma_m d\Sigma$$
$$dM' = \sigma'_m d\Sigma'$$

$\begin{array}{c}+\\-\end{array}$ fictitious magnetic charges

$$V_m(P) = \frac{\mu_0}{4\pi} \left[\int_\Sigma \frac{\sigma_m d\Sigma}{r} + \int_{\Sigma'} \frac{\sigma'_m d\Sigma'}{r'} \right].$$

As the distance between the two surfaces of the double layer is very small, both integrals can be considered on the median surface S. For the same reason, $1/r'$ can be approximate as

$$\frac{1}{r'} \simeq \frac{1}{r} + \mathbf{l} \cdot \nabla \left(\frac{1}{r} \right),$$

such that

$$V_m(P) = \frac{\mu_0}{4\pi} \int_S \sigma_m \left(\frac{1}{r} - \frac{1}{r'} \right) dS = -\frac{\mu_0}{4\pi} \int_S \sigma_m \mathbf{l} \cdot \nabla \left(\frac{1}{r} \right) dS.$$

Noting that

$$-\mathbf{n} \cdot \nabla \left(\frac{1}{r} \right) dS = \frac{\mathbf{n} \cdot \mathbf{r}}{r^3} dS,$$

we find the potential $V_m(P)$ (see also (2.36)):

$$V_m(P) \equiv V_m^{double\ layer} = \frac{\mu_0}{4\pi} \int_S \frac{\mathbf{n} \cdot \mathbf{r}}{r^3} \tau_m \, dS, \qquad (2.37)$$

with the notation $\sigma_m l = \tau_m$, where τ_m is called the *moment* or the *power* of the magnetic double layer. Since

$$\frac{\mathbf{n} \cdot \mathbf{r}}{r^3} dS = \frac{dS \cos \alpha}{r^2} = d\Omega,$$

we obtain

$$V_m = \frac{\mu_0}{4\pi} \int_\Omega \tau_m \, d\Omega, \qquad (2.38)$$

where Ω is the solid angle subtended by the double layer from the observation point P of potential V_m. This solid angle and the fictitious magnetic charges situated on that side of the double layer which is "seen" from the observation point P should have the same sign (Fig. 2.5). If the double layer is homogeneous, then

$$V_m = \frac{\mu_0}{4\pi} \tau_m \, \Omega. \qquad (2.39)$$

We find the power of the magnetic double layer using the fact that the density of fictitious magnetic charge is $d\sigma_m = \frac{d\mathcal{M}}{dS}$ and taking into account the relation (2.31):

$$\tau_m = \sigma_m l = \frac{l \, d\mathcal{M}}{dS} = \frac{l \, dm}{l \, dS} = \frac{l I \, dS}{l \, dS} = I,$$

i.e. the power of a magnetic double layer equals the current I which circulates through the closed loop representing the contour of the double layer "generated" by that current (for example, the closed loop Γ in Fig. 2.5).

The nature of the quantity τ_m can be easily found also by using dimensional analysis. Thus, from the third and the last relation in (2.36) we have

$$\left[\frac{E^{double\ layer}}{B^{double\ layer}}\right]_{SI} = \left[\frac{\tau_e}{\tau_m}\right]_{SI} \left[\frac{1}{\epsilon_0 \mu_0}\right]_{SI}.$$

From here, taking into account (4.5) and (4.23), follows that

$$[\tau_m]_{SI} = [\tau_e]_{SI}\, [c]_{SI}\,,$$

where c is the speed of light in vacuum. Since $\tau_e = \sigma l$, where σ is the density of electric charge on one surface of an electric double layer, and l is the distance between the two charges of the electric dipole, we can write

$$[\tau_m]_{SI} = [\sigma l]_{SI}\, [c]_{SI} = \frac{1C}{1s} = 1A,$$

meaning that the power τ_m of a magnetic double layer has to be an electric current.

2.1.5 Ampère's Circuital Law

Let us calculate the circulation of the field $\mathbf{B}$ along a closed contour (path) C, around a wire carrying the current I (Fig. 2.6). If the observer is at a fixed origin O, while the field is determined at the point $P(\mathbf{r})$, the Biot–Savart law (2.26) reads

$$\mathbf{B}(\mathbf{r}) = \frac{\mu_0}{4\pi} \int_{V'} \frac{\mathbf{j}(\mathbf{r}') \times (\mathbf{r} - \mathbf{r}')}{|\mathbf{r} - \mathbf{r}'|^3} d\tau', \qquad (2.40)$$

Fig. 2.6 A current I
surrounded by the contour C
(the *arrows* show the
orientation of the curve).

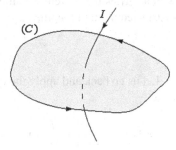

where V' is the volume of the three dimensional domain D' occupied by the sources of the magnetic field, i.e. by the current density $\mathbf{j}(\mathbf{r}')$.

As the operator ∇ acts only on vector and scalar functions which depend on $\mathbf{r}$, we observe that

$$\frac{\mathbf{r} - \mathbf{r}'}{|\mathbf{r} - \mathbf{r}'|^3} = -\nabla\left(\frac{1}{|\mathbf{r} - \mathbf{r}'|}\right),$$

consequently

$$\nabla \times \left(\frac{\mathbf{j}(\mathbf{r}')}{|\mathbf{r} - \mathbf{r}'|}\right) = \frac{1}{|\mathbf{r} - \mathbf{r}'|}\nabla \times \mathbf{j}(\mathbf{r}') - \mathbf{j}(\mathbf{r}') \times \nabla\left(\frac{1}{|\mathbf{r} - \mathbf{r}'|}\right).$$

The first term on the r.h.s. of the last relation is null as $\mathbf{j}(\mathbf{r}')$ does not depend on $\mathbf{r}$, and thus we can express (2.40) as

$$\mathbf{B}(\mathbf{r}) = \frac{\mu_0}{4\pi}\int_{V'}\nabla \times \left(\frac{\mathbf{j}(\mathbf{r}')}{|\mathbf{r} - \mathbf{r}'|}\right)d\tau' = \nabla \times \left[\frac{\mu_0}{4\pi}\int_{V'}\frac{\mathbf{j}(\mathbf{r}')}{|\mathbf{r} - \mathbf{r}'|}d\tau'\right]. \quad (2.41)$$

Formula (2.41) can be recast into the form

$$\mathbf{B}(\mathbf{r}) = \nabla \times \mathbf{A}(\mathbf{r}), \quad (2.42)$$

where

$$\mathbf{A}(\mathbf{r}) = \frac{\mu_0}{4\pi}\int_{V'}\frac{\mathbf{j}(\mathbf{r}')}{|\mathbf{r} - \mathbf{r}'|}d\tau'. \quad (2.43)$$

The vector field $\mathbf{A}(\mathbf{r})$ is called the *vector potential* of the stationary vector field $\mathbf{B}(\mathbf{r})$. If the distribution of currents is both in volume and on surface, by analogy with the discussion in Chap. 1, we may write

$$\mathbf{A}(\mathbf{r}) = \frac{\mu_0}{4\pi}\int_{V'}\frac{\mathbf{j}(\mathbf{r}')}{|\mathbf{r} - \mathbf{r}'|}d\tau' + \frac{\mu_0}{4\pi}\int_{S'}\frac{\mathbf{i}(\mathbf{r}')}{|\mathbf{r} - \mathbf{r}'|}dS', \quad (2.44)$$

where S' is any open surface bounded by the closed curve (of elemental length $d\mathbf{l}$) which carries the current $\mathbf{i}$, while $\mathbf{i}(\mathbf{r})$ is the *linear current density*, i.e. the current distributed per unit length,

$$I = \int_1^2 \mathbf{i} \cdot d\mathbf{l}.$$

Let us go back and apply the operator curl $(\nabla\times)$ to (2.41). In view of (A.51), we have

$$\nabla \times \mathbf{B}(\mathbf{r}) = \frac{\mu_0}{4\pi}\left[\nabla\int_{V'}\mathbf{j}(\mathbf{r}') \cdot \nabla\left(\frac{1}{|\mathbf{r} - \mathbf{r}'|}\right)d\tau - \int_{V'}\mathbf{j}(\mathbf{r}')\Delta\left(\frac{1}{|\mathbf{r} - \mathbf{r}'|}\right)d\tau'\right].$$

Using the expressions (1.41) and (1.45),

$$\Delta\left(\frac{1}{4\pi|\mathbf{r}-\mathbf{r}'|}\right) = -\delta(\mathbf{r}-\mathbf{r}') \quad \text{and} \quad \nabla\left(\frac{1}{|\mathbf{r}-\mathbf{r}'|}\right) = -\nabla'\left(\frac{1}{|\mathbf{r}-\mathbf{r}'|}\right),$$

and then integrating by parts the first term in the square bracket and extending the integration domain over the whole space,

$$\nabla \times \mathbf{B}(\mathbf{r}) = \frac{\mu_0}{4\pi}\nabla\int\frac{1}{|\mathbf{r}-\mathbf{r}'|}\nabla'\cdot\mathbf{j}(\mathbf{r}')d\tau' + \mu_0\int\mathbf{j}(\mathbf{r}')\,\delta(\mathbf{r}-\mathbf{r}')d\tau'.$$

Since our field is stationary, the first term is null by virtue of the continuity equation. Using the sifting property of Dirac's δ-distribution in the second term, we obtain finally

$$\nabla \times \mathbf{B}(\mathbf{r}) = \mu_0\mathbf{j}(\mathbf{r}), \tag{2.45}$$

which is the *differential form of Ampère's law*. Integrating this relation over an open surface S bounded by the integration contour C surrounding the current, we can write

$$\int_S \nabla \times \mathbf{B}(\mathbf{r}) \cdot d\mathbf{S} = \mu_0\int_S \mathbf{j}\cdot d\mathbf{S} = \mu_0 I,$$

or, using the Stokes theorem (A.34),

$$\oint_C \mathbf{B}\cdot d\mathbf{l} = \mu_0 I, \tag{2.46}$$

which is the *integral form of Ampère's law*. The current I above represents the total current passing through the surface S enclosed by the curve C, which includes both free and bound currents. We shall revisit this issue in Sect. 2.2.1.

We conclude that the circulation of the field $\mathbf{B}$ along a closed contour, surrounding the current I, depends on the current intensity only. (It can be shown that, if the contour does not surround the current, the circulation is zero).

2.1.6 Vector Potential of the Field of a Stationary Current

Taking the divergence of (2.42), and using (A.48), we find that

$$\nabla\cdot\mathbf{B} = 0 \quad \text{or} \quad \nabla\cdot\mathbf{H} = 0. \tag{2.47}$$

Therefore, the field $\mathbf{B}$ is a *solenoidal field* (without sources). The lines of such a field are closed curves. This result can be also obtained by means of Gauss's law for a system of N fictitious magnetic charges,

$$\oint \mathbf{B} \cdot d\mathbf{S} = \mu_0 \sum_{i=1}^{N} \mathcal{M}_i = 0, \tag{2.48}$$

and of the divergence theorem.

Note that the vector potential $\mathbf{A}$ is not uniquely defined by (2.42). Indeed, if instead of $\mathbf{A}$ one takes

$$\mathbf{A}' = \mathbf{A} + \nabla \psi, \tag{2.49}$$

where $\psi = \psi(\mathbf{r})$ is an arbitrary differentiable function of coordinates, one obtains the same field

$$\mathbf{B}' = \nabla \times \mathbf{A}' = \nabla \times \mathbf{A} + \nabla \times (\nabla \psi) = \nabla \times \mathbf{A} = \mathbf{B}. \tag{2.50}$$

The relation (2.49) is known as *gauge transformation* of the vector potential, while the fact that both $\mathbf{A}$ and $\mathbf{A}'$ lead to the same field $\mathbf{B}$ is called *gauge invariance* of the field. Therefore, to be uniquely defined, the field $\mathbf{A}$ must obey a supplementary condition, for example

$$\nabla \cdot \mathbf{A} = \chi, \tag{2.51}$$

where χ is a given function.

By choosing $\chi = 0$ in (2.51) and using (2.42), one obtains

$$\Delta \mathbf{A} = -\mu_0 \mathbf{j}. \tag{2.52}$$

This particular choice of the vector potential is known as *Coulomb gauge* (see Sect. 3.9). Equation (2.52) is a second-order partial differential equation, analogous to Poisson's equation of electrostatics (1.37). The solution of this equation is given by (2.44) being, in its turn, analogous to the scalar potential formula of a continuous distribution of electric charges (1.29).

2.1.7 Energy of the Magnetic Field of Stationary Currents

Let us calculate first the potential energy of a circuit C traveled by a stationary current I, and placed in a magnetic field of induction $\mathbf{B}$. The field acts on the current by the Laplace force (see Eq. (2.28)),

$$\mathbf{F} = I \oint d\mathbf{l} \times \mathbf{B}.$$

Due to this force, the circuit C will suffer an elementary displacement $d\mathbf{s}$ as in Fig. 2.7 and, as a result, it will perform the elementary mechanical work

Fig. 2.7 Displacement of an electric circuit C placed in the magnetic field **B**.

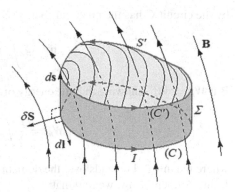

$$dL = \mathbf{F} \cdot d\mathbf{s} = I \oint \mathbf{B} \cdot (d\mathbf{s} \times d\mathbf{l}).$$

But $d\mathbf{s} \times d\mathbf{l} = -\delta\mathbf{S}$ is an oriented element of the surface Σ, swept out by the contour C as a result of displacement, so that

$$dL = -I \int_{\Sigma} \mathbf{B} \cdot \delta\mathbf{S}.$$

The surface Σ can be expressed as the difference between the surface S bounded by C and passing through C', and the surface S' bounded by C' which – except for Σ – coincides with S. Hence,

$$\delta L = -I \left[\int_{S} \mathbf{B} \cdot d\mathbf{S} - \int_{S'} \mathbf{B} \cdot d\mathbf{S} \right] = -I\,\delta \int_{S} \mathbf{B} \cdot d\mathbf{S},$$

or, by means of the Stokes theorem (A.34),

$$\delta L = -I\,\delta \oint_{C} \mathbf{A} \cdot d\mathbf{l}.$$

The potential energy of the circuit C in the field **B** is then

$$W_m = I \oint_{C} \mathbf{A} \cdot d\mathbf{l}. \tag{2.53}$$

If the circuit is not linear, we can pass to an integral over the domain occupied by the current using the substitution (see Eq. (2.2)) $I\,d\mathbf{l} \to \mathbf{j}\,d\tau$, which means

$$W_m = \int_{V} \mathbf{A} \cdot \mathbf{j}\,d\tau. \tag{2.54}$$

Recalling the procedure leading to the energy of the electrostatic field, one observes that **B**, in its turn, is created by a current of density $\mathbf{j}'$ which in the field $\mathbf{B}'$ generated by the circuit C has the potential energy

$$W'_m = \int_V \mathbf{A} \cdot \mathbf{j}' d\tau.$$

But $W_m = W'_m$, showing that the energy of the system is

$$W_m = \frac{1}{2} \int_V \mathbf{A} \cdot \mathbf{j} \, d\tau, \tag{2.55}$$

where the integral extends over the domain occupied by both currents.
Using Ampère's law, we find that

$$W_m = \frac{1}{2} \int_V \mathbf{A} \cdot \nabla \times \mathbf{H} \, d\tau = \frac{1}{2} \int_V \nabla \cdot (\mathbf{H} \times \mathbf{A}) \, d\tau + \frac{1}{2} \int_V \mathbf{H} \cdot \nabla \times \mathbf{A} \, d\tau.$$

With the help of the divergence theorem (A.32), the first integral in the r.h.s of
the above formula can be transformed into a surface integral. When we extend the
integration domain over the whole space, this integral goes to zero as the fields vanish
at infinity, and we finally obtain

$$W_m = \frac{1}{2} \int \mathbf{H} \cdot \mathbf{B} \, d\tau, \tag{2.56}$$

analogous to the formula (1.65), which expresses the energy of the electrostatic field.

2.1.8   Magnetic Multipoles

The definition of the magnetic multipole is analogous to that of the electric multipole.
A magnetic multipole is, therefore, a system of magnetic dipoles situated at small
mutual distances.

In the following, we shall calculate the vector potential for various magnetic
multipoles. Using the procedure given in Sect. 1.1.11, one expands in series $\frac{1}{|\mathbf{r} - \mathbf{r}'|}$
(see Eq. (1.80)). The component A_i of (2.43) can then be cast into the form

$$\begin{aligned}
A_i(\mathbf{r}) &= \frac{\mu_0}{4\pi} \left[\frac{1}{r} \int_{V'} j_i(\mathbf{r}') d\tau' + \frac{1}{r^3} \mathbf{r} \cdot \int_{V'} \mathbf{r}' \, j_i(\mathbf{r}') d\tau' \right. \\
&\quad \left. + \frac{1}{6} \frac{\partial^2}{\partial x_j \partial x_k} \left(\frac{1}{r} \right) \int_{V'} \left(3x'_j x'_k - r'^2 \delta_{jk} \right) j_i(\mathbf{r}') d\tau' + \dots \right] \\
&= A_i^{(0)} + A_i^{(1)} + A_i^{(2)} + \dots
\end{aligned} \tag{2.57}$$

Here $\mathbf{A}^{(0)}$, $\mathbf{A}^{(1)}$, $\mathbf{A}^{(2)}$, etc. are the vector potentials of the monopole, dipole, quadrupole, octupole, ..., respectively. In a condensed form, (2.57) can be written as

$$A_i(\mathbf{r}) = \frac{\mu_0}{4\pi} \sum_{l=0}^{\infty} \frac{(-1)^l}{l!} \int_{V'} j_i(\mathbf{r}') \left(\mathbf{r}' \cdot \nabla\right)^{(l)} \left(\frac{1}{r}\right) d\tau'. \tag{2.58}$$

In what follows, we shall calculate the first two terms of (2.57). We start by showing that, if $f(\mathbf{r}')$ and $g(\mathbf{r}')$ are two well-behaved functions (i.e. continuous and differentiable), then

$$\int \left[f\mathbf{j}(\mathbf{r}') \cdot \nabla' g + g\,\mathbf{j}(\mathbf{r}') \cdot \nabla' f \right] d\tau' = 0. \tag{2.59}$$

Integrating by parts the second term of (2.59), we have

$$\int f\mathbf{j} \cdot \nabla' g \, d\tau' + \int \nabla' \cdot (g f \mathbf{j}) \, d\tau' - \int f\mathbf{j} \cdot \nabla' g \, d\tau' = 0,$$

where we used the fact that the currents are stationary and localized, i.e. $\nabla' \cdot \mathbf{j}(\mathbf{r}') = 0$. The middle integral goes to zero when extending the integration domain over the whole space and this completes the proof.

By choosing $f = 1$, $g = x_i'$ in (2.59), it follows that

$$\int \mathbf{j}(\mathbf{r}')d\tau' = 0, \quad \text{i.e.} \quad \mathbf{A}^{(0)} = 0, \tag{2.60}$$

which is expected, since magnetic monopoles have not been observed.

Next, consider $f = x_i'$, $g = x_k'$ in (2.59), which leads to

$$\int \left(x_i' \, j_k + x_k' \, j_i \right) d\tau' = 0.$$

Using this formula and after some manipulations of the terms, we obtain

$$\begin{aligned}
A_i^{(1)} &= \frac{\mu_0}{4\pi} \left[\frac{1}{r^3} x_k \int x_k' \, j_i \, d\tau' \right] = \frac{\mu_0}{4\pi} \frac{1}{r^3} \left[\frac{1}{2} x_k \int \left(j_i x_k' - j_k x_i' \right) d\tau' \right] \\
&= \frac{\mu_0}{4\pi} \frac{1}{r^3} \left[\frac{1}{2} \epsilon_{kil} \, x_k \int (\mathbf{r}' \times \mathbf{j})_l \, d\tau' \right] \\
&= \frac{\mu_0}{4\pi} \frac{1}{r^3} \left[\epsilon_{ilk} \left(\frac{1}{2} \int \mathbf{r}' \times \mathbf{j} \, d\tau' \right)_l x_k \right] \\
&= \frac{\mu_0}{4\pi} \frac{1}{r^3} \left[\left(\frac{1}{2} \int \mathbf{r}' \times \mathbf{j} \, d\tau' \right) \times \mathbf{r} \right]_i.
\end{aligned}$$

By definition, the quantity

$$\mathbf{m} = \frac{1}{2} \int \mathbf{r}' \times \mathbf{j}(\mathbf{r}') \, d\tau' \tag{2.61}$$

is the *magnetic dipole moment* of the distribution of stationary currents. Thus,

$$\mathbf{A}^{(1)}(\mathbf{r}) = \frac{\mu_0}{4\pi} \frac{1}{r^3} (\mathbf{m} \times \mathbf{r}).\tag{2.62}$$

The field $\mathbf{B}^{(1)}$ of the magnetic dipole (first-order multipole) is found by means of (2.42) and (A.44):

$$\begin{aligned}
\mathbf{B}^{(1)} &= \frac{\mu_0}{4\pi} \nabla \times \left(\frac{\mathbf{m} \times \mathbf{r}}{r^3} \right) \\
&= \frac{\mu_0}{4\pi} \left[\frac{1}{r^3} \nabla \times (\mathbf{m} \times \mathbf{r}) + \nabla \left(\frac{1}{r^3} \right) \times (\mathbf{m} \times \mathbf{r}) \right],
\end{aligned}$$

and finally

$$\mathbf{B}^{(1)} = \frac{\mu_0}{4\pi} \left[\frac{3(\mathbf{m} \cdot \mathbf{r})\mathbf{r}}{r^5} - \frac{\mathbf{m}}{r^3} \right].\tag{2.63}$$

This formula is analogous to the expression of the field of an electric dipole. If by $\hat{\mathbf{r}}$ we denote the unit vector of $\mathbf{r}$,

$$\mathbf{B}^{(1)} = \frac{\mu_0}{4\pi r^3} [\, 3(\mathbf{m} \cdot \hat{\mathbf{r}})\hat{\mathbf{r}} - \mathbf{m}].\tag{2.64}$$

The same result is found if, by similarity to the treatment of the electrostatic field, one defines the *scalar potential* of the magnetostatic field as (see Eq. (2.32)):

$$V_m = \frac{\mu_0}{4\pi} \frac{\mathbf{m} \cdot \mathbf{r}}{r^3},\tag{2.65}$$

and one calculates $\mathbf{B}^{(1)}$ by formula

$$\mathbf{B}^{(1)} = -\nabla V_m.\tag{2.66}$$

At first sight, (2.66) contradicts Ampère's law: according to (2.66), $\nabla \times \mathbf{B} = 0$, while according to Ampère's law, $\nabla \times \mathbf{B} = \mu_0 \mathbf{j} \neq 0$. In fact, there is no contradiction: in the formula of Ampère's law the circulation is taken along a closed path surrounding the currents, while (2.66) is calculated far from (and outside) the circuit, where $\oint \mathbf{B} \cdot d\mathbf{l} = 0$. In the discussion to follow, we shall use only the vector potential.

Observations:

(a) If the circuit is plane, the magnetic moment $\mathbf{m}$ is orthogonal to the surface of the circuit. Substituting $\mathbf{j}\, d\tau$ by $I\, d\mathbf{l}$, we obtain

$$\mathbf{m} = \frac{1}{2} I \oint \mathbf{r}' \times d\mathbf{l}' = I\mathbf{S}, \tag{2.67}$$

where $\mathbf{S}$ is the plane area bounded by the circuit, whatever its form.

(b) Using (2.58), we can write the vector potential of a distribution of magnetic dipoles in a condensed form:

$$\mathbf{A}(\mathbf{r}) = \frac{\mu_0}{4\pi} \sum_{l=0}^{\infty} \frac{(-1)^l}{(l+1)!} \mathcal{M}_{i_1 \dots i_l} \, \partial_{i_1} \dots \partial_{i_l} \left(\frac{1}{r}\right), \tag{2.68}$$

where

$$\mathcal{M}_{i_1 \dots i_l} = (l+1) \int_{V'} x'_{i_1} \dots x'_{i_l} \mathbf{j}(\mathbf{r}') \, d\tau' \tag{2.69}$$

is the 2^l-th order *multipolar magnetic moment tensor*.

(c) Although magnetic monopoles have not been observed, their existence is not forbidden by any law of Nature. On the contrary, various quantum theories predict the existence of magnetic monopoles. However, the calculated masses of monopoles are extremely large, such that their production in the laboratory at the presently attainable energies is not possible. The theoretical research on magnetic monopoles is very active. In 2008, it was reported that quasi-particles in materials called spin ices resembled in behaviour magnetic monopoles (but of course such monopoles can not be isolated from the spin ice).

2.2 Magnetostatic Field in Magnetized Media

2.2.1 Polarized Magnetic Media

It has been experimentally shown that any substance, when introduced in a magnetic field, is *magnetically polarized*. As proven by Ampère, the magnetization can be explained by the equivalence between molecular currents and magnetic dipoles.

To study the magnetic polarization one introduces some quantities and definitions similar to those encountered in the case of the dielectric polarization (see Sect. 1.2). One of these is the *magnetization intensity*, or *magnetic polarization vector*,

$$\mathbf{M} = \lim_{\Delta\tau \to 0} \frac{\Delta \mathbf{m}}{\Delta\tau} = \frac{d\mathbf{m}}{d\tau}, \tag{2.70}$$

where $\mathbf{m}$ is the magnetic dipole moment.

According to (2.62), the vector potential of a continuous distribution of magnetic dipoles is

$$\mathbf{A}(\mathbf{r}) = -\frac{\mu_0}{4\pi} \int_{V'} d\mathbf{m} \times \nabla \left(\frac{1}{|\mathbf{r} - \mathbf{r}'|} \right) = \frac{\mu_0}{4\pi} \int_{V'} \mathbf{M}(\mathbf{r}') \times \nabla' \left(\frac{1}{|\mathbf{r} - \mathbf{r}'|} \right) d\tau'.$$

Using (A.39) and (A.44), it follows that

$$\begin{aligned}
\mathbf{A}(\mathbf{r}) &= -\frac{\mu_0}{4\pi} \int_{V'} \nabla' \times \left(\frac{\mathbf{M}(\mathbf{r}')}{|\mathbf{r} - \mathbf{r}'|} \right) d\tau' + \frac{\mu_0}{4\pi} \int_{V'} \frac{\nabla' \times \mathbf{M}(\mathbf{r}')}{|\mathbf{r} - \mathbf{r}'|} d\tau' \\
&= \frac{\mu_0}{4\pi} \oint_{S'} \frac{\mathbf{M} \times \mathbf{n}'}{|\mathbf{r} - \mathbf{r}'|} dS' + \frac{\mu_0}{4\pi} \int_{V'} \frac{\nabla' \times \mathbf{M}(\mathbf{r}')}{|\mathbf{r} - \mathbf{r}'|} d\tau'.
\end{aligned} \tag{2.71}$$

Comparing now (2.71) and (2.44), one observes that the two relations are formally identical, therefore we may take

$$\mathbf{j}_m = \nabla \times \mathbf{M}, \qquad \mathbf{i}_m = \mathbf{M} \times \mathbf{n}'. \tag{2.72}$$

Consequently, from the point of view of magnetic external actions, a continuous distribution of magnetic multipoles behaves as a fictitious distribution of currents, located both in the volume and on the surface. These currents are equivalent to the bound charges encountered in electrostatics and are due to the orbital motion of electrons in the atoms.

Recall that in the formula expressing Ampère's law (2.45) the current represents the sum of free and bound currents. The bound currents are precisely those corresponding to the magnetic polarization. Thus,

$$\nabla \times \frac{\mathbf{B}}{\mu_0} = \mathbf{j}_{free} + \mathbf{j}_m = \mathbf{j}_{free} + \nabla \times \mathbf{M},$$

or

$$\nabla \times \mathbf{H} = \mathbf{j}_{free}, \tag{2.73}$$

where, by definition,

$$\mathbf{H} = \frac{\mathbf{B}}{\mu_0} - \mathbf{M} \tag{2.74}$$

is the *magnetic field intensity* in the medium under consideration. We would like to emphasize that the curl of $\mathbf{H}$ is determined solely by the free current, which makes Ampère' law in the form (2.73) (together with the definition (2.74)) extremely important in solving magnetostatics problems.

Unlike Ampère's law, the equation $\nabla \cdot \mathbf{B} = 0$ remains unchanged in a medium. This is obvious, since *any* magnetic field is source-free. Anticipating, the equations (2.45), (2.47), and (2.73) are at the basis of Maxwell's electromagnetic field theory.

The phenomenon of magnetization appears due to the bound magnetic charges, which are due to the motion of elementary charged particles in atoms and molecules. Similarly to the electric polarization, the magnetization $\mathbf{M}$ is microscopically defined by

$$\mathbf{M} = N\boldsymbol{m}, \tag{2.75}$$

where N is the number of elementary currents (or particles bearing a magnetic moment m) per unit volume.

Consider an electron on its orbit around the nucleus. Associated with this motion is a *dipole orbital moment*

$$\boldsymbol{m}_L = \frac{e}{2m}\mathbf{L}, \tag{2.76}$$

where e is the electron charge, m is its mass, and $\mathbf{L}$ is the electron's orbital angular momentum. Moreover, the electron has also an intrinsic angular momentum or spin $\mathbf{S}$ which leads to a magnetic moment. This is a quantum effect:

$$\boldsymbol{m}_S = \frac{eg_s}{m}\mathbf{S}, \tag{2.77}$$

where $S = \pm\hbar/2$ ($\hbar$ is the reduced Planck constant) and $g_s \sim 2.002$ is the gyromagnetic ratio of the electron. The customary unit for elementary magnetic momenta is the *Bohr magneton*,

$$m_B = \frac{e\hbar}{2m} = 9.274 \times 10^{-24} \text{J} \cdot \text{T}^{-1}.$$

The total magnetic moment of the electron is

$$\boldsymbol{m} = \frac{e}{2m}(\mathbf{L} + 2\mathbf{S}).$$

A similar expression to (2.77) gives the proton magnetic moment, but as its mass M_p is much greater than m ($M_p \sim 1840\,m$), the proton has a magnetic moment three orders of magnitude smaller than the electron. In the low energy (non-relativistic) case of electron motion inside an atom placed in a magnetic field, two independent magnetic moments appear, orbital and spin, interacting with each other and contributing to the total energy of the electron in the atom. These two contributions play an important role in determining the magnetic properties of matter.

2.2.2 Types of Magnetizable Media

Using the definition (2.74), one can establish the relation between the magnetization intensity $\mathbf{M}$ and the field $\mathbf{H}$. Homogeneous and isotropic media are characterized by the constitutive relation

$$\mathbf{B} = \mu\mathbf{H},$$

where $\mu = \mu_0 \mu_r$ is the (*absolute*) *magnetic permeability* of the considered medium, while μ_r is its *relative magnetic permeability*. Thus,

$$\mathbf{M} = (\mu_r - 1)\mathbf{H} = \chi \mathbf{H}, \qquad (2.78)$$

where

$$\chi = \mu_r - 1 \qquad (2.79)$$

is called the *magnetic susceptibility* of the medium. Obviously,

$$\mu = \mu_0(1 + \chi). \qquad (2.80)$$

In isotropic, non-homogeneous media, the coefficients μ and χ depend on the point:

$$\mathbf{B} = \mu(x, y, z)\,\mathbf{H}, \quad \mathbf{M} = \chi(x, y, z)\,\mathbf{H}. \qquad (2.81)$$

In anisotropic media, the relations between $\mathbf{B}, \mathbf{M}$, and $\mathbf{H}$, have tensorial character:

$$B_i = \mu_{ik} H_k; \quad M_i = [(\mu_r)_{ik} - \delta_{ik}]\, H_k = \chi_{ik} H_k, \qquad (2.82)$$

where μ_{ik} and χ_{ik} are the (*absolute*) *magnetic permeability tensor* and the *magnetic susceptibility tensor*, respectively.

2.2.2.1 Diamagnetism and Paramagnetism

Substances characterized by a small susceptibility, $|\chi| \ll 1$, are called *diamagnetic* if $\chi < 0$, while those with $\chi > 0$ are called *paramagnetic*. The magnetization in diamagnetic and paramagnetic media depends linearly on the applied magnetic field.

In *diamagnetic* media, as a result of the action of the external field $\mathbf{H}$, a magnetization $\mathbf{M}$ of opposite direction to $\mathbf{H}$ is created, and the magnetic susceptibility is therefore negative. This arises because the external field induces elementary currents in the substance. The external field also acts independently on the spin magnetic moments, but the net effect of this action on the system of electrons of a diamagnetic medium is negligible, as compared to the effect produced by the orbital motion of the charges.

According to Lenz's law, the field created by the moving charges is opposed to the applied external field $\mathbf{H}$. Inside the substance, the magnetic field decreases and the medium expels the magnetic lines of force. The magnetic lines of force within a diamagnetic substance have opposite direction to the external field $\mathbf{H}$, and the net result of embedding a diamagnetic substance in a magnetic field $\mathbf{H}$ is the expulsion of the lines of force (Fig. 2.8a).

The phenomenon of diamagnetism is particularly important in superconductors.

In some substances there exist permanent magnetic dipoles associated with the electron spin. Under the action of an external magnetic field $\mathbf{H}$, the magnetic dipoles

Fig. 2.8 (**a**) A diamagnetic body placed in a magnetic field expels the lines of force. (**b**) A paramagnetic body placed in a magnetic field attracts the lines of force and tends to concentrate them inside it.

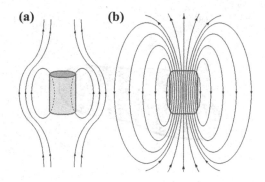

line up parallel to the field. This effect is referred to as *paramagnetism*. Not all the dipoles are aligned parallel to the field, since the ordering action of the magnetic field is opposed by the disordering action of thermal motions, which increases with temperature.

In a paramagnetic medium, the lines of force of the external field **H** tend to concentrate inside the substance (Fig. 2.8b). If the magnetic field is switched off, the dipoles become disordered again and the substance does not retain magnetic properties.

2.2.2.2 Ferromagnetism

Some media show a non-linear relation between **M** and **H**, such as

$$M = \chi_1 H + \chi_2 H^2 + \chi_3 H^3 + \cdots . \tag{2.83}$$

This type of connection appears, for example, in the case of *ferromagnetic* substances. Here, χ (which is positive) and μ are not constants, but depend on **H**, and typically have high values. The curve $\mu = \mu(H)$ is experimentally determined. Ferromagnetic substances can present a magnetic polarization $\mathbf{M}_0 \neq 0$ even when **H** is absent (permanent magnetization):

$$\mathbf{M} = \mathbf{M}_0 + \chi \mathbf{H}. \tag{2.84}$$

In ferromagnetic substances there is a spontaneous tendency for parallel neighbouring spins to couple. This is a purely quantum effect. When an external magnetic field **H** is applied to a ferromagnetic substance, the substance acquires a macroscopic magnetization **M** parallel to the field. When the external field **H** is removed, the substance preserves some magnetization **M** and behaves as a permanent dipole, like a common magnet.

In a ferromagnetic substance there are elementary regions or domains with spontaneous magnetization (see Fig. 2.9a). Ferromagnetic materials exhibit the phenomenon of *hysteresis*, which is depicted in Fig. 2.9b. If an external magnetic field **H** is applied, the domains align themselves with **H** up to some maximum value called

(a)   **(b)**

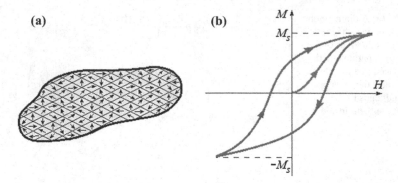

Fig. 2.9 (**a**) In a ferromagnetic material there are domains with spontaneous magnetization. Under the action of an external field, the magnetic moments of the domains line up parallel to the applied field. If the external field is turned off, the ferromagnetic substance retains the acquired magnetization and behaves as a permanent dipole. (**b**) Hysteresis cycle for a ferromagnet.

saturation magnetization M_S (above which the magnetization does not increase anymore, if we increase the applied magnetic field). If H decreases, the ferromagnet maintains some magnetization, and even when the external field H becomes zero, part of the alignment is retained, as a memory, and the sample would stay magnetized indefinitely. To demagnetize the ferromagnet, it would be necessary to apply a magnetic field in the opposite direction. For a large enough negative field $-H$, we can reach a negative saturation magnetization $-M_S$. The change in magnetization from $-M_S$ to M_S follows a similar path to the previous one, from negative to positive magnetization, closing the cycle of hysteresis. (The element of memory in a hard disk drive is based on this effect).

If the temperature is increased, the ferromagnetic property disappears at some temperature T_c characteristic of each ferromagnetic substance. For $T > T_c$ the material behaves like a paramagnet. This critical temperature T_c is called the Curie temperature, in honour of Pierre Curie (1859–1906).

As the temperature decreases, the magnetic susceptibility of a ferromagnetic substance varies, becoming infinite at the Curie temperature T_c. Furthermore, a spontaneous magnetization M appears in the substance even in the absence of an external field H. Here occurs a phenomenon called *second order phase transition*. The ferromagnetic substance reaches its minimal energy, or ground state, when all dipoles are oriented in one direction, at a nonzero value of its magnetization.

2.2.3 Jump Conditions for the Components of the Fields H and B

The relations describing the behaviour of the normal and tangent components of the fields H and B, when passing through interfaces which separate media with different

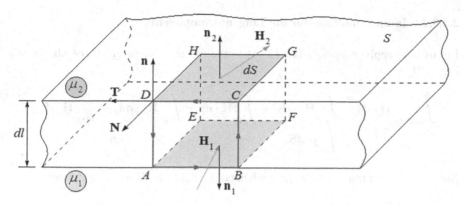

Fig. 2.10 An elementary prism $ABCDEFGH$ used to determine the jump conditions for the normal and tangent components of the fields **B** and **H**.

permeabilities are obtained using a procedure very similar to the one described in Sect. 1.2.4.

Consider two homogeneous and isotropic magnetizable media of permeabilities μ_1 and μ_2, separated by an interface in which we assume that there are electric currents. Let $\mathbf{H}_1$, $\mathbf{B}_1$ and $\mathbf{H}_2$, $\mathbf{B}_2$ be the corresponding fields in the two media. To obtain the desired relations we shall use Fig. 2.10.

2.2.3.1 Jump Conditions for the Normal Components

The flux of **B** through the elementary prism $ABCDEFGH$ is

$$d\Phi_m = \mathbf{B} \cdot d\mathbf{S} = d\Phi_1 + d\Phi_2 + d\Phi_l = (B_{2n} - B_{1n})\, dS' + d\Phi_l = 0.$$

Since

$$\lim_{dl \to 0} d\Phi_l = 0,$$

it follows that

$$B_{2n} - B_{1n} = 0, \tag{2.85}$$

as well as

$$\mu_2 H_{2n} - \mu_1 H_{1n} = 0, \tag{2.86}$$

meaning that, if $\mu_1 \neq \mu_2$, the normal component of the field **H** varies discontinuously (suffers a *refraction*).

2.2.3.2 Jump Conditions for the Tangent Components

Let us now apply Ampère's law (2.46) to the closed contour $ABCDA$ shown in Fig. 2.10:

$$\oint_{ABCDA} \mathbf{H} \cdot d\mathbf{s} = \int_{AB} \mathbf{H}_1 \cdot d\mathbf{s}_1 + \int_{BC} \mathbf{H} \cdot d\mathbf{s} + \int_{CD} \mathbf{H}_2 \cdot d\mathbf{s}_2 + \int_{DA} \mathbf{H} \cdot d\mathbf{s}$$
$$= \int_{dS} \mathbf{j} \cdot d\mathbf{S}.$$

Since the current must remain finite when taking the limit $dl \to 0$, one obtains

$$\int (\mathbf{H}_2 - \mathbf{H}_1) \cdot \mathbf{T} \, ds = \lim_{dl \to 0} \int \mathbf{j} \cdot d\mathbf{S} = \lim_{dl \to 0} \int j_N dS = \int i_N ds,$$

where i_N is the component of the *linear density of current* along the direction of the unit vector $\mathbf{N}$. Therefore, we have

$$H_{2T} - H_{1T} = i_N, \tag{2.87}$$

or, in vector form,

$$\mathbf{n} \times (\mathbf{H}_2 - \mathbf{H}_1) = \mathbf{i},$$

as well as

$$\frac{B_{2T}}{\mu_2} - \frac{B_{1T}}{\mu_1} = i_N. \tag{2.88}$$

These relations show that the tangent component of $\mathbf{H}$ and $\mathbf{B}$ present a discontinuous variation (if $i_N \neq 0$). There is no jump if $\mu_1 = \mu_2$ and $i_N = 0$.

2.3 Solved Problems

Problem 1. Show that the magnetostatic uniform field $\mathbf{B}_0$ admits as vector potential $\mathbf{A} = \frac{1}{2}\mathbf{B}_0 \times \mathbf{r}$.

Solution. Using the Levi-Civita symbol ϵ_{ijk}, we have

$$\nabla \times \mathbf{A} = \frac{1}{2}\nabla \times (\mathbf{B}_0 \times \mathbf{r}) = \frac{1}{2}\epsilon_{ijk}\partial_j(\mathbf{B}_0 \times \mathbf{r})_k \mathbf{u}_i = \frac{1}{2}\epsilon_{ijk}\partial_j(\epsilon_{klm}B_{0l}x_m)\mathbf{u}_i$$
$$= \frac{1}{2}\epsilon_{ijk}\epsilon_{klm}B_{0l}(\partial_j x_m)\mathbf{u}_i = \frac{1}{2}\epsilon_{ijk}\epsilon_{klm}B_{0l}\delta_{jm}\mathbf{u}_i = \frac{1}{2}\epsilon_{imk}\epsilon_{klm}B_{0l}\mathbf{u}_i$$
$$= \frac{1}{2}\epsilon_{imk}\epsilon_{lmk}B_{0l}\mathbf{u}_i = \frac{1}{2}\, 2\,\delta_{il}B_{0l}\mathbf{u}_i = B_{0i}\,\mathbf{u}_i = \mathbf{B}_0.$$

Fig. 2.11 Schematic
representation of the infinite,
rectilinear current I, given in
Problem 2.

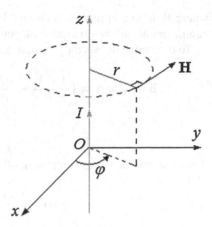

Problem 2. Find the scalar and vector potentials corresponding to a magnetostatic field created by an infinite, rectilinear current of intensity I.

Solution. Since

$$\nabla \times \mathbf{H} = \nabla \times (\nabla V_m) = 0 \tag{2.89}$$

anywhere around the current, it follows that one can define a scalar potential V_m such that

$$\mathbf{H} = -\nabla V_m. \tag{2.90}$$

The geometry of the problem suggests the use of cylindrical coordinates r, φ, z, with the z-axis oriented along the current. The magnetic field lines are then circles orthogonal to the wire, with centres on the wire, while the magnitude of the field is constant along a field line (Fig. 2.11). Applying Ampère's law (2.46) in the form $\oint \mathbf{H} \cdot d\mathbf{l} = I$, we have

$$H_r = H_z = 0, \quad H_\varphi = \frac{I}{2\pi r}. \tag{2.91}$$

On the other hand, projecting (2.90) on the coordinate axes, one obtains

$$H_r = -\frac{\partial V_m}{\partial r},$$

$$H_\varphi = -\frac{1}{r}\frac{\partial V_m}{\partial \varphi}, \tag{2.92}$$

$$H_z = -\frac{\partial V_m}{\partial z}.$$

The magnetostatic potential is then

$$V_m = -\frac{I}{2\pi}\varphi + V_0, \tag{2.93}$$

where V_0 is an arbitrary real constant. Equation (2.93) shows that the magnetostatic equipotential surfaces are meridian planes.

To determine the vector potential **A** we use (2.42):

$$\mathbf{B} = \nabla \times \mathbf{A} = \left(\frac{1}{r} \frac{\partial A_z}{\partial \varphi} - \frac{\partial A_\varphi}{\partial z} \right) \mathbf{e}_r + \left(\frac{\partial A_r}{\partial z} - \frac{\partial A_z}{\partial r} \right) \mathbf{e}_\varphi$$
$$+ \left[\frac{1}{r} \frac{\partial (r A_\varphi)}{\partial r} - \frac{\partial A_r}{\partial \varphi} \right] \mathbf{e}_z. \quad (2.94)$$

Since the only non-zero component of the current is oriented along the z-axis, from (2.43), i.e.

$$\mathbf{A}(\mathbf{r}) = \frac{\mu_0}{4\pi} \int_{V'} \frac{\mathbf{j}(\mathbf{r}')}{|\mathbf{r} - \mathbf{r}'|} d\tau'.$$

we find that $A_r = A_\varphi = 0$. Consequently,

$$B_r = \frac{1}{r} \frac{\partial A_z}{\partial \varphi},$$
$$B_\varphi = -\frac{\partial A_z}{\partial r}, \quad (2.95)$$
$$B_z = 0,$$

and, in view of (2.91), we finally obtain

$$A_z = -\frac{\mu_0 I}{2\pi} \ln r + A_0, \quad (2.96)$$

where A_0 is an arbitrary real constant.

Problem 3. Determine the field **B** produced in vacuum by a stationary, rectilinear current of length $2L$ and intensity I.

Solution. This time we choose a Cartesian coordinate system, with the z-axis along the wire and the origin at its middle (Fig. 2.12).

At the point P defined by the radius vector **r**, the vector potential is

$$\mathbf{A}(\mathbf{r}) = \frac{\mu_0}{4\pi} \int \frac{\mathbf{j}(\mathbf{r}') d\tau'}{|\mathbf{r} - \mathbf{r}'|}. \quad (2.97)$$

As

$$\mathbf{j} d\tau = \mathbf{j} S \, dl = j S \, d\mathbf{l} = I \, d\mathbf{l},$$

it follows that

$$\mathbf{A}(\mathbf{r}) = \frac{\mu_0 I}{4\pi} \int_{-L}^{+L} \frac{d\mathbf{l}}{r}, \quad (2.98)$$

Fig. 2.12 The geometry of the finite current in Problem 3.

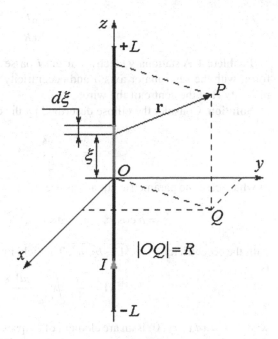

which, projected on the coordinate axes, reads

$$A_x = A_y = 0, \quad A_z = \frac{\mu_0 I}{4\pi} \int_{-L}^{+L} \frac{d\xi}{r}. \tag{2.99}$$

To perform the integration in (2.99) we make the change of variable $z - \xi = u$. Since $r = \sqrt{R^2 + u^2}, d\xi = -du$, we have

$$A_z = \frac{\mu_0 I}{4\pi} \int_{z-L}^{z+L} \frac{du}{\sqrt{R^2 + u^2}} = \frac{\mu_0 I}{4\pi} \left(\text{arcsinh} \frac{z+L}{R} - \text{arcsinh} \frac{z-L}{R} \right). \tag{2.100}$$

Due to the cylindrical symmetry of the problem, we find immediately (see Appendix D):

$$B_r = B_z = 0, \quad B_\varphi = -\frac{\partial A_z}{\partial R},$$

and, performing the derivatives,

$$B_\varphi = \frac{\mu_0 I}{4\pi R} \left[\frac{z+L}{\sqrt{R^2 + (z+L)^2}} - \frac{z-L}{\sqrt{R^2 + (z-L)^2}} \right]. \tag{2.101}$$

Remark that, for $L \to \infty$ and $z \neq 0$, (2.101) leads to the expected formula,

$$B_\varphi = \frac{\mu_0 I}{2\pi R}.$$

Problem 4. A stationary electric current I passes through a thin wire of an elliptic form, with the semi-major axis a and eccentricity e. Determine the magnetic field intensity $\mathbf{H}$ at the centre of the wire.

Solution. Consider the ellipse described by the canonical equation

$$\frac{x^2}{a^2} + \frac{y^2}{b^2} = 1,$$

in which case the parametric equations are

$$x = a\cos\varphi, \quad y = b\sin\varphi, \quad \varphi \in [-\pi, +\pi], \tag{2.102}$$

with the eccentricity $e = \sqrt{1 - b^2/a^2}$. The field intensity at the centre of the ellipse is

$$\mathbf{H} = \frac{I}{4\pi} \oint_{ellipse} \frac{d\mathbf{l} \times \mathbf{r}}{r^3}, \tag{2.103}$$

where $d\mathbf{l} = (dx, dy, 0)$ is an arc element of ellipse and $\mathbf{r} = (-x, -y, 0)$ is the radius vector of the centre O with respect to $d\mathbf{l}$ (Fig. 2.13). Therefore

$$d\mathbf{l} \times \mathbf{r} = \begin{vmatrix} \mathbf{i} & \mathbf{j} & \mathbf{k} \\ dx & dy & 0 \\ -x & -y & 0 \end{vmatrix} = (x\,dy - y\,dx)\mathbf{k},$$

showing that the only non-zero component of $\mathbf{H}$ at O is orthogonal to the surface of the ellipse (i.e. oriented along the z-axis):

$$H_z = \mathbf{H} \cdot \mathbf{k} = \frac{I}{4\pi} \oint_{ellipse} \frac{x\,dy - y\,dx}{r^3}. \tag{2.104}$$

Fig. 2.13 A stationary electric current I passing through a thin wire of an elliptic form (Problem 4).

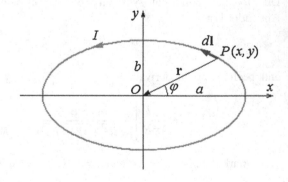

Using (2.102), we may write

$$r^2 = x^2 + y^2 = a^2(1 - \sin^2 \varphi) + b^2 \sin^2 \varphi$$
$$= a^2 \left[1 - \left(\frac{a^2 - b^2}{a^2} \right) \sin^2 \varphi \right] = a^2(1 - e^2 \sin^2 \varphi), \qquad (2.105)$$

and thus (2.104) becomes

$$H_z = \frac{I}{4\pi} \int_{-\pi}^{+\pi} \frac{[(-b \sin \varphi)(-a \sin \varphi) + (a \cos \varphi)(b \cos \varphi)] d\varphi}{a^3 (1 - e^2 \sin^2 \varphi)^{3/2}} \qquad (2.106)$$

$$= \frac{bI}{4\pi a^2} \int_{-\pi}^{\pi} \frac{d\varphi}{(1 - e^2 \sin^2 \varphi)^{3/2}} = \frac{I\sqrt{1-e^2}}{4\pi a} \int_{-\pi}^{\pi} \frac{d\varphi}{(1 - e^2 \sin^2 \varphi)^{3/2}}$$

$$= \frac{I\sqrt{1-e^2}}{2\pi a} \int_{0}^{\pi} \frac{d\varphi}{(1 - e^2 \sin^2 \varphi)^{3/2}} = \frac{I\sqrt{1-e^2}}{\pi a} \int_{0}^{\pi/2} \frac{d\varphi}{(1 - e^2 \sin^2 \varphi)^{3/2}}.$$

Here we used the facts that the integrand is an even function and the integration interval is symmetric about the origin. Also, we took into account that $\sin^2 \varphi$ is a positive periodical function (the period is π). To calculate the integral (2.106) one uses the change of variable

$$1 - e^2 \sin^2 \varphi = \frac{1 - e^2}{1 - e^2 \sin^2 \alpha}. \qquad (2.107)$$

Then

$$\sin \varphi = \frac{\cos \alpha}{\sqrt{1 - e^2 \sin^2 \alpha}}, \qquad \cos \varphi = \sqrt{1 - e^2} \frac{\sin \alpha}{\sqrt{1 - e^2 \sin^2 \alpha}},$$

and

$$\cos \varphi d\varphi = \frac{-\sin \alpha \sqrt{1 - e^2 \sin^2 \alpha} \, d\alpha + \frac{e^2 \sin \alpha \cos^2 \alpha \, d\alpha}{\sqrt{1 - e^2 \sin^2 \alpha}}}{1 - e^2 \sin^2 \alpha}$$

$$= \frac{-(1 - e^2) \sin \alpha}{(1 - e^2 \sin^2 \alpha)^{3/2}} d\alpha.$$

The last formula leads to

$$d\varphi = \frac{-\sin \alpha (1 - e^2)}{\cos \varphi (1 - e^2 \sin^2 \alpha)^{3/2}} d\alpha = -\frac{\sqrt{1 - e^2}}{1 - e^2 \sin^2 \alpha} d\alpha.$$

Returning to (2.106), the z-component of the field $\mathbf{H}$ is

$$H_z = -\frac{I\sqrt{1-e^2}}{\pi a} \int_{\pi/2}^{0} \frac{(1-e^2\sin^2\alpha)^{3/2}\sqrt{1-e^2}\,d\alpha}{(1-e^2)^{3/2}(1-e^2\sin^2\alpha)} \qquad (2.108)$$

$$= \frac{I}{\pi a\sqrt{1-e^2}} \int_{0}^{\pi/2} \sqrt{1-e^2\sin^2\alpha}\,d\alpha = \frac{I\,E(e)}{\pi a\sqrt{1-e^2}} = \frac{I\,E(e)}{\pi b},$$

where

$$E(e) = \int_{0}^{\pi/2} \sqrt{1-e^2\sin^2\alpha}\,d\alpha$$

is Legendre's trigonometric form of the complete elliptic integral of the second kind.

To conclude, the magnetic field intensity produced by the current in the wire is orthogonal to the plane of the wire, its magnitude being given by (2.108).

Problem 5. Calculate the vector potential $\mathbf{A}$ of the field produced by a current I passing through a circular loop of radius R.

Solution. It is convenient to use a cylindrical coordinate system ρ, φ, z, with the origin at the centre of the loop, and the z-axis orthogonal to its surface (Fig. 2.14). Let $P(\rho, \varphi, z)$ be the point where $\mathbf{A}$ is determined, P' – its orthogonal projection on the plane of the loop, Q – an actual point on the loop, and $I\,d\mathbf{l}$ – an elementary current with the origin at Q. The triangle $PP'Q$ is a right triangle, with the right angle at P', while the unit vectors $\mathbf{e}_\rho, \mathbf{e}_\varphi, \mathbf{e}_z$ are shown in Fig. 2.14.

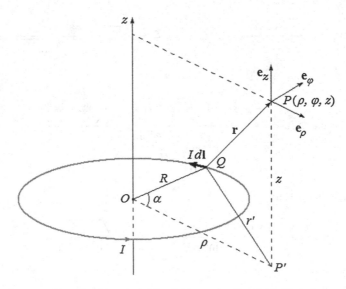

Fig. 2.14 A current I passing through the circular loop of radius R (Problem 5).

Fig. 2.15 An auxiliary geometrical construction for the Problem 5.

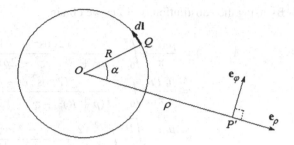

The vector potential is defined by

$$\mathbf{A} = \frac{\mu_0}{4\pi} I \oint_{loop} \frac{d\mathbf{l}}{r}, \tag{2.109}$$

where

$$r^2 = r'^2 + z^2 = R^2 + \rho^2 - 2\rho R \cos\alpha + z^2. \tag{2.110}$$

The components of $\mathbf{A}$ in cylindrical coordinates are obtained by performing the dot product between $\mathbf{A}$ and the unit vectors $\mathbf{e}_\rho$, $\mathbf{e}_\varphi$, $\mathbf{e}_z$. Since $d\mathbf{l} \perp \mathbf{e}_z$, we have

$$A_z = \mathbf{A} \cdot \mathbf{e}_z = 0. \tag{2.111}$$

From Fig. 2.15 we note that

$$d\mathbf{l} \cdot \mathbf{e}_\rho = dl \cos\left(\alpha + \frac{\pi}{2}\right) = dl \sin\alpha, \quad dl = R \, d\alpha,$$

such that

$$A_\rho = \mathbf{A} \cdot \mathbf{e}_\rho = \frac{\mu_0 I R}{4\pi} \int_{-\pi}^{+\pi} \frac{\sin\alpha \, d\alpha}{r} = 0, \tag{2.112}$$

because the integrand is an odd function of α, and the integration interval is symmetric with respect to the origin O.

The only non-zero component of $\mathbf{A}$ is A_φ:

$$A_\varphi = \mathbf{A} \cdot \mathbf{e}_\varphi = \frac{\mu_0 I}{4\pi} \oint_{loop} \frac{d\mathbf{l} \cdot \mathbf{e}_\varphi}{r} = \frac{\mu_0 I R}{4\pi} \int_{-\pi}^{+\pi} \frac{\cos\alpha \, d\alpha}{r}$$

$$= \frac{\mu_0 I R}{2\pi} \int_0^{+\pi} \frac{\cos\alpha \, d\alpha}{r}. \tag{2.113}$$

By using the substitution $\alpha = 2\psi$, we obtain

$$
\begin{aligned}
A_\varphi(\rho, z) &= \frac{\mu_0 I R}{\pi} \int_0^{\pi/2} \frac{(2\cos^2\psi - 1)d\psi}{(\rho^2 + R^2 + z^2 - 2\rho R \cos 2\psi)^{1/2}} \\
&= \frac{\mu_0 I R}{\pi} \int_0^{\pi/2} \frac{(2\cos^2\psi - 1)d\psi}{\left[(\rho + R)^2 + z^2 - 4\rho R \cos^2\psi\right]^{1/2}} \\
&= \frac{\mu_0 I \xi}{2\pi} \sqrt{\frac{R}{\rho}} \int_0^{\pi/2} \frac{(2\cos^2\psi - 1)d\psi}{(1 - \xi^2 \cos^2\psi)^{1/2}} = \frac{\mu_0 I \xi}{2\pi} \sqrt{\frac{R}{\rho}} \, \mathcal{I}(\xi),
\end{aligned}
\tag{2.114}
$$

where

$$
\xi^2 = \frac{4\rho R}{(\rho + R)^2 + z^2} \quad \text{and} \quad \mathcal{I}(\xi) = \int_0^{\pi/2} \frac{(2\cos^2\psi - 1)d\psi}{(1 - \xi^2 \cos^2\psi)^{1/2}}.
$$

Remark that

$$
\xi^2 \le \frac{4\rho R}{(\rho + R)^2} = 1 - \left(\frac{\rho - R}{\rho + R}\right)^2 \le 1.
$$

meaning that $0 \le \xi \le 1$.

Formula (2.114) can be expressed in terms of Legendre's trigonometric form of complete elliptic integrals of the first and second kind:

$$
K(\xi) = \int_0^{\pi/2} \frac{d\psi}{\sqrt{1 - \xi^2 \sin^2\psi}},
\tag{2.115}
$$

$$
E(\xi) = \int_0^{\pi/2} \sqrt{1 - \xi^2 \sin^2\psi}\, d\psi, \qquad 0 \le \xi \le 1.
$$

Using these auxiliary formulae, let us now go back to (2.114) and perform the integration. This can be done by a suitable change of variable, namely $\cos\psi = \sin\theta$. Then

$$
\begin{aligned}
\mathcal{I}(\xi) &= \int_0^{\pi/2} \frac{\left(2\cos^2\psi - 1\right)d\psi}{\left(1 - \xi^2\cos^2\psi\right)^{1/2}} = \int_0^{\pi/2} \frac{\left(2\sin^2\theta - 1\right)d\theta}{\left(1 - \xi^2\sin^2\theta\right)^{1/2}} \\
&= -\frac{2}{\xi^2} \int_0^{\pi/2} \frac{-\xi^2\sin^2\theta\, d\theta}{\sqrt{1 - \xi^2\sin^2\theta}} - \int_0^{\pi/2} \frac{d\theta}{\sqrt{1 - \xi^2\sin^2\theta}} \\
&= -\frac{2}{\xi^2} \int_0^{\pi/2} \frac{\left(1 - \xi^2\sin^2\theta - 1\right)d\theta}{\sqrt{1 - \xi^2\sin^2\theta}} - \int_0^{\pi/2} \frac{d\theta}{\sqrt{1 - \xi^2\sin^2\theta}}
\end{aligned}
$$

$$= -\frac{2}{\xi^2} \int_0^{\pi/2} \sqrt{1 - \xi^2 \sin^2 \theta} \, d\theta + \left(\frac{2}{\xi^2} - 1\right) \int_0^{\pi/2} \frac{d\theta}{\sqrt{1 - \xi^2 \sin^2 \theta}}$$

$$= -\frac{2}{\xi^2} E(\xi) + \left(\frac{2}{\xi^2} - 1\right) K(\xi).$$

Hence,

$$A_\varphi(\rho, z) = \frac{\mu_0 I \xi}{2\pi} \sqrt{\frac{R}{\rho}} \int_0^{\pi/2} \frac{(2\cos^2 \psi - 1) \, d\psi}{(1 - \xi^2 \cos^2 \psi)^{1/2}}$$

$$= \frac{\mu_0 I \xi}{2\pi} \sqrt{\frac{R}{\rho}} \left[-\frac{2}{\xi^2} E(\xi) + \left(\frac{2}{\xi^2} - 1\right) K(\xi)\right],$$

or, after some arrangement of terms,

$$A_\varphi(\rho, z) = \frac{\mu_0 I}{\pi \xi} \sqrt{\frac{R}{\rho}} \left[\left(1 - \frac{\xi^2}{2}\right) K(\xi) - E(\xi)\right]. \qquad (2.116)$$

2.4 Proposed Problems

1. Find the stationary current distribution in a metallic rectangular plate with the sides a, b, thickness h, and electric conductivity λ. The electrodes through which the potential difference is applied to the plate are placed on two opposite sides.
2. Find the stationary current distribution in a metallic circular plate of thickness h, radius a, and electric conductivity λ. The electrodes through which the potential difference is applied to the plate are placed at two diametrically opposed points.
3. Show that the force of interaction between two circuits C_1 and C_2, traveled by the stationary currents I_1 and I_2, satisfies the action and reaction principle, while the force of interaction between two elements of C_1 and C_2 does not obey this law. (Hint: Derive the formula (2.30) and interpret it in terms of action and reaction principle.)
4. Show that the magnetic lines of force of a linear, plane circuit of an arbitrary form, situated in vacuum, are curves symmetric with respect to the plane of the circuit.
5. Determine the magnetic moment of a non-magnetic, homogeneously filled sphere with charge Q, uniformly rotating ($\omega =$ const.) around an axis passing through its centre. The radius of the sphere is R.
6. A stationary electric current I travels through an n-sided convex regular polygon, whose circumscribed circle has radius R. Calculate the magnetic field both at

Fig. 2.16 The electric
circuit closing through the
pendulum in Problem 10.

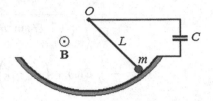

the centre of the polygon, and at some point of the polygon axis, situated at a
distance z relative to its centre.

7. Determine the magnetic field produced by an infinite electric current I, situated
   at a distance a with respect to the plane surface of an iron block of permeability
   μ.

8. Determine the shape of the lines of the magnetic field produced by a stationary
   electric current I passing through a circular loop of radius R.

9. Show that far from the loop in Problem 8, the vector potential is given by

$$\mathbf{A}_\varphi(\mathbf{r}) = \frac{\mu_0}{4\pi} \frac{\mathbf{m} \times \mathbf{r}}{r^3},$$

where $\mathbf{r}$ is the radius-vector (with respect to the centre of the loop) of the point
where the field is determined, and $\mathbf{m}$ is the magnetic moment of the loop.

10. A mathematical pendulum of length L whose rod is made of a conducting mate-
    rial moves so that its inferior end slides without friction on a metallic support in
    the shape of an arc of circle. The rod is rigid, of negligible mass, and it moves in
    a static magnetic field of induction $\mathbf{B}$, orthogonal to the plane of the pendulum.
    The bob of the pendulum is a point mass m. The metallic support and the fix
    end of the pendulum are connected through an ideal capacitor of capacity C, the
    electric circuit closing through the rod, as depicted in Fig. 2.16.

    Show that, neglecting the resistance and the inductance of the circuit, the period
    of oscillation of the pendulum is

$$T = 2\pi \sqrt{\frac{L}{g} \left(1 + \frac{CB^2L^2}{4m}\right)}.$$

Chapter 3
The Electromagnetic Field

3.1 Maxwell's Equations in Vacuum

The study of time-variable electric and magnetic fields showed the strong interdependence between them: a variable electric field produces a magnetic field and vice-versa.

Based on Ampère's and Faraday's research, James Clerk Maxwell (1831–1879) elaborated the theory of the *electromagnetic field*. As shown in his fundamental work, *A Treatise on Electricity and Magnetism* (1873), the variable electric and magnetic fields are not only interdependent, but also carriers of energy.

The fundamental principles of Maxwell's electromagnetic theory are contained in a system of linear partial differential equations, known as *Maxwell's equations*. Two of these equations show the time variation of electric and magnetic fields, being deduced by generalizing Ampère's law for non-stationary regime and Faraday's law of electromagnetic induction. Let us deduce and discuss these two equations.

3.1.1 Maxwell–Ampère Equation

Let us consider Ampère's equation in the form (2.73), $\nabla \times \mathbf{H} = \mathbf{j}$, where on the r.h.s. we have the free (not bound) current, since we wish to write the Maxwell–Ampère equation in terms of free currents, i.e. the currents which we can turn on and off at will. By applying the operator divergence to (2.73), it follows that

$$\nabla \cdot \mathbf{j} = 0.$$

On the other hand, the equation of continuity (2.15) produces

$$\nabla \cdot \mathbf{j} = -\partial \rho / \partial t.$$

© Springer-Verlag Berlin Heidelberg 2016
M. Chaichian et al., *Electrodynamics*, DOI 10.1007/978-3-642-17381-3_3

Consequently, while in stationary regime these two equations coincide, in a non-stationary situation ($\partial \rho / \partial t \neq 0$) these equations seem to contradict each other. This apparent inconsistency was brilliantly solved by Maxwell, by introducing the notion of *displacement current*.

The current density $\mathbf{j}$ appearing in the equation of continuity is due to the motion of charges and generally contains conduction and convection currents. In contrast, the current appearing in Ampère's law is subject to the condition $\nabla \cdot \mathbf{j} = 0$, i.e. it is a *closed* current. This means that in Ampère's equation should appear not only conduction and convection currents, but also an additional current which "closes the circuit".

Maxwell postulated the validity of Ampère's law in non-stationary regime, provided that $\mathbf{j}$ is composed of two parts:

$$\mathbf{j} = \mathbf{j}_{cond} + \mathbf{j}_{disp},$$

where $\mathbf{j}_{disp}$ is termed *displacement current*.[1] Ampère's equation then reads

$$\nabla \times \mathbf{H} = \mathbf{j}_{cond} + \mathbf{j}_{disp}. \tag{3.1}$$

Taking the divergence of this equation, one obtains $\nabla \cdot \mathbf{j}_{cond} + \nabla \cdot \mathbf{j}_{disp} = 0$, which means

$$\nabla \cdot \mathbf{j}_{disp} = -\nabla \cdot \mathbf{j}_{cond} = \frac{\partial \rho}{\partial t},$$

or, in view of Gauss's theorem (1.96) and taking into account that the operators ∇ and $\partial / \partial t$ are independent,

$$\nabla \cdot \mathbf{j}_{disp} = \frac{\partial}{\partial t}(\nabla \cdot \mathbf{D}) = \nabla \cdot \left(\frac{\partial \mathbf{D}}{\partial t} \right).$$

The simplest solution of the last equation is

$$\mathbf{j}_{disp} = \frac{\partial \mathbf{D}}{\partial t}, \tag{3.2}$$

and (3.1) becomes

$$\nabla \times \mathbf{H} = \mathbf{j} + \frac{\partial \mathbf{D}}{\partial t}, \tag{3.3}$$

known as the *Maxwell–Ampère equation*. Remark that in (3.3) we abandoned the subindex "cond" attached to $\mathbf{j}$ in (3.1); we shall use this notation in the discussion to follow. Equation (3.3) is valid in any polarizable medium, with $\mathbf{j}$ denoting the current density produced by the motion of free charges.

---

[1]The name *displacement current* comes from the fact that this current is produced by the time variation of the electric displacement $\mathbf{D}$, see Eq. (3.2).

In vacuum, $\mathbf{H} = \mathbf{B}/\mu_0$ and $\mathbf{D} = \epsilon_0 \mathbf{E}$, and (3.3) becomes

$$\nabla \times \mathbf{B} = \mu_0 \mathbf{j} + \frac{1}{c^2} \frac{\partial \mathbf{E}}{\partial t}, \tag{3.4}$$

where $c = 1/\sqrt{\epsilon_0 \mu_0}$ is the speed of light in vacuum. Actually, the form of the Eq. (3.4) is valid in any polarizable medium, as long as the current on the r.h.s. includes also the contribution of the bound electric and magnetic charges:

$$\nabla \times \mathbf{B} = \mu_0 \mathbf{J} + \frac{1}{c^2} \frac{\partial \mathbf{E}}{\partial t}, \tag{3.5}$$

where

$$\mathbf{J} = \mathbf{j} + \nabla \times \mathbf{M} + \frac{\partial \mathbf{P}}{\partial t}. \tag{3.6}$$

The displacement currents can also exist in vacuum. They are produced not by the motion of charges, but by the time variation of the electric field $\mathbf{E}$.

Equation (3.4) expresses the fundamental fact that *the time variation of an electric field produces a variable magnetic field*.

3.1.2 Maxwell–Faraday Equation

In 1831, Michael Faraday (1791–1867) discovered the phenomenon of *electromagnetic induction*. The integral form of *Faraday's law of induction* reads

$$\mathcal{E} = \oint_C \mathbf{E} \cdot d\mathbf{l} = -\frac{d}{dt} \int_S \mathbf{B} \cdot d\mathbf{S} = -\frac{d\Phi_m}{dt}, \tag{3.7}$$

where $\mathcal{E}$ is the *electromotive force*, S is an open surface bounded by a simple (or Jordanian[2]) closed contour C (Fig. 3.1), both being time-independent, $\Phi_m = \int_S \mathbf{B} \cdot d\mathbf{S}$ is the magnetic flux through the surface S, while $\mathbf{E}$ is the electric field induced in the circuit C by the changing magnetic flux Φ_m. The SI unit for magnetic flux is the *weber*, with the symbol Wb, named in honour of Wilhelm Eduard Weber (1804–1891). The weber is the magnetic flux that, linking a circuit of one turn, would produce in it an electromotive force of 1 volt if it were reduced to zero at a uniform rate in 1 second.

The flux Φ_m can be varied by changing the magnetic induction $\mathbf{B}$ or by changing the shape, or orientation, or position of the circuit.

Electromagnetic induction was discovered, independently and almost at the same time, by Joseph Henry (1797–1878). The quantitative expression (3.7) of Faraday's law was given by Franz Neumann (1798–1895) in 1845.

---

[2]A contour C is called *simple* or *Jordanian* if $x_i(t_1) = x_i(t_2)$ only for $t_1 = t_2$, where $x_i = x_i(t)$, $t_1 \leq t \leq t_2$, $(i = 1, 2, 3)$ are the parametric equations of the contour. In other words, a contour is called Jordanian if it does not intersect itself.

Fig. 3.1 A closed circuit in
a variable magnetic field.
According to Faraday's law
of induction, a current is
induced in the circuit.

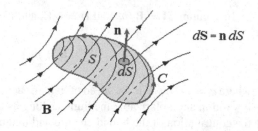

The sign "minus" in (3.7) expresses the *Lenz rule: the magnetic flux produced by
the induced current opposes the flux variation that generated it*. Note that in a variable
regime characterized by $\oint_C \mathbf{E} \cdot d\mathbf{l} \neq 0$, the induced electric field *is not conservative*.

Maxwell generalized the induction law (3.7) admitting its universal validity, which
means that the law is valid even if C is an arbitrary closed curve (i.e. not necessarily
a conductor), in a medium or even in vacuum.

The theory presented in this chapter is based on the assumption that the medium
and the observer are at rest with respect to each other. Mathematically, this assumption
is expressed by the fact that the fields $\mathbf{E}$ and $\mathbf{B}$ do not *explicitly* depend on the
coordinates x, y, z. Consequently, we may write

$$\frac{d}{dt} \int_S \mathbf{B} \cdot d\mathbf{S} = \int_S \frac{\partial \mathbf{B}}{\partial t} \cdot d\mathbf{S}.$$

By means of the Stokes theorem (A.34), Eq. (3.7) leads to

$$\int_S \left(\nabla \times \mathbf{E} + \frac{\partial \mathbf{B}}{\partial t} \right) \cdot d\mathbf{S} = 0,$$

or, since the surface S is arbitrary,

$$\nabla \times \mathbf{E} = -\frac{\partial \mathbf{B}}{\partial t}. \tag{3.8}$$

This is the *Maxwell–Faraday equation*. It expresses the fact that *a time variable
magnetic field gives rise to a (spatial) variable electric field, both in material media
and/or in vacuum*.

Equations (3.3) and (3.8) are not enough to describe the space-time variation of
the fields $\mathbf{E}$ and $\mathbf{B}$. To complete the system, we must add the differential forms of
Gauss's theorem (1.96) and (2.48), describing the nature of the fields (with or without
sources).

We can conclude that the electromagnetic field in any medium, including vacuum,
is described by the following system of linear partial differential equations of the
first order

$$\nabla \times \mathbf{H} = \mathbf{j} + \frac{\partial \mathbf{D}}{\partial t},$$

$$\nabla \times \mathbf{E} = -\frac{\partial \mathbf{B}}{\partial t},$$

$$\nabla \cdot \mathbf{B} = 0, \qquad (3.9)$$

$$\nabla \cdot \mathbf{D} = \rho.$$

We emphasize that in this manner of writing the Maxwell equations, $\mathbf{j}$ represents the free current density and ρ – the free charge density. In order to solve them, one needs as well the constitutive relations $\mathbf{D}(\mathbf{E})$ and $\mathbf{H}(\mathbf{B})$, which for the vacuum read

$$\mathbf{D} = \epsilon_0 \mathbf{E}, \quad \mathbf{H} = \mathbf{B}/\mu_0. \qquad (3.10)$$

Equivalently, Maxwell's equations can be written only in terms of the fields $\mathbf{E}$ and $\mathbf{B}$,

$$\nabla \times \mathbf{B} = \mu_0 \mathbf{J} + \epsilon_0 \mu_0 \frac{\partial \mathbf{E}}{\partial t},$$

$$\nabla \times \mathbf{E} = -\frac{\partial \mathbf{B}}{\partial t},$$

$$\nabla \cdot \mathbf{B} = 0, \qquad (3.11)$$

$$\nabla \cdot \mathbf{E} = \frac{\rho_{total}}{\epsilon_0},$$

but in this case $\mathbf{J}$ and ρ_{total} represent the densities of total current (free and bound) and total charge (free and bound), respectively.

Moreover, for a complete description of the electromagnetic field, we have to define the energy of the electromagnetic field as

$$W_{em} = \frac{1}{2} \int_V (\mathbf{E} \cdot \mathbf{D} + \mathbf{H} \cdot \mathbf{B}) \, d\tau, \qquad (3.12)$$

which in vacuum becomes

$$W_{em} = \frac{1}{2}\epsilon_0 \int_V E^2 d\tau + \frac{1}{2\mu_0} \int_V B^2 d\tau. \qquad (3.13)$$

Maxwell's equations can be grouped as follows:

(a) Equations $(3.11)_{1,2}$ show the space-time variation of the field components $\mathbf{E}, \mathbf{B}$. They are called *evolution equations*. The remaining equations $(3.11)_{3,4}$ are *condition equations* (see Sect. 3.2).
(b) Equations $(3.11)_{1,4}$ are called *source equations*, while $(3.11)_{2,3}$ are *source-free equations*. The names are obvious.

Maxwell's system of equations expresses the fundamental principles of the theory of electromagnetic field. According to these equations, the variable electric and magnetic fields cannot be separated; they represent a unity, called by Maxwell *electromagnetic field*, while the electrostatic and magnetostatic fields appear as special cases of the electromagnetic field.

3.2   Maxwell's Equations for Polarizable Media

The discovery of the electron by J.J. Thomson (1856–1940) in 1895 (recall that Maxwell died in 1879) led to some conclusions that contradict the concept of electric charge continuity embraced by Maxwell's theory. At the end of the 19th century, the Dutch physicist Hendrik Antoon Lorentz (1853–1928) elaborated the microscopic electrodynamics, taking into account the atomic structure of matter. In this theory, the material media are regarded as a set of charged particles, electrons and atomic nuclei, moving in vacuum.

According to Lorentz, the charged particles in a material give rise to a microscopic electromagnetic field $\mathbf{e}, \mathbf{b}$, while the *macroscopic* fields $\mathbf{E}, \mathbf{D}, \mathbf{H}, \mathbf{B}$ are obtained by taking the space and time average of the corresponding microscopic quantities.

The interactions between microscopic charges and microscopic fields take place in vacuum. Lorentz postulated the *microscopic field equations* as being Maxwell's equations for vacuum:

$$\frac{1}{\mu_0} \nabla \times \mathbf{b} = \mathbf{j}_{micro} + \epsilon_0 \frac{\partial \mathbf{e}}{\partial t},$$

$$\nabla \times \mathbf{e} = -\frac{\partial \mathbf{b}}{\partial t},$$

$$\nabla \cdot \mathbf{b} = 0, \tag{3.14}$$

$$\epsilon_0 \nabla \cdot \mathbf{e} = \rho_{micro},$$

where $\mathbf{j}_{micro}$ and ρ_{micro} are the microscopic current and charge densities, respectively. Equations (3.14) are called *Maxwell–Lorentz equations*.

The microscopic electromagnetic field is characterized by a rapid variation over small distances, of the order of the atomic dimensions (approx. 10^{-10} m). The measuring instruments cannot follow these variations, so that what we can detect are only the *mean values* of these quantities, taken over large space and time intervals as compared to the microscopic ones. These average values are the *macroscopic* fields. Besides, in a medium can exist both *free* charges and currents ($\rho_{free}, \mathbf{j}_{free}$), and *bound* charges and currents (microscopic circuits) ($\rho_{bound}, \mathbf{j}_{bound}$). The free charges (currents) can become bound charges (currents) and vice-versa, depending on the intensity of the external electric and/or magnetic fields.

Under these assumptions, the obvious question which arises is: how large has to be the space and time domain of averaging? To answer this question, let us consider a microscopic quantity $\Psi(x, y, z, t)$ as being any of $\mathbf{e}, \mathbf{b}, \rho_{micro}$ or $\mathbf{j}_{micro}$. The space-time *mean value* of this quantity is by definition

$$\langle \Psi(x, y, z, t) \rangle = \frac{1}{\Delta V}\frac{1}{\Delta t} \int \Psi(x+\xi, y+\eta, z+\zeta, t+\theta)\, d\xi\, d\eta\, d\zeta\, d\theta, \quad (3.15)$$

where ΔV is a spatial vicinity of an arbitrary point $P(x, y, z)$ of the polarizable medium, and Δt is a temporal vicinity of the instant t, while the integral (3.15) is taken over the domain $\Delta V \Delta t$.

To be uniquely determined, the mean value (3.15) must be independent of the choice of ΔV and Δt. In other words, on the r.h.s. of (3.15) we mean, in fact, the limit of this expression when the product $\Delta V \Delta t$ becomes infinitely small (tends to zero). But the mathematical procedure of performing the limit is restricted by the possibility of having no particle in ΔV, meaning $\rho = 0$. To solve this problem, Lorentz defined the notion of *physically infinitesimal* domain. This is either a volume ΔV, a duration Δt, or a surface ΔS etc., each of them satisfying two conditions:

(a) it is large as compared to the space-time inhomogeneities of the microscopic quantities;
(b) it is small as compared to the space-time inhomogeneities of the macroscopic quantities, detectable by the measuring instruments.

In other words, the domain is chosen large enough to contain so many particles that the behaviour of an individual particle has a negligible effect on the average of a quantity, but it is taken small enough to let the average of a quantity follow all the changes that are observable at the macroscopic level.

For example, if 10^{-4} m is the sensitivity of the instrument and 10^{-10} m the dimension of an atom, then a line element of 10^{-7} m can be conceived as physically infinitesimal. So, if we take the limit of (3.15) in view of the above assumptions, we have

$$\langle \Psi(x, y, z, t) \rangle = \frac{1}{V}\frac{1}{\tau} \int_D \int_{\theta=-\frac{\tau}{2}}^{\theta=+\frac{\tau}{2}} \Psi(x+\xi, y+\eta, z+\zeta, t+\theta)\, d\xi\, d\eta\, d\zeta\, d\theta, \quad (3.16)$$

where D is a three-dimensional physically infinitesimal spatial domain of volume $V = \int d\xi\, d\eta\, d\zeta$, and τ – a physically infinitesimal duration.

Regarding the order of magnitude of the physically infinitesimal time interval, we have to consider the fact that the periods of the temporal microscopic fluctuations are situated in a relatively large domain of variation, going from 10^{-13} s for nuclear vibrations, up to 10^{-17} s for the orbital motion of the electrons. For example, if the sensitivity of the measurement device is 10^{-6} s, then an interval of 10^{-10} s is an infinitesimal time interval.

By *macroscopic quantity* we therefore mean the average value of the corresponding microscopic quantity over a physically infinitesimal domain:

$$\Psi_{macro} = \langle \Psi_{micro} \rangle.$$

Since the integration in (3.16) is independent of x, y, z, t, we can write

$$\nabla\langle\Psi\rangle = \langle\nabla\Psi\rangle, \quad \frac{\partial}{\partial t}\langle\Psi\rangle = \left\langle\frac{\partial\Psi}{\partial t}\right\rangle, \text{ etc.} \qquad (3.17)$$

Observations:

(a) The Maxwell–Lorentz equations are written only in terms of the field components **e** and **b**. Since in vacuum the polarization phenomenon does not exist (all field sources are included in ρ_{micro} and $\mathbf{j}_{micro}$), the macroscopic field components **D** and **H** do not have microscopic correspondents **d** and **h**.

(b) Rigorously speaking, at the macroscopic level, the result of averaging the microscopic distribution of molecular electric charge is a system of multipoles. In the considerations to follow we shall take into account only the first-order approximation, that is only the first-order multipoles (i.e. dipoles). In other words, the quadrupolar, octupolar, etc. macroscopic densities of electric charge shall be neglected, as being very small quantities. We shall limit the averaging of microscopic current densities to the non-relativistic approximation. In addition, all higher-order terms (as compared to the first-order terms, representing the macroscopic magnetization) appearing as a result of averaging, shall be neglected as being small. This operation is allowed because the molecular velocities are small (such as thermal velocities in gases, or oscillation velocities in the crystal nets of solid bodies); moreover, there are random fluctuations in velocities, so that their macroscopic mean value is very close to zero. The only exception is the global motion of the medium as a whole; this case is studied separately in Sect. 3.12.

(c) Even if at microscopic level we deal with atoms and molecules, which should be described by means of the quantum formalism, we shall use an eminently classic description. There is a double reason for this choice: on the one hand, a classical description is easier and simpler (but not simplistic); on the other, a quantum approach would replace the classical quantities with the corresponding probabilistic ones, specific to quantum mechanics, but the qualitative reasoning would remain the same.

(d) Finally, one last observation refers to the fact that spatial averaging is the only necessary. As we shall see in the following, within the first-order approximation the Lorentz approach does not imply concrete performing of any averaging, spatial or temporal. In fact, the time variable is irrelevant for systems in statistical equilibrium. The time averaging is not necessary because in a spatially small infinite there are so many electrons/nuclei, that upon spatial averaging the microscopic time fluctuations of the medium are completely erased. Indeed, the time variations of the microscopic fields, for any microscopic system at statistical equilibrium and in normal external conditions, are totally non-correlated on distances comparable to the dimensions of a spatially small infinity.

We are now prepared to perform the Lorentzian averaging of the Maxwell–Lorentz equations (3.14). By definition, we take

$$\langle\mathbf{e}\rangle = \mathbf{E} \quad \text{and} \quad \langle\mathbf{b}\rangle = \mathbf{B}. \qquad (3.18)$$

3.2.1 Source-free Equations

Using the definitions (3.18) and the properties (3.17), we have

$$\langle \nabla \times e \rangle = \nabla \times \langle e \rangle = \nabla \times \mathbf{E},$$
$$\left\langle \frac{\partial b}{\partial t} \right\rangle = \frac{\partial}{\partial t} \langle b \rangle = \frac{\partial \mathbf{B}}{\partial t},$$
$$\langle \nabla \cdot b \rangle = \nabla \cdot \langle b \rangle = \nabla \cdot \mathbf{B}.$$

Taking the average of $(3.14)_{2,3}$ in which we plug the above results, we find the corresponding Maxwell's equations for the macroscopic fields:

$$\nabla \times \mathbf{E} = -\frac{\partial \mathbf{B}}{\partial t},$$
$$\nabla \cdot \mathbf{B} = 0. \tag{3.19}$$

3.2.2 Equations with Sources

Let us start by averaging Eq. $(3.14)_4$:

$$\langle \epsilon_0 \nabla \cdot e \rangle = \epsilon_0 \langle \nabla \cdot e \rangle = \epsilon_0 \nabla \cdot \langle e \rangle = \epsilon_0 \nabla \cdot \mathbf{E} = \langle \rho_{micro} \rangle. \tag{3.20}$$

To perform the average of ρ_{micro}, we have to take into account the fact that it comprises two terms:

(a) ρ_{free}, corresponding to the free charges (electrons in metals, ions in gases and solutions, etc.);

(b) ρ_{bound}, due to the bound charges that are subjected to infinitesimally small displacements under the action of the field and form dipoles (electrons in atoms, ions in crystal nets and in neutral molecules, etc.).

Hence,

$$\langle \rho_{micro} \rangle = \langle \rho_{free} \rangle + \langle \rho_{bound} \rangle. \tag{3.21}$$

The quantity $\langle \rho_{free} \rangle$ is the macroscopic charge density as determined by the measurement instruments:

$$\langle \rho_{free} \rangle = \rho. \tag{3.22}$$

Regarding the bound charges, recall that, in the first-order approximation, they appear as elementary dipoles formed (and/or oriented) by the external electric field. The electric moment of a continuous dipole distribution is (see Eq. (1.82)):

$$\mathbf{p} = \int_{V'} \rho(\mathbf{r}')\,\mathbf{r}'d\tau'. \tag{3.23}$$

The charge density ρ in the last relation is a macroscopic quantity and appears as a result of the formation of dipoles, therefore we can write

$$\mathbf{p} = \int_{V'} \langle \rho_{bound} \rangle\,\mathbf{r}'d\tau'. \tag{3.24}$$

On the other hand, the definition (1.88) of the electric polarization intensity $\mathbf{P}$ allows us to write

$$p_i = \int_{V'} P_i(\mathbf{r}')\,d\tau' = \int_{V'} P_k\,\delta_{ik}\,d\tau' = \int_{V'} P_k\,\frac{\partial x_i'}{\partial x_k'}\,d\tau'$$

$$= \int_{V'} \frac{\partial}{\partial x_k'}(P_k x_i')\,d\tau' - \int_{V'} x_i'\,\nabla'\cdot\mathbf{P}\,d\tau'.$$

Applying now the divergence theorem to the first term on the r.h.s., and extending the integration domain over the whole space, we have

$$\int \frac{\partial}{\partial x_k'}(P_k x_i')\,d\tau' = \oint P_k\,x_i'\,dS_k' = 0,$$

because there are no sources at infinity. Thus,

$$\mathbf{p} = -\int \mathbf{r}'\,\nabla'\cdot\mathbf{P}\,d\tau'. \tag{3.25}$$

Combining now (3.24) and (3.25), we find

$$\int \left(\langle \rho_{bound} \rangle + \nabla'\cdot\mathbf{P} \right)\mathbf{r}'d\tau' = 0.$$

Since the dimensions and shape of the body are arbitrary, we arrive at

$$\langle \rho_{bound} \rangle = -\nabla'\cdot\mathbf{P} = -\mathrm{div}\,\mathbf{P}, \tag{3.26}$$

which is an already known result (see Eq. (1.92)). Substituting (3.22) and (3.26) in (3.21), then in (3.20), we obtain the expected equation

$$\nabla\cdot(\epsilon_0\mathbf{E} + \mathbf{P}) = \rho,$$

or

$$\nabla \cdot \mathbf{D} = \rho, \tag{3.27}$$

where

$$\mathbf{D} \equiv \epsilon_0 \mathbf{E} + \mathbf{P} \tag{3.28}$$

is the *electric induction*, or *electric displacement*. Obviously, in vacuum $\mathbf{P} = 0$, i.e. $\mathbf{D} = \epsilon_0 \mathbf{E}$.

Next, let us apply Lorentz's averaging procedure to the Maxwell–Ampère equation $(3.14)_1$. We have

$$\left\langle \frac{1}{\mu_0} \nabla \times \mathbf{b} \right\rangle = \langle \mathbf{j}_{micro} \rangle + \left\langle \epsilon_0 \frac{\partial \mathbf{e}}{\partial t} \right\rangle,$$

or, by virtue of (3.17),

$$\frac{1}{\mu_0} \nabla \times \mathbf{B} = \langle \mathbf{j}_{micro} \rangle + \epsilon_0 \frac{\partial \mathbf{E}}{\partial t}. \tag{3.29}$$

As was the case with the charge density, the microscopic current density $\mathbf{j}_{micro}$ is also composed of two parts:

$$\mathbf{j}_{micro} = \mathbf{j}_{free} + \mathbf{j}_{bound},$$

where:

(a) $\mathbf{j}_{free}$ is produced by the motion of free charges (free electrons in metals, ions in gases and electrolytes, etc.);
(b) $\mathbf{j}_{bound}$ is produced by the bound charges. In its turn, $\mathbf{j}_{bound}$ consists of two terms: $\mathbf{j}_{bound} = \mathbf{j}_{molec} + \mathbf{j}_{pol}$, where:
(i) $\mathbf{j}_{molec}$ is the density of molecular microscopic currents, produced by the orbital motion of electrons in atoms;
(ii) $\mathbf{j}_{pol}$ stands for the polarization microscopic current densities, which in the first-order approximation are due to the relative displacement of the poles in dipoles, or, in other words, to the polarization of molecules.

We may then write

$$\langle \mathbf{j}_{micro} \rangle = \langle \mathbf{j}_{free} \rangle + \langle \mathbf{j}_{molec} \rangle + \langle \mathbf{j}_{pol} \rangle. \tag{3.30}$$

The quantity $\langle \mathbf{j}_{free} \rangle$ is the density $\mathbf{j}$ of macroscopic current determined by the measurement instruments:

$$\langle \mathbf{j}_{free} \rangle = \langle (\rho \mathbf{v})_{free} \rangle = \langle \rho_{free} \mathbf{v}_{free} \rangle = \mathbf{j}, \tag{3.31}$$

where $\mathbf{v}_{free}$ is the velocity of the free charges with respect to a fixed observer. The current density $\mathbf{j}$ is generally formed of a conduction and a convection part: $\mathbf{j} = \mathbf{j}_{cond} + \mathbf{j}_{conv}$. In metals, $\mathbf{j}_{conv} = 0$, such that $\mathbf{j} = \mathbf{j}_{cond}$.

The current density $\mathbf{j}_{molec}$ is created by the motion of electrons in atoms, meaning that the atoms behave like elementary magnets (magnetic sheets). By the definition (2.61), the dipole magnetic moment of a continuous distribution of such dipoles is

$$\mathbf{m} = \frac{1}{2} \int_{V'} \mathbf{r}' \times \langle \mathbf{j}(\mathbf{r}')_{molec} \rangle \, d\tau'. \tag{3.32}$$

On the other hand, the definition (2.70) gives

$$\mathbf{m} = \int_{V'} \mathbf{M}(\mathbf{r}') \, d\tau'. \tag{3.33}$$

To compare the last two relations we use (A.46) and write

$$\nabla'(\mathbf{r}' \cdot \mathbf{M}) = \mathbf{r}' \times (\nabla' \times \mathbf{M}) + (\mathbf{r}' \cdot \nabla')\mathbf{M} + (\mathbf{M} \cdot \nabla')\mathbf{r}'.$$

In a Cartesian reference frame, the last term can be written as

$$(\mathbf{M} \cdot \nabla')\mathbf{r}' = M_i \frac{\partial}{\partial x_i'}(x_k' \mathbf{u}_k') = M_i \delta_{ik} \mathbf{u}_k' = M_i \mathbf{u}_i' = \mathbf{M},$$

leading to

$$\int_{V'} \mathbf{M}(\mathbf{r}') \, d\tau' = \int_{V'} \nabla'(\mathbf{r}' \cdot \mathbf{M}) \, d\tau'$$
$$- \int_{V'} \mathbf{r}' \times (\nabla' \times \mathbf{M}) \, d\tau' - \int_{V'} (\mathbf{r}' \cdot \nabla')\mathbf{M} \, d\tau'. \tag{3.34}$$

Applying (A.37) and extending the integration domain over the whole space, one observes that

$$\int \nabla'(\mathbf{r}' \cdot \mathbf{M}) \, d\tau' = \oint (\mathbf{r}' \cdot \mathbf{M}) \, d\mathbf{S}' = 0,$$

because at infinity there are neither charges, nor currents. This procedure is legitimate, since the electric charges are localized and the fields fall off very fast at a distance from the charges.

Let us consider the x_i-component of the last integral in (3.34):

$$-\int x_k' \frac{\partial M_i}{\partial x_k'} \, d\tau' = -\int \frac{\partial}{\partial x_k'}(x_k' M_i) \, d\tau' + \int M_i \frac{\partial x_k'}{\partial x_k'} \, d\tau'$$
$$= -\oint x_k' M_i \, dS_k' + 3\int M_i \, d\tau' = 3\int M_i \, d\tau',$$

since the surface integral vanishes for the already known reasons. In this case (3.34) leads to

$$\mathbf{m} = \int_{V'} \mathbf{M} \, d\tau' = \frac{1}{2} \int_{V'} \mathbf{r}' \times (\nabla' \times \mathbf{M}) \, d\tau'. \tag{3.35}$$

Comparing now (3.32) and (3.35), we find (since the dimensions and shape of the body are arbitrary):

$$\langle \mathbf{j}_{molec} \rangle = \nabla' \times \mathbf{M} = \text{curl } \mathbf{M}. \tag{3.36}$$

To calculate the last term of (3.30) we remember that, in the first-order approximation, $\mathbf{j}_{pol}$ is due to the time variation of the distance l between the dipole charges, so that

$$\langle \mathbf{j}_{pol} \rangle = ne\dot{\mathbf{l}} = \frac{\partial}{\partial t}(ne\mathbf{l}) = \frac{\partial \mathbf{P}}{\partial t}, \tag{3.37}$$

because

$$ne\mathbf{l} = \frac{N\mathbf{p}}{V} = \mathbf{P},$$

where we denoted by e the charge of one electron. Then we can write

$$\langle \mathbf{j}_{bound} \rangle = \nabla \times \mathbf{M} + \frac{\partial \mathbf{P}}{\partial t}. \tag{3.38}$$

It is worthwhile to observe that the last relation follows as an immediate consequence of averaging the equation of continuity written for bound charges in the first-order approximation, when $\langle \rho_{bound} \rangle = \frac{\partial P}{\partial t}$. Indeed,

$$\nabla \cdot (\langle \mathbf{j}_{bound} \rangle) = -\frac{\partial}{\partial t} \langle \rho_{bound} \rangle = \frac{\partial}{\partial t}(\nabla \cdot \mathbf{P}) = \nabla \cdot \frac{\partial \mathbf{P}}{\partial t},$$

which proves the statement.

Using (3.31) and (3.38) in (3.30), we find

$$\langle \mathbf{j}_{micro} \rangle = \mathbf{j} + \nabla \times \mathbf{M} + \frac{\partial \mathbf{P}}{\partial t}, \tag{3.39}$$

and (3.29) becomes

$$\nabla \times \left(\frac{\mathbf{B}}{\mu_0} - \mathbf{M} \right) = \mathbf{j} + \frac{\partial}{\partial t}(\epsilon_0 \mathbf{E} + \mathbf{P}). \tag{3.40}$$

Using the familiar notations

$$\frac{\mathbf{B}}{\mu_0} - \mathbf{M} = \mathbf{H},$$
$$\epsilon_0 \mathbf{E} + \mathbf{P} = \mathbf{D}, \tag{3.41}$$

we finally obtain

$$\nabla \times \mathbf{H} = \mathbf{j} + \frac{\partial \mathbf{D}}{\partial t}. \tag{3.42}$$

Observations:

(a) We shall continue to call the equations established in this section *Maxwell's equations*, without ignoring Lorentz's contribution to the explanation of the phenomena in polarizable media;

(b) We remind the reader that our study concerns only *linear media*;

(c) Taking the divergence of the source-free equation $(3.19)_1$, we have

$$\nabla \cdot (\nabla \times \mathbf{E}) = -\frac{\partial}{\partial t}(\nabla \cdot \mathbf{B}) = 0,$$

meaning that the source-free equations are not completely independent. As one can see, equation $\nabla \cdot \mathbf{B} = 0$ appears as an *initial condition* for the equation $\nabla \times \mathbf{E} = -\frac{\partial \mathbf{B}}{\partial t}$. Similarly, taking the divergence of the source equation (3.42), we find

$$\nabla \cdot (\nabla \times \mathbf{H}) = -\frac{\partial \rho}{\partial t} + \frac{\partial}{\partial t}(\nabla \cdot \mathbf{D}) = \frac{\partial}{\partial t}(\nabla \cdot \mathbf{D} - \rho) = 0.$$

Therefore, the source equations, in their turn, are not completely independent, and equation $\nabla \cdot \mathbf{D} = \rho$ acts as an initial condition for the equation $\nabla \times \mathbf{H} = \mathbf{j} + \frac{\partial \mathbf{D}}{\partial t}$. For these reasons we called Maxwell's equations $\nabla \cdot \mathbf{B} = 0$, and $\nabla \cdot \mathbf{D} = \rho$ *equations of condition*;

(d) Since strictly independent are only Maxwell's equations of evolution (six equations when written in components), while the number of unknown quantities is 16 (E_i, D_i, H_i, B_i, j_i, ρ, with $i = 1, 2, 3$), the system has to be completed with 10 more equations. Nine of them are the *constitutive relations*:

$$\mathbf{D} = \mathbf{D}(\mathbf{E}), \quad \mathbf{B} = \mathbf{B}(\mathbf{H}), \quad \mathbf{j} = \mathbf{j}(\mathbf{E}).$$

In linear, homogeneous, and isotropic media these relations are

$$\mathbf{D} = \epsilon \mathbf{E}, \quad \mathbf{B} = \mu \mathbf{H}, \quad \mathbf{j} = \lambda(\mathbf{E} + \mathbf{E}^{ext}).$$

The tenth equation is the equation of continuity. We then have a system of 16 linear partial differential equations with 16 unknown quantities. To connect the field and mechanical quantities, one postulates the expression of the electromagnetic energy (3.12).

Maxwell's equations, together with the constitutive relations, the equation of continuity, and the definition of electromagnetic energy (if no discontinuity surfaces are present, i.e. in infinite media) represent the *fundamental axioms of Maxwell's phenomenological electrodynamics*.

3.3 Jump Conditions

We want to specify, for the beginning, that the jump conditions (1.103), (1.105), (2.85), and (2.87), obtained separately for the electrostatic and magnetostatic fields, remain unchanged in a variable regime. Let us write them out, at the same time indicating the integral fundamental law from which they emerge:

$$\oint \mathbf{D} \cdot d\mathbf{S} = Q \quad \rightarrow \quad D_{2n} - D_{1n} = \sigma,$$

$$\oint \mathbf{E} \cdot d\mathbf{l} = 0 \quad \rightarrow \quad E_{2T} - E_{1T} = 0,$$

$$\oint \mathbf{B} \cdot d\mathbf{S} = 0 \quad \rightarrow \quad B_{2n} - B_{1n} = 0, \tag{3.43}$$

$$\oint \mathbf{H} \cdot d\mathbf{l} = I \quad \rightarrow \quad H_{2T} - H_{1T} = i_N.$$

To check up these assertions, we first observe that Eqs. $(3.43)_1$ and $(3.43)_3$ have been obtained using Gauss's theorem for the fields $\mathbf{D}$ and $\mathbf{B}$, respectively, that do not change in a variable regime.

To ascertain the Eq. $(3.43)_2$, let us integrate Maxwell's equation $(3.19)_1$ over a fixed surface S bounded by the contour C:

$$\int_S (\nabla \times \mathbf{E}) \cdot d\mathbf{S} = \oint_C \mathbf{E} \cdot d\mathbf{l} = - \int_S \left(\frac{\partial \mathbf{B}}{\partial t} \right) \cdot d\mathbf{S} = -\frac{d}{dt} \int_S \mathbf{B} \cdot d\mathbf{S}.$$

Using Fig. 1.11, one calculates the circulation on the closed contour $ABCDA$ by means of a procedure analogous to that given in Sect. 1.2.4. In the limit $dl \rightarrow 0$, the surface dS vanishes, therefore $\int_{dS} \mathbf{B} \cdot d\mathbf{S} \rightarrow 0$. We are then left with $(3.43)_2$.

Finally, to justify the Eq. $(3.43)_4$ let us integrate Maxwell's equation (3.42) on the above defined fixed surface S. We have

$$\int_S \nabla \times \mathbf{H} \cdot d\mathbf{S} = \oint \mathbf{H} \cdot d\mathbf{l} = \int_S \mathbf{j} \cdot d\mathbf{S} + \frac{d}{dt} \int_S \mathbf{D} \cdot d\mathbf{S}.$$

Choosing again $ABCDA$ as the integration contour, and observing that $\lim_{dl \rightarrow 0} \int_S \mathbf{D} \cdot d\mathbf{S} = 0$, we arrive at the jump relation $(3.43)_4$.

Let us now write again the jump relations, also showing the corresponding Maxwell's equations (in their partial differential form):

$$\nabla \cdot \mathbf{D} = \rho \qquad \rightarrow \quad D_{2n} - D_{1n} = \sigma,$$

$$\nabla \times \mathbf{E} = -\frac{\partial \mathbf{B}}{\partial t} \qquad \rightarrow \quad E_{2T} - E_{1T} = 0,$$

$$\nabla \cdot \mathbf{B} = 0 \qquad \rightarrow \quad B_{2n} - B_{1n} = 0, \qquad (3.44)$$

$$\nabla \times \mathbf{H} = \mathbf{j} + \frac{\partial \mathbf{D}}{\partial t} \qquad \rightarrow \quad H_{2T} - H_{1T} = i_N.$$

These equations are applied whenever the field quantities $\mathbf{E}$, $\mathbf{D}$, $\mathbf{H}$, $\mathbf{B}$ suffer refractions when passing through sheets separating media with different material constants ϵ, μ, and λ. Consequently, for problems involving the existance of such surfaces, Maxwell's equations must be supplemented with the jump conditions (3.44).

3.4   Electromagnetic Field Energy. Poynting's Theorem

As we mentioned in Sect. 3.1, one postulates the energy of the electromagnetic field as being

$$W_{em} = \frac{1}{2} \int_V (\mathbf{E} \cdot \mathbf{D} + \mathbf{H} \cdot \mathbf{B}) \, d\tau, \qquad (3.45)$$

where the integration is performed over a volume V, bounded by the surface S.

Our purpose is to evaluate the time variation of the energy W_{em}. Taking the total time derivative of (3.45), we have

$$\frac{dW_{em}}{dt} = \frac{1}{2} \int_V \frac{\partial}{\partial t} (\mathbf{E} \cdot \mathbf{D} + \mathbf{H} \cdot \mathbf{B}) \, d\tau.$$

Since the medium is linear, homogeneous, and isotropic (this assumption has been previously made), the constitutive relations are

$$\mathbf{D} = \epsilon \mathbf{E}, \quad \mathbf{B} = \mu \mathbf{H}, \quad \mathbf{j} = \lambda(\mathbf{E} + \mathbf{E}^{ext}). \qquad (3.46)$$

Then, we can write

$$\frac{dW_{em}}{dt} = \frac{1}{2} \int_V \frac{\partial}{\partial t} \left(\epsilon \mathbf{E}^2 + \mu \mathbf{H}^2 \right) d\tau$$

$$= \int_V \left(\epsilon \mathbf{E} \cdot \frac{\partial \mathbf{E}}{\partial t} + \mu \mathbf{H} \cdot \frac{\partial \mathbf{H}}{\partial t} \right) d\tau = \int_V \left(\mathbf{E} \cdot \frac{\partial \mathbf{D}}{\partial t} + \mathbf{H} \cdot \frac{\partial \mathbf{B}}{\partial t} \right) d\tau.$$

Using Maxwell's evolution equations $\nabla \times \mathbf{H} = \mathbf{j} + \frac{\partial \mathbf{D}}{\partial t}$, $\nabla \times \mathbf{E} = -\frac{\partial \mathbf{B}}{\partial t}$, we find

$$\frac{dW_{em}}{dt} = \int_V [\mathbf{E} \cdot (\nabla \times \mathbf{H}) - \mathbf{H} \cdot (\nabla \times \mathbf{E})]\, d\tau - \int_V \mathbf{j} \cdot \mathbf{E}\, d\tau$$

$$= -\int_V \nabla \cdot (\mathbf{E} \times \mathbf{H})\, d\tau - \int_V \mathbf{j} \cdot \mathbf{E}\, d\tau,$$

or, using the divergence theorem,

$$\frac{dW_{em}}{dt} = -\oint_S (\mathbf{E} \times \mathbf{H}) \cdot d\mathbf{S} - \int_V \mathbf{j} \cdot \mathbf{E}\, d\tau. \tag{3.47}$$

Denoting *Poynting's vector*

$$\mathbf{\Pi} = \mathbf{E} \times \mathbf{H} \tag{3.48}$$

and using $(3.46)_3$, we finally obtain

$$\frac{dW_{em}}{dt} = -\oint_S \mathbf{\Pi} \cdot d\mathbf{S} - \int_V \frac{\mathbf{j}^2}{\lambda}\, d\tau + \int_V \mathbf{j} \cdot \mathbf{E}^{ext} d\tau. \tag{3.49}$$

This relation expresses the energy conservation law in electromagnetic processes and is known as *Poynting's theorem*, after the English physicist John Poynting (1852–1914) who pubished it in 1884. The first term on the r.h.s. is the flux of the electromagnetic energy passing through the surface S in unit time. This shows that Poynting's vector $\mathbf{\Pi}$ signifies *the radiant flux, crossing a unit area in unit time, orthogonal to* $\mathbf{\Pi}$. In other words, the normal component Π_n of Poynting's vector represents *the power density dissipated as electromagnetic radiation*. The second term in (3.49) represents the Joule heat dissipated in unit time and unit volume by the conduction currents, while the last term gives the mechanical work done on conduction currents, in unit time, by the external electromotive forces.

Thus, Poynting's theorem can be formulated as follows: *the electromagnetic energy is dissipated into electromagnetic radiation and Joule heat, being recovered from the external sources (if they exist).*

Observing that

$$\frac{dW_{mec}}{dt} = \int_V \mathbf{j} \cdot \mathbf{E}\, d\tau$$

is the mechanical work done in unit time by the field $\mathbf{E}$ on the currents $\mathbf{j}$, and neglecting the external sources, we can give an alternative formulation of Poynting's theorem:

$$\frac{d}{dt}(W_{em} + W_{mec}) = -\oint_S \mathbf{\Pi} \cdot d\mathbf{S}. \tag{3.50}$$

According to (3.50), if the system is closed (there is no radiation), the sum $W = W_{mec} + W_{em}$ is conserved.

By definition, the quantity

$$w_{em} = \frac{1}{2}(\mathbf{E} \cdot \mathbf{D} + \mathbf{H} \cdot \mathbf{B}) \qquad (3.51)$$

is the *electromagnetic energy density*, and (3.47) can also be written as

$$\frac{\partial w_{em}}{\partial t} + \nabla \cdot \mathbf{\Pi} = -\mathbf{j} \cdot \mathbf{E}.$$

If $\mathbf{j} \cdot \mathbf{E} = 0$ (for example, if $\lambda = 0$), the last equation reduces to

$$\frac{\partial w_{em}}{\partial t} + \nabla \cdot \mathbf{\Pi} = 0, \qquad (3.52)$$

which is an equation of continuity.

> *Observation*:
> There exist mechanisms of dissipation of electromagnetic energy which we have not considered. This is the case, for example, when the electromagnetic energy is dissipated as a result of *hysteresis*.

3.5   Uniqueness of the Solutions of Maxwell's Equations

Poynting's theorem is an essential ingredient in finding the boundary conditions under which Maxwell's equations (supplemented with constitutive relations, jump conditions, and the definition of the electromagnetic energy), have a unique solution. Let us prove the following

Theorem. If at the initial moment $t = 0$ are given $\mathbf{E}(\mathbf{r}, 0)$, $\mathbf{H}(\mathbf{r}, 0)$ at any point of the spatial domain D, of volume V, bounded by the surface S, while on the boundary S is given either $E_T(\mathbf{r}, t)$ or $H_T(\mathbf{r}, t)$ for $0 \leq t \leq t_1$, then the electromagnetic field is uniquely determined by Maxwell's equations at any time $t_1 > t$.

The proof is based on *reductio ad absurdum*. Let us assume that there exist at least two solutions of Maxwell's equations, denoted by $\mathbf{E}_1$, $\mathbf{H}_1$ and $\mathbf{E}_2$, $\mathbf{H}_2$. The linearity of Maxwell's equations implies the superposition property, meaning that the field $\mathbf{E}'$, $\mathbf{H}'$, given by

$$\mathbf{E}' = \mathbf{E}_1 - \mathbf{E}_2,$$
$$\mathbf{H}' = \mathbf{H}_1 - \mathbf{H}_2,$$

is also a solution of Maxwell's equations, satisfying both the initial and the boundary conditions:

$$\mathbf{E}'(\mathbf{r}, 0) = \mathbf{E}_1(\mathbf{r}, 0) - \mathbf{E}_2(\mathbf{r}, 0) = 0,$$
$$\mathbf{H}'(\mathbf{r}, 0) = \mathbf{H}_1(\mathbf{r}, 0) - \mathbf{H}_2(\mathbf{r}, 0) = 0, \qquad (3.53)$$

with

$$\text{either} \quad E'_T(\mathbf{r}, t) = 0 \quad \text{or} \quad H'_T(\mathbf{r}, t) = 0, \quad \text{for} \ 0 \leq t \leq t_1. \tag{3.54}$$

It should be emphasized that, if $\mathbf{E}_1$, $\mathbf{H}_1$ and $\mathbf{E}_2$, $\mathbf{H}_2$ are solutions of Maxwell's equations with sources, then $\mathbf{E}'$, $\mathbf{H}'$ is a solution of the sourceless Maxwell's equations, consequently it satisfies the sourceless Poynting's theorem:

$$\frac{d \, W'_{em}}{dt} = \frac{1}{2} \int_V \frac{\partial}{\partial t} \left(\epsilon E'^2 + \mu H'^2 \right) = - \oint_S (\mathbf{E}' \times \mathbf{H}') \cdot d\mathbf{S} \tag{3.55}$$

One observes that, on the one hand,

$$W'_{em} = \frac{1}{2} \int_V \left(\epsilon E'^2 + \mu H'^2 \right) d\tau \geq 0$$

with

$$W'_{em} = 0 \quad \text{at} \ \ t = 0, \tag{3.56}$$

due to (3.53). On the other hand, we have at any time $0 \leq t \leq t_1$

$$P'_{rad}(t) = \oint (\mathbf{E}' \times \mathbf{H}') \cdot \mathbf{n} \, dS = 0, \tag{3.57}$$

since, due to the boundary conditions (3.54),

$$(\mathbf{E}' \times \mathbf{H}') \cdot \mathbf{n} = (\mathbf{n} \times \mathbf{E}') \cdot \mathbf{H}' = (\mathbf{H}' \times \mathbf{n}) \cdot \mathbf{E}' = 0.$$

(Recall that $\mathbf{n} \times \mathbf{E}'$ picks up the tangent component of $\mathbf{E}'$ and similarly $\mathbf{H}' \times \mathbf{n}$ picks up the tangent component of $\mathbf{H}'$.) Consequently, W'_{em} is constant on the time interval $0 \leq t \leq t_1$ and, having in view (3.58), we infer that

$$W'_{em}(t) = 0 \quad \text{for} \ \ 0 \leq t \leq t_1. \tag{3.58}$$

This is possible if and only if $\mathbf{E}_1(\mathbf{r}, t) = \mathbf{E}_2(\mathbf{r}, t)$ and $\mathbf{H}_1(\mathbf{r}, t) = \mathbf{H}_2(\mathbf{r}, t)$, which proves the theorem.

Remark that the crucial step of the proof was to show that the radiated power $P'_{rad}(t)$ vanishes at any moment, which was possible due to the specific boundary conditions involving the tangent components of the fields. In other words, Poynting's theorem indicates unequivocally the type of boundary conditions needed to prove the uniqueness of the solutions of Maxwell's equations. When $V \to \infty$, uniqueness is guaranteed if the fields go to zero fast enough for $r \to 0$, such that the r.h.s. of (3.55) vanishes.

3.6   Electromagnetic Momentum. Momentum Theorem

The electromagnetic field $(\mathbf{E}, \mathbf{B})$ acts on a charge q moving in the field with the force

$$\mathbf{F}_{em} = q(\mathbf{E} + \mathbf{v} \times \mathbf{B}), \tag{3.59}$$

called *electromagnetic force*. If the charge q is continuously distributed in a volume V, bounded by the surface S, then the electromagnetic force density is

$$\mathbf{f}_{em} = \rho(\mathbf{E} + \mathbf{v} \times \mathbf{B}) = \rho\mathbf{E} + \mathbf{j} \times \mathbf{B},$$

where ρ is the charge density, and $\mathbf{j}$ – the conduction current density. The equation of motion of the charge $q = \int_V \rho \, d\tau$ in the electromagnetic field $(\mathbf{E}, \mathbf{B})$ is then

$$\mathbf{F}_{mec} = \frac{d\mathbf{P}_{mec}}{dt} = \int_V (\rho\mathbf{E} + \mathbf{j} \times \mathbf{B}) \, d\tau. \tag{3.60}$$

Using Maxwell's equations

$$\nabla \cdot \mathbf{D} = \rho,$$

$$\nabla \times \mathbf{H} = \mathbf{j} + \frac{\partial \mathbf{D}}{\partial t},$$

$$\nabla \times \mathbf{E} = -\frac{\partial \mathbf{B}}{\partial t},$$

we shall write (3.60) in terms of fields only. Thus, we have

$$\rho\mathbf{E} + \mathbf{j} \times \mathbf{B} = \mathbf{E}\nabla \cdot \mathbf{D} + (\nabla \times \mathbf{H}) \times \mathbf{B} - \frac{\partial \mathbf{D}}{\partial t} \times \mathbf{B}$$

$$= \mathbf{E}\nabla \cdot \mathbf{D} - \mathbf{D} \times (\nabla \times \mathbf{E}) - \mathbf{B} \times (\nabla \times \mathbf{H}) - \frac{\partial}{\partial t}(\mathbf{D} \times \mathbf{B}).$$

Adding the null term $\mathbf{H}\nabla \cdot \mathbf{B}$ on the r.h.s. and integrating over the fixed volume V, we find

$$\frac{d}{dt}\mathbf{P}_{mec} + \frac{d}{dt}\int_V \mathbf{D} \times \mathbf{B} \, d\tau \tag{3.61}$$

$$= \int_V [\mathbf{E}\nabla \cdot \mathbf{D} - \mathbf{D} \times (\nabla \times \mathbf{E}) + \mathbf{H}\nabla \cdot \mathbf{B} - \mathbf{B} \times (\nabla \times \mathbf{H})] \, d\tau.$$

The integral on the l.h.s. of (3.61) has the dimension of momentum. Let us denote it by $\mathbf{P}_{em}$ and call this quantity *momentum of the electromagnetic field*, or, shorter, *electromagnetic momentum*:

$$\mathbf{P}_{em} = \int_V \mathbf{D} \times \mathbf{B} \, d\tau. \tag{3.62}$$

Its volume density is

$$\mathbf{p}_{em} = \mathbf{D} \times \mathbf{B},$$

and in vacuum it has the expression

$$\mathbf{p}_{em} = \epsilon_0 \mu_0 \, \mathbf{E} \times \mathbf{H} = \frac{1}{c^2} \mathbf{\Pi}, \tag{3.63}$$

$\mathbf{\Pi}$ being Poynting's vector (3.48).

The r.h.s. of (3.61) can be written in a more convenient form. Thus, the x_i-component of the first two terms is

$$\epsilon \left(E_i \frac{\partial E_k}{\partial x_k} - \varepsilon_{ijk} \varepsilon_{klm} E_j \frac{\partial E_m}{\partial x_l} \right) = \epsilon \left[E_i \frac{\partial E_k}{\partial x_k} - (\delta_{il}\delta_{jm} - \delta_{im}\delta_{jl}) E_j \frac{\partial E_m}{\partial x_l} \right]$$

$$= \epsilon \left[E_i \frac{\partial E_k}{\partial x_k} + E_l \frac{\partial E_i}{\partial x_l} - \frac{1}{2} \frac{\partial}{\partial x_i} (E_m E_m) \right]$$

$$= \frac{\partial}{\partial x_k} \left(E_i D_k - \frac{1}{2} \mathbf{E} \cdot \mathbf{D} \, \delta_{ik} \right).$$

The quantity in parentheses is nothing else but Maxwell's electric stress tensor, defined in formula (1.176) in Chap. 1 for the more general case of the variation of permittivity with respect to the density. Denoting

$$T_{ik}^{(e)} = \frac{1}{2} \mathbf{E} \cdot \mathbf{D} \, \delta_{ik} - E_i D_k, \tag{3.64}$$

we have

$$[\mathbf{E} \, \nabla \cdot \mathbf{D} - \mathbf{D} \times (\nabla \times \mathbf{E})]_i = -\frac{\partial}{\partial x_k} T_{ik}^{(e)}.$$

A similar calculation for the magnetic part gives

$$[\mathbf{H} \, \nabla \cdot \mathbf{B} - \mathbf{B} \times (\nabla \times \mathbf{H})]_i = -\frac{\partial}{\partial x_k} T_{ik}^{(m)},$$

with

$$T_{ik}^{(m)} = \frac{1}{2} \mathbf{H} \cdot \mathbf{B} \, \delta_{ik} - H_i B_k. \tag{3.65}$$

The x_i-component of the vectorial equation (3.61) is then

$$\frac{d}{dt} (\mathbf{P}_{mec} + \mathbf{P}_{em})_i = -\int_V \frac{\partial T_{ik}}{\partial x_k} \, d\tau = -\oint_S T_{ik} \, dS_k, \tag{3.66}$$

where

$$T_{ik} = T_{ik}^{(e)} + T_{ik}^{(m)} = \frac{1}{2}(\mathbf{E} \cdot \mathbf{D} + \mathbf{H} \cdot \mathbf{B})\, \delta_{ik} - E_i D_k - H_i B_k \qquad (3.67)$$

is *Maxwell's electromagnetic stress tensor*.

The term "stress" comes from Maxwell's concept of the *æther* as being a perfectly elastic medium, which gets deformed under the action of the electromagnetic field, in other words the existence of the electromagnetic forces leads to the appearance of elastic tensions which deform the æther. Thus, Maxwell's electromagnetic stress tensor is analogous to the *elastic stress tensor*, which expresses the deformations in elastic media. Equation (3.66) can also be written in a vector form, which is

$$\frac{d}{dt}(\mathbf{P}_{mec} + \mathbf{P}_{em}) = - \oint_S \boldsymbol{T} \cdot d\mathbf{S} = - \oint_S \boldsymbol{T} \cdot \mathbf{n}\, dS, \qquad (3.68)$$

where the *dyadic form* of the tensor T_{ik} has been used. Therefore, Maxwell's stress tensor represents the momentum flux density, per unit time, of the electromagnetic field. In other words, the *electromagnetic energy flux, of density* $\boldsymbol{\Pi}$, *is accompanied by an electromagnetic momentum flux, of density* $\boldsymbol{T}$.

If the integration boundary surface goes to infinity, or if on S we have both $\mathbf{E} = 0$, and $\mathbf{H} = 0$, then it follows from (3.68) that the momentum is conserved:

$$\mathbf{P}_{mec} + \mathbf{P}_{em} = \text{const.}$$

All these observations lead to the conclusion that formula (3.68) expresses the law of conservation of momentum in electromagnetic processes.

3.7 Electromagnetic Angular Momentum. Angular Momentum Theorem

Taking advantage of the fact that Maxwell's stress tensor is symmetric and using relations (3.60) and (3.66), let us write the balance of the forces acting on a unit volume of the spatial fixed domain D:

$$\mathbf{f}_{em} + \frac{\partial \mathbf{p}_{em}}{\partial t} = - \frac{\partial \mathbf{T}_l}{\partial x_l}, \qquad (3.69)$$

where $\mathbf{T}$ is the dyadic representation of the tensor T_{ij}, $i, j = 1, 2, 3$, i.e.

$$\mathbf{T} = T_{xx}\mathbf{ii} + T_{xy}\mathbf{ij} + T_{xz}\mathbf{ik} + T_{yx}\mathbf{ji} + T_{yy}\mathbf{jj} + T_{yz}\mathbf{jk} + T_{zx}\mathbf{ki} + T_{zy}\mathbf{kj} + T_{zz}\mathbf{kk}.$$

Multiplying this equation by the radius-vector $\mathbf{r}$ of an arbitrary point $P(\mathbf{r})$ in D and then integrating over D, we have

$$\int_V \mathbf{r} \times \mathbf{f}_{em}\, d\tau + \frac{d}{dt} \int_V \mathbf{r} \times \mathbf{p}_{em}\, d\tau = -\int_V \frac{\partial}{\partial x_l}(\mathbf{r} \times \mathbf{T}_l)\, d\tau = -\oint_S \mathbf{r} \times \mathbf{T}_l\, dS_l,$$
(3.70)

where we used the fact that $\mathbf{r}$ does not depend on time as the domain D is fixed, while

$$\mathbf{r} \times \frac{\partial \mathbf{T}_l}{\partial x_l} = \frac{\partial}{\partial x_l}(\mathbf{r} \times \mathbf{T}_l) - \mathbf{u}_l \times \mathbf{T}_l = \frac{\partial}{\partial x_l}(\mathbf{r} \times \mathbf{T}_l),$$

because $\mathbf{u}_l$ and $\mathbf{T}_l$ have the same direction.

The first term on the l.h.s. of (3.70) is the moment of the electromagnetic force. According to the angular momentum theorem, this equals the total time derivative of the mechanical angular momentum $\mathbf{L}_{mec}$. The second integral on the l.h.s. is the angular momentum of the electromagnetic field $\mathbf{L}_{em}$. Thus we can write

$$\frac{d}{dt}(\mathbf{L}_{mec} + \mathbf{L}_{em}) = -\oint_S \mathbf{r} \times \mathbf{T}_l\, dS_l,$$
(3.71)

or, as projected on the x_i-direction,

$$\frac{d}{dt}(\mathbf{L}_{mec} + \mathbf{L}_{em})_i = -\oint_S L_{il}\, dS_l,$$
(3.72)

where

$$L_{il} = \varepsilon_{ijk} x_j T_{lk} = \varepsilon_{ijk} x_j T_{kl}$$
(3.73)

is a tensor signifying the *flux density of the angular momentum* of the electromagnetic field.

If the integration surface in (3.72) covers the whole space, or $\mathbf{E}|_S = 0$ and $\mathbf{H}|_S = 0$ simultaneously, then (3.72) leads to

$$\mathbf{L}_{mec} + \mathbf{L}_{em} = \text{const.}$$

Thus, Eq. (3.71) expresses the *law of conservation of angular momentum* in electromagnetic processes.

3.8 Electrodynamic Potentials

To know the electromagnetic field at a point situated either in vacuum, or in a homogeneous and isotropic medium whose material constants are given, one has to know six quantities: the components of the vector fields $\mathbf{E}$ and $\mathbf{B}$.

This problem is simplified by expressing the field in terms of its potentials. The electrostatic field and the magnetic field of stationary currents can be written in terms of scalar (V) and vector ($\mathbf{A}$) potentials, respectively. In a variable regime the electric and magnetic fields are connected, and this leads to a change in the relations between the fields and their potentials. Let us find these new relationships.

To this end, we use Maxwell's source-free equations

$$\nabla \cdot \mathbf{B} = 0,$$
$$\nabla \times \mathbf{E} = -\frac{\partial \mathbf{B}}{\partial t}. \tag{3.74}$$

As we already know, the first equation gives

$$\mathbf{B} = \nabla \times \mathbf{A}. \tag{3.75}$$

Introducing (3.75) into (3.74)$_2$, we have

$$\nabla \times \mathbf{E} = -\frac{\partial}{\partial t}(\nabla \times \mathbf{A}) = -\nabla \times \left(\frac{\partial \mathbf{A}}{\partial t}\right) \; \Rightarrow \; \nabla \times \left(\mathbf{E} + \frac{\partial \mathbf{A}}{\partial t}\right) = 0,$$

or, since the curl of a gradient is zero ($\nabla \times \nabla \varphi = 0$),

$$\mathbf{E} = -\nabla V - \frac{\partial \mathbf{A}}{\partial t}. \tag{3.76}$$

The functions $\mathbf{A}(\mathbf{r}, t)$ and $V(\mathbf{r}, t)$ are twice differentiable. They are called *electrodynamic (electromagnetic) potentials* and play an essential role in the electromagnetic field theory. For the moment we only mention that the problem of determining the electromagnetic field (six real quantities) reduces to the knowledge of electrodynamic potentials (four real components). This representation will prove its usefullness in the theory of the propagation of the electromagnetic field and even more in the relativistically covariant approach to electromagnetism.

3.9 Differential Equations for the Electrodynamic Potentials

Let us introduce (3.75) and (3.76) into Maxwell's source equations

$$\nabla \times \mathbf{B} = \mu \mathbf{j} + \epsilon \mu \frac{\partial \mathbf{E}}{\partial t},$$
$$\nabla \cdot \mathbf{E} = \frac{\rho}{\epsilon}, \tag{3.77}$$

written for a homogeneous and isotropic medium, characterized by the constitutive relations $\mathbf{D} = \epsilon \mathbf{E}$ and $\mathbf{B} = \mu \mathbf{H}$. Using (A.51), we have

$$\nabla \times (\nabla \times \mathbf{A}) = \nabla(\nabla \cdot \mathbf{A}) - \Delta \mathbf{A}$$
$$= \mu \mathbf{j} - \epsilon \mu \nabla \left(\frac{\partial V}{\partial t} \right) - \epsilon \mu \frac{\partial^2 \mathbf{A}}{\partial t^2},$$

or

$$\Delta \mathbf{A} - \epsilon \mu \frac{\partial^2 \mathbf{A}}{\partial t^2} = -\mu \mathbf{j} + \nabla \left(\nabla \cdot \mathbf{A} + \epsilon \mu \frac{\partial V}{\partial t} \right). \tag{3.78}$$

Using a similar procedure, we obtain

$$\nabla \cdot \left(-\nabla V - \frac{\partial \mathbf{A}}{\partial t} \right) = -\Delta V - \frac{\partial}{\partial t} (\nabla \cdot \mathbf{A}) = \frac{\rho}{\epsilon},$$

or, by adding the term $\epsilon \mu \, \partial^2 V / \partial t^2$ to both sides,

$$\Delta V - \epsilon \mu \frac{\partial^2 V}{\partial t^2} = -\frac{\rho}{\epsilon} - \frac{\partial}{\partial t} \left(\nabla \cdot \mathbf{A} + \epsilon \mu \frac{\partial V}{\partial t} \right). \tag{3.79}$$

Equations (3.78) and (3.79) form a system of four second order partial differential equations for A_i, $i = 1, 2, 3$, and V. Recall from Chap. 2 (see (2.51)), that the vector potential $\mathbf{A}(\mathbf{r}, t)$ is fixed if one imposes $\nabla \cdot \mathbf{A} = \chi$, where $\chi = \chi(\mathbf{r}, t)$ is a given function. Equations (3.78) and (3.79) suggest as a possible choice $\chi = -\epsilon \mu \frac{\partial V}{\partial t}$, or

$$\nabla \cdot \mathbf{A} + \epsilon \mu \frac{\partial V}{\partial t} = 0, \tag{3.80}$$

known as *Lorenz condition*, after the Dutch physicist Ludvig Lorenz (1829–1891), who proposed it in 1867. In this case, Eqs. (3.78) and (3.79) acquire the simpler form

$$\Delta \mathbf{A} - \epsilon \mu \frac{\partial^2 \mathbf{A}}{\partial t^2} = -\mu \mathbf{j}, \tag{3.81}$$

$$\Delta V - \epsilon \mu \frac{\partial^2 V}{\partial t^2} = -\frac{\rho}{\epsilon}, \tag{3.82}$$

and, more importantly, they become uncoupled. Using the Lorenz condition, the two electrodynamic potentials satisfy the same type of equation: a non-homogeneous wave propagation equation. This fact is very important for the relativistically covariant canonical quantization theory of the electromagnetic field.

Once $\mathbf{A} = \mathbf{A}(\mathbf{r}, t)$ and $V = V(\mathbf{r}, t)$ are obtained as solution of (3.81) and (3.82), the electromagnetic field can be determined by means of (3.75) and (3.76). We shall dedicate to this problem a special section in Chap. 4.

In some cases it is more advantageous to choose

$$\nabla \cdot \mathbf{A} = \chi = 0, \tag{3.83}$$

known as the *Coulomb condition*. Equations (3.78) and (3.79) then become

$$\Delta \mathbf{A} - \epsilon\mu \frac{\partial^2 \mathbf{A}}{\partial t^2} = -\mu \mathbf{j} + \epsilon\mu \nabla \left(\frac{\partial V}{\partial t} \right),$$

$$\Delta V = -\frac{\rho}{\epsilon}.$$

The last equation is formally identical to Poisson's equation (1.37). Recall that the solution of this equation is

$$V(\mathbf{r}, t) = \frac{1}{4\pi\epsilon} \int_{V'} \frac{\rho(\mathbf{r}', t)}{|\mathbf{r} - \mathbf{r}'|} \, d\tau'.$$

Using this expression, as well as the equation of continuity

$$\frac{\partial \rho}{\partial t} = -\nabla' \cdot \mathbf{j}(\mathbf{r}', t),$$

one obtains the following equation for $\mathbf{A}(\mathbf{r}', t)$:

$$\Delta \mathbf{A} - \epsilon\mu \frac{\partial^2 \mathbf{A}}{\partial t^2} = -\mu \mathbf{j} - \frac{\mu}{4\pi} \nabla \int_{V'} \frac{\nabla' \cdot \mathbf{j}(\mathbf{r}', t)}{|\mathbf{r} - \mathbf{r}'|} \, d\tau'. \tag{3.84}$$

To interpret the r.h.s. of (3.84), we shall establish an identity satisfied by *any* vector function, for example $\mathbf{j}(\mathbf{r}, t)$. We have

$$\nabla \times \left[\nabla \times \int_{V'} \frac{\mathbf{j}(\mathbf{r}', t)}{|\mathbf{r} - \mathbf{r}'|} \, d\tau' \right] = \nabla \left[\nabla \cdot \int_{V'} \frac{\mathbf{j}(\mathbf{r}', t)}{|\mathbf{r} - \mathbf{r}'|} \, d\tau' \right] - \Delta \int_{V'} \frac{\mathbf{j}(\mathbf{r}', t)}{|\mathbf{r} - \mathbf{r}'|} \, d\tau',$$

or, in view of (1.41), (1.45), and the divergence theorem,

$$\nabla \times \left[\nabla \times \int_{V'} \frac{\mathbf{j}(\mathbf{r}', t)}{|\mathbf{r} - \mathbf{r}'|} \, d\tau' \right]$$

$$= -\nabla \int_{V'} \mathbf{j}(\mathbf{r}', t) \cdot \nabla' \left(\frac{1}{|\mathbf{r} - \mathbf{r}'|} \right) d\tau' + \int_{V'} \mathbf{j}(\mathbf{r}', t) \left[4\pi \delta(\mathbf{r} - \mathbf{r}') \right] d\tau'$$

$$= \nabla \int_{V'} \frac{\nabla' \cdot \mathbf{j}(\mathbf{r}', t)}{|\mathbf{r} - \mathbf{r}'|} \, d\tau' + 4\pi \mathbf{j}(\mathbf{r}, t),$$

where a surface integral has been eliminated by the usual procedure. We then obtain

$$\mathbf{j}(\mathbf{r}, t) = \frac{1}{4\pi} \nabla \times \left[\nabla \times \int_{V'} \frac{\mathbf{j}(\mathbf{r}', t)}{|\mathbf{r} - \mathbf{r}'|} d\tau' \right] - \frac{1}{4\pi} \nabla \int_{V'} \frac{\nabla' \cdot \mathbf{j}(\mathbf{r}', t)}{|\mathbf{r} - \mathbf{r}'|} d\tau'$$

$$= \mathbf{j}_{trans} + \mathbf{j}_{long}, \tag{3.85}$$

meaning that $\mathbf{j}$ has been separated into two vector components, one *transversal* and the other *longitudinal*, so that

$$\nabla \cdot \mathbf{j}_{trans} = 0, \quad \nabla \times \mathbf{j}_{long} = 0,$$

i.e. $\mathbf{j}_{trans}$ is a *solenoidal* current, while $\mathbf{j}_{long}$ is an *irrotational* one.

The terms "longitudinal" and "transversal" are connected with the possible polarizations of a plane wave, characterized by either $\nabla \times \mathbf{a} = 0$, or $\nabla \cdot \mathbf{a} = 0$, where $\mathbf{a} = \mathbf{a}(\mathbf{r}, t)$ is a vector field. Then, by means of (3.85), Eq. (3.84) finally reads

$$\Delta \mathbf{A} - \epsilon\mu \frac{\partial^2 \mathbf{A}}{\partial t^2} = -\mu(\mathbf{j} - \mathbf{j}_{long}) = -\mu \mathbf{j}_{trans}. \tag{3.86}$$

Thus, if Coulomb's condition is chosen, then $\mathbf{j}_{trans}$ plays the role of *source* for the vector potential $\mathbf{A}$. Equation (3.86) is used to find $\mathbf{A}$, then the field $(\mathbf{E}, \mathbf{B})$ is determined by means of (3.75) and (3.76).

3.9.1   Gauge Transformations

The vector and scalar potentials $(\mathbf{A}, V)$ are not uniquely determined by the Eqs. (3.75) and (3.76). Indeed, if we choose

$$\mathbf{A}' = \mathbf{A} + \nabla\psi, \tag{3.87}$$

where ψ is an arbitrary differentiable function of position and time, we obtain the same field $\mathbf{B}$:

$$\mathbf{B}' = \nabla \times \mathbf{A}' = \nabla \times \mathbf{A} + \nabla \times (\nabla\psi) = \nabla \times \mathbf{A} = \mathbf{B}.$$

Introducing (3.87) into (3.76), we find

$$\mathbf{E} = -\nabla V - \frac{\partial}{\partial t}(\mathbf{A}' - \nabla\psi) = -\nabla \left(V - \frac{\partial\psi}{\partial t} \right) - \frac{\partial \mathbf{A}'}{\partial t} \equiv \mathbf{E}'.$$

Obviously, to obtain the same field, i.e. to have $\mathbf{E}' = \mathbf{E}$, we have to take

$$V' = V - \frac{\partial\psi}{\partial t}. \tag{3.88}$$

The transformations (3.87) and (3.88) leave unchanged the electromagnetic field. The property of the electromagnetic field to remain unchanged while transformations (3.87), (3.88) are performed is called *gauge invariance* (*Eichinvarianz* in German, *invariance de jauge* in French, etc.), while (3.87) and (3.88) are named *gauge transformations*.

The freedom offered by the gauge transformations allows us to choose those electrodynamic potentials which uncouple the second-order partial differential equations satisfied by these potentials. Indeed, it is always possible to choose the electrodynamic potentials in such a way as to satisfy the Lorenz condition. Let us assume that the pair of potentials V and $\mathbf{A}$ satisfy the coupled equations (3.78) and (3.79), but do not satisfy the Lorenz condition, and let us perform a gauge transformation to a new set of potentials V' and $\mathbf{A}'$ which satisfy the Lorenz condition:

$$0 = \nabla \cdot \mathbf{A}' + \epsilon\mu\frac{\partial V'}{\partial t} = \nabla \cdot (\mathbf{A} + \nabla\psi) + \epsilon\mu\frac{\partial}{\partial t}\left(V - \frac{\partial\psi}{\partial t}\right)$$

$$= \nabla \cdot \mathbf{A} + \epsilon\mu\frac{\partial V}{\partial t} + \Delta\psi - \epsilon\mu\frac{\partial^2\psi}{\partial t^2}.$$

Thus, if the gauge function ψ satisfies the relation

$$\Delta\psi - \epsilon\mu\frac{\partial^2\psi}{\partial t^2} = -\left(\nabla \cdot \mathbf{A} + \epsilon\mu\frac{\partial V}{\partial t}\right),$$

then V' and $\mathbf{A}'$ will satisfy both the Lorenz condition and the non-homogeneous, decoupled, wave equations.

The gauge transformation whose gauge function satisfies the homogeneous wave equation

$$\Delta\psi - \epsilon\mu\frac{\partial^2\psi}{\partial t^2} = 0$$

is a *restricted gauge transformation*. It maintains the Lorenz condition for the new set of potentials, on condition that the initial potentials have already satisfied it. The potentials connected by such restricted gauge transformation are called *Lorenz gauge potentials*.

The Lorenz gauge is independent of the choice of coordinate system (which makes possible a straight generalization to the relativistic case) and, in addition, allows a manifestly covariant canonical quantization of the electromagnetic field. The disadvantage of this type of quantization resides in the necessity to introduce unphysical degrees of freedom whose effects are further eliminated by means of some relatively complicated procedures. Due to this inconvenience, with the price of the loss of the manifest relativistic covariance of the theory, sometimes the *Coulomb gauge* (also called *radiation gauge* or *transversal gauge*) is preferred instead. In this gauge, we have

$$\nabla \cdot \mathbf{A} = 0,$$

while the scalar potential V is a solution of the Poisson equation (and not of the inhomogeneous wave equation (3.82)). The scalar potential V is, therefore, precisely the instantaneous Coulomb potential

$$V(\mathbf{r}, t) = \frac{1}{4\pi\epsilon} \int_{V'} \frac{\rho(\mathbf{r}', t)}{|\mathbf{r} - \mathbf{r}'|} \, d\tau'.$$

For this reason, this type of gauge is called "Coulomb gauge". As we have seen, in this case the vector potential $\mathbf{A}$ satisfies the inhomogeneous wave equation in which the term responsible for inhomogeneity (i.e. the source-term in the equation for $\mathbf{A}$) is expressed only by means of the transversal component of the density current, $\mathbf{j}_{trans}$. Consequently, the Coulomb gauge is also called "transversal gauge". The third name of this gauge, "radiation gauge", is given by the fact that the transverse radiation fields are essentially determined by the vector potential $\mathbf{A}$ – remark that even the field $\mathbf{E}$ involves the change in time of the vector potential $\mathbf{A}$, not only the gradient of the instantaneous Coulomb potential. In other words, although the inobservable scalar potential at a given time depends on the charge density distribution at that moment, the propagation of the measurable electric field is not instantaneous. Due to its simplicity, the Coulomb gauge is mostly preferred when sources are absent.

The gauge invariance is preserved in the quantum theory of electromagnetism, called *quantum electrodynamics*, or QED. Maxwell's gauge theory is the prototype of *gauge theories* formulated at different stages and in various contexts mainly by Hermann Weyl (1885–1955), Chen-Ning Yang (b. 1922), and Robert Mills (1927–1999).

3.10 Different Types of Electrodynamic Potentials

If the medium is homogeneous and isotropic, free of charges ($\rho = 0$) and conduction currents ($\mathbf{j} = 0$), the electromagnetic field can be expressed in terms of some other potentials, such as *antipotentials* or the *Hertz vector*.

3.10.1 Antipotentials

Maxwell's source equations, written for $\rho = 0$, $\mathbf{j} = 0$, are

$$\nabla \cdot \mathbf{D} = 0,$$
$$\nabla \times \mathbf{H} = \frac{\partial \mathbf{D}}{\partial t}. \tag{3.89}$$

The form of these equations is similar to those we have already used to define the electromagnetic field in terms of the potentials $\mathbf{A}, V$. This fact suggests to write $\mathbf{D}$

and **H** as

$$\mathbf{D} = \epsilon \mathbf{E} = -\nabla \times \mathbf{A}^*,$$
$$\mathbf{H} = \frac{1}{\mu} \mathbf{B} = -\nabla V^* - \frac{\partial \mathbf{A}^*}{\partial t},$$

which give

$$\mathbf{E} = -\frac{1}{\epsilon} \nabla \times \mathbf{A}^*,$$
$$\mathbf{B} = -\mu \left(\nabla V^* + \frac{\partial \mathbf{A}^*}{\partial t} \right). \tag{3.90}$$

It is easy to prove that the potentials $\mathbf{A}^*$, V^* satisfy the homogeneous second order partial differential equations

$$\Delta \mathbf{A}^* - \epsilon \mu \frac{\partial^2 \mathbf{A}^*}{\partial t^2} = 0,$$
$$\Delta V^* - \epsilon \mu \frac{\partial^2 V^*}{\partial t^2} = 0, \tag{3.91}$$

if the Lorenz-type condition

$$\nabla \cdot \mathbf{A}^* + \epsilon \mu \frac{\partial V^*}{\partial t} = 0 \tag{3.92}$$

is fulfilled. The potentials $\mathbf{A}^*$ and V^* are (improperly) called *antipotentials* and their usage is less frequent.

3.10.2   Hertz's Vector Potential

Heinrich Hertz (1857–1894) showed that, instead of the potentials $\mathbf{A}(\mathbf{r}, t)$ and $V(\mathbf{r}, t)$ can be used a single vector field $\mathbf{Z}(\mathbf{r}, t)$, known as the *Hertz vector*. This is defined by observing that the Lorenz condition (3.80) is identically satisfied if we choose, on the one hand

$$\mathbf{A} = \epsilon \mu \frac{\partial \mathbf{Z}}{\partial t}, \tag{3.93}$$

and on the other

$$V = -\nabla \cdot \mathbf{Z}. \tag{3.94}$$

Being a potential for the potentials $\mathbf{A}$ and V, the Hertz vector appears as a "super-potential". The electromagnetic field $\mathbf{E}$, $\mathbf{B}$ is then expressed in terms of $\mathbf{Z}$ by

$$\mathbf{E} = -\nabla V - \frac{\partial \mathbf{A}}{\partial t} = \nabla(\nabla \cdot \mathbf{Z}) - \epsilon\mu\frac{\partial^2 \mathbf{Z}}{\partial t^2},$$

$$\mathbf{B} = \nabla \times \mathbf{A} = \epsilon\mu\nabla \times \left(\frac{\partial \mathbf{Z}}{\partial t}\right). \tag{3.95}$$

The partial differential equation satisfied by $\mathbf{Z}$ is found using Maxwell's equation

$$\frac{1}{\mu}\nabla \times \mathbf{B} = \epsilon\frac{\partial \mathbf{E}}{\partial t}, \tag{3.96}$$

then (3.95), and finally integrating with respect to time. This gives

$$\Delta\mathbf{Z} - \epsilon\mu\frac{\partial^2 \mathbf{Z}}{\partial t^2} = 0. \tag{3.97}$$

A similar reasoning allows us to define $\mathbf{Z}$ in terms of the antipotentials $\mathbf{A}^*$, V^*. Choosing

$$\mathbf{A}^* = \epsilon\mu\frac{\partial \mathbf{Z}^*}{\partial t}, \quad V^* = -\nabla \cdot \mathbf{Z}^* \tag{3.98}$$

and using (3.90), one easily finds

$$\mathbf{D} = -\nabla \times \mathbf{A}^* = -\epsilon\mu\nabla \times \left(\frac{\partial \mathbf{Z}^*}{\partial t}\right),$$

$$\mathbf{H} = -\nabla V^* - \frac{\partial \mathbf{A}^*}{\partial t} = \nabla(\nabla \cdot \mathbf{Z}^*) - \epsilon\mu\frac{\partial^2 \mathbf{Z}^*}{\partial t^2}, \tag{3.99}$$

while the partial differential equation satisfied by $\mathbf{Z}^*$ (the *Hertz antipotential*) is

$$\Delta\mathbf{Z}^* - \epsilon\mu\frac{\partial^2 \mathbf{Z}^*}{\partial t^2} = 0.$$

The physical significance of the potentials $\mathbf{Z}$ and $\mathbf{Z}^*$ is closely related to the electric and magnetic polarization vectors $\mathbf{P}$ and $\mathbf{M}$, respectively. Let us admit the existence of an electric polarization ($\mathbf{P}_0 \neq 0$) and a magnetic polarization ($\mathbf{M}_0 \neq 0$) when the external polarizing fields are absent. Using (1.100) and (2.84), we have

$$\mathbf{P} = \mathbf{P}_0 + \alpha\mathbf{E} = \mathbf{P}_0 + \epsilon_0(\epsilon_r - 1)\mathbf{E},$$

$$\mathbf{M} = \mathbf{M}_0 + \chi\mathbf{H} = \mathbf{M}_0 + (\mu_r - 1)\mathbf{H},$$

leading to

$$\mathbf{D} = \epsilon_0\mathbf{E} + \mathbf{P} = \epsilon\mathbf{E} + \mathbf{P}_0,$$
$$\mathbf{B} = \mu_0(\mathbf{H} + \mathbf{M}) = \mu\mathbf{H} + \mu_0\mathbf{M}_0.$$

Let us consider two special cases:

(a) $\mathbf{M}_0 = 0$. Integrating with respect to time Maxwell's equation

$$\nabla \times \mathbf{H} = \frac{\partial\mathbf{D}}{\partial t},$$

we then obtain as the simplest solution

$$\Delta\mathbf{Z} - \epsilon\mu\frac{\partial^2\mathbf{Z}}{\partial t^2} = -\frac{1}{\epsilon_0}\mathbf{P}_0, \qquad (3.100)$$

showing that $\mathbf{P}_0$ is the *source* of the vector field $\mathbf{Z}$. For this reason, the Hertz vector $\mathbf{Z}$ is also called *potential of electric polarization*.

(b) $\mathbf{P}_0 = 0$. This time we consider Maxwell's equation

$$\nabla \times \mathbf{E} = -\frac{\partial\mathbf{B}}{\partial t}$$

and obtain

$$\frac{1}{\epsilon}\nabla \times \mathbf{D} = -\mu\frac{\partial\mathbf{H}}{\partial t} - \mu_0\frac{\partial\mathbf{H}_0}{\partial t}.$$

Expressing $\mathbf{D}$ and $\mathbf{H}$ in terms of $\mathbf{Z}^*$ (see (3.99)) and integrating with respect to time, one obtains as the simplest solution

$$\Delta\mathbf{Z}^* - \epsilon\mu\frac{\partial^2\mathbf{Z}^*}{\partial t^2} = -\frac{\mu_0}{\mu}\mathbf{M}_0. \qquad (3.101)$$

Since $\mathbf{M}_0$ is the source of the field $\mathbf{Z}^*$, the latter is also called *potential of magnetic polarization*.

3.11   Electrodynamic Potentials and the Analytical Derivation of Some Fundamental Equations

The formalism of analytical mechanics is very powerful in the derivation of some fundamental equations in field theory, in particular in the electromagnetic field theory. The expressions of the fields $\mathbf{E}$ and $\mathbf{B}$ in terms of the potentials $\mathbf{A}$ and V allow one to apply the analytical formalism in the derivation of the equations of motion for both discrete and continuous systems, i.e. point charges and fields, respectively.

3.11.1 Analytical Derivation of the Equation of Motion of a Point Charge in an External Electromagnetic Field

Let us consider a point charge (e.g. an electron), of mass m and charge e, moving with the velocity $\mathbf{v}$ in the electromagnetic field $\mathbf{E}$ and $\mathbf{B}$. The Lagrangian of such a system is composed of two terms, the kinetic part, customarily denoted by L_0, and the interaction part (between charge and the field), L_{int}:

$$L = L_0 + L_{int} = \frac{m}{2}\dot{x}_k\dot{x}_k - eV + e\,\dot{x}_k A_k, \qquad (3.102)$$

where Einstein's summation convention has been used. The differential equations of motion of the particle are Lagrange's equations of second kind, written for natural systems (the force is conservative, in the general sense[3]):

$$\frac{d}{dt}\left(\frac{\partial L}{\partial \dot{q}_j}\right) - \frac{\partial L}{\partial q_j} = 0, \quad j = 1, 2, \ldots, n, \qquad (3.103)$$

where q_j are generalized coordinates and $\dot{q}_j$ the generalized velocities. In our case, since the particle is not subjected to any constraints, as generalized coordinates we can choose the Cartesian coordinates:

$$q_i = x_i, \quad \dot{q}_i = \dot{x}_i = v_i, \quad i = 1, 2, 3.$$

Thus, we have

$$p_i = \frac{\partial L}{\partial \dot{x}_i} = m\dot{x}_k\delta_{ik} + eA_k\delta_{ik} = m\dot{x}_i + eA_i,$$

$$\frac{d}{dt}\left(\frac{\partial L}{\partial \dot{x}_j}\right) = m\ddot{x}_i + e\frac{dA_i}{dt} = m\ddot{x}_i + e\left(\frac{\partial A_i}{\partial t} + v_k\frac{\partial A_i}{\partial x_k}\right),$$

$$\frac{\partial L}{\partial x_i} = -e\frac{\partial V}{\partial x_i} + ev_k\frac{\partial A_k}{\partial x_i}.$$

Introducing these results into (3.103), we obtain

$$m\ddot{x}_i = -e\frac{\partial V}{\partial x_i} - e\frac{\partial A_i}{\partial t} + ev_k\left(\frac{\partial A_k}{\partial x_i} - \frac{\partial A_i}{\partial x_k}\right).$$

It can be easily verified that the expression in parentheses on the r.h.s. is a second order antisymmetric tensor, denoted by F_{ik}, defined on the three-dimensional Euclidean

---

[3]The potential of the interaction between a point charge and the electromagnetic field is a *generalized potential*, i.e. a potential which depends not only on the generalized coordinates $q_i = q_i(t)$ and time t, but also on the generalized velocities $\dot{q}_i = \dot{q}_i(t)$, i.e. $V = V(q, \dot{q}, t)$.

space E_3. If we denote by B_s the components of the dual pseudovector associated to F_{ik}, we have (see (A.60)):

$$F_{ik} = \frac{\partial A_k}{\partial x_i} - \frac{\partial A_i}{\partial x_k} = \varepsilon_{ikm} B_m, \qquad i, k, m = 1, 2, 3.$$

Hence

$$B_s = \frac{1}{2} \varepsilon_{sik} F_{ik} = \frac{1}{2} \varepsilon_{sik} \left(\frac{\partial A_k}{\partial x_i} - \frac{\partial A_i}{\partial x_k} \right) = \varepsilon_{sik} \partial_i A_k = (\nabla \times \mathbf{A})_s.$$

The equation of motion then becomes

$$m\ddot{x}_i = e \left(-\frac{\partial V}{\partial x_i} - \frac{\partial A_i}{\partial t} \right) + e \varepsilon_{iks} v_k B_s, \tag{3.104}$$

and finally, in view of (3.75) and (3.76):

$$m\ddot{x}_i = e(\mathbf{E} + \mathbf{v} \times \mathbf{B})_i, \qquad m\ddot{\mathbf{r}} = e(\mathbf{E} + \mathbf{v} \times \mathbf{B}), \tag{3.105}$$

which is the well-known equation of motion of a point charge in an external electro-magnetic field.

3.11.2   Analytical Derivation of Maxwell's Equations

The electromagnetic field is treated analytically as a continuous systems, i.e. as a system with an infinity of degrees of freedom, corresponding to each point in space. The components of the electromagnetic field are the generalized coordinates. Consequently, from each point in space one has a contribution to the total Lagrangian of the system. Let us denote by $\mathcal{L}(\mathbf{r})$ the contribution to the Lagrangian from a vicinity of the point defined by the radius vector $\mathbf{r}$. Then the total Lagrangian will be given by the volume integral

$$L = \int \mathcal{L}(\mathbf{r}) d\tau.$$

Thus, $\mathcal{L}$ is nothing else but the density of Langrangian. The study of continous media starts always from the construction of the Lagrangian density $\mathcal{L}$ of the studied model. Once the Lagrangian density is established, the evolution in time and space of the system is found by solving the second order partial differential equations called Euler–Lagrange equations:

$$\frac{\partial \mathcal{L}}{\partial \varphi^{(s)}} - \frac{\partial}{\partial x_J} \left(\frac{\partial \mathcal{L}}{\partial \varphi^{(s)}_{,J}} \right) = 0, \qquad s = 1, 2, \ldots, h, \quad J = 1, 2, \ldots, n, \tag{3.106}$$

where $\mathcal{L}(x_j, \varphi^{(s)}, \varphi^{(s)}_{,j})$ is the Lagrangian density, x_J labels the space-time points, $\varphi^{(s)}$ are the components of the fields, i.e. the generalized coordinates (variational parameters), and we use the notation

$$\varphi^{(s)}_{,J} = \frac{\partial \varphi^{(s)}}{\partial x_J} = \partial_J \varphi^{(s)}.$$

Choosing

$$x_1 = x, \ x_2 = y, \ x_3 = z, \ x_4 = t,$$

the system (3.106) becomes

$$\frac{\partial \mathcal{L}}{\partial \varphi^{(s)}} - \frac{\partial}{\partial x_i}\left(\frac{\partial \mathcal{L}}{\partial \varphi^{(s)}_{,i}}\right) - \frac{\partial}{\partial t}\left(\frac{\partial \mathcal{L}}{\partial \varphi^{(s)}_{,t}}\right) = 0, \quad i = 1, 2, 3. \tag{3.107}$$

The Lagrangian density $\mathcal{L}$ is composed of two terms: a kinetic term $\mathcal{L}_0$, which is formally the same as the Lagrangian density of the free field, and a part $\mathcal{L}_{int}$ that expresses the interaction between the field and the sources:

$$\mathcal{L} = \mathcal{L}_0 + \mathcal{L}_{int}.$$

Supposing that the interactions take place in a homogeneous and isotropic medium, it can be shown that the Lagrangian density of the electromagnetic field is

$$\mathcal{L}_0 = \frac{1}{2}\epsilon \mathbf{E}^2 - \frac{1}{2\mu}\mathbf{B}^2. \tag{3.108}$$

This expression is relativistically invariant and was found in 1900 by Joseph Larmor (1857–1942) (see Part II).

The interaction Lagrangian density $\mathcal{L}_{int}$ is found by dividing L_{int} from (3.102) by the infinitesimal volume $\Delta\tau$ where the charge e is distributed:

$$\mathcal{L}_{int} = \frac{e}{\Delta\tau}(-V + \mathbf{v} \cdot \mathbf{A}) = -\rho V + \mathbf{j} \cdot \mathbf{A}. \tag{3.109}$$

The Lagrangian density of the field, in the presence of the sources, is then

$$\mathcal{L} = \frac{1}{2}\epsilon E_m E_m - \frac{1}{2\mu}B_m B_m - \rho V + j_m A_m. \tag{3.110}$$

The variables $\varphi^{(s)}$ are, in our case, A_i, $i = 1, 2, 3$ and V, therefore we shall use relations (3.75), (3.76) between the field and its potentials:

$$E_m = -V_{,m} - A_{m,t},$$
$$B_m = \varepsilon_{msj} A_{j,s}. \tag{3.111}$$

Choosing, for the beginning, $\varphi^{(1,2,3)} \equiv A_k$, $k = 1, 2, 3$ in (3.107), we have

$$\frac{\partial \mathcal{L}}{\partial A_k} - \frac{\partial}{\partial x_i}\left(\frac{\partial \mathcal{L}}{\partial A_{k,i}}\right) - \frac{\partial}{\partial t}\left(\frac{\partial \mathcal{L}}{\partial A_{k,t}}\right) = 0, \quad i, k = 1, 2, 3. \tag{3.112}$$

Performing the calculations, we successively obtain

$$\frac{\partial \mathcal{L}}{\partial A_k} = j_m \delta_{mk} = j_k,$$

$$\frac{\partial \mathcal{L}}{\partial A_{k,i}} = \frac{\partial \mathcal{L}}{\partial B_m}\frac{\partial B_m}{\partial A_{k,i}} = -\frac{1}{\mu}B_m \varepsilon_{msj}\delta_{jk}\delta_{si} = -\frac{1}{\mu}\varepsilon_{ikm}B_m,$$

$$\frac{\partial}{\partial x_i}\left(\frac{\partial \mathcal{L}}{\partial A_{k,i}}\right) = -\frac{1}{\mu}\varepsilon_{ikm}B_{m,i} = \frac{1}{\mu}(\nabla \times \mathbf{B})_k = (\nabla \times \mathbf{H})_k,$$

$$\frac{\partial \mathcal{L}}{\partial A_{k,t}} = \frac{\partial \mathcal{L}}{\partial E_m}\frac{\partial E_m}{\partial A_{k,t}} = \epsilon E_m(-\delta_{mk}) = -\epsilon E_k = -D_k,$$

$$\frac{\partial}{\partial t}\left(\frac{\partial \mathcal{L}}{\partial A_{k,t}}\right) = -\frac{\partial D_k}{\partial t}.$$

Introducing these results into (3.112), we find

$$(\nabla \times \mathbf{H})_k = j_k + \frac{\partial D_k}{\partial t},$$

which is the x_k-component of Maxwell's source equation (3.42). The last step is achieved by choosing $\varphi^{(4)} = V$ in (3.107). We then obtain a single equation,

$$\frac{\partial \mathcal{L}}{\partial V} - \frac{\partial}{\partial x_i}\left(\frac{\partial \mathcal{L}}{\partial V_{,i}}\right) - \frac{\partial}{\partial t}\left(\frac{\partial \mathcal{L}}{\partial V_{,t}}\right) = 0, \quad i = 1, 2, 3. \tag{3.113}$$

Similar calculations as above give

$$\frac{\partial \mathcal{L}}{\partial V} = -\rho,$$

$$\frac{\partial \mathcal{L}}{\partial V_{,i}} = \frac{\partial \mathcal{L}}{\partial E_m}\frac{\partial E_m}{\partial V_{,i}} = \epsilon E_m(-\delta_{im}) = -\epsilon E_i = -D_i,$$

$$\frac{\partial}{\partial x_i}\left(\frac{\partial \mathcal{L}}{\partial V_{,i}}\right) = -D_{i,i},$$

$$\frac{\partial \mathcal{L}}{\partial V_{,t}} = 0,$$

and (3.113) leads to

$$D_{i,i} = \nabla \cdot \mathbf{D} = \rho,$$

which is the remaining Maxwell's source equation.

Observation:

In the above example we have used the source-free Maxwell's equations to define the electromagnetic field in terms of its potentials and, by applying the analytic formalism, we have deduced the other group, i.e. the Maxwell equations with sources. The problem may also be posed the other way around: knowing Maxwell's equations with sources, determine the electromagnetic field in terms of $\mathbf{A}$ and V, as in formulas (3.111), which finally lead to the source-free Maxwell equations. To this end, we construct the following Lagrangian density:

$$\mathcal{L} = \frac{1}{2}\epsilon\mathbf{E}^2 - \frac{1}{2\mu}\mathbf{B}^2 - V\left(\epsilon\,\nabla\cdot\mathbf{E} - \rho\right) + \mathbf{A}\cdot\left(\frac{1}{\mu}\nabla\times\mathbf{B} - \epsilon\frac{\partial\mathbf{E}}{\partial t} - \mathbf{j}\right). \quad (3.114)$$

Here the variational parameters are the field components E_i, B_i, $i = 1, 2, 3$, while the source equations have been used as *constraints* for these components, introduced by the *Lagrange multipliers* $V(\mathbf{r}, t)$ and $\mathbf{A}(\mathbf{r}, t)$. Choosing $\varphi^{(1,2,3)} = E_k$, $k = 1, 2, 3$ and $\varphi^{(4,5,6)} = B_k$, $k = 1, 2, 3$, we find $(3.111)_1$ and $(3.111)_2$, respectively. Using then the procedure described above, the source-free equations are obtained immediately.

3.12 Electromagnetic Field Equations for Moving Media

The theory presented up to this point is based on the assumption that the medium where the electromagnetic phenomena take place is *at rest* with respect to the observer; in other words, all measurements are done in the reference system attached to the medium. Under these circumstances we have deduced, among other results, Maxwell's system of equations.

The formulation of the system of equations describing the electromagnetic field for moving media can be unitarily treated only in the framework of the theory of relativity. In this section, nevertheless, we shall present a pre-relativistic approach to this problem, based on the mechanics of continuous media and valid only for *small velocities as compared to the speed of light in vacuum*. This theory was developed by Heinrich Hertz and Hendrik Antoon Lorentz. It is of a special importance, since the special theory of relativity, as we shall see in the second part of the book, appeared and was developed *within* the framework of the electrodynamics of moving media.

3.12.1 Source-free Equations

The motion of the medium affects only the equations of evolution. Indeed, if in a certain reference frame the field $\mathbf{B}$ is solenoidal, it will retain this property in any reference frame which is in a uniform motion of translation with respect to the first (inertial frame). In other words, the equation $\nabla \cdot \mathbf{B} = 0$ will keep its form, no matter in which inertial frame it is considered. As we have already specified, we assume that the medium moves uniformly with the velocity $|\mathbf{v}| \ll c$.

Let $C(t)$ be a closed contour connected to the body, and $S(t)$ a surface bounded by C. The equation expressing the time variation of $\mathbf{B}$ is then given by Faraday's law of electromagnetic induction:

$$\oint_C \mathbf{E} \cdot d\mathbf{l} = -\frac{d}{dt} \int_S \mathbf{B} \cdot d\mathbf{S} = -\frac{d\Phi_m}{dt}. \tag{3.115}$$

Since the surface is variable in time, one has to know the rule for performing the time-derivative of the magnetic flux in (3.115). To this end we shall apply the formalism of the mechanics of continuous media.

Assume that at the initial time $t = t_0$ the medium occupies the spatial domain D_0, of volume V_0, bounded by the surface S_0. These elements define the so-called *configuration* of the medium at the initial moment t_0. Next, assume that at the moment $t > t_0$ the same medium (therefore the same number of particles) fills up a domain D, of volume V, bounded by the surface S. Thus, if $\mathbf{r}_0$ is the initial radius vector of a particle (molecule) P_0 of D_0, and $\mathbf{r}$ is the radius vector of *the same* particle in D at the time t, we note that $\mathbf{r}$ is a function of $\mathbf{r}_0$ and t:

$$\mathbf{r} = \mathbf{r}(\mathbf{r}_0, t), \quad x_i = x_i(x_j^0, t), \quad i, j = 1, 2, 3. \tag{3.116}$$

For $\mathbf{r}_0$ fixed and t variable, these equations define the *trajectory* of the particle under observation, while for t fixed, $\mathbf{r}$ describes the domain D when $\mathbf{r}_0$ describes the domain D_0. In the latter case Eq. (3.116) defines an application of the Euclidean space onto itself. We call *material manifold* any manifold (curve, surface, volume) formed by particles of the medium. For example, the parametric equations of a material surface S_0 in D_0 is

$$x_i^0 = x_i^0(q_1, q_2), \quad i = 1, 2, 3, \tag{3.117}$$

while the equations of its *image* S in D is

$$x_i = x_i[\mathbf{r}_0(q_1, q_2), t] = x_i(q_1, q_2, t). \tag{3.118}$$

Since, by hypothesis, the particles keep their individuality (to a point P_0 of D_0 corresponds a single point P of D and reciprocally), from the mathematical point of view this is an example of a *one-to-one correspondence*, expressed by

$$J = \det\left(\frac{\partial x_i}{\partial x_j^0}\right) = \frac{\partial(x_1, x_2, x_3)}{\partial(x_1^0, x_2^0, x_3^0)} \neq 0, \tag{3.119}$$

where J is the functional determinant (Jacobian) of the transformation (3.116) (t being fixed).

By means of (3.119), we can prove that the image of a material manifold is a manifold of the same order. For example, if the material manifold is a surface, we have

$$\frac{\partial x_i}{\partial q_\alpha} = \frac{\partial x_i}{\partial x_j^0}\frac{\partial x_j^0}{\partial q_\alpha}, \quad i, j = 1, 2, 3, \quad \alpha = 1, 2.$$

Thus, according to (3.119):

$$\mathrm{rank}\left\|\frac{\partial x_i}{\partial q_\alpha}\right\| = \mathrm{rank}\left\|\frac{\partial x_j^0}{\partial q_\alpha}\right\| = 2.$$

Based on these preliminary observations, let us consider the total time derivative

$$\frac{d}{dt}\int_S \mathbf{a}\cdot d\mathbf{S}, \tag{3.120}$$

where $S(t)$ is a material surface and $\mathbf{a}(\mathbf{r}, t)$ – an arbitrary vector field. To perform the integration in (3.120), one has to calculate the time derivative of the surface element $d\mathbf{S}$. To this end, we shall use the parametric representation and write $d\mathbf{S}$ as the cross product of the vectors $\frac{\partial \mathbf{r}}{\partial q_1}dq_1$ and $\frac{\partial \mathbf{r}}{\partial q_2}dq_2$, tangent to the curves $q_1 = \lambda_1$ and $q_2 = \lambda_2$, respectively:

$$d\mathbf{S} = \frac{\partial \mathbf{r}}{\partial q_1} \times \frac{\partial \mathbf{r}}{\partial q_2}dq_1 dq_2,$$

$$dS_i = \varepsilon_{ijk}\frac{\partial x_j}{\partial q_1}\frac{\partial x_k}{\partial q_2}dq_1 dq_2, \quad i, j, k = 1, 2, 3. \tag{3.121}$$

The determinant of any 3×3 matrix A can be written as

$$\det A = \varepsilon_{ijk}A_{1i}A_{2j}A_{3k}, \quad i, j, k = 1, 2, 3,$$

in which case is valid also the relation

$$\varepsilon_{lmn}(\det A) = \varepsilon_{ijk}A_{li}A_{mj}A_{nk}, \quad i, j, k, l, m, n = 1, 2, 3.$$

Since in our case $\det A = J$, we have

$$J\varepsilon_{lmn} = \varepsilon_{ijk}\frac{\partial x_i}{\partial x_l^0}\frac{\partial x_j}{\partial x_m^0}\frac{\partial x_k}{\partial x_n^0}, \quad i, j, k, l, m, n = 1, 2, 3. \tag{3.122}$$

Let us write (3.121) in the form

$$dS_i = \varepsilon_{ijk} \frac{\partial x_j}{\partial x_m^0} \frac{\partial x_k}{\partial x_n^0} \frac{\partial x_m^0}{\partial q_1} \frac{\partial x_n^0}{\partial q_2} dq_1 dq_2, \quad i, j, k, m, n = 1, 2, 3,$$

and multiply it by $\frac{\partial x_i}{\partial x_l^0} \frac{\partial x_l^0}{\partial x_k} = \delta_{ik}$, $i, k, l = 1, 2, 3$. The result is

$$dS_k = J \frac{\partial x_l^0}{\partial x_k} dS_l^0, \quad k, l = 1, 2, 3. \tag{3.123}$$

The derivative (3.120) implies the calculation of $\frac{d}{dt} \left(\frac{\partial x_l^0}{\partial x_k} \right)$, $k, l = 1, 2, 3$. To this end, we take an arbitrary elementary variation $dx_l^0 = \frac{\partial x_l^0}{\partial x_k} dx_k$, $k, l = 1, 2, 3$ and calculate (taking into account that dx_l^0 does not vary in time):

$$0 = \frac{d}{dt}(dx_l^0) = \frac{d}{dt} \left(\frac{\partial x_l^0}{\partial x_k} \right) dx_k + \frac{\partial x_l^0}{\partial x_k} dv_k, \quad v_k = \frac{dx_k}{dt}, \quad k, l = 1, 2, 3.$$

In the last relation we replace dv_k by $\frac{\partial v_k}{\partial x_i} dx_i$, $i, k = 1, 2, 3$ and interchange the summation indices. Since dx_k is arbitrary, this yields

$$\frac{d}{dt} \left(\frac{\partial x_l^0}{\partial x_k} \right) = -\frac{\partial x_l^0}{\partial x_i} \frac{\partial v_i}{\partial x_k}, \quad i, k, l = 1, 2, 3. \tag{3.124}$$

We are now prepared to perform the time derivative of the surface element (3.123). Using (3.124) and Euler's formula (2.11), which we re-write here for convenience:

$$\frac{dJ}{dt} = J \nabla \cdot \mathbf{v},$$

we have

$$\frac{d}{dt}(dS_i) = \left[J (\nabla \cdot \mathbf{v}) \frac{\partial x_l}{\partial x_i} - J \frac{\partial x_l^0}{\partial x_k} \frac{\partial v_k}{\partial x_i} \right] dS_l^0$$

$$= (\nabla \cdot \mathbf{v}) dS_i - \frac{\partial v_k}{\partial x_i} dS_k, \quad i, k, l = 1, 2, 3.$$

The derivative (3.120) then becomes

$$\frac{d}{dt} \int_S a_i \, dS_i = \int_S \left\{ \frac{da_i}{dt} dS_i + a_i \left[(\nabla \cdot \mathbf{v}) dS_i - \frac{\partial v_k}{\partial x_i} dS_k \right] \right\}$$

$$= \int_S \left[\frac{d\mathbf{a}}{dt} + \mathbf{a} \nabla \cdot \mathbf{v} - (\mathbf{a} \cdot \nabla)\mathbf{v} \right] \cdot d\mathbf{S}, \quad i, k = 1, 2, 3.$$

The last two terms can be manipulated by means of the vector relation (see (A.47)):

$$\nabla \times (\mathbf{a} \times \mathbf{v}) = \mathbf{a}\nabla \cdot \mathbf{v} - \mathbf{v}\nabla \cdot \mathbf{a} + (\mathbf{v} \cdot \nabla)\mathbf{a} - (\mathbf{a} \cdot \nabla)\mathbf{v},$$

while the substantial time derivative $d\mathbf{a}/dt$ is given by

$$\frac{d\mathbf{a}}{dt} = \frac{\partial \mathbf{a}}{\partial t} + (\mathbf{v} \cdot \nabla)\mathbf{a}.$$

One then finally obtains

$$\frac{d}{dt} \int_S \mathbf{a} \cdot d\mathbf{S} = \int_S \frac{D\mathbf{a}}{Dt} \cdot d\mathbf{S} = \int_S \left[\frac{\partial \mathbf{a}}{\partial t} + \nabla \times (\mathbf{a} \times \mathbf{v}) + \mathbf{v}\nabla \cdot \mathbf{a} \right] \cdot d\mathbf{S}. \quad (3.125)$$

The notation

$$\frac{D\mathbf{a}}{Dt} = \frac{\partial \mathbf{a}}{\partial t} + \nabla \times (\mathbf{a} \times \mathbf{v}) + \mathbf{v}\nabla \cdot \mathbf{a}$$

belongs to Lorentz, being called *Lorentzian derivative*.

Let us apply now the above results to the electromagnetic field theory. We take $\mathbf{a}(\mathbf{r}, t) \equiv \mathbf{B}(\mathbf{r}, t)$, which means $\nabla \cdot \mathbf{a} = \nabla \cdot \mathbf{B} = 0$, so that

$$\frac{d}{dt} \int_S \mathbf{B} \cdot d\mathbf{S} = \int_S \left[\frac{\partial \mathbf{B}}{\partial t} + \nabla \times (\mathbf{B} \times \mathbf{v}) \right] \cdot d\mathbf{S}. \quad (3.126)$$

Denoting by $\mathbf{E}'$ the electric field measured by an observer which is at rest with respect to the moving medium, let us write the induction law (3.115) in the form

$$\oint_C \mathbf{E}' \cdot d\mathbf{l} = -\int_S \left[\frac{\partial \mathbf{B}}{\partial t} + \nabla \times (\mathbf{B} \times \mathbf{v}) \right] \cdot d\mathbf{S}, \quad (3.127)$$

or, by means of the Stokes–Ampère theorem and a convenient arrangement of the terms,

$$\nabla \times (\mathbf{E}' - \mathbf{v} \times \mathbf{B}) = -\frac{\partial \mathbf{B}}{\partial t}. \quad (3.128)$$

The vector $\mathbf{E}' - \mathbf{v} \times \mathbf{B}$ is nothing else but the field $\mathbf{E}$, measured by a stationary observer, therefore we can write

$$\nabla \times \mathbf{E} = -\frac{\partial \mathbf{B}}{\partial t}, \quad (3.129)$$

as well as

$$\mathbf{E} = \mathbf{E}' - \mathbf{E}'', \quad (3.130)$$

where $\mathbf{E}'' \equiv \mathbf{v} \times \mathbf{B}$ was termed by Lorentz *effective electric field*.

Thus we came to the conclusion that Maxwell's equation expressing the induction law does not change its form when passing from a medium at rest to a uniformly moving medium. The relation (3.130) is also written as

$$\mathbf{E}' = \mathbf{E} + \mathbf{v} \times \mathbf{B}, \tag{3.131}$$

i.e. the field $\mathbf{E}'$, measured by the mobile observer, is the sum of $\mathbf{E}$, measured by the stationary observer, and the *effective field* $\mathbf{E}''$, that appears as a result of the displacement of the medium. We shall encounter again Eq. (3.131) when discussing the relativistic approach to electrodynamics.

Observations:

(a) We emphasize, for very good reasons, that the velocity of the medium is much smaller than the speed of light in vacuum.
(b) Relation (3.128) also says that the two observers determine *the same magnetic field*: $\mathbf{B}' = \mathbf{B}$.
(c) The three terms appearing in the integral (3.125) have the following significance: the first term expresses the time variation of the flux of the vector $\mathbf{a}$ due to the time variation of $\mathbf{a}$; the second gives the flux variation through the surface bounded by the moving contour; the third appears as a result of the surface S passing through an inhomogeneous region of the field.

3.12.2 Source Equations

Recall that by averaging the Maxwell–Lorentz microscopic equations we obtained (see Sect. 3.2):

$$\nabla \times \frac{\mathbf{B}}{\mu_0} = \langle \mathbf{j}_{micro} \rangle + \epsilon_0 \frac{\partial \mathbf{E}}{\partial t},$$
$$\nabla \cdot (\epsilon_0 \mathbf{E}) = \langle \rho_{micro} \rangle, \tag{3.132}$$

and found

$$\langle \mathbf{j}_{micro} \rangle = \mathbf{j} + \nabla \times \mathbf{M} + \frac{\partial \mathbf{P}}{\partial t},$$
$$\langle \rho_{micro} \rangle = \rho - \nabla \cdot \mathbf{P}, \tag{3.133}$$

leading to the following source equations for media at rest:

$$\nabla \times \left(\frac{\mathbf{B}}{\mu_0} - \mathbf{M} \right) = \mathbf{j} + \frac{\partial}{\partial t}(\epsilon_0 \mathbf{E} + \mathbf{P}),$$
$$\nabla \cdot (\epsilon_0 \mathbf{E} + \mathbf{P}) = \rho. \tag{3.134}$$

If the medium is moving, Eq. $(3.134)_2$ will remain unchanged, while the derivation of some terms in $(3.134)_1$ must be reconsidered. First, besides the conduction current we have to consider the convection current

$$\langle \rho_{micro} \rangle \, \mathbf{v} = (\rho - \nabla \cdot \mathbf{P}) \, \mathbf{v}. \tag{3.135}$$

Secondly, the currents produced by the time variation of $\mathbf{P}$ as a result of the motion of the medium are now given by the Lorentzian derivative of $\mathbf{P}$, which means that the term $\partial \mathbf{P} / \partial t$ in (3.133) has to be replaced by

$$\frac{D\mathbf{P}}{Dt} = \frac{\partial \mathbf{P}}{\partial t} + \nabla \times (\mathbf{P} \times \mathbf{v}) + \mathbf{v} \nabla \cdot \mathbf{P} \tag{3.136}$$

This conclusion can also be drawn from the fact that, since the medium is moving,

$$\langle \mathbf{j}_{pol} \rangle = n \, e \langle \dot{\mathbf{l}} \rangle = \frac{D}{Dt}(n \, e \langle \mathbf{l} \rangle) = \frac{D\mathbf{P}}{Dt}.$$

Therefore, instead of Eq. $(3.134)_1$ we have to write

$$\nabla \times \mathbf{H} = \mathbf{j} + \rho \mathbf{v} + \nabla \times (\mathbf{P} \times \mathbf{v}) + \frac{\partial \mathbf{D}}{\partial t}, \tag{3.137}$$

where the fields $\mathbf{H}$, $\mathbf{D}$ have the usual significance:

$$\mathbf{H} = \frac{\mathbf{B}}{\mu_0} - \mathbf{M},$$
$$\mathbf{D} = \epsilon_0 \mathbf{E} + \mathbf{P}. \tag{3.138}$$

Let us write the source equations for two particular cases.

3.12.2.1 The Moving Medium Is a Non-polarizabile Conducting Fluid ($\mathbf{P} = 0$, $\mathbf{M} = 0$)

The source equations read

$$\frac{1}{\mu_0} \nabla \times \mathbf{B} = \mathbf{j} + \rho \mathbf{v} + \epsilon_0 \frac{\partial \mathbf{E}}{\partial t},$$
$$\epsilon_0 \nabla \cdot \mathbf{E} = \rho. \tag{3.139}$$

As shown by M.G. Calkin,[4] Eq. (3.139) can be written in a symmetric form, similar to that of the source-free equations:

---

[4]M.G. Calkin, *An Action Principle for Magnetohydrodynamics*, Can. J. Phys., **41**, 1963, p. 2241.

$$\nabla \times \left(\frac{\mathbf{B}}{\mu_0} - \boldsymbol{P} \right) = \frac{\partial}{\partial t}(\epsilon_0 \mathbf{E} + \boldsymbol{P}),$$

$$\nabla \cdot (\epsilon_0 \mathbf{E} + \boldsymbol{P}) = 0, \tag{3.140}$$

by means of the vector field $\boldsymbol{P}$, called *pseudo-polarization* and defined by

$$\mathbf{j} = \frac{\partial \boldsymbol{P}}{\partial t} + \nabla \times (\boldsymbol{P} \times \mathbf{v}) + \mathbf{v}\nabla \cdot \boldsymbol{P},$$

$$\rho = -\nabla \cdot \boldsymbol{P}. \tag{3.141}$$

This formalism is justified by the fact that the equation of continuity

$$\frac{\partial \rho}{\partial t} + \nabla \cdot (\mathbf{j} + \rho \mathbf{v}) = 0 \tag{3.142}$$

is identically satisfied by (3.141). Since the field described by the Eq. (3.140) is, formally, source-free, these can serve to define the electromagnetic field $(\mathbf{E}, \mathbf{B})$ in terms of the *generalized antipotentials* $\boldsymbol{M}$ and ψ as

$$\mathbf{E} = \frac{1}{\epsilon_0}(\nabla \times \boldsymbol{M} - \boldsymbol{P}),$$

$$\mathbf{B} = \mu_0 \left(\nabla \psi + \boldsymbol{P} \times \mathbf{v} + \frac{\partial \boldsymbol{M}}{\partial t} \right). \tag{3.143}$$

Equations (3.143) generalize the definition of the electromagnetic field in terms of the antipotentials (3.90). (The sign does not have any significance).

The generalized antipotentials are useful in the study of moving charged fluids (magnetofluid dynamics, plasma theory, etc.).

3.12.2.2   The Moving Medium Is a Polarized ($\mathbf{P} \neq 0$) Dielectric ($\mathbf{M} = 0$), without Sources ($\mathbf{j} = 0$, $\rho = 0$)

In this case, the source equations become

$$\nabla \times \left(\frac{\mathbf{B}}{\mu_0} - \mathbf{P} \times \mathbf{v} \right) = \frac{\partial \mathbf{D}}{\partial t},$$

$$\nabla \cdot \mathbf{D} = 0. \tag{3.144}$$

Comparing $(3.144)_1$ and (3.134) one observes that a polarized dielectric behaves, from the macroscopic point of view, as a *magnetized body*, possessing the magnetization intensity

$$\mathbf{M}_d = \mathbf{P} \times \mathbf{v}.$$

The Hertz–Lorentz theory was experimentally confirmed by Wilhelm Röntgen (1845–1923) in 1888. He investigated the magnetic effect accompanying the motion of a polarized dielectric. The current density $\nabla \times (\mathbf{P} \times \mathbf{v})$, corresponding to this type of magnetization, is called *Röntgen current*. Röntgen's studies were resumed in 1903 by Alexander Eichenwald (1863–1944), and in 1904 by Harold A. Wilson (1874–1964), the latter performing the polarization of a dielectric cylinder, while rotating in a magnetic field parallel to the cylinder axis. In 1903–1904, Nicolae Vasilescu-Karpen (1870–1964) proved experimentally, with a high precision, that the magnetic field produced by the convection current is the same with the effect produced by the conduction current, in any conditions. All these facts show that a moving polarized dielectric gives rise to a magnetic field of the same nature as the field produced by permanent magnets.

Summarizing the previous discussion, let us write the system of equations of the electromagnetic field for slowly moving media, as given by Lorentz:

$$\nabla \times \mathbf{E} = -\frac{\partial \mathbf{B}}{\partial t},$$
$$\nabla \cdot \mathbf{B} = 0,$$
$$\nabla \times \mathbf{H} = \mathbf{j} + \rho \mathbf{v} + \nabla \times (\mathbf{P} \times \mathbf{v}) + \frac{\partial \mathbf{D}}{\partial t},$$
$$\nabla \cdot \mathbf{D} = \rho, \tag{3.145}$$

where by ρ and $\mathbf{j}$ we mean the free charge density and the conduction current density, respectively. Equations (3.145) have to be completed with constitutive relations and jump conditions.

The differential form of Ohm's law for moving media, due to (3.131), becomes

$$\mathbf{j} = \lambda(\mathbf{E} + \mathbf{v} \times \mathbf{B}). \tag{3.146}$$

Observation:

In the derivation of Eqs. (3.145) Lorentz neglected the possibility of variation of the magnetic polarization of moving media. Subsequent research and, first of all, the theory of relativity, showed the connection between the two types of polarization and gave the appropriate form of the relations between fields. This issue will be discussed in Part II of the book.

3.13 Solved Problems

Problem 1. Determine the expression (3.102) of the Lagrangian used in Sect. 3.11 to find the equation of motion of a point charge in an external electromagnetic field.

Solution. In the framework of the Lagrangian formalism, the differential equations of motion are generally written as

$$\frac{d}{dt}\left(\frac{\partial T}{\partial \dot{q}_k}\right) - \frac{\partial T}{\partial q_k} = Q_k, \qquad k = 1, 2, \ldots, n, \tag{3.147}$$

where $q_1, q_2, \ldots, q_n$ are the generalized coordinates, $T(q, \dot{q}, t)$ is the kinetic energy, and Q_k, $k = 1, 2, \ldots, n$ are the associated generalized forces. If the applied force comes from a generalized potential, linear in $\dot{q}_k$, of the form

$$U(q, \dot{q}, t) = \sum_{k=1}^{n} C_k \dot{q}_k + U_0 = U_1 + U_0, \quad C_k = C_k(q, t),$$

Eqs. (3.147) can be put in a more convenient form. To this end, one adds to both sides of (3.147) the expression

$$\frac{\partial U}{\partial q_k} - \frac{d}{dt}\left(\frac{\partial U}{\partial \dot{q}_k}\right),$$

obtaining

$$\frac{d}{dt}\left[\frac{\partial (T - U)}{\partial \dot{q}_k}\right] - \frac{\partial (T - U)}{\partial q_k} = Q_k - \frac{d}{dt}\left(\frac{\partial U}{\partial \dot{q}_k}\right) + \frac{\partial U}{\partial q_k}.$$

If one may write

$$Q_k = \frac{d}{dt}\left(\frac{\partial U}{\partial \dot{q}_k}\right) - \frac{\partial U}{\partial q_k}, \tag{3.148}$$

and define the Lagrangian of the system as

$$L(q, \dot{q}, t) = T(q, \dot{q}, t) - U(q, \dot{q}, t), \tag{3.149}$$

then we get the usual form of the Lagrange equations:

$$\frac{d}{dt}\left(\frac{\partial L}{\partial \dot{q}_k}\right) - \frac{\partial L}{\partial q_k} = 0. \tag{3.150}$$

Our Lagrangian falls into this general description. Since there are no definite laws to construct a Lagrangian, we should mention that the Lagrangian associated with a physical system is not unique, but it has to obey certain rules. First, it must contain all the quantities necessary to describe the system and, finally, to lead to the desired equation(s) of motion. Secondly, it has to be as simple as possible. Thirdly, it must have the dimension of energy. Fourthly, it has to be invariant with respect to a set of space-time transformations (see Part II). Since there are no constraints on the motion of the particle, it has three degrees of freedom. Thus, we can choose as general coordinates its Cartesian coordinates, and as general velocities – its usual velocity components:

$$q_k = x_k, \quad \dot{q}_k = \dot{x}_k = v_k, \quad k = 1, 2, 3.$$

In the Lagrangian formalism, the general coordinates and general velocities are considered independent quantities.

In the absence of the magnetic field, the Lagrangian is

$$L = T - U = T - eV,$$

where V is the potential of the electric field. To consider the contribution of the magnetic field, we use the above mentioned properties and observe that, in order to be a scalar, the Lagrangian must contain the vector potential $\mathbf{A}$ only as a dot product with another vector. There are three meaningful possibilities: $\mathbf{A} \cdot \mathbf{A}, \mathbf{A} \cdot \dot{\mathbf{r}}$ and $\mathbf{A} \cdot \ddot{\mathbf{r}}$. The first possibility would give a term proportional to $\mathbf{B}^2$ in the equation of motion, and since such a term does not exist, the first possibility is excluded. The third case implies the existence of a term containing the derivative of acceleration in the equation of motion; this option is also out of the question, since customarily the equations of motion are of the second order. Therefore, it remains only the second solution, and the Lagrangian acquires the form (see (3.102)):

$$L = \frac{1}{2}m v^2 - eV + e\mathbf{A} \cdot \mathbf{v}. \tag{3.151}$$

Problem 2. Starting from the Lagrangian (3.151), find the equation of motion of a charged particle in the electromagnetic field $\mathbf{E}$ and $\mathbf{B}$, using the Hamiltonian formalism.

Solution. Let us write the Lagrangian (3.151) in the form

$$L = \frac{1}{2}m v_j v_j - eV + e v_j A_j, \quad j = 1, 2, 3, \tag{3.152}$$

where Einstein's summation convention has been used. The first step in the Hamiltonian formalism is to determine Hamilton's function $H(q, p, t)$, where p stands for the generalized momenta

$$p_j = \frac{\partial L}{\partial \dot{q}_j} = mv_j + eA_j. \tag{3.153}$$

The Hamiltonian of the system is then

$$H = p_j \dot{q}_j - L = v_j(mv_j + eA_j) - \frac{1}{2}m v_j v_j + eV - e v_j A_j = \frac{1}{2}mv_j v_j + eV.$$

Using (3.153) we shall express the Hamiltonian in terms of coordinates and canonically conjugated momenta:

$$H = \frac{1}{2}(p_j - eA_j)(p_j - eA_j) + eV. \tag{3.154}$$

Next step is to write Hamilton's canonical equations

$$\dot{x}_k = \frac{\partial H}{\partial p_k},$$
$$\dot{p}_k = -\frac{\partial H}{\partial x_k}. \tag{3.155}$$

For the Hamiltonian (3.154) these equations become

$$\dot{x}_k = \frac{1}{m}(p_k - eA_k),$$
$$\dot{p}_k = \frac{e}{m}(p_j - eA_j)\frac{\partial A_j}{\partial x_k} - e\frac{\partial V}{\partial x_k}. \tag{3.156}$$

Taking the total time derivative of $(3.156)_1$, and using $(3.156)_2$, we find the equation of motion in the form

$$\ddot{x}_k = -e\frac{\partial V}{\partial x_k} - e\frac{\partial A_k}{\partial t} + ev_j\left(\frac{\partial A_j}{\partial x_k} - \frac{\partial A_k}{\partial x_j}\right), \quad j, k = 1, 2, 3,$$

or, in terms of the electric and magnetic fields,

$$m\ddot{x}_k = eE_k + e(\mathbf{v} \times \mathbf{B})_k, \quad k = 1, 2, 3, \tag{3.157}$$

as we have expected. The last equation has been already obtain in Sect. 3.11 by means of the Lagrangian formalism, and the final results coincide.

Problem 3. Find the transformation of the Lagrangian of a charged particle, moving in an external electromagnetic field, as a result of a gauge transformation of the potentials.

Solution. Let us consider $L(V, \mathbf{A})$ and $L'(V', \mathbf{A}')$, where (see (3.87), (3.88)):

$$\mathbf{A}' = \mathbf{A} + \nabla\psi,$$
$$V' = V - \frac{\partial\psi}{\partial t}, \tag{3.158}$$

$\psi(\mathbf{r}, t)$ being a differentiable function. Then, we have

$$L = \frac{1}{2}mv^2 - eV + e(\mathbf{v} \cdot \mathbf{A}),$$

and, in view of (3.158):

$$L = \frac{1}{2}mv^2 - e\left(V' + \frac{\partial \psi}{\partial t}\right) + e\mathbf{v} \cdot (\mathbf{A}' - \nabla\psi) = L' - e\mathbf{v} \cdot \nabla\psi - e\frac{\partial \psi}{\partial t}$$

$$= L' - e\left[\frac{\partial \psi}{\partial t} + (\mathbf{v} \cdot \nabla)\psi\right] = L' - e\frac{d\psi}{dt} = L' - \frac{d}{dt}(e\psi). \qquad (3.159)$$

But two Lagrangian functions, which differ from each other by a term which is a total time derivative of an arbitrary scalar function of coordinates and time, are *equivalent*. In other words, they give the same description of the motion. In conclusion, the gauge transformation (3.158) does not affect the equations of motion.

Problem 4. The Lagrangian of a charged particle, subject to an electromagnetic force is (see (3.151)):

$$L_{em} = \frac{1}{2}m\,v^2 - eV + e\mathbf{v} \cdot \mathbf{A}, \qquad (3.160)$$

while the Lagrangian of a particle of mass m, in motion with respect to a non-inertial frame S' (see Fig. 3.2) is

$$L_{mech} = \frac{1}{2}m\,|\mathbf{v}_r|^2 + \frac{1}{2}m\,|\boldsymbol{\omega} \times \mathbf{r}'|^2 + m\,\mathbf{v}_r \cdot (\boldsymbol{\omega} \times \mathbf{r}') - m\,\mathbf{a}_0 \cdot \mathbf{r}' - W_{potential}, \qquad (3.161)$$

where $\boldsymbol{\omega}$ is the instantaneous vector of rotation, $\mathbf{v}_r$ – the relative velocity, and $W_{potential}$ – the potential energy of the particle.

a) Using the correspondences $m \leftrightarrow e$ si $\boldsymbol{\omega} \leftrightarrow \nabla \times \mathbf{A}'$, show that the equation of motion of a massive and electrically neutral particle in a non-inertial system can be obtained from the Lagrangian of a charged particle moving under the action of a force which derives from a generalized potential (depending on velocities), where the generalized potential energy is of the form

$$W' = m\,(V' - 2\mathbf{v} \cdot \mathbf{A}'),$$

Fig. 3.2 The two referentials: S – the inertial reference frame (IRF) and S' – the non-inertial reference frame (NIRF) for Problem 4.

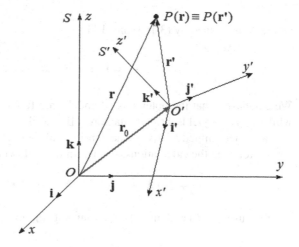

with

$$V' = \mathbf{r} \cdot \mathbf{a}_0 - \frac{1}{2}|\boldsymbol{\omega} \times \mathbf{r}|^2 \quad \text{and} \quad \mathbf{A}' = \frac{1}{2}\boldsymbol{\omega} \times \mathbf{r},$$

and the magnetic field is considered constant and homogeneous.

b) Show that the "new fields" $\mathbf{E}'$ and $\mathbf{B}'$ (corresponding to the potentials V' si $\mathbf{A}'$ defined above) satisfy the sourceless Maxwell equations.

Solution. Starting from the Lagrangian (3.161) one obtains the equation of motion of the particle with respect to the non-inertial frame as the Lagrange equation:

$$m\,\mathbf{a}_r = \mathbf{F} - m\,\mathbf{a}_0 - m\,\dot{\boldsymbol{\omega}} \times \mathbf{r}' - m\,\boldsymbol{\omega} \times (\boldsymbol{\omega} \times \mathbf{r}') - 2m\,\boldsymbol{\omega} \times \mathbf{v}_r. \tag{3.162}$$

The analogy between the two systems whose Lagrangians are (3.160) and (3.161) appears at various levels, as follows. First, one observes that both the Coriolis force $\mathbf{F}_c = 2m\,\mathbf{v}_r \times \boldsymbol{\omega}$ and the Lorentz force are *gyroscopic* (their power is zero). Since we deal with a motion in a non-inertial frame, we are allowed to use $\mathbf{r}$, $\mathbf{v}$, instead of $\mathbf{r}'$, $\mathbf{v}_r$, respectively. Secondly, each Lagrangian contains two velocity-dependent terms. Thirdly, if the field $\mathbf{B}$ is constant and homogeneous, then it is easy to show (see Problem 1, Chap. 2) that $\mathbf{A}$ can be written as $\mathbf{A} = \frac{1}{2}\mathbf{B} \times \mathbf{r}$. This means that the terms $m\,\mathbf{v} \cdot (\boldsymbol{\omega} \times \mathbf{r})$ and $e\,\mathbf{v} \cdot \mathbf{A}$ are equivalent, if we set the correspondence $m \leftrightarrow e$, and choose

$$\mathbf{A}' = \frac{1}{2}\boldsymbol{\omega} \times \mathbf{r}, \tag{3.163}$$

i.e.

$$\boldsymbol{\omega} = \nabla \times \mathbf{A}'. \tag{3.164}$$

This analogy leads to the following Lagrangian of the particle, relative to the non-inertial frame:

$$L = \frac{1}{2}m\,|\mathbf{v}|^2 - m\,V' + 2m\,\mathbf{v} \cdot \mathbf{A}' - V(\mathbf{r}), \tag{3.165}$$

where $\mathbf{A}'$ is defined by (3.163) and V' by

$$V' = \mathbf{r} \cdot \mathbf{a}_0 - \frac{1}{2}|\boldsymbol{\omega} \times \mathbf{r}|^2. \tag{3.166}$$

We emphasize that the potentials $\mathbf{A}'$ and V' are functions of coordinates and time, while $W_{potential}$ yields the potential force. By the above settled convention, $\dot{\mathbf{r}} = \mathbf{v}$ and $\ddot{\mathbf{r}} = \mathbf{a}$ are determined relative to the non-inertial frame. Using the Lagrange equations and performing the calculations, we obtain the following equation of motion:

$$m\,\ddot{\mathbf{r}} + 2m\,\mathbf{A}' + m\,\nabla V' - 2m\,\nabla(\mathbf{r} \cdot \mathbf{A}') + \nabla W = 0,$$

or, if we use (A.46) and make some re-arrangements of the terms,

$$m\,\ddot{\mathbf{r}} = m\,(\mathbf{E'} + \mathbf{v} \times \mathbf{B'}) + \mathbf{F},\qquad(3.167)$$

where

$$\mathbf{E'} = -\nabla V' - \frac{\partial}{\partial t}(2\mathbf{A'}),$$
$$\mathbf{B'} = \nabla \times (2\mathbf{A'}).\qquad(3.168)$$

Equation (3.167) shows that the terms $\mathbf{E'}$ and $\mathbf{v} \times \mathbf{B'}$ have the units of an acceleration. It also shows that the particle moves in an inertial force field, defined by the potentials $\mathbf{A'}$, V', and in an applied force field $\mathbf{F} = -\nabla V$ as well. If the frame S' becomes inertial, and $\mathbf{F} = 0$, then $\mathbf{a}_0 = 0$, $\boldsymbol{\omega} = 0$, and Eq. (3.167) yields $\ddot{\mathbf{r}} = 0$, as expected. We reach the same result if the charged particle is neither accelerated by the electric field $\mathbf{E}$ nor rotated by the magnetic field $\mathbf{B}$.

We can still go further with this analogy, and observe that the fields $\mathbf{E'}$ and $\mathbf{B'}$ satisfy the source-free Maxwell equations,

$$\nabla \times \mathbf{E'} = -\frac{\partial \mathbf{B'}}{\partial t},$$
$$\nabla \cdot \mathbf{B'} = 0.\qquad(3.169)$$

We can then conclude that the study of a massive ($m \neq 0$), non-charged particle, moving in a non-inertial frame, can be accomplished by using the same Lagrangian as for a charged particle, moving in a velocity-dependent force field, the generalized potential energy being given by

$$W = m\,(V' - 2\mathbf{v} \cdot \mathbf{A'}).\qquad(3.170)$$

Problem 5. Using the Hamiltonian formalism, study the motion of a charged particle in a stationary magnetic field with axial symmetry.

Solution. Since the field is stationary, the vector potential $\mathbf{A}$ does not explicitly depend on time and the magnetic field is given by

$$\mathbf{B} = \nabla \times \mathbf{A}.\qquad(3.171)$$

The geometry of the problem indicates to choose a cylindrical system of coordinates ρ, φ, z, with the z-axis directed along the symmetry axis, and assume that $A_\rho = 0$, $A_\varphi \neq 0$, $A_z = 0$. Since by definition of the field with axial symmetry $B_\varphi = 0$, the non-zero components of $\mathbf{B}$ are

$$B_\rho = -\frac{\partial A_\varphi}{\partial z},$$
$$B_z = \frac{1}{\rho}\frac{\partial}{\partial \rho}\,(\rho A_\varphi).$$

Observing that **A** is independent of φ, we may write the Hamiltonian as

$$H = \frac{1}{2m}\left\{ p_\rho^2 + \frac{1}{\rho^2}\left[p_\varphi - e\rho A_\varphi(\rho, z) \right]^2 + p_z^2 \right\}. \tag{3.172}$$

Hamilton's equations for the conjugate variables φ and p_φ are

$$\dot{\varphi} = \frac{\partial H}{\partial p_\varphi} = \frac{1}{m\rho^2}(p_\varphi - e\rho A_\varphi),$$

$$\dot{p}_\varphi = -\frac{\partial H}{\partial \varphi} = 0.$$

The second equation leads to the first integral

$$p_\varphi = m\rho^2\dot{\varphi} + e\rho A_\varphi = \text{const.} \tag{3.173}$$

Since H does not explicitly depend on time, there exists also the energy first integral

$$E = \frac{1}{2}m(\dot{\rho}^2 + \rho^2\dot{\varphi}^2 + \dot{z}^2) = \text{const.} \tag{3.174}$$

The differential equations of motion for ρ and z are then

$$m(\ddot{\rho} - \rho\dot{\varphi}^2) = e\dot{\rho}B_z,$$

$$m\ddot{z} = -e\rho\dot{\varphi}B_\rho. \tag{3.175}$$

Assume now that **B** is constant and homogeneous. Then, according to the result of Problem 1, Chap. 2,

$$A_\varphi = \frac{1}{2}(\mathbf{B} \times \mathbf{r})_\varphi = \frac{1}{2}\rho B_z,$$

and the first integral (3.173) reads

$$m\rho^2\dot{\varphi} + \frac{1}{2}e\rho^2 B_z = C \text{ (const.)}, \tag{3.176}$$

known as *Busch's relation*, after the German physicist Hans Busch (1884–1973). If at the initial moment $t = 0$ the particle is at the origin O of the reference frame, where $\rho = 0$ and $v_\varphi = \rho\dot{\varphi} = 0$, then $C = 0$ and (3.176) yields

$$\dot{\varphi} = -\frac{eB_z}{2m}, \tag{3.177}$$

meaning that the particle performs a motion of precession around the z-axis. This effect is applied in the construction of magnetic focusing devices, called *magnetic lenses*. Substituting (3.177) into the energy first integral (3.174) (and noting that for

$\mathbf{B} = (0, 0, B_z)$, the second equation in (3.175) leads to $\dot{z} = \text{const.}$), one obtains the following second-order differential equation:

$$\ddot{\rho} + \left(\frac{eB_z}{2m}\right)^2 \rho = 0. \qquad (3.178)$$

Assume that the particle moves close to the z-axis, while the magnetic field acts only over a small portion of the beam (paraxial beam). The components of the velocity along ρ and φ are then negligible as compared to the component along z. As $\dot{z} = v = \text{const.}$, we have

$$\ddot{\rho} = \dot{z}^2 \frac{d^2\rho}{dz^2} = v^2 \frac{d^2\rho}{dz^2},$$

and (3.178) yields

$$\frac{d^2\rho}{dz^2} + \left(\frac{eB_z}{2mv}\right)^2 \rho = 0. \qquad (3.179)$$

Since $d^2\rho/dz^2 < 0$, the magnetic lens is *convergent*, independently of the sign of charged particles.

The *electric lens* is based on a similar focusing principle. Both electric and magnetic lenses are used in electronic microscopy, television devices, etc.

3.14 Proposed Problems

1. Write Maxwell's equations (3.9), as well as the relations between fields and potentials, in spherical and cylindrical coordinates.
2. Show that, if in a homogeneous conducting medium one can neglect the displacement current as compared with the conduction current, then the density $\mathbf{j}$ of the conduction current satisfies the equations

$$\nabla \cdot \mathbf{j} = 0, \qquad \Delta\mathbf{j} = \lambda\mu \frac{\partial \mathbf{j}}{\partial t},$$

where λ is the electric conductivity. Also, show that the fields $\mathbf{E}$, $\mathbf{D}$, $\mathbf{H}$, $\mathbf{B}$ satisfy analogous relations.
3. Given the Lagrangian density of the electromagnetic field in the presence of sources (3.110), find Maxwell's equations by using the Hamiltonian formalism for continuous media.
4. Using the Hamiltonian approach, determine the energy of a dipole in an external, constant, and homogeneous electromagnetic field.
5. Show that Maxwell's equations (3.9) are not covariant (do not keep their form) under the Galilei–Newton group of transformations:

$$\mathbf{r}' = \mathbf{r} - \mathbf{V}t,$$
$$t' = t.$$

Later on, we shall see that they are covariant under the Lorentz group of transformations, which represents the symmetry of special relativity.

6. A flux of charged particles is radially and uniformly ejected from the surface
   of an empty sphere, towards the outside. The result is a radial current, having
   the same magnitude in all directions. Let R be the radius of the sphere, $Q(r)$
   the charge inside a given spherical surface of radius $r > R$, $\mathbf{j}(r)$ the density
   of the radial current, and $\mathbf{E}(r)$ the intensity of the electric field at any point
   of the spherical surface of radius $r > R$. Determine $\mathbf{j}(r)$ in terms of $Q(r)$ and
   $\mathbf{E}(r)$. Using Maxwell's equations, find also the magnetic field produced by the
   currents.

7. Show that the vector potential $\mathbf{A}$, in the Coulomb gauge, can be derived from a
   superpotential in such a way that the Eq. (3.86) and the Coulomb condition (3.83)
   are both satisfied, the superpotential having to satisfy only a wave equation. (Hint:
   one takes $\mathbf{A} = \nabla \times \boldsymbol{\mathcal{Z}}$.)

8. Derive Maxwell's equations in Hamiltonian formalism, when no sources are
   present, by using the complex function $\psi_k = E_k - i\, B_k$, where i is the imaginary
   unit.

9. An electric charge q is uniformly distributed inside a cone whose height h equals
   the radius of its basis. The cone rotates around its symmetry axis with the constant
   angular velocity ω. A particle whose internal magnetic moment $\mathbf{m}$ is tilted by
   the angle α with respect to the axis, is placed at the tip of the cone. Show that
   the magnetic energy of interaction between the particle and the cone is given by
   the expression

$$W = \frac{3(4 - 3\sqrt{2})}{2} \frac{\mu_0}{4\pi} \frac{m\omega q}{h} \cos \alpha.$$

10. A bounded distribution of electric currents $\mathbf{j}$ produces in vacuum a field whose
    total energy is W_0. In a magnetizable (non-ferromagnetic) medium, the same dis-
    tribution gives rise to a field whose total energy is W. Show that the contribution
    of the medium to the energy is

$$\delta W = W - W_0 = \frac{1}{2} \int \mathbf{M} \cdot \mathbf{B}_0 \, d\tau,$$

where $\mathbf{M}$ is the magnetization of the medium, and $\mathbf{B}_0$ is the initial value in
vacuum of the magnetic induction.

Chapter 4
Electromagnetic Waves

4.1 Conductors, Semiconductors, Dielectrics

The electromagnetic field propagates both in vacuum and in media as *electromagnetic waves*. In this respect, it is necessary to establish a classification criterion of material media regarding the propagation phenomenon of electromagnetic waves.

Consider a linear, homogeneous, and isotropic medium, characterized by the permittivity ϵ, the permeability μ, and the conductivity λ. If the medium is fixed relative to the observer, the constitutive equation are:

$$\mathbf{D} = \epsilon \mathbf{E}, \quad \mathbf{B} = \mu \mathbf{H}, \quad \mathbf{j} = \lambda \mathbf{E}.$$

In this case, the sum of the conduction and displacement current densities arising in Maxwell's equation (3.42) is:

$$\mathbf{j}_{tot} = \mathbf{j}_{cond} + \mathbf{j}_{disp} = \lambda \mathbf{E} + \epsilon \frac{\partial \mathbf{E}}{\partial t}.$$

Let us define the ratio

$$\beta = \frac{|\mathbf{j}_{cond}|}{|\mathbf{j}_{disp}|}.$$

Depending on the value of this ratio, one can distinguish three cases:

(a) $\beta \gg 1$. Such a situation is found in case of *conductors*, where $\mathbf{j}_{disp}$ is negligible with respect to $\mathbf{j}_{cond}$;
(b) $\beta \ll 1$. This property characterizes the *dielectric* media;
(c) $\beta \simeq 1$, which means that both currents have to be considered. This is the case of *semiconductors*.

© Springer-Verlag Berlin Heidelberg 2016
M. Chaichian et al., *Electrodynamics*, DOI 10.1007/978-3-642-17381-3_4

Assuming that there is no spatial charge ($\rho = 0$), consider now Maxwell's evolution equations

$$\frac{1}{\mu} \nabla \times \mathbf{B} = \lambda \mathbf{E} + \epsilon \frac{\partial \mathbf{E}}{\partial t},$$

$$\nabla \times \mathbf{E} = -\frac{\partial \mathbf{B}}{\partial t}.$$

Taking the curl of both equations and using Maxwell's condition equations $\nabla \cdot \mathbf{B} = 0$ and $\nabla \cdot \mathbf{E} = 0$, one obtains:

$$\Delta \mathbf{E} - \epsilon \mu \frac{\partial^2 \mathbf{E}}{\partial t^2} = \lambda \mu \frac{\partial \mathbf{E}}{\partial t},$$

$$\Delta \mathbf{B} - \epsilon \mu \frac{\partial^2 \mathbf{B}}{\partial t^2} = \lambda \mu \frac{\partial \mathbf{B}}{\partial t}. \tag{4.1}$$

According to the above discussion, these equations describe the propagation of the electromagnetic waves in semiconductive media. The dielectrics and conductors will then appear as particular cases of semiconductors, by neglecting either the term $\epsilon \mu\, \partial^2 \mathbf{E}/\partial t^2$ (conductors), or the term $\lambda \mu\, \partial \mathbf{E}/\partial t$ (dielectrics).

For given λ, ϵ, and μ, the type of a material depends, from the point of view of wave propagation, on the frequency of the incident radiation. Let us consider the electric field varying periodically according to the law $\mathbf{E} = \mathbf{E}_0 \exp(iwt)$. In this case,

$$\beta = \frac{|\mathbf{j}_{cond}|}{|\mathbf{j}_{disp}|} = \frac{\lambda}{\epsilon \omega}.$$

Take, as an example, the *silicon* (Si), characterized by $\lambda/\epsilon \simeq 5 \times 10^7$ Hz. Here are three values for β for three different frequencies (since we do not take into account the dispersion, the figures are only indicative):

$$\omega = 10^3 \text{ Hz} \rightarrow \beta = 5 \times 10^4 \qquad \text{(metal-type)},$$
$$\omega = 10^{11} \text{ Hz} \rightarrow \beta = 5 \times 10^{-4} \qquad \text{(dielectric-type)},$$
$$\omega = 5 \times 10^7 \text{ Hz} \rightarrow \beta = 1 \qquad \text{(semiconductor-type)}.$$

Therefore, the type of one and the same material depends on the frequency of the incident radiation. To avoid this complication, we shall consider the type of material as being exclusively given by its intrinsic parameters (permittivity, permeability, conductivity, etc.).

We shall begin with the analysis of electromagnetic wave propagation in dielectric-type media. This is the simplest case and allows us to settle some notions with general validity.

4.2 Propagation of Electromagnetic Waves in Dielectric Media

Consider a homogeneous and isotropic, non-dissipative and non-dispersive dielectric medium, free of charge and conduction currents. Maxwell's equations describing the electromagnetic field in such a medium are

$$
\nabla \times \mathbf{H} - \epsilon \frac{\partial \mathbf{E}}{\partial t} = 0,
$$
$$
\nabla \cdot \mathbf{H} = 0,
$$
$$
\nabla \times \mathbf{E} + \mu \frac{\partial \mathbf{H}}{\partial t} = 0, \qquad (4.2)
$$
$$
\nabla \cdot \mathbf{E} = 0.
$$

Taking the curl of the evolution equations and using the condition equations and formula (A.51), we are led to the following two homogeneous second order partial differential equations:

$$
\epsilon \mu \frac{\partial^2 \mathbf{E}}{\partial t^2} - \Delta \mathbf{E} = 0,
$$
$$
\epsilon \mu \frac{\partial^2 \mathbf{H}}{\partial t^2} - \Delta \mathbf{H} = 0. \qquad (4.3)
$$

This means that both $\mathbf{E}$ and $\mathbf{H}$ satisfy the d'Alembert wave equation

$$
\frac{1}{u^2} \frac{\partial^2 \mathbf{f}}{\partial t^2} - \Delta \mathbf{f} = 0, \qquad \mathbf{f} = \mathbf{f}(\mathbf{r}, t). \qquad (4.4)
$$

Comparing (4.3) and (4.4), we find

$$
u = \frac{1}{\sqrt{\epsilon \mu}}; \quad \text{in vacuum } u_0 = \frac{1}{\sqrt{\epsilon_0 \mu_0}} = c. \qquad (4.5)
$$

Plugging the numerical values of ϵ_0 and μ_0 into (4.5), we obtain

$$
u_0 = c \simeq 3 \times 10^8 \, \text{m} \cdot \text{s}^{-1},
$$

We conclude that the electromagnetic field, characterized by the vectors $\mathbf{E}$ and $\mathbf{H}$, propagates in dielectric media as *waves*. Their (phase) velocity in vacuum equals the speed of light in vacuum.

Assume the x-axis as the direction of propagation, and take $\mathbf{f} = \mathbf{f}(x, t)$. D'Alembert's equation (4.4) then becomes

$$
\frac{1}{u^2} \frac{\partial^2 \mathbf{f}}{\partial t^2} - \frac{\partial^2 \mathbf{f}}{\partial x^2} = 0. \qquad (4.6)
$$

To integrate this equation, one uses the substitution

$$\xi = x - ut,$$
$$\eta = x + ut,$$

which gives

$$\frac{\partial^2 \mathbf{f}}{\partial \xi \partial \eta} = 0.$$

Integrating, we have

$$\frac{\partial \mathbf{f}}{\partial \xi} = \mathbf{F}(\xi), \quad \mathbf{f} = \int \mathbf{F}(\xi) d\xi + \mathbf{g}_2(\eta),$$

and, finally,

$$\mathbf{f}(x, t) = \mathbf{g}_1(\xi) + \mathbf{g}_2(\eta) = \mathbf{g}_1(x - ut) + \mathbf{g}_2(x + ut), \tag{4.7}$$

where $\mathbf{g}_1(x - ut)$ and $\mathbf{g}_2(x + ut)$ are two arbitrary functions. The solution $\mathbf{g}_1$ gives the propagation in the positive direction of the x–axis, while $\mathbf{g}_2$ gives the propagation in the negative direction. The first corresponds to a *progressive wave* and the second to a *regressive wave*. We can write also

$$\mathbf{f}(x, t) = \mathbf{f}_1\left(t - \frac{x}{u}\right) + \mathbf{f}_2\left(t + \frac{x}{u}\right). \tag{4.8}$$

In the theory of wave propagation, only the progressive wave is practically used. It is chosen as

$$\mathbf{f}_{p1}(x, t) = \mathbf{f}_0 \cos \omega \left(t - \frac{x}{u}\right) = \mathbf{f}_0 \cos(\omega t - kx), \tag{4.9}$$

where $\mathbf{f}_0$ is the *amplitude* of the wave, $k = 2\pi/\lambda$ – the *wave number*, λ – the *wavelength*, while the argument $(\omega t - kx)$ of the trigonometric function is the *phase of the wave*.

Another solution can be

$$\mathbf{f}_{p2}(x, t) = \mathbf{f}_0 \sin(\omega t - kx). \tag{4.10}$$

According to the superposition principle, the combination

$$\mathbf{f}(x, t) = \mathbf{f}_{p1}(x, t) + i\mathbf{f}_{p2}(x, t)$$
$$= \mathbf{f}_0[\cos(\omega t - kx) + i \sin(\omega t - kx)] = \mathbf{f}_0 e^{i(\omega t - kx)} \tag{4.11}$$

is also a solution of the wave equation (4.6).

If the wave propagates in an arbitrary direction, defined by the unit vector $s(\alpha, \beta, \gamma)$, where α, β, γ are the directional cosines of the normal to the wavefront, then the *wave vector* is

$$\mathbf{k} = k\mathbf{s} = \frac{2\pi}{\lambda}\mathbf{s}, \tag{4.12}$$

and the general form of (4.11) is

$$\mathbf{f}(\mathbf{r}, t) = \mathbf{f}_0\, e^{i(\omega t - \mathbf{k} \cdot \mathbf{r})}. \tag{4.13}$$

The solution (4.13) is the *equation of a monochromatic plane wave*. The term *mono-chromatic* shows that the frequency of the wave is always the same, while *plane* means that at every moment the equal phase surfaces are planes. Indeed, the locus of the points which, at a given moment ($t = $ const.), are characterized by the same phase, is given by

$$\omega t - \mathbf{k} \cdot \mathbf{r} = \omega t - \frac{2\pi}{\lambda}(\alpha x + \beta y + \gamma z) = \text{const.}, \tag{4.14}$$

which is the *normal equation* of a plane, called *phase plane of the wave*.

In conclusion, the solutions of the partial differential equations (4.3) are

$$\mathbf{E}(\mathbf{r}, t) = \mathbf{E}_0\, e^{i(\omega t - \mathbf{k} \cdot \mathbf{r})},$$
$$\mathbf{H}(\mathbf{r}, t) = \mathbf{H}_0\, e^{i(\omega t - \mathbf{k} \cdot \mathbf{r})}, \tag{4.15}$$

and describe the *plane monochromatic electromagnetic waves*. (The *amplitudes* $\mathbf{E}_0$ and $\mathbf{H}_0$ can be either real, or complex.)

4.2.1 Spherical Waves

A wave is characterized, among other properties, by the shape of its *wavefront* (i.e. of the equal-phase surface). By this criterion, a wave can be: plane, spherical, cylindrical, etc. In a homogeneous and isotropic medium, a point source gives rise to *spherical waves*. When studying the propagation of a spherical wave, one has to distinguish the shape (curvature) of the wave front very near to the source (say, at 100 m), and very far from it (say, at 1000 km). Since the spherical waves play an important role in the study of wave propagation in various media, let us determine the law of propagation of such a wave.

We shall use again the d'Alembert wave equation (4.4) where, this time, the Laplacian has to be written in spherical coordinates. Assuming the wave to be mono-chromatic, we choose the solution in the form

$$\mathbf{f}(\mathbf{r}, t) = \mathbf{f}_0\, \varphi(r)e^{i\omega t}.$$

Then the wave equation (4.4) yields

$$\Delta \mathbf{f} = \mathbf{f}_0 e^{i\omega t} \Delta \varphi(r), \qquad -\frac{1}{u^2} \frac{\partial^2 \mathbf{f}}{\partial t^2} = \frac{\omega^2}{u^2} \mathbf{f}_0 \varphi(r) e^{i\omega t},$$

hence

$$\Delta \varphi(r) + \frac{\omega^2}{u^2} \varphi(r) = \Delta_r \varphi(r) + \frac{\omega^2}{u^2} \varphi(r) = 0,$$

or

$$\frac{d^2 \varphi}{dr^2} + \frac{2}{r} \frac{d\varphi}{dr} + k^2 \varphi = 0, \tag{4.16}$$

where

$$k = \frac{\omega}{u} = \frac{2\pi}{\lambda} \tag{4.17}$$

is the wave number. Using the substitution $\varphi = \psi/r$, one obtains the equation

$$\frac{d^2 \psi}{dr^2} + k^2 \psi = 0,$$

with the solution (for the progressive wave) $\psi = A e^{-ikr}$, which gives

$$\varphi(r) = \frac{A}{r} e^{-ikr},$$

so that

$$\mathbf{f}(\mathbf{r}, t) = \frac{\mathbf{f}_0}{r} e^{i(\omega t - kr)}. \tag{4.18}$$

It is easily seen that the equal-phase surfaces are given by $r = \text{const.}$, representing a system of concentric spheres, with their common centre at the point source.

Thus, a spherical, monochromatic, electromagnetic wave is expressed by

$$\mathbf{E}(\mathbf{r}, t) = \frac{\mathbf{X}_0}{r} e^{i(\omega t - kr)},$$

$$\mathbf{H}(\mathbf{r}, t) = \frac{\mathbf{Y}_0}{r} e^{i(\omega t - kr)}. \tag{4.19}$$

Note that, this time, the amplitudes $\mathbf{X}_0, \mathbf{Y}_0$ do not have dimensions of electric or magnetic field, respectively.

4.2.2 Transversality of Electromagnetic Waves

Consider an electromagnetic plane wave, described by the Eq. (4.15). Introducing $\mathbf{E}$ and $\mathbf{H}$ into Maxwell's equations $(4.2)_{2,4}$, we find

$$\nabla \cdot \mathbf{E} = \mathbf{E}_0 \cdot \nabla e^{i(\omega t - \mathbf{k} \cdot \mathbf{r})} = -i\mathbf{k} \cdot \mathbf{E}_0 e^{i(\omega t - \mathbf{k} \cdot \mathbf{r})} = -i\mathbf{k} \cdot \mathbf{E} = 0,$$
$$\nabla \cdot \mathbf{H} = \mathbf{H}_0 \cdot \nabla e^{i(\omega t - \mathbf{k} \cdot \mathbf{r})} = -i\mathbf{k} \cdot \mathbf{H}_0 e^{i(\omega t - \mathbf{k} \cdot \mathbf{r})} = -i\mathbf{k} \cdot \mathbf{H} = 0, \qquad (4.20)$$

expressing the fact that the vector fields $\mathbf{E}$ and $\mathbf{H}$ are *orthogonal to the propagation direction of the wave*. Using the same procedure in the remaining Maxwell's equations $(4.2)_{1,3}$, we obtain

$$\nabla \times \mathbf{E} = \nabla e^{i(\omega t - \mathbf{k} \cdot \mathbf{r})} \times \mathbf{E}_0 = -\mu \mathbf{H}_0 \frac{\partial}{\partial t} \left[e^{i(\omega t - \mathbf{k} \cdot \mathbf{r})} \right],$$
$$\nabla \times \mathbf{H} = \nabla e^{i(\omega t - \mathbf{k} \cdot \mathbf{r})} \times \mathbf{H}_0 = \epsilon \mathbf{E}_0 \frac{\partial}{\partial t} \left[e^{i(\omega t - \mathbf{k} \cdot \mathbf{r})} \right].$$

Performing the derivative and taking into account the relation

$$\frac{\omega}{k} = \frac{\lambda}{T} = u = \frac{1}{\sqrt{\epsilon \mu}},$$

we find

$$\mathbf{E} = \sqrt{\frac{\mu}{\epsilon}} \, (\mathbf{H} \times \mathbf{s}),$$
$$\mathbf{H} = \sqrt{\frac{\epsilon}{\mu}} \, (\mathbf{s} \times \mathbf{E}). \qquad (4.21)$$

These two relations are equivalent. Taking the dot product of $(4.21)_1$ by $\mathbf{H}$, and/or of $(4.21)_2$ by $\mathbf{E}$, we obtain the same result

$$\mathbf{E} \cdot \mathbf{H} = 0, \qquad (4.22)$$

meaning that $\mathbf{E}$ and $\mathbf{H}$ are *reciprocally orthogonal*.

Summarizing our results, we conclude that the vectors $\mathbf{E}$, $\mathbf{H}$, and $\mathbf{s}$ form a right trihedron, and, therefore, in homogeneous and isotropic dielectrics, the electromagnetic field propagates as *transverse waves*.

Expressions (4.21) also give the relation between the magnitudes of $\mathbf{E}$ and $\mathbf{H}$:

$$|\mathbf{E}| = \sqrt{\frac{\mu}{\epsilon}} |\mathbf{H}|. \qquad (4.23)$$

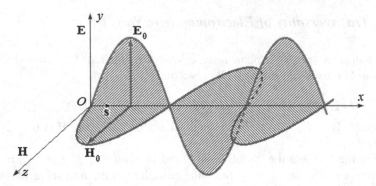

Fig. 4.1 Transversality of a plane electromagnetic wave propagating through a homogeneous and isotropic dielectric medium.

This relation shows that $|\mathbf{E}|$ and $|\mathbf{H}|$ are *proportional*. In other words, the two vector fields oscillate *in phase* or *synchronously*.

Choosing $\mathbf{E}$ to oscillate in the xOy-plane and $\mathbf{H}$ in the plane zOx, one can represent a progressive, plane electromagnetic wave as shown in Fig. 4.1.

This analysis makes it possible to calculate several quantities which characterize the electromagnetic field in dielectrics. For example, since

$$\epsilon E^2 = \mu H^2 \quad \text{and} \quad w_{em} = 2w_e = 2w_m, \tag{4.24}$$

where w is the energy density, we find that the energy carried by the electromagnetic wave is equally distributed between the electric and magnetic components.

In its turn, Poynting's vector is

$$\mathbf{\Pi} = \mathbf{E} \times \mathbf{H} = EH\,\mathbf{s},$$
$$|\mathbf{\Pi}| = EH = \sqrt{\frac{\epsilon}{\mu}}\,E^2 = \sqrt{\frac{\mu}{\epsilon}}\,H^2, \tag{4.25}$$

while the *momentum density of the electromagnetic field* is

$$\mathbf{p}_{em} = \epsilon\mu\,(\mathbf{E} \times \mathbf{H}) = \epsilon\mu\mathbf{\Pi},$$
$$|\mathbf{p}_{em}| = \epsilon\sqrt{\epsilon\mu}\,E^2 = \mu\sqrt{\epsilon\mu}\,H^2 = \sqrt{\epsilon\mu}\,w_{em}. \tag{4.26}$$

Since in vacuum $\sqrt{\epsilon_0\mu_0} = 1/c$, we have in that case

$$|\mathbf{p}_{em}| = \frac{1}{c}\,w_{em}. \tag{4.27}$$

4.2.3 *Electromagnetic Theory of Light*

In 1865, Maxwell realized that between the electromagnetic waves, predicted by his theory, and the light waves exists a close analogy. Both types of waves:

a)   are transversal;
b)   satisfy d'Alembert's homogeneous equations;
c)   propagate with the same velocity in vacuum.

The ratio between the velocity of light in vacuum, and the phase velocity of light in a transparent medium is called the *refractive index*, n, of that medium. According to Maxwell's theory, we should have (see (4.5)):

$$n = \frac{c}{u} = \sqrt{\epsilon_r \mu_r} , \tag{4.28}$$

known as the *Maxwell relation*. In transparent media with $\mu_r \simeq 1$, the refractive index becomes

$$n = \sqrt{\epsilon_r} . \tag{4.29}$$

This relation is verified for gases, transparent crystals[1] and certain liquids (toluene, benzene, etc.). There are, however, some considerable mismatches between Maxwell's theory and experimental data. For example, in the case of water, the results are

$$n_{theor} = \sqrt{\epsilon_r} = \sqrt{81} = 9, \quad n_{exp} = 1.3.$$

These deviations come from the fact that Maxwell's theory does not take into consideration the phenomenon of *dispersion*, i.e. the variation of ϵ_r and μ_r (and, implicitly, of n) with the frequency of the incident radiation. The variation of the permittivity and permeability of the medium with frequency can only be explained by taking into account the discontinuous, atomic structure of substance. Such a theory was given by Lorentz.

Maxwell's theory was a great success, in spite of being incomplete. It predicted the existence of the electromagnetic field and electromagnetic waves, leading to the unification of optical and electromagnetic phenomena, for many years considered as being independent.

---

[1] According to Max Planck, "Ludwig Boltzmann studied particularly Maxwell's asserted relation between the refractive index and the dielectric constant, and verified it completely by extremely careful experiments on various substances, especially on gases." (*James Clerk Maxwell, A Commemoration Volume 1831–1931*, With Essays By J.J. Thomson, Max Planck, Albert Einstein and others, Cambridge University Press, 1931)

4.3   Polarization of the Electromagnetic Waves

Let $\mathbf{u}(\mathbf{r}, t)$ be the vector defining an oriented wave. In Cartesian coordinates,

$$\mathbf{u}(\mathbf{r}, t) = \mathbf{u}_x + \mathbf{u}_y + \mathbf{u}_z = u_x\mathbf{i} + u_y\mathbf{j} + u_z\mathbf{k}.$$

Choosing the z-axis as the direction of propagation and examining Fig. 4.2, one can write

$$\mathbf{u}(\mathbf{r}, t) = \mathbf{u}_\perp(\mathbf{r}, t) + \mathbf{u}_\|(\mathbf{r}, t), \qquad\qquad (4.30)$$

where $\mathbf{u}_\perp = \mathbf{u}_x + \mathbf{u}_y$ is orthogonal to the direction of propagation, and $\mathbf{u}_\| = \mathbf{u}_z$ is parallel to it. Here, $\mathbf{u}_\perp$ is called transverse component and $\mathbf{u}_\|$ – longitudinal component of the field.

A transverse wave is a wave for which, at every moment, $\mathbf{u}_\| = 0$, while a longitudinal wave is characterized by $\mathbf{u}_\perp = 0$. There are also mixed waves, in which the field oscillates both transversally and longitudinally.

In the case of a *transverse wave*, the direction defined by $\mathbf{u}_\perp$ is called *direction of polarization*, while the plane defined by $\mathbf{u}_\perp$ and the propagation direction of the wave is the *plane of polarization*. The plane containing $\mathbf{u}_\perp$, orthogonal to the direction of propagation (i.e. to the plane of polarization), is called *plane of oscillation*.

The polarization of a transverse wave can be:

a) *linear*, if the oscillations of $\mathbf{u}_\perp$ take place along a fixed direction. In this case, the arrow of $\mathbf{u}_\perp$ describes a straight line;

b) *plane*, if the arrow of $\mathbf{u}_\perp$ describes a curve in the plane of oscillation. Depending on the shape of the curve, the plane polarization can be:
   (*i*) *elliptical* (*right* or *left*);
   (*ii*) *circular* (*right* or *left*), as a particular case of the elliptic polarization.

Under certain circumstances, the elliptic polarization can become linear. Let us point out some characteristics of the elliptic polarization. We can write $\mathbf{u}_\perp(x, t)$ as

Fig. 4.2 A convenient decomposition of the vector **u**, defining an oriented wave.

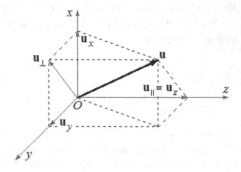

$$\mathbf{u}_\perp(z, t) = \mathbf{i}\, u_x(z, t) + \mathbf{j}\, u_y(z, t)$$
$$= \mathbf{i}\, u_{0x} e^{i\alpha} e^{i(\omega t - kz)} + \mathbf{j}\, u_{0y} e^{i\beta} e^{i(\omega t - kz)}, \tag{4.31}$$

where u_{0x} and u_{0y} are real amplitudes, while α and β are the phase differences of the components. With the notation $\Phi = \omega t - kz$, we write the real part of the components $\mathbf{u}_x$ and $\mathbf{u}_y$ as

$$\text{Re}\, u_x = u_{0x}\, \cos(\Phi + \alpha),$$
$$\text{Re}\, u_y = u_{0y}\, \cos(\Phi + \beta).$$

Expanding the trigonometric functions and denoting $\varphi = \cos \Phi$, we have

$$\frac{\text{Re}\, u_x}{u_{0x}} = \varphi \cos \alpha - \sqrt{1 - \varphi^2}\, \sin \alpha,$$
$$\frac{\text{Re}\, u_y}{u_{0y}} = \varphi \cos \beta - \sqrt{1 - \varphi^2}\, \sin \beta.$$

These relations are used to extract φ and $\sqrt{1 - \varphi^2}$:

$$\varphi = \frac{\frac{\text{Re}\, u_x}{u_{0x}} \sin \beta - \frac{\text{Re}\, u_y}{u_{0y}} \sin \alpha}{\sin \theta},$$
$$\sqrt{1 - \varphi^2} = \frac{\frac{\text{Re}\, u_x}{u_{0x}} \cos \beta - \frac{\text{Re}\, u_y}{u_{0y}} \cos \alpha}{\sin \theta},$$

where $\theta = \beta - \alpha$. Squaring and adding, we arrive at

$$\left(\frac{\text{Re}\, u_x}{u_{0x}}\right)^2 + \left(\frac{\text{Re}\, u_y}{u_{0y}}\right)^2 - 2\frac{(\text{Re}\, u_x)(\text{Re}\, u_y)}{u_{0x} u_{0y}} \cos \theta = \sin^2 \theta.$$

This is the equation of an ellipse, whose axes are rotated by an angle (that depends on the value of θ) relative to the plane yOz, and inscribed in the rectangle of sides $2u_{0x}$ and $2u_{0y}$. The eccentricity, the direction of the axes, and the orientation of the ellipse all depend on the value θ of the phase difference. The possibilities are the following:

(i) $0 < \theta < \pi$ – the orientation of the ellipse is counterclockwise (left-handed elliptic polarization, or positive helicity);

(ii) $\pi < \theta < 2\pi$ – the orientation of the ellipse is clockwise (right-handed elliptic polarization, or negative helicity);

(iii) $\theta = \pm\frac{\pi}{2}$, $u_{0x} = u_{0y} = u_0$ – the ellipse becomes a circle, inscribed in a square of side $2u_0$ (circular left-handed/right-handed polarization);

(iv) $\theta = (2n + 1)\frac{\pi}{2}$ – the axes of the ellipse (circle) coincide with the coordinate axes;

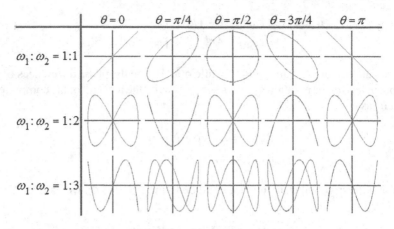

Fig. 4.3 Some examples of Lissajous figures.

(v) $\theta = n\pi$ – the ellipse (circle) degenerates into a straight line, and the polarization becomes linear. For example, if $\theta = 0$ or $\theta = \pi$, the corresponding straight lines are $\mathrm{Re}\, u_x = \pm \frac{u_{0x}}{u_{0y}} (\mathrm{Re}\, u_y)$.

If the two components u_x and u_y have different frequencies, such that ω_1/ω_2 is a rational number, then the tip of the vector $\mathbf{u}_\perp$ describes some complicated closed curves, called *Lissajous figures* or *Bowditch curves*, after the American mathematician Nathaniel Bowditch (1773–1838), who studied them in 1815, and the French physicist Jules Lissajous (1822–1880), who embarked on their detailed analysis in 1857 (see Fig. 4.3). If the ratio ω_1/ω_2 is irrational, the curves will be open.

Assume a plane monochromatic electromagnetic wave, given by (4.15), in a homogeneous and isotropic medium, and let $\mathbf{k} = k\,\mathbf{s}$ be its direction of propagation. The wave transversality, expressed by (see (4.21)):

$$\mathbf{B} = \sqrt{\epsilon\mu}\,\mathbf{s} \times \mathbf{E}, \tag{4.32}$$

suggests the definition of three reciprocally orthogonal vectors $\mathbf{e}_1$, $\mathbf{e}_2$, $\mathbf{s}$, so that the amplitudes $\mathbf{E}_0$ and $\mathbf{B}_0$ are given either by

$$\mathbf{E}_0 = \mathbf{e}_1\, E_{01}, \qquad \mathbf{B}_0 = \mathbf{e}_2\sqrt{\epsilon\mu}\, E_{01}, \tag{4.33}$$

or by

$$\mathbf{E}_0 = \mathbf{e}_2\, E_{02}, \qquad \mathbf{B}_0 = -\mathbf{e}_1\sqrt{\epsilon\mu}\, E_{02}. \tag{4.34}$$

The wave described by (4.33) has the vector $\mathbf{E}$ permanently oriented in direction $\mathbf{e}_1$, while in the case of the wave described by (4.34), $\mathbf{E}$ is always oriented along $\mathbf{e}_2$. The first wave is linearly polarized in the direction $\mathbf{e}_1$, and the second in the

Fig. 4.4 Orientation of the field vectors $\mathbf{E}_1$ and $\mathbf{E}_2$ in a polarized wave.

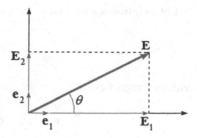

direction $\mathbf{e}_2$. The two waves can be combined to give a plane wave, propagating in the $\mathbf{k} = k\,\mathbf{s}$ direction, whose most general form is

$$\mathbf{E}(\mathbf{r}, t) = (\mathbf{e}_1 E_{01} + \mathbf{e}_2 E_{02})e^{i(\omega t - \mathbf{k}\cdot\mathbf{r})}. \tag{4.35}$$

The amplitudes E_{01} and E_{02} are, in general, complex quantities:

$$E_{01} = |E_{01}|\,e^{i\alpha}, \qquad E_{02} = |E_{02}|\,e^{i\beta}.$$

If E_{01} and E_{02} are in phase, then (4.35) describes a *linearly polarized wave*; its polarization vector makes with $\mathbf{e}_1$ the angle

$$\theta = \arctan\frac{E_{02}}{E_{01}},$$

and has the modulus $E_0 = \sqrt{E_{01}^2 + E_{02}^2}$ (see Fig. 4.4). If E_{01} and E_{02} have different phases, the wave (4.35) is *elliptically polarized*. In particular, if $|E_{01}| = |E_{02}| = E_0$ (real), while the phase difference is $\pi/2$, then (4.35) gives

$$\mathbf{E}(\mathbf{r}, t) = (\mathbf{e}_1 \pm i\mathbf{e}_2)E_0\,e^{i(\omega t - \mathbf{k}\cdot\mathbf{r})}. \tag{4.36}$$

Let us now consider $\mathbf{e}_1$ and $\mathbf{e}_2$ in the directions of the x- and y-axes, respectively. The real parts of the components of $\mathbf{E}$ are then

$$E_x(\mathbf{r}, t) = E_0\,\cos(\omega t - kz),$$
$$E_y(\mathbf{r}, t) = \mp E_0\,\sin(\omega t - kz). \tag{4.37}$$

At a fixed point in space, (4.36) shows that the vector $\mathbf{E}$ has constant modulus, while its tip describes a circle, with the frequency $\nu = \omega/2\pi$, in the xOy-plane. In the case $\mathbf{e}_1 + i\mathbf{e}_2$, an observer placed so that the wave propagates towards him detects a counterclockwise rotation of $\mathbf{E}$ (left-handed circular polarization, or positive helicity), while in the case $\mathbf{e}_1 - i\mathbf{e}_2$ the rotation of $\mathbf{E}$ is clockwise (right-handed circular polarization, or negative helicity).

Let us introduce the unit orthogonal vectors

$$\mathbf{e}_\pm = \frac{1}{\sqrt{2}}(\mathbf{e}_1 \pm i\mathbf{e}_2),$$

with the properties

$$\mathbf{e}_\pm^* \cdot \mathbf{e}_\pm = 1, \quad \mathbf{e}_\pm^* \cdot \mathbf{e}_\mp = 0, \quad \mathbf{e}_\pm^* \cdot \mathbf{k} = 0.$$

Then (4.35) becomes equivalent to

$$\mathbf{E}(\mathbf{r}, t) = (\mathbf{e}_+ E_{0+} + \mathbf{e}_- E_{0-})e^{i(\omega t - \mathbf{k}\cdot\mathbf{r})}, \tag{4.38}$$

where E_+ and E_- are complex amplitudes. If E_{0+} and E_{0-} have different moduli but the same phase, then (4.38) describes an *elliptically polarized wave*, with the principal axes of the ellipse oriented along $\mathbf{e}_1$ and $\mathbf{e}_2$. Denoting $E_-/E_+ \equiv r$, the ratio between the major and the minor axes of the ellipse is

$$\left|\frac{1+r}{1-r}\right|.$$

If there exists a phase difference between amplitudes, i.e. $E_-/E_+ = r\,e^{i\theta}$, then the axes of the ellipse described by $\mathbf{E}$ are rotated by $\theta/2$ (see Fig. 4.5). For $r = 1$, we recover the case of linear polarization.

Fig. 4.5 An elliptically polarized wave, with the oscillation plane xOy.

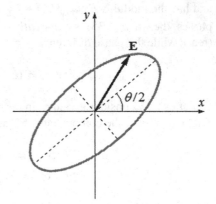

4.4 Reflection and Refraction of Plane Electromagnetic Waves

Consider two different, homogeneous and isotropic, transparent dielectric media, characterized by permittivities ϵ_1 and ϵ_2 ($\epsilon_1 \neq \epsilon_2$) and permeabilities $\mu_1 = \mu_2$, separated by the plane surface S, like in Fig. 4.6.

The experiment shows that the incident beam SI is partially reflected and refracted at the point I. To distinguish them from each other, we attach an index to each wave: 1 – for the incident wave, 2 – for the refracted wave, and 3 – for the reflected wave. Suppose that the field $\mathbf{E}_1$ is known. We shall decompose its amplitude $\mathbf{E}_{01}$ into a component parallel to the y-axis, denoted by E_{01}^y, and a component orthogonal to the direction of propagation of the incident wave and situated in the zOx-plane, denoted by $E_{01}^\perp$. The same significance have the amplitudes E_{02}^y and $E_{02}^\perp$ for the refracted wave, and E_{03}^y and $E_{03}^\perp$ for the reflected one.

If the incident wave is plane and monochromatic, the field $\mathbf{E}_1$ can be written as

$$\mathbf{E}_1 = \mathbf{E}_{01} e^{i(\omega t - \mathbf{k}_1 \cdot \mathbf{r})} = \mathbf{E}_{01} \exp\left[i\omega \left(t - \frac{\mathbf{s}_1 \cdot \mathbf{r}}{u_1} \right) \right]. \tag{4.39}$$

Let us denote the vector components as follows:

$$\mathbf{E}_1 = (X_1, Y_1, Z_1), \quad \mathbf{s}_1 = (l_1, m_1, n_1),$$
$$\mathbf{E}_{01} = (E_{01}^\perp \cos\varphi, E_{01}^y, -E_{01}^\perp \sin\varphi).$$

Since

$$l_1 = \sin\varphi, \quad m_1 = 0, \quad n_1 = \cos\varphi,$$
$$u_1 = u_3 = \frac{1}{\sqrt{\epsilon_1 \mu_1}}, \quad u_2 = \frac{1}{\sqrt{\epsilon_2 \mu_2}},$$

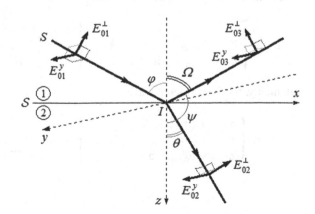

Fig. 4.6 Reflection and refraction of a plane electromagnetic wave.

we can write

$$X_1 = E_{01}^{\perp} \cos \varphi \exp \left\{ i\omega \left[t - \sqrt{\epsilon_1 \mu_1} \left(x \sin \varphi + z \cos \varphi \right) \right] \right\}.$$

Expressing the last two components of the field $\mathbf{E}_1$ and summarizing, we have

$$\mathbf{E}_1 \equiv \begin{cases} X_1 = E_{01}^{\perp} \eta_1 \cos \varphi\,, \\ Y_1 = E_{01}^{y} \eta_1\,, \\ Z_1 = -E_{01}^{\perp} \eta_1 \sin \varphi, \end{cases} \tag{4.40}$$

where

$$\eta_1 = \exp \left[i\omega \left(t - \frac{\mathbf{s}_1 \cdot \mathbf{r}}{u_1} \right) \right] = \exp \left\{ i\omega \left[t - \sqrt{\epsilon_1 \mu_1} \left(x \sin \varphi + z \cos \varphi \right) \right] \right\}$$

is the periodical part of the incident wave.

Next, let $\mathbf{B}_1 = (\alpha_1, \beta_1, \gamma_1)$ be the magnetic induction vector in the incident beam. Using (4.21), we can write

$$\mathbf{B}_1 = \mu_1 \mathbf{H}_1 = \sqrt{\epsilon_1 \mu_1}\ \mathbf{s}_1 \times \mathbf{E}_1. \tag{4.41}$$

The x-component of (4.41) is

$$B_{1x} = \alpha_1 = \sqrt{\epsilon_1 \mu_1}\ (m_1 Z_1 - n_1 Y_1) = -\sqrt{\epsilon_1 \mu_1} E_{01}^{y}\ \eta_1 \cos \varphi.$$

Calculating the other two components, then grouping the terms, one obtains

$$\mathbf{B}_1 \equiv \begin{cases} \alpha_1 = -\sqrt{\epsilon_1 \mu_1}\ E_{01}^{y} \eta_1 \cos \varphi\,, \\ \beta_1 = \sqrt{\epsilon_1 \mu_1}\ E_{01}^{\perp} \eta_1\,, \\ \gamma_1 = \sqrt{\epsilon_1 \mu_1}\ E_{01}^{y} \eta_1 \sin \varphi. \end{cases} \tag{4.42}$$

In a similar way, we find for the reflected wave

$$\mathbf{E}_3 \equiv \begin{cases} X_3 = E_{03}^{\perp} \eta_3 \cos \psi\,, \\ Y_3 = E_{03}^{y} \eta_3\,, \\ Z_3 = -E_{03}^{\perp} \eta_3 \sin \psi, \end{cases} \qquad \mathbf{B}_3 \equiv \begin{cases} \alpha_3 = -\sqrt{\epsilon_1 \mu_1}\ E_{03}^{y} \eta_3 \cos \psi\,, \\ \beta_3 = \sqrt{\epsilon_1 \mu_1}\ E_{03}^{\perp} \eta_3\,, \\ \gamma_3 = \sqrt{\epsilon_1 \mu_1}\ E_{03}^{y} \eta_3 \sin \psi, \end{cases} \tag{4.43}$$

and, finally, for the refracted wave

$$\mathbf{E}_2 \equiv \begin{cases} X_2 = E_{02}^{\perp} \eta_2 \cos \theta\,, \\ Y_2 = E_{02}^{y} \eta_2\,, \\ Z_2 = -E_{02}^{\perp} \eta_2 \sin \theta, \end{cases} \qquad \mathbf{B}_2 \equiv \begin{cases} \alpha_2 = -\sqrt{\epsilon_2 \mu_2}\ E_{02}^{y} \eta_2 \cos \theta\,, \\ \beta_2 = \sqrt{\epsilon_2 \mu_2}\ E_{02}^{\perp} \eta_2\,, \\ \gamma_2 = \sqrt{\epsilon_2 \mu_2}\ E_{02}^{y} \eta_2 \sin \theta. \end{cases} \tag{4.44}$$

4.4.1 Laws of Reflection and Refraction

Now we shall find the relations between the angles φ, Ω, θ, and ψ shown in Fig. 4.6. To this end, we use the boundary conditions: on the plane separation surface $z = 0$, the incident, reflected, and refracted waves must have the same phase:

$$\eta_1 = \eta_2 = \eta_3,$$

or, since $\mu_1 = \mu_2$,

$$\sqrt{\epsilon_1} \sin \varphi = \sqrt{\epsilon_2} \sin \theta = \sqrt{\epsilon_1} \sin \psi. \tag{4.45}$$

Hence,

(a) $\sin \varphi = \sin \psi = \sin(\pi - \Omega)$, which means

$$\varphi = \Omega, \tag{4.46}$$

i.e. *the angles of incidence and of reflection are equal*. This is the *reflection law*;
(b) By definition, $n_2/n_1 \equiv n_{21}$ is the *relative refractive index* of medium 2 with respect to medium 1. In view of Maxwell's relation (4.28), we then have

$$\frac{\sin \varphi}{\sin \theta} = \sqrt{\frac{\epsilon_2}{\epsilon_1}} = \frac{n_2}{n_1} = n_{21}, \tag{4.47}$$

known as the *refraction law* or the *Snellius–Descartes law*, after the Dutch astronomer Willebrord Snellius (1580–1626) and the French philosopher and mathematician René Descartes (1596–1650), although the law was discovered as early as 984 by the Persian mathematician Ibn Sahl (c. 940–1000). One can distinguish two cases:
(i) $\epsilon_2 > \epsilon_1$, implying $\sin \varphi > \sin \theta$. Since $0 \leq \varphi, \theta \leq \frac{\pi}{2}$, and the sinus function is monotonically increasing on this interval, we conclude that, when passing to a more refringent medium ($n_{21} > 1$), the refracted wave bends towards the normal;
(ii) $\epsilon_2 < \epsilon_1$. This implies $\sin \varphi < \sin \theta$, therefore $\varphi < \theta$, meaning that, when entering a less refringent medium, the refracted wave bends away from the normal.

There exists a specific incident angle, called *limit angle* and denoted φ_l, for which $\theta = \theta_l = \frac{\pi}{2}$. Then (4.47) gives

$$\sin \varphi_l = n_{21} \sin \theta_l = n_{21} < 1.$$

To keep constant the ratio n_{21}, for $\varphi > \varphi_l$ we should have $\sin \theta > \sin \theta_l$. In reality, $\sin \theta$ $\left(\theta > \frac{\pi}{2}\right)$ decreases when φ increases. Since this is not possible, we conclude that for $\varphi > \varphi_l$ there is no refraction, the incident ray being totally reflected on the separation surface. This phenomenon is called *total (internal) reflection*.

4.4.2   Fresnel's Formulas

Equations (4.40)–(4.44) allow us to find some relations between the amplitudes of reflected and refracted waves, on the one hand, and the amplitude of the incident wave, on the other hand. To this end, we use the boundary conditions: on the separation surface, the normal and tangent components of the fields $\mathbf{E}$ and $\mathbf{H}$ satisfy the jump conditions (3.43). Since there are no sources on the plane surface S ($\mathbf{j}|_S = 0$, $\rho|_S = 0$), these equations are written as

$$D_{2n} - D_{1n} = 0,$$
$$B_{2n} - B_{1n} = 0,$$
$$E_{2T} - E_{1T} = 0,$$
$$H_{2T} - H_{1T} = 0.$$

In our case, both $\mathbf{E}$ and $\mathbf{H}$ have two tangent components (in the Ox and Oy directions), and a normal component (in the Oz direction). Recalling that $\mu_1 = \mu_2$, we have

$$\mathbf{E}: \begin{cases} E_{1T} + E_{3T} = E_{2T}, \\ \epsilon_1(E_{1n} + E_{3n}) = \epsilon_2 E_{2n}, \end{cases} \qquad \mathbf{H}: \begin{cases} H_{1T} + H_{3T} = H_{2T}, \\ H_{1n} + H_{3n} = H_{2n}. \end{cases}$$

On the separation surface the phases are equal ($\eta_1 = \eta_2 = \eta_3$), so that

$$\begin{cases} X_1 + X_3 = X_2, \\ Y_1 + Y_3 = Y_2, \\ \epsilon_1(Z_1 + Z_3) = \epsilon_2 Z_2, \end{cases} \qquad \begin{cases} \alpha_1 + \alpha_3 = \alpha_2, \\ \beta_1 + \beta_3 = \beta_2, \\ \gamma_1 + \gamma_3 = \gamma_2, \end{cases}$$

or, if we use (4.40)–(4.44),

$$\mathbf{E}: \begin{cases} E_{01}^\perp \cos\varphi + E_{03}^\perp \cos\psi = E_{02}^\perp \cos\theta, \\ E_{01}^y + E_{03}^y = E_{02}^y, \\ \epsilon_1(E_{01}^\perp \sin\varphi + E_{03}^\perp \sin\psi) = \epsilon_2 E_{02}^\perp \sin\theta, \end{cases}$$

$$\tag{4.48}$$

$$\mathbf{B}: \begin{cases} E_{01}^y \sqrt{\epsilon_1} \cos\varphi + E_{03}^y \sqrt{\epsilon_1} \cos\psi = E_{02}^y \sqrt{\epsilon_2} \cos\theta, \\ E_{01}^\perp \sqrt{\epsilon_1} + E_{03}^\perp \sqrt{\epsilon_1} = E_{02}^\perp \sqrt{\epsilon_2}, \\ E_{01}^y \sqrt{\epsilon_1} \sin\varphi + E_{03}^y \sqrt{\epsilon_1} \sin\psi = E_{02}^y \sqrt{\epsilon_2} \sin\theta. \end{cases}$$

According to (4.45), Eq. (4.48)$_6$ reduces to (4.48)$_2$, and (4.48)$_3$ to (4.48)$_5$, so that only four of the relations (4.48) are independent. Then, we are left with

$$(E_{01}^{\perp} - E_{03}^{\perp}) \cos\varphi = E_{02}^{\perp} \cos\theta,$$
$$E_{01}^{y} + E_{03}^{y} = E_{02}^{y},$$
$$\sqrt{\epsilon_1}\,(E_{01}^{y} - E_{03}^{y}) \cos\varphi = \sqrt{\epsilon_2}\,E_{02}^{y} \cos\theta,$$
$$\sqrt{\epsilon_1}\,(E_{01}^{\perp} + E_{03}^{\perp}) = \sqrt{\epsilon_2}\,E_{02}^{\perp}. \tag{4.49}$$

If the angles φ and θ, and permittivities ϵ_1 and ϵ_2 are given, then Eq. (4.49) serve to determine $E_{02}^{\perp}$, E_{02}^{y}, $E_{03}^{\perp}$, and E_{03}^{y} in terms of $E_{01}^{\perp}$ and E_{01}^{y}:

$$E_{02}^{\perp} = E_{01}^{\perp} \frac{2\sqrt{\epsilon_1}\cos\varphi}{\sqrt{\epsilon_1}\cos\theta + \sqrt{\epsilon_2}\cos\varphi},$$

$$E_{02}^{y} = E_{01}^{y} \frac{2\sqrt{\epsilon_1}\cos\varphi}{\sqrt{\epsilon_1}\cos\varphi + \sqrt{\epsilon_2}\cos\theta},$$

$$E_{03}^{\perp} = E_{01}^{\perp} \frac{\sqrt{\epsilon_2}\cos\varphi - \sqrt{\epsilon_1}\cos\theta}{\sqrt{\epsilon_2}\cos\varphi + \sqrt{\epsilon_1}\cos\theta}, \tag{4.50}$$

$$E_{03}^{y} = E_{01}^{y} \frac{\sqrt{\epsilon_1}\cos\varphi - \sqrt{\epsilon_2}\cos\theta}{\sqrt{\epsilon_1}\cos\varphi + \sqrt{\epsilon_2}\cos\theta},$$

known as *Fresnel's formulas.*

It is not difficult to show that, making use of the refraction law

$$\sqrt{\epsilon_1}\sin\varphi = \sqrt{\epsilon_2}\sin\theta,$$

we can write (4.50) in a form which does not contain ϵ_1 and ϵ_2 anymore:

$$E_{02}^{\perp} = E_{01}^{\perp} \frac{2\sin\theta\cos\varphi}{\sin(\varphi+\theta)\cos(\varphi-\theta)},$$

$$E_{02}^{y} = E_{01}^{y} \frac{2\sin\theta\cos\varphi}{\sin(\varphi+\theta)},$$

$$E_{03}^{\perp} = E_{01}^{\perp} \frac{\tan(\varphi-\theta)}{\tan(\varphi+\theta)}, \tag{4.51}$$

$$E_{03}^{y} = -E_{01}^{y} \frac{\sin(\varphi-\theta)}{\sin(\varphi+\theta)}.$$

Let us discuss the following five cases:

(a) If $E_{01}^{y} \neq 0$ and $\varphi \neq 0$, it follows that $E_{03}^{y} \neq 0$. In particular, the case $\varphi = 0$ (normal incidence) implies $\theta = 0$ and $(4.50)_4$, for $\epsilon_1 \neq \epsilon_2$, gives

$$E_{03}^{y} = E_{01}^{y} \frac{\sqrt{\epsilon_1} - \sqrt{\epsilon_2}}{\sqrt{\epsilon_1} + \sqrt{\epsilon_2}} \neq 0,$$

therefore the component of the reflected wave orthogonal to the plane of incidence never disappears. In other words, by reflection an electromagnetic wave

can never be polarized perpendicular to the plane of incidence. We have, in
general, a *partial polarization* in the plane of incidence;

(b) Formula $(4.51)_3$ shows that the amplitude $E_{03}^{\perp}$ decreases when φ increases (starting from 0). There is a certain angle of incidence $\varphi = \varphi_B$, called *Brewster angle*,
for which $\varphi + \theta = \pi/2$. In this case $E_{03}^{\perp} = 0$, because $\tan(\pi/2) = \infty$, so that
only $E_{03}^y \neq 0$ remains in the reflected wave, its oscillations being perpendicular
to the plane of incidence. This means that the reflected wave is totally linearly
polarized in the plane of incidence;

(c) Let us consider the ratio $E_{03}^y/E_{03}^{\perp}$. Using $(4.51)_{3,4}$, we have

$$\frac{E_{03}^y}{E_{03}^{\perp}} = -\frac{E_{01}^y}{E_{01}^{\perp}} \frac{\cos(\varphi - \theta)}{\cos(\varphi + \theta)}.$$

Choosing $0 < \theta < \varphi < \pi/2$, one obtains

$$\left|\frac{\cos(\varphi - \theta)}{\cos(\varphi + \theta)}\right| > 1 \Rightarrow \left|\frac{E_{03}^y}{E_{03}^{\perp}}\right| > \left|\frac{E_{01}^y}{E_{01}^{\perp}}\right|.$$

Suppose that the reflected wave reflects again, the amplitudes being $E_{03}^{\prime\perp}$ and $E_{03}^{\prime y}$,
respectively. A similar reasoning then leads us to the conclusion that

$$\left|\frac{E_{03}^{\prime y}}{E_{03}^{\prime\perp}}\right| > \left|\frac{E_{03}^y}{E_{03}^{\perp}}\right| > \left|\frac{E_{01}^y}{E_{01}^{\perp}}\right|, \quad \text{etc.}$$

The increase in the ratio $|E_{03}^y/E_{03}^{\perp}|$ happens on account of the decrease in $E_{03}^{\perp}$,
and not of the increase in E_{03}^y. This means that, by multiple reflections, $E_{03}^{\perp}$
vanishes. Therefore we find again a linear polarization, in the plane of incidence,
by *multiple reflections*;

(d) Let us now consider the refracted wave. Using $(4.51)_{1,2}$, we have

$$\frac{E_{02}^y}{E_{02}^{\perp}} = \frac{E_{01}^y}{E_{01}^{\perp}} \cos(\varphi - \theta).$$

Since $|\cos(\varphi - \theta)| < 1$, it follows that $|E_{02}^y/E_{02}^{\perp}| < |E_{01}^y/E_{01}^{\perp}|$, and, if the wave
is refracted again,

$$\left|\frac{E_{02}^{\prime y}}{E_{02}^{\prime\perp}}\right| < \left|\frac{E_{02}^y}{E_{02}^{\perp}}\right| < \left|\frac{E_{01}^y}{E_{01}^{\perp}}\right|, \quad \text{etc.}$$

Thus, by multiple refractions, the ratio $|E_{02}^y/E_{02}^{\perp}| \to 0$. Since $E_{02}^{\perp}$ cannot go
to infinity, this means $E_{02}^y \to 0$. The refracted wave is thus *linearly polarized,
perpendicular to the plane of incidence* (the oscillations take place in the plane
of incidence);

(e) Let us now calculate the *reflection* and *transmission coefficients* R, respectively T. To this end, we define the *complex Poynting's vector*

$$\tilde{\Pi} = \frac{1}{2}(\mathbf{E} \times \mathbf{H}^*), \tag{4.52}$$

where the symbol * means complex conjugation.

Using this definition, let us first show that the average flux density of electromagnetic energy emitted per unit time is the *real part of the normal component of* $\tilde{\Pi}$:

$$\langle \Phi_{em} \rangle = \frac{1}{2} \left[\text{Re} \left(\mathbf{E} \times \mathbf{H}^* \right) \right] \cdot \mathbf{n}, \tag{4.53}$$

where $\langle \rangle$ signifies the average over a period τ. To show this, let us take

$$E = \text{Re} \left[E_0 e^{i\varphi} e^{i(\omega t - kx)} \right],$$
$$H = \text{Re} \left[H_0 e^{i\varphi'} e^{i(\omega t - kx)} \right],$$

or

$$E = E_0 \cos(\Phi + \varphi),$$
$$H = H_0 \cos(\Phi + \varphi'), \tag{4.54}$$

where the amplitudes E_0 and H_0 are real, while $\Phi = \omega t - kx$.
Consider now the product

$$EH = E_0 H_0 \cos(\Phi + \varphi) \cos(\Phi + \varphi').$$

Using the trigonometric formula

$$\cos \alpha \cos \beta = \frac{1}{2} \left[\cos(\alpha + \beta) + \cos(\alpha - \beta) \right]$$

one finds $\alpha = 2\Phi + \varphi + \varphi'$, $\beta = \varphi - \varphi'$, which give

$$EH = \frac{1}{2} E_0 H_0 \left[\cos(2\Phi + \varphi + \varphi') + \cos(\varphi - \varphi') \right].$$

The average of this expression over a period is

$$\langle EH \rangle = \frac{1}{2} E_0 H_0 \frac{1}{\tau} \int_0^\tau \left[\cos(2\Phi + \varphi + \varphi') + \cos(\varphi - \varphi') \right] dt$$
$$= \frac{1}{2} E_0 H_0 \cos(\varphi - \varphi').$$

On the other hand,

$$\text{Re}(EH^*) = \text{Re}\left[E_0 H_0\, e^{i(\varphi - \varphi')}\right] = E_0 H_0\, \cos(\varphi - \varphi').$$

The last two relations imply

$$\langle EH \rangle = \frac{1}{2}\,\text{Re}(EH^*),$$

in which case

$$\langle \mathbf{E} \times \mathbf{H} \rangle = \langle \mathbf{\Pi} \rangle = \frac{1}{2}\,\text{Re}(\mathbf{E} \times \mathbf{H}^*) \tag{4.55}$$

is also true. Taking the dot product of (4.55) with $\mathbf{n}$ (the unit vector of the normal to the surface S that bounds the volume V in which the electromagnetic field is considered), one finds (4.53).

Using $(4.21)_2$, we have

$$\tilde{\mathbf{\Pi}} = \frac{1}{2}(\mathbf{E} \times \mathbf{H}^*) = \frac{1}{2}\sqrt{\frac{\epsilon}{\mu}}\,\mathbf{E} \times (\mathbf{s} \times \mathbf{E}^*)$$

$$= \frac{1}{2}\sqrt{\frac{\epsilon}{\mu}}\,(\mathbf{E} \cdot \mathbf{E}^*)\,\mathbf{s} = \frac{1}{2}\sqrt{\frac{\epsilon}{\mu}}\,|\mathbf{E}_0|^2\,\mathbf{s}.$$

Thus,

$$\langle \Phi_{em} \rangle = \frac{1}{2}\sqrt{\frac{\epsilon}{\mu}}\,|\mathbf{E}_0|^2\,\mathbf{s} \cdot \mathbf{n}. \tag{4.56}$$

Using this preliminary analysis, let us now come back to the main subject of this case. By definition, the ratio between the modulus of the average energy flux of the reflected electromagnetic wave and the average energy flux of the incident electromagnetic wave is called *reflection coefficient* (or *reflection power*):

$$R = \frac{|\langle \Phi_{em} \rangle_3|}{|\langle \Phi_{em} \rangle_1|}. \tag{4.57}$$

The *transmission coefficient* is defined as:

$$T = \frac{|\langle \Phi_{em} \rangle_2|}{|\langle \Phi_{em} \rangle_1|}. \tag{4.58}$$

Using (4.56) and observing that $|\mathbf{n} \cdot \mathbf{s}_1| = \cos\varphi$, $|\mathbf{n} \cdot \mathbf{s}_2| = \cos\theta$, $|\mathbf{n} \cdot \mathbf{s}_3| = \cos\varphi$, one obtains

$$R = \frac{|E_{03}|^2}{|E_{01}|^2}, \qquad T = \frac{|E_{02}|^2}{|E_{01}|^2}\,\frac{\sqrt{\epsilon_2}}{\sqrt{\epsilon_1}}\,\frac{\cos\theta}{\cos\varphi}. \tag{4.59}$$

The law of conservation of energy implies

$$\mathbf{n} \cdot \tilde{\mathbf{\Pi}}_1 = \mathbf{n} \cdot (\tilde{\mathbf{\Pi}}_2 + \tilde{\mathbf{\Pi}}_3),$$

or

$$\sqrt{\epsilon_1}\, |E_{01}|^2 \cos\varphi = \sqrt{\epsilon_1}\, |E_{03}|^2 \cos\varphi + \sqrt{\epsilon_2}\, |E_{02}|^2 \cos\theta. \qquad (4.60)$$

Using (4.59) and (4.60), we find

$$R + T = 1.$$

Since $0 \le R \le 1$ and $0 \le T \le 1$, this relation gives rise to a probabilistic interpretation: R is the *probability of reflection* of the incident wave, while T is the *probability of transmission* of the incident wave. Since their sum equals one, it is *certain* that at least one of the events will happen.

Fresnel's formulas (4.50) allow us to verify the relation $R + T = 1$ in the case of normal incidence ($\varphi = \theta = 0$). Indeed,

$$E_{03}^{\perp} = E_{01}^{\perp} \frac{\sqrt{\epsilon_2} - \sqrt{\epsilon_1}}{\sqrt{\epsilon_2} + \sqrt{\epsilon_1}} = E_{01}^{\perp} \frac{n-1}{n+1}, \qquad E_{02}^{\perp} = E_{01}^{\perp} \frac{2}{n+1},$$

which yields

$$R = \frac{(n-1)^2}{(n+1)^2}, \quad T = \frac{4n}{(n+1)^2}, \quad R + T = 1.$$

In case of total reflection ($\theta = \pi/2$), relation (4.60) gives $|E_{03}|^2 = |E_{01}|^2$, and, by virtue of (4.59), $R = 1$ and $T = 0$, which is obvious.

To conclude, Fresnel's formulas are of major importance, since they:

(a) give a unitary explanation of the phenomena connected to the reflection and refraction of electromagnetic waves;
(b) describe the polarization by reflection and refraction of the electromagnetic waves;
(c) allow the determination of the relative intensity of reflected and refracted waves in terms of the refractive index.

4.5 Propagation of Electromagnetic Waves in Massive Conductors. Skin Effect

Suppose that the plane xOy coincides with the plane surface of a massive metallic conductor, that occupies the whole half-space $z > 0$ (Fig. 4.7). Consider a plane electromagnetic wave propagating in the z-direction (normal incidence), the vectors

Fig. 4.7 Propagation of an electromagnetic wave in a massive conductor.

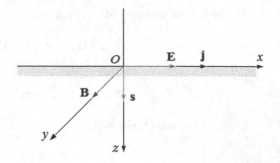

$\mathbf{E}(\mathbf{r}, t)$ and $\mathbf{B}(\mathbf{r}, t)$ being chosen as shown in Fig. 4.7. The phenomenon that happens when the wave falls on the metal is described by Maxwell's equations in which, according to the assumptions made in the beginning of this chapter, the displacement current is negligible compared to the conduction current. If we take the incident wave as

$$\mathbf{E}(\mathbf{r}, t) = \mathbf{E}(\mathbf{r})e^{i\omega t}, \quad \mathbf{B}(\mathbf{r}, t) = \mathbf{B}(\mathbf{r})e^{i\omega t}, \tag{4.61}$$

then

$$\frac{\partial \mathbf{B}}{\partial t} = i\omega \mathbf{B} = \frac{1}{i\omega}\frac{\partial^2 \mathbf{B}}{\partial t^2}. \tag{4.62}$$

On the other hand, taking the curl of Maxwell's equation

$$\frac{1}{\mu}\nabla \times \mathbf{B} = \mathbf{j} = \lambda \mathbf{E},$$

then using (4.62) and the results given in Sect. 4.1, we find

$$\tilde{\epsilon}\mu \frac{\partial^2 \mathbf{B}}{\partial t^2} - \Delta \mathbf{B} = 0, \tag{4.63}$$

where we denoted

$$\tilde{\epsilon} = -i\,\lambda/\omega. \tag{4.64}$$

Thus, the wave equation for $\mathbf{B}$ is the homogeneous d'Alembert-type equation (4.3), the only (but essential) difference being the fact that now *the permittivity is imaginary*. To an imaginary (in general, complex) permittivity corresponds a complex phase velocity

$$\tilde{u} = \frac{1}{\sqrt{\tilde{\epsilon}\mu}}, \tag{4.65}$$

as well as a complex wave number

$$\tilde{k} = \frac{\omega}{\tilde{u}} = k_1 - i k_2 = \omega\sqrt{1 - i\frac{\lambda}{\omega}\mu}. \tag{4.66}$$

Separating the real and imaginary parts, we find $k_1^2 = k_2^2 = \omega\lambda\mu/2$. Choosing the sign + for solutions, we have

$$\tilde{k} = k(1 - i) = \frac{1 - i}{\delta}, \tag{4.67}$$

where $k = k_1 = k_2 = \sqrt{\lambda\mu\omega/2}$, and

$$\delta = \sqrt{\frac{2}{\lambda\mu\omega}} \tag{4.68}$$

is called the *penetration depth*. Since the propagation occurs in the z-direction, the field $\mathbf{B}(z)$ is obtained as a solution of the equation

$$\frac{d^2\mathbf{B}(z)}{dz^2} + \tilde{k}^2\mathbf{B}(z) = 0.$$

Choosing the progressive wave, the solution is

$$\mathbf{B}(z) = \mathbf{B}_0 e^{-i\tilde{k}z}.$$

Having in view (4.61), we find the field $\mathbf{B}(z, t)$:

$$\mathbf{B}(z, t) = \mathbf{B}_0 e^{i(\omega t - \tilde{k}z)} = \mathbf{B}_0 e^{-\frac{z}{\delta}} e^{i(\omega t - \frac{z}{\delta})}. \tag{4.69}$$

Since the wave is transverse, the field $\mathbf{E}(z, t)$ is expressed by a relation of the form $(4.21)_1$:

$$\mathbf{E}(z, t) = \sqrt{\frac{\mu}{\tilde{\epsilon}}}\,\mathbf{H} \times \mathbf{s} = \tilde{u}\,\mathbf{B} \times \mathbf{s} = \frac{1 + i}{\lambda\mu\delta}\,\mathbf{B} \times \mathbf{s}$$

$$= \frac{1 + i}{\lambda\mu\delta}\,(\mathbf{B}_0 \times \mathbf{s})\,e^{-\frac{z}{\delta}} e^{i(\omega t - \frac{z}{\delta})}. \tag{4.70}$$

According to our convention, the field $\mathbf{E}$ is oriented along the x-axis and $\mathbf{B}$ along the y-axis, so that we may denote $E \equiv E_x$ and $B \equiv B_y$. Since $E = (1 + i)B/\lambda\mu\delta$, the real parts of the two fields are:

$$\text{Re}\,E = \frac{B_0}{\lambda\mu\delta}e^{-\frac{z}{\delta}}\left[\cos\left(\omega t - \frac{z}{\delta}\right) - \sin\left(\omega t - \frac{z}{\delta}\right)\right]$$

$$= \frac{\sqrt{2}B_0}{\lambda\mu\delta}e^{-\frac{z}{\delta}}\cos\left(\omega t - \frac{z}{\delta} + \frac{\pi}{4}\right), \tag{4.71}$$

$$\text{Re}\,B = B_0 e^{-\frac{z}{\delta}}\cos\left(\omega t - \frac{z}{\delta}\right). \tag{4.72}$$

Comparing (4.71) with (4.72), we notice that the phase difference between the fields **E** and **B** of the wave propagating through the conductor is $\pi/4$. Since $\mathbf{j} = \lambda\mathbf{E}$, the phase difference between **B** and **j** is also $\pi/4$. The factor $\exp(-z/\delta)$ shows that the two fields are attenuated; also the current density decays exponentially starting with $z = 0$ (the surface of the conductor). The smaller δ is, i.e. the greater λ, μ, ω are, the greater the wave attenuation is. For given λ and μ, the penetration depth decreases with increasing wave frequency. This means that the *high frequency currents* are localized in a thin layer at the surface of the conductor. The wave only penetrates a few wavelengths into a good conductor before decaying away. The phenomenon is known as *skin effect*, and δ is also termed *skin depth*.

The experiment confirms this theoretical investigation. Consider a typical metallic conductor such as copper, whose electrical conductivity at room temperature is about $6 \times 10^7\,(\Omega\cdot\text{m})^{-1}$. Copper, therefore, acts as a good conductor for all electromagnetic waves of frequency below about 10^{18} Hz. The skin-depth in copper for such waves is

$$\delta = \sqrt{\frac{2}{\lambda\mu\omega}} = \frac{6}{\sqrt{\omega(\text{Hz})}}\,\text{cm}.$$

It follows that the skin-depth is about 6 cm at 1 Hz, but only about 2 mm at 1 kHz.

As another example, the conductivity of sea water is about $\lambda \simeq 5\,(\Omega\cdot\text{m})^{-1}$. This conductivity is high enough for the sea water to behave as a good conductor for frequencies $\omega < 10^9$ Hz. The skin depth at 1 MHz ($\lambda \sim 2$ km) is about 0.2 m, while at 1 kHz is about 7 m. This effect is important for submarines: to stay away from the surface of the sea, they have to communicate at very low frequencies (under 100 Hz), which means λ greater than 10^4 km, but such wavelengths are produced only by very large antennas.

Observation:

In the case of a conductor, the ratio between the magnetic and the electric components of the electromagnetic energy is

$$\frac{w_m}{w_e} \sim \left(\frac{\lambda}{\omega\epsilon}\right)^2 = \beta^2. \tag{4.73}$$

Since metals are characterized by $\beta \gg 1$ (see Sect. 4.1), it follows that in conductors, unlike dielectrics, the energy is not equally distributed between the two components, but $w_m \gg w_e$.

4.6 Propagation of Electromagnetic Waves in Semiconductors

As previously mentioned in Sect. 4.1, the propagation of electromagnetic waves in semiconductors is described by Eq. (4.1), which we repeat here for convenience:

$$\epsilon\mu \frac{\partial^2 \mathbf{E}}{\partial t^2} - \Delta\mathbf{E} = -\lambda\mu \frac{\partial \mathbf{E}}{\partial t},$$

$$\epsilon\mu \frac{\partial^2 \mathbf{B}}{\partial t^2} - \Delta\mathbf{B} = -\lambda\mu \frac{\partial \mathbf{B}}{\partial t}.$$

Consider a monochromatic wave, characterized by the pulsation ω; simple calculations along the lines of the previous section lead to

$$\tilde{\epsilon}\mu \frac{\partial^2 \mathbf{E}}{\partial t^2} - \Delta\mathbf{E} = 0,$$

$$\tilde{\epsilon}\mu \frac{\partial^2 \mathbf{B}}{\partial t^2} - \Delta\mathbf{B} = 0, \tag{4.74}$$

where

$$\tilde{\epsilon} = \epsilon - i\frac{\lambda}{\omega} \tag{4.75}$$

is the *complex permittivity*.

It follows that, formally, the wave equation in semiconductors is similar to the case of dielectrics, except for the complex permittivity. This implies a complex phase velocity $\tilde{u} = 1/\sqrt{\tilde{\epsilon}\mu}$, and, consequently, a complex wave number

$$\tilde{k} = \frac{\omega}{\tilde{u}} = \omega\sqrt{\tilde{\epsilon}\mu} = \omega\sqrt{\mu\left(\epsilon - i\frac{\lambda}{\omega}\right)} = k_1 - ik_2. \tag{4.76}$$

To find the real and imaginary parts of the complex wave number $\tilde{k}$ in terms of ϵ, λ, μ, and ω, we square (4.76), then identify the real and imaginary parts on both sides, thus obtaining:

$$k_1^2 - k_2^2 = \epsilon\mu\omega^2, \quad k_1 k_2 = \frac{1}{2}\lambda\mu\omega.$$

The solutions of this system are

$$k_1^2 = \frac{\epsilon\mu\omega^2}{2}\left[\sqrt{1 + \frac{\lambda^2}{\omega^2\epsilon^2}} + 1\right],$$

$$k_2^2 = \frac{\epsilon\mu\omega^2}{2}\left[\sqrt{1 + \frac{\lambda^2}{\omega^2\epsilon^2}} - 1\right]. \tag{4.77}$$

Thus

$$\tilde{k} = k_1 - i\,k_2 = \sqrt{k_1^2 + k_2^2}\,e^{i\varphi}, \quad \tan\varphi = \frac{k_2}{k_1}. \tag{4.78}$$

Let us now exploit the analogy between Eqs. (4.74) and (4.3). As for dielectrics, the solutions are

$$(\mathbf{E}, \mathbf{B}) = (\mathbf{E}_0, \mathbf{B}_0)e^{i(\omega t - \tilde{k}\,\mathbf{s}\cdot\mathbf{r})} = (\mathbf{E}_0, \mathbf{B}_0)e^{-k_2\mathbf{s}\cdot\mathbf{r}}e^{i(\omega t - \mathbf{k}_1\cdot\mathbf{r})}. \tag{4.79}$$

By means of (4.20) and (4.21), we also have

$$\mathbf{s}\cdot\mathbf{E} = 0,$$
$$\mathbf{s}\cdot\mathbf{H} = 0, \tag{4.80}$$
$$\mathbf{H} = \sqrt{\frac{\tilde{\epsilon}}{\mu}}\,\mathbf{s}\times\mathbf{E} = \frac{\tilde{k}}{\mu\omega}\,\mathbf{s}\times\mathbf{E},$$

and, by virtue of (4.78) and (4.79):

$$\mathbf{H} = \frac{\sqrt{k_1^2 + k_2^2}}{\mu\omega}\,\mathbf{s}\times\mathbf{E}_0\,e^{-k_2\mathbf{s}\cdot\mathbf{r}}\,e^{i\varphi}\,e^{i(\omega t - \mathbf{k}_1\cdot\mathbf{r})}. \tag{4.81}$$

These results show that in semiconductor-type media the electromagnetic field propagates as *plane, transverse, damped waves*, with a phase difference φ between **E** and **H**. The attenuation of the wave is due to its absorption by the medium, expressed by the factor $\exp(-k_2\,\mathbf{s}\cdot\mathbf{r})$, which contains the *imaginary part* of the complex wave number. If $\lambda = 0$, (4.77) gives $k_2 = 0$, in which case there is no absorption, the two fields are in phase, and we find the case of dielectrics.

Observations:

(a) The presence of a complex permittivity $\tilde{\epsilon}$ yields a complex refractive index

$$\tilde{n} = \frac{c}{\tilde{u}} = \frac{c}{\omega}\tilde{k} = n_1 - i\,n_2. \tag{4.82}$$

Generally speaking, the refractive index varies with the wavelength, meaning that the semiconductors are not only absorbent, but also *dispersive*. The phenomenon of dispersion will be separately studied;

(b) For $\beta \ll 1$ (dielectrics), expanding in series the radical in (4.77), we have

$$\sqrt{1 + \frac{\lambda^2}{\omega^2 \epsilon^2}} \simeq 1 + \frac{1}{2} \frac{\lambda^2}{\omega^2 \epsilon^2}, \tag{4.83}$$

so that

$$k_1^2 \simeq \epsilon \mu \omega^2 \left(1 + \frac{1}{4} \frac{\lambda^2}{\omega^2 \epsilon^2}\right), \quad k_2^2 \simeq \frac{1}{4} \frac{\mu \lambda^2}{\epsilon}, \tag{4.84}$$

meaning that the attenuation is independent of ω. If $\lambda \to 0$, we get the already known (and expected) result

$$k_2 = 0, \quad k_1 = k = \omega \sqrt{\epsilon \mu}; \tag{4.85}$$

(c) In the case of conductors ($\beta \gg 1$), we can write

$$\sqrt{1 + \frac{\lambda^2}{\omega^2 \epsilon^2}} \simeq \frac{\lambda}{\epsilon \omega}, \tag{4.86}$$

and (4.77) yields the familiar result

$$k_1^2 = k_2^2 = \frac{1}{2} \omega \lambda \mu. \tag{4.87}$$

Thus, the conductor and dielectric media can be studied as two limit cases of semiconductors, corresponding to $\beta \gg 1$ and $\beta \ll 1$, respectively.

4.7 Propagation of Electromagnetic Waves in Anisotropic Media

The anisotropic media (e.g. crystals) have different properties in different directions, and, consequently, the permittivity and permeability are tensor quantities. In this case, the constitutive relations are written as

$$D_i = \epsilon_{ik} E_k, \quad B_i = \mu_{ik} H_k, \quad i, k = 1, 2, 3. \tag{4.88}$$

In the following, we shall limit ourselves to dielectric-type crystals, transparent for electromagnetic waves, with $\mu_r \simeq 1$. Since

$$\mathbf{E} \cdot \mathbf{D} = E_i D_i = E_k D_k,$$

we have

$$E_i(\epsilon_{ik} E_k) = E_k(\epsilon_{ki} E_i),$$

meaning that the permittivity tensor is symmetric:

$$\epsilon_{ik} = \epsilon_{ki}.$$

In a similar way it can be shown that

$$\mu_{ik} = \mu_{ki}.$$

4.7.1 Fresnel's Ellipsoid

Assume that the magnetic characteristics of a generic crystal are close to those of the vacuum. An immediate consequence of this assumption is that $W_e \gg W_m$. The density of the electric component of electromagnetic energy is then

$$2\,w_e = \epsilon_{ik}E_iE_k = \sum_i \epsilon_{ii}E_i^2 + 2\sum_{i>k}\epsilon_{ik}E_iE_k \quad (i,k = 1,2,3).$$

Dividing by $2w_e$ and denoting $X_i = E_i/\sqrt{2w_e}$, we have

$$\sum_i \epsilon_{ii}X_i^2 + 2\sum_{i>k}\epsilon_{ik}X_iX_k = 1. \tag{4.89}$$

This is the equation of a quadric, namely an ellipsoid (all $\epsilon_{ik} > 0$). One can conveniently rotate the coordinate axes $(x_i \to X_i')$, so that the new axes X_i' coincide with the principal axes of the ellipsoid. In this case the permittivity tensor becomes diagonal, and Eq. (4.89) reduces to the canonical form

$$\sum_i \epsilon_{ii}'X_i'^2 = 1.$$

Denoting

$$\epsilon_{ii}' \equiv \epsilon_i', \qquad \sqrt{\epsilon_i'} \equiv \frac{1}{a_i'}, \tag{4.90}$$

we find

$$\sum_i \frac{X_i'^2}{a_i'^2} = \frac{X'^2}{a'^2} + \frac{Y'^2}{b'^2} + \frac{Z'^2}{c'^2} = 1, \tag{4.91}$$

where we put $a_1' = a'$, $a_2' = b'$, and $a_3' = c'$. The components ϵ_i', $i = 1, 2, 3$ of the permittivity tensor, reduced to the diagonal form, are the medium permittivities along the principal axes of the ellipsoid, and are called *principal permittivities*, while the ellipsoid (4.91) is known as *Fresnel's ellipsoid*. Dropping the "prime" index, we write (4.91) as

$$\epsilon_x X^2 + \epsilon_y Y^2 + \epsilon_z Z^2 = 1,$$

or, taking into account that $n_i^2 = (\epsilon_r)_i = \epsilon_i/\epsilon_0$, $i = 1, 2, 3$,

$$\epsilon_0 \left(n_x^2 X^2 + n_y^2 Y^2 + n_z^2 Z^2\right) = 1. \tag{4.92}$$

Since along the principal axes of Fresnel's ellipsoid we have

$$D_x = \epsilon_x E_x, \quad D_y = \epsilon_y E_y, \quad D_z = \epsilon_z E_z, \tag{4.93}$$

we may conclude that, if the direction of the field **E** coincides with the direction of one of the principal axes of Fresnel's ellipsoid, the field **D** has the same direction. For any other direction of **E**, the two fields **E** and **D** are no more collinear.

4.7.2 Fresnel's Law of Velocities for Electromagnetic Waves

We shall now turn our attention towards establishing a relation between the velocity and the direction of propagation of an electromagnetic wave in an anisotropic medium. Assume that in a crystal characterized by $\mathbf{j} = 0$, $\rho = 0$, $\mu_r \simeq 1$ propagates a plane, monochromatic, electromagnetic wave, given by

$$\begin{aligned} \mathbf{E} &= \mathbf{E}_0 \, e^{i(\omega t - \mathbf{k} \cdot \mathbf{r})}, \\ \mathbf{D} &= \mathbf{D}_0 \, e^{i(\omega t - \mathbf{k} \cdot \mathbf{r})}, \\ \mathbf{H} &= \mathbf{H}_0 \, e^{i(\omega t - \mathbf{k} \cdot \mathbf{r})}. \end{aligned} \tag{4.94}$$

If the direction of propagation of the wave is defined by the unit vector $\mathbf{s} = (\alpha, \beta, \gamma)$, then the phase of the wave will be

$$\omega t - \mathbf{k} \cdot \mathbf{r} = \omega \left(t - \frac{\alpha x + \beta y + \gamma z}{u}\right) = \omega \left(t - \frac{\alpha_k x_k}{u}\right),$$

with the notations $\alpha_1 = \alpha$, $\alpha_2 = \beta$, $\alpha_3 = \gamma$.

We shall use a convenient form of Maxwell equations, namely

$$\begin{aligned} \nabla \times \mathbf{H} &= \frac{\partial \mathbf{D}}{\partial t}, \\ \nabla \times \mathbf{E} &= -\mu \frac{\partial \mathbf{H}}{\partial t}, \\ \nabla \cdot \mathbf{H} &= 0, \\ \nabla \cdot \mathbf{D} &= 0. \end{aligned} \tag{4.95}$$

Supposing that the permittivity tensor ϵ_{ik} is diagonal, we compress the relations (4.93) and write them in the form

$$D_i = \epsilon_i E_i \quad \text{(no summation)}, \quad i = 1, 2, 3. \tag{4.96}$$

Denoting by l_i, $i = 1, 2, 3$ the direction cosines of the field $\mathbf{D}$, we have

$$D_i = l_i |\mathbf{D}| = l_i D.$$

Using $(4.94)_2$ in $(4.95)_4$, we obtain

$$\mathbf{D} \cdot \mathbf{s} = \alpha_k l_k D = 0. \tag{4.97}$$

This shows that $\mathbf{D}$ is orthogonal to the direction of propagation. However, due to (4.96), the vectors $\mathbf{E}$ and $\mathbf{D}$ are not parallel and $\mathbf{E}$ is not orthogonal to $\mathbf{s}$.

As μ_r has the same value in all directions, we have $\mathbf{H} \parallel \mathbf{B}$. Moreover, from $(4.95)_3$ follows that $\mathbf{H} \perp \mathbf{s}$, while $(4.95)_{1,2}$ yield $\mathbf{H} \perp \mathbf{D}$ and $\mathbf{H} \perp \mathbf{E}$. Putting all this information together, we deduce that $\mathbf{E}$, $\mathbf{D}$, and $\mathbf{s}$ lie all in one plane, and $\mathbf{H}$ and $\mathbf{B}$ are orthogonal to this plane. Remark that in this case, the Poynting vector $\mathbf{\Pi} = \mathbf{E} \times \mathbf{H}$ is not parallel to the direction of the wave vector $\mathbf{k}$.

Let us now turn to determining the law of velocities for the anisotropic medium. Equation $(4.95)_{1,2}$ can be combined to give

$$\mu \frac{\partial^2 \mathbf{D}}{\partial t^2} = \Delta \mathbf{E} - \nabla(\nabla \cdot \mathbf{E}). \tag{4.98}$$

The x_i-component of this equation is

$$\mu \frac{\partial^2 D_i}{\partial t^2} = \frac{1}{\epsilon_i} \Delta D_i - \frac{\partial}{\partial x_i} \left(\frac{1}{\epsilon_k} \frac{\partial D_k}{\partial x_k} \right),$$

with summation over the index k and no summation over i. This can also be written as

$$\mu l_i \frac{\partial^2 D}{\partial t^2} = \frac{l_i}{\epsilon_i} \Delta D - \frac{\partial}{\partial x_i} \left(\frac{l_k}{\epsilon_k} \frac{\partial D}{\partial x_k} \right),$$

or, in view of $(4.94)_2$

$$\mu l_i = \frac{l_i}{\epsilon_i u^2} - \frac{\alpha_i \alpha_k}{\epsilon_k u^2} l_k.$$

Denoting

$$\frac{1}{\mu \epsilon_i} = v_i^2, \quad \frac{l_k \alpha_k}{\mu \epsilon_k} = K^2, \tag{4.99}$$

we find

$$l_i = -\frac{\alpha_i}{u^2 - v_i^2} K^2.$$

Now we multiply this equation by α_i and perform the summation over i. Using (4.97), we obtain

$$\sum_{i=1}^{3} \frac{\alpha_i^2}{u^2 - v_i^2} = 0,$$

or, by denoting $v_1 = a$, $v_2 = b$, $v_3 = c$,

$$\frac{\alpha^2}{u^2 - a^2} + \frac{\beta^2}{u^2 - b^2} + \frac{\gamma^2}{u^2 - c^2} = 0, \qquad (4.100)$$

which is *Fresnel's law of velocities*, or *Fresnel's formula*.

To find the significance of a, b, and c let us successively suppose that the wave propagates along the Ox, Oy, and Oz axes. For example, when propagation takes place in the x-direction we have $\alpha = 1$, $\beta = \gamma = 0$, and (4.100) yields

$$\left(u^2 - b^2\right)\left(u^2 - c^2\right) = 0,$$

which means either $u^2 = b^2$ or $u^2 = c^2$. This can also be written as $u_{1x}^2 = b^2$, $u_{2x}^2 = c^2$. In a similar way we find $u_{1y}^2 = a^2$, $u_{2y}^2 = c^2$, and $u_{1z}^2 = a^2$, $u_{2z}^2 = b^2$. We therefore conclude that a, b, and c signify the velocities of the electromagnetic wave along the principal axes of Fresnel's ellipsoid. They are called *principal velocities*.

Fresnel's formula is a biquadratic algebraic equation in u with real coefficients, therefore it admits two roots u_1^2 and u_2^2. Thus, for a given direction of propagation **s** of the incident electromagnetic wave, there are two distinct absolute values for the velocity in a crystal, meaning two waves with different velocities u_1 and u_2. Let us analyze these velocities, admitting, for instance, that $a > b > c$.

a) Suppose, first, that the wave propagates in the yOz-plane ($\alpha = 0$). Then the velocity law (4.100) gives

$$\left(u^2 - a^2\right)\left[\beta^2(u^2 - c^2) + \gamma^2(u^2 - b^2)\right] = 0,$$

which leads to

$$u_1^2 - a^2 = 0,$$
$$\beta^2(u_2^2 - c^2) + \gamma^2(u_2^2 - b^2) = 0.$$

As $\alpha = 0$, we have $\beta^2 + \gamma^2 = 1$, and the last relation yields

$$u_2^2 = \beta^2 c^2 + \gamma^2 b^2. \qquad (4.101)$$

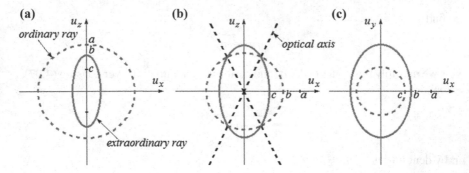

Fig. 4.8 Propagation of an electromagnetic wave in an anisotropic medium. Note that the axes are the principal axes of Fresnel's ellipsoid.

This means that, in a crystal, the electromagnetic wave suffers the phenomenon of *birefringence*. It is decomposed into two waves (rays):

(*i*) the *ordinary ray* (or *O-ray*), travelling with the velocity $|u_1| = a$ in all directions (since it does not depend on β and γ), and

(*ii*) the *extraordinary ray* (or *E-ray*), with the velocity $|u_2| = \sqrt{\beta^2 c^2 + \gamma^2 b^2}$, which depends on the direction of propagation through the direction cosines β and γ.

In the plane of propagation (*yOz*-plane) the first represents a circle, and the latter – an ellipse. The intersections of the ellipse with the coordinate axes are found from (4.101), by setting $\gamma = 0$, leading to $u_{2y} = \pm c$, respectively $\beta = 0$, leading to $u_{2z} = \pm b$ (see Fig. 4.8a). As $a > b > c$, we note that $u_1 \neq u_2$ for any direction of propagation in the *yOz*-plane.

b) Let us now assume that the direction of the wave propagation lies in the *zOx*-plane ($\beta = 0$). Then (4.100) leads to

$$\left(u^2 - b^2\right)\left[\alpha^2(u^2 - c^2) + \gamma^2(u^2 - a^2)\right] = 0,$$

with the solutions

$$u_1^2 = b^2,$$

$$u_2^2 = \alpha^2 c^2 + \gamma^2 a^2 \Rightarrow \begin{cases} u_{2x} = \pm c, \\ u_{2z} = \pm a. \end{cases}$$

There are two directions in the *zOx*-plane corresponding to $u_1 = u_2$ (see Fig. 4.8b). These two directions define the *optical axes* of the crystal. A crystal possessing a single optical axis is called *uniaxial*, and that having two optical axes – *biaxial*. A crystal may have maximum two optical axes.

Fig. 4.9 Geometrical representation of the velocities of ordinary and extraordinary rays along different directions.

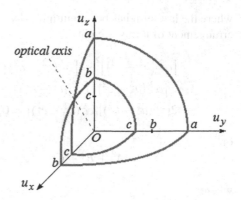

c) As the third possible situation, suppose that the direction of propagation is situated in the xOy-plane. In this case, (4.100) leads to

$$(u^2 - c^2)\left[\alpha^2(u^2 - b^2) + \beta^2(u^2 - a^2)\right] = 0,$$

whose solutions are

$$u_1^2 = c^2,$$
$$u_2^2 = \alpha^2 b^2 + \beta^2 a^2 \Rightarrow \begin{cases} u_{2x} = \pm b, \\ u_{2y} = \pm a, \end{cases}$$

as shown in Fig. 4.8c. Again, as in the first case, there are no optical axes in the xOy-plane.

The three geometric representations of Fig. 4.8 can be unified in a single picture as shown in Fig. 4.9. This gives the spatial geometric representation of the velocities of the ordinary and extraordinary rays along different directions.

Let us now determine the direction cosines α, β, and γ of the optical axis of the crystal in terms of principal velocities a, b, c. To this end, we write Fresnel's velocity law as

$$u^4 - \left[\alpha^2\left(b^2 + c^2\right) + \beta^2\left(a^2 + c^2\right) + \gamma^2\left(a^2 + b^2\right)\right]u^2$$
$$+ \alpha^2 b^2 c^2 + \beta^2 a^2 c^2 + \gamma^2 a^2 b^2 = 0.$$

Using the general formula for solving biquadratic equation, in order to have no distinct roots ($u_1^2 = u_2^2$), which would correspond to an optical axis, we must have

$$\left[\alpha^2\left(b^2 + c^2\right) + \beta^2\left(a^2 + c^2\right) + \gamma^2\left(a^2 + b^2\right)\right]^2$$
$$- 4(\alpha^2 + \beta^2 + \gamma^2)(\alpha^2 b^2 c^2 + \beta^2 a^2 c^2 + \gamma^2 a^2 b^2) = 0,$$

where the last term has been multiplied by $1 = \alpha^2 + \beta^2 + \gamma^2$. Making a convenient arrangement of terms, we obtain

$$
\begin{aligned}
&\left[\alpha^2 \left(b^2 - c^2\right)\right]^2 + \left[\beta^2 \left(c^2 - a^2\right)\right]^2 + \left[\gamma^2 \left(a^2 - b^2\right)\right]^2 \\
&- 2\left[\alpha^2 \left(b^2 - c^2\right)\right]\left[\beta^2 \left(c^2 - a^2\right)\right] - 2\left[\beta^2 \left(c^2 - a^2\right)\right]\left[\gamma^2 \left(a^2 - b^2\right)\right] \\
&- 2[\gamma^2(a^2 - b^2)][\alpha^2(b^2 - c^2)] = 0,
\end{aligned}
$$

or

$$
A^2 + B^2 + C^2 - 2AB - 2BC - 2CA = 0,
$$

where

$$
\begin{aligned}
A &\equiv \alpha^2 \left(b^2 - c^2\right), \\
B &\equiv \beta^2 (c^2 - a^2), \\
C &\equiv \gamma^2 (a^2 - b^2).
\end{aligned}
\tag{4.102}
$$

The last equation can also be written as

$$
(A + B - C)^2 - 4AB = 0.
\tag{4.103}
$$

Maintaining the supposition $a > b > c$ (see Figs. 4.8 and 4.9), we have $A > 0, B < 0$, and $C > 0$, which means $-4AB > 0$. Then (4.103) yields

$$
A + B - C = 0,
$$
$$
AB = 0.
\tag{4.104}
$$

If $A = 0$, it follows from $(4.104)_1$ that $B = C$. Since B and C have opposite signs, the only possibility is $B = C = 0$. But $A = B = C = 0$ means $a = b = c$, which corresponds to an isotropic medium. Therefore, to have an anisotropic medium, we must have $A \neq 0$, which implies $B = 0$, while $(4.104)_1$ gives $A = C$, leading to the system

$$
\beta^2(c^2 - a^2) = 0,
$$
$$
\alpha^2(b^2 - c^2) = \gamma^2(a^2 - b^2).
\tag{4.105}
$$

Since, by hypothesis, $a > b > c$, the possibility $c = a$ would presume $a = b = c$, corresponding to an isotropic medium. Excluding this possibility, we are left with $\beta = 0$. By means of the relations $\alpha^2 + \gamma^2 = 1$ and $(4.105)_2$, it then follows that

$$
\alpha = \pm\sqrt{\frac{a^2 - b^2}{a^2 - c^2}},
$$

$$
\gamma = \pm\sqrt{\frac{b^2 - c^2}{a^2 - c^2}}.
\tag{4.106}
$$

Thus, the principal optical axis lies in the plane xOz (as we already know) and its direction parameters are defined by (4.106). Since $a > b > c$, this parameters are real.

Let $\mathbf{E}_1$, $\mathbf{D}_1$ be the field vectors attached to the ordinary ray, and $\mathbf{E}_2$, $\mathbf{D}_2$ those corresponding to the extraordinary ray. Using (4.94) and (4.95), we have

$$\mathbf{D} = \frac{1}{u}\mathbf{H} \times \mathbf{s},$$

$$\mathbf{H} = \frac{1}{\mu u}\mathbf{s} \times \mathbf{E},$$

or, by eliminating $\mathbf{H}$,

$$\mathbf{D} = \frac{1}{\mu u^2}[\mathbf{E} - (\mathbf{s} \cdot \mathbf{E})\,\mathbf{s}]. \tag{4.107}$$

Let us now transcribe this relation for the two rays:

$$\mu u_1^2\mathbf{D}_1 = \mathbf{E}_1 - (\mathbf{s} \cdot \mathbf{E}_1)\,\mathbf{s},$$
$$\mu u_2^2\mathbf{D}_2 = \mathbf{E}_2 - (\mathbf{s} \cdot \mathbf{E}_2)\,\mathbf{s},$$

then multiply them by $\mathbf{D}_2$ and $\mathbf{D}_1$, respectively, and finally subtract one from the other. The result is

$$\mu\left(u_1^2 - u_2^2\right)\mathbf{D}_1 \cdot \mathbf{D}_2 = \mathbf{E}_1 \cdot \mathbf{D}_2 - \mathbf{E}_2 \cdot \mathbf{D}_1 = 0,$$

because

$$\mathbf{E}_1 \cdot \mathbf{D}_2 = E_{1i}D_{2i} = \epsilon_i E_{1i}E_{2i},$$
$$\mathbf{E}_2 \cdot \mathbf{D}_1 = E_{2i}D_{1i} = \epsilon_i E_{2i}E_{1i}.$$

Therefore, if $u_1 \neq u_2$, we have $\mathbf{D}_1 \perp \mathbf{D}_2$, meaning that the vectors $\mathbf{D}_1$ and $\mathbf{D}_2$ are reciprocally orthogonal and oscillate in perpendicular planes. The plane containing the optical axis and the incident ray is called *principal plane* of the crystal, or *principal section*.

We conclude that when propagating in a crystal, the incident ray splits into two rays, polarized in planes perpendicular to each other: the ordinary ray, polarized in the plane of principal section, and the extraordinary ray, which is polarized perpendicularly to this plane. The two rays have different velocity of propagation, therefore they experience different refraction indices, and thus they refract at different angles when entering the anisotropic medium.

4.8 Dispersion of Electromagnetic Waves

Let $f(x)$ be a function defined on the whole real axis. By definition, the *Fourier transform* of $f(x)$ is

$$F(k) = \frac{1}{\sqrt{2\pi}} \int_{-\infty}^{+\infty} f(x)e^{-ikx}\, dx. \tag{4.108}$$

The function $F(k)$ is the *spectral distribution* of $f(x)$, or, in short, the *spectrum* of $f(x)$. The inverse operation

$$f(x) = \frac{1}{\sqrt{2\pi}} \int_{-\infty}^{+\infty} F(k)e^{ikx}\, dk, \tag{4.109}$$

is the *inverse Fourier transform*. Replacing formally $F(k)$ in (4.109), and changing x to x' in (4.108), we have

$$f(x) = \int_{-\infty}^{+\infty} f(x')\left[\frac{1}{2\pi}\int_{-\infty}^{+\infty} e^{ik(x-x')}dk\right]dx'. \tag{4.110}$$

Comparing (4.110) with the relation expressing the sifting property of Dirac δ function (see (E.7)),

$$f(x) = \int_{-\infty}^{+\infty} f(x')\delta(x-x')dx', \tag{4.111}$$

we find the Fourier representation of the Dirac δ function:

$$\delta(x-x') = \frac{1}{2\pi}\int_{-\infty}^{+\infty} e^{ik(x-x')}\, dk. \tag{4.112}$$

Let, now, $f(x,t)$ be the solution of the propagation equation in the x-direction (see (4.6)), of a plane monochromatic wave. The Fourier representation of $f(x,t)$, at any fix t, is

$$f(x,t) = \frac{1}{\sqrt{2\pi}} \int_{-\infty}^{+\infty} F(k,t)e^{ikx}\, dk. \tag{4.113}$$

Plugging this expression into the wave equation

$$\frac{\partial^2 f}{\partial x^2} - \frac{1}{u^2}\frac{\partial^2 f}{\partial t^2} = 0,$$

we find

$$\frac{1}{\sqrt{2\pi}} \int_{-\infty}^{+\infty} \left[k^2 F(k,t) + \frac{1}{u^2}\frac{\partial^2 F(k,t)}{\partial t^2}\right]e^{ikx}dk = 0.$$

If the Fourier transform of some function is zero, this means that the function itself is zero, hence

$$\frac{\partial^2 F(k, t)}{\partial t^2} + u^2 k^2 F(k, t) = 0. \tag{4.114}$$

The general solution of this differential equation is

$$F(k, t) = A(k)e^{-ikut} + B(k)e^{ikut}, \tag{4.115}$$

and $f(x, t)$ is thus expressed in Fourier series as

$$f(x, t) = \frac{1}{\sqrt{2\pi}} \int_{-\infty}^{+\infty} \left[A(k)e^{ik(x-ut)} + B(k)e^{ik(x+ut)} \right] dk.$$

The first term corresponds to a progressive wave, while the second represents a regressive wave. According to our previous convention, we shall use only the first term, i.e.

$$f(x, t) = \frac{1}{\sqrt{2\pi}} \int_{-\infty}^{+\infty} A(k)e^{ik(x-ut)} \, dk, \tag{4.116}$$

where u is the phase velocity and k – the wave number, so that

$$ku = \frac{2\pi}{\lambda} u = 2\pi \nu = \omega.$$

The solution (4.116) can be interpreted as a superposition of an infinite number of waves of amplitudes $A(k)$, all propagating in the x-direction and having a continuous frequency spectrum.

Suppose that at time $t = 0$, the function $f(x, 0) = g(x)$ is known:

$$f(x, 0) = g(x) = \frac{1}{\sqrt{2\pi}} \int_{-\infty}^{+\infty} A(k)e^{ikx} \, dk.$$

The spectrum $A(k)$ is obtained by the inverse Fourier transform,

$$A(k) = \frac{1}{\sqrt{2\pi}} \int_{-\infty}^{+\infty} g(x')e^{-ikx'} \, dx',$$

and (4.116) becomes

$$f(x, t) = \frac{1}{\sqrt{2\pi}} \int_{-\infty}^{+\infty} g(x') \left[\frac{1}{2\pi} \int_{-\infty}^{+\infty} e^{ik(x-x'-ut)} dk \right] dx'$$

$$= \int_{-\infty}^{+\infty} g(x')\delta(x - x' - ut)dx' = g(x - ut),$$

which is nothing else but the term corresponding to a progressive wave in the general solution (4.7). In the analysis above we did not consider the dispersion of the medium.

In three dimensions, the passage from the space of radius vectors to that of wave vectors is accomplished by the transformation

$$f(\mathbf{r}, t) = \frac{1}{(2\pi)^{3/2}} \int_{-\infty}^{+\infty} A(\mathbf{k}) \, e^{i(\mathbf{k}\cdot\mathbf{r} - \omega t)} d\mathbf{k}, \qquad (4.117)$$

where $d\mathbf{k} = dk_x dk_y dk_z$.

Formula (4.117) is also valid for the electromagnetic field vectors, for example

$$\mathbf{E}(\mathbf{r}, t) = \frac{1}{(2\pi)^{3/2}} \int_{-\infty}^{+\infty} \tilde{\mathbf{E}}(\mathbf{k}) \, e^{i(\mathbf{k}\cdot\mathbf{r} - \omega t)} d\mathbf{k}. \qquad (4.118)$$

Observation:

The Fourier transforms (4.108) and (4.109) are inverse of each other, therefore the choice of sign in the exponential function in (4.108), as well as in (4.118), is arbitrary. This is also shown by the fact that $\mathrm{Re}\{\exp[i(\mathbf{k}\cdot\mathbf{r} - \omega t)]\}$ is an even function.

4.8.1 Phase Velocity and Group Velocity

The analyses in the present chapter have been based on the hypothesis of propagation of electromagnetic perturbations as plane, monochromatic waves, through linear and homogeneous media, meaning that the wave function (i.e. the solution of the wave equation) is proportional to $\exp[i(\mathbf{k}\cdot\mathbf{r} - \omega t)]$, where $\mathbf{k}$ and ω are independent of $\mathbf{r}$ and t. We have also assumed that two waves may exist simultaneously, independent of each other.

In reality, the angular frequency ω is a function of the wave vector: $\omega = \omega(\mathbf{k})$. This relation is the *dispersion equation*, and characterizes each medium in which the wave propagation takes place. In isotropic media all directions are equivalent, and ω depends only on $|\mathbf{k}| = k$, i.e. $\omega = \omega(k)$. For example, the frequency of a light wave in a linear dielectric is $\omega = k\,u = k\,c/n = k\,c/\sqrt{\epsilon_r \mu_r}$.

The phase $\mathbf{k}\cdot\mathbf{r} - \omega t$ can be written as

$$\mathbf{k} \cdot \left(\mathbf{r} - \frac{\omega}{k} t\,\mathbf{s} \right),$$

where $\mathbf{s}$ is the unit vector of $\mathbf{k}$. The locus of the points of constant phase, termed *wave front*, is given by the equation

$$\mathbf{r} - \frac{\omega}{k}\mathbf{s}\,t = \mathbf{r}_0 = \text{const.},$$

or

$$\mathbf{r} = \mathbf{u}_p\, t + \mathbf{r}_0,$$

The vector quantity

$$\mathbf{u}_p = \frac{\omega}{k}\,\mathbf{s} \qquad (4.119)$$

is called *phase velocity*.

The model of plane monochromatic wave, used by us as a working frame, corresponds only approximately to the reality. Indeed, such an ideal wave has no spatial and/or temporal limits, while its characteristic elements (frequency, amplitude, wavelength) are permanently constant. In order to transmit and receive a signal, this has to be limited in space and variable in amplitude or/and frequency. The signals emitted by a source of electromagnetic waves are in fact *groups of waves*, or *wave packets*, obtained by superposition of waves with frequencies close to each other. In a dispersive medium, the phase velocity is not the same for all the frequencies which compose the real wave. As a result, the components of the wave propagate with different velocities and the phase differences between the components change in time, resulting in a loss of the original coherence. Moreover, in a dispersive medium the velocity of energy propagation by a real wave can differ sometimes quite a lot from the average phase velocity of the component waves.

Let us elucidate the meaning of a wave packet. Consider a wave propagating in the x–direction. Suppose that the signal is characterized by the function $\varphi(x)$, whose spectral decomposition is:

$$\varphi(x) = \frac{1}{\sqrt{2\pi}} \int_{-\infty}^{+\infty} A(k)e^{ikx}dk.$$

Assume that the perturbation is localized around a position x_0, being formed of a set of wave numbers distributed around some value k_0 (Fig. 4.10a and b).

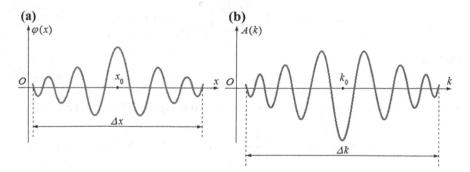

Fig. 4.10 Graphical representations of: (**a**) the perturbation $\varphi(x)$ as a function of the distance x; (**b**) the Fourier transform $A(k)$ as a function of $k = |\mathbf{k}|$, where $\mathbf{k}$ is the wave vector.

Let Δx and Δk be the extensions of the wave packet in the coordinate space and in the wave vector space, respectively. In this case, a wave e^{ikx} will suffer a phase variation $k\,\Delta x$, from one end of the packet to the other, while a wave $e^{i(k+\Delta k)x}$ will be subject to a phase change $(k+\Delta k)\Delta x$. Due to the superposition of components with different wave numbers, at the ends of the packet $\varphi(x) \simeq 0$. Therefore, the two *in-phase positions* give the difference

$$(k + \Delta k)\Delta x - k\Delta x = \Delta k \Delta x \simeq 2\pi\,.$$

In three dimensions, according to (4.117), we have

$$\varphi(\mathbf{r}, t) = \frac{1}{(2\pi)^{3/2}} \int_{-\infty}^{+\infty} A(\mathbf{k})\, e^{i[\mathbf{k}\cdot\mathbf{r} - \omega(\mathbf{k})t]}d\mathbf{k}, \qquad (4.120)$$

where $\omega(\mathbf{k})$ is supposed to be a real function (there is no absorption). If the values of $\mathbf{k}$ are distributed within a narrow interval around an arbitrary $\mathbf{k}_0$, and the variation of ω with respect to $\mathbf{k}$ in that interval is not very large, then $\omega(\mathbf{k})$ can be expanded in series around this value:

$$\omega(\mathbf{k}) = \omega(\mathbf{k}_0) + (\mathbf{k} - \mathbf{k}_0)_i \left[\left(\frac{d\omega}{d\mathbf{k}}\right)_i\right]_{\mathbf{k}=\mathbf{k}_0} + \dots, \quad i = 1, 2, 3, \qquad (4.121)$$

where the gradient of ω in the $\mathbf{k}$-space was formally written as $d\omega/d\mathbf{k}$, meaning

$$\frac{d\omega}{d\mathbf{k}} \equiv \frac{\partial\omega}{\partial k_1}\hat{k}_1 + \frac{\partial\omega}{\partial k_2}\hat{k}_2 + \frac{\partial\omega}{\partial k_3}\hat{k}_3,$$

$\hat{k}_i$, $i = 1, 2, 3$ being the three unit vectors of the corresponding axes in the $\mathbf{k}$-space.

Suppose that the third term in the series expansion (4.121) is much smaller than the second one, so that we keep only two terms. Denoting

$$\omega(\mathbf{k}_0) = \omega_0 \quad \text{and} \quad \left[\left(\frac{d\omega}{d\mathbf{k}}\right)_i\right]_{\mathbf{k}=\mathbf{k}_0} = (\mathbf{u}_g)_i, \qquad (4.122)$$

we have

$$\omega(\mathbf{k}) = \omega_0 + \mathbf{u}_g \cdot (\mathbf{k} - \mathbf{k}_0). \qquad (4.123)$$

Writing $\mathbf{k} \cdot \mathbf{r} = \mathbf{k}_0 \cdot \mathbf{r} + (\mathbf{k} - \mathbf{k}_0) \cdot \mathbf{r}$ in (4.120) and using (4.123), we obtain

$$\varphi(\mathbf{r}, t) = e^{i(\mathbf{k}_0 \cdot \mathbf{r} - \omega_0 t)} \frac{1}{(2\pi)^{3/2}} \int A(\mathbf{k}) e^{i(\mathbf{k}-\mathbf{k}_0)\cdot(\mathbf{r}-\mathbf{u}_g t)}d\mathbf{k}. \qquad (4.124)$$

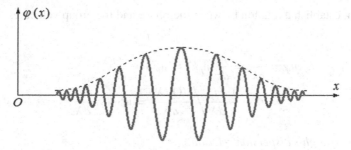

Fig. 4.11 A modulated wave and the group envelope.

This wave is called *modulated*, i.e. composed of many frequencies. Such a wave can be regarded as the result of multiplying the phase factor $e^{i(\mathbf{k}_0\cdot\mathbf{r}-\omega_0 t)}$ by a perturbation of the form (4.120) (an envelope-amplitude) that depends on space and time only through the combination $\mathbf{r} - \mathbf{u}_g t$:

$$\varphi(\mathbf{r}, t) = \eta(\mathbf{r} - \mathbf{u}_g t)e^{i(\mathbf{k}_0\cdot\mathbf{r}-\omega_0 t)}.$$

The shape of the waves, as well as their envelope, are illustrated in Fig. 4.11. The group envelope is constant along the trajectories $\mathbf{r} - \mathbf{u}_g t = \mathbf{r}_0 = \text{const.}$, or

$$\mathbf{r} = \mathbf{r}_0 + \mathbf{u}_g t. \tag{4.125}$$

This means that, excepting a phase factor, the real wave (or the wave packet) propagates practically undistorted, with the *group velocity*

$$\mathbf{u}_g = \nabla_{\mathbf{k}}\, \omega, \tag{4.126}$$

where $\nabla_{\mathbf{k}}$ signifies the gradient in the $\mathbf{k}$-space, formally written as $\frac{d\omega}{d\mathbf{k}}$.

The analysis above is valid in the case of a wave packet with a rather narrow spectrum of frequencies or propagating in a weakly dispersive medium, in which the frequency varies slowly with the wave number (under these circumstances we are allowed to ignore the higher order terms in the series expansion (4.121)). The general case of a strongly dispersive medium or of a sharply localized wave packet, which contains a very large number of component frequencies, is much more complex and beyond our present scope.

In isotropic media $\mathbf{u}_g = (d\omega/dk)\mathbf{s}$, while in vacuum

$$\omega = ck = c(\mathbf{k}\cdot\mathbf{k})^{1/2} \quad \rightarrow \quad u_p = u_g = c.$$

One can establish a relation between the phase and the group velocities. Indeed, we have

$$u_g = \frac{d\omega}{dk} = \frac{d}{dk}(u_p k) = u_p + k\frac{du_p}{dk}$$

$$= u_p + \frac{1}{\lambda}\frac{du_p}{d\lambda}\Big/\frac{d(1/\lambda)}{d\lambda} = u_p - \lambda\frac{du_p}{d\lambda}, \qquad (4.127)$$

known as *Rayleigh's dispersion relation*.

Application

Consider a group of light waves propagating along the x-axis, and assume that the wave numbers of the constitutive waves are contained in the interval $k_0 - \frac{\Delta k}{2} \le k \le k_0 + \frac{\Delta k}{2}$, with $|\Delta k| \ll k_0$. We also assume that the amplitudes of the component waves are equal to each other, for any k. Since in the light wave the effect on matter of the electric field $\mathbf{E}$ is overwhelming compared to that of the magnetic field $\mathbf{B}$, we shall characterize the wave packet (according to formula (4.124)) by:

$$\mathbf{E}(x, t) = e^{i(k_0 x - \omega_0 t)}\frac{\mathbf{E}_0}{\sqrt{2\pi}}\int_{k_0 - \frac{\Delta k}{2}}^{k_0 + \frac{\Delta k}{2}} e^{i(k-k_0)(x-u_g t)}dk,$$

where $\mathbf{E}_0 = \mathbf{E}(k_0)$. Using the substitution $k - k_0 = \xi$, we have

$$\mathbf{E}(x, t) = e^{i(k_0 x - \omega_0 t)}\frac{\mathbf{E}_0}{\sqrt{2\pi}}\int_{-\Delta k/2}^{+\Delta k/2} e^{i\xi(x-u_g t)}d\xi.$$

The integral is easily worked out and the result is

$$\mathbf{E}(x, t) = \frac{2\mathbf{E}_0}{\sqrt{2\pi}}e^{i(k_0 x - \omega_0 t)}\frac{\sin\left[(x - u_g t)\frac{\Delta k}{2}\right]}{x - u_g t}. \qquad (4.128)$$

In this expression the first factor, $\sqrt{\frac{2}{\pi}}\mathbf{E}_0 e^{i(k_0 x - \omega_0 t)}$, represents a spatially homogeneous wave, propagating with the average "carrier" frequency ω_0. The amplitude of the resulting wave varies from one point to another due to the second factor,

$$\frac{\sin\left[(x - u_g t)\frac{\Delta k}{2}\right]}{(x - u_g t)\frac{\Delta k}{2}}\frac{\Delta k}{2} \equiv g\left[(x - u_g t)\frac{\Delta k}{2}\right] \equiv g(\alpha). \qquad (4.129)$$

This factor is sharply peaked at $x = u_g t$, as $\lim_{\alpha \to 0}\frac{\sin\alpha}{\alpha} = 1$. This maximum is not fixed, but propagates with the velocity u_g, which is the group velocity (Fig. 4.12).

The width of the principal maximum, in units of α, equals π, which means that its space extension, Δx, at a certain moment, can be obtained from $\Delta\alpha = \frac{\Delta k}{2}\Delta x \simeq \pi$, that is

$$\Delta k\,\Delta x \simeq 2\pi. \qquad (4.130)$$

Fig. 4.12 Graphical representation of the function $g(\alpha)$, the envelope of a wave group.

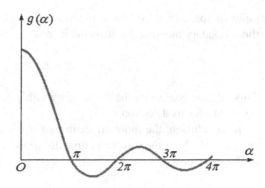

Fig. 4.13 Space localization of an electromagnetic signal.

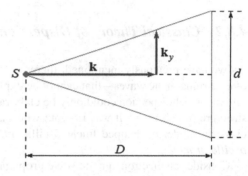

If the secondary maxima are also considered, then

$$\Delta k\, \Delta x \geq 2\pi. \tag{4.131}$$

This relation furnishes information regarding the space localization of an electromagnetic signal. Take, for example, $\Delta x = d$ (the transverse width of the signal), $\Delta k = 2k_y$, and D the distance from source (see Fig. 4.13). Then,

$$k_y = k\frac{d}{2D} = \frac{\pi d}{\lambda D},$$

in which case (4.131) gives $2k_y D \geq 2\pi$. This yields

$$d \geq \sqrt{\lambda D}.$$

Note that the transverse width of the signal is directly proportional to the square root of the wavelength. That is why the controlled transmission of electromagnetic radiation (e.g. in a radar) is performed by means of short wavelength radio waves or *microwaves* ($\lambda \sim 10^{-3} - 10^{-2}\,\text{m}$).

Following a similar reasoning, the inequality (4.131) can also be written for the pair of variables $\Delta\omega$ and Δt: the time duration Δt of a signal, received at a fixed point of space ($x = $ const.), is determined by the relation $\Delta \alpha = \frac{\Delta \omega}{2} \Delta t \simeq \pi$, or, if the secondary maxima are also considered,

$$\Delta \omega \, \Delta t \geq 2\pi. \tag{4.132}$$

This relation conditions the frequency width $\Delta \omega$ of a signal and the time interval Δt necessary for its detection.

In conclusion, the more concentrated in space a wave packet is, the larger the interval Δk. Also, the shorter its time duration is, the larger the frequency dispersion of the wave group.

4.8.2 Classical Theory of Dispersion

As we have previously mentioned, there are some phenomena – as the dispersion of electromagnetic waves – that cannot be explained by Maxwell's theory. An accurate theory of dispersion could only be elaborated by considering the discontinuous structure of substance. It was Lorentz who elaborated such a theory, based on the classical model of damped linear oscillators. The approach is known as *Lorentz oscillator model*.

Consider an electromagnetic wave propagating in an electrically neutral, linear, homogeneous, and isotropic medium. Let N be the number of atoms per unit volume, each atom having an electron of charge $q = -e$ (where $e = +1.602 \times 10^{-19}$C) and mass m, bound by a positive nucleus. Each electron is subject to three types of force:

(i) the quasi-elastic force $-m\omega_0^2 \mathbf{r}$, where ω_0 is the natural frequency of the oscillator;
(ii) the damping force $-m\gamma\dot{\mathbf{r}}$, equivalent to a friction force, due to the proper field of the electron;
(iii) the electromagnetic force $\mathbf{F}_{em} = -e(\mathbf{E} + \mathbf{v} \times \mathbf{B})$.

Since our model is non-relativistic ($v \ll c$), the magnetic component of this force is negligible as compared to the electric component. Supposing $\epsilon \sim \epsilon_0$ and $\mu \sim \mu_0$ (as for gases), we have indeed

$$|\mathbf{v} \times \mathbf{B}| = v\mu_0 H \sin(\widehat{\mathbf{v}, \mathbf{H}}) = \frac{v}{c^2} \frac{H}{\epsilon} \sin(\widehat{\mathbf{v}, \mathbf{H}}) \ll |\mathbf{E}|.$$

The equation of motion of the electron therefore is

$$m\ddot{\mathbf{r}} = -m\omega_0^2 \mathbf{r} - m\gamma\dot{\mathbf{r}} - e\mathbf{E},$$

or

$$\ddot{\mathbf{r}} + \gamma\dot{\mathbf{r}} + \omega_0^2 \mathbf{r} = -\frac{e}{m}\mathbf{E}. \tag{4.133}$$

To integrate this equation, we seek the solution in the form

$$\mathbf{r} = \mathbf{r}_0 e^{-i\omega t}. \tag{4.134}$$

Since the wave is plane, that is

$$\mathbf{E} = \mathbf{E}_0 e^{i(\mathbf{k}\cdot\mathbf{r}-\omega t)}, \quad \mathbf{B} = \mathbf{B}_0 e^{i(\mathbf{k}\cdot\mathbf{r}-\omega t)},$$

we easily find

$$\mathbf{r}_0 = \frac{e}{m} \frac{1}{\omega^2 - \omega_0^2 + i\omega\gamma} \mathbf{E}_0 e^{i\mathbf{k}\cdot\mathbf{r}}$$

and

$$\mathbf{r} = \frac{e}{m} \frac{1}{\omega^2 - \omega_0^2 + i\omega\gamma} \mathbf{E}. \tag{4.135}$$

Hence, we obtain the electron velocity $\mathbf{v} = \dot{\mathbf{r}} = -i\omega\mathbf{r}$, which permits us to write the conduction current density

$$\mathbf{j} = Nq\mathbf{v} = -Ne\mathbf{v} = iNe\omega\mathbf{r} = i\frac{Ne^2}{m} \frac{\omega}{\omega^2 - \omega_0^2 + i\omega\gamma} \mathbf{E}. \tag{4.136}$$

Since we have considered a semiconductor-type medium, the complex permittivity is given by (4.75). However, due to the choice of the phase of the field vectors, this time we have

$$\tilde{\epsilon} = \epsilon + i\frac{\lambda}{\omega}. \tag{4.137}$$

Using Ohm's law, $\mathbf{j} = \lambda\mathbf{E}$, from (4.136) we obtain

$$\lambda = i\frac{Ne^2}{m} \frac{\omega}{\omega^2 - \omega_0^2 + i\omega\gamma}, \tag{4.138}$$

which leads to

$$\tilde{\epsilon} = \epsilon\left(1 + \frac{Ne^2}{m\epsilon} \frac{1}{\omega_0^2 - \omega^2 - i\omega\gamma}\right). \tag{4.139}$$

The refractive index of the studied medium is then

$$\tilde{n} = \frac{c}{\tilde{u}} = c\sqrt{\tilde{\epsilon}\mu} \simeq \sqrt{\frac{\tilde{\epsilon}}{\epsilon_0}} \simeq \left(1 + \frac{Ne^2}{m\epsilon_0} \frac{1}{\omega_0^2 - \omega^2 - i\omega\gamma}\right)^{1/2}, \tag{4.140}$$

where we assumed $\mu_r \simeq 1$, and in (4.139) we considered the real part of the permittivity to be ϵ_0. Relation (4.140),

$$\tilde{n} = \left(1 + \frac{Ne^2}{m\epsilon_0} \frac{1}{\omega_0^2 - \omega^2 - i\omega\gamma}\right)^{1/2},$$

is called the *dispersion equation* for the medium under consideration.

Since the refractive index is complex, it is convenient to write it as

$$\tilde{n} = n_1 + in_2,$$

which gives

$$(\tilde{n})^2 = n_1^2 - n_2^2 + 2in_1n_2 = 1 + \frac{Ne^2}{m\epsilon_0} \frac{\omega_0^2 - \omega^2 + i\omega\gamma}{(\omega_0^2 - \omega^2)^2 + \omega^2\gamma^2},$$

or, by separating the real and imaginary parts,

$$n_1^2 - n_2^2 = 1 + \frac{Ne^2}{m\epsilon_0} \frac{\omega_0^2 - \omega^2}{(\omega_0^2 - \omega^2)^2 + \omega^2\gamma^2},$$

$$2n_1n_2 = \frac{Ne^2}{m\epsilon_0} \frac{\omega\gamma}{(\omega_0^2 - \omega^2)^2 + \omega^2\gamma^2}. \qquad (4.141)$$

Relations (4.141) have been obtained for N oscillators with the natural frequency ω_0. For $\mathcal{N}$ types of oscillators, these formulas are written as follows

$$n_1^2 - n_2^2 = 1 + \frac{e^2}{m\epsilon_0} \sum_{i=1}^{N} \frac{N_i(\omega_i^2 - \omega^2)}{(\omega_i^2 - \omega^2)^2 + \omega^2\gamma^2},$$

$$2n_1n_2 = \frac{e^2}{m\epsilon_0} \sum_{i=1}^{N} \frac{N_i\omega\gamma}{(\omega_i^2 - \omega^2)^2 + \omega^2\gamma^2}, \qquad (4.142)$$

where ω_i is the natural frequency of the i-type oscillators, N_i being their volume density.

It is obvious that a complex refraction index implies a complex wave number, $\tilde{k} = k_1 + i k_2$. The following discussion is therefore valid for both n and k. Keeping in mind that the propagation takes place along the x-axis and the field vector **E** is proportional to $\exp(ikx)$, one can distinguish three cases:

(i) $k_1 = 0$: the wave number is purely imaginary, and the wave is rapidly attenuated, its amplitude dropping precipitously with no propagation in the medium. This is the case of an *evanescent wave*;

(ii) $k_2 = 0$: the wave propagates without damping;

(iii) $k_1, k_2 \neq 0$: the amplitude of the wave decreases exponentially with x. The larger k_2, the greater the attenuation of the wave.

One can conclude that the imaginary part k_2 of $\tilde{k}$ describes the *absorption* of the wave in the medium, due to the energy dissipated as a result of damping of the electron by its own electromagnetic field.

As an immediate application, suppose our medium to be a neutral plasma (e.g. the ionosphere or a metal), with N electrons per unit volume. Neglecting the quasielastic ($\omega_0 = 0$) and friction ($\gamma = 0$) forces, relation (4.140) reduces to

$$n^2 = 1 - \frac{Ne^2}{m\epsilon_0}\frac{1}{\omega^2} = 1 - \frac{\omega_{pl}^2}{\omega^2} < 1, \qquad (4.143)$$

where

$$\omega_{pl}^2 = \frac{Ne^2}{m\epsilon_0}$$

is called *plasma frequency*. Formula (4.143) is useful, for example, when studying reflection and refraction of X-rays at the surface of a metal (neutral plasma). If φ and θ are the incidence and the refraction angles, respectively, we have

$$\frac{\sin^2 \varphi}{\sin^2 \theta} = n^2 = 1 - \frac{\omega_{pl}^2}{\omega^2}.$$

If the X-ray frequency ω is smaller than ω_{pl}, the refraction index becomes purely imaginary, which means that the X-ray beam is *totally reflected*.

Let us go back and apply the equation of dispersion (4.140) to the case of gases and/or vapors. Since the density of polarized atoms is small, the square root can be expanded in series; retaining only the first two terms of the expansion, we have

$$\tilde{n} = n_1 + i n_2 \simeq 1 + \frac{Ne^2}{2m\epsilon_0}\frac{1}{\omega_0^2 - \omega^2 - i\omega\gamma}, \qquad (4.144)$$

which yields

$$n_1 - 1 = f_1(\omega) = \frac{Ne^2}{2m\epsilon_0}\frac{\omega_0^2 - \omega^2}{(\omega_0^2 - \omega^2)^2 + \omega^2\gamma^2},$$

$$n_2 = f_2(\omega) = \frac{Ne^2}{2m\epsilon_0}\frac{\omega\gamma}{(\omega_0^2 - \omega^2)^2 + \omega^2\gamma^2}. \qquad (4.145)$$

The graphical representation of the functions $f_1(\omega)$ and $f_2(\omega)$ (see Fig. 4.14) allows us to draw some important conclusions. For the resonance frequency $\omega = \omega_0$, the function $f_1(\omega)$ vanishes, while $f_2(\omega)$ exhibits a sharp peak. Since the imaginary part of the complex refractive index (or, equivalently, of the wave number) gives the absorption, the medium is *opaque* for the incident wave in the vicinity of ω_0.

It is not difficult to find the extrema of $f_1(\omega)$, by equalizing to zero its derivative with respect to ω. This function admits a maximum for $\omega = \omega_{max} = \omega_0 - \frac{1}{2}\gamma$, and

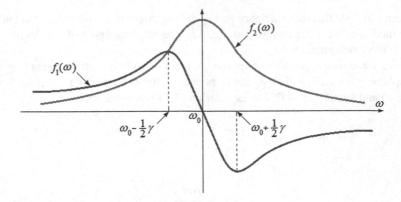

Fig. 4.14 *Normal* and *anomal* dispersion.

a minimum for $\omega = \omega_{min} = \omega_0 + \frac{1}{2}\gamma$. Thus, in the frequency intervals $\omega < \omega_0 - \frac{1}{2}\gamma$ and $\omega > \omega_0 + \frac{1}{2}\gamma$ the real part of the refractive index increases as ω increases, and this is the case of *normal dispersion*. In the frequency interval $\omega_0 - \frac{1}{2}\gamma < \omega < \omega_0 + \frac{1}{2}\gamma$, that is, close to the resonance frequency, the refractive index decreases as ω increases, in which case we have *anomal dispersion*.

We can conclude that a substance has a limited transparency domain for the electromagnetic waves, and the width of this interval depends on the frequency of the incident waves. This conclusion contradicts Maxwell's theory, regarding the transparency of dielectric media for *all* electromagnetic waves, irrespective of their frequencies.

Lorentz's theory, based on the simplistic model of atomic oscillators, turns out to be valid only in some special cases, such as vapors and ideal gases (this is the reason why we considered above $\epsilon_r \simeq 1$ and $\mu_r \simeq 1$). Nevertheless, its fundamental ideas remain valid.

Observation:
Rayleigh's dispersion law (4.127) can also be written as

$$u_g = \frac{1}{\left(\frac{dk}{d\omega}\right)} = \frac{1}{\frac{n}{c} + \frac{\omega}{c}\frac{dn}{d\omega}} = \frac{c}{n + \omega\frac{dn}{d\omega}}.$$

For frequencies ω corresponding to normal dispersion, the derivative $dn/d\omega$ is positive. In case of media with $n > 1$, we then have

$$u_g < u_p < c.$$

In the vicinity of the resonance frequency ω_0, the derivative $dn/d\omega$ is negative (see Fig. 4.14), which might be interpreted as $u_g > c$. However, in this region ω varies very rapidly with respect to **k**, and the approximation (4.121) is no more valid.

Consequently, in the nonlinear regime, the group velocity is no more relevant for such a situation as velocity of propagation of signal.

In its turn, the phase velocity u_p may be greater than c. As an example, let us use the dispersion relation (4.143) for a neutral plasma (e.g. the ionosphere):

$$\omega^2 = \omega_{pl}^2 + \omega^2 n^2 = \omega_{pl}^2 + c^2 k^2$$

and calculate

$$u_g = \frac{d\omega}{dk} = c^2 \frac{k}{\omega} < c, \quad u_p = \frac{\omega}{k} = \sqrt{c^2 + \frac{\omega_{pl}^2}{k^2}} > c,$$

since $ck < \omega$. We also have $u_p u_g = c^2$.

We therefore arrived at a notable result: the phase velocity of an electromagnetic wave *can* exceed the speed of light in vacuum. As we shall show in Part II of the book, this result does not contradict the special theory of relativity.

4.8.3 Kramers–Kronig Dispersion Relations

A more rigorous formulation of dispersion theory was given by Ralph Kronig (1904–1995) and Hans A. Kramers (1894–1952) in the third decade of the 20th century. Nowadays, their results are extended and derived based on the theory of complex functions. The Kramers–Kronig relations connect the real and imaginary parts of any complex function which is analytic in the upper half plane.

Let us start by recalling Cauchy's formula. If $f(z)$ is a holomorphic function on an open set containing the domain $\mathcal{D}$ of the complex plane, bounded by the closed simple curve C, then for any $z \in \mathcal{D}$ is valid Cauchy's formula:

$$f(z) = \frac{1}{2\pi i} \oint_C \frac{f(\zeta)\, d\zeta}{\zeta - z}. \tag{4.146}$$

As the function which interests us is $\tilde{n} = \tilde{n}(\omega)$, and this is analytic in the whole upper half-plane, we shall choose the contour C composed of the axis of real ζ and a half-circle of infinite radius ($\operatorname{Im} f(\zeta) > 0$). If the function $|f(\zeta)|$ tends to zero sufficiently fast at infinity (such a condition is satisfied by the physical quantity represented by the generic function f), the contribution to the integral from the half-circle vanishes, and the Cauchy integral can be written as

$$f(z) = \frac{1}{2\pi i} \int_{-\infty}^{+\infty} \frac{f(\zeta)\, d\zeta}{\zeta - z}, \tag{4.147}$$

where now z is an arbitrary point in the upper half-plane, and the integral is taken on the real axis. Clearly, the integrand in (4.147) has a pole on the real axis, at $\zeta = z$. (If $f(\zeta) \neq 0$, $\lim_{\zeta \to z} \left(\frac{f(\zeta)}{\zeta - z} \right) = \infty$, and the integral (4.147) is called in this case *singular*). For such integrals, which can generally be defined on a rectifiable curve γ, one defines the *Cauchy principal value* by the relation

$$P \int_\gamma \frac{f(\zeta)\, d\zeta}{\zeta - z} = \lim_{\varepsilon \to 0} \int_{\gamma - \gamma_\varepsilon} \frac{f(\zeta)\, d\zeta}{\zeta - z}, \qquad (4.148)$$

when the limit exists and it is finite. In (4.148), γ_ε is the arc of the curve γ situated inside the circle with the centre at z and of radius ε (see Fig. 4.15). In our case, the curve γ is the real axis, and the arc γ_ε can be chosen as a half-circle of radius ε, centred at z, such that the pole is contained inside the integration contour (see (Fig. 4.16)).

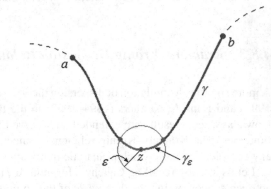

Fig. 4.15 Graphical representation of the arc of curve γ_ε appearing in the definition of the Cauchy principal value.

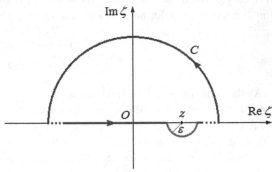

Fig. 4.16 Choice of the integration contour leading to the dispersion relations.

Having in view the above considerations, we can express $f(z)$ as

$$f(z) = \frac{1}{2\pi i} \lim_{\varepsilon \to 0} \left[\int_{-\infty}^{z-\varepsilon} \frac{f(\zeta)\,d\zeta}{\zeta - z} \right.$$
$$\left. + \int_{z+\varepsilon}^{+\infty} \frac{f(\zeta)\,d\zeta}{\zeta - z} \right] + \frac{1}{2\pi i} \lim_{\varepsilon \to 0} \int_{z-\varepsilon}^{z+\varepsilon} \frac{f(\zeta)\,d\zeta}{\zeta - z}. \tag{4.149}$$

According to (4.148), the limit of the square bracket is the Cauchy principal value of the integral at $\zeta = z$,

$$\lim_{\varepsilon \to 0} \left[\int_{-\infty}^{z-\varepsilon} \frac{f(\zeta)\,d\zeta}{\zeta - z} + \int_{z+\varepsilon}^{+\infty} \frac{f(\zeta)\,d\zeta}{\zeta - z} \right] = P \int_{-\infty}^{+\infty} \frac{f(\zeta)\,d\zeta}{\zeta - z}. \tag{4.150}$$

The last integral of (4.149) is easily worked out by using the parameterization $\zeta - z = \varepsilon\, e^{i\theta}$. Then,

$$\lim_{\varepsilon \to 0} \int_{z-\varepsilon}^{z+\varepsilon} \frac{f(\zeta)\,d\zeta}{\zeta - z} = \lim_{\varepsilon \to 0} \int_{\pi}^{2\pi} \frac{f\left(z + \varepsilon e^{i\theta}\right)}{\varepsilon e^{i\theta}} \left(i\,\theta\, \varepsilon\, e^{i\theta}\right) d\theta = \pi\, i f(z). \tag{4.151}$$

Putting together (4.149)–(4.151), we find

$$f(z) = \frac{1}{\pi i} P \int_{-\infty}^{+\infty} \frac{f(\zeta)\,d\zeta}{\zeta - z}. \tag{4.152}$$

Writing $f(\zeta)$ in the complex form as

$$f(\zeta) = \text{Re}(\zeta) + i\,\text{Im}(\zeta),$$

where $\text{Re}(\zeta)$ and $\text{Im}(\zeta)$ are real functions, then (4.152) yields

$$\text{Re}(z) = \frac{1}{\pi} P \int_{-\infty}^{+\infty} \frac{\text{Im}(\zeta)d\zeta}{\zeta - z},$$

$$\text{Im}(z) = -\frac{1}{\pi} P \int_{-\infty}^{+\infty} \frac{\text{Re}(\zeta)d\zeta}{\zeta - z}. \tag{4.153}$$

To apply these relations in the case of dispersion, the function $f(z)$ becomes $\tilde{n}(\omega)$, with

$$\tilde{n}(\omega) = n_1(\omega) + i\, n_2(\omega), \tag{4.154}$$

where $n_1(\omega)$ and $n_2(\omega)$ are defined by (4.145). Then

$$n_1(\omega_0) = 1 + \frac{1}{\pi}P\int_{-\infty}^{+\infty}\frac{n_2(\omega)\,d\omega}{\omega - \omega_0},$$

$$n_2(\omega_0) = -\frac{1}{\pi}P\int_{-\infty}^{+\infty}\frac{[n_1(\omega) - 1]\,d\omega}{\omega - \omega_0}. \tag{4.155}$$

These formulas are called *dispersion relations*, or *Kramers–Kronig relations*. They were derived independently by Ralph Kronig in 1926 and by Hans Kramers in 1927, and they are important not only for electrodynamics, but also in some other branches of physics, for example, in elementary particle physics or solid state physics.

Since in classical electrodynamics negative frequencies have no physical meaning, the dispersion relations (4.155) can be written as integrals only over positive frequencies. This is possible due to the symmetry properties of the real and imaginary parts of the complex refraction index $\tilde{n}$: the real part, $n_1(\omega)$, is an even function of ω, while the imaginary part, $n_2(\omega)$, is an odd function. This follows from the property

$$\tilde{n}(-\omega) = \tilde{n}^*(\omega^*).$$

This relation can be easily deduced from the representation[2] of $\tilde{n}(\omega)$ as

$$\tilde{n}(\omega) = \left(1 + \int_0^\infty G(\tau)\,e^{-i\omega\tau}d\tau\right)^{1/2}.$$

In this expression was used the kernel[3] $G(\tau)$ of the integral giving the most general linear, spatially local, and causal relation between **D** and **E** in a uniform isotropic medium,

$$\mathbf{D}(\mathbf{r}, t) = \epsilon_0\mathbf{E}(\mathbf{r}, t) + \epsilon_0\int_0^\infty G(\tau)\mathbf{E}(\mathbf{r}, t - \tau)d\tau.$$

One uses as well the fact that **E** and **D** (and implicitly $G(\tau)$) are real, measurable quantities.

Returning to the refraction index and using the above considerations, let us show that we can write the relations (4.155) in the form

---

[2]This expression is valid assuming that the relative permeability μ_r of the considered medium is very close to unity, $\mu_r \simeq 1$.

[3]In fact, the kernel $G(\tau)$ is the Fourier transform of the electric susceptibility of the medium, $\chi_e = \frac{\epsilon(\omega)}{\epsilon_0} - 1$:

$$G(\tau) = \frac{1}{2\pi}\int_{-\infty}^{+\infty}\chi_e(\omega)\,e^{-i\omega\tau}d\omega.$$

$$n_1(\omega_0) = 1 + \frac{2}{\pi} P \int_0^\infty \frac{\omega\, n_2(\omega)\, d\omega}{\omega^2 - \omega_0^2},$$

$$n_2(\omega_0) = -\frac{2\omega}{\pi} P \int_0^\infty \frac{[n_1(\omega) - 1]\, d\omega}{\omega^2 - \omega_0^2}. \tag{4.156}$$

We shall prove only the first relation (4.156), since the second can be derived analogously. Taking into account that $n_2(-\omega) = -n_2(\omega)$, from $(4.155)_1$ we have

$$n_1(\omega_0) = 1 + \frac{1}{\pi} P \int_{-\infty}^{+\infty} \frac{n_2(\omega)\, d\omega}{\omega - \omega_0}$$

$$= 1 + \frac{1}{\pi} P \left[\int_0^\infty \frac{n_2(\omega)\, d\omega}{\omega - \omega_0} + \int_{-\infty}^0 \frac{n_2(\omega)\, d\omega}{\omega - \omega_0} \right]$$

$$= 1 + \frac{1}{\pi} P \left[\int_0^\infty \frac{n_2(\omega)\, d\omega}{\omega - \omega_0} + \int_0^\infty \frac{n_2(-\omega)\, d\omega}{-\omega - \omega_0} \right]$$

$$= 1 + \frac{1}{\pi} P \int_0^\infty n_2(\omega) \left(\frac{1}{\omega - \omega_0} + \frac{1}{\omega + \omega_0} \right) d\omega$$

$$= 1 + \frac{2}{\pi} P \int_0^\infty \frac{\omega\, n_2(\omega)\, d\omega}{\omega^2 - \omega_0^2}.$$

Usually, the imaginary part of the refraction index, $n_2(\omega)$, is determined experimentally from the absorption spectra, and the real part is subsequently derived from $(4.156)_1$. Moreover, the dispersion relations (4.155) or (4.156) contain as well the connection between the absorption and anomalous dispersion presented in Fig. 4.14.

The Kramers–Kronig relations show that the dispersion of electromagnetic waves must be accompanied by absorption. Indeed, if there is no absorption ($n_2 = 0$), (4.155) gives $n_1 = n = 1$, the phase velocity is

$$u_p = \frac{\omega}{k} = \frac{c}{n} = c,$$

and we have *no dispersion*.

4.8.4 Dispersion in Crystals

Let us resume the analysis of the electromagnetic wave propagation in anisotropic media, briefly discussed in Sect. 4.7, in order to emphasize some important characteristics of dispersion in such media. Limiting ourselves to light waves, let us write the phase of a plane wave as

$$\varphi(\mathbf{r}, t) = -\omega t + \mathbf{k} \cdot \mathbf{r} = -\omega t + k\, \mathbf{s} \cdot \mathbf{r} = -\omega t + \frac{\omega}{c} \psi(x, y, z).$$

The function ψ is called *eikonal* (from the Greek *eikon*, meaning image), being a common concept in ray optics and particle mechanics. The gradient of the eikonal is

$$\nabla\psi = \mathbf{n}, \quad \text{where} \quad \mathbf{k} = \frac{\omega}{c}\mathbf{n}. \qquad (4.157)$$

The magnitude of $\mathbf{n}$ is the refractive index of the (transparent) medium. In an isotropic medium $n = |\mathbf{n}|$ depends on frequency, but in anisotropic media like crystals n depends on both frequency and direction.

To establish the equation that explains the "shape" of light rays, we start from *Fermat's principle*: along the ray trajectory between two points A and B, the integral

$$\psi = \int_A^B \mathbf{n} \cdot d\mathbf{s} = \int_A^B n\, ds$$

is minimum. This means that the variation of ψ vanishes:

$$\delta\psi = \int_A^B [\delta n(ds) + n\,\delta(ds)] = 0. \qquad (4.158)$$

Let $\boldsymbol{\tau} = d\mathbf{r}/ds$ be the unit vector of the tangent to the ray. Then we can write

$$\delta n = \delta\mathbf{r} \cdot \nabla n, \quad \delta(ds) = \boldsymbol{\tau} \cdot d(\delta\mathbf{r}).$$

Using these relations and integrating by parts (4.158), one obtains

$$\delta\psi = \int_A^B \left[\left(\nabla n - \frac{d(n\boldsymbol{\tau})}{ds} \right) \cdot \delta\mathbf{r} \right] ds = 0,$$

which leads to

$$\frac{d(n\boldsymbol{\tau})}{ds} = \nabla n.$$

As $dn/ds = \boldsymbol{\tau} \cdot \nabla n$, we have

$$\frac{d\boldsymbol{\tau}}{ds} = \frac{1}{n}[\nabla n - \boldsymbol{\tau}(\boldsymbol{\tau} \cdot \nabla n)]. \qquad (4.159)$$

An interesting conclusion is obtained by using the Serret–Frenet frame. Here, in addition to $\boldsymbol{\tau}$, one introduces the unit vector of the principal normal, $\boldsymbol{\nu}$, orthogonal to $\boldsymbol{\tau}$ and defined by

$$\frac{d\boldsymbol{\tau}}{ds} = \frac{1}{\rho}\boldsymbol{\nu},$$

where ρ is the radius of curvature, and $1/\rho$ is the curvature of the ray at the chosen point. So we have

$$\frac{1}{\rho} = \nu \cdot \frac{\nabla n}{n}, \qquad (4.160)$$

saying that the curvature of the ray depends on the refractive index: *the curvature increases with n.*

The propagation velocity of the light ray is the group velocity $\mathbf{u}_g = d\omega/d\mathbf{k}$, its orientation being given by the unit vector τ.

Let us now determine the dispersion law of electromagnetic radiation in anisotropic media. We start from Eq. (4.107):

$$\mathbf{D} = \epsilon \, [\mathbf{E} - (\mathbf{s} \cdot \mathbf{E})\mathbf{s}], \quad D_i = \epsilon \, [E_i - (\mathbf{s} \cdot \mathbf{E})s_i] = \epsilon_{ik} E_k.$$

Since $\epsilon/\epsilon_0 = \epsilon_r \sim n^2$ and $\epsilon_{ik} = \epsilon_0(\epsilon_r)_{ik}$, we have

$$n^2 \, [E_i - (\mathbf{s} \cdot \mathbf{E})s_i] = (\epsilon_r)_{ik} E_k.$$

But $n\mathbf{s} = \mathbf{n}$, hence

$$n^2 E_i - (n_k E_k) \, n_i = (\epsilon_r)_{ik} E_k,$$

or

$$\left[n^2 \delta_{ik} - n_i n_k - (\epsilon_r)_{ik} \right] E_k = 0. \qquad (4.161)$$

The system of three equations (4.161) has nontrivial solutions if the determinant of the matrix of coefficients vanishes:

$$\left| n^2 \delta_{ik} - n_i n_k - (\epsilon_r)_{ik} \right| = 0. \qquad (4.162)$$

If x, y, and z are the principal directions of the tensor $(\epsilon_r)_{ik}$, and $(\epsilon_r)_x$, $(\epsilon_r)_y$, and $(\epsilon_r)_z$ are the principal relative permittivities, then expanding the determinant (4.162) we arrive at the following equation:

$$
\begin{aligned}
n^2 & \left[(\epsilon_r)_x n_x^2 + (\epsilon_r)_y n_y^2 + (\epsilon_r)_z n_z^2 \right] - \left\{ n_x^2 (\epsilon_r)_x [(\epsilon_r)_y + (\epsilon_r)_z] \right. \\
& + n_y^2 (\epsilon_r)_y [(\epsilon_r)_x + (\epsilon_r)_z] + n_z^2 (\epsilon_r)_z [(\epsilon_r)_x + (\epsilon_r)_y] \} \\
& + (\epsilon_r)_x (\epsilon_r)_y (\epsilon_r)_z = 0.
\end{aligned}
\qquad (4.163)
$$

This equation gives the implicit form of the dispersion law: the dependence of the wave vector on frequency. For a given direction of $\mathbf{n}$ and also of $\mathbf{k}$, it is a quadratic equation in n^2 with real coefficients. This means that to each direction $\mathbf{n}$ correspond, in general, two distinct absolute values of the wave vector.

Assuming $(\epsilon_r)_i$, $i = 1, 2, 3$ to be constants, and taking n_i, $i = 1, 2, 3$ as coordinates, Eq. (4.163) represents a surface, called the *wave vector surface*, or the *surface of normals.*

Remark that Eq. (4.163) is nothing else but a different form of Fresnel's equation encountered in Sect. 4.7, emphasizing the phenomenon of dispersion.

The direction of light rays is given by the group velocity vector $\mathbf{u}_g = d\omega/d\mathbf{k}$. In isotropic media, its direction coincides with that of $\mathbf{k}$; in anisotropic media, their directions do not coincide. Let $\boldsymbol{\sigma}$ be a vector oriented in the $\mathbf{u}_g$ direction, called the *ray vector*, and defined such that

$$\mathbf{n} \cdot \boldsymbol{\sigma} = 1. \tag{4.164}$$

Consider the divergent beam of monochromatic light emitted by a point source. The eikonal is

$$\psi = \int \mathbf{n} \cdot d\mathbf{s} = \int \frac{\mathbf{n} \cdot \boldsymbol{\sigma}}{\sigma} ds = \int \frac{ds}{\sigma}.$$

In a homogeneous medium, σ is constant along the ray, that is $\psi = L/\sigma$, where L is the length of the ray segment. If on each ray of the beam one takes a segment proportional to σ, one obtains a surface, whose points are all in phase. This is the *ray surface*.

Equation (4.163) can be written in a condensed form as $f(k_x, k_y, k_z) = 0$. We may write

$$\frac{\partial \omega}{\partial k_i} = \frac{\partial f}{\partial k_i} \Big/ \frac{\partial f}{\partial \omega},$$

meaning that the group velocity components are proportional to the derivatives $\partial f/\partial k_i$, therefore to $\partial f/\partial n_i$. Since the vector $df/d\mathbf{n}$ is oriented along the normal to the surface $f = 0$, we conclude that the direction of $\boldsymbol{\sigma}$, at any point, is given by the normal to the wave vector surface at that point.

4.9 Propagation of Electromagnetic Waves in Waveguides

A *waveguide* is a structure used for the efficient transfer of electromagnetic power. Waveguides differ in their geometry, their names coming from the shape of the cross section: circular, rectangular, elliptic, etc. Waveguides can be constructed to carry waves over a wide portion of the electromagnetic spectrum, but are especially useful in the microwave and optical frequency ranges. Depending on the frequency, they can be constructed from either conductive or dielectric materials. The first waveguide was imagined by J.J. Thomson in 1893 and experimentally verified by O.J. Lodge (1851–1940) in 1894; the mathematical analysis of the propagating modes within a hollow metal cylinder was first performed by Rayleigh in 1897.

Waveguides used at optical frequencies are typically dielectric waveguides, structures in which a dielectric material with high permittivity, and thus high index of refraction, is surrounded by a material with lower permittivity. These devices guide optical waves by total internal reflection, as in *optical fibers*.

Waveguides are widely utilized in electronics, nuclear physics, safety engineering, etc.

There are several methods of theoretical approach to electromagnetic waves transmission through waveguides. In the following analysis, we shall use the simplest procedure, by a direct application of Maxwell's equation. At the end of the section we shall briefly mention an alternative method.

4.9.1 Rectangular Waveguides

Suppose that the cross section of the waveguide is a rectangle, with length a and width b. The internal walls of the pipe are considered to possess a very high conductivity ($\lambda \to \infty$), and the dielectric is supposed to be air ($\epsilon \sim \epsilon_0, \mu \sim \mu_0$).

The equations describing the electromagnetic field inside the waveguide are Maxwell's equations, supplemented with jump conditions and the constitutive equations. We write Maxwell's equations in the form:

$$\nabla \times \mathbf{H} = \epsilon_0 \frac{\partial \mathbf{E}}{\partial t},$$

$$\nabla \cdot \mathbf{B} = 0,$$

$$\nabla \times \mathbf{E} = -\mu_0 \frac{\partial \mathbf{H}}{\partial t},$$

$$\nabla \cdot \mathbf{E} = 0,$$

then take the curl of the equations of evolution and find

$$\epsilon_0 \mu_0 \frac{\partial^2 \mathbf{E}}{\partial t^2} - \Delta \mathbf{E} = 0,$$

$$\epsilon_0 \mu_0 \frac{\partial^2 \mathbf{H}}{\partial t^2} - \Delta \mathbf{H} = 0. \tag{4.165}$$

The shape of the cross section prompts us to use Cartesian coordinates. If the z-axis is oriented along the pipe axis, the solutions of the equations (4.165) are

$$\mathbf{E}(x, y, z) = \mathbf{E}(x, y)e^{i(\omega t - k_z z)},$$

$$\mathbf{H}(x, y, z) = \mathbf{H}(x, y)e^{i(\omega t - k_z z)}. \tag{4.166}$$

Observing that $\partial/\partial t \to i\omega$ and $\Delta_z \to -k_z^2$, Eqs. (4.165) yield

$$\left(\Delta_2 + \alpha^2\right) \mathbf{E}(x, y) = 0,$$

$$\left(\Delta_2 + \alpha^2\right) \mathbf{H}(x, y) = 0, \tag{4.167}$$

where

$$\alpha^2 = \frac{\omega^2}{c^2} - k_z^2 = k^2 - k_z^2, \tag{4.168}$$

and

$$\Delta_2 = \frac{\partial^2}{\partial x^2} + \frac{\partial^2}{\partial y^2}$$

is the *two-dimensional Laplacian*.

The solutions of the equations of propagation (4.167) have to obey the *boundary conditions* (see (3.43)):

$$D_{2n} - D_{1n} = \sigma,$$
$$E_{2T} - E_{1T} = 0,$$
$$B_{2n} - B_{1n} = 0,$$
$$H_{2T} - H_{1T} = i_N.$$

Choosing the *xz*-plane as being one of the internal metallic walls, and the unit vectors **N**, **T**, and **n** as shown in Fig. 4.17, the second and the third jump conditions give

$$(\mathbf{E} - \mathbf{E}_c) \cdot \mathbf{n} = 0,$$
$$(\mathbf{B} - \mathbf{B}_c) \cdot \mathbf{n} = 0, \tag{4.169}$$

where $\mathbf{E}_c$ and $\mathbf{B}_c$ are the field vectors on the internal walls of the waveguide.

Since the wall is highly conductive, according to Ohm's law, $\mathbf{j} = \lambda\mathbf{E}$, for $\lambda \to \infty$ and $\mathbf{j}$ finite, we must have $\mathbf{E}_c = 0$, which implies $\mathbf{B}_c = 0$. Then we are left with

$$E_T|_S = 0 \to E_z|_S = 0, \quad \text{or} \quad E_x|_S = 0,$$
$$H_n|_S = 0, \quad \text{or} \quad \frac{\partial H_z}{\partial n}\bigg|_S = 0. \tag{4.170}$$

Fig. 4.17 Geometry of a rectangular waveguide with the axis along the *z*-direction.

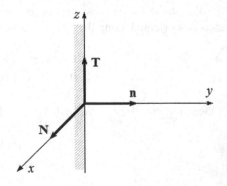

The last relation has been obtained by taking the x-component of Maxwell's equation $\nabla \times \mathbf{H} = \epsilon_0 \frac{\partial \mathbf{E}}{\partial t}$, and using it at the waveguide wall.

Combining the general solutions with the boundary conditions, we can have different types (modes) of waves, both longitudinal and transverse. The transverse waves, which is our case, are classified into: TM (Transverse Magnetic), TE (Transverse Electric), TEM (Transverse ElectroMagnetic), and hybrid modes (which have both electric and magnetic field components in the direction of propagation). In the following we shall discuss the first two modes:

(a) TM (or E) modes have no magnetic field in the direction of propagation: $H_z = 0$ everywhere, with the boundary condition $E_z|_S = 0$;

(b) TE (or M) modes have no electric field in the direction of propagation: $E_z = 0$ everywhere, with the boundary condition $\left. \frac{\partial H_z}{\partial n} \right|_S = 0$.

4.9.1.1 Transverse Magnetic (TM) Modes

We start by projecting Maxwell's evolution equations on x and y directions:

$$\frac{\partial E_z}{\partial y} - \frac{\partial E_y}{\partial z} = -i\omega\mu_0 H_x,$$

$$\frac{\partial E_x}{\partial z} - \frac{\partial E_z}{\partial x} = -i\omega\mu_0 H_y,$$

$$\frac{\partial H_z}{\partial y} - \frac{\partial H_y}{\partial z} = i\omega\epsilon_0 E_x,$$

$$\frac{\partial H_x}{\partial z} - \frac{\partial H_z}{\partial x} = i\omega\epsilon_0 E_y,$$

or, since $\partial/\partial z \rightarrow -ik_z$, $H_z = 0$,

$$\frac{\partial E_z}{\partial y} + ik_z E_y = -i\omega\mu_0 H_x,$$

$$ik_z E_x + \frac{\partial E_z}{\partial x} = i\omega\mu_0 H_y,$$

$$k_z H_y = \epsilon_0 \omega E_x,$$

$$-k_z H_x = \epsilon_0 \omega E_y.$$

One observes that the field components can be expressed in terms of partial derivatives of E_z. Using (4.168), one obtains

$$E_x = -\frac{ik_z}{\alpha^2}\frac{\partial E_z}{\partial x},$$

$$E_y = -\frac{ik_z}{\alpha^2}\frac{\partial E_z}{\partial y},$$

$$H_x = \frac{i\epsilon_0\omega}{\alpha^2}\frac{\partial E_z}{\partial y}, \qquad\qquad (4.171)$$

$$H_y = -\frac{i\epsilon_0\omega}{\alpha^2}\frac{\partial E_z}{\partial x},$$

$$H_z = 0.$$

The field component E_z is determined as solution of the z-projection of $(4.167)_1$:

$$\left(\Delta_2 + \alpha^2\right) E_z(x, y) = 0.$$

To separate the variables, we take

$$E_z(x, y) = X(x)\, Y(y), \qquad\qquad (4.172)$$

which leads to

$$X'' + \alpha_1^2 X = 0,$$
$$Y'' + \alpha_2^2 Y = 0, \qquad\qquad (4.173)$$

where

$$\alpha_1^2 + \alpha_2^2 = \alpha^2. \qquad\qquad (4.174)$$

The solutions of (4.173) are of the form

$$X(x) = A\,\sin\alpha_1 x + B\,\cos\alpha_1 x, \quad Y(y) = C\,\sin\alpha_2 y + D\,\cos\alpha_2 y.$$

The constants are determined by using the boundary condition $E_z|_S = 0$, which means $E_z = 0$ for $x = 0$, $x = a$, $y = 0$, and $y = b$. We find $B = D = 0$, as well as

$$\alpha_1 = \frac{n_1\pi}{a}, \qquad \alpha_2 = \frac{n_2\pi}{b}. \qquad\qquad (4.175)$$

Thus,

$$E_z(x, y) = E_0 \sin\frac{n_1\pi}{a}x \sin\frac{n_2\pi}{b}y,$$

and the complete solution is

$$E_z(x, y, z, t) = E_0 \sin\frac{n_1\pi}{a}x \sin\frac{n_2\pi}{b}y e^{i(\omega t - k_z z)}. \qquad\qquad (4.176)$$

As the last step, we use (4.171) to obtain all the other field components in the TM wave:

$$E_x = -\frac{ik_z}{\alpha^2} E_0 \frac{n_1\pi}{a} \cos\frac{n_1\pi}{a}x \sin\frac{n_2\pi}{b}y e^{i(\omega t - k_z z)},$$

$$E_y = -\frac{ik_z}{\alpha^2} E_0 \frac{n_2\pi}{b} \sin\frac{n_1\pi}{a}x \cos\frac{n_2\pi}{b}y e^{i(\omega t - k_z z)},$$

$$E_z = E_0 \sin\frac{n_1\pi}{a}x \sin\frac{n_2\pi}{b}y \, e^{i(\omega t - k_z z)}, \qquad (4.177)$$

$$H_x = \frac{i\omega\epsilon_0}{\alpha^2} E_0 \frac{n_2\pi}{b} \sin\frac{n_1\pi}{a}x \cos\frac{n_2\pi}{b}y e^{i(\omega t - k_z z)},$$

$$H_y = -\frac{i\omega\epsilon_0}{\alpha^2} E_0 \frac{n_1\pi}{a} \cos\frac{n_1\pi}{a}x \sin\frac{n_2\pi}{b}y e^{i(\omega t - k_z z)},$$

$$H_z = 0.$$

4.9.1.2 Transverse Electric (TE) Modes

In this case the solution is found by using the same procedure, except for the fact that the field components are expressed in terms of partial derivatives of H_z, while the boundary conditions are of the form $\frac{\partial H_z}{\partial n}\big|_S = 0$, leading to

$$\frac{dX}{dx} = 0, \text{ for } x = 0 \text{ and } x = a,$$

$$\frac{dY}{dy} = 0, \text{ for } y = 0 \text{ and } y = b.$$

We leave to the reader the derivation of the field components for the TE modes, which will read as follows:

$$E_x = \frac{i\omega\mu_0}{\alpha^2} H_0 \frac{n_2\pi}{b} \cos\frac{n_1\pi}{a}x \sin\frac{n_2\pi}{b}y \, e^{i(\omega t - k_z z)},$$

$$E_y = -\frac{i\omega\mu_0}{\alpha^2} H_0 \frac{n_1\pi}{a} \sin\frac{n_1\pi}{a}x \cos\frac{n_2\pi}{b}y \, e^{i(\omega t - k_z z)},$$

$$E_z = 0, \qquad (4.178)$$

$$H_x = \frac{ik_z}{\alpha^2} H_0 \frac{n_1\pi}{a} \sin\frac{n_1\pi}{a}x \cos\frac{n_2\pi}{b}y \, e^{i(\omega t - k_z z)},$$

$$H_y = \frac{ik_z}{\alpha^2} H_0 \frac{n_2\pi}{b} \cos\frac{n_1\pi}{a}x \sin\frac{n_2\pi}{b}y \, e^{i(\omega t - k_z z)},$$

$$H_z = H_0 \cos\frac{n_1\pi}{a}x \cos\frac{n_2\pi}{b}y \, e^{i(\omega t - k_z z)}.$$

Discussion:

(a) With $\alpha = \omega_c/c$, (4.168) yields

$$k_z^2 = \frac{\omega^2}{c^2} - \frac{\omega_c^2}{c^2}, \qquad k_z = \frac{1}{c}\sqrt{\omega^2 - \omega_c^2}. \tag{4.179}$$

The wave propagates without absorption (damping) only if $k_z^2 > 0$, which implies $\omega \geq \omega_c = \omega_{min}$. Therefore, the wave must have a minimum frequency, called *critical*, or *cut-off frequency*, in order to propagate through the waveguide:

$$\omega_c = \omega_{min} = 2\pi\nu_{min} = 2\pi\,\frac{c}{\lambda_{max}}. \tag{4.180}$$

If $\omega < \omega_c$, the wave is attenuated by absorption and does not propagate. Thus, the waveguide plays the role of a *frequency filter*.

(b) Combining (4.168) with (4.164) it follows that

$$\alpha^2 = \alpha_1^2 + \alpha_2^2 = \pi^2 \left(\frac{n_1^2}{a^2} + \frac{n_2^2}{b^2} \right) = \frac{\omega_c^2}{c^2},$$

which gives

$$\lambda_c = \lambda_{max} = \frac{2\pi c}{\omega_c} = \frac{2}{\sqrt{\frac{n_1^2}{a^2} + \frac{n_2^2}{b^2}}}. \tag{4.181}$$

The cut-off wavelength is usually denoted by two indices, corresponding to the values of n_1 and n_2 that make $\lambda = \lambda_{max}$. In the case of TM modes, the smallest values of n_1 and n_2 are $n_1 = n_2 = 1$ (the value zero is not possible, since it would annul the wave). Thus,

$$\lambda_{max} = \lambda_{11} = \frac{2}{\sqrt{\frac{1}{a^2} + \frac{1}{b^2}}}. \tag{4.182}$$

If $a = b$, this means $\lambda_{11} = a\sqrt{2}$, being of the order of the side of the rectangular cross section.

The cut-off frequency for the TE modes is found by analyzing the solution (4.178), in particular H_z'. Unlike the previous case, the value zero for n_1 and n_2 is possible, and we find

$$\lambda_{10} = 2a, \qquad \lambda_{01} = 2b. \tag{4.183}$$

We note that the propagation of electromagnetic waves through waveguides is strongly connected to the technique of production of high frequency waves ($\sim$ GHz). Such waves are called *microwaves*.

(c) The phase and group velocities of the wave propagating through the waveguide are

$$u_p = \frac{\omega}{k_z} = \frac{k}{k_z} c > c,$$

$$u_g = \frac{d\omega}{dk_z} = \frac{ck_z}{\sqrt{\alpha^2 + k_z^2}} = \frac{c^2 k_z}{\omega} = \frac{k_z}{k} c < c.$$

Obviously $u_p u_g = c^2$, as expected.

(d) An important quantity to be determined is the *energy carried by the wave* in the waveguide. We choose a TM-mode and use the complex Poynting's vector $\mathbf{\Pi}$, defined in Sect. 4.4.2. From (4.53), the average of the electromagnetic flux density for one period is

$$\langle \Phi_{em} \rangle = \frac{1}{2} \left[\mathrm{Re}(\mathbf{E} \times \mathbf{H}^*) \right] \cdot \mathbf{n}$$

$$= \frac{1}{2} \mathrm{Re}(\mathbf{E} \times \mathbf{H}^*)_z = \frac{1}{2} \mathrm{Re} \left(E_x H_y^* - E_y H_x^* \right),$$

and, by means of (4.171),

$$\langle \Phi_{em} \rangle = \frac{1}{2} \frac{k_z \epsilon_0 \omega}{\alpha^4} \left(\frac{\partial E_z}{\partial x} \frac{\partial E_z^*}{\partial x} + \frac{\partial E_z}{\partial y} \frac{\partial E_z^*}{\partial y} \right)$$

$$= \frac{1}{2} \frac{k_z \epsilon_0 \omega}{\alpha^4} (\nabla_2 E_z) \cdot (\nabla_2 E_z^*),$$

where ∇_2 is the "nabla" operator in the xOy-plane. The electromagnetic energy flux through the cross section S of the guide is

$$\Phi_{em} = \int_S \langle \Phi_{em} \rangle dS = \frac{1}{2} \frac{k_z \epsilon_0 \omega}{\alpha^4} \int_S (\nabla_2 E_z) \cdot (\nabla_2 E_z^*) \, dS.$$

Integrating by parts, we find

$$\int_S (\nabla_2 E_z) \cdot (\nabla_2 E_z^*) dS = \oint_C E_z^* \frac{\partial E_z}{\partial n} dl - \int_S E_z^* \Delta_2 E_z dS,$$

where C is the contour of the cross section. Since $E_z|_S = 0$ and $\Delta_2 E_z = -\alpha^2 E_z$, the contour integral vanishes and we are left with

$$\Phi_{em} = \frac{1}{2} \frac{k_z \epsilon_0 \omega}{\alpha^2} \int |E_z|^2 dS.$$

The energy per unit length of the waveguide is

$$\frac{W_{em}}{l} = \Phi_{em} \frac{t}{l} = \frac{\Phi_{em}}{u_z} = \frac{1}{2} \frac{k_z \epsilon_0 \omega}{\alpha^2 u_z} \int_S |E_z|^2 dS.$$

In vacuum, for the cut-off frequency, we find

$$\frac{W_{em}}{l} = \frac{1}{2}\epsilon_0 \int_S |E_z|^2 dS.$$

A similar result is obtained for the TE-mode:

$$\frac{W_{em}}{l} = \frac{1}{2}\mu_0 \int_S |H_z|^2 dS.$$

4.9.2 Circular Waveguides

The natural choice in this case is to work in cylindrical coordinates ρ, φ, z. Choosing again the z-axis oriented along the waveguide axis, the solutions of Eq. (4.165), written in cylindrical coordinates, are

$$\mathbf{E}(\rho, \varphi, z) = \mathbf{E}(\rho, \varphi)e^{i(\omega t - k_z z)},$$
$$\mathbf{H}(\rho, \varphi, z) = \mathbf{H}(\rho, \varphi)e^{i(\omega t - k_z z)}. \qquad (4.184)$$

Introducing (4.184) into (4.165), we find again (4.167), where the notation (4.168) remains valid, while the two-dimensional Laplacian Δ_2 this time is expressed in cylindrical coordinates:

$$\Delta_2 = \frac{\partial^2}{\partial \rho^2} + \frac{1}{\rho}\frac{\partial}{\partial \rho} + \frac{1}{\rho^2}\frac{\partial^2}{\partial \varphi^2}.$$

Using the method of separation of variables, we take as solution for the TM wave

$$E_z(\rho, \varphi) = F(\rho)\, \Phi(\varphi),$$

which leads to

$$\rho^2 \frac{F''}{F} + \rho\frac{F'}{F} + \alpha^2 \rho^2 = -\frac{\Phi''}{\Phi} \equiv p,$$

where p is a constant. The solution we are looking for must be periodical, with period 2π. Therefore, the solution of the equation

$$\Phi'' + p\,\Phi = 0,$$

which is $\Phi \sim e^{\pm i\sqrt{p}\,\varphi}$, requires $\sqrt{p} = m$ to be integer. We then have

$$F'' + \frac{1}{\rho}F' + \left(\alpha^2 - \frac{m^2}{\rho^2}\right)F = 0,$$

or, by a suitable change to a new variable $\rho_1 = \alpha\rho$:

$$\frac{d^2F}{d\rho_1^2} + \frac{1}{\rho_1}\frac{dF}{d\rho_1} + \left(1 - \frac{m^2}{\rho_1^2}\right) = 0. \tag{4.185}$$

As we know (see (1.141)), this is a Bessel-type equation. The only solution of this equation which satisfies the physical conditions is the Bessel function of integer order m, $J_m(\rho_1) = J_m(\alpha\rho)$.

Thus, the field components for the TM modes are

$$E_\rho = -ik_z\alpha J'_m(\alpha\rho) \frac{\sin}{\cos}(m\varphi)e^{i(\omega t - k_z z)},$$

$$E_\varphi = -ik_z\frac{m}{\rho}J_m(\alpha\rho) \frac{\cos}{-\sin}(m\varphi)e^{i(\omega t - k_z z)},$$

$$E_z = \alpha^2 J_m(\alpha\rho) \frac{\sin}{\cos}(m\varphi)e^{i(\omega t - k_z z)}, \tag{4.186}$$

$$H_\rho = \frac{ik^2}{\mu\omega\rho} m J_m(\alpha\rho) \frac{\cos}{-\sin}(m\varphi)e^{i(\omega t - k_z z)},$$

$$H_\varphi = -\frac{ik^2}{\mu\omega} \alpha J'_m(\alpha\rho) \frac{\sin}{\cos}(m\varphi)e^{i(\omega t - k_z z)},$$

$$H_z = 0.$$

Using the same procedure, one finds the field components for the TE modes:

$$E_\rho = \frac{i\omega\mu}{\rho} m J_m(\alpha\rho) \frac{\cos}{-\sin}(m\varphi)e^{i(\omega t - k_z z)},$$

$$E_\varphi = -i\omega\mu\alpha J'_m(\alpha\rho) \frac{\sin}{\cos}(m\varphi)e^{i(\omega t - k_z z)},$$

$$E_z = 0,$$

$$H_\rho = ik_z\alpha J'_m(\alpha\rho) \frac{\sin}{\cos}(m\varphi)e^{i(\omega t - k_z z)}, \tag{4.187}$$

$$H_\varphi = \frac{ik_z m}{\rho} J_m(\alpha\rho) \frac{\cos}{-\sin}(m\varphi)e^{i(\omega t - k_z z)},$$

$$H_z = -\alpha^2 J_m(\alpha\rho) \frac{\sin}{\cos}(m\varphi)e^{i(\omega t - k_z z)},$$

where $J'_m(\alpha\rho) = J'_m(\rho_1) = dJ_m/d\rho_1$.

Since $\mathbf{E}$ is oriented along the normal to the waveguide wall (ρ-direction), this implies for the tangent components (see Fig. 4.18):

$$E_z|_{\rho=R} = E_\varphi|_{\rho=R} = 0,$$

Fig. 4.18 Geometry of a
circular waveguide.

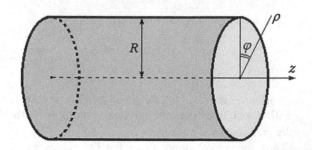

which leads to the equations

$$J_m(\alpha R) = 0, \quad \text{for TM modes,}$$
$$J'_m(\alpha R) = 0, \quad \text{for TE modes.} \qquad (4.188)$$

Let us denote by $\mu_1^{(m)}, \mu_2^{(m)}, \ldots, \mu_i^{(m)}, \ldots$ the zeros of the J_m function in $(4.188)_1$, and by $\nu_1^{(m)}, \nu_2^{(m)}, \ldots, \nu_i^{(m)}, \ldots$ the zeros of the J'_m function in $(4.188)_2$. We find

$$\alpha_{mi} \equiv \alpha_i = \frac{\mu_i^{(m)}}{R} \quad \text{for TM modes,}$$

$$\alpha'_{mi} \equiv \alpha'_i = \frac{\nu_i^{(m)}}{R} \quad \text{for TE modes.} \qquad (4.189)$$

Thus, we have obtained a double series of values α, each being characterized by two indices, for which the propagation in the waveguide is possible. For a given k, that is for a given frequency, there exists the following double series with possible values for k_z:

$$k_z^2 = k^2 - \alpha_{mi}^2 \text{ (TM mode), } \quad \text{or } k_z^2 = k^2 - \alpha'^2_{mi} \text{ (TE mode).} \qquad (4.190)$$

To each value of k_z corresponds a wavelength and a phase velocity of the wave given by

$$\lambda = \frac{2\pi}{k_z}, \quad u_p = \frac{\omega}{k_z}.$$

According to the dispersion relation in vacuum $\omega = c\,k$, it then follows that

$$u_p = \frac{k}{k_z} c > c. \qquad (4.191)$$

To calculate the group velocity, we give up the indices of α and obtain

$$u_g = \frac{d\omega}{dk_z} = \frac{c\,k_z}{\sqrt{k_z^2 + \alpha^2}} = \frac{c^2 k_z}{\omega} = \frac{k_z}{k} c < c,$$

which leads to the same result as for rectangular waveguides,

$$u_p \, u_g = c^2.$$

In order to propagate *without damping*, the wave must satisfy

$$k_z^2 = \frac{1}{c^2}(\omega^2 - \omega_c^2) > 0,$$

where ω_c is the minimum (cut-off) frequency, expressing the border between transmission and absorption. We may write

$$\omega_c = 2\pi \nu_{min} = c \, \alpha_{min} = c \, \frac{\mu_1^{(0)}}{R},$$

where $\mu_1^{(0)} = 2.4$ is the zeroth-order root of Eq. (4.188)$_1$. Thus,

$$\nu_{min} = \frac{2.4\,c}{2\pi R}, \qquad \lambda_{max} = \frac{c}{\nu_{min}} = \frac{2\pi R}{2.4}. \tag{4.192}$$

For example, if $R = 10$ cm, an approximative calculation gives

$$\nu \geq \nu_{min} \simeq 10^9 \, \text{Hz}, \qquad \lambda \leq \lambda_{max} \simeq 30 \, \text{cm}.$$

This frequency (wavelength) corresponds to *microwaves*. These are electromagnetic waves with wavelengths ranging from as long as one meter down to as short as one millimeter, or equivalently, with frequencies between 300 MHz (0.3 GHz) and 300 GHz. This is a broad definition including both ultra-high frequencies (UHF), 0.3–3 GHz and extremely high frequencies (EHF) (millimeter waves), 30–300 GHz. The technical problems of production and transmission are beyond the scope of this book.

4.9.3 Borgnis' Method

An alternative approach to the theoretical study of electromagnetic waves propagation through waveguides was offered by Fritz Edward Borgnis. Below we shall sketch the idea of the method, and the intermediate calculations are left to the reader.

Borgnis' method starts with the electromagnetic field equations (4.2), written in curvilinear orthogonal coordinates q_1, q_2, q_3 (see Appendix D), with the choice

$$h_1 = 1, \qquad \frac{h_2}{h_3} \text{ independent of } q_1.$$

We remark that these two conditions are fulfilled in the cases treated in Sects. 4.9.1 and 4.9.2. Supposing the field vectors to be given by

$$\mathbf{E} = \mathbf{E}_0 \, e^{i\omega t},$$
$$\mathbf{H} = \mathbf{H}_0 \, e^{i\omega t}, \tag{4.193}$$

Borgnis showed that Maxwell's equations (in number of eight when written in components) can be reduced to a single equation, where a single function $U(q_1, q_2, q_3)$, called *Borgnis' function*, takes the place of the field variables E_i, B_i. This can be done either by the substitution

$$E_1 = k^2 U + \frac{\partial^2 U}{\partial q_1^2}, \quad E_2 = \frac{1}{h_2} \frac{\partial^2 U}{\partial q_1 \partial q_2}, \quad E_3 = \frac{1}{h_3} \frac{\partial^2 U}{\partial q_1 \partial q_3},$$

$$H_1 = 0, \quad H_2 = \frac{ik^2}{\mu \omega h_3} \frac{\partial U}{\partial q_3}, \quad H_3 = -\frac{ik^2}{\mu \omega h_2} \frac{\partial U}{\partial q_2}, \tag{4.194}$$

or by the substitution

$$E_1 = 0, \quad E_2 = \frac{i\omega \mu}{h_3} \frac{\partial U'}{\partial q_3}, \quad E_3 = -\frac{i\omega \mu}{h_2} \frac{\partial U'}{\partial q_2},$$

$$H_1 = -k^2 U' - \frac{\partial^2 U'}{\partial q_1^2}, \quad H_2' = -\frac{1}{h_2} \frac{\partial^2 U'}{\partial q_1 \partial q_2}, \quad H_3' = -\frac{1}{h_3} \frac{\partial^2 U'}{\partial q_1 \partial q_3}. \tag{4.195}$$

It is not difficult to show that these choices lead to

$$\frac{\partial^2 U}{\partial q_1^2} + \frac{1}{h_2 h_3} \left[\frac{\partial}{\partial q_2} \left(\frac{h_3}{h_2} \frac{\partial U}{\partial q_2} \right) + \frac{\partial}{\partial q_3} \left(\frac{h_2}{h_3} \frac{\partial U}{\partial q_3} \right) \right] + k^2 U = 0, \tag{4.196}$$

known as *Borgnis' equation*. The two types of solutions (4.194) and (4.195) give the already known TM and TE modes. Taking the z-axis along the waveguide axis, Borgnis' function can be written as

$$U(q_1, q_2, q_3, t) = U_0(q_2, q_3) e^{i(\omega t - k_z z)}.$$

The next step is to impose boundary conditions and find the corresponding solutions.

Borgnis' approach contains the results presented in Sects. 4.9.1 and 4.9.2 as particular cases, however the substitutions (4.194) and (4.195) are not obvious and it takes time to find them.

Observation:
Since the waves transmitted by waveguides are *collimated*, the microwave receiver must "see" the emitting source.

A special type of waveguides, mostly used for guiding light waves, are the so-called *optical fibers*. Light is kept in the core of the optical fiber by *total internal reflection*. This causes the fiber to act as a waveguide. Fibers which support many propagation paths or transverse modes are called *multi-mode fibers* (MMF), while those which can only support a single mode are called *single-mode fibers* (SMF). Optical fibers are used for a variety of applications, including communications at long distances and sensors and fiber lasers.

4.10 Electromagnetic Radiation

4.10.1 Solutions of the Electrodynamic Potential Equations

When discussing electrostatic and magnetostatic multipole systems, we defined the concepts of scalar and vector potentials (see (1.79) and (2.43)):

$$V(\mathbf{r}) = \frac{1}{4\pi\epsilon} \int_{V'} \frac{\rho(\mathbf{r}')d\tau'}{|\mathbf{r} - \mathbf{r}'|}, \quad \mathbf{A}(\mathbf{r}) = \frac{\mu}{4\pi} \int_{V'} \frac{\mathbf{j}(\mathbf{r}')d\tau'}{|\mathbf{r} - \mathbf{r}'|}, \quad (4.197)$$

where the integrals are taken on the volume V' of the three-dimensional domain D' of the sources (see Fig. 4.19). In the formulas above, $\mathbf{r}$ is the radius vector of an arbitrary point, $P(\mathbf{r})$, of the domain D, where the *effect* is observed (where the electromagnetic field produced by the sources is detected). The fact that sources are considered stationary is mathematically expressed by their explicit time independence: $\rho = \rho(\mathbf{r})$ and $\mathbf{j} = \mathbf{j}(\mathbf{r})$.

In the general case, the sources are variable both in space as well as in time, $\rho = \rho(\mathbf{r}', t')$ and $\mathbf{j} = \mathbf{j}(\mathbf{r}', t')$ (where t' is the *time of the sources*), which represent the *cause* of the electromagnetic field. The principle of causality requires that the *time of the effect*, t, be different from the time of sources, t', because the electromag-

Fig. 4.19 The domain D' of the sources (the cause) and the domain D of the fields (the effect).

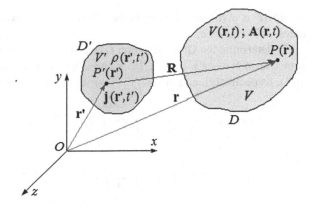

netic perturbation needs a certain finite time (exactly the difference $\Delta t = t - t'$) to propagate from $P'(\mathbf{r}')$ to $P(\mathbf{r})$. In the following, when the sources are not involved in the calculations, but only generically mentioned, we shall drop the index $'$ (prime), writing simply $\rho = \rho(\mathbf{r}, t)$ and $\mathbf{j} = \mathbf{j}(\mathbf{r}, t)$.

In a non-stationary regime, the electrodynamic potentials are solutions of the non-homogeneous d'Alembert equations, in the Lorenz gauge (see (3.81) and (3.82)):

$$\epsilon\mu \frac{\partial^2 V}{\partial t^2} - \Delta V = \frac{\rho}{\epsilon},$$

$$\epsilon\mu \frac{\partial^2 \mathbf{A}}{\partial t^2} - \Delta \mathbf{A} = \mu\mathbf{j}, \qquad (4.198)$$

where the sources ρ and $\mathbf{j}$ are functions of both position and time, $\rho = \rho(\mathbf{r}, t)$ and $\mathbf{j} = \mathbf{j}(\mathbf{r}, t)$.

The purpose of this section is to find the solutions of Eq. (4.198) by using the *Green's function method* (see Appendix F). As a first observation, since Eqs. (4.198) have a similar form, we may write them as a single equation

$$\Box\psi = f(\mathbf{r}, t), \qquad (4.199)$$

where $\Box = \frac{1}{u^2}\frac{\partial^2}{\partial t^2} - \Delta$ is the *d'Alembert operator*, also called *d'Alembertian* or *wave operator* ($u = 1/\sqrt{\epsilon\mu} = c/n$ is the phase velocity of the electromagnetic perturbation which propagates through the non-dispersive medium having the refraction index $n = \sqrt{\epsilon_r\mu_r}$), while $f(\mathbf{r}, t)$ is either $\rho(\mathbf{r}, t)/\epsilon$ or $\mu\mathbf{j}(\mathbf{r}, t)$.

The corresponding Green function is defined as the solution of the equation

$$\Box G(\mathbf{r}, t; \mathbf{r}', t') = \delta(\mathbf{r} - \mathbf{r}')\delta(t - t'). \qquad (4.200)$$

If we succeed to determine G, then the solution of (4.199) is written as

$$\psi(\mathbf{r}, t) = \int d\mathbf{r}' \int dt' G(\mathbf{r}, t; \mathbf{r}', t') f(\mathbf{r}', t'), \qquad (4.201)$$

where $d\mathbf{r}'$ is a volume element in the domain D' (for example, in Cartesian coordinates, $d\mathbf{r}' = dx'dy'dz'$).

To determine the Green function G, we shall Fourier transform it, as well as the source term $\delta(\mathbf{r} - \mathbf{r}')\delta(t - t')$. Let $\mathbf{k}$ and ω be the new variables (corresponding to $\mathbf{r}$ and t, respectively), and $g(\mathbf{k}, \omega)$ the Fourier transform of the Green function. Then

$$G(\mathbf{r}, t; \mathbf{r}', t') = \frac{1}{(2\pi)^2} \int d\mathbf{k} \int d\omega\, e^{i\mathbf{k}\cdot(\mathbf{r}-\mathbf{r}')} e^{-i\omega(t-t')} g(\mathbf{k}, \omega) \qquad (4.202)$$

and

$$\delta(\mathbf{r} - \mathbf{r}')\delta(t - t') = \frac{1}{(2\pi)^4} \int d\mathbf{k} \int d\omega\, e^{i\mathbf{k}\cdot(\mathbf{r}-\mathbf{r}')} e^{-i\omega(t-t')}, \qquad (4.203)$$

where $d\mathbf{k}$ is a volume element in the $\mathbf{k}$-space (for example, $d\mathbf{k} = dk_x dk_y dk_z$), and the integrals are taken from $-\infty$ to $+\infty$, if not specified otherwise. Denoting $\mathbf{r} - \mathbf{r}' \equiv \mathbf{R}$ and $t - t' \equiv \tau$ and introducing (4.202) into (4.200), we have

$$(2\pi)^2 g(\mathbf{k}, \omega) \,\Box\, \left(e^{i\mathbf{k}\cdot\mathbf{R}} e^{-i\omega\tau} \right) = e^{i(\mathbf{k}\cdot\mathbf{R} - \omega\tau)}. \tag{4.204}$$

Since

$$\Box \left[e^{i\mathbf{k}\cdot\mathbf{R}} e^{-i\omega\tau} \right] = \left(\frac{1}{u^2} \frac{\partial^2}{\partial t^2} - \Delta \right) \left[e^{i\mathbf{k}\cdot(\mathbf{r}-\mathbf{r}')} e^{-i\omega(t-t')} \right]$$
$$= \left(k^2 - \frac{\omega^2}{u^2} \right) e^{i(\mathbf{k}\cdot\mathbf{R} - \omega\tau)},$$

from (4.204) it follows that

$$g(\mathbf{k}, \omega) = \frac{1}{(2\pi)^2} \frac{1}{k^2 - \frac{\omega^2}{u^2}}, \tag{4.205}$$

and the Green function (4.202) becomes

$$G(\mathbf{r}, t; \mathbf{r}', t') = \frac{1}{(2\pi)^4} \int d\mathbf{k} \int d\omega \, \frac{e^{i(\mathbf{k}\cdot\mathbf{R} - \omega\tau)}}{k^2 - \frac{\omega^2}{u^2}}. \tag{4.206}$$

The existence of the poles at $\omega = \pm uk$ in the denominator makes the integral (4.206) divergent. To perform the integration, we shall analytically continue the integrand in the complex ω-plane and treat the problem as a contour integral of a function of the complex variable ω and use the residue theorem. In the following, we shall use this method to calculate the Green function $G(\mathbf{r}, t; \mathbf{r}', t')$ for an infinite domain, assuming the medium to be nondispersive ($k = \omega/u$).

The two simple poles of the integrand in

$$\int_{-\infty}^{+\infty} \frac{e^{i(\mathbf{k}\cdot\mathbf{R} - \omega\tau)}}{k^2 - \frac{\omega^2}{u^2}} d\omega \tag{4.207}$$

are on the real axis, symmetrically with respect to the origin (see Fig. 4.20). By choosing various integration contours in the complex plane we shall obtain Green functions with different properties.

In Fig. 4.21 are given two examples of the most important contours, denoted by r and a. These contours can be closed at infinity by half-circles in the lower or upper half-planes, depending on the sign of $\tau = t - t'$ on the boundary.

Retarded Green Function

Let us consider first the case when $\tau > 0$. In this situation, the exponential $\exp(-i\omega\tau)$ grows unlimitedly in the upper half-plane, and for $\text{Im}\,\omega \to \infty$ it becomes infinite.

Fig. 4.20 The two simple poles of the integrand in (4.207).

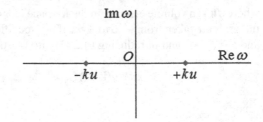

Fig. 4.21 Two integration contours for the integral (4.207), physically significant in the theory of the electromagnetic radiation.

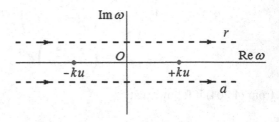

If we wish to have no contribution to the integral from the half-circle that closes the contour at infinity, then we cannot choose the half-circle in the upper half-plane, but in the lower half-plane, where it is easy to see that the integrand becomes zero for $\tau > 0$. In Fig. 4.22 is represented the contour r which satisfies the above requirements.

According to the residue theorem, the integral with respect to ω on the contour $r \equiv C_-$ is given by the sum of the residues at the two poles which are enclosed by the integration contour:

$$\oint_{C_-} \frac{e^{-i\omega\tau}}{k^2 - \frac{\omega^2}{u^2}} \, d\omega \equiv \oint_{C_-} w(\omega) d\omega = -2\pi i \sum_{j=1}^{2} \text{Res}\{w, \omega_j\} \qquad (4.208)$$

$$= -2\pi i \left(\text{Res}\{w, \omega_1 = -uk\} + \text{Res}\{w, \omega_2 = uk\} \right),$$

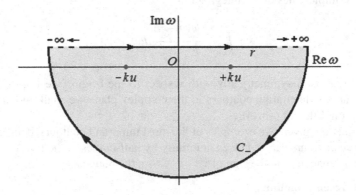

Fig. 4.22 Integration contour for the retarded (causal) Green function.

where $\text{Res}\{w, \omega_j\}$, $j = 1, 2$ are the residues of the function

$$w(\omega) = \frac{e^{-i\omega\tau}}{k^2 - \frac{\omega^2}{u^2}}$$

at the poles $\omega_j = \pm uk$. The minus sign in front of the sum accounts for the fact that the contour $C_- \equiv (r)$ is negatively oriented (clockwise).

For simple poles, as it is our case, the formula for calculating the residues is

$$\text{Res}\{w, \omega_j\} = \lim_{\omega \to \omega_j} (\omega - \omega_j)w(\omega), \quad j = 1, 2. \tag{4.209}$$

Hence,

$$\text{Res}\{w, \omega_1 = -ku\} = \frac{u}{2}\frac{e^{iku\tau}}{k}, \qquad \text{Res}\{w, \omega_2 = ku\} = -\frac{u}{2}\frac{e^{-iku\tau}}{k},$$

and plugging these results into (4.208),

$$\oint_{(r)} w(\omega)d\omega = \frac{2\pi u}{k}\sin ku\tau.$$

Thus, the Green function $G(\mathbf{r}, t; \mathbf{r}', t')$, given by (4.206), becomes

$$G(\mathbf{r}, t; \mathbf{r}', t') = \frac{1}{(2\pi)^4}\int d\mathbf{k} \int \frac{e^{i(\mathbf{k}\cdot\mathbf{R}-\omega\tau)}}{k^2 - \frac{\omega^2}{u^2}}d\omega = \frac{u}{(2\pi)^3}\int \frac{\sin ku\tau}{k}e^{i\mathbf{k}\cdot\mathbf{R}}d\mathbf{k}.$$
$$\tag{4.210}$$

The integral over $\mathbf{k}$ becomes simpler in spherical coordinates, where we have $d\mathbf{k} = k^2 \sin\theta \, dkd\theta \, d\varphi$ and $\mathbf{k} \cdot \mathbf{R} = kR\cos\theta$. We obtain

$$G(\mathbf{r}, t; \mathbf{r}', t') = \frac{u}{(2\pi)^3}\int_0^\infty dk \left(k\sin ku\tau \int_0^\pi e^{ikR\cos\theta}\sin\theta \, d\theta\right)\int_0^{2\pi}d\varphi$$

$$= \frac{u}{(2\pi)^2}\int_0^\infty \left(k\sin ku\tau \int_{-kR}^{+kR}\frac{e^{i\xi}}{kR}d\xi\right)dk$$

$$= \frac{2u}{(2\pi)^2 R}\int_0^\infty \sin ku\tau \sin kR \, dk$$

$$= \frac{u}{(2\pi)^2 R}\int_0^\infty [\cos(kR - ku\tau) - \cos(kR + ku\tau)] \, dk$$

$$= \frac{u}{2(2\pi)^2 R}\int_{-\infty}^{+\infty} [\cos(kR - ku\tau) - \cos(kR + ku\tau)] \, dk$$

$$= \frac{u}{2(2\pi)^2 R}\int_{-\infty}^{+\infty} \Big\{ [\cos(kR - ku\tau) + i\sin(kR - ku\tau)]$$

$$- [\cos(kR + ku\tau) + i \sin(kR + ku\tau)] \Big\} \, dk$$

$$= \frac{u}{2(2\pi)^2 R} \left(\int_{-\infty}^{+\infty} e^{ik(R-u\tau)} \, dk - \int_{-\infty}^{+\infty} e^{ik(R+u\tau)} \, dk \right)$$

$$= \frac{u}{2(2\pi)^2 R} 2\pi \left[\delta(R - u\tau) - \delta(R + u\tau) \right]$$

$$= \frac{u}{4\pi R} \delta(R - u\tau) = \frac{1}{4\pi R} \delta\left(\tau - \frac{R}{u} \right), \qquad (4.211)$$

for which we have used, successively:

(1) change of variable $kR \cos\theta = \xi$;
(2) trigonometric formula $\sin\alpha \sin\beta = \frac{1}{2} [\cos(\alpha - \beta) - \cos(\alpha + \beta)]$;
(3) the fact that cosine is an even function:
     $\int_0^\infty \cos[kf(R, \tau)] dk = \frac{1}{2} \int_{-\infty}^{+\infty} \cos[kf(R, \tau)] dk$;
(4) the fact that sine is an odd function: $\int_{-\infty}^{+\infty} \sin[kf(R, \tau)] dk = 0$;
(5) Euler's formula: $\cos x \pm i \sin x = e^{\pm ix}$;
(6) the Fourier representation of the delta function: $\delta(x) = \frac{1}{2\pi} \int_{-\infty}^{+\infty} e^{ikx} dk$;
(7) $\delta(R + u\tau) = 0$, because $R > 0$, $u > 0$ and $\tau > 0 \Rightarrow R + u\tau > 0$;
(8) the property of the delta function that $\forall a > 0$, $\delta(ax) = \frac{\delta(x)}{a}$.

Reverting to the initial notation, for $\tau > 0$ and the integration contour r closing in the lower half-plane, we obtained the *retarded Green function*:

$$G(\mathbf{r}, t; \mathbf{r}', t') \Big[\equiv G_r(\mathbf{r}, t; \mathbf{r}', t') \Big] = \frac{1}{4\pi |\mathbf{r} - \mathbf{r}'|} \delta\left(t - t' - \frac{|\mathbf{r} - \mathbf{r}'|}{u} \right). \qquad (4.212)$$

This Green function is also called *causal*, because the time of observation, t, is always later than the time of the source, t'. In other words, the relation (4.212) expresses the fact that the effect observed at time t at the point $\mathbf{r}$ is due to the *cause* (disturbance) which took place at the point $\mathbf{r}'$, at the earlier time $t' = t - R/u < t$. The time difference $\Delta t = t - t' = R/u$, where $R = |\mathbf{r} - \mathbf{r}'|$ is the distance between the points corresponding to the cause and effect, and $u = c/n$ is the phase velocity of the electromagnetic perturbation, is nothing but the time necessary to the signal to propagate between the two points.

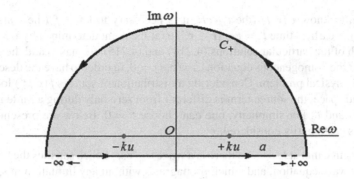

Fig. 4.23 Integration contour for the advanced Green function.

Advanced Green Function

When $\tau = t - t' < 0$, an analogous line of reasoning leads to the choice of the integration contour a as in Fig. 4.23. Consequently, one obtains the *advanced Green function*:

$$G(\mathbf{r}, t; \mathbf{r}', t')\Big[\equiv G_a(\mathbf{r}, t; \mathbf{r}', t') \Big] = \frac{1}{4\pi|\mathbf{r} - \mathbf{r}'|} \delta\left(t - t' + \frac{|\mathbf{r} - \mathbf{r}'|}{u}\right). \quad (4.213)$$

Plugging (4.212) and (4.213) into (4.201) and integrating over t', one obtains the particular solutions of the inhomogeneous equation (4.199):

$$\psi_r(\mathbf{r}, t) = \int d\mathbf{r}' \int dt' G_r(\mathbf{r}, t; \mathbf{r}', t') f(\mathbf{r}', t')$$

$$= \frac{1}{4\pi} \int d\mathbf{r}' \int \frac{\delta\left(t - t' - \frac{|\mathbf{r}-\mathbf{r}'|}{u}\right)}{|\mathbf{r} - \mathbf{r}'|} f(\mathbf{r}', t')\, dt' \quad (4.214)$$

$$= \frac{1}{4\pi} \int \frac{f\left(\mathbf{r}', t - \frac{|\mathbf{r}-\mathbf{r}'|}{u}\right)}{|\mathbf{r} - \mathbf{r}'|}\, d\mathbf{r}',$$

as well as

$$\psi_a(\mathbf{r}, t) = \int d\mathbf{r}' \int dt' G_a(\mathbf{r}, t; \mathbf{r}', t') f(\mathbf{r}', t')$$

$$= \frac{1}{4\pi} \int d\mathbf{r}' \int \frac{\delta\left(t - t' + \frac{|\mathbf{r}-\mathbf{r}'|}{u}\right)}{|\mathbf{r} - \mathbf{r}'|} f(\mathbf{r}', t')\, dt' \quad (4.215)$$

$$= \frac{1}{4\pi} \int \frac{f\left(\mathbf{r}', t + \frac{|\mathbf{r}-\mathbf{r}'|}{u}\right)}{|\mathbf{r} - \mathbf{r}'|}\, d\mathbf{r}'.$$

Therefore, to know $\psi(\mathbf{r}, t)$ (the *effect*), it is necessary to know f (the *cause*) at a retarded (i.e. earlier) time $t' = t - |\mathbf{r} - \mathbf{r}'|/u$, at the point determined by $\mathbf{r}'$.

To each of the particular solutions (4.214) and (4.215) one has to add the general solution of the homogeneous equation $\Box\, \psi(\mathbf{r}, t) = 0$, in order to have the description of a given physical problem. Considering a distribution of sources $f(\mathbf{r}', t')$ localized in time and space, the source term is different from zero only during a finite interval of time around t'. For simplicity, one can choose $t' = 0$. Below are presented two limit cases customarily considered:

(*1*) We assume that at $t \to -\infty$ there is a wave $\psi_{in}(\mathbf{r}, t)$ which satisfies the homogeneous wave equation, and which propagrates without any limitation in space or time. At a given moment, the source is turned on and it starts to generate its own waves. In this situation, the general solution of the inhomogeneous equation is

$$\psi(\mathbf{r}, t) = \psi_{in}(\mathbf{r}, t) + \int d\mathbf{r}' \int dt' G_r(\mathbf{r}, t; \mathbf{r}', t') f(\mathbf{r}', t')$$

$$= \psi_{in}(\mathbf{r}, t) + \int \frac{f\left(\mathbf{r}', t - \frac{|\mathbf{r}-\mathbf{r}'|}{u}\right)}{4\pi|\mathbf{r} - \mathbf{r}'|} d\mathbf{r}'. \tag{4.216}$$

The retarded Green function $G_r(\mathbf{r}, t; \mathbf{r}', t')$ ensures that before turning on the source there is no contribution from the integral, but only the wave $\psi_{in}(\mathbf{r}, t)$;

(*2*) For very late moments of time, $t \to +\infty$, the wave is again a solution, $\psi_{out}(\mathbf{r}, t)$, of the same homogeneous wave equation. In this case, the complete solution of the inhomogeneous equation is

$$\psi(\mathbf{r}, t) = \psi_{out}(\mathbf{r}, t) + \int d\mathbf{r}' \int dt' G_a(\mathbf{r}, t; \mathbf{r}', t') f(\mathbf{r}', t')$$

$$= \psi_{out}(\mathbf{r}, t) + \int \frac{f\left(\mathbf{r}', t + \frac{|\mathbf{r}-\mathbf{r}'|}{u}\right)}{4\pi|\mathbf{r} - \mathbf{r}'|} d\mathbf{r}'. \tag{4.217}$$

Now it is the advanced Green function, $G_a(\mathbf{r}, t; \mathbf{r}', t')$, which ensures that after turning off the source, its signals do not appear explicitly in the solution, being included by assumption in the solution of the homogeneous equation, $\psi_{out}(\mathbf{r}, t)$.

The most common physical situation is described by (4.216) with $\psi_{in}(\mathbf{r}, t) = 0$:

$$\psi(\mathbf{r}, t) = \int \frac{1}{4\pi|\mathbf{r} - \mathbf{r}'|} f\left(\mathbf{r}', t - \frac{|\mathbf{r} - \mathbf{r}'|}{u}\right) d\mathbf{r}'. \tag{4.218}$$

If here we replace the function f by ρ/ϵ or $\mu\mathbf{j}$, respectively, then $\psi(\mathbf{r}, t)$ will stand for the scalar potential $V(\mathbf{r}, t)$ or the vector potential $\mathbf{A}(\mathbf{r}, t)$, respectively:

$$V(\mathbf{r}, t) = \frac{1}{4\pi\epsilon} \int_{V'} \frac{1}{|\mathbf{r} - \mathbf{r}'|} \rho\left(\mathbf{r}', t - \frac{|\mathbf{r} - \mathbf{r}'|}{u}\right) d\mathbf{r}',$$

$$\mathbf{A}(\mathbf{r}, t) = \frac{\mu}{4\pi} \int_{V'} \frac{1}{|\mathbf{r} - \mathbf{r}'|} \mathbf{j}\left(\mathbf{r}', t - \frac{|\mathbf{r} - \mathbf{r}'|}{u}\right) d\mathbf{r}', \qquad (4.219)$$

which are called *retarded electrodynamic potentials*. Once the potentials are known, the electromagnetic field can be determined. Thus, to know the electromagnetic field $\mathbf{E}$ and $\mathbf{B}$ at the point $\mathbf{r}$ and time t, one must know the *sources* ρ and $\mathbf{j}$, at the point $\mathbf{r}'$ and retarded time $t' = t - |\mathbf{r} - \mathbf{r}'|/u$.

The quasi-stationary regime is characterized by $|\mathbf{r} - \mathbf{r}'|/u \ll T$, where T is the period of the electromagnetic oscillation. For example, if $\nu = 1/T = 50\,\mathrm{Hz}$ (industrial alternating current), then, assuming $n \simeq 1$ we have $R \ll cT = 6 \times 10^3\,\mathrm{km}$, which means that low frequency currents are practically quasi-stationary. Thus, the electrodynamic potentials describing quasi-stationary electromagnetic phenomena become

$$V(\mathbf{r}, t) = \frac{1}{4\pi\epsilon} \int_{V'} \frac{\rho(\mathbf{r}', t)}{|\mathbf{r} - \mathbf{r}'|} d\mathbf{r}',$$

$$\mathbf{A}(\mathbf{r}, t) = \frac{\mu}{4\pi} \int_{V'} \frac{\mathbf{j}(\mathbf{r}', t)}{|\mathbf{r} - \mathbf{r}'|} d\mathbf{r}'. \qquad (4.220)$$

Finally, in a purely stationary regime the sources ρ and $\mathbf{j}$ do not depend on time explicitly, and one obtains the relations (1.79) and (2.43) of the stationary potentials:

$$V(\mathbf{r}) = \frac{1}{4\pi\epsilon} \int_{V'} \frac{\rho(\mathbf{r}')}{|\mathbf{r} - \mathbf{r}'|} d\mathbf{r}',$$

$$\mathbf{A}(\mathbf{r}) = \frac{\mu}{4\pi} \int_{V'} \frac{\mathbf{j}(\mathbf{r}')}{|\mathbf{r} - \mathbf{r}'|} d\mathbf{r}'.$$

The solutions of Maxwell's equations have been obtained in the Lorenz gauge. We may now convince ourselves that the *Lorenz gauge condition* (see (3.80)),

$$\nabla \cdot \mathbf{A} + \epsilon\mu \frac{\partial V}{\partial t} = 0 \qquad (4.221)$$

is satisfied indeed by the solutions. As this condition must be satisfied at the observation point $P(\mathbf{r})$ (see Fig. 4.19), at the moment t, we have

$$\frac{\partial V}{\partial t} = \frac{\partial}{\partial t} V(\mathbf{r}, t) = \frac{1}{4\pi\epsilon} \frac{\partial}{\partial t} \left[\int_{V'} \frac{1}{|\mathbf{r} - \mathbf{r}'|} \rho\left(\mathbf{r}', t - \frac{|\mathbf{r} - \mathbf{r}'|}{u}\right) d\mathbf{r}' \right]$$

$$= \frac{1}{4\pi\epsilon} \int_{V'} \frac{1}{R} \frac{\partial \rho\left(\mathbf{r}', t - \frac{R}{u}\right)}{\partial\left(t - \frac{R}{u}\right)} \frac{\partial\left(t - \frac{R}{u}\right)}{\partial t} d\mathbf{r}' = \frac{1}{4\pi\epsilon} \int_{V'} \frac{1}{R} \frac{\partial \rho\left(\mathbf{r}', t - \frac{R}{u}\right)}{\partial\left(t - \frac{R}{u}\right)} d\mathbf{r}'.$$

We also observe that

$$\nabla \cdot \mathbf{j} = \nabla \cdot \mathbf{j}\left(\mathbf{r}', t - \frac{R}{u}\right) = \frac{\partial \mathbf{j}}{\partial \left(t - \frac{R}{u}\right)} \cdot \nabla \left(t - \frac{R}{u}\right)$$

$$= -\frac{1}{u} \frac{\partial \mathbf{j}}{\partial \left(t - \frac{R}{u}\right)} \cdot \nabla R = \frac{1}{u} \frac{\partial \mathbf{j}}{\partial \left(t - \frac{R}{u}\right)} \cdot \nabla' R,$$

$$\nabla' \cdot \mathbf{j} = (\nabla \cdot \mathbf{j})_{t - \frac{R}{u} = \text{const.}} - \frac{1}{u} \frac{\partial \mathbf{j}}{\partial \left(t - \frac{R}{u}\right)} \cdot \nabla' R,$$

leading to

$$\nabla \cdot \mathbf{j} = (\nabla' \cdot \mathbf{j})_{t - \frac{R}{u} = \text{const.}} - \nabla' \cdot \mathbf{j}.$$

By using this result, if one applies the divergence theorem and disregards the vanishing term, the divergence of $\mathbf{A}$ (given by $(4.219)_2$):

$$\nabla \cdot \mathbf{A}(\mathbf{r}, t) = \frac{\mu}{4\pi} \nabla \cdot \left[\int_{V'} \frac{1}{|\mathbf{r} - \mathbf{r}'|} \mathbf{j}\left(\mathbf{r}', t - \frac{|\mathbf{r} - \mathbf{r}'|}{u}\right) d\mathbf{r}' \right]$$

$$\equiv \frac{\mu}{4\pi} \nabla \cdot \left[\int_{V'} \frac{\mathbf{j}}{R} d\mathbf{r}' \right] = \frac{\mu}{4\pi} \int_{V'} \left[\frac{1}{R} \nabla \cdot \mathbf{j} + \mathbf{j} \cdot \nabla \left(\frac{1}{R}\right) \right] d\mathbf{r}'$$

$$= \frac{\mu}{4\pi} \int_{V'} \frac{1}{R} (\nabla' \cdot \mathbf{j})_{t - \frac{R}{u} = \text{const.}} d\mathbf{r}' - \frac{\mu}{4\pi} \int_{V'} \nabla' \cdot \left(\frac{\mathbf{j}}{R}\right) d\mathbf{r}'$$

$$= \frac{\mu}{4\pi} \int_{V'} \frac{1}{R} (\nabla' \cdot \mathbf{j})_{t - \frac{R}{u} = \text{const.}} d\mathbf{r}' - \oint_{S'} \frac{1}{R} \mathbf{j} \cdot d\mathbf{S}'$$

can be written as

$$\nabla \cdot \mathbf{A} = \frac{\mu}{4\pi} \int_{V'} \frac{1}{R} (\nabla' \cdot \mathbf{j})_{t - \frac{R}{u} = \text{const.}} d\mathbf{r}'.$$

Introducing all these results into the Lorenz condition (4.221), we have

$$\nabla \cdot \mathbf{A} + \epsilon \mu \frac{\partial V}{\partial t} = \frac{\mu}{4\pi} \int_{V'} \frac{1}{R} \left[\frac{\partial \rho}{\partial \left(t - \frac{R}{u}\right)} + (\nabla' \cdot \mathbf{j})_{t - \frac{R}{u} = \text{const.}} \right] d\mathbf{r}' = 0,$$

which finally leads to

$$\frac{\partial \rho}{\partial \left(t - \frac{R}{u}\right)} + (\nabla' \cdot \mathbf{j})_{t - \frac{R}{u} = \text{const.}} = 0. \qquad (4.222)$$

This is nothing else but the *equation of continuity*, valid at the moment $t' = t - \frac{R}{u}$ (the time of the sources), at an arbitrary point $P'(\mathbf{r}')$ from the domain of the sources D' (see Fig. 4.19).

Thus, the Lorenz gauge condition is satisfied by the obtained potentials, as it translates into the equation of continuity for the electric charge. However, the Lorenz gauge condition is not equivalent to the equation of continuity. For example, if the solutions of Maxwell's equations are found in another gauge, say the Coulomb gauge, and we plug the solutions into the Lorenz gauge condition, the latter will not be satisfied. However, the equation of continuity for electric charge will always be valid, irrespective of which gauge condition we are using. The equation of continuity expresses a physical law, which is the conservation of electric charge, while gauge fixing conditions do not have any physical significance, being just some relations by which we pick up a certain form for the potentials out of an infinity of physically equivalent possibilities.

4.10.2 Liénard–Wiechert Potentials

Let us determine the potentials $V(\mathbf{r}, t)$ and $\mathbf{A}(\mathbf{r}, t)$ corresponding to the field produced by a point charge moving in vacuum. It is supposed that the motion of the charge is known. Denote by e the charge of the particle and by $\mathbf{x}(t)$ its trajectory. The velocity of the particle can be written as

$$\frac{d\mathbf{x}}{dt} = c\boldsymbol{\beta}(t), \tag{4.223}$$

where c is the speed of light in vacuum. The charge density $\rho(\mathbf{r}, t)$ and the current density $\mathbf{j}(\mathbf{r}, t)$ are expressed by means of the Dirac delta function:

$$\rho(\mathbf{r}, t) = e\delta[\mathbf{r} - \mathbf{x}(t)],$$
$$\mathbf{j}(\mathbf{r}, t) = ec\boldsymbol{\beta}(t)\delta[\mathbf{r} - \mathbf{x}(t)]. \tag{4.224}$$

Let us start with the scalar potential V. According to $(4.219)_1$ (written for the vacuum) and $(4.224)_1$, this is

$$V(\mathbf{r}, t) = \frac{e}{4\pi\epsilon_0} \int \frac{\delta\left[\mathbf{r}' - \mathbf{x}\left(t - \frac{|\mathbf{r}-\mathbf{r}'|}{c}\right)\right]}{|\mathbf{r} - \mathbf{r}'|} \, d\mathbf{r}'.$$

The delta function in the integrand can be written as

$$\delta\left[\mathbf{r}' - \mathbf{x}\left(t - \frac{|\mathbf{r} - \mathbf{r}'|}{c}\right)\right] = \int \delta[\mathbf{r}' - \mathbf{x}(t')]\delta\left(t' - t + \frac{|\mathbf{r} - \mathbf{r}'|}{c}\right) dt',$$

such that

$$V(\mathbf{r}, t) = \frac{e}{4\pi\epsilon_0} \int dt' \int \frac{\delta[\mathbf{r}' - \mathbf{x}(t')]\delta\left(t' - t + \frac{|\mathbf{r}-\mathbf{r}'|}{c}\right)}{|\mathbf{r} - \mathbf{r}'|} d\mathbf{r}'$$

$$= \frac{e}{4\pi\epsilon_0} \int dt' \frac{1}{|\mathbf{r} - \mathbf{x}(t')|} \delta\left(t' - t + \frac{|\mathbf{r} - \mathbf{x}(t')|}{c}\right). \qquad (4.225)$$

On the other hand, according to the properties of delta function (see Appendix E),

$$\int g(t')\delta[f(t') - t]dt' = \left[\frac{g(t')}{df/dt'}\right]_{f(t')=t}. \qquad (4.226)$$

Since in our case

$$g(t') = \frac{1}{|\mathbf{r} - \mathbf{x}(t')|}, \qquad f(t') = t' + \frac{|\mathbf{r} - \mathbf{x}(t')|}{c},$$

and

$$\frac{df}{dt'} = 1 + \frac{1}{c}\frac{d}{dt'}\left\{\left[r_i - x_i(t')\right]\left[r_i - x_i(t')\right]\right\}^{1/2}$$

$$= 1 + \frac{1}{c}\frac{r_i - x_i(t')}{|\mathbf{r} - \mathbf{x}(t')|}\left(-\frac{dx_i}{dt'}\right) = 1 - \frac{r_i - x_i(t')}{|\mathbf{r} - \mathbf{x}(t')|}\beta_i = 1 - \mathbf{n}\cdot\boldsymbol{\beta},$$

where $\mathbf{n}$ is the unit vector of the direction $\mathbf{r} - \mathbf{x}(t')$, we finally obtain

$$V(\mathbf{r}, t) = \frac{e}{4\pi\epsilon_0}\left[\frac{1}{|\mathbf{r} - \mathbf{x}|(1 - \mathbf{n}\cdot\boldsymbol{\beta})}\right]_{t=t'+\frac{|\mathbf{r}-\mathbf{x}(t')|}{c}}, \qquad (4.227)$$

where $\mathbf{n}$, $\boldsymbol{\beta}$, and $\mathbf{x}$ are functions of t'.

In a similar way, one finds for the vector potential

$$\mathbf{A}(\mathbf{r}, t) = \frac{\mu_0\, e\, c}{4\pi}\left[\frac{\boldsymbol{\beta}}{|\mathbf{r} - \mathbf{x}|(1 - \mathbf{n}\cdot\boldsymbol{\beta})}\right]_{t=t'+\frac{|\mathbf{r}-\mathbf{x}(t')|}{c}}. \qquad (4.228)$$

These expressions were found in part by Alfred-Marie Liénard (1869–1958) in 1898 and independently by Emil Johann Wiechert (1861–1928) in 1900. They are known as *Liénard–Wiechert potentials*. These potentials describe relativistically correctly the time-varying classical electromagnetic field of a point charge in arbitrary motion.

Thus we realize that *it is not correct* to express the potential (say, scalar) of the field produced by a number of *moving* electrons by the formula

$$V(\mathbf{r}, t) = \frac{1}{4\pi\epsilon}\int_{V'}\frac{\rho(\mathbf{r}', t')}{|\mathbf{r} - \mathbf{r}'|}d\tau' = \frac{1}{4\pi\epsilon}\frac{q}{R},$$

because the microscopic charge density depends on the retarded time t', which in its turn depends on $|\mathbf{r} - \mathbf{r}'|$.

Turning now towards the Hertz potential $\mathbf{Z}$, its source is the polarization $\mathbf{P}$ (see (3.100)). Since Eq. (3.100) is perfectly analogous to the equations satisfied by $\mathbf{A}$ and V, the solution

$$\mathbf{Z}(\mathbf{r}, t) = \frac{1}{4\pi\epsilon} \int_{V'} \frac{\mathbf{P}\left(\mathbf{r}', t - \frac{|\mathbf{r}-\mathbf{r}'|}{u}\right)}{|\mathbf{r} - \mathbf{r}'|} \, d\mathbf{r}', \tag{4.229}$$

is the *retarded Hertz vector (potential)*.

4.11 Potentials of a Time-Variable Continuous Charge Distribution

Consider a continuous charge distribution placed in air ($\epsilon \sim \epsilon_0$, $\mu \sim \mu_0$), the sources $\mathbf{j}$ and ρ varying periodically with time according to the laws

$$\rho\left(\mathbf{r}', t - \frac{R}{c}\right) = \rho(\mathbf{r}') e^{-i\omega(t-\frac{R}{c})} = \rho(\mathbf{r}') e^{-i\omega t} e^{ikR},$$

$$\mathbf{j}\left(\mathbf{r}', t - \frac{R}{c}\right) = \mathbf{j}(\mathbf{r}') e^{-i\omega(t-\frac{R}{c})} = \mathbf{j}(\mathbf{r}') e^{-i\omega t} e^{ikR}, \tag{4.230}$$

where $R = |\mathbf{r} - \mathbf{r}'|$. The corresponding retarded potentials (4.219) are

$$V(\mathbf{r}, t) = \frac{1}{4\pi\epsilon_0} e^{-i\omega t} \int_V \frac{1}{R} \rho(\mathbf{r}') e^{ikR} \, d\tau',$$

$$\mathbf{A}(\mathbf{r}, t) = \frac{\mu_0}{4\pi} e^{-i\omega t} \int_V \frac{1}{R} \mathbf{j}(\mathbf{r}') e^{ikR} \, d\tau', \tag{4.231}$$

where obviously $d\tau' \equiv d\mathbf{r}'$.

Given the potentials $\mathbf{A}$ and V, one can determine the electromagnetic field components $\mathbf{E}$ and $\mathbf{B}$ at the point $P(\mathbf{r})$, at time t, through (3.75) and (3.76). The field distribution depends upon the distance between the sources and the point P. The sources $\rho(\mathbf{r}', t')$ and $\mathbf{j}(\mathbf{r}', t')$ are located in a spatial region whose dimensions are of the order $d \ll \lambda$, where λ is the wavelength of the electromagnetic radiation. Depending on the distance from the region of the sources to the observation point $P(\mathbf{r})$, one can distinguish three zones of interest:

(i) *near zone* (or *static*), characterized by $d \ll r \ll \lambda$;
(ii) *intermediate* (or *induction*) *zone*, defined by $(d \ll r \simeq \lambda)$;
(iii) *far* (or *radiation*) *zone*, where $(d \ll \lambda \ll r)$.

We shall focus on the propagation phenomena in the *radiation zone*, and all mathematical approximations will be made accordingly. Moreover, since in the region where there are no sources Maxwell's equation $\nabla \times \mathbf{B} = \mu_0 \mathbf{j} + \epsilon_0 \mu_0 \frac{\partial \mathbf{E}}{\partial t}$ becomes $\nabla \times \mathbf{B} = \epsilon_0 \mu_0 \frac{\partial \mathbf{E}}{\partial t} = \frac{1}{c^2} \frac{\partial \mathbf{E}}{\partial t}$, and the time-dependence is given by the exponential $\exp(-i\omega t)$, we have

$$\frac{\partial \mathbf{E}}{\partial t} = c^2 \nabla \times \mathbf{B} \quad \Leftrightarrow \quad -i\omega \mathbf{E} = c^2 \nabla \times \mathbf{B} \quad \Rightarrow \quad \mathbf{E} = \frac{ic^2}{\omega} \nabla \times \mathbf{B} = \frac{ic}{k} \nabla \times \mathbf{B}.$$

As a result, both $\mathbf{B}$ (through the relation $\mathbf{B} = \nabla \times \mathbf{A}$), and $\mathbf{E}$ (through the above equation, $\mathbf{E} = \frac{ic}{k} \nabla \times \nabla \times \mathbf{A}$) are determined only by the retarded vector potential $\mathbf{A}(\mathbf{r}, t)$, which reads, in Lorenz gauge,

$$\mathbf{A}(\mathbf{r}, t) = \frac{\mu_0}{4\pi} \int d\mathbf{r}' \int dt' \frac{\mathbf{j}(\mathbf{r}', t')}{|\mathbf{r} - \mathbf{r}'|} \delta \left(t - t' - \frac{|\mathbf{r} - \mathbf{r}'|}{c} \right),$$

as long as there are no boundary surfaces. The delta function ensures the causal behaviour of fields and, if we assume a time-variation of the source $\mathbf{j}$ as in $(4.230)_2$, then $\mathbf{A}(\mathbf{r}, t)$ is given by $(4.231)_2$. In the following we shall analyze only this case.

Before discussing various types of radiation (electric dipole, magnetic dipole, electric quadrupole, etc.), let us examine briefly the electric fields generated by time-variable point-like sources. The scalar potential $V(\mathbf{r}, t)$ is given by the relation

$$V(\mathbf{r}, t) = \frac{1}{4\pi\epsilon_0} \int d\mathbf{r}' \int dt' \frac{\rho(\mathbf{r}', t')}{|\mathbf{r} - \mathbf{r}'|} \delta \left(t - t' - \frac{|\mathbf{r} - \mathbf{r}'|}{c} \right)$$

$$= \frac{1}{4\pi\epsilon_0} \int d\mathbf{r}' \int dt' \frac{\rho(\mathbf{r}', t')}{|\mathbf{r} - \mathbf{r}'|} \delta \left(t' + \frac{|\mathbf{r} - \mathbf{r}'|}{c} - t \right).$$

The contribution of the monopole is obtained by replacing under the integral $|\mathbf{r} - \mathbf{r}'|$ by $|\mathbf{r}| \cong R$, with the result

$$V_{monopole}(\mathbf{r}, t) = \frac{q \left(t' = t - \frac{R}{c} \right)}{4\pi\epsilon_0 R},$$

where $q(t')$ is the total charge of the source. Since the electric charge is conserved and a localized source is by definition a source into which and out of which charge does not flow, it follows that the total charge q is time independent. Thus, the electric monopole contribution to the potentials and fields of a localized source is necessarily *static*. The time-harmonic fields (i.e. fields whose time variation is of the type $e^{-i\omega t}$) do not have monopole terms.

Let us return to the multipole fields of the lowest order. Since these fields can be calculated by using only the vector potential, as we have shown earlier, in the following we shall omit to refer explicitly to the scalar potential $V(\mathbf{r}, t)$.

The series expansion of the factor $\frac{1}{R}$ which appears in the integrand of (4.231) gives

$$\frac{1}{R} = \frac{1}{|\mathbf{r} - \mathbf{r}'|} = \frac{1}{r} - x_i' \frac{\partial}{\partial x_i}\left(\frac{1}{r}\right) + \dots \cong \frac{1}{r} + \frac{x_i' x_i}{r^3} = \frac{1}{r}\left(1 + \frac{\mathbf{r}' \cdot \mathbf{s}}{r}\right),$$

leading to

$$R \cong \frac{r}{1 + \frac{\mathbf{r}' \cdot \mathbf{s}}{r}} \cong r\left(1 - \frac{\mathbf{r}' \cdot \mathbf{s}}{r}\right)$$

as well as

$$e^{-ik\mathbf{r}' \cdot \mathbf{s}} \cong 1 - ik\mathbf{r}' \cdot \mathbf{s},$$

$$\frac{e^{-ik\mathbf{r}' \cdot \mathbf{s}}}{R} \cong \frac{1}{r}(1 - ik\mathbf{r}' \cdot \mathbf{s})\left(1 + \frac{\mathbf{r}' \cdot \mathbf{s}}{r}\right) \cong \frac{1}{r}\left[1 + \left(\frac{1}{r} - ik\right)(\mathbf{r}' \cdot \mathbf{s})\right],$$

where $\mathbf{s}$ is the unit vector, with the components $(\alpha_1, \alpha_2, \alpha_3)$, in the direction of $\mathbf{r}$. With these results, the vector potential becomes

$$\mathbf{A}(\mathbf{r}, t) = \frac{\mu_0}{4\pi} \frac{e^{i(kr - \omega t)}}{r}\left[\int_{V'} \mathbf{j}(\mathbf{r}')d\mathbf{r}' + \left(\frac{1}{r} - ik\right)\int_{V'} \mathbf{j}(\mathbf{r}')(\mathbf{r}' \cdot \mathbf{s})d\mathbf{r}' + \dots\right]$$

$$= \sum_{i=1}^{\infty} \mathbf{A}^{(i)}(\mathbf{r}, t). \tag{4.232}$$

The first term of the series (4.232) represents the vector potential of the field produced by an *oscillating electric dipole*:

$$\mathbf{A}^{(1)}(\mathbf{r}, t) = \frac{\mu_0}{4\pi} \frac{e^{i(kr - \omega t)}}{r}\int_{V'} \mathbf{j}(\mathbf{r}')d\mathbf{r}'. \tag{4.233}$$

In the stationary case, such a concept is meaningless, as the integral in (4.233) vanishes; this is the simplest mathematical justification in classical non-relativistic electrodynamics for the non-existence of the magnetic monopole.

To find the significance of the second term of the series (4.232),

$$\mathbf{A}^{(2)}(\mathbf{r}, t) = \frac{\mu_0}{4\pi} \frac{e^{i(kr - \omega t)}}{r}\left(\frac{1}{r} - ik\right)\int_{V'} \mathbf{j}(\mathbf{r}')(\mathbf{r}' \cdot \mathbf{s})d\mathbf{r}', \tag{4.234}$$

we shall transcribe the integrand in the form

$$\mathbf{j}(\mathbf{r}')(\mathbf{r}' \cdot \mathbf{s}) \equiv \mathbf{j}(\mathbf{r}' \cdot \mathbf{s}) = \frac{1}{2}\left[(\mathbf{s} \cdot \mathbf{r}')\mathbf{j} - (\mathbf{s} \cdot \mathbf{j})\mathbf{r}'\right] + \frac{1}{2}\left[(\mathbf{s} \cdot \mathbf{r}')\mathbf{j} + (\mathbf{s} \cdot \mathbf{j})\mathbf{r}'\right], \tag{4.235}$$

which suggests to write $\mathbf{A}^{(2)}$ as a sum of two terms: $\mathbf{A}^{(2)} = \mathbf{A}_1^{(2)} + \mathbf{A}_2^{(2)}$, associated with the antisymmetric and the symmetric part of $\mathbf{A}^{(2)}$, respectively. We note that

$$\frac{1}{2}\Big[(\mathbf{s} \cdot \mathbf{r}')\mathbf{j} - (\mathbf{s} \cdot \mathbf{j})\mathbf{r}'\Big] = \frac{1}{2}(\mathbf{r}' \times \mathbf{j}) \times \mathbf{s}$$

and

$$\left(\frac{1}{2}\int_{V'} \mathbf{r}' \times \mathbf{j}\, d\mathbf{r}'\right) \times \mathbf{s} = \mathbf{m} \times \mathbf{s}, \tag{4.236}$$

where $\mathbf{m} = \frac{1}{2}\int_{V'} \mathbf{r}' \times \mathbf{j}\, d\mathbf{r}'$ is the magnetic dipole moment of a continuous distribution of currents (magnetic sources) (see (2.61)). As a result,

$$\mathbf{A}_1^{(2)}(\mathbf{r}, t) = \frac{\mu_0}{4\pi} \frac{e^{i(kr-\omega t)}}{r} \left(\frac{1}{r} - ik\right)(\mathbf{m} \times \mathbf{s}), \tag{4.237}$$

showing that $\mathbf{A}_1^{(2)}(\mathbf{r}, t)$ represents the vector potential of the field produced by a *magnetic dipole*. Remark that, if the regime were stationary ($k \to 0$, $\omega \to 0$), we would fall back on formula (2.62):

$$\mathbf{A}(\mathbf{r}) = \frac{\mu_0}{4\pi}\frac{\mathbf{m} \times \mathbf{s}}{r^2} = \frac{\mu_0}{4\pi}\frac{\mathbf{m} \times \mathbf{r}}{r^3}.$$

Next, let us find the significance of the symmetric part of $\mathbf{A}(\mathbf{r}, t)$,

$$\mathbf{A}_2^{(2)}(\mathbf{r}, t) = \frac{\mu_0}{8\pi} \frac{e^{i(kr-\omega t)}}{r} \left(\frac{1}{r} - ik\right) \int_{V'} \Big[(\mathbf{s} \cdot \mathbf{r}')\mathbf{j} + (\mathbf{s} \cdot \mathbf{j})\mathbf{r}'\Big] d\mathbf{r}'. \tag{4.238}$$

Assuming that the sources are localized in a region of space whose characteristic dimensions are much smaller than the distance from observer and sources to the origin of the coordinate system (see Fig. 4.24), $d \ll \lambda \ll r$, the projection on the x_i-direction of the expression in the right-hand side of (4.238) is

$$\int_{V'} \Big[(\mathbf{s} \cdot \mathbf{r}')\mathbf{j} + (\mathbf{s} \cdot \mathbf{j})\mathbf{r}'\Big]_i d\mathbf{r}' = \alpha_k \int_{V'} (j_i x_k' + j_k x_i')\, d\mathbf{r}'$$

$$= \alpha_k \int_{V'} \left(x_k' j_m \frac{\partial x_i'}{\partial x_m'} + x_i' j_m \frac{\partial x_k'}{\partial x_m'}\right) d\mathbf{r}'$$

$$= \alpha_k \int_{V'} \left[\frac{\partial}{\partial x_m'}(2 j_m x_i' x_k') - x_i' \frac{\partial}{\partial x_m'}(j_m x_k') - x_k' \frac{\partial}{\partial x_m'}(j_m x_i')\right] d\mathbf{r}'$$

$$= 2\alpha_k \int_{V'} \frac{\partial}{\partial x_m'}(x_i' x_k' j_m)\, d\mathbf{r}' - 2\alpha_k \int_{V'} x_i' x_k' \nabla' \cdot \mathbf{j}\, d\mathbf{r}'$$

$$- \alpha_k \int_{V'} (x_i' j_k + x_k' j_i)\, d\mathbf{r}' = 2\alpha_k \oint_{S'} j_m x_i' x_k'\, dS_m$$

Fig. 4.24 Sources localized within a small region, $|\mathbf{r}'| \ll |\mathbf{r}| \cong |\mathbf{R}|$.

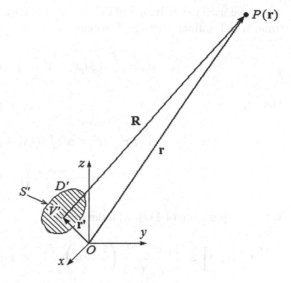

$$- 2\alpha_k \int_{V'} x_i' x_k' \nabla' \cdot \mathbf{j} \, d\mathbf{r}' - \alpha_k \int_{V'} (j_i x_k' + j_k x_i') d\mathbf{r}'$$

$$= -2\alpha_k \int_{V'} x_i' x_k' \nabla' \cdot \mathbf{j} \, d\mathbf{r}' - \alpha_k \int_{V'} (j_i x_k' + j_k x_i') d\mathbf{r}',$$

where the divergence theorem has been used and we took into account the fact that the sources are localized (consequently, at infinity there are no sources).

Thus, we obtain

$$\alpha_k \int_{V'} (j_i x_k' + j_k x_i') d\mathbf{r}' = -2\alpha_k \int_{V'} x_i' x_k' \nabla' \cdot \mathbf{j} \, d\mathbf{r}' - \alpha_k \int_{V'} (j_i x_k' + j_k x_i') d\mathbf{r}',$$

hence

$$\int_{V'} (j_i x_k' + j_k x_i') d\mathbf{r}' = - \int_{V'} x_i' x_k' \nabla' \cdot \mathbf{j} \, d\mathbf{r}'.$$

In view of the continuity equation and recalling the time-dependence of the electric charge density as $e^{-i\omega t}$ (leading to $\nabla' \cdot \mathbf{j} = -\partial \rho / \partial t = i\omega \rho$), we can further write:

$$\int_{V'} (j_i x_k' + j_k x_i') d\mathbf{r}' = - \int_{V'} x_i' x_k' \nabla' \cdot \mathbf{j} \, d\mathbf{r}' = -i\omega \int_{V'} x_i' x_k' \rho(\mathbf{r}') d\mathbf{r}'$$

$$= -\frac{i\omega}{3} \int_{V'} 3 x_i' x_k' \rho(\mathbf{r}') d\mathbf{r}' = -\frac{i\omega}{3} \int_{V'} \left(3 x_i' x_k' - r'^2 \delta_{ik} \right) \rho(\mathbf{r}') d\mathbf{r}'$$

$$- \frac{i\omega}{3} \int_{V'} r'^2 \rho(\mathbf{r}') \delta_{ik} \, d\mathbf{r}' = -\frac{i\omega}{3} p_{ik} - \frac{i\omega}{3} \int_{V'} r'^2 \rho(\mathbf{r}') \delta_{ik} \, d\mathbf{r}',$$

where we used the definition of the tensor of electric quadrupole moment of a continuous and stationary charge distribution,

$$p_{ik} = \int_{V'} \left(3x_i'x_k' - r'^2\delta_{ik}\right) \rho(\mathbf{r}')d\mathbf{r}'.$$

Hence,

$$\int_{V'} \left[(\mathbf{s} \cdot \mathbf{r}')\mathbf{j} + (\mathbf{s} \cdot \mathbf{j})\mathbf{r}'\right]_i d\mathbf{r}' = \alpha_k \int_{V'} (j_i x_k' + j_k x_i')d\mathbf{r}'$$

$$= -\frac{i\omega\alpha_k}{3}\left(p_{ik} + \int_{V'} r'^2 \rho(\mathbf{r}')\delta_{ik} \, d\mathbf{r}'\right),$$

which we plug into (4.238), with the result:

$$\left[\mathbf{A}_2^{(2)}(\mathbf{r}, t)\right]_i = \frac{\mu_0}{8\pi} \frac{e^{i(kr-\omega t)}}{r} \left(\frac{1}{r} - ik\right) \int_{V'} \left[(\mathbf{s} \cdot \mathbf{r}')\mathbf{j} + (\mathbf{s} \cdot \mathbf{j})\mathbf{r}'\right]_i d\mathbf{r}'$$

$$= \frac{\mu_0}{8\pi}\left(-\frac{i\omega}{3}\right) \frac{e^{i(kr-\omega t)}}{r}\left(\frac{1}{r} - ik\right) p_{ik}\alpha_k + \frac{\mu_0}{8\pi}\left(-\frac{i\omega}{3}\right) \frac{e^{i(kr-\omega t)}}{r}$$

$$\times \left(\frac{1}{r} - ik\right) \alpha_k \delta_{ik} \int_{V'} r'^2 \rho(\mathbf{r}') \, d\mathbf{r}' = \frac{i\mu_0\omega}{24\pi} \frac{e^{i(kr-\omega t)}}{r}\left(\frac{1}{r} - ik\right)$$

$$\times p_{ik}\alpha_k - \frac{i\mu_0\omega}{24\pi} \frac{e^{i(kr-\omega t)}}{r}\left(\frac{1}{r} - ik\right) \alpha_i \int_{V'} r'^2 \rho(\mathbf{r}') \, d\mathbf{r}'$$

$$= \frac{i\mu_0\omega}{24\pi} \frac{e^{i(kr-\omega t)}}{r}\left(\frac{1}{r} - ik\right) p_{ik}\alpha_k.$$

In obtaining the above expression we omitted the term

$$\frac{-i\mu_0\omega}{24\pi} \frac{e^{i(kr-\omega t)}}{r}\left(\frac{1}{r} - ik\right)\left(\int_{V'} r'^2 \rho(\mathbf{r}') \, d\mathbf{r}'\right) \alpha_i$$

as this is proportional to the unit vector $\mathbf{s}$ along the vector $\mathbf{r}$, and the field does not change if one adds to the vector potential an arbitrary vector proportional to $\mathbf{s}$. Thus, the symmetric part of $\mathbf{A}^{(2)}(\mathbf{r}, t)$ is proportional to the electric quadrupole moment.

The higher order terms ($\mathbf{A}^{(3)}$, $\mathbf{A}^{(4)}$, ...) of the series (4.232) lead to all the components of the vector potential of this type of charge distribution (magnetic quadrupole, electric octupole, magnetic octupole radiation, etc.). Thus, the multipole analysis brings with each new term a higher degree of complexity of the electromagnetic radiation field produced by an arbitrary electric and magnetic charge distribution. If the dimensions d of the source are small as compared to the wavelength, given that r' is of the order of d, and kd is small compared to the unity by assumption, the contribution of the higher order terms in the expansion of $\mathbf{A}(\mathbf{r}, t)$ decreases fast with the order of the multipole.

4.11.1 Electric Dipole Radiation

In the following we shall calculate the electromagnetic field $\mathbf{E}$, $\mathbf{B}$ generated by an oscillating electric dipole at an arbitrary point situated in the radiation zone, characterized by $d \ll \lambda \ll r$. We shall perform the calculations using two different methods.

4.11.1.1 Method of Vector Potential $\mathbf{A}(\mathbf{r}, t)$

We re-write the expression of the electromagnetic field in terms of the vector potential:

$$\mathbf{E} = \frac{ic}{k} \nabla \times \nabla \times \mathbf{A},$$

$$\mathbf{B} = \nabla \times \mathbf{A}, \tag{4.239}$$

where $\mathbf{A}(\mathbf{r}, t)$, for the case of the oscillating electric dipole, is given by (4.233):

$$\mathbf{A}(\mathbf{r}, t) = \frac{\mu_0}{4\pi} \frac{e^{i(kr-\omega t)}}{r} \int_{V'} \mathbf{j}(\mathbf{r}') d\mathbf{r}'.$$

The x_i-component of the integral in the above relation is

$$\int_{V'} j_i(\mathbf{r}') \, d\mathbf{r}' = \int_{V'} j_k(\mathbf{r}') \frac{\partial x_i'}{\partial x_k'} d\mathbf{r}' = \int_{V'} \frac{\partial}{\partial x_k'} \left(j_k(\mathbf{r}') x_i' \right) d\mathbf{r}' - \int_{V'} x_i' \frac{\partial j_k(\mathbf{r}')}{\partial x_k'} d\mathbf{r}'$$

$$= \oint_{S'} j_k(\mathbf{r}') x_i' dS_k - \int_{V'} x_i' \frac{\partial j_k(\mathbf{r}')}{\partial x_k'} d\mathbf{r}' = -\int_{V'} x_i' \nabla' \cdot \mathbf{j}(\mathbf{r}') \, d\mathbf{r}',$$

where the surface integral vanishes, since the sources are localized and there are no sources at infinity. As a result, in vectorial form we can write

$$\int_{V'} \mathbf{j}(\mathbf{r}') \, d\mathbf{r}' = -\int_{V'} \mathbf{r}' \, \nabla' \cdot \mathbf{j}(\mathbf{r}') \, d\mathbf{r}' = -i\omega \int_{V'} \mathbf{r}' \rho(\mathbf{r}') \, d\mathbf{r}', \tag{4.240}$$

where we used the fact that, due to the time-dependence of the sources of the form $e^{-i\omega t}$, we have $\nabla' \cdot \mathbf{j} = -\partial \rho / \partial t = i\omega \rho$. According to the definition of the electric dipole moment of a distribution of stationary electric charges, the last integral in the relation (4.240) is $\int_{V'} \mathbf{r}' \rho(\mathbf{r}') \, d\mathbf{r}' = \mathbf{p}_0$, such that the vector potential $\mathbf{A}(\mathbf{r}, t)$ given by (4.233) becomes

$$\mathbf{A}(\mathbf{r}, t) = -i\omega \frac{\mu_0}{4\pi} \frac{\mathbf{p}_0 \, e^{i(kr-\omega t)}}{r} = \frac{\mu_0}{4\pi} \left(\frac{-i\omega \mathbf{p}}{r} \right) = \frac{\mu_0}{4\pi} \frac{\dot{\mathbf{p}}}{r}. \tag{4.241}$$

We denoted above the moment of the oscillating electric dipole by $\mathbf{p} = \mathbf{p}_0 \, e^{i(kr-\omega t)}$ and we used the fact that $\dot{\mathbf{p}} = d\mathbf{p}/dt = -i\omega\mathbf{p}$.

The magnetic component of the electromagnetic field created by the dipole is given by $(4.239)_2$:

$$\mathbf{B} = \nabla \times \mathbf{A} = \nabla \times \left(\frac{-i\omega\mu_0}{4\pi} \frac{\mathbf{p}_0 \, e^{i(kr-\omega t)}}{r} \right) = -\frac{i\omega\mu_0}{4\pi} \frac{e^{i(kr-\omega t)}}{r} (\nabla \times \mathbf{p}_0)$$

$$-\frac{i\omega\mu_0}{4\pi} \nabla \left(\frac{e^{i(kr-\omega t)}}{r} \right) \times \mathbf{p}_0 = -\frac{i\omega\mu_0}{4\pi} \left(-\frac{\mathbf{r}}{r^3} e^{i(kr-\omega t)} + \frac{i\mathbf{k}}{r} e^{i(kr-\omega t)} \right)$$

$$\times \mathbf{p}_0 = \frac{i\omega\mu_0}{4\pi} \nabla \left(\frac{ik}{r} - \frac{1}{r^2} \right) \mathbf{p} \times \mathbf{s} \cong \frac{\mu_0}{4\pi cr} (-\omega^2 \mathbf{p}) \times \mathbf{s} = \frac{\mu_0}{4\pi cr} \ddot{\mathbf{p}} \times \mathbf{s},$$

where we used the appropriate approximations for the radiation zone ($r \gg \lambda \gg d$), i.e. we neglected all the terms with powers of r bigger than 2 in the denominator.

For the electric component of the field we have

$$\mathbf{E} = \frac{ic}{k} \nabla \times \nabla \times \mathbf{A} \cong -\frac{ic}{k} \frac{\mu_0 \omega^2}{4\pi c} \nabla \times \left(\frac{\mathbf{p} \times \mathbf{s}}{r} \right) = -\frac{i\omega}{4\pi\epsilon_0 c} \nabla \times \left(\frac{\mathbf{p} \times \mathbf{s}}{r} \right)$$

$$= -\frac{i\omega}{4\pi\epsilon_0 cr} \left[\mathbf{p} \nabla \cdot \mathbf{s} - \mathbf{s}\nabla \cdot \mathbf{p} + (\mathbf{s} \cdot \nabla)\mathbf{p} - (\mathbf{p} \cdot \nabla)\mathbf{s} \right] + \frac{i\omega}{4\pi\epsilon_0 c} \frac{\mathbf{s} \times (\mathbf{p} \times \mathbf{s})}{r^2}$$

$$\cong -\frac{i\omega}{4\pi\epsilon_0 cr} \left[\mathbf{p} \nabla \cdot \mathbf{s} - \mathbf{s}\nabla \cdot \mathbf{p} + (\mathbf{s} \cdot \nabla)\mathbf{p} - (\mathbf{p} \cdot \nabla)\mathbf{s} \right] = -\frac{i\omega}{4\pi\epsilon_0 cr}$$

$$\times \left[\frac{2\mathbf{p}}{r} - ik\,\mathbf{s}(\mathbf{s} \cdot \mathbf{p}) + ik\,(\mathbf{s} \cdot \mathbf{s})\mathbf{p} - \frac{\mathbf{p}}{r^2} + \frac{\mathbf{s}(\mathbf{s} \cdot \mathbf{p})}{r} \right] \cong -\frac{(i\omega)(ik)}{4\pi\epsilon_0 cr}$$

$$\times \left[\mathbf{p}(\mathbf{s} \cdot \mathbf{s}) - \mathbf{s}(\mathbf{s} \cdot \mathbf{p}) \right] = -\frac{\omega^2}{4\pi\epsilon_0 c^2 r} \mathbf{s} \times (\mathbf{s} \times \mathbf{p}) = \frac{1}{4\pi\epsilon_0 c^2} \frac{\mathbf{s} \times (\mathbf{s} \times \ddot{\mathbf{p}})}{r},$$

where we used again the approximation $\mathcal{O}(r^{-2}) = 0$. Thus, in the radiation zone, the field of the oscillating electric dipole is given by

$$\mathbf{E} = \frac{1}{4\pi\epsilon_0 c^2} \frac{\mathbf{s} \times (\mathbf{s} \times \ddot{\mathbf{p}})}{r},$$

$$\mathbf{B} = \frac{\mu_0}{4\pi c} \frac{\ddot{\mathbf{p}} \times \mathbf{s}}{r}. \tag{4.242}$$

Equation (4.233) shows that in the wave zone the vector potential behaves like an emerging spherical wave, and relations (4.242) show that $\mathbf{E}$ and $\mathbf{B}$ are orthogonal to the radius-vector and fall off like r^{-1}, in other words they correspond to radiative fields. For these reasons it it preferable to work in spherical coordinates r, θ, φ (in the wave zone, the wave front can be considered spherical). We shall consider the

Fig. 4.25 Orientation of the electric dipole moment **p** with respect to the axes of a Cartesian reference system.

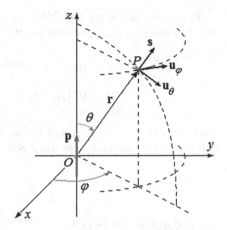

dipole oriented along the z-axis (see Fig. 4.25) and we shall re-write the fields in Eq. (4.242) as

$$\mathbf{E} = \frac{1}{4\pi\epsilon_0 c^2 r} \left[(\mathbf{s} \cdot \ddot{\mathbf{p}})\mathbf{s} - \ddot{\mathbf{p}} \right],$$

$$\mathbf{H} = \frac{\mathbf{u}_\varphi}{4\pi c\, r} \ddot{p} \sin\theta. \tag{4.243}$$

The components of the electromagnetic field are then

$$E_r = \mathbf{E} \cdot \mathbf{s} = 0, \quad E_\theta = \mathbf{E} \cdot \mathbf{u}_\theta = \frac{\ddot{p}\sin\theta}{4\pi\epsilon_0 c^2 r}, \quad E_\varphi = \mathbf{E} \cdot \mathbf{u}_\varphi = 0,$$

$$H_r = \mathbf{H} \cdot \mathbf{s} = 0, \quad H_\theta = \mathbf{H} \cdot \mathbf{u}_\theta = 0, \quad H_\varphi = \mathbf{H} \cdot \mathbf{u}_\varphi = \frac{\ddot{p}\sin\theta}{4\pi c\, r}. \tag{4.244}$$

These relations serve, among other things, to calculate the average power radiated by the oscillating electric dipole (regarded as an antenna), in other words the electromagnetic energy which passes in the unit time through a surface S located in the wave zone.

Defining the complex Poynting vector as

$$\tilde{\mathbf{\Pi}} = \frac{1}{2}\mathbf{E} \times \mathbf{H}^*, \tag{4.245}$$

where "$*$" represents complex conjugation, one can straightforwardly show that the average of the density flux of electromagnetic energy emitted in the unit time is the real part of the normal component of the vector $\tilde{\mathbf{\Pi}}$, i.e.

$$\langle \Phi_{em} \rangle = \frac{1}{2}\left[\text{Re}(\mathbf{E} \times \mathbf{H}^*) \right] \cdot \mathbf{n}, \tag{4.246}$$

where $\langle\ \rangle$ represents the average over one period, and in the wave zone $\mathbf{n} \cong \mathbf{s}$.

To demonstrate relation (4.246) one starts from the expressions of the fields $|\mathbf{E}|$ and $|\mathbf{H}|$, which, for calculational simplicity, we can choose as

$$E = \mathrm{Re}\left[E_0\, e^{i\varphi}\, e^{i(kx-\omega t)}\right] = E_0 \cos(\theta + \varphi),$$

$$H = \mathrm{Re}\left[H_0\, e^{i\varphi'}\, e^{i(kx-\omega t)}\right] = H_0 \cos(\theta + \varphi'),$$

where $\theta \equiv kx - \omega t$, and E_0 and H_0 are real amplitudes. Then, the average power radiated by the oscillating electric dipole is

$$\langle P_{rad}\rangle = \oint_S \langle \Phi_{em}\rangle dS = \oint_S (\mathrm{Re}\,\tilde{\mathbf{\Pi}}) \cdot \mathbf{s}\, dS. \qquad (4.247)$$

The calculations can be done

(*a*) either in Cartesian coordinates, using (4.242), in which case

$$\tilde{\mathbf{\Pi}} = \frac{1}{2}\mathbf{E} \times \mathbf{H}^* = \frac{1}{32\pi^2\epsilon_0 c^3 r^2}[\mathbf{s} \times (\mathbf{s} \times \ddot{\mathbf{p}})] \times (\ddot{\mathbf{p}} \times \mathbf{s})^*$$

$$= \frac{\mathbf{s}}{32\pi^2\epsilon_0 c^3 r^2}\left[|\ddot{\mathbf{p}}|^2 - (\mathbf{s} \cdot \ddot{\mathbf{p}})(\mathbf{s} \cdot \ddot{\mathbf{p}}^*)\right], \qquad (4.248)$$

(*b*) or in spherical coordinates, using (4.244).

If we adopt the first approach, then

$$(\mathrm{Re}\,\tilde{\mathbf{\Pi}}) \cdot \mathbf{s} = \frac{1}{32\pi^2\epsilon_0 c^3 r^2}\mathrm{Re}\left[|\ddot{\mathbf{p}}|^2 - (\mathbf{s} \cdot \ddot{\mathbf{p}})(\mathbf{s} \cdot \ddot{\mathbf{p}}^*)\right]$$

$$= \frac{|\ddot{\mathbf{p}}|^2 \sin^2\theta}{32\pi^2\epsilon_0 c^3 r^2} = \frac{\omega^4 p_0^2 \sin^2\theta}{32\pi^2\epsilon_0 c^3 r^2}$$

and taking into account $dS = r^2 \sin\theta\, d\theta\, d\varphi$, we find

$$\langle P_{rad}\rangle = \oint_S (\mathrm{Re}\,\tilde{\mathbf{\Pi}}) \cdot \mathbf{s}\, dS = \frac{\omega^4 p_0^2 r^2\theta}{32\pi^2\epsilon_0 c^3 r^2}\int_0^\pi \sin^3\theta d\theta \int_0^{2\pi} d\varphi = \frac{\omega^4 p_0^2}{12\pi\epsilon_0 c^3}.$$

Since in vacuum $\omega = 2\pi c/\lambda$, we finally obtain

$$\langle P_{rad}\rangle = \frac{4\pi^3 p_0^2 c}{3\epsilon_0}\frac{1}{\lambda^4}. \qquad (4.249)$$

Equation (4.249) shows that the average power radiated by the oscillating dipole varies inversely proportional to the fourth power of the wavelength. As a result, the smaller the wavelength of the emitted waves, the larger the electromagnetic energy which reaches a receiver in unit time. For example, the broadcasts of faraway radio stations are much better received in the shortwave bands than in the medium or longwave bands. For the same reason, mobile telephony uses microwaves for signal transmission.

Relation (4.249) is known as the *Rayleigh scattering law*. Based on it, one can explain why the colour of the sky is blue. Lord Rayleigh approached quantitatively the subject of the sky colour for the first time in the work published in *Phil. Mag.*, **XLI** 107–120, 274–279 (1871), which later became famous. The scattering of the Sun light by the atmosphere is due almost entirely to the oscillations of electric dipoles, since the magnetic moments of most of the atmospheric gases are negligible as compared to their electric dipole moments. According to the λ^{-4} scattering law, the red wavelength is scattered the least while violet is scattered the most. The light received away from the direction of the incident beam contains more high-frequency components (blue–violet) as compared to the incident beam. The light transmitted along the direction of the incident beam is richer in low-frequency components (yellow–red) and has a diminished intensity. This makes the sky to appear overall blue and bright and the Sun to be yellow, as viewed from the Earth. (It should be pointed out that, as seen from the space, the sky is black and the Sun is white, as it would be expected without the atmospheric scattering of the light.) Actually, the problem is more complicated, due to the influence of other factors, like the atmospheric instability, the water vapours (which have strong absorption bands in infrared), the ozone (which absorbs the violet wavelength), the other molecular species and the dust, the gradients of temperature, etc.

Observation:

Since $\mathbf{p} = q\mathbf{l}$, we also have $\ddot{\mathbf{p}} = q\ddot{\mathbf{l}}$, meaning that according to $(\mathrm{Re}\,\tilde{\mathbf{\Pi}}) \cdot \mathbf{s} = \frac{|\ddot{\mathbf{p}}|^2 \sin^2 \theta}{32\pi^2 \epsilon_0 c^3 r^2}$ and (4.247), the average power of dipole radiation is proportional to the acceleration of the electric charges. In other words, only an accelerated charged system can produce electromagnetic waves.

The planetary model of the atom was conceived in 1911 by Ernest Rutherford (1871–1937) as such a system. The deficiency of this model is the inconsistency of its build-up. Due to a continuous emission of radiation, the electron should "fall" on the nucleus. In fact, the atoms are very stable systems, and there is no relation between the electron rotation frequency in this model and the frequency of emitted radiation. Therefore, at this level of matter organization, the classical electrodynamics laws cannot offer a correct explanation of the observations. The model was refined by Niels Bohr (1885–1962) in 1913, by the introduction of quantum postulates which state that the electrons revolve around the atomic nucleus on certain stationary orbits, without radiating electromagnetic energy. The energy is gained or lost by the electrons only when they jump from one stationary orbit to another. This was the first quantum model of the atom and, although it was soon superseded by accurate quantum mechanical models, it remains a cornerstone of the history of physics in the conceptual leap from the classical to the quantum approach.

4.11.1.2 Method of Hertz's Vector

As shown in Chap. 3, Hertz's vector satisfies Eq. (3.100), which is written in vacuum as

$$\epsilon_0 \mu_0 \frac{\partial^2 \mathbf{Z}}{\partial t^2} - \Delta \mathbf{Z} = \frac{1}{\epsilon_0} \mathbf{P}, \tag{4.250}$$

and whose solution is the retarded potential (4.229):

$$\mathbf{Z}(\mathbf{r}, t) = \frac{1}{4\pi\epsilon_0} \int_{V'} \frac{\mathbf{P}\left(\mathbf{r}', t - \frac{|\mathbf{r}-\mathbf{r}'|}{c}\right)}{|\mathbf{r} - \mathbf{r}'|} \, d\mathbf{r}' = \frac{1}{4\pi\epsilon_0} \int_{V'} \frac{\mathbf{P}\left(\mathbf{r}', t - \frac{R}{c}\right)}{R} \, d\mathbf{r}'. \tag{4.251}$$

Considering a time-periodical variation of the source of the form

$$\mathbf{P}\left(\mathbf{r}', t - \frac{R}{c}\right) = \mathbf{P}(\mathbf{r}') \, e^{-i\omega(t-R/c)} = \mathbf{P}(\mathbf{r}') \, e^{-i\omega t} \, e^{ikR}, \tag{4.252}$$

the Hertz vector (4.251) becomes

$$\mathbf{Z}(\mathbf{r}, t) = \frac{1}{4\pi\epsilon_0} \int_{V'} \frac{\mathbf{P}(\mathbf{r}') \, e^{-i\omega t} \, e^{ikR}}{R} \, d\mathbf{r}' = \frac{e^{-i\omega t}}{4\pi\epsilon_0} \int_{V'} \frac{\mathbf{P}(\mathbf{r}') \, e^{ikR}}{R} \, d\mathbf{r}'. \tag{4.253}$$

In the first approximation ($R \simeq r$), this is

$$\mathbf{Z}(\mathbf{r}, t) = \frac{e^{-i\omega t}}{4\pi\epsilon_0} \int_{V'} \frac{\mathbf{P}(\mathbf{r}') \, e^{ikR}}{R} \, d\mathbf{r}' \cong \frac{e^{i(kr-\omega t)}}{4\pi\epsilon_0 r} \int_{V'} \mathbf{P}(\mathbf{r}') \, d\mathbf{r}'$$

$$= \frac{\mathbf{p}_0}{4\pi\epsilon_0 r} \, e^{i(kr-\omega t)} = \frac{1}{4\pi\epsilon_0} \frac{\mathbf{p}}{r}, \tag{4.254}$$

where we made the notation $\mathbf{p}_0 = \int_{V'} \mathbf{P}(\mathbf{r}') \, d\mathbf{r}'$, while the oscillating electric dipole moment is $\mathbf{p} = \mathbf{p}_0 \, e^{i(kr-\omega t)}$.

Using Eq. (3.95) for vacuum, the electromagnetic field $(\mathbf{E}, \mathbf{B})$ is expressed by means of the Hertz potential as

$$\mathbf{E} = \nabla(\nabla \cdot \mathbf{Z}) - \epsilon_0 \mu_0 \frac{\partial^2 \mathbf{Z}}{\partial t^2},$$

$$\mathbf{B} = \epsilon_0 \mu_0 \nabla \times \left(\frac{\partial \mathbf{Z}}{\partial t}\right).$$

Let us find first $\mathbf{E}$:

$$\mathbf{E} = \nabla(\nabla \cdot \mathbf{Z}) - \epsilon_0 \mu_0 \frac{\partial^2 \mathbf{Z}}{\partial t^2} = \frac{1}{4\pi\epsilon_0} \nabla \left[\nabla \cdot \left(\frac{\mathbf{p}}{r} \right) \right] - \frac{1}{4\pi\epsilon_0 c^2 r} \frac{\partial^2 \mathbf{p}}{\partial t^2}$$

$$= \frac{1}{4\pi\epsilon_0} \nabla \left[\frac{ik(\mathbf{s} \cdot \mathbf{p})}{r} - \frac{\mathbf{p} \cdot \mathbf{s}}{r^2} \right] + \frac{\omega^2 \mathbf{p}}{4\pi\epsilon_0 c^2 r} \cong \frac{ik}{4\pi\epsilon_0} \nabla \left(\frac{\mathbf{s} \cdot \mathbf{p}}{r} \right) - \frac{-\omega^2 \mathbf{p}}{4\pi\epsilon_0 c^2 r}$$

$$= \frac{ik}{4\pi\epsilon_0} \left[(\mathbf{s} \cdot \mathbf{p}) \nabla \left(\frac{1}{r} \right) + \frac{1}{r} \nabla (\mathbf{s} \cdot \mathbf{p}) \right] - \frac{\ddot{\mathbf{p}}}{4\pi\epsilon_0 c^2 r} \cong ik \frac{\nabla(\mathbf{s} \cdot \mathbf{p})}{4\pi\epsilon_0 r} - \frac{\ddot{\mathbf{p}}}{4\pi\epsilon_0 c^2 r}$$

$$= \frac{1}{4\pi\epsilon_0 c^2 r} \left[(\mathbf{s} \cdot \ddot{\mathbf{p}}) \mathbf{s} - \ddot{\mathbf{p}} \right].$$

The magnetic component of the field is

$$\mathbf{B} = \epsilon_0 \mu_0 \nabla \times \left(\frac{\partial \mathbf{Z}}{\partial t} \right) = \frac{1}{4\pi\epsilon_0} \epsilon_0 \mu_0 \nabla \times \left[\frac{\partial}{\partial t} \left(\frac{\mathbf{p}}{r} \right) \right] = -\frac{i\omega\mu_0}{4\pi} \nabla \times \left(\frac{\mathbf{p}}{r} \right)$$

$$= -\frac{i\omega\mu_0}{4\pi} \left[\frac{ik\,\mathbf{s} \times \mathbf{p}}{r} - \frac{\mathbf{s} \times \mathbf{p}}{r^2} \right] \cong \frac{k\omega\mu_0}{4\pi r} \mathbf{s} \times \mathbf{p}.$$

In the above calculations we have used everywhere the approximation $\mathcal{O}(r^{-2}) = 0$, which is valid in the wave zone.

Summarizing, we obtained for the electromagnetic field $(\mathbf{E}, \mathbf{B})$ the expressions

$$\mathbf{E} = \frac{1}{4\pi\epsilon_0 c^2 r} \left[(\mathbf{s} \cdot \ddot{\mathbf{p}}) \mathbf{s} - \ddot{\mathbf{p}} \right] = \frac{1}{4\pi\epsilon_0 c^2} \frac{\mathbf{s} \times (\mathbf{s} \times \ddot{\mathbf{p}})}{r},$$

$$\mathbf{B} = \frac{k\omega\mu_0}{4\pi r} \mathbf{s} \times \mathbf{p} = \frac{\mu_0}{4\pi c} \frac{\ddot{\mathbf{p}} \times \mathbf{s}}{r},$$

where we took into account the fact that $\ddot{\mathbf{p}} = -\omega^2 \mathbf{p}$. As expected, we re-obtained the relations (4.242).

4.11.2 The Centre-Fed Thin Linear Antenna

In this section we focus on a simple radiative system, but very useful in practice, since by a convenient arrangement of such simple systems, with suitably chosen current phases, more complex combinations – with a large angular distribution of radiative power – can be obtained.

4.11.2.1 The Sinusoidal Current Approximation

We shall determine the angular distribution of the time average of the power radiated in a unit of solid angle by a thin, linear antenna of length d, excited at its centre, assuming that the current is sinusoidally distributed along the antenna.

Let us consider that the antenna is oriented along the z-axis, as in Fig. 4.26. The average flux density of the electromagnetic energy emitted per unit time (in other words, the time average of the power radiated per unit surface) is the real part of the normal component of the complex Poynting vector $\tilde{\boldsymbol{\Pi}}$:

$$\langle \Phi_{em} \rangle \equiv \frac{dP_{rad}}{dS} = \operatorname{Re} \tilde{\Pi}_n = \frac{1}{2}\operatorname{Re}\left[\mathbf{n}\cdot(\mathbf{E}\times\mathbf{H}^*)\right] = \frac{1}{2}\operatorname{Re}\left[\mathbf{s}\cdot(\mathbf{E}\times\mathbf{H}^*)\right], \quad (4.255)$$

where $\langle\ \rangle$ represents the average value over a period of time, and $\mathbf{s}$ is the unit vector along the direction of $\mathbf{r}$ (which is orthogonal to the surface S and determines the solid angle Ω). This relation follows immediately if one uses the relation

$$\langle \boldsymbol{\Pi} \rangle \equiv \langle \mathbf{E}\times\mathbf{H} \rangle = \operatorname{Re}\tilde{\boldsymbol{\Pi}} = \frac{1}{2}\operatorname{Re}\left(\mathbf{E}\times\mathbf{H}^*\right), \quad (4.256)$$

where $\boldsymbol{\Pi} = \mathbf{E}\times\mathbf{H} = EH\,\mathbf{s}$ is the Poynting vector, signifying the radiant flux per unit time and unit surface orthogonal to $\mathbf{s}$ (the unit vector of the direction of propagation of the wave in the far zone). In relation (4.255) the average brackets for power have been omitted, but the significance of the expression is the time average of the power radiated per unit surface. This notation convention will be used everywhere from now on. Thus, we write the time average of the power radiated by the antenna per unit solid angle as

Fig. 4.26 Orientation of the linear antenna with respect to a Cartesian reference frame.

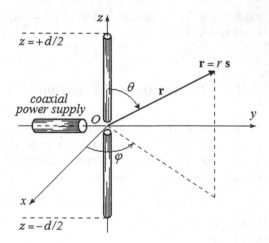

$$\frac{dP_{rad}}{d\Omega} = \text{Re}\left(r^2 \tilde{\Pi}_n\right) = \frac{1}{2}\text{Re}\left[r^2 \, \mathbf{s} \cdot (\mathbf{E} \times \mathbf{H}^*)\right]. \tag{4.257}$$

In order to calculate $dP_{rad}/d\Omega$, the components $\mathbf{E}$ and $\mathbf{H}$ of the field emitted by antenna have to be first determined. Supposing that the antenna is situated in vacuum, the magnetic field is given by the relation

$$\mathbf{H} = \frac{1}{\mu_0} \nabla \times \mathbf{A},$$

while $\mathbf{E}$ can be determined by means of Maxwell's equation, with the assumption that no conduction currents are present in the zone where the field is calculated:

$$\mu_0 \nabla \times \mathbf{H} = \frac{1}{c^2}\frac{\partial \mathbf{E}}{\partial t}. \tag{4.258}$$

Outside the source we consider a periodical time variation for the field $\mathbf{E}$ of the form $\mathbf{E}(\mathbf{r}, t) = \mathbf{E}_0(\mathbf{r})\, e^{-i\omega t}$. Then from (4.258) it follows that

$$\mathbf{E} = \frac{ic^2}{\omega} \nabla \times \mathbf{B} = \frac{i}{\omega \epsilon_0} \nabla \times \mathbf{H} = \frac{i}{k}\sqrt{\frac{\mu_0}{\epsilon_0}} \nabla \times \mathbf{H}. \tag{4.259}$$

Consequently, to determine $dP_{rad}/d\Omega$ it is only necessary to know the vector potential $\mathbf{A}$, since

$$\frac{dP_{rad}}{d\Omega} = \frac{1}{2}\text{Re}\left[r^2 \, \mathbf{s} \cdot (\mathbf{E} \times \mathbf{H}^*)\right]$$

$$= \frac{1}{2}\text{Re}\left\{\frac{i\,r^2}{k}\sqrt{\frac{\mu_0}{\epsilon_0}}\left[(\nabla \times \mathbf{H}) \times \mathbf{H}^*\right] \cdot \mathbf{s}\right\}, \tag{4.260}$$

while the magnetic field $\mathbf{H}$ (and, implicitly, $\mathbf{H}^*$) are directly expressed in terms of $\mathbf{A}$ which, if no significant boundary surfaces are present, is the retarded (causal) potential:

$$\mathbf{A}(\mathbf{r}, t) = \frac{\mu_0}{4\pi}\int d\mathbf{r}' \int dt' \, \frac{\mathbf{j}(\mathbf{r}', t')}{|\mathbf{r} - \mathbf{r}'|}\, \delta\left(t' + \frac{|\mathbf{r} - \mathbf{r}'|}{c} - t\right). \tag{4.261}$$

For a system of charges and time variable currents one can perform a Fourier analysis of the time dependence, and thus each component can be separately handled; consequently, there is no loss of generality if the potentials, fields, and radiation of a system of localized charges and currents, as in our case, is described by a sinusoidal time variation of the form

$$\mathbf{j}(\mathbf{r}, t) = \mathbf{j}(\mathbf{r})\, e^{-i\omega t}.$$

Plugging the above expression into (4.261), $\mathbf{A}(\mathbf{r}, t)$ becomes

$$\mathbf{A}(\mathbf{r}, t) = \frac{\mu_0}{4\pi} e^{-i\omega t} \int_{V'} \mathbf{j}(\mathbf{r}') \frac{e^{ik|\mathbf{r}-\mathbf{r}'|}}{|\mathbf{r}-\mathbf{r}'|} d\mathbf{r}', \qquad (4.262)$$

with the integral taken over the spatial domain of volume V', where the sources are distributed; in the present case, since the antenna is considered to be thin, this domain is represented by a cylinder of length d, with infinitely small transversal section.

In the wave zone ($kr \gg 1$), the exponential $e^{ik|\mathbf{r}-\mathbf{r}'|}$ oscillates very fast and determines the behaviour of the vector potential $\mathbf{A}(\mathbf{r}, t)$. In this region we use the approximation $|\mathbf{r} - \mathbf{r}'| \cong r - \mathbf{s} \cdot \mathbf{r}'$. This relation is valid also for $r \gg l$, where l is the dimension of the source, independently of the value of the product kr. Therefore, this approximation is adequate even for the (static) near-zone, which is characterized by $d \ll r \ll \lambda$, as well as for the intermediate (induction) zone, defined by $d \ll r \sim \lambda$, where $\lambda = 2\pi c/\omega$ is the wavelength of the electromagnetic radiation. Moreover, if only the main term is required, then the inverse of the distance appearing in (4.262) can be simply replaced by r, and the vector potential becomes

$$\mathbf{A}(\mathbf{r}, t) = \frac{\mu_0}{4\pi} \frac{e^{i(kr-\omega t)}}{r} \int_{V'} \mathbf{j}(\mathbf{r}') e^{-ik(\mathbf{s}\cdot\mathbf{r}')} d\mathbf{r}'. \qquad (4.263)$$

This formula shows that in the far zone the vector potential behaves like an emergent spherical wave, with an angle-dependent coefficient. Besides, it can be easily shown that the relations $\mathbf{H} = \mu_0^{-1} \nabla \times \mathbf{A}$ and $\mathbf{E} = ik^{-1}\sqrt{\mu_0/\epsilon_0} \nabla \times \mathbf{H}$ yield fields orthogonal to the radius vector r and decreasing like $|\mathbf{r}|^{-1}$, i.e. radiative fields.

If the damping due to radiation is neglected and the antenna is thin enough, then it can be considered that along the antenna the current varies sinusoidally in space and time, the wave number being $k = \omega/c$. In addition, the current is symmetrically distributed in the two sides of the antenna and it vanishes at its ends. These properties are mathematically modeled by the expression

$$\mathbf{j}(\mathbf{r}) = I \sin\left(\frac{kd}{2} - k|z|\right) \delta(x) \, \delta(y) \, \mathbf{u}_3, \qquad (4.264)$$

where $\mathbf{u}_3 \equiv \hat{z}$ is the unit vector of the Oz-axis. The delta functions restrict the flow of the current only along the Oz-axis. If $kd \geq \pi$, then I is the maximum value of the current, while at the origin of the coordinate system (i.e. the excitation point of the antenna) the current has the constant value $I \sin(kd/2)$.

Introducing (4.264) into (4.263), it follows that in the radiation zone the vector potential $\mathbf{A}(\mathbf{r}, t)$ is oriented along the Oz-axis and has the form

$$
\begin{aligned}
\mathbf{A}(\mathbf{r}, t) &= \frac{\mu_0}{4\pi} \frac{e^{i(kr-\omega t)}}{r} \int_{V'} \mathbf{j}(\mathbf{r}') e^{-ik(\mathbf{s}\cdot\mathbf{r}')} d\mathbf{r}' \\
&= \mathbf{u}_3 \frac{\mu_0 I e^{ikr} e^{-i\omega t}}{4\pi r} \int_{-\infty}^{+\infty} \delta(x') dx' \int_{-\infty}^{+\infty} \delta(y') dy' \\
&\quad \times \int_{-d/2}^{+d/2} \sin\left(\frac{kd}{2} - k|z'|\right) e^{-ikz' \cos\theta} dz' \\
&= \mathbf{u}_3 \frac{\mu_0 I e^{ikr} e^{-i\omega t}}{4\pi r} \int_{-d/2}^{+d/2} \sin\left(\frac{kd}{2} - k|z'|\right) e^{-ikz' \cos\theta} dz',
\end{aligned}
\tag{4.265}
$$

where the integration domain over x' and y' has been extended from $-\infty$ to $+\infty$. In order to calculate the integral

$$
J \equiv \int_{-d/2}^{+d/2} \sin\left(\frac{kd}{2} - k|z'|\right) e^{-ikz' \cos\theta} dz',
$$

we consider the auxiliary integral

$$
K \equiv \int_{-d/2}^{+d/2} \cos\left(\frac{kd}{2} - k|z'|\right) e^{-ikz' \cos\theta} dz'
$$

and form the complex conjugated combinations $C_1 = K - iJ$ and $C_2 = K + iJ$, which are easier to calculate, the integral K being then given by $J = \frac{C_2 - C_1}{2i}$.

Using Euler's formula, one finds

$$
\begin{aligned}
C_1 &= \int_{-d/2}^{+d/2} e^{-i\left(\frac{kd}{2} - k|z'|\right)} e^{-ikz' \cos\theta} dz' \\
&= \frac{i}{k} \frac{e^{-i\frac{kd}{2}}}{\sin^2\theta} \left[2 - 2e^{i\frac{kd}{2}} \cos\left(\frac{kd}{2} \cos\theta\right) + 2ie^{i\frac{kd}{2}} \cos\theta \sin\left(\frac{kd}{2} \cos\theta\right) \right],
\end{aligned}
$$

and

$$
\begin{aligned}
C_2 &= \int_{-d/2}^{+d/2} e^{i\left(\frac{kd}{2} - k|z'|\right)} e^{-ikz' \cos\theta} dz' \\
&= \frac{-i}{k} \frac{e^{i\frac{kd}{2}}}{\sin^2\theta} \left[2 - 2e^{-i\frac{kd}{2}} \cos\left(\frac{kd}{2} \cos\theta\right) - 2ie^{-i\frac{kd}{2}} \cos\theta \sin\left(\frac{kd}{2} \cos\theta\right) \right].
\end{aligned}
$$

Then, we have

$$
J = \int_{-d/2}^{+d/2} \sin\left(\frac{kd}{2} - k|z'|\right) e^{-ikz'\cos\theta} dz' = \frac{C_2 - C_1}{2i}
$$

$$
= \frac{2}{k\sin^2\theta}\left[\cos\left(\frac{kd}{2}\cos\theta\right) - \cos\frac{kd}{2}\right]. \tag{4.266}
$$

Introducing this result into (4.265), one obtains the expression for the vector potential $\mathbf{A}(\mathbf{r}, t)$,

$$
\mathbf{A}(\mathbf{r}, t) = A_0 \frac{e^{ikr}}{r} \mathbf{u}_3, \tag{4.267}
$$

where

$$
A_0 = \frac{\mu_0 I}{2\pi k} \frac{\cos\left(\frac{kd}{2}\cos\theta\right) - \cos\frac{kd}{2}}{\sin^2\theta} e^{-i\omega t}. \tag{4.268}
$$

To obtain the angular distribution of the power radiated by the antenna, $dP_{rad}/d\Omega$, using relation (4.260), one must calculate the quantity $\left[(\nabla \times \mathbf{H}) \times \mathbf{H}^*\right] \cdot \mathbf{s}$, where $\mathbf{H} = \mu_0^{-1}\nabla \times \mathbf{A}$. We have

$$
\mathbf{H} = \mu_0^{-1}\nabla \times \mathbf{A} = \mu_0^{-1}A_0\nabla \times \left(\frac{e^{ikr}}{r}\mathbf{u}_3\right) = \mu_0^{-1}A_0 e^{ikr}\left(\frac{ik}{r} - \frac{1}{r^2}\right)\frac{\mathbf{r}}{r} \times \mathbf{u}_3. \tag{4.269}
$$

In the wave zone this relation acquires the asymptotic form (ignoring the $\mathcal{O}(r^{-2})$ terms)

$$
\mathbf{H} = \frac{iI}{2\pi r} \frac{\cos\left(\frac{kd}{2}\cos\theta\right) - \cos\frac{kd}{2}}{\sin^2\theta} e^{-i(\omega t - kr)}(\mathbf{s} \times \mathbf{u}_3), \tag{4.270}
$$

or, in a simpler form,

$$
\mathbf{H} = \frac{ik}{\mu_0}\mathbf{s} \times \mathbf{A}, \tag{4.271}
$$

which yields

$$
|\mathbf{H}| = \sqrt{\mathbf{H} \cdot \mathbf{H}^*} = \frac{I}{2\pi r} \frac{\cos\left(\frac{kd}{2}\cos\theta\right) - \cos\frac{kd}{2}}{\sin\theta}. \tag{4.272}
$$

For $\nabla \times \mathbf{H}$ one finds

$$
\nabla \times \mathbf{H} = \nabla \times \left[\mu_0^{-1}A_0 e^{ikr}\left(\frac{ik}{r} - \frac{1}{r^2}\right)\frac{\mathbf{r}}{r} \times \mathbf{u}_3\right] = \frac{kI\, e^{-i(\omega t - kr)}}{2\pi r}
$$

$$
\times \frac{\cos\left(\frac{kd}{2}\cos\theta\right) - \cos\frac{kd}{2}}{\sin^2\theta}(\mathbf{s} \times \mathbf{u}_3) \times \mathbf{s} + \frac{I\, e^{-i(\omega t - kr)}}{\pi k}
$$

$$
\times \frac{\cos\left(\frac{kd}{2}\cos\theta\right) - \cos\frac{kd}{2}}{\sin^2\theta}\left(\frac{ik}{r^2} - \frac{1}{r^3}\right)(\mathbf{s} \times \mathbf{u}_3) \times \mathbf{s}. \tag{4.273}
$$

Thus, in the radiative zone, the electric field takes the asymptotic form

$$\mathbf{E} = \frac{i}{k} \sqrt{\frac{\mu_0}{\epsilon_0}} \nabla \times \mathbf{H} = \frac{iIe^{-i(\omega t - kr)}}{2\pi r} \frac{\cos\left(\frac{kd}{2}\cos\theta\right) - \cos\frac{kd}{2}}{\sin^2\theta} \sqrt{\frac{\mu_0}{\epsilon_0}} (\mathbf{s} \times \mathbf{u}_3) \times \mathbf{s}$$

$$= \sqrt{\frac{\mu_0}{\epsilon_0}} \mathbf{H} \times \mathbf{s}. \tag{4.274}$$

Putting together the previous results, for the wave zone we can write

$$\left[(\nabla \times \mathbf{H}) \times \mathbf{H}^*\right] \cdot \mathbf{s} = \left\{ \left[\frac{k\, Ie^{-i(\omega t - kr)}}{2\pi r} \frac{\cos\left(\frac{kd}{2}\cos\theta\right) - \cos\frac{kd}{2}}{\sin^2\theta} \right. \right.$$

$$\times (\mathbf{s} \times \mathbf{u}_3) \times \mathbf{s}] \times \left[\frac{i\, Ie^{-i(\omega t - kr)}}{2\pi r} \frac{\cos\left(\frac{kd}{2}\cos\theta\right) - \cos\frac{kd}{2}}{\sin^2\theta} \mathbf{s} \times \mathbf{u}_3 \right]^* \right\} \cdot \mathbf{n}$$

$$= -\frac{ikI^2}{4\pi^2 r^2} \left| \frac{\cos\left(\frac{kd}{2}\cos\theta\right) - \cos\frac{kd}{2}}{\sin^2\theta} \right|^2 \sin^2\theta.$$

Introducing this expression into (4.260), we find

$$\frac{dP_{rad}}{d\Omega} = \frac{I^2 Z_0}{8\pi^2} \left| \frac{\cos\left(\frac{kd}{2}\cos\theta\right) - \cos\frac{kd}{2}}{\sin\theta} \right|^2, \tag{4.275}$$

where $Z_0 = \sqrt{\mu_0/\epsilon_0} \cong 376.7\,\Omega$ is the *impedance of the free space*.

The electric vector is oriented along the component of $\mathbf{A}$ orthogonal to $\mathbf{s}$ (that is, the direction of the vector $(\mathbf{s} \times \mathbf{u}_3) \times \mathbf{s}$, or, equivalently, the direction of the vector $\mathbf{H} \times \mathbf{s}$). Consequently, the polarization of the radiation lies in the plane defined by the antenna and the radius vector of the observation point.

The angular distribution (4.275) depends, obviously, on the value of the product kd. It can be easily shown that, for large wavelengths, this result reduces to the one corresponding to the dipole, that is

$$\frac{dP_{rad}}{d\Omega} = \frac{I^2 Z_0}{32\pi^2} \left(\frac{kd}{2}\right)^4 \sin^2\theta = \frac{\pi^2 I^2 d^4}{32} Z_0 \sin^2\theta \frac{1}{\lambda^4}, \tag{4.276}$$

i.e. the Rayleigh scattering law $(dP_{rad}/d\Omega \sim \lambda^{-4})$. Indeed, for $kd \ll 1$ we have

$$\cos\left(\frac{kd}{2}\right) \cong 1 - \frac{k^2 d^2}{8},$$

$$\cos\left(\frac{kd}{2}\cos\theta\right) \cong 1 - \frac{k^2 d^2 \cos^2\theta}{8}$$

and then

$$\left| \frac{\cos\left(\frac{kd}{2}\cos\theta\right) - \cos\frac{kd}{2}}{\sin\theta} \right|^2 \cong \left| \frac{\frac{k^2 d^2}{8}(1 - \cos^2\theta)}{\sin\theta} \right|^2 = \frac{k^4 d^4}{64}\sin^2\theta.$$

Thus, (4.275) leads to (4.276).

For the special values $kd = \pi$ and $kd = 2\pi$, corresponding to the length of the antenna being equal to one half or two halves of the wavelength ($d = \lambda/2$ and $d = \lambda$, respectively) of the current oscillating along the antenna, the angular distributions are

$$\frac{dP_{rad}}{d\Omega} = \frac{I^2 Z_0}{8\pi^2} \frac{\cos^2\left(\frac{\pi}{2}\cos\theta\right)}{\sin^2\theta} \times \begin{cases} 1, & kd = \pi \quad \left(d = \frac{\lambda}{2}\right); \\ 4\cos^2\left(\frac{\pi}{2}\cos\theta\right), & kd = 2\pi \quad (d = \lambda). \end{cases}$$

$$(4.277)$$

These angular distributions are graphically represented in Fig. 4.27a for the half-wave antenna and Fig. 4.27b for the full-wave antenna. In Fig. 4.27c both distributions are drawn on the same figure, for comparison.

Fig. 4.27 Dipole angular distributions for a thin, linear, centrally-fed antenna.

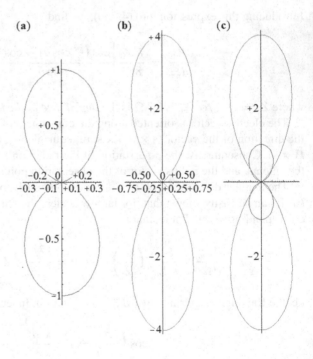

The distribution for the half-wave antenna is very similar to a simple dipole picture, while the distribution of the full-wave antenna is much sharper. The full-wave antenna distribution can be imagined as the coherent superposition of the fields corresponding to two half-wave antennas, one above the other, both of them being excited in phase. The intensity at $\theta = \pi/2$, where the waves add algebraically, is four times bigger than that of a half-wave antenna. Far from $\theta = \pi/2$, the amplitudes interfere giving a sharper graph. Using a convenient arrangement of basic antennas, such as half-wave antennas, with the phases of the currents suitably chosen, one can form by coherent superposition arbitrary radiative figures.

The total power emitted by the antenna is

$$P_{rad} = \int dP_{rad} = \iint \frac{dP_{rad}}{d\Omega} \, d\Omega = \frac{I^2 Z_0}{8\pi^2} \int_0^\pi \left| \frac{\cos\left(\frac{kd}{2}\cos\theta\right) - \cos\frac{kd}{2}}{\sin\theta} \right|^2$$

$$\times \sin\theta \, d\theta \int_0^{2\pi} d\varphi = \frac{I^2 Z_0}{4\pi} \left\{ \gamma - 2\cos^2\left(\frac{kd}{2}\right) \mathrm{Ci}(kd) + \ln(kd) \right.$$

$$+ \frac{1}{2}\left[\gamma + \mathrm{Ci}(2kd) + \ln\left(\frac{kd}{2}\right)\right]\cos(kd)$$

$$\left. + \frac{1}{2}\left[\mathrm{Si}(2kd) - 2\,\mathrm{Si}(kd)\right]\sin(kd) \right\}, \tag{4.278}$$

where

$$\gamma = \lim_{n\to\infty}\left[\left(\sum_{k=1}^\infty \frac{1}{k}\right) - \ln n\right] \cong 0.577216$$

is the *Euler–Mascheroni constant* (sometimes simply called the *Euler constant*), while $\mathrm{Si}\,z = \int_0^z \frac{\sin t}{t}\,dt$ and $\mathrm{Ci}\,z = -\int_z^\infty \frac{\cos t}{t}\,dt$ are the special functions *sine integral* and *cosine integral*, respectively.

We note that, for a fixed excitation current, the total radiated power depends on frequency through the function

$$F(k) = F\left(\frac{2\pi\nu}{c}\right) = \gamma - 2\cos^2\left(\frac{kd}{2}\right)\mathrm{Ci}(kd) + \ln(kd)$$

$$+ \frac{1}{2}\left[\gamma + \mathrm{Ci}(2kd) + \ln\left(\frac{kd}{2}\right)\right]\cos(kd) \tag{4.279}$$

$$+ \frac{1}{2}\left[\mathrm{Si}(2kd) - 2\,\mathrm{Si}(kd)\right]\sin(kd).$$

The relative extremum points of the function $F(k)$ can be determined by solving the equation

$$
0 = \frac{dF(k)}{dk} = \frac{1}{k} - \frac{2}{k}\cos(kd)\cos^2\left(\frac{kd}{2}\right) + \frac{\cos(kd)}{2k}[1+\cos(2kd)] + 2d
$$
$$
\times \cos\left(\frac{kd}{2}\right)\mathrm{Ci}(kd)\sin\left(\frac{kd}{2}\right) - \frac{d\sin(kd)}{2}\left[\gamma + \mathrm{Ci}(2kd) + \ln\left(\frac{kd}{2}\right)\right]
$$
$$
+ \frac{\sin(kd)}{2k}[\sin(2kd) - 2\sin(kd)] + \frac{d\cos(kd)}{2}[\mathrm{Si}(2kd) - 2\,\mathrm{Si}(kd)].
$$

For a typical antenna with $d = 2 \times 10^{-1}$m, the first solution of the above equation (the first relative maximum) is $k_{M_1} \cong 28.1617\,\mathrm{m}^{-1}$, corresponding to the frequency $\nu_{M_1} \cong 1.345\,\mathrm{GHz}$ and to the wavelength $\lambda_{M_1} \cong 22.31\,\mathrm{cm}$.

Figure 4.28 shows the dependence of the total power radiated by the antenna on the wave number $k = 2\pi\nu/c$, for the value $I = 10\,\mathrm{mA}$ of the excitation current. This representation corresponds to the interval $k \in [(2\pi/3) \times 10^{-6},\ 10\pi]$ for the wave number, or, equivalently, the frequency interval $\nu \in [0.1\,\mathrm{kHz},\ 1.5\,\mathrm{GHz}]$, or, still, the wavelength interval $\lambda \in [20\,\mathrm{cm},\ 3000\,\mathrm{km}]$.

Using the power series expansion of the sine and cosine functions, as well as of the special functions sine integral

$$
\mathrm{Si}\,x = x - \frac{x^3}{18} + \frac{x^5}{600} - \frac{x^7}{35280} + \mathcal{O}(x^9)
$$

and cosine integral

$$
\mathrm{Ci}\,x = \gamma + \ln x - \frac{x^2}{4} + \frac{x^4}{96} - \frac{x^6}{4320} + \mathcal{O}(x^8),
$$

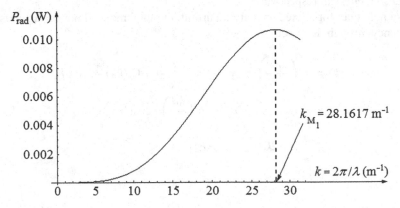

Fig. 4.28 Dependence of the total power radiated by a thin, linear, centrally-fed antenna on the wave number $k = |\mathbf{k}|$, for $k \in [(2\pi/3) \times 10^{-6},\ 10\pi]$.

for a fixed excitation current and small frequency values (corresponding to large wavelengths, $kd \ll 1$), the radiated power increases with frequency to approximately the fourth power. Indeed, we have

$$F(k) = \frac{k^4 d^4}{48} + \mathcal{O}\left[(kd)^6\right] \cong \frac{k^4 d^4}{48},$$

showing that in the range of large wavelengths this type of antenna behaves like an electric dipole radiator.

Due to the contribution of the multipoles of higher order, the approximation $F(k) \cong k^4 d^4/48$ (in other words, $\mathcal{O}[(kd)^6] = 0$) is valid only in the limit of large wavelengths. This fact is displayed in Fig. 4.29, showing the dependencies

$$P_{rad}^{exact} = P_{rad}^{exact}(k) = \frac{I^2 Z_0}{4\pi} F(k)$$

and

$$P_{rad}^{dipole\,type} \equiv P_{rad}^{dtype}(k) = \frac{I^2 Z_0}{4\pi} \frac{k^4 d^4}{48}.$$

Remark that the two curves overlap with a good approximation only for frequencies not higher than 477 MHz, or, equivalently, wavelengths not lower than $\lambda \simeq 62.8\,\text{cm}$. The percentage error in calculating the total power radiated by the antenna when taking into consideration only the dipole-type component (i.e., if the contribution of the higher-order dipoles is neglected) is graphically represented in Fig. 4.30 as a function of the wave vector modulus $k = |\mathbf{k}|$:

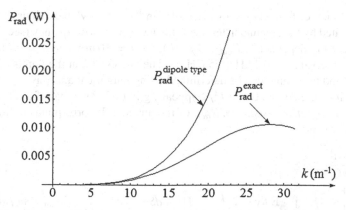

Fig. 4.29 Contribution of higher order multipoles to the total power radiated by a thin, linear, centrally-fed antenna.

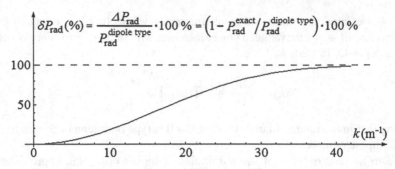

Fig. 4.30 Dependence of the percentage relative error of the total power radiated by a thin, linear, centrally-excited antenna on the wave number modulus $k = |\mathbf{k}|$.

$$\frac{P_{rad}^{dipole\ type} - P_{rad}^{exact}}{P_{rad}^{dipole\ type}} \cdot 100\,\% \equiv \frac{P_{rad}^{dtype} - P_{rad}}{P_{rad}^{dtype}} \cdot 100\,\%$$

$$= \left\{ 1 - \frac{48}{k^4 d^4} \left([\mathrm{Si}(2kd) - 2\,\mathrm{Si}(kd)] \frac{\sin(kd)}{2} \right.\right.$$

$$- 2\cos^2 \left(\frac{kd}{2}\right) \mathrm{Ci}(kd) + \gamma + \ln(kd)$$

$$\left.\left. + \frac{1}{2} \left[\gamma + \ln\left(\frac{kd}{2}\right) + \mathrm{Ci}(2kd) \right] \cos(kd) \right) \right\} \cdot 100\,\%.$$

As can be observed from Fig. 4.30, at least within the frequency domain corresponding to the interval $k \in [0,\ 40]\,\mathrm{m}^{-1}$, the larger the wavelength, the smaller this relative "error" is.

Going to larger frequencies, one remarks an "oscillatory" behaviour of the total power radiated by the antenna in terms of the wave number/frequency (see Fig. 4.31, representing the dependence $P_{rad} = P_{rad}(k)$, for $I = 10\,\mathrm{mA}$ and $d = 20\,\mathrm{cm}$ in the frequency interval $\nu \in [0.1\,\mathrm{kHz},\ 55\,\mathrm{GHz}]$). The values of k at the points of relative minimum and maximum can be determined using numerical analysis.

The unit of the coefficient of $I^2/2$ appearing in (4.278) is that of a resistance; it is called the *radiative resistance*, R_{rad}, of the antenna. It corresponds to the second term in the relation

$$R = \frac{1}{|I_e|^2} \tag{4.280}$$

$$\times \left\{ \mathrm{Re} \int_{(V)} \mathbf{j}^* \cdot \mathbf{E}\, d\mathbf{r} + 2 \oint_{(S-S_e)} \tilde{\Pi} \cdot \mathbf{s}\, dS + 4\omega\, \mathrm{Im} \int_{(V)} (w_{mag} - w_{el}) d\mathbf{r} \right\},$$

expressing the real part of the entrance impedance of an arbitrary, linear, passive electromagnetic system, with two terminals. This relation can be deduced by means of the complex Poynting theorem

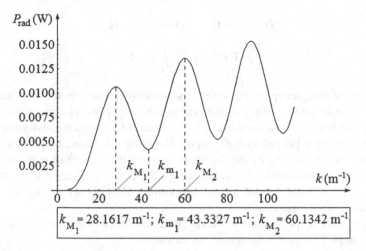

Fig. 4.31 The total power radiated by a thin, linear, centrally-fed antenna, as a function of the wave number modulus, $k = |\mathbf{k}|$, for an extended frequency interval.

$$\frac{1}{2} \int_V \mathbf{j}^* \cdot \mathbf{E} \, d\mathbf{r} + 2i\omega \int_V (w_{el} - w_{mag}) \, d\mathbf{r} + \oint_S \tilde{\Pi} \cdot \mathbf{s} \, dS = 0, \qquad (4.281)$$

written for harmonic fields, i.e. fields with the time dependence of the form $e^{-i\omega t}$, i.e.,

$$\mathbf{E}(\mathbf{r}, t) = \mathrm{Re}(\mathbf{E}(\mathbf{r})e^{i\omega t}) = \frac{1}{2} \left[\mathbf{E}(\mathbf{r})e^{i\omega t} + \mathbf{E}^*(\mathbf{r})e^{i\omega t} \right].$$

In the relations (4.280) and (4.281), V denotes the volume of the three-dimensional spatial domain occupied by the passive electromagnetic system, S is the surface bounding the volume V, I_e is the entrance current (complex quantity), S_e is the "entrance surface" through which the transfer of the entrance power takes place (usually, it is the section of the entrance coaxial line), $w_{el} = \frac{1}{4}(\mathbf{E} \cdot \mathbf{D}^*)$ and $w_{mag} = \frac{1}{4}(\mathbf{B} \cdot \mathbf{H}^*)$ are the (complex) harmonic electric and magnetic energy densities, respectively, $\tilde{\Pi}$ is the complex Poynting's vector, and $\frac{1}{2}\int_V \mathbf{j}^* \cdot \mathbf{E} \, d\mathbf{r}$ is the integral (corresponding to formula $\int_V \mathbf{j} \cdot \mathbf{E} \, d\mathbf{r}$, valid for real quantities and signifying the power of conversion of electromagnetic energy into some other forms of energy: mechanical, thermal, etc.), whose real part represents the average velocity of performing the mechanical work by the fields in the volume V.

The factor $1/2$ which appears in the definition of the radiative resistance (which, as we previously mentioned, is not the coefficient of I^2 – as it could appear at the first sight – but of $I^2/2$, in the formula of the total radiative power) is due to the complex character of harmonic fields. The most obvious effect – as we previously mentioned – consists in the replacement of the real Poynting's vector with the complex one, and of the real mechanical work performed in unit time by the field $\mathbf{E}$ on currents $\mathbf{j}$, by the complex correspondent

$$\Pi = \mathbf{E} \times \mathbf{H} \to \tilde{\Pi} = \frac{1}{2}\mathbf{E} \times \mathbf{H}^*,$$

$$\int_V \mathbf{j} \cdot \mathbf{E} \, d\mathbf{r} \to \frac{1}{2} \int_V \mathbf{j}^* \cdot \mathbf{E} \, d\mathbf{r}.$$

The use of complex quantities offers a significant advantage: it allows an easier identification and, at the same time, tracking of the active from reactive quantities. For example, the real coils and capacitors have both an active and a reactive component. Another example is offered by the complex Poynting's theorem: the real part of the mathematical relation expressing the theorem gives the conservation of energy for the time-averaged quantities, while the imaginary part is connected to the reactive energy (stocked in the system) and its alternating "flow" between various reactive components of the circuit.

Coming back to the problem of the radiative resistance of the antenna, in view of the relation (4.278) and its definition, we can write

$$
\begin{aligned}
R_{rad} = \frac{Z_0}{2\pi} &\left\{ \gamma - 2\cos^2\left(\frac{kd}{2}\right) \mathrm{Ci}(kd) + \ln(kd) \right. \\
&+ \frac{1}{2}\left[\gamma + \mathrm{Ci}(2kd) + \ln\left(\frac{kd}{2}\right) \right] \cos(kd) \\
&\left. + \frac{1}{2}\left[\mathrm{Si}(2kd) - 2\,\mathrm{Si}(kd) \right] \sin(kd) \right\}.
\end{aligned}
\tag{4.282}
$$

Just as in the case of the angular distribution of the power radiated by the antenna, here also there exist two cases of interest, namely:

(i) "half-wave" antenna, for which $d = \frac{\lambda}{2}$ ($kd = \pi$) and
(ii) "full-wave" antenna, for which $d = \lambda$ ($kd = 2\pi$).

In these cases, we have

$$
\begin{aligned}
R_{rad}^{half-wave} &= \frac{Z_0}{2\pi}\left\{ \gamma + \ln\pi - \frac{1}{2}\left[\gamma + \mathrm{Ci}(2\pi) + \ln\left(\frac{\pi}{2}\right) \right] \right\} \\
&\cong 1.22\,\frac{Z_0}{2\pi} \cong 73.14\,\Omega
\end{aligned}
\tag{4.283}
$$

and, respectively,

$$
\begin{aligned}
R_{rad}^{full-wave} &= \frac{Z_0}{2\pi}\left\{ \gamma - 2\,\mathrm{Ci}(2\pi) + \ln(2\pi) + \frac{1}{2}\left[\gamma + \mathrm{Ci}(4\pi) + \ln\pi \right] \right\} \\
&\cong 3.32\,\frac{Z_0}{2\pi} \cong 199.05\,\Omega.
\end{aligned}
\tag{4.284}
$$

As can easily be observed, in these two particular cases the radiative resistance of the antenna does not depend on frequency.

4.11.2.2   The Method of Multi-polar Expansion

Supposing that the antenna is excited in such a way that a sinusoidal current performs a complete wavelength oscillation as in Fig. 4.32, the current distribution along the antenna is

$$\mathbf{j}(\mathbf{r}) = I_0 \sin(kz)\delta(x)\delta(y)\hat{\mathbf{z}}. \tag{4.285}$$

If the dimensions of the source are small compared to the wavelength of the electromagnetic radiation, it is convenient to expand the integral (4.263) in powers of k:

$$\lim_{kr\to\infty} \mathbf{A}(\mathbf{r}) = \frac{\mu_0}{4\pi} \frac{e^{i(kr-\omega t)}}{r} \int_{V'} \mathbf{j}(\mathbf{r}') \, e^{-ik(\mathbf{s}\cdot\mathbf{r}')} \, d\mathbf{r}'$$

$$= \frac{\mu_0}{4\pi} \frac{e^{i(kr-\omega t)}}{r} \sum_{n=0}^{\infty} \frac{(-ik)^n}{n!} \int_{V'} \mathbf{j}(\mathbf{r}') \, (\mathbf{s}\cdot\mathbf{r}')^n \, d\mathbf{r}'.$$

For $n = 0$, the *electric dipole* contribution is obtained (see (4.240)), that is

$$\mathbf{A}_{el\ dipole} = \frac{\mu_0}{4\pi} \frac{e^{i(kr-\omega t)}}{r} \int \mathbf{j}(\mathbf{r}') \, d\mathbf{r}' = \frac{\mu_0 I_0}{4\pi} \frac{e^{i(kr-\omega t)}}{r}\hat{\mathbf{z}}$$

$$\times \int \delta(x') \, dx' \int \delta(y') \, dy' \int_{-d/2}^{+d/2} \sin(kz') \, dz' \tag{4.286}$$

$$= \frac{\mu_0 I_0}{4\pi} \frac{e^{i(kr-\omega t)}}{r}\hat{\mathbf{z}} \int_{-d/2}^{+d/2} \sin(kz') \, dz' = 0.$$

Thus, in this particular case, the electric dipole contribution vanishes.

For $n = 1$, the *magnetic dipole* and *electric quadrupole* contributions are obtained (see relation (4.235), as well as the related discussions). According to relation (4.235), the magnetic dipole contribution is

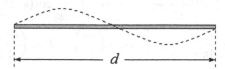

Fig. 4.32 Excitation of a thin, linear, centrally-fed antenna with a sinusoidal current performing a complete wavelength oscillation.

$$\mathbf{A}_{mag\ dipole} = -\frac{\mu_0}{4\pi} \frac{e^{i(kr-\omega t)}}{r} \frac{ik}{2} \int \left[(\mathbf{s} \cdot \mathbf{r}')\mathbf{j} - (\mathbf{s} \cdot \mathbf{j})\mathbf{r}' \right] d\mathbf{r}'$$

$$= -\frac{\mu_0}{4\pi} \frac{e^{i(kr-\omega t)}}{r} \frac{ik}{2} \int (\mathbf{r}' \times \mathbf{j}) \times \mathbf{s}\, d\mathbf{r}' \tag{4.287}$$

$$= \frac{\mu_0 I_0}{4\pi} \frac{e^{i(kr-\omega t)}}{r} \frac{ik}{2} \mathbf{s} \times \int \delta(x')\delta(y') \sin(kz')\, (\mathbf{r}' \times \hat{\mathbf{z}})\, dx'dy'dz'$$

$$= \frac{\mu_0 I_0}{4\pi} \frac{e^{i(kr-\omega t)}}{r} \frac{ik}{2} (\mathbf{s} \times \hat{\mathbf{x}}) \int y'\delta(y')dy' \int_{-d/2}^{+d/2} \sin(kz')\, dz'$$

$$- \frac{\mu_0 I_0}{4\pi} \frac{e^{i(kr-\omega t)}}{r} \frac{ik}{2} (\mathbf{s} \times \hat{\mathbf{y}}) \int x'\delta(x')\, dx' \int_{-d/2}^{+d/2} \sin(kz')\, dz' = 0,$$

because the integrals over x' and y' can be extended over the whole real axis, in which case both integrals are null. As a result, in this particular case the magnetic dipole contribution is also null.

Finally, according to the same relation (4.235), the electric quadrupole contribution is given by

$$\mathbf{A}_{el\ quadrupole} = -\frac{\mu_0}{4\pi} \frac{e^{i(kr-\omega t)}}{r} \frac{ik}{2} \int \left[(\mathbf{s} \cdot \mathbf{r}')\mathbf{j} - (\mathbf{s} \cdot \mathbf{j})\mathbf{r}' \right] d\mathbf{r}'$$

$$= -\hat{\mathbf{z}} \frac{\mu_0 I_0}{4\pi} \frac{e^{i(kr-\omega t)}}{r} \frac{ik}{2} \int \left[z' \cos\theta \sin(kz') + z' \cos\theta \sin(kz') \right]$$

$$\times \delta(x')\delta(y')dx'dy'dz' = -\hat{\mathbf{z}} \frac{\mu_0 I_0}{4\pi} \frac{e^{i(kr-\omega t)}}{r} ik \cos\theta$$

$$\times \int \delta(x')dx' \int \delta(y')dy' \int_{-d/2}^{+d/2} z' \sin(kz')dz' \tag{4.288}$$

$$= e^{i(kr-\omega t)} \frac{\mu_0 I_0}{4\pi} \frac{ik\hat{\mathbf{z}} \cos\theta}{r} \int_{-d/2}^{+d/2} \frac{\partial}{\partial k}[\cos(kz')]\, dz'$$

$$= e^{i(kr-\omega t)} \frac{\mu_0 I_0}{4\pi} \frac{ik\hat{\mathbf{z}} \cos\theta}{r} \frac{\partial}{\partial k} \left[\int_{-d/2}^{+d/2} \cos(kz')dz' \right]$$

$$= i\hat{\mathbf{z}} \frac{\mu_0 I_0}{4\pi} \frac{e^{i(kr-\omega t)}}{kr} \left[kd \cos\left(\frac{kd}{2}\right) - 2\sin\left(\frac{kd}{2}\right) \right] \cos\theta.$$

The temporal average of the power radiated by the antenna per unit solid angle is

$$\frac{dP_{rad}}{d\Omega} = \mathrm{Re}\left(r^2 \tilde{\Pi}_n \right) = \frac{1}{2}\mathrm{Re}\left[r^2 \mathbf{s} \cdot (\mathbf{E} \times \mathbf{H}^*) \right] = \frac{r^2}{2\mu_0}\mathrm{Re}\left[(\mathbf{E} \times \mathbf{B}^*) \cdot \mathbf{s} \right]. \tag{4.289}$$

In the wave zone ($\mathcal{O}(r^{-2}) = 0$), the magnetic component of the field is

$$
\begin{aligned}
\mathbf{B} = \nabla \times \mathbf{A} = \nabla \times & \left\{ i\hat{\mathbf{z}} \frac{\mu_0 I_0}{4\pi} \frac{e^{i(kr-\omega t)}}{kr} \left[kd \cos\left(\frac{kd}{2}\right) - 2\sin\left(\frac{kd}{2}\right) \right] \cos\theta \right\} \\
= & \ i\frac{\mu_0 I_0}{4\pi k} \left[kd\cos\left(\frac{kd}{2}\right) - 2\sin\left(\frac{kd}{2}\right) \right] \cos\theta \ \nabla\left(\frac{e^{i(kr-\omega t)}}{r}\right) \times \hat{\mathbf{z}} \\
= & \ i\frac{\mu_0 I_0}{4\pi k} \cos\theta \left[kd\cos\left(\frac{kd}{2}\right) - 2\sin\left(\frac{kd}{2}\right) \right] \frac{e^{i(krt-\omega t)}}{r} \left[ik - \frac{1}{r} \right] (\mathbf{s} \times \hat{\mathbf{z}}) \\
\cong & \ -\mathbf{u}_\varphi \frac{\mu_0 I_0}{4\pi} \left[kd\cos\left(\frac{kd}{2}\right) - 2\sin\left(\frac{kd}{2}\right) \right] \frac{e^{i(kr-\omega t)}}{r} \sin\theta \cos\theta, \qquad (4.290)
\end{aligned}
$$

and the electric component is given by

$$
\mathbf{E} = \frac{i}{k}\sqrt{\frac{\mu_0}{\epsilon_0}} \ \nabla \times \mathbf{H},
$$

which, as a result of some calculations similar to those leading to relation (4.274), within the same approximation $\mathcal{O}(r^{-2}) = 0$, becomes

$$
\mathbf{E} = \sqrt{\frac{\mu_0}{\epsilon_0}} \ \mathbf{H} \times \mathbf{s} = c\,\mathbf{B} \times \mathbf{s}.
$$

Then, using relation (4.290),

$$
\begin{aligned}
\mathrm{Re}\left[(\mathbf{E} \times \mathbf{B}^*) \cdot \mathbf{s} \right] = c\,\mathrm{Re}&\left\{ [(\mathbf{B} \times \mathbf{s}) \times \mathbf{B}^*] \cdot \mathbf{s} \right\} = c\,\left(|\mathbf{B}|^2 - |\mathbf{s} \cdot \mathbf{B}|^2 \right) = c\,|\mathbf{B}|^2 \\
= & \frac{c\mu_0^2 I_0^2}{16\pi^2 r^2} \left[kd\cos\left(\frac{kd}{2}\right) - 2\sin\left(\frac{kd}{2}\right) \right]^2 \sin^2\theta \cos^2\theta.
\end{aligned}
$$

Under these conditions, the angular distribution of the power radiated by the antenna is

$$
\begin{aligned}
\frac{dP_{rad}}{d\Omega} = \mathrm{Re}\left(r^2 \tilde{\Pi}_n \right) = & \ \frac{1}{2}\mathrm{Re}\left[r^2 \mathbf{s} \cdot (\mathbf{E} \times \mathbf{H}^*) \right] \qquad (4.291) \\
= & \ \frac{c\mu_0 I_0^2}{32\pi^2} \left[kd\cos\left(\frac{kd}{2}\right) - 2\sin\left(\frac{kd}{2}\right) \right]^2 \sin^2\theta \cos^2\theta.
\end{aligned}
$$

It can be easily shown that, in the limit of large wavelengths ($kd \ll 1$), this result leads to an oscillatory, spheroidal distribution of electric charge, representing one of the simplest examples of quadrupolar radiative source:

$$
\frac{dP_{rad}}{d\Omega} = \frac{c\mu_0 I_0^2}{32\pi^2} \frac{k^6 d^6}{144} \sin^2\theta \cos^2\theta = \frac{ck^6 Q_0^2}{512\epsilon_0 \pi^2} \sin^2\theta \cos^2\theta. \qquad (4.292)
$$

The quadrupolar momentum tensor corresponding to this symmetric charge distribution is diagonal; the elements of the tensor are

$$Q_{33} = Q_0, \quad Q_{11} = Q_{22} = -\frac{1}{2}Q_0,$$

where

$$Q_0 = \frac{I_0 d^3}{3c}.$$

Indeed, for $kd \ll 1$ we have $\cos\left(\frac{kd}{2}\right) \cong 1 - \frac{k^2 d^2}{8}$, as well as $\sin\left(\frac{kd}{2}\right) \cong \frac{kd}{2} - \frac{k^3 d^3}{48}$, therefore

$$\left[kd\cos\left(\frac{kd}{2}\right) - 2\sin\left(\frac{kd}{2}\right) \right]^2 \cong \left[kd\left(1 - \frac{k^2 d^2}{8}\right) - 2\left(\frac{kd}{2} - \frac{k^3 d^3}{48}\right) \right]^2$$
$$= \frac{k^6 d^6}{144}.$$

Introducing this relation into (4.291), we arrive at (4.292).

For the special values $kd = \pi$ and $kd = 2\pi$, corresponding to one half-wave ($d = \lambda/2$) and two half-waves ($d = \lambda$) of the current oscillating along the antenna, the angular distributions are

$$\frac{dP_{rad}}{d\Omega} = \frac{c\mu_0 I_0^2}{32\pi^2} \sin^2 2\theta \times \begin{cases} 1, & kd = \pi \quad \left(d = \frac{\lambda}{2}\right); \\[2mm] \pi^2, & kd = 2\pi \quad (d = \lambda). \end{cases} \tag{4.293}$$

Relations (4.293) show that both the half-wave and the full-wave antennas have the same angular distribution of radiated power. This distribution is graphically represented in Fig. 4.33. The figure has four lobes, with maxima for $\theta = \pi/4$ and $\theta = 3\pi/4$. The total power radiated by such a quadrupole is

$$P_{rad} = \int dP_{rad} = \iint \frac{dP_{rad}}{d\Omega} \, d\Omega = \frac{c\mu_0 I_0^2}{32\pi^2} \left[kd\cos\left(\frac{kd}{2}\right) - 2\sin\left(\frac{kd}{2}\right) \right]^2$$
$$\times \int_0^\pi \sin^2\theta \cos^2\theta \sin\theta \, d\theta \int_0^{2\pi} d\varphi \tag{4.294}$$
$$= \frac{c\mu_0 I_0^2}{30\pi} \left[\frac{kd}{2}\cos\left(\frac{kd}{2}\right) - \sin\left(\frac{kd}{2}\right) \right]^2.$$

Remark that, in the general case, the total radiated power depends on the frequency.

The radiative resistance of the antenna is the coefficient of $I_0^2/2$ appearing in the expression (4.294) of the power, that is

Fig. 4.33 Quadrupole angular distribution for a thin, linear, centrally-fed antenna.

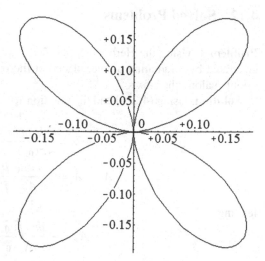

$$R_{rad} = R_{rad}(k) = \frac{c\mu_0}{15\pi}\left[\frac{kd}{2}\cos\left(\frac{kd}{2}\right) - \sin\left(\frac{kd}{2}\right)\right]^2. \tag{4.295}$$

For the special values $kd = \pi$ and $kd = 2\pi$, the total radiated power is

$$P_{rad} = \frac{c\mu_0 I_0^2}{30\pi} \times \begin{cases} 1, & kd = \pi \quad (d = \frac{\lambda}{2}); \\ \pi^2, & kd = 2\pi \quad (d = \lambda), \end{cases} \tag{4.296}$$

and depends only on the current I_0 traveling through the antenna. In this case, the radiative resistance in the two cases is

$$R_{rad} = \frac{c\mu_0}{15\pi} \times \begin{cases} 1, & kd = \pi, \ d = \frac{\lambda}{2}; \\ \pi^2, & kd = 2\pi, \ d = \lambda \end{cases} = \frac{Z_0}{15\pi} \times \begin{cases} 1, & kd = \pi, \ d = \frac{\lambda}{2}; \\ \pi^2, & kd = 2\pi, \ d = \lambda \end{cases}$$

$$\cong \begin{cases} 8\,\Omega, & kd = \pi, \quad d = \frac{\lambda}{2}; \\ 78.96\,\Omega, & kd = 2\pi, \quad d = \lambda. \end{cases} \tag{4.297}$$

Note that there is a considerable difference between the values of the radiative resistance of the "half-wave" and the "full-wave" antennas when we use the model which provides a "closed"-form result compared to the multipolar expansion method. This outcome can be easily understood, since in the latter case only the first two terms of the infinite series corresponding to the multipolar expansion (4.286) have been considered. Practically, the values given by the relation (4.297) represent only the contribution of the electric quadrupole momentum (which is the lowest order momentum whose contribution to the vector potential is not null).

4.12   Solved Problems

Problem 1. Using the Hertz potential $\mathbf{Z}(\mathbf{r}, t)$, determine the electromagnetic field generated by a radiating dipole, placed at the origin of the coordinate system and oriented along the z-axis.

  Solution. Using (4.229) and the notation $|\mathbf{p}| = p_z \equiv p$, we have

$$Z_z \equiv Z = \frac{1}{4\pi\epsilon_0} \frac{p_0 e^{i(kr-\omega t)}}{r},$$

$$A_z \equiv A = -\frac{i\omega\mu_0}{4\pi} \frac{p}{r}, \qquad (4.298)$$

leading to

$$A = -\frac{ik}{4\pi}\sqrt{\frac{\mu_0}{\epsilon_0}} \frac{p}{r}. \qquad (4.299)$$

We also have

$$V = -\frac{1}{4\pi\epsilon_0}\mathbf{p}_0 \cdot \nabla\left(\frac{e^{i(kr-\omega t)}}{r}\right),$$

or, by neglecting the terms proportional to r^{-n} $(n \geq 2)$,

$$V \simeq \frac{p\cos\theta}{4\pi\epsilon_0 r}\left(\frac{1}{r} - ik\right). \qquad (4.300)$$

Since the geometry of the problem suggests to use spherical coordinates, one has to write the components of the vector $\mathbf{A}$ in such coordinates, namely

$$A_r = A\cos\theta,$$
$$A_\theta = -A\sin\theta, \qquad (4.301)$$
$$A_\varphi = 0.$$

  Now we are able to determine the electromagnetic field at some point P, far enough from the source to fulfill the condition $\mathcal{O}(r^{-2}) = 0$. Taking only the real part of the field components, we have

$$E_r = -\frac{\partial V}{\partial r} - \frac{\partial A_r}{\partial t} = -\frac{p\cos\theta}{4\pi\epsilon_0}\left(\frac{ik}{r^2} - \frac{2}{r^3}\right) - \frac{ikp\cos\theta}{4\pi\epsilon_0 r}\left(\frac{1}{r} - ik\right) \simeq 0,$$

$$E_\theta = -\frac{1}{r}\frac{\partial V}{\partial \theta} - \frac{\partial A_\theta}{\partial t} = \frac{p\sin\theta}{4\pi\epsilon_0 r^2}\left(\frac{1}{r} - ik\right) - \frac{\omega k p\sin\theta}{4\pi r}\sqrt{\frac{\mu_0}{\epsilon_0}}$$

$$\simeq -\frac{\omega^2\sin\theta}{4\pi\epsilon_0 c^2 r}p_0\cos(kr - \omega t), \qquad (4.302)$$

$$E_\varphi = -\frac{1}{r\sin\theta}\frac{\partial V}{\partial \varphi} = 0.$$

In a similar way are calculated the components of the magnetic field **B**:

$$B_r = \frac{1}{r^2 \sin\theta} \left[\frac{\partial}{\partial\theta}(\sin\theta A_\varphi) - \frac{\partial}{\partial\varphi}(rA_\theta) \right] = 0,$$

$$B_\theta = \frac{1}{r \sin\theta} \left[\frac{\partial}{\partial\varphi}(A_r) - \frac{\partial}{\partial r}(r \sin\theta A_\varphi) \right] = 0, \tag{4.303}$$

$$B_\varphi = \frac{1}{r} \left[\frac{\partial}{\partial r}(r A_\theta) - \frac{\partial}{\partial\theta}(A_r) \right] = -\frac{\omega^2 \mu_0 \sin\theta}{4\pi rc} p_0 \cos(kr - \omega t).$$

Note that
$$|E_\theta| = c|B_\varphi|. \tag{4.304}$$

Observation:

There is an alternative method to reach the same result. Using (3.95) and (4.298), we have

$$\nabla \cdot \mathbf{Z} = i\mathbf{k} \cdot \mathbf{Z} = ikZ \cos\theta,$$
$$\nabla(\nabla \cdot \mathbf{Z}) = -k^2 \cos\theta \, \mathbf{s},$$
$$\frac{\partial \mathbf{Z}}{\partial t} = i\omega \mathbf{Z},$$
$$\frac{\partial^2 \mathbf{Z}}{\partial t^2} = -\omega^2 \mathbf{Z},$$

which allow us to write

$$\mathbf{E} = \nabla(\nabla \cdot \mathbf{Z}) - \epsilon_0 \mu_0 \frac{\partial^2 \mathbf{Z}}{\partial t^2} = -k^2 Z \cos\theta \, \mathbf{s} + k^2 Z \mathbf{u}_z,$$

$$\mathbf{B} = \epsilon_0 \mu_0 \nabla \times \left(\frac{\partial \mathbf{Z}}{\partial t} \right) = \epsilon_0 \mu_0 i\mathbf{k} \times (-i\omega \mathbf{Z}) = -\frac{k^2}{c} Z \sin\theta \, \mathbf{u}_\varphi.$$

The field components are then

$$E_r = 0, \quad E_\varphi = 0, \quad E_\theta = -k^2 Z \sin\theta,$$
$$B_r = 0, \quad B_\theta = 0, \quad B_\varphi = -\frac{k^2}{c} Z \sin\theta.$$

If now Z is expressed in terms of p, we find (4.302) and (4.303), as expected.

Problem 2. An atom radiates electromagnetic waves and remains in an excited state during the time interval τ. The time dependence of the electric field emitted by the atom is given by

$$E(t) = E_0 \, e^{-\frac{t}{\tau} - i\omega_0 t}.$$

Determine the halfwidth of the spectral line emitted by the atom, which is given by the expression

$$I\left(\omega_0 \pm \frac{\Delta\omega}{2}\right) = \frac{1}{2}I(\omega_0),\tag{4.305}$$

where $I(\omega)$ represents the intensity of the emitted radiation.

Solution. Before emission, both the magnetic and the electric components of the field are obviously null. Since the magnitude of the magnetic component of the emitted electromagnetic field is much smaller than the electric component, we shall further consider only the latter.

Taking into consideration these preliminary observations, we may write the time dependence of the emitted field as

$$E(t) = \begin{cases} 0, & \text{for } t < 0; \\ E_0\, e^{-\frac{t}{\tau} - i\omega_0 t}, & \text{for } t > 0. \end{cases}\tag{4.306}$$

Due to the fact that the dependence on time and ω is continuous, the Fourier expansion of the field is

$$E(t) = \frac{1}{\sqrt{2\pi}} \int_{-\infty}^{+\infty} E(\omega) e^{-i\omega t}\, d\omega,\tag{4.307}$$

where $E(\omega)$ is the Fourier transform of $E(t)$:

$$\begin{aligned}
E(\omega) &= \frac{1}{\sqrt{2\pi}} \int_{-\infty}^{+\infty} E(t)\, e^{i\omega t}\, dt = \frac{1}{\sqrt{2\pi}} \int_{-\infty}^{0} 0 \cdot e^{i\omega t} dt \\
&+ \frac{1}{\sqrt{2\pi}} \int_{0}^{+\infty} E_0\, e^{-\frac{t}{\tau} - i\omega_0 t}\, e^{i\omega t} dt \\
&= \frac{E_0}{\sqrt{2\pi}} \int_{0}^{+\infty} e^{-t\left[\frac{1}{\tau} + i(\omega_0 - \omega)\right]} dt \\
&= \frac{E_0}{\sqrt{2\pi}} \left[\frac{-1}{\frac{1}{\tau} + i(\omega_0 - \omega)} e^{-t\left[\frac{1}{\tau} + i(\omega_0 - \omega)\right]} \right]_{t=0}^{t=+\infty} \\
&= \frac{E_0}{\sqrt{2\pi}\left[\frac{1}{\tau} + i(\omega_0 - \omega)\right]}.
\end{aligned}$$

The intensity of the emitted radiation $I(\omega)$ is, by definition,

$$I(\omega) \sim |E(\omega)|^2 = \frac{E_0^2}{2\pi} \frac{1}{\left|\frac{1}{\tau} + i(\omega_0 - \omega)\right|^2} = \frac{E_0^2}{2\pi} \frac{1}{\frac{1}{\tau^2} + (\omega_0 - \omega)^2},$$

or

$$I(\omega) = \frac{I_0}{\frac{1}{\tau^2} + (\omega_0 - \omega)^2}.\tag{4.308}$$

Here I_0 comes from

$$I_0 = \int_{-\infty}^{+\infty} I(\omega)d\omega,$$

but its exact expression is not important for this problem.

Using (4.305) in which we plug the expression (4.308), we find

$$\frac{I_0}{\frac{1}{\tau^2} + \left[\omega_0 - \left(\omega_0 \pm \frac{\Delta\omega}{2}\right)\right]^2} = \frac{1}{2}\frac{I_0}{\frac{1}{\tau^2}} = \frac{I_0\tau^2}{2},$$

leading to

$$\frac{1}{\tau^2} + \left[\omega_0 - \left(\omega_0 \pm \frac{\Delta\omega}{2}\right)\right]^2 = \frac{2}{\tau^2},$$

which finally gives

$$\Delta\omega = \frac{2}{\tau}. \tag{4.309}$$

Problem 3. Decompose in plane waves the electromagnetic field generated by an electron, moving uniformly along a straight line in vacuum.

Solution. Let $\mathbf{v}$ be the velocity and e the modulus of the charge of the electron, ρ – the charge density, and $\mathbf{A}(\mathbf{r}, t)$, $V(\mathbf{r}, t)$ – the potentials associated to the field of the electron. Choosing as origin of a reference system the position of the electron at the time $t = 0$, we can write the charge density of the physical system under consideration by means of the Dirac delta function

$$\rho = e\delta(\mathbf{r} - \mathbf{v}t). \tag{4.310}$$

The inhomogeneous d'Alembert-type equations satisfied by $\mathbf{A}(\mathbf{r}, t)$ and $V(\mathbf{r}, t)$ are, respectively,

$$\Delta\mathbf{A} - \frac{1}{c^2}\frac{\partial^2\mathbf{A}}{\partial t^2} = -\mu_0 e\,\mathbf{v}\,\delta(\mathbf{r} - \mathbf{v}t), \tag{4.311}$$

$$\Delta V - \frac{1}{c^2}\frac{\partial^2 V}{\partial t^2} = -\frac{e}{\epsilon_0}\delta(\mathbf{r} - \mathbf{v}t). \tag{4.312}$$

We note that

$$\mathbf{A} = \frac{V}{c^2}\mathbf{v}, \tag{4.313}$$

meaning that we may study only one potential, for example V. Keeping t unchanged and denoting by $V_{\mathbf{k}}(t)$ the Fourier transform of $V(\mathbf{r}, t)$, we have

$$V(\mathbf{r}, t) = \frac{1}{(2\pi)^{3/2}}\int V_{\mathbf{k}}(t)\,e^{i\mathbf{k}\cdot\mathbf{r}}d\mathbf{k}. \tag{4.314}$$

Now we introduce (4.314) into (4.312) and write on the right-hand side the Fourier transform of $\delta(\mathbf{r} - \mathbf{v}t)$:

$$\frac{1}{(2\pi)^{3/2}} \int \left[-k^2 V_{\mathbf{k}}(t) - \frac{1}{c^2} \frac{\partial^2 V_{\mathbf{k}}(t)}{\partial t^2} \right] e^{i\mathbf{k}\cdot\mathbf{r}} d\mathbf{k} = -\frac{e}{(2\pi)^3 \epsilon_0} \int e^{i\mathbf{k}\cdot(\mathbf{r}-\mathbf{v}t)} d\mathbf{k},$$

finding the equation for the Fourier transform $V_{\mathbf{k}}(t)$:

$$\frac{\partial^2 V_{\mathbf{k}}(t)}{\partial t^2} + c^2 k^2 V_{\mathbf{k}}(t) = \frac{c^2 e}{(2\pi)^{3/2}\epsilon_0} e^{-i\mathbf{k}\cdot\mathbf{v}t}. \tag{4.315}$$

We seek a solution of this equation of the form

$$V_{\mathbf{k}}(t) = V_{\mathbf{k}}^{(0)} e^{-i\mathbf{k}\cdot\mathbf{v}t}, \tag{4.316}$$

which we plug into (4.315), with the result

$$V_{\mathbf{k}}(t) = \frac{e}{(2\pi)^3 \epsilon_0} \frac{e^{-i\mathbf{k}\cdot\mathbf{v}t}}{k^2 - \left(\frac{\mathbf{k}\cdot\mathbf{v}}{c}\right)^2}. \tag{4.317}$$

Using this expression, we obtain both $V(\mathbf{r}, t)$ from (4.314):

$$V(\mathbf{r}, t) = \frac{e}{(2\pi)^3 \epsilon_0} \int \frac{e^{i\mathbf{k}\cdot(\mathbf{r}-\mathbf{v}t)}}{k^2 - \left(\frac{\mathbf{k}\cdot\mathbf{v}}{c}\right)^2} d\mathbf{k}, \tag{4.318}$$

and $\mathbf{A}(\mathbf{r}, t)$ from (4.313):

$$\mathbf{A}(\mathbf{r}, t) = \frac{e\,\mathbf{v}}{(2\pi)^3 \epsilon_0 c^2} \int \frac{e^{i\mathbf{k}\cdot(\mathbf{r}-\mathbf{v}t)}}{k^2 - \left(\frac{\mathbf{k}\cdot\mathbf{v}}{c}\right)^2} d\mathbf{k}. \tag{4.319}$$

Summarizing, the Fourier transforms of the scalar and vector potentials are

$$V_{\mathbf{k}}(\mathbf{r}, t) = \frac{e}{(2\pi)^3 \epsilon_0} \frac{e^{i\mathbf{k}\cdot(\mathbf{r}-\mathbf{v}t)}}{k^2 - \left(\frac{\mathbf{k}\cdot\mathbf{v}}{c}\right)^2}, \tag{4.320}$$

and

$$\mathbf{A}_{\mathbf{k}}(\mathbf{r}, t) = \frac{\mathbf{v}}{c^2} V_{\mathbf{k}}(\mathbf{r}, t). \tag{4.321}$$

In this representation, the electromagnetic field associated to the electron is given by

$$\mathbf{e}_{\mathbf{k}} = -\nabla V_{\mathbf{k}} - \frac{\partial \mathbf{A}_{\mathbf{k}}}{\partial t} = -\nabla V_{\mathbf{k}} - \frac{\mathbf{v}}{c^2} \frac{\partial V_{\mathbf{k}}}{\partial t}, \tag{4.322}$$

and

$$\mathbf{b_k} = \nabla \times \mathbf{A_k} = -\frac{1}{c^2}\mathbf{v} \times \nabla V_k. \tag{4.323}$$

Using (4.320) and (4.321) and carrying out the calculation, we finally obtain

$$\mathbf{e_k} = -\frac{ie}{(2\pi)^3\epsilon_0}\frac{\mathbf{k} - \frac{1}{c^2}\mathbf{v}(\mathbf{k} \cdot \mathbf{v})}{k^2 - \left(\frac{\mathbf{k}\cdot\mathbf{v}}{c}\right)^2}e^{i\mathbf{k}\cdot(\mathbf{r}-\mathbf{v}t)}, \tag{4.324}$$

and

$$\mathbf{b_k} = \frac{ie\mu_0}{(2\pi)^3}\frac{\mathbf{k}\times\mathbf{v}}{k^2 - \left(\frac{\mathbf{k}\cdot\mathbf{v}}{c}\right)^2}e^{i\mathbf{k}\cdot(\mathbf{r}-\mathbf{v}t)}, \tag{4.325}$$

Note that, unlike the free plane waves which are purely transverse, the electric component $\mathbf{e_k}$ has also a longitudinal component, whose modulus is

$$|\mathbf{e}_k^{long}| = C\frac{\mathbf{k}\cdot\mathbf{s} - \frac{(\mathbf{v}\cdot\mathbf{s})(\mathbf{v}\cdot\mathbf{k})}{c^2}}{k^2 - \left(\frac{\mathbf{k}\cdot\mathbf{v}}{c}\right)^2} = \frac{C}{k}, \tag{4.326}$$

where C is a constant, and $\mathbf{s}$ is the unit vector of $\mathbf{k}$. Then,

$$\mathbf{e}_k^{long} = \frac{C}{k}\mathbf{s} = -\frac{ie}{(2\pi)^3\epsilon_0}\frac{\mathbf{k}}{k^2}e^{i\mathbf{k}\cdot(\mathbf{r}-\mathbf{v}t)}. \tag{4.327}$$

One also observes that

$$\mathbf{b}_k^{long} = 0. \tag{4.328}$$

If $\mathbf{v} \to 0$, $\mathbf{e}_k^{trans} \to 0$, and we fall back on the electrostatic field case.

Problem 4. An anisotropic dielectric, with N_0 electrons per unit volume, is placed in a combination of electric and magnetic fields, such that the electric component varies periodically with time, while the magnetic component $\mathbf{B_0}$ is uniform and homogeneous. Determine the relative permittivity tensor $(\epsilon_r)_{ik}$, $i, k = 1, 2, 3$ of the dielectric. It is assumed that all electrons have the same natural frequency ω_0, while the damping (friction) force is neglected.

Solution. The electron obeys the equation of motion

$$m\frac{d^2\mathbf{r}}{dt^2} = -m\omega_0^2\mathbf{r} + e\mathbf{E} + e\mathbf{v}\times\mathbf{B_0} = -m\omega_0^2\mathbf{r} + e\mathbf{E_0}\,e^{-i\omega t} + e\frac{d\mathbf{r}}{dt}\times\mathbf{B_0}. \tag{4.329}$$

Taking the z-axis along $\mathbf{B_0}$, we get

$$m\frac{d^2\mathbf{r}}{dt^2} = -m\omega_0^2\mathbf{r} + e\mathbf{E_0}\,e^{-i\omega t} + e\frac{dy}{dt}B_0\mathbf{i} - e\frac{dx}{dt}B_0\mathbf{j}, \tag{4.330}$$

288 4 Electromagnetic Waves

where $\mathbf{i}$ and $\mathbf{j}$ are the unit vectors of the x- and y-axes, respectively. This equation, projected on the axes, gives

$$m\frac{d^2x}{dt^2} = -m\omega_0^2 x + eE_x + e\frac{dy}{dt}B_0,$$

$$m\frac{d^2y}{dt^2} = -m\omega_0^2 y + eE_y - e\frac{dx}{dt}B_0, \tag{4.331}$$

$$m\frac{d^2z}{dt^2} = -m\omega_0^2 z + eE_z.$$

The calculation is simplified if one introduces the pair of coordinates

$$\xi = -\frac{1}{\sqrt{2}}(x+iy),$$

$$\eta = \frac{1}{\sqrt{2}}(x-iy), \tag{4.332}$$

in the xOy-plane, orthogonal to $\mathbf{B}$. To re-write (4.331) in the new variables ξ, η, z, we first multiply $(4.331)_1$ by $-1/\sqrt{2}$ and $(4.331)_2$ by $-i/\sqrt{2}$, then add the resulting equations side by side. The result is

$$\frac{d^2\xi}{dt^2} + 2i\Omega\frac{d\xi}{dt} + \omega_0^2\xi = \frac{e}{m}E_+, \tag{4.333}$$

where

$$\Omega = \frac{eB_0}{2m},$$

$$E_+ = -(1/\sqrt{2})(E_x + iE_y).$$

Further, we multiply $(4.331)_1$ by $1/\sqrt{2}$ and $(4.331)_2$ by $-i/\sqrt{2}$, then add the results side by side. This yields the equation for η:

$$\frac{d^2\eta}{dt^2} - 2i\Omega\frac{d\eta}{dt} + \omega_0^2\eta = \frac{e}{m}E_-, \tag{4.334}$$

where $E_- = \frac{1}{\sqrt{2}}(E_x - iE_y)$. The last equation $(4.331)_3$ remains unchanged,

$$\frac{d^2z}{dt^2} + \omega_0^2 z = \frac{e}{m}E_z. \tag{4.335}$$

For each Eqs. (4.333)–(4.335), the solution is sought in the form (*variable*) = (*amplitude*) $e^{-i\omega t}$. Introducing these solutions into the equations, one obtains the amplitude for each variable. The solutions are then

$$\xi = \frac{e}{m} \frac{E_+}{\omega_0^2 - \omega^2 + 2\omega\Omega},$$ (4.336)

$$\eta = \frac{e}{m} \frac{E_-}{\omega_0^2 - \omega^2 - 2\omega\Omega},$$ (4.337)

$$z = \frac{e}{m} \frac{E_z}{\omega_0^2 - \omega^2}.$$ (4.338)

Returning now to the old variables x, y, z, we first observe that

$$x = -\frac{1}{\sqrt{2}}(\xi - \eta),$$

$$y = \frac{i}{\sqrt{2}}(\xi + \eta),$$

$$z = z.$$

The desired expressions are obtained by some simple manipulations of (4.336)–(4.338). For example, to find x one multiplies (4.336) by $-1/\sqrt{2}$ and (4.337) by $1/\sqrt{2}$, then add side by side. A similar procedure is used to determine y. The results are

$$x = \frac{e}{m} \frac{aE_x - ibE_y}{a^2 - b^2},$$

$$y = \frac{e}{m} \frac{aE_y + ibE_x}{a^2 - b^2},$$ (4.339)

$$z = \frac{e}{m} \frac{E_z}{a},$$

where $a = \omega_0^2 - \omega^2$ and $b = 2\omega\Omega$.

Within this simple model of dielectric polarization, the dipole moment induced in a single atom/molecule is $\mathbf{p} = q\mathbf{r}$. The dipole moment per unit volume then is $\mathbf{P} = N_0 \mathbf{p} = e N_0 \mathbf{r}$, where N_0 is the number of dipoles (electron-ion pairs) per unit volume, i.e. the electron number density. We assume, therefore, that each atomic/molecular system consists of a pair ion-electron, which is an elementary dipole. In this case, the electric induction is

$$\mathbf{D} = \epsilon_0 \mathbf{E} + \mathbf{P} = \epsilon_0 \mathbf{E} + e N_0 \mathbf{r}.$$ (4.340)

Taking the components of (4.340) and using (4.339), one obtains

$$D_x = \epsilon_0 E_x + eN_0 x = \epsilon_0 E_x + eN_0 \frac{e}{m} \frac{aE_x - ibE_y}{a^2 - b^2}$$

$$= \left[\epsilon_0 + \frac{ae^2 N_0}{m(a^2 - b^2)} \right] E_x - \frac{ibe^2 N_0}{m(a^2 - b^2)} E_y, \tag{4.341}$$

$$D_y = \epsilon_0 E_y + eN_0 y = \epsilon_0 E_y + eN_0 \frac{e}{m} \frac{aE_y + ibE_x}{a^2 - b^2}$$

$$= \left[\epsilon_0 + \frac{ae^2 N_0}{m(a^2 - b^2)} \right] E_y + \frac{ibe^2 N_0}{m(a^2 - b^2)} E_x, \tag{4.342}$$

$$D_z = \epsilon_0 E + eN_0 z = \epsilon_0 E_z + eN_0 \frac{e}{m} \frac{E_z}{a} = \left(\epsilon_0 + \frac{e^2 N_0}{ma} \right) E_z. \tag{4.343}$$

Comparing (4.341)–(4.343) with the general tensor relations

$$D_i = \epsilon_{ik} E_k = \epsilon_0 (\epsilon_r)_{ik} E_k, \quad i, k = 1, 2, 3,$$

one finds the components of the relative permittivity tensor

$$(\epsilon_r)_{ik} = \begin{pmatrix} 1 + \frac{ae^2 N_0}{\epsilon_0 m(a^2 - b^2)} & -\frac{ibe^2 N_0}{\epsilon_0 m(a^2 - b^2)} & 0 \\ \frac{ibe^2 N_0}{\epsilon_0 m(a^2 - b^2)} & 1 + \frac{ae^2 N_0}{\epsilon_0 m(a^2 - b^2)} & 0 \\ 0 & 0 & 1 + \frac{e^2 N_0}{\epsilon_0 m a} \end{pmatrix}. \tag{4.344}$$

Problem 5. Determine the electromagnetic field associated with an electron performing an arbitrary (accelerated, but non-relativistic) motion.

Solution. The electromagnetic field associated with the moving electron is given by

$$\mathbf{E} = -\nabla V - \frac{\partial \mathbf{A}}{\partial t},$$

$$\mathbf{B} = \nabla \times \mathbf{A}, \tag{4.345}$$

where

$$V(\mathbf{r}, t) = \left[\frac{e}{4\pi\epsilon_0 |\mathbf{r} - \mathbf{x}| (1 - \mathbf{n} \cdot \boldsymbol{\beta})} \right]_{t = t' + \frac{|\mathbf{r} - \mathbf{x}(t')|}{c}}, \tag{4.346}$$

$$\mathbf{A}(\mathbf{r}, t) = \left[\frac{\mu_0 e c \boldsymbol{\beta}}{4\pi |\mathbf{r} - \mathbf{x}| (1 - \mathbf{n} \cdot \boldsymbol{\beta})} \right]_{t = t' + \frac{|\mathbf{r} - \mathbf{x}(t')|}{c}} \tag{4.347}$$

are the *Liénard–Wiechert potentials*. Here $\mathbf{x} = \mathbf{x}(t')$ is the parametric equation of the electron trajectory, $\mathbf{r}$ is the radius-vector of the observation point, $\mathbf{n}$ is the unit vector of the direction $\mathbf{r} - \mathbf{x}$, and $c\boldsymbol{\beta} = d\mathbf{x}/dt'$, with $\mathbf{v} = d\mathbf{x}/dt'$, is the velocity of the electron on its trajectory. If $P(\mathbf{r})$ is the observation point and $P'(\mathbf{x})$ the actual

Fig. 4.34 The relative position P' of the electron in motion, with respect to the fixed observation point P.

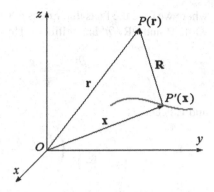

position of the electron (see Fig. 4.34), then the electromagnetic signal emitted by the electron at time t' will arrive at the observation point P at the later time

$$t = t' + \frac{|\mathbf{r} - \mathbf{x}(t')|}{c} = t' + \frac{R(t')}{c}, \qquad R = |\mathbf{r} - \mathbf{x}|.$$

For convenience, from now on we shall omit the "index" $t = t' + \frac{R}{c}$, but keep in mind that all quantities on the r.h.s. are considered at the time $t' = t - \frac{R}{c}$. Since $\mathbf{A} = \frac{\mathbf{v}}{c^2} V$ (see also (4.313)), we can write

$$
\begin{aligned}
\mathbf{E} &= -\frac{\partial V}{\partial \mathbf{r}} - \frac{\partial \mathbf{A}}{\partial t} = -\left(\frac{\partial}{\partial \mathbf{r}} + \frac{\mathbf{v}}{c^2}\frac{\partial}{\partial t}\right) V - \frac{V}{c^2}\frac{\partial \mathbf{v}}{\partial t} \\
&= -\frac{e}{4\pi\epsilon_0}\left(\frac{\partial}{\partial \mathbf{r}} + \frac{\mathbf{v}}{c^2}\frac{\partial}{\partial t}\right)\left(R - \frac{\mathbf{R}\cdot\mathbf{v}}{c}\right)^{-1} - \frac{V}{c^2}\frac{\partial \mathbf{v}}{\partial t'}\frac{\partial t'}{\partial t} \\
&= \frac{e}{4\pi\epsilon_0\left(R - \frac{\mathbf{R}\cdot\mathbf{v}}{c}\right)^2}\left(\frac{\partial}{\partial \mathbf{r}} + \frac{\mathbf{v}}{c^2}\frac{\partial}{\partial t}\right)\left(R - \frac{\mathbf{R}\cdot\mathbf{v}}{c}\right) - \frac{\mathbf{a}V}{c^2}\frac{\partial t'}{\partial t},
\end{aligned}
\qquad (4.348)
$$

where $\mathbf{a} = \partial\mathbf{v}/\partial t'$ is the acceleration of the electron.

The derivative $\partial t'/\partial t$ is found using $R = c(t - t')$ and observing that

$$\frac{\partial t'}{\partial t} = \frac{c}{c + \frac{\partial R}{\partial t'}},$$

$$\frac{\partial R^2}{\partial t'} = 2\,\mathbf{R}\cdot\frac{\partial}{\partial t'}(\mathbf{r} - \mathbf{x}) = -2\,\mathbf{R}\cdot\mathbf{v},$$

$$2\,\mathbf{R}\cdot\frac{\partial \mathbf{R}}{\partial t'} = 2R\frac{\partial R}{\partial t'} = -2\,\mathbf{R}\cdot\mathbf{v},$$

$$\frac{\partial R}{\partial t'} = -\frac{\mathbf{R}\cdot\mathbf{v}}{R},$$

where we used the facts that $\mathbf{r}$ does not explicitly depend on t', $\mathbf{v} = d\mathbf{x}/dt' = \partial\mathbf{x}/\partial t'$, while $\mathbf{R}$ and $\partial\mathbf{R}/\partial t'$ are collinear. Thus,

$$\frac{\partial t'}{\partial t} = \frac{c}{c + \frac{\partial R}{\partial t'}} = \frac{c}{c - \frac{\mathbf{R}\cdot\mathbf{v}}{R}} = \frac{R}{R - \frac{\mathbf{R}\cdot\mathbf{v}}{c}}$$

and $\mathbf{E}$ becomes

$$\mathbf{E} = \frac{e}{4\pi\epsilon_0 \left(R - \frac{\mathbf{R}\cdot\mathbf{v}}{c}\right)^2} \left(\frac{\partial}{\partial\mathbf{r}} + \frac{\mathbf{v}}{c^2}\frac{\partial}{\partial t}\right)\left(R - \frac{\mathbf{R}\cdot\mathbf{v}}{c}\right) - \frac{e R \mathbf{a}}{4\pi\epsilon_0 c^2 \left(R - \frac{\mathbf{R}\cdot\mathbf{v}}{c}\right)^2}.$$
(4.349)

The four derivatives appearing in (4.349) are calculated as follows. Since x, y, z, t are independent variables, while R can be expressed in terms of both the coordinates of the observation point x, y, z, t and those of the electron x', y', z', t', we have

$$\frac{\partial t'}{\partial\mathbf{r}} = \frac{\partial}{\partial\mathbf{r}}\left(t - \frac{R}{c}\right) = -\frac{1}{c}\frac{\partial R}{\partial\mathbf{r}},$$

$$\frac{\partial R}{\partial\mathbf{r}} = \frac{\partial R}{\partial x_i'}\frac{\partial x_i'}{\partial\mathbf{r}} + \frac{\partial R}{\partial t'}\frac{\partial t'}{\partial\mathbf{r}}, \quad i = 1,2,3; \quad x_1 = x, x_2 = y, x_3 = z,$$

$$\frac{\partial R}{\partial x_i'}\frac{\partial x_i'}{\partial\mathbf{r}} = \frac{\partial R}{\partial x_k}\frac{\partial x_k}{\partial x_i'}\frac{\partial x_m}{\partial\mathbf{r}}\frac{\partial x_i'}{\partial x_m} = \frac{\partial R}{\partial x_k}\frac{\partial x_m}{\partial\mathbf{r}}\delta_{km}$$

$$= \frac{\partial R}{\partial x_k}\frac{\partial x_k}{\partial\mathbf{r}} = \nabla R = \frac{\mathbf{R}}{R}, \quad i,k,m = 1,2,3.$$

Therefore,

$$\frac{\partial R}{\partial\mathbf{r}} = \frac{\mathbf{R}}{R} + \frac{\partial R}{\partial t'}\frac{\partial t'}{\partial\mathbf{r}} = \frac{\mathbf{R}}{R} - \frac{\mathbf{R}\cdot\mathbf{v}}{R}\frac{\partial t'}{\partial\mathbf{r}},$$

as well as

$$\frac{\partial t'}{\partial\mathbf{r}} = -\frac{1}{c}\frac{\partial R}{\partial\mathbf{r}} = \frac{\mathbf{R}\cdot\mathbf{v}}{cR}\frac{\partial t'}{\partial\mathbf{r}} - \frac{\mathbf{R}}{cR},$$

leading to

$$\frac{\partial t'}{\partial\mathbf{r}} = -\frac{\mathbf{R}}{cR}\frac{1}{1 - \frac{\mathbf{R}\cdot\mathbf{v}}{cR}} = -\frac{\mathbf{R}}{c\left(R - \frac{\mathbf{R}\cdot\mathbf{v}}{c}\right)},$$

so that

$$\frac{\partial R}{\partial\mathbf{r}} = -c\frac{\partial t'}{\partial\mathbf{r}} = \frac{\mathbf{R}}{R - \frac{\mathbf{R}\cdot\mathbf{v}}{c}}.$$
(4.350)

Next derivative, $\partial R/\partial t$, is easily obtained using the previous calculations:

$$\frac{\partial R}{\partial t} = c\left(1 - \frac{\partial t'}{\partial t}\right) = c\left(1 - \frac{R}{R - \frac{\mathbf{R}\cdot\mathbf{v}}{c}}\right) = -\frac{\mathbf{R}\cdot\mathbf{v}}{R - \frac{\mathbf{R}\cdot\mathbf{v}}{c}}.$$
(4.351)

The third derivative is

$$\frac{\partial}{\partial r}(\mathbf{R} \cdot \mathbf{v}) = \frac{\partial}{\partial r}[(\mathbf{r} - \mathbf{x}) \cdot \mathbf{v}] = \mathbf{v} + \mathbf{r} \cdot \frac{\partial \mathbf{v}}{\partial r} - \frac{\partial \mathbf{x}}{\partial r} \cdot \mathbf{v} - \mathbf{x} \cdot \frac{\partial \mathbf{v}}{\partial r}. \qquad (4.352)$$

But

$$\frac{\partial \mathbf{v}}{\partial r} = \frac{\partial \mathbf{v}}{\partial t'} \frac{\partial t'}{\partial r} = \mathbf{a} \frac{\partial t'}{\partial r} = -\frac{R\mathbf{a}}{c \left(R - \frac{\mathbf{R} \cdot \mathbf{v}}{c}\right)},$$

$$\frac{\partial \mathbf{x}}{\partial r} = \frac{\partial \mathbf{x}}{\partial t'} \frac{\partial t'}{\partial r} = -\frac{R}{c \left(R - \frac{\mathbf{R} \cdot \mathbf{v}}{c}\right)} \mathbf{v},$$

so that

$$\frac{\partial}{\partial r}(\mathbf{R} \cdot \mathbf{v}) = \mathbf{v} + \mathbf{r} \cdot \left[-\frac{\mathbf{a}R}{c \left(R - \frac{\mathbf{R} \cdot \mathbf{v}}{c}\right)}\right] - \left[-\frac{R}{c \left(R - \frac{\mathbf{R} \cdot \mathbf{v}}{c}\right)} \mathbf{v}\right] \cdot \mathbf{v}$$

$$- \mathbf{x} \cdot \left[-\frac{\mathbf{a}R}{c \left(R - \frac{\mathbf{R} \cdot \mathbf{v}}{c}\right)}\right] = \mathbf{v} - (\mathbf{r} - \mathbf{x}) \cdot \frac{\mathbf{a}R}{c \left(R - \frac{\mathbf{R} \cdot \mathbf{v}}{c}\right)} + \frac{v^2 R}{c \left(R - \frac{\mathbf{R} \cdot \mathbf{v}}{c}\right)}$$

$$= \mathbf{v} + \frac{v^2 - \mathbf{R} \cdot \mathbf{a}}{c \left(R - \frac{\mathbf{R} \cdot \mathbf{v}}{c}\right)} \mathbf{R}. \qquad (4.353)$$

The last derivative gives

$$\frac{\partial}{\partial t}(\mathbf{R} \cdot \mathbf{v}) = \frac{\partial}{\partial t}[(\mathbf{r} - \mathbf{x}) \cdot \mathbf{v}] = \frac{\partial \mathbf{r}}{\partial t} \cdot \mathbf{v} + \mathbf{r} \cdot \frac{\partial \mathbf{v}}{\partial t} - \frac{\partial \mathbf{x}}{\partial t} \cdot \mathbf{v} - \mathbf{x} \cdot \frac{\partial \mathbf{v}}{\partial t}.$$

Since $\mathbf{r}$ does not explicitly depend on t' we have

$$\frac{\partial \mathbf{r}}{\partial t} = \frac{\partial \mathbf{r}}{\partial t'} \frac{\partial t'}{\partial t} = 0,$$

$$\frac{\partial \mathbf{x}}{\partial t} = \frac{\partial \mathbf{x}}{\partial t'} \frac{\partial t'}{\partial t} = \mathbf{v} \frac{R}{R - \frac{\mathbf{R} \cdot \mathbf{v}}{c}},$$

$$\frac{\partial \mathbf{v}}{\partial t} = \frac{\partial \mathbf{v}}{\partial t'} \frac{\partial t'}{\partial t} = \mathbf{a} \frac{R}{R - \frac{\mathbf{R} \cdot \mathbf{v}}{c}},$$

and then

$$\frac{\partial}{\partial t}(\mathbf{R} \cdot \mathbf{v}) = (\mathbf{r} - \mathbf{x}) \cdot \mathbf{a} \frac{R}{R - \frac{\mathbf{R} \cdot \mathbf{v}}{c}} - \frac{v^2 R}{R - \frac{\mathbf{R} \cdot \mathbf{v}}{c}} = \frac{\mathbf{R} \cdot \mathbf{a} - v^2}{R - \frac{\mathbf{R} \cdot \mathbf{v}}{c}} R. \qquad (4.354)$$

Introducing now the derivatives (4.350), (4.351), (4.353), and (4.354) into (4.349), we successively have

$$
\mathbf{E} = \frac{e}{4\pi\epsilon_0 \left(R - \frac{\mathbf{R}\cdot\mathbf{v}}{c}\right)^2} \left\{ \frac{\mathbf{R}}{R - \frac{\mathbf{R}\cdot\mathbf{v}}{c}} - \frac{\mathbf{v}}{c^2}\frac{\mathbf{R}\cdot\mathbf{v}}{R - \frac{\mathbf{R}\cdot\mathbf{v}}{c}} - \frac{1}{c}\left[\mathbf{v} + \frac{v^2 - \mathbf{R}\cdot\mathbf{a}}{c\left(R - \frac{\mathbf{R}\cdot\mathbf{v}}{c}\right)}\mathbf{R}\right] \right.
$$

$$
\left. - \frac{\mathbf{v}}{c^2}\frac{\mathbf{R}\cdot\mathbf{a} - v^2}{R - \frac{\mathbf{R}\cdot\mathbf{v}}{c}}\mathbf{R} \right\} - \frac{e R \mathbf{a}}{4\pi\epsilon_0 c^2 \left(R - \frac{\mathbf{R}\cdot\mathbf{v}}{c}\right)^2} = \frac{e}{4\pi\epsilon_0 \left(R - \frac{\mathbf{R}\cdot\mathbf{v}}{c}\right)^3}
$$

$$
\times \left\{ \left[\left(1 - \frac{v^2}{c^2}\right) + \frac{\mathbf{R}\cdot\mathbf{a}}{c^2}\right]\left(\mathbf{R} - \frac{R\mathbf{v}}{c}\right) - \frac{R\mathbf{a}}{c^2}\left(R - \frac{\mathbf{R}\cdot\mathbf{v}}{c}\right) \right\}.
$$

Taking $\mathbf{A} \equiv \mathbf{R}$, $\mathbf{B} \equiv \mathbf{R} - \frac{R\mathbf{v}}{c}$, $\mathbf{C} \equiv \mathbf{a}$ and using the vector formula (A.13), we finally obtain

$$
\mathbf{E} = \frac{e}{4\pi\epsilon_0} \frac{\left(1 - \frac{v^2}{c^2}\right)\left(\mathbf{R} - \frac{R\mathbf{v}}{c}\right) + \frac{\mathbf{R}}{c^2} \times \left[\left(\mathbf{R} - \frac{R\mathbf{v}}{c}\right) \times \mathbf{a}\right]}{\left(R - \frac{\mathbf{R}\cdot\mathbf{v}}{c}\right)^3} = \mathbf{E}_v + \mathbf{E}_a, \qquad (4.355)
$$

where

$$
\mathbf{E}_v = \frac{e}{4\pi\epsilon_0} \frac{\left(1 - \frac{v^2}{c^2}\right)\left(\mathbf{R} - \frac{R\mathbf{v}}{c}\right)}{\left(R - \frac{\mathbf{R}\cdot\mathbf{v}}{c}\right)^3}
$$

is the electric component of the electromagnetic field created by the moving electron which depends on the velocity only, and

$$
\mathbf{E}_a = \frac{\mu_0}{4\pi} \frac{e\mathbf{R} \times \left[\left(\mathbf{R} - \frac{R\mathbf{v}}{c}\right) \times \mathbf{a}\right]}{\left(R - \frac{\mathbf{R}\cdot\mathbf{v}}{c}\right)^3}
$$

is the component depending on both the velocity and the acceleration of the electron.

Next, let us calculate the magnetic component $\mathbf{B}$:

$$
\mathbf{B} = \nabla \times \mathbf{A} = \nabla \times \left(\frac{\mathbf{v}V}{c^2}\right) = \frac{1}{c^2}(\nabla V) \times \mathbf{v} + \frac{V}{c^2}\nabla \times \mathbf{v}. \qquad (4.356)
$$

We have

$$
\nabla V = \frac{\partial V}{\partial \mathbf{r}} = -\frac{e}{4\pi\epsilon_0}\frac{1}{\left(R - \frac{\mathbf{R}\cdot\mathbf{v}}{c}\right)^2}\frac{\partial}{\partial \mathbf{r}}\left(R - \frac{\mathbf{R}\cdot\mathbf{v}}{c}\right)
$$

$$
= -\frac{e}{4\pi\epsilon_0}\frac{1}{\left(R - \frac{\mathbf{R}\cdot\mathbf{v}}{c}\right)^2}\left\{ \frac{\mathbf{R}}{R - \frac{\mathbf{R}\cdot\mathbf{v}}{c}} - \frac{1}{c}\left[\mathbf{v} + \frac{v^2 - \mathbf{R}\cdot\mathbf{a}}{c\left(R - \frac{\mathbf{R}\cdot\mathbf{v}}{c}\right)}\mathbf{R}\right] \right\}
$$

$$
= -\frac{e}{4\pi\epsilon_0}\frac{\mathbf{R}\left(1 - \frac{v^2}{c^2} + \frac{\mathbf{R}\cdot\mathbf{a}}{c^2}\right) - \frac{\mathbf{v}}{c}\left(R - \frac{\mathbf{R}\cdot\mathbf{v}}{c}\right)}{\left(R - \frac{\mathbf{R}\cdot\mathbf{v}}{c}\right)^3}, \qquad (4.357)
$$

as well as

$$\nabla \times \mathbf{v} = \epsilon_{ijk} \frac{\partial v_k}{\partial x_j} \mathbf{u}_i = \epsilon_{ijk} \frac{\partial v_k}{\partial t'} \frac{\partial t'}{\partial x_j} \mathbf{u}_i = \epsilon_{ijk} a_k \left(-\frac{1}{c} \frac{R_j}{R - \frac{\mathbf{R} \cdot \mathbf{v}}{c}} \right) \mathbf{u}_i$$

$$= -\frac{1}{c} \frac{\mathbf{R} \times \mathbf{a}}{R - \frac{\mathbf{R} \cdot \mathbf{v}}{c}}, \tag{4.358}$$

where we have used the x_j-projection of the relation

$$\frac{\partial t'}{\partial \mathbf{r}} = -\frac{\mathbf{R}}{c} \frac{1}{R - \frac{\mathbf{R} \cdot \mathbf{v}}{c}}.$$

With (4.357) and (4.358), (4.356) becomes

$$\mathbf{B} = -\frac{1}{c^2} \frac{e}{4\pi\epsilon_0} \frac{\mathbf{R} \left(1 - \frac{v^2}{c^2} + \frac{\mathbf{R} \cdot \mathbf{a}}{c^2} \right) - \frac{\mathbf{v}}{c} \left(R - \frac{\mathbf{R} \cdot \mathbf{v}}{c} \right)}{\left(R - \frac{\mathbf{R} \cdot \mathbf{v}}{c} \right)^3} \times \mathbf{v} + \frac{V}{c^2} \left(-\frac{1}{c} \frac{\mathbf{R} \times \mathbf{a}}{R - \frac{\mathbf{R} \cdot \mathbf{v}}{c}} \right)$$

$$= -\frac{e}{4\pi\epsilon_0 c^2} \frac{\left(1 - \frac{v^2}{c^2} + \frac{\mathbf{R} \cdot \mathbf{a}}{c^2} \right) (\mathbf{R} \times \mathbf{v}) + \left(R - \frac{\mathbf{R} \cdot \mathbf{v}}{c} \right) \frac{\mathbf{R} \times \mathbf{a}}{c}}{\left(R - \frac{\mathbf{R} \cdot \mathbf{v}}{c} \right)^3}. \tag{4.359}$$

The electric and magnetic components of the electromagnetic field created by the moving electron are given by (4.355) and (4.359). If the electron is moving uniformly ($\mathbf{a} = 0$) along a straight line, and $|\mathbf{v}| \ll c$, not only the ratio v^2/c^2, but also v/c can be neglected, and we are left with the field of an "inertial" electron,

$$\mathbf{E} = \frac{e}{4\pi\epsilon_0} \frac{\mathbf{R}}{R^3},$$

$$\mathbf{B} = \frac{\mu_0 e}{4\pi} \frac{\mathbf{v} \times \mathbf{R}}{R^3}. \tag{4.360}$$

As an application, let us calculate the energy of the non-relativistic electron, conceived as a sphere of radius R_0, uniformly charged with electricity of superficial density $\sigma = e/4\pi R_0^2$:

$$W = \frac{\epsilon_0}{2} \int_V E^2 d\mathbf{r} + \frac{1}{2\mu_0} \int_V B^2 d\mathbf{r} = \frac{e^2}{32\pi^2\epsilon_0} \int_{R_0}^{\infty} \int_0^{\pi} \int_0^{2\pi} \frac{1}{r^2} \sin\theta \, dr \, d\theta \, d\varphi$$

$$+ \frac{\mu_0^2 e^2 v^2}{32\pi^2} \int_{R_0}^{\infty} \int_0^{\pi} \int_0^{2\pi} \frac{1}{r^2} \sin^3\theta \, dr \, d\theta \, d\varphi$$

$$= \frac{e^2}{8\pi\epsilon_0 R_0} \left(1 + \frac{2}{3} \frac{v^2}{c^2} \right) \simeq \frac{e^2}{8\pi\epsilon_0 R_0}, \tag{4.361}$$

which is very close to the energy of the electron at rest. We may conclude that the non-relativistic, uniform motion of the electron does not affect very much its energy, or its classical radius, whose order of magnitude is 10^{-15} m. This spatial dimension shows, at the same time, the *applicability limits of classical electrodynamics for the electron*.

4.13 Proposed Problems

1. Given the vector potential $\mathbf{A} = \mathbf{A}_0 \, e^{i(\omega t - \mathbf{k} \cdot \mathbf{r})}$, and the conditions $V = 0$, $\nabla \cdot \mathbf{A} = 0$, $\mathbf{j} = 0$ and $\rho = 0$,
   (a) Prove the transversality of the electromagnetic waves in a homogeneous and isotropic medium;
   (b) Determine the magnitude and orientation of Poynting's vector.

2. On a plane surface separating two dielectric media falls a plane wave, the angle of incidence being φ and the angle of refraction being θ. The wave is polarized, the electric vector field of the incident wave making the angle γ_i with the plane of incidence. Determine the angles γ_{rfl} and γ_{rft} between the electric vector and the plane of incidence in both the reflected and refracted waves. Show that, if $\gamma_i = 0$, $\pi/2$, there is no change in these angles as compared to the incident wave.

3. Two dielectric plates of refractive index n are separated by an air layer of refractive index 1 and thickness d. Determine the ratio between the amplitude of the wave transmitted in one plate, and the amplitude of the wave incident on the layer from the other plate.

4. Determine the electric and magnetic energy densities of a plane wave propagating in the direction $\mathbf{s}$, in an anisotropic dielectric medium.

5. Consider a metal with N charge carriers per unit volume, each having charge e and mass m and satisfying the equation of motion

$$m\frac{d\mathbf{v}}{dt} + \frac{m}{\tau}\mathbf{v} = e\,\mathbf{E},$$

   where τ is the relaxation time (the time constant of an exponential return of the system to equilibrium after disturbance). Find
   (a) The complex electric conductivity $\tilde{\lambda}$;
   (b) The dispersion equations for the conductivity;
   (c) The connection between $\mathrm{Re}(\lambda)$ and the energy loss by Joule effect.

6. Determine the wavelength λ_{max} of a TE mode propagating in a rectangular waveguide, of sides a and b, the waveguide being filled with a semiconducting medium.

7. Calculate the phase velocity of the magnetic field generated by an electric oscillating dipole, situated in vacuum.

8. A linear quadrupole is composed of the charges $-q$, $2q$, $-q$ placed on z-axis at the points $-a$, 0, a, respectively. The quadrupole moment is $P = P_0 \sin \omega t$,

where $P_0 = a^2 q$. Determine the field generated by the quadrupole at some point $r \gg a$.

9. A plane polarized electromagnetic wave of frequency ω falls normally on a flat surface of a medium with $\epsilon \neq 1$, $\mu = 1$, $\lambda \neq 0$. Calculate the amplitude and phase of the reflected wave relative to the incident wave and discuss the limiting cases of good and poor conductors.

10. Find the reflection coefficients corresponding to a perpendicular/parallel polarization with respect to the plane of incidence, if the angle of incidence φ_i is close to the total reflection angle φ_l ($\varphi_l - \varphi_i = \delta \ll 1$).

Chapter 5
Elements of Magnetofluid Dynamics

In the fourth decade of the 20th century appeared a new branch of physical sciences, as a border discipline between the electromagnetic field theory, on the one side, and the fluid mechanics, on the other, known as *magnetofluid dynamics*, or *magnetohydrodynamics – MHD*. By its object and applications, this discipline can be included in the larger framework of *plasma physics*.

Plasma is conceived as a mixture of electrons, ions, photons, neutral atoms, and molecules. The density of negative charges is close to that of the positive charges, so that, on the whole, plasma is quasineutral. In some cases, the "electron gas" in metals can be studied using methods of plasma physics.

If no external electric or/and magnetic field acts on the plasma, then it behaves like an ordinary fluid (liquid, gas). In the presence of a field, the interaction with the fluid leads to modifications of the dimensions and shape of the fluid.

There are two fundamental methods of study of plasma:

(a) *Statistical method*, which uses the statistical physics formalism, in particular the kinetic theory of gases, Boltzmann and Fokker–Planck equations, etc.
(b) *Magnetohydrodynamical method*, which conceives plasma as a conducting, ionized fluid, subject to the action of an external electromagnetic field. The equations describing the behaviour of such a model are fluid mechanics equations, combined with the equations of the electromagnetic field.

The plasma state is somewhat uncommon on Earth, and can be found in some special cases: aurora borealis, lightning, ball lightning, etc. On the contrary, in the Universe the condensed state of matter is an exception and more than 99 % of matter is in plasma state.

The basic source of solar energy is given by thermonuclear fusion reactions of light atoms (hydrogen, deuterium, tritium) at very high temperatures ($10^6 - 10^8$ K). Under these physical circumstances, the matter is in a hot plasma state. Fundamental research and advanced techniques are directed towards obtaining similar reactions in

© Springer-Verlag Berlin Heidelberg 2016
M. Chaichian et al., *Electrodynamics*, DOI 10.1007/978-3-642-17381-3_5

laboratory conditions. Over the years, starting with the 1950s, various devices have been invented and built, like the *stellarator, tokamak, magnetic traps*, etc.

To continuously produce controlled thermonuclear reactions, it is necessary for plasma to simultaneously satisfy three conditions: a minimum temperature ($\sim 10^8$ K), a minimum concentration ($\sim 10^{15}$ nuclei/cm^3), and a minimum time of stability (fractions of seconds). This triple criterion has already been fulfilled in laboratory conditions and it is hoped that the need of energy of our civilization will be (at least) partially covered through the achievements in hot plasma research.

Applications of plasma physics have been also developed in some other directions, such as: construction of magnetohydrodynamical generators, plasma burners, plasma chemistry, plasma deposition, plasma surface modification, etc.

In the following we shall present a summary of the most interesting phenomena of plasma physics, in the framework of the magnetohydrodynamical approach.

5.1 Basic Equations of Magnetofluid Dynamics

The basis of magnetofluid dynamics were established around 1930, by the astrophysicists who were studying the origin of the solar and star energy. Initially, research was directed towards investigating the solar magnetism, the origin of the solar spots, stellar structure, as well as the magnetic storms accompanying the solar prominences. In this respect, we quote the pioneering work of Joseph Larmor (1857–1942), Thomas George Cowling (1906–1990), and Sydney Chapman (1888–1970). During the 1930s began the laboratory research of phenomena that appear in a conducting fluid submitted to the action of a magnetic field.

Magnetofluid dynamics became an independent science when the Swedish physicist Hannes Olof Gösta Alfvén (1908–1995) settled the fundamental system of equations describing the behaviour of a magnetofluid in an external field. In 1970 he was awarded the Nobel Prize in Physics for his work "on magnetohydrodynamics with fruitful applications in different parts of plasma physics".

We consider a model consisting of a conducting fluid, of conductivity λ, moving with the velocity **v** in the external electromagnetic field (**E**, **B**). The mass density of the fluid is ρ, the electric charge density is ρ_e, and the static pressure of the fluid is p. We also assume that the displacement current $\partial \mathbf{D}/\partial t$ is much smaller than **j**, which is the sum of conduction and convection current densities. Maxwell's equations are then

$$\nabla \times \mathbf{E} = -\frac{\partial \mathbf{B}}{\partial t},$$
$$\nabla \cdot \mathbf{B} = 0,$$
$$\nabla \times \frac{\mathbf{B}}{\mu} = \mathbf{j}, \tag{5.1}$$
$$\nabla \cdot \mathbf{D} = \rho_e.$$

To these equations we have to add Ohm's law for a medium whose velocity is much smaller than the speed of light (see (3.146)):

$$\mathbf{j} = \lambda \mathbf{E}' = \lambda(\mathbf{E} + \mathbf{v} \times \mathbf{B}). \tag{5.2}$$

Using Maxwell's equation $(5.1)_3$ and Ohm's law, we find

$$\frac{1}{\mu} \nabla \times (\nabla \times \mathbf{B}) = \nabla \times \mathbf{j} = \lambda \nabla \times \mathbf{E} + \lambda \nabla \times (\mathbf{v} \times \mathbf{B}).$$

Using $(5.1)_1$ in the above formula we obtain

$$\frac{\partial \mathbf{B}}{\partial t} = \nabla \times (\mathbf{v} \times \mathbf{B}) + \nu_m \Delta \mathbf{B}, \tag{5.3}$$

where the quantity $\nu_m = (\lambda \mu)^{-1}$ is called the *coefficient of magnetic viscosity*. The name comes from the similarity between Eq. (5.3) and the vortex equation of fluid dynamics, where instead of $\mathbf{B}$ appears the *vorticity*

$$\boldsymbol{\Omega} = \frac{1}{2} (\nabla \times \mathbf{v}).$$

Equation (5.3) is the *induction equation*.

The equation of motion of the magnetofluid is obtained by adding the magnetic force density

$$\mathbf{f}_{em} = \rho_e(\mathbf{E} + \mathbf{v} \times \mathbf{B}) \simeq \rho_e \mathbf{v} \times \mathbf{B} = \mathbf{j} \times \mathbf{B},$$

to the equation of motion of a *perfectly viscous*, or *Newtonian fluid*, known from fluid mechanics. Thus, the equation of motion of our model of magnetofluid is

$$\rho \mathbf{a} = \rho \mathbf{F} - \nabla p + (\xi + \eta) \nabla \theta' + \eta \Delta \mathbf{v} + \mathbf{j} \times \mathbf{B}, \tag{5.4}$$

where ρ is the mass density, ξ and η are the coefficients of dynamical viscosity, $\mathbf{F}$ is the non-electromagnetic (e.g. the gravitational) force per unit mass, and $\theta' = \nabla \cdot \mathbf{v}$.

Another essential equation of the model is the equation of mass conservation, written as the equation of continuity

$$\frac{\partial \rho}{\partial t} + \nabla \cdot (\rho \mathbf{v}) = 0. \tag{5.5}$$

Replacing ρ by ρ_e, one obtains the equation of charge conservation.

To these equations we have to add the *energy equation*. This equation says that in the unit volume of fluid and per unit time, due to viscosity, thermal conductivity, and Joule loss, the heat $\rho T \, ds/dt$ is dissipated, that is

$$\rho T \left[\frac{\partial s}{\partial t} + (\mathbf{v} \cdot \nabla) s \right] = T'_{ik} \frac{\partial v_i}{\partial x_k} + \nabla \cdot (\kappa \nabla T) + \mathbf{j} \cdot \mathbf{E}, \qquad (5.6)$$

where s is the entropy per unit mass of fluid, κ is the thermal conductivity coefficient, while T'_{ik} is the *viscous stress tensor*, defined by

$$T_{ik} = - p \, \delta_{ik} + T'_{ik},$$

where T_{ik} is the *stress tensor*. In the case of homogeneous and isotropic fluids, T'_{ik} is given by

$$T'_{ik} = \xi \, \theta' \, \delta_{ik} + 2\eta' \, e'_{ik},$$

where

$$e'_{ik} = \frac{1}{2} \left(\frac{\partial v_k}{\partial x_i} + \frac{\partial v_i}{\partial x_k} \right)$$

is the *velocity of deformation tensor*.

Since the above equations are not enough to characterize the system from the thermodynamic point of view, we have to add also the *equation of state*, usually written as

$$f(p, \, \rho, \, T) = 0. \qquad (5.7)$$

Equations (5.3)–(5.7), together with (5.1)₄ form the *fundamental system of equations* of magnetofluid dynamics. If the fluid is perfect and infinitely conducting (perfect fluids have no shear stresses, viscosity, or heat conduction), then $\xi = \eta = 0, \lambda \to \infty$ (or $\nu_m \to 0$), and the motion is *isentropic* ($s = $ const.); in other words, there are no dissipative processes, and the fundamental system of equations reduces to

$$\frac{\partial \mathbf{B}}{\partial t} = \nabla \times (\mathbf{v} \times \mathbf{B}),$$
$$\epsilon \nabla \cdot \mathbf{E} = \rho_e,$$
$$\rho \mathbf{a} = \rho \mathbf{F} - \nabla p + \mathbf{j} \times \mathbf{B},$$
$$\frac{\partial \rho}{\partial t} + \nabla \cdot (\rho \mathbf{v}) = 0, \qquad (5.8)$$
$$\frac{ds}{dt} = 0,$$
$$f(p, \rho, T) = 0.$$

Using this system of equations, we shall survey some of the most interesting applications.

5.2   Freezing-In of Magnetic Field Lines

Consider a perfect magnetofluid ($\lambda \to \infty$), moving with the velocity $\mathbf{v}$ in the magnetic field $\mathbf{B}$, and expand the r.h.s. of Eq. (5.8)$_1$:

$$\frac{\partial \mathbf{B}}{\partial t} = - \mathbf{B} \nabla \cdot \mathbf{v} + (\mathbf{B} \cdot \nabla)\mathbf{v} - (\mathbf{v} \cdot \nabla)\mathbf{B},$$

or

$$\frac{d\mathbf{B}}{dt} = - \mathbf{B} \nabla \cdot \mathbf{v} + (\mathbf{B} \cdot \nabla)\mathbf{v}. \tag{5.9}$$

Multiplying (5.9) by $1/\rho$ and using the equation of continuity (5.5), we have

$$\frac{1}{\rho}\frac{d\mathbf{B}}{dt} = \frac{\mathbf{B}}{\rho^2}\frac{\partial \rho}{\partial t} + \frac{\mathbf{B}}{\rho^2}\mathbf{v} \cdot \nabla \mathbf{v} + \left(\frac{\mathbf{B}}{\rho} \cdot \nabla\right)\mathbf{v}$$

$$= \frac{\mathbf{B}}{\rho^2}\frac{d\rho}{dt} + \left(\frac{\mathbf{B}}{\rho} \cdot \nabla\right)\mathbf{v},$$

or

$$\frac{d}{dt}\left(\frac{\mathbf{B}}{\rho}\right) = \left(\frac{\mathbf{B}}{\rho} \cdot \nabla\right)\mathbf{v}, \tag{5.10}$$

which is very similar to *Beltrami's diffusion equation* of fluid mechanics.

   To extract the physical significance of Eq. (5.10), let us consider two current lines Γ and Γ', so that all the fluid particles which are on Γ at the time t, will be situated on Γ' at the time $t + dt$. Let $\delta \mathbf{l}$ be an element of Γ, and $\delta \mathbf{l}'$ an element of Γ', composed of the same number of particles as $\delta \mathbf{l}$ (Fig. 5.1). The conservation of the current lines implies

$$\delta \mathbf{l}' - \delta \mathbf{l} = \frac{d}{dt}(\delta \mathbf{l})\, dt. \tag{5.11}$$

Fig. 5.1 Geometrical construction auxiliary to the freezing-in of magnetic field lines.

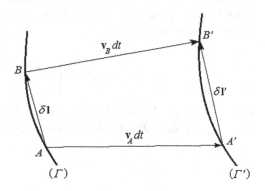

If $\mathbf{v}_A$ and $\mathbf{v}_B$ are the displacement velocities of the points A and B belonging to Γ (with respect to an orthogonal Cartesian frame), within the time interval dt, we can write

$$\delta\mathbf{l}' = \delta\mathbf{l} + \mathbf{v}_B dt - \mathbf{v}_A dt = \delta\mathbf{l} + \mathbf{v}(\mathbf{r} + \delta\mathbf{l})dt - \mathbf{v}_A dt$$
$$= \delta\mathbf{l} + [\mathbf{v}_A + (\delta\mathbf{l} \cdot \nabla)\mathbf{v}_A + ...] - \mathbf{v}_A dt \simeq \delta\mathbf{l} + (\delta\mathbf{l} \cdot \nabla)\mathbf{v}_A dt,$$

or

$$\frac{d}{dt}(\delta\mathbf{l}) \simeq (\delta\mathbf{l} \cdot \nabla)\mathbf{v}. \tag{5.12}$$

Comparing Eqs. (5.10) and (5.12) we realize that they have a similar form, meaning that the time variation of the vectors $\delta\mathbf{l}$ and $\mathbf{B}/\rho$ obey the same law. Consequently, the conservation of the current lines of the fluid implies the conservation of the lines of the field $\mathbf{B}$ (in fact, of a vector collinear with $\mathbf{B}$). This latter vector moves together with the fluid, as if being "stuck" into the current lines. The phenomenon is known as the *freezing-in of magnetic field lines*.

An alternative formulation of this conclusion is that the magnetic induction flux Φ_m crossing a surface $S(t)$ bounded by a closed contour composed by fluid particles (material contour), is conserved. Indeed, recalling the theory developed in Sect. 3.12 on the electrodynamics of moving media, we have

$$\frac{d\Phi_m}{dt} = \frac{d}{dt}\int_S \mathbf{B} \cdot d\mathbf{S} = \int_S \left[\frac{\partial \mathbf{B}}{\partial t} - \nabla \times (\mathbf{v} \times \mathbf{B})\right] \cdot d\mathbf{S} = 0,$$

meaning that $\Phi_m = \int_S \mathbf{B} \cdot d\mathbf{S} = $ const.

5.3 Magnetohydrodynamic Waves

According to the mechanics of continuous media, in a homogeneous medium (e.g. air) the small perturbations propagate as compression waves called *sound waves*. The speed of these waves is

$$v_s = \left[\left(\frac{\partial p}{\partial \rho}\right)_S\right]^{1/2}, \tag{5.13}$$

where p is the mechanical pressure, ρ is the density of the medium, and the index S shows that the derivative is taken at constant entropy (isentropic process).

In case of a conducting medium, besides the mechanical pressure p there will be present a magnetic pressure $B^2/2\mu$ (see next section). Thus we expect to have both sound waves, whose phase velocity is given by (5.13), and some specific waves characteristic for conducting media, that propagate with the velocity (see further):

$$v \sim \frac{B}{\sqrt{\mu\rho}}. \tag{5.14}$$

To determine the nature and properties of these waves, we consider the model of a compressible, homogeneous, conducting fluid, placed in a constant and uniform magnetic field $\mathbf{B}_0$. We also suppose that we may neglect the energy dissipation by Joule effect, viscosity, and thermal conductivity. In this case, (5.6) gives $s = \text{const.}$, so that the process may be considered isentropic.

To study the propagation of small oscillations in such a medium, we use the equations (see (5.8)):

$$\frac{\partial \mathbf{B}}{\partial t} - \nabla \times (\mathbf{v} \times \mathbf{B}) = 0,$$

$$\rho \left[\frac{\partial \mathbf{v}}{\partial t} + (\mathbf{v} \cdot \nabla)\mathbf{v} \right] = -\nabla p - \frac{1}{\mu}\mathbf{B} \times (\nabla \times \mathbf{B}), \tag{5.15}$$

$$\frac{\partial \rho}{\partial t} + \nabla \cdot (\rho \mathbf{v}) = 0,$$

as well as the law of conservation of entropy $(5.8)_5$, and the state equation $(5.8)_6$.

Suppose that

$$\mathbf{B} = \mathbf{B}_0 + \mathbf{B}'(\mathbf{r}, t),$$

$$\rho = \rho_0 + \rho'(\mathbf{r}, t), \tag{5.16}$$

$$\mathbf{v} = \mathbf{v}'(\mathbf{r}, t),$$

where $\mathbf{B}_0$, ρ_0 are the equilibrium, constant quantities, while $\mathbf{B}'$, ρ' are small perturbations from equilibrium. The equilibrium velocity is considered zero, and $\mathbf{v}'$ is also a small quantity, of the same order of magnitude as $\mathbf{B}'$ and ρ'. Introducing (5.16) into (5.15) and keeping only the first order quantities, we arrive at the following system of *linear partial differential equations*

$$\frac{\partial \mathbf{B}'}{\partial t} - \nabla \times (\mathbf{v}' \times \mathbf{B}_0) = 0,$$

$$\rho_0 \frac{\partial \mathbf{v}'}{\partial t} + v_s^2 \nabla \rho' + \frac{1}{\mu}\mathbf{B}_0 \times (\nabla \times \mathbf{B}') = 0, \tag{5.17}$$

$$\frac{\partial \rho'}{\partial t} + \rho_0 \nabla \cdot \mathbf{v}' = 0.$$

We take now the time derivative of $(5.17)_2$ and use the other two equations. Thus, we find the following equation in $\mathbf{v}'$:

$$\rho_0 \frac{\partial^2 \mathbf{v}'}{\partial t^2} - v_s^2 \rho_0 \nabla(\nabla \cdot \mathbf{v}') + \frac{1}{\mu} \mathbf{B}_0 \times \{\nabla \times [\nabla \times (\mathbf{v}' \times \mathbf{B}_0)]\} = 0.$$

If one introduces the *Alfvén velocity*

$$\mathbf{v}_A = \frac{\mathbf{B}_0}{\sqrt{\mu \rho_0}}, \qquad (5.18)$$

the last equation can also be written as

$$\rho_0 \frac{\partial^2 \mathbf{v}'}{\partial t^2} - v_s^2 \nabla(\nabla \cdot \mathbf{v}') + \mathbf{v}_A \times \{\nabla \times [\nabla \times (\mathbf{v}' \times \mathbf{v}_A)]\} = 0. \qquad (5.19)$$

Suppose that $\mathbf{v}'$ propagates as a plane wave of the form

$$\mathbf{v}'(\mathbf{r}, t) = \mathbf{v}'_0 \, e^{i(\mathbf{k} \cdot \mathbf{r} - \omega t)}. \qquad (5.20)$$

The last term of (5.19) then yields:

$$
\begin{aligned}
\mathbf{v}_A &\times \{\nabla \times [-\mathbf{v}_A \nabla \cdot \mathbf{v}' + (\mathbf{v}_A \cdot \nabla)\mathbf{v}']\} \\
&= \mathbf{v}_A \times \{-[\nabla(\nabla \cdot \mathbf{v}')] \times \mathbf{v}_A + (\mathbf{v}_A \cdot \nabla)\nabla \times \mathbf{v}'\} \\
&= \mathbf{v}_A \times \{-[\mathbf{k}(\mathbf{k} \cdot \mathbf{v}')] \times \mathbf{v}_A + (\mathbf{v}_A \cdot \mathbf{k})(\mathbf{k} \times \mathbf{v}')\} \\
&= v_A^2 \mathbf{k}(\mathbf{k} \cdot \mathbf{v}') - (\mathbf{v}_A \cdot \mathbf{k})(\mathbf{k} \cdot \mathbf{v}')\mathbf{v}_A - (\mathbf{v}_A \cdot \mathbf{k})[(\mathbf{v}_A \cdot \mathbf{v}')\mathbf{k} - (\mathbf{v}_A \cdot \mathbf{k})\mathbf{v}'],
\end{aligned}
$$

and from (5.19) we obtain

$$
\begin{aligned}
&- \omega^2 \mathbf{v}'_0 + (v_s^2 + v_A^2)(\mathbf{k} \cdot \mathbf{v}'_0)\mathbf{k} \\
&\quad + (\mathbf{v}_A \cdot \mathbf{k})[(\mathbf{v}_A \cdot \mathbf{k})\mathbf{v}'_0 - (\mathbf{v}_A \cdot \mathbf{v}'_0)\mathbf{k} - (\mathbf{k} \cdot \mathbf{v}'_0)\mathbf{v}_A] = 0. \qquad (5.21)
\end{aligned}
$$

If $\mathbf{v}_A \perp \mathbf{k}$, the last term (containing the square bracket) vanishes and one obtains as solution for $\mathbf{v}'_0$ a wave called *longitudinal magnetoacoustic*, whose phase velocity is

$$u_{long} = \sqrt{v_s^2 + v_A^2}. \qquad (5.22)$$

If $\mathbf{v}_A \parallel \mathbf{k}$, Eq. (5.21) reduces to

$$\left(v_A^2 k^2 - \omega^2\right) \mathbf{v}'_0 + \left(\frac{v_s^2}{v_A^2} - 1\right) k^2 (\mathbf{v}_A \cdot \mathbf{v}'_0)\mathbf{v}_A = 0. \qquad (5.23)$$

In this case, two types of periodical motion are possible: an ordinary longitudinal wave ($\mathbf{v}_0' \parallel \mathbf{k}$, $\mathbf{v}_A$) with phase velocity v_s, and a transversal wave, whose phase velocity is the Alfvén velocity $\mathbf{v}_A$. This *Alfvén wave* is a pure magnetodynamical phenomenon, that depends only on the magnetic field and the fluid density. Under laboratory conditions we have $v_A \ll v_s$, irrespective of the intensity of the magnetic field; under cosmic conditions, due to small densities, v_A can reach considerable values. Let us illustrate this point numerically: taking the density of the solar photosphere to be $\rho \sim 2.2 \times 10^{-4}\,\mathrm{kg \cdot m^{-3}}$ and the solar magnetic field at the surface of the Sun about 2×10^{-4} T (it is much larger around the solar spots), we find $v_A = 12$ km $\cdot$ s^{-1}. For comparison, the sound velocity in photosphere is ~ 10 km $\cdot$ s^{-1}.

The magnetic fields corresponding to the three types of wave defined above can be determined from $(5.17)_1$:

$$
\mathbf{B}' = \begin{cases}
\frac{k}{\omega} v' \mathbf{B}_0, & \text{for } \mathbf{k} \perp \mathbf{B}_0, \\[2mm]
0, & \text{for } \mathbf{k} \parallel \mathbf{B}_0 \text{ (longit.)}, \\[2mm]
-\frac{k}{\omega} v' \mathbf{B}_0, & \text{for } \mathbf{k} \parallel \mathbf{B}_0 \text{ (transv.)},
\end{cases}
\tag{5.24}
$$

where we used the fact that $\mathbf{v}_A$ and $\mathbf{B}_0$ are collinear. The magnetoacoustic wave, moving orthogonally to $\mathbf{v}_0$, produces compressions and rarefactions of the magnetic field lines, without changing their direction (Fig. 5.2a), while the Alfvén wave propagates parallel to $\mathbf{B}_0$ and makes the field lines oscillate laterally, and back and forth (Fig. 5.2b). In each case, since the magnetic viscosity is zero, the magnetic field lines are "frozen" in the conducting fluid.

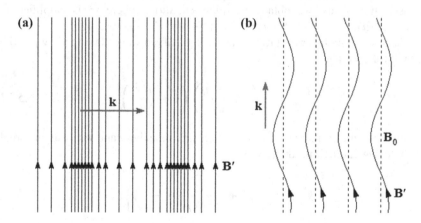

Fig. 5.2 The behaviour of: (**a**) the magnetoacoustic wave; (**b**) the Alfvèn wave.

If we take into account the losses due to viscosity and Joule effect, the first two equations (5.17) must be replaced by

$$\frac{\partial \mathbf{B}'}{\partial t} - \nabla \times (\mathbf{v}' \times \mathbf{B}_0) = \frac{1}{\lambda \mu} \Delta \mathbf{B}',$$

$$\rho_0 \frac{\partial \mathbf{v}'}{\partial t} + v_s^2 \nabla \rho' + \frac{1}{\mu} \mathbf{B}_0 \times (\nabla \times \mathbf{B}') = \eta \Delta \mathbf{v}'. \tag{5.25}$$

To analyze the change produced by adding the two dissipative terms, we assume that $\mathbf{B}'$ and $\mathbf{v}'$ propagate as plane waves of the form (5.20). Thus, we have

$$\frac{\partial \mathbf{B}'}{\partial t} - \frac{1}{\lambda \mu} \Delta \mathbf{B}' = -i\omega \left(1 + \frac{k^2}{\omega \lambda \mu}\right) \mathbf{B}' = \left(1 + \frac{ik^2}{\omega \lambda \mu}\right) \frac{\partial \mathbf{B}'}{\partial t},$$

$$\rho_0 \frac{\partial \mathbf{v}'}{\partial t} - \eta \Delta \mathbf{v}' = \rho_0 \left(1 + \frac{ik^2 \eta}{\omega \rho_0}\right) \frac{\partial \mathbf{v}'}{\partial t}.$$

Therefore, Eqs. (5.25) are equivalent to

$$\frac{\partial \mathbf{B}'}{\partial t} = \frac{1}{\left(1 + \frac{ik^2}{\omega \lambda \mu}\right)} \nabla \times (\mathbf{v}' \times \mathbf{B}_0),$$

$$\rho_0 \frac{\partial \mathbf{v}'}{\partial t} = -\frac{1}{\left(1 + \frac{ik^2 \eta}{\omega \rho_0}\right)} \left[v_s^2 \nabla \rho' + \frac{1}{\mu} \mathbf{B}_0 \times (\nabla \times \mathbf{B}')\right]. \tag{5.26}$$

Resuming the arguments leading to Eq. (5.21), but this time taking into account the dissipative terms, we obtain a similar equation, where v_s^2 is amplified by $(1 + ik^2/\omega\lambda\mu)$, and ω^2 by $(1 + ik^2/\omega\lambda\mu)(1 + ik^2\eta/\omega\rho_0)$.

Let us now apply this result to the case of the Alfvén wave. Taking $\mathbf{v}' \cdot \mathbf{v}_A = 0$ in (5.23), it follows that

$$k^2 v_A^2 = \omega^2 \left(1 + \frac{ik^2 \eta}{\omega \rho_0}\right)\left(1 + \frac{ik^2}{\lambda \mu \omega}\right). \tag{5.27}$$

Assuming that the corrective terms due to electric resistivity and viscosity are small enough to neglect their product, we have

$$k^2 v_A^2 \simeq \omega^2 \left[1 + \frac{ik^2}{\omega}\left(\frac{\eta}{\rho_0} + \frac{1}{\lambda \mu}\right)\right],$$

which yields

$$k^2 = \frac{\omega^2}{v_A^2 - i\omega\left(\frac{\eta}{\rho_0} + \frac{1}{\lambda\mu}\right)} = \frac{\omega^2\left[v_A^2 + i\omega\left(\frac{\eta}{\rho_0} + \frac{1}{\lambda\mu}\right)\right]}{v_A^4 + \omega^2\left(\frac{\eta}{\rho_0} + \frac{1}{\lambda\mu}\right)^2}$$

$$\simeq \frac{\omega^2}{v_A^2}\left[1 + \frac{i\omega}{v_A^2}\left(\frac{\eta}{\rho_0} + \frac{1}{\lambda\mu}\right)\right],$$

finally giving the *dispersion relation* for our case:

$$k \simeq \frac{\omega}{v_A} + i\frac{\omega^2}{2v_A^3}\left(\frac{\eta}{\rho_0} + \frac{1}{\lambda\mu}\right). \tag{5.28}$$

Since the wave number is complex, the wave is attenuated. The damping increases with frequency, but decreases even faster with the magnetic field. If the dissipation is considerable, that is, if

$$\frac{k^2\eta}{\omega\rho_0} \gg 1, \qquad \frac{k^2}{\lambda\mu\omega} \gg 1,$$

then k^2 in (5.27) becomes negative, meaning that the wave number is purely imaginary, and the wave attenuates very rapidly, irrespective of the magnitude of $\mathbf{B}_0$.

5.4 Some Problems of Magnetohydrostatics

In the absence of external sources, the presence of a *static* magnetic field in a fluid at rest is possible only if the fluid conductivity is infinite (see Sect. 5.2), otherwise the magnetic energy would dissipate as Joule heat.

Magnetohydrostatics is the study of the equilibrium conditions of an infinitely conducting fluid, under the action of (hydrostatic) pressure, as well as Lorentz and gravitational forces. If the fluid is at rest ($\mathbf{v} = 0$), the equation of motion (5.8)$_3$ reduces to

$$- \nabla p + \frac{1}{\mu}(\nabla \times \mathbf{B}) \times \mathbf{B} + \rho\mathbf{g} = 0, \tag{5.29}$$

where $\rho\mathbf{g}$ is the gravitational force density. If the gravitational force is negligible as compared to the other two forces, the state of equilibrium implies $\mathbf{B}$ and $\mathbf{j} = \frac{1}{\mu}\nabla \times \mathbf{B}$ orthogonal to ∇p, which means that $\mathbf{B}$ and $\mathbf{j}$ are placed on the constant pressure surface $p = \text{const}$.

In the following we shall present two applications of this formalism, one encountered in laboratory practice, and the other in astrophysics.

5.4.1   Magnetic Thermal Insulation. The Pinch Effect

We consider again an ideal magnetofluid (no dissipative effects) at rest ($\mathbf{v} = 0$). Neglecting the gravitational forces, Eq. (5.29) gives

$$\nabla p = \frac{1}{\mu}(\nabla \times \mathbf{B}) \times \mathbf{B}.$$

Since (see (A.46))

$$\mathbf{B} \times (\nabla \times \mathbf{B}) = \frac{1}{2}\nabla B^2 - (\mathbf{B} \cdot \nabla)\mathbf{B},$$

we can write

$$\nabla \left(p + \frac{1}{2\mu}B^2 \right) = \frac{1}{\mu}(\mathbf{B} \cdot \nabla)\mathbf{B}. \tag{5.30}$$

It follows that the dimension of the quantity $p_m = B^2/2\mu$ is that of a pressure; it is called the *magnetic pressure*. In some simple geometric dispositions, like that of a one-component field, the r.h.s. of (5.30) vanishes and we are left with

$$p + \frac{1}{2\mu}B^2 = \text{const.} \tag{5.31}$$

Let us choose the magnetic field $\mathbf{B}$ on the Ox-axis. In this case $\nabla \cdot \mathbf{B} = 0$ leads to $\partial B/\partial x = 0$, and

$$(\mathbf{B} \cdot \nabla)\mathbf{B} = \left(B \frac{\partial}{\partial x} \right)(B\,\mathbf{i}) = B \frac{\partial B}{\partial x}\mathbf{i} = 0.$$

According to (5.31), any increase in mechanical pressure must be compensated by a lowering of the magnetic pressure, and vice-versa. If the mechanical pressure decreases close to zero in some region, the magnetic pressure has to increase very much in that region if we want to confine the fluid. This is the principle of the *pinch effect*: the confinement of a conducting fluid by its own plasma field, created by electric currents inside the fluid. This effect is very important in thermonuclear research, and we shall briefly discuss below this application.

Consider an infinite cylinder of our conducting medium, with its axis oriented along the Oz-direction (see Fig. 5.3), and let $j_z = j(r)$ be the axial current density, and $B_\theta = B(r)$ the azimuthal magnetic field created by $j(r)$. Our problem is to find a possible equilibrium state, in which the fluid is confined in a cylinder of radius R, under the action of its own magnetic field $\mathbf{B}$.

The geometry of the problem requires us to work in cylindrical coordinates r, θ, z. Beforehand, we observe that, denoting by $\mathbf{u}_r, \mathbf{u}_\theta, \mathbf{k}$ the unit vectors of the cylindrical frame directions, the r.h.s of (5.30) can be written as

$$(\mathbf{B} \cdot \nabla)\mathbf{B} = \left(B_\theta \frac{1}{r}\frac{\partial}{\partial \theta} \right)(B_\theta\,\mathbf{u}_\theta) = \frac{B^2}{r}\frac{d\mathbf{u}_\theta}{d\theta} = -\frac{B^2}{r}\mathbf{u}_r.$$

Fig. 5.3 The column of
conducting fluid in
cylindrical coordinates.

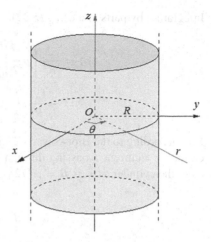

Projecting (5.20) on the r-direction, one obtains

$$\frac{d}{dr}\left(p+\frac{B^2}{3\mu}\right)=-\frac{B^2}{\mu r},$$

or

$$\frac{dp}{dr}=-\frac{1}{2\mu r^2}\frac{d}{dr}\left(r^2 B^2\right), \tag{5.32}$$

with the solution

$$p(r)=p(0)-\frac{1}{2\mu}\int_0^r \frac{1}{r^2}\frac{d}{dr}\left(r^2 B^2\right)dr, \tag{5.33}$$

where $p(0)$ is the pressure of the fluid for $r = 0$. If the fluid is confined inside the
cylindrical column $r \le R$, at the limit $r = R$ the pressure p decreases to zero, so
that

$$p(0)=\frac{1}{2\mu}\int_0^R \frac{1}{r^2}\frac{d}{dr}\left(r^2 B^2\right)dr, \tag{5.34}$$

and (5.33) becomes

$$p(r)=\frac{1}{2\mu}\int_r^R \frac{1}{r^2}\frac{d}{dr}\left(r^2 B^2\right)dr, \tag{5.35}$$

The *average pressure* $\langle p \rangle$ inside the cylinder can be expressed in terms of the
current intensity traveling through the fluid. By definition,

$$\langle p \rangle=\frac{\int\limits_0^R r\,p(r)dr}{\int\limits_0^R r\,dr}=\frac{2}{R^2}\int\limits_0^R r\,p(r)\,dr.$$

Integrating by parts and using (5.32), we find

$$\langle p \rangle = \frac{2}{R^2} \left[p \frac{r^2}{2} \Big|_0^R - \frac{1}{2} \int_0^R r^2 \, dp \right]$$

$$= \frac{1}{2\mu R^2} \int_0^R d\left(r^2 B^2\right) = \frac{1}{2\mu} B^2 \Big|_{r=R}. \tag{5.36}$$

But, according to the Biot–Savart–Laplace law, the field of magnetic induction generated by a current I passing through an infinite, rectilinear wire, at the distance r from the cylinder axis, is $B = \mu I / 2\pi r$, and we finally obtain

$$\langle p \rangle = \frac{\mu I^2}{8\pi^2 R^2}. \tag{5.37}$$

One observes that, according to (5.36), the average pressure equals the pressure at the surface of the cylinder of fluid.

In thermonuclear experiments one works with hot plasmas at temperatures of about 10^8 K and densities of about 10^{15} particle/cm^3. These figures correspond to pressures $p = nkT \sim 14$ atm. To confine such plasmas, if we assume $B|_R = 20$ kGs, one must have $I \sim (9 \times 10^4 \, R)$ A, with R in cm.

To determine the radial variation of pressure, we assume that $j(r)$ is a constant for $r < R$. The field B at some point inside the plasma column is then

$$B(r) = \frac{\mu I}{2\pi R^2} r, \quad r < R.$$

With this relation, (5.35) gives

$$p(r) = \frac{\mu I^2}{4\pi^2 R^2} \left(1 - \frac{r^2}{R^2}\right), \tag{5.38}$$

indicating that the pressure depends parabolically on r.

The r-dependence of the pressure, magnetic induction, and current density is shown in Fig. 5.4.

Fig. 5.4 The r-dependence of the pressure, magnetic field, and current density in the *pinch effect*.

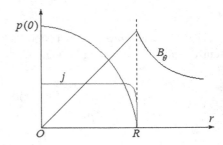

5.4.2 Force-Free Fields

A special interest in astrophysical observations is presented by those magnetic fields that do not produce any mechanical action on the conducting medium, i.e. those characterized by

$$(\nabla \times \mathbf{B}) \times \mathbf{B} = 0, \tag{5.39}$$

which means

$$\nabla \times \mathbf{B} = \alpha(r)\mathbf{B}. \tag{5.40}$$

These fields are called *force-free fields*. For example, such a field is the relatively strong magnetic field in a region where the pressure of the ionized fluid (plasma) is very low, as above solar spots. Indeed, since the temperature above solar spots is smaller than the temperature of the surrounding solar surface, according to $p = nkT$ it results that the pressure is also smaller.

   To justify the equilibrium of the solar spots, we therefore have to explain why the surrounding plasma cannot penetrate the spots. First, we shall prove that a force-free field (with $\alpha = $ const.) represents, in a closed system, a state of minimum energy. We begin by showing that the integral

$$I_1 = \int_V \mathbf{A} \cdot \mathbf{B} \, d\tau = \int_V \mathbf{A} \cdot \nabla \times \mathbf{A} \, d\tau, \tag{5.41}$$

where V is a fixed volume, is a constant of motion (first integral) for the system of equations (5.8). We have

$$\frac{d}{dt} \int_V \mathbf{A} \cdot \nabla \times \mathbf{A} \, d\tau = \int_V \frac{\partial \mathbf{A}}{\partial t} \cdot \nabla \times \mathbf{A} \, d\tau + \int_V \mathbf{A} \cdot \nabla \times \frac{\partial \mathbf{A}}{\partial t} \, d\tau.$$

But, according to $(5.8)_1$,

$$\frac{\partial}{\partial t}(\nabla \times \mathbf{A}) = \nabla \times \left(\frac{\partial \mathbf{A}}{\partial t}\right) = \nabla \times [\mathbf{v} \times (\nabla \times \mathbf{A})],$$

or, by a convenient choice of the gauge,

$$\frac{\partial \mathbf{A}}{\partial t} = \mathbf{v} \times (\nabla \times \mathbf{A}). \tag{5.42}$$

This yields

$$\frac{\partial \mathbf{A}}{\partial t} \cdot (\nabla \times \mathbf{A}) = 0, \tag{5.43}$$

so that

$$\frac{d}{dt} \int_V \mathbf{A} \cdot \nabla \times \mathbf{A} \, d\tau = \int_V \frac{\partial \mathbf{A}}{\partial t} \cdot \nabla \times \mathbf{A} \, d\tau + \int_V \nabla \cdot \left(\mathbf{A} \times \frac{\partial \mathbf{A}}{\partial t} \right) d\tau.$$

The first integral vanishes according to (5.43). The same happens with the second integral due to the divergence theorem

$$\int_V \nabla \cdot \left(\mathbf{A} \times \frac{\partial \mathbf{A}}{\partial t} \right) d\tau = \oint_S \left(\mathbf{A} \times \frac{\partial \mathbf{A}}{\partial t} \right) \cdot d\mathbf{S} = 0, \qquad (5.44)$$

because the system is closed, which means that the internal motion cannot affect the vector potential outside the system, and, since $\mathbf{A}$ is supposed to be a continuous function, the quantity $\partial \mathbf{A}/\partial t$ must vanish on the boundary surface S.

Next, let us calculate the extremum of the magnetic energy

$$W_m = \frac{1}{2\mu} \int_V B^2 d\tau, \qquad (5.45)$$

subject to the first integral (5.41). This is a *constrained extremum problem* which can generally be formulated as follows: given a constraint equation of the form $g(x, y, z) = 0$ and a function $f(x, y, z)$, find the extremum values of $f(x, y, z)$, with the restriction that the point (x, y, z) must be a solution of the constraint equation $g(x, y, z) = 0$. The problem can also be expressed this way: find the extremum of a functional, if an integral has a given value.

To solve our constraint problem, we multiply the constraint (5.41) by a Lagrangian multiplier β and add it to (5.45). The result is the functional

$$\Phi = \int_V \left[\frac{1}{2\mu} (\nabla \times \mathbf{A})^2 - \beta \mathbf{A} \cdot (\nabla \times \mathbf{A}) \right] d\tau. \qquad (5.46)$$

The necessary conditions of existence of an extremum for the functional (5.46) are given by the Euler–Lagrange equations

$$\frac{\partial f}{\partial A_i} - \frac{\partial}{\partial x_k} \left(\frac{\partial f}{\partial A_{i,k}} \right) - \frac{\partial}{\partial t} \left(\frac{\partial f}{\partial A_{i,t}} \right) = 0, \qquad (5.47)$$

where

$$f = \frac{1}{2\mu} (\nabla \times \mathbf{A})^2 - \beta \mathbf{A} \cdot (\nabla \times \mathbf{A}),$$

$$A_{i,k} = \frac{\partial A_i}{\partial x_k},$$

$$A_{i,t} = \frac{\partial A_i}{\partial t}.$$

Performing the derivatives, one finds

$$\frac{\partial f}{\partial A_i} = -\beta (\nabla \times \mathbf{A})_i = -\beta B_i,$$

$$\frac{\partial f}{\partial A_{i,k}} = \frac{1}{\mu} \epsilon_{mki} B_m - \beta \epsilon_{mki} A_m,$$

$$\frac{\partial}{\partial x_k} \left(\frac{\partial f}{\partial A_{i,k}} \right) = \frac{1}{\mu} \epsilon_{mki} B_{m,k} - \beta \epsilon_{mki} A_{m,k} = -\frac{1}{\mu} (\nabla \times \mathbf{B})_i + \beta B_i.$$

Replacing these results into (5.47) and denoting $2\mu\beta \equiv \alpha$, we finally obtain

$$\nabla \times \mathbf{B} = \alpha \mathbf{B},$$

meaning that the magnetic energy has an extremum if the field $\mathbf{B}$ is *force-free*. (From the physical point of view, this extremum can only be a minimum.) It follows, on the one hand, that such a field is relatively stable against small perturbations, and on the other, that a system where the magnetic forces are dominant and in which there exists a mechanism of dissipation, the *force-free fields* (with $\alpha = $ const.) are *final natural configurations*.

To conclude, if there are no Joule losses, but the motion can be damped due to viscosity or some other causes, the most probable final states are those corresponding to *force-free fields*.

5.5 Solved Problems

Problem 1. Using the representation of the electromagnetic field in terms of the generalized antipotentials $\mathcal{M}$, ψ (see (3.143)), determine in the Lagrangian formalism the equation of motion of a compressible, non-viscous, infinitely conducting fluid, which performs an isentropic motion in the external electromagnetic field $(\mathbf{E}, \mathbf{B})$.

Solution. As we have seen, the electromagnetic field $(\mathbf{E}, \mathbf{B})$ can be expressed in terms of the generalized antipotentials $\mathcal{M}$, ψ as

$$\mathbf{E} = \frac{1}{\epsilon_0} (\nabla \times \mathcal{M} - \mathcal{P}),$$

$$\mathbf{B} = \mu_0 \left(\nabla\psi + \mathcal{P} \times \mathbf{v} + \frac{\partial \mathcal{M}}{\partial t} \right). \tag{5.48}$$

The quantities describing the behaviour of the fluid are connected by the equation of continuity

$$\frac{d\rho}{dt} + \rho \nabla \cdot \mathbf{v} = 0, \tag{5.49}$$

as well as the fundamental equation of thermodynamics for quasistatic, reversible processes (the entropy equation):

$$T\,ds = d\varepsilon + pd\left(\frac{1}{\rho}\right),\tag{5.50}$$

where s is the entropy and ε is the internal energy, both taken for unit mass. Since the motion of the fluid is isentropic,

$$\frac{ds}{dt} = 0.\tag{5.51}$$

To determine the equation of motion, one first needs to construct the Lagrangian density. Recall that the choice of the Lagrangian function is not unique; therefore a suitable Lagrangian density has to be simple, but contain all the characteristic physical quantities. To construct a suitable Lagrangian, we shall use the method of *Lagrangian multipliers* (see Sect. 5.4). In mathematical optimization, this method provides a strategy for finding the extremum of a function subject to constraints. For example, consider the variational problem: maximize/minimize $f(x, y, z)$, subject to the constraint $g(x, y, z) = C$. We then introduce a new variable λ, called Lagrangian multiplier, and study the Lagrange function defined by

$$\Lambda(x, y, z, \lambda) = f(x, y, z) + \lambda\,(g(x, y, z) - C) \equiv f(x, y, z) + \lambda g(x, y, z).$$

This method can be extend to more than one Lagrangian multiplier, depending on the concrete problem. In our case, for example, Eqs. (5.49) and (5.51) play the role of constrains, and our choice for the Lagrangian density is

$$\mathcal{L} = \frac{1}{2\mu_0}B^2 - \frac{1}{2}\epsilon_0 E^2 + \frac{1}{2}\rho v^2 - \rho\varepsilon - \alpha\left(\frac{d\rho}{dt} + \rho\nabla\cdot\mathbf{v}\right) - \rho\beta\frac{ds}{dt}.\tag{5.52}$$

Here all quantities, including the Lagrangian multipliers α and β, are functions of the space coordinates and time. Denoting, generically, the space-time coordinates by x_γ, $\gamma = 1, 2, 3, 4$, and the variational parameters by σ_i, $i = 1, 2, \ldots, 6$, the Euler–Lagrange equations read

$$(\mathcal{L})_i \equiv \frac{\partial\mathcal{L}}{\partial\sigma_i} - \frac{\partial}{\partial x_\gamma}\left(\frac{\partial\mathcal{L}}{\partial\sigma_{i,\gamma}}\right) = 0,\quad \gamma = 1, 2, 3, 4.\tag{5.53}$$

Expressing the electromagnetic field $(\mathbf{E}, \mathbf{B})$ in terms of the antipotentials $\mathcal{M}$, ψ, and taking $\mathcal{M}$, ψ, $\mathcal{P}$, $\mathbf{v}$, ρ, and s as variational parameters, one obtains

$$(\mathcal{L})_{\mathcal{M}} = -\frac{\partial \mathbf{B}}{\partial t} - \nabla \times \mathbf{E} = 0,$$

$$(\mathcal{L})_{\psi} = \nabla \cdot \mathbf{B} = 0,$$

$$(\mathcal{L})_{\mathcal{P}} = \mathbf{E} + \mathbf{v} \times \mathbf{B} = 0,$$

$$(\mathcal{L})_{\mathbf{v}} = \rho \mathbf{v} + \mathbf{B} \times \mathcal{P} + \rho \nabla \alpha - \beta \rho \nabla s = 0, \qquad (5.54)$$

$$(\mathcal{L})_{\rho} = \frac{1}{2} v^2 - \varepsilon - \frac{p}{\rho} + \mathbf{v} \nabla \alpha + \frac{\partial \alpha}{\partial t} = 0,$$

$$(\mathcal{L})_s = \mathbf{v} \cdot \nabla \beta + \frac{\partial \beta}{\partial t} - T = 0.$$

Thus, we have found Maxwell's source-free equations $(5.54)_{1,2}$, as well as Ohm's law for infinite conductivity $(5.54)_3$. Relation $(5.54)_4$ is a generalization of the representation of the velocity field $\mathbf{v}$ in terms of the *Clebsch potentials* which, in our case, are α, β, and ρ.

To find the equation of motion of the magnetofluid, we eliminate the Lagrangian multipliers α and β from the equations $(5.54)_{4,5,6}$. Let us first take the gradient of $(5.54)_5$. Using $(5.54)_4$, we have

$$\nabla(\beta \mathbf{v} \cdot \nabla s) + \nabla \left(\frac{\partial \alpha}{\partial t} \right) - \nabla \left(\frac{v^2}{2} \right) = \nabla \left[\varepsilon + \frac{p}{\rho} + \frac{1}{\rho} \mathbf{v} \cdot (\mathbf{B} \times \mathcal{P}) \right]. \qquad (5.55)$$

Next, we use (5.51), $(5.54)_6$, as well as the vector formula (see Appendix A):

$$\nabla(\mathbf{A} \cdot \mathbf{B}) = \mathbf{A} \times (\nabla \times \mathbf{B}) + \mathbf{B} \times (\nabla \times \mathbf{A}) + (\mathbf{A} \cdot \nabla)\mathbf{B} + (\mathbf{B} \cdot \nabla)\mathbf{A},$$

and find

$$\frac{\partial \mathbf{v}}{\partial t} + (\mathbf{v} \cdot \nabla)\mathbf{v} = -\frac{1}{\rho} \nabla p + \frac{\partial}{\partial t} \left(\frac{1}{\rho} \mathbf{B} \times \mathcal{P} \right)$$

$$+ \mathbf{v} \times \left[\nabla \times \left(\frac{1}{\rho} \mathbf{B} \times \mathcal{P} \right) \right] + \nabla \left[\frac{1}{\rho} \mathbf{v} \cdot (\mathcal{P} \times \mathbf{B}) \right]. \qquad (5.56)$$

As a last step, we take advantage of the vector identity

$$\nabla[\mathbf{A} \cdot (\mathbf{B} \times \mathbf{C})] = \mathbf{A} \times [\nabla \times (\mathbf{B} \times \mathbf{C})] + \mathbf{B} \times [\nabla \times (\mathbf{C} \times \mathbf{A})]$$

$$+ \mathbf{C} \times [\nabla \times (\mathbf{A} \times \mathbf{B})] - (\mathbf{A} \times \mathbf{B})\nabla \cdot \mathbf{C}$$

$$- (\mathbf{B} \times \mathbf{C})\nabla \cdot \mathbf{A} - (\mathbf{C} \times \mathbf{A})\nabla \cdot \mathbf{B},$$

and observe that some terms disappear in (5.56). Therefore, we arrive at the required equation of motion

$$\rho \left[\frac{\partial \mathbf{v}}{\partial t} + (\mathbf{v} \cdot \nabla)\mathbf{v} \right] = -\nabla p + \mathbf{j} \times \mathbf{B}. \qquad (5.57)$$

Problem 2. The equation of motion of an electronic plasma, in which the hydrostatic pressure is neglected, but the collisions between electrons are considered, is written as

$$\frac{\partial \mathbf{v}}{\partial t} + (\mathbf{v} \cdot \nabla)\mathbf{v} = \frac{e}{m}(\mathbf{E} + \mathbf{v} \times \mathbf{B}) - \nu\mathbf{v}, \tag{5.58}$$

where ν is the collision frequency. Show that in the presence of static, uniform electric and magnetic fields $(\mathbf{E}, \mathbf{B})$, Ohm's law is

$$j_i = \lambda_{ik}E_k, \quad i, k = 1, 2, 3, \tag{5.59}$$

where the conductivity tensor λ_{ik} is given by

$$\lambda_{ik} = \frac{\epsilon_0\omega_p^2}{\nu\left(1 + \frac{4\omega_B^2}{\nu^2}\right)} \begin{pmatrix} 1 & \frac{2\omega_B}{\nu} & 0 \\ -\frac{2\omega_B}{\nu} & 1 & 0 \\ 0 & 0 & 1 + \frac{4\omega_B^2}{\nu^2} \end{pmatrix}. \tag{5.60}$$

Here ω_p is the electronic plasma frequency, ω_B – the Larmor precession frequency, while $\mathbf{B}$ is oriented along the z-axis. It is assumed that $\epsilon \sim \epsilon_0$ and $\mu \sim \mu_0$.

Solution. We multiply (5.58) by the electric charge density ρ_e and use the relation $\mathbf{j} = \rho_e\mathbf{v}$, with the result

$$\rho_e\frac{d\mathbf{v}}{dt} = \frac{e\rho_e}{m}\mathbf{E} + \frac{e}{m}\mathbf{j} \times \mathbf{B} - \nu\mathbf{j}. \tag{5.61}$$

To calculate the l.h.s. of (5.61) we use the equation $\nabla \times \mathbf{B} = \mu_0\mathbf{j} = \mu_0\rho_e\mathbf{v}$. In view of the equation of continuity, and recalling that the magnetic field is static $(d\mathbf{B}/dt = 0)$, we can write

$$\frac{d\mathbf{v}}{dt} = \frac{d}{dt}\left(\frac{1}{\mu_0\rho_e}\nabla \times \mathbf{B}\right) = \frac{1}{\mu_0}\frac{\rho_e\nabla \times \frac{d\mathbf{B}}{dt} - \frac{d\rho_e}{dt}\nabla \times \mathbf{B}}{\rho_e^2},$$

and

$$\rho_e\frac{d\mathbf{v}}{dt} = \frac{1}{\mu_0}\nabla \times \frac{d\mathbf{B}}{dt} - \frac{\nabla \times \mathbf{B}}{\mu_0}\frac{1}{\rho_e}\frac{d\rho_e}{dt}$$
$$= -\frac{\nabla \times \mathbf{B}}{\mu_0}\frac{1}{\rho_e}\frac{d\rho_e}{dt} = -\mathbf{j}\frac{1}{\rho_e}\frac{d\rho_e}{dt} = \mathbf{j}\nabla \cdot \mathbf{v}.$$

In the linear approximation, we may take $\rho_0 \simeq \rho_{e0}$, where $\rho_{e0} = e\,n_0$ is the equilibrium electronic charge density. Then we may write

$$\nabla \cdot \mathbf{v} = \nabla \cdot \left(\frac{1}{\mu_0\rho_e}\nabla \times \mathbf{B}\right) \simeq \nabla \cdot \left(\frac{1}{\mu_0 n_0 e}\nabla \times \mathbf{B}\right) = \frac{1}{\mu_0 n_0 e}[\nabla \cdot (\nabla \times \mathbf{B})] = 0,$$

and (5.61) becomes

$$\epsilon_0 \omega_p^2 \mathbf{E} + \frac{e}{m} \mathbf{j} \times \mathbf{B} - \nu \mathbf{j} = 0, \qquad (5.62)$$

where $\omega_p^2 = \frac{n_0 e^2}{m \epsilon_0}$ is the *plasma frequency*. Introducing the notation

$$\omega_B = \frac{e B_z}{2m} \simeq \frac{e B_0}{2m}$$

for the *Larmor precession frequency*, and recalling that $\mathbf{B} = (0, 0, B_0)$, we project on axes the vector equation (5.62) and obtain

$$\begin{aligned}
\epsilon_0 \omega_p^2 E_x + 2\omega_B j_y - \nu j_x &= 0, \\
\epsilon_0 \omega_p^2 E_y + 2\omega_B j_x - \nu j_y &= 0, \\
\epsilon_0 \omega_p^2 E_z - \nu j_z &= 0.
\end{aligned} \qquad (5.63)$$

Equation (5.63)₃ gives

$$j_z = \frac{\epsilon_0 \omega_p^2}{\nu} E_z,$$

which can also be cast into the form

$$\begin{aligned}
j_z &= 0 \cdot E_x + 0 \cdot E_y + \frac{\epsilon_0 \omega_p^2}{\nu} E_z \\
&= \frac{\epsilon_0 \omega_p^2}{\nu \left(1 + \frac{4\omega_B^2}{\nu^2}\right)} \left[0 \cdot E_x + 0 \cdot E_y + \left(1 + \frac{4\omega_B^2}{\nu^2}\right) E_z \right].
\end{aligned} \qquad (5.64)$$

The first two equations (5.63) are used to find j_x and j_y. We have

$$j_x = \frac{\epsilon_0 \nu \omega_p^2 E_x + 2\epsilon_0 \omega_B \omega_p^2 E_y}{\nu^2 + 4\omega_B^2},$$

or

$$\begin{aligned}
j_x &= \frac{\epsilon_0 \omega_p^2}{\nu \left(1 + \frac{4\omega_B^2}{\nu^2}\right)} E_x + \frac{\epsilon_0 \omega_p^2}{\nu \left(1 + \frac{4\omega_B^2}{\nu^2}\right)} \frac{2\omega_B}{\nu} E_y + 0 \cdot E_z \\
&= \frac{\epsilon_0 \omega_p^2}{\nu \left(1 + \frac{4\omega_B^2}{\nu^2}\right)} \left(1 \cdot E_x + \frac{2\omega_B}{\nu} E_y + 0 \cdot E_z \right).
\end{aligned} \qquad (5.65)$$

The component j_y is found by means of $(5.63)_1$ and (5.65):

$$j_y = \frac{\nu j_x - \epsilon_0 \omega_p^2 E_x}{2\omega_B} = \frac{-4\omega_B^2 \omega_p^2 E_x + 2\nu \omega_B \omega_p^2 E_y}{\omega_B(\nu^2 + 4\omega_B^2)} \epsilon_0,$$

or

$$j_y = -\epsilon_0 \frac{\omega_p^2}{\nu\left(1 + \frac{4\omega_B^2}{\nu^2}\right)} \frac{2\omega_B}{\nu} E_x + \frac{\epsilon_0 \omega_p^2}{\nu\left(1 + \frac{4\omega_B^2}{\nu^2}\right)} E_y + 0 \cdot E_z$$

$$= \frac{\epsilon_0 \omega_p^2}{\nu\left(1 + \frac{4\omega_B^2}{\nu^2}\right)} \left(-\frac{2\omega_B}{\nu} E_x + 1 \cdot E_y + 0 \cdot E_z\right). \tag{5.66}$$

Equations (5.64)–(5.66) can be unified as

$$j_i = \lambda_{ik} E_k,$$

where

$$\lambda_{ik} = \frac{\epsilon_0 \omega_p^2}{\nu\left(1 + \frac{4\omega_B^2}{\nu^2}\right)} \begin{pmatrix} 1 & \frac{2\omega_B}{\nu} & 0 \\ -\frac{2\omega_B}{\nu} & 1 & 0 \\ 0 & 0 & 1 + \frac{4\omega_B^2}{\nu^2} \end{pmatrix}$$

is the *electric conductivity tensor* (5.60).

Problem 3. Resume the previous problem assuming that, in addition to the magnetic field $\mathbf{B} = (0, 0, B)$, at the time $t = 0$ is instantly applied an electric field $\mathbf{E} = (E, 0, 0)$. If $\mathbf{j} = 0$ at the time $t = 0$, determine the time dependence of the current density components.

Solution. Suppose, first, that the convective term $(\mathbf{v} \cdot \nabla)\mathbf{v}$ arising in the equation of motion (5.58) is small as compared to the first term (linearized equation). The components of the equation of motion (5.58) are then

$$\frac{\partial v_x}{\partial t} = \frac{eE}{m} + \frac{eB}{m} v_y - \nu v_x,$$

$$\frac{\partial v_y}{\partial t} = -\frac{eB}{m} v_x - \nu v_y, \tag{5.67}$$

$$\frac{\partial v_z}{\partial t} = -\nu v_z.$$

Integrating Eq. $(5.67)_3$, one obtains

$$v_z = C_1(x, y, z) e^{-\nu t}. \tag{5.68}$$

Taking the derivative with respect to time of $(5.67)_1$ and using $(5.67)_2$, we have

$$\frac{\partial^2 v_x}{\partial t^2} = w'_B \frac{\partial v_y}{\partial t} - \nu \frac{\partial v_x}{\partial t} = -w'^2_B v_x - \nu w'_B v_y - \nu \frac{\partial v_x}{\partial t}$$

$$= -w'^2_B v_x - \nu \left(\frac{\partial v_x}{\partial t} + \nu v_x - \frac{eE}{m} \right) - \nu \frac{\partial v_x}{\partial t},$$

where $w'_B = eB/m = 2(eB/2m) = 2w_B$. Thus,

$$\frac{\partial^2 v_x}{\partial t^2} + 2\nu \frac{\partial v_x}{\partial t} + \left(\nu^2 + w'^2_B \right) v_x = \frac{\nu e E}{m}. \tag{5.69}$$

This is a linear inhomogeneous second order differential equation, with constant coefficients. The general solution of this equation is the sum of the general solution of the corresponding homogeneous equation with a particular solution of the inhomogeneous equation, which can be determined by the method of variation of constants.

The homogeneous equation attached to Eq. (5.69) is

$$\frac{\partial^2 v_x}{\partial t^2} + 2\nu \frac{\partial v_x}{\partial t} + \left(\nu^2 + w'^2_B \right) v_x = 0.$$

Its characteristic equation

$$r^2 + 2\nu r + (\nu^2 + w'^2_B) = 0$$

has two complex conjugate solutions

$$r_{1,2} = -\nu \pm i\, w'_B.$$

The general solution of the homogeneous equation therefore is

$$v_{x0} = C'_2(x, y, z) e^{-\nu t} e^{i w'_B t} + C'_3(x, y, z) e^{-\nu t} e^{-i w'_B t}. \tag{5.70}$$

Let $v_{xp} = K_p = $ const. be a particular solution of the inhomogeneous equation (5.69). Imposing on this solution to verify (5.69), we find

$$K_p = \frac{\nu e E}{m(\nu^2 + w'^2_B)}.$$

The general solution of (5.69) then is

$$v_x = v_{x0} + v_{xp}$$

$$= C'_2(x, y, z) e^{-\nu t} e^{i w'_B t} + C'_3(x, y, z) e^{-\nu t} e^{-i w'_B t} + \frac{\nu e E}{m\left(\nu^2 + w'^2_B \right)},$$

or, in an alternative form,

$$v_x = \left[C_2(x, y, z)\sin\omega'_B t + C_3(x, y, z)\cos\omega'_B t\right] e^{-\nu t} + \frac{\nu e E}{m\left(\nu^2 + {\omega'_B}^2\right)},$$

(5.71)

with $C_2 = i(C'_2 - C'_3)$ and $C_3 = C'_2 + C'_3$.
The component v_y of the velocity results from $(5.67)_1$:

$$v_y = \frac{1}{\omega'_B}\frac{\partial v_x}{\partial t} + \frac{\nu v_x}{\omega'_B} - \frac{eE}{m\omega'_B}$$

(5.72)

$$= \left[C_2(x, y, z)\cos\omega'_B t - C_3(x, y, z)\sin\omega'_B t\right] e^{-\nu t} - \frac{eE\omega'_B}{m(\nu^2 + {\omega'_B}^2)}.$$

Consider now the simplest case when C_1, C_2, and C_3 are true constants. Using again the approximation $\rho_0 \simeq \rho_{e0} = en_0$, the time dependence of the current density components $j_i = \rho_e v_i$, $i = 1, 2, 3$ is given by

$$j_x = en_0\left[C_2\sin\omega'_B t + C_3\cos\omega'_B t\right] e^{-\nu t} + \frac{n_0 e^2 E\nu}{m\left(\nu^2 + {\omega'_B}^2\right)},$$

$$j_y = en_0\left[C_2\cos\omega'_B t - C_3\sin\omega'_B t\right] e^{-\nu t} - \frac{n_0 e^2 E\omega'_B}{m\left(\nu^2 + {\omega'_B}^2\right)},$$

(5.73)

$$j_z = en_0 C_1 e^{-\nu t}.$$

The three constants C_i, $i = 1, 2, 3$ are determined using the initial conditions, $j_i(t = 0) = 0$, leading to three algebraic equations

$$0 = en_0 C_1,$$

$$0 = en_0 C_2 - \frac{n_0 e^2 E\omega'_B}{m\left(\nu^2 + {\omega'_B}^2\right)},$$

$$0 = en_0 C_3 + \frac{n_0 e^2 E\nu}{m\left(\nu^2 + {\omega'_B}^2\right)},$$

with the solution

$$C_1 = 0,$$

$$C_2 = \frac{eE\omega'_B}{m(\nu^2 + {\omega'_B}^2)},$$

(5.74)

$$C_3 = -\frac{eE\nu}{m\left(\nu^2 + {\omega'_B}^2\right)}.$$

Introducing (5.74) into (5.73), one finally obtains the time dependence of the current density components:

$$
j_x = e n_0 \left[\frac{e E \omega'_B}{m \left(\nu^2 + {\omega'_B}^2 \right)} \sin \omega'_B t - \frac{e E \nu}{m \left(\nu^2 + {\omega'_B}^2 \right)} \cos \omega'_B t \right] e^{-\nu t}
$$
$$
+ \frac{n_0 \nu e^2 E}{m \left(\nu^2 + {\omega'_B}^2 \right)},
$$

$$
j_y = e n_0 \left[\frac{e E \omega'_B}{m \left(\nu^2 + {\omega'_B}^2 \right)} \cos \omega'_B t + \frac{e E \nu}{m \left(\nu^2 + {\omega'_B}^2 \right)} \sin \omega'_B t \right] e^{-\nu t}
$$
$$
- \frac{n_0 \omega'_B e^2 E}{m (\nu^2 + {\omega'_B}^2)},
$$
$$
j_z = 0,
$$

or, in a more condensed form,

$$
j_x = \frac{n_0 e^2 E}{m \left(\nu^2 + {\omega'_B}^2 \right)} \left[\omega'_B \sin \omega'_B t + \nu (e^{\nu t} - \cos \omega'_B t) \right] e^{-\nu t},
$$
$$
j_y = \frac{n_0 e^2 E}{m \left(\nu^2 + {\omega'_B}^2 \right)} \left[\nu \sin \omega'_B t - \omega'_B (e^{\nu t} - \cos \omega'_B t) \right] e^{-\nu t}, \qquad (5.75)
$$
$$
j_z = 0.
$$

Problem 4. A plasma is placed in the static and homogeneous magnetic field $\mathbf{B}_0$.

(a) Assuming that the plasma is a rarefied electronic gas, determine the motion of the electrons subject to a plane, monochromatic wave propagating in the direction of $\mathbf{B}_0$, and find the corresponding dispersion relation.

(b) Determine the group velocity of the electromagnetic wave, and discuss the particular case $n - 1 \ll 1$.

Solution

(a) Suppose that $\mathbf{B}_0$ is oriented along the z-axis. Since the electronic gas is rarefied, the plasma electrons can be considered as being free, so that the quasi-elastic force between electrons and nuclei may be neglected. In addition, we also neglect the damping force, due to the field produced by the electrons, as well as the electric component of the Lorentz force as compared to the magnetic component. The equation of motion of an electron of plasma then is

$$
m \frac{d^2 \mathbf{r}}{dt^2} = e \mathbf{E} + e \frac{d \mathbf{r}}{dt} \times \mathbf{B}_0, \qquad (5.76)
$$

where $\mathbf{E}$ is the electric component of the electromagnetic wave propagating through plasma, and $\mathbf{B}_0$ the external magnetic field in which the plasma is placed. Expanding the cross product in (5.76), we have

$$m\frac{d^2\mathbf{r}}{dt^2} = e\mathbf{E} + e\dot{y}B_0\mathbf{i} - e\dot{x}B_0\mathbf{j}, \tag{5.77}$$

with $\mathbf{i}, \mathbf{j}, \mathbf{k}$ being the unit vectors of the coordinate axes. Since there are no z-components of $\mathbf{E}$ and $\mathbf{v} \times \mathbf{B}_0$, we next consider only the x- and y-components of the equation of motion, which are

$$\ddot{x} = \frac{e}{m}E_x + \frac{eB_0}{m}\dot{y},$$
$$\ddot{y} = \frac{e}{m}E_y - \frac{eB_0}{m}\dot{x}. \tag{5.78}$$

Let us now make the change of variables

$$u = x + iy,$$
$$\mathcal{E} = E_x + iE_y. \tag{5.79}$$

Multiplying (5.78)$_2$ by the imaginary unit i and adding it to (5.78)$_1$, one obtains

$$\ddot{u} + i\omega_0\dot{u} = \frac{e}{m}\mathcal{E}, \quad \text{where} \quad \omega_0 = \frac{eB_0}{m}. \tag{5.80}$$

On the other hand, as stated in the problem, the direction of propagation of the plane monochromatic wave and the direction of $\mathbf{B}_0$ coincide. Since the electromagnetic wave is transversal, the electric component $\mathbf{E}$ of the wave lies in a plane orthogonal to the direction of propagation which, in our case, is the xOy-plane, so that $E_z = 0$. Recalling that $B_z = B_0 = \text{const.}$, the x- and y-projections of Maxwell's equations

$$\nabla \times \mathbf{E} = -\frac{\partial \mathbf{B}}{\partial t},$$
$$\nabla \times \mathbf{B} = \mu_0\mathbf{j} + \frac{1}{c^2}\frac{\partial \mathbf{E}}{\partial t},$$

become

$$-\frac{\partial E_y}{\partial z} = -\frac{\partial B_x}{\partial t}, \quad \frac{\partial E_x}{\partial z} = -\frac{\partial B_y}{\partial t}, \tag{5.81}$$

$$-\frac{\partial B_y}{\partial z} = \mu_0 j_x + \frac{1}{c^2}\frac{\partial E_x}{\partial t}, \quad \frac{\partial B_x}{\partial z} = \mu_0 j_y + \frac{1}{c^2}\frac{\partial E_y}{\partial t}. \tag{5.82}$$

We now use again the change of variables (5.79) and, in addition, denote

$$\mathcal{B} \equiv B_x + i B_y,$$
$$\mathcal{J} \equiv j_x + i j_y = Ne\dot{x} + iNe\dot{y} = Ne(\dot{x} + i\dot{y}) = Ne\dot{u},$$

N being the density of the plasma electrons. Since the plasma is rarefied, the electrons may be considered as free particles. A convenient multiplication by the imaginary unit i of (5.81)$_2$ and (5.82)$_2$, then addition to (5.81)$_1$ and (5.82)$_1$, give

$$i\frac{\partial \mathcal{E}}{\partial z} = -\frac{\partial \mathcal{B}}{\partial t}, \quad i\frac{\partial \mathcal{B}}{\partial z} = \mu_0 \mathcal{J} + \frac{1}{c^2}\frac{\partial \mathcal{E}}{\partial t}.$$

Thus, we arrive at the following system of three coupled differential equations for the variables u, $\mathcal{E}$, and $\mathcal{B}$:

$$\ddot{u} + i\omega_0\dot{u} = \frac{e}{m}\mathcal{E},$$
$$\frac{\partial \mathcal{E}}{\partial z} - i\frac{\partial \mathcal{B}}{\partial t} = 0, \tag{5.83}$$
$$\frac{\partial \mathcal{B}}{\partial z} + \frac{i}{c^2}\frac{\partial \mathcal{E}}{\partial t} = -i\mu_0 Ne\dot{u}.$$

Having in view the formulation of the problem, we search the unknown quantities u, $\mathcal{E}$, and $\mathcal{B}$ as plane waves of the same frequency and the same wave number (monochromatic waves), all propagating in the z-direction (direction of $\mathbf{B}_0$):

$$u = U\, e^{\pm i(kz - \omega t)},$$
$$\mathcal{E} = V\, e^{\pm i(kz - \omega t)}, \tag{5.84}$$
$$\mathcal{B} = W\, e^{\pm i(kz - \omega t)}.$$

Then, the three equations (5.83) lead to

$$-\omega^2 U + i\omega_0(\mp i\omega U) = \frac{e}{m}V,$$
$$\pm ikV \mp \omega W = 0,$$
$$\pm ikW + \frac{i}{c^2}(\mp i\omega V) = -i\mu_0 Ne(\mp i\omega U),$$

or, in a more organized form,

$$\begin{cases} \left(-\omega^2 \pm \omega\omega_0\right)U - \frac{e}{m}V + 0 \cdot W = 0 \\[2mm] 0 \cdot U \qquad\qquad +ikV - \omega W = 0 \\[2mm] \mu_0 Ne\omega U \qquad\quad +\frac{\omega}{c^2}V + ikW = 0. \end{cases} \tag{5.85}$$

In order to obtain non-trivial solutions of (5.85), one must have

$$
\begin{vmatrix}
-\omega^2 \pm \omega\omega_0 & -\frac{e}{m} & 0 \\
0 & ik & -\omega \\
\mu_0 N e\omega & \frac{\omega}{c^2} & ik
\end{vmatrix} = 0,
$$

that is

$$
\left(1 - \frac{k^2 c^2}{\omega^2}\right)\left(\omega^2 \mp \omega\,\omega_0\right) = \frac{Ne^2}{\epsilon_0 m}.
$$

This yields

$$
\frac{k^2 c^2}{\omega^2}\left(= \frac{c^2}{v_{ph}^2} = n^2\right) = 1 - \frac{\frac{Ne^2}{m\epsilon_0}}{\omega^2 \mp \omega\omega_0},
$$

and, finally,

$$
n^2 = 1 - \frac{\omega_p^2}{\omega^2 \mp \omega\omega_0}, \tag{5.86}
$$

where $v_{ph} = \omega/k$ is the phase velocity of the wave, and $\omega_p^2 = \frac{Ne^2}{m\epsilon_0}$ is the squared plasma frequency.

To find the equation of motion of the electrons, one must determine u. Taking the ratio of $(5.84)_1$ and $(5.84)_2$, we have

$$
u = \frac{U}{V}\mathcal{E} = \frac{\frac{e}{m}}{-\omega^2 \pm \omega\omega_0}\mathcal{E} = -\frac{e\mathcal{E}}{m\omega(\omega \mp \omega_0)}. \tag{5.87}
$$

Recalling that $u = x + iy$, $\mathcal{E} = E_x + iE_y$, and separating the real and imaginary parts, we arrive at

$$
x = -\frac{eE_x}{m\omega(\omega \mp \omega_0)},
$$

$$
y = -\frac{eE_x}{m\omega(\omega \mp \omega_0)},
$$

or, in the vector form,

$$
\mathbf{r} = -\frac{e}{m}\frac{\mathbf{E}}{\omega^2 \mp \omega\frac{eB_0}{m}}. \tag{5.88}
$$

Observation:
If $\mathbf{B}_0 = 0$, the relations (5.86) and (5.88) lead to

$$
\epsilon_r(\simeq n^2) = 1 - \frac{\omega_p^2}{\omega^2} \quad (\mu_r \sim 1), \quad \mathbf{r} = -\frac{e\mathbf{E}}{m\omega^2},
$$

as expected.

(b) According to the definition of the group velocity, $v_g = d\omega/dk$, we can write

$$\frac{1}{v_g} = \frac{dk}{d\omega} = \frac{d}{d\omega}\left(\frac{n\omega}{c}\right) = \frac{n}{c} + \frac{\omega}{c}\frac{dn}{d\omega}. \qquad (5.89)$$

Since the density N of electrons (in fact, of the polarized atoms) is small by hypothesis, we may approximate (see (5.86)):

$$n = \sqrt{1 - \frac{\omega_p^2}{\omega(\omega \mp \omega_0)}} \simeq 1 - \frac{\omega_p^2}{2\omega(\omega \mp \omega_0)},$$

so that

$$\frac{dn}{d\omega} = \frac{\omega_p^2}{2}\frac{2\omega \mp \omega_0}{\omega^2(\omega \mp \omega_0)^2},$$

and (5.89) gives

$$\frac{1}{v_g} = \frac{1}{c}\left[1 - \frac{\omega_p^2}{2\omega(\omega \mp \omega_0)}\right]$$

$$+ \frac{\omega\omega_p^2}{2c}\frac{2\omega \mp \omega_0}{\omega^2(\omega \mp \omega_0)^2} = \frac{1}{c}\left[1 + \frac{\omega_p^2}{2(\omega \mp \omega_0)^2}\right],$$

which finally leads to

$$v_g = c\,\frac{1}{1 + \frac{1}{2}\left(\frac{\omega_p}{\omega \mp \omega_0}\right)^2} < c. \qquad (5.90)$$

Problem 5. A viscous, homogeneous, and incompressible conducting fluid is placed between the conducting planes $z = 0$ and $z = d$. The plane $z = d$ moves along the x-axis with a constant velocity v_0. If along the z-axis acts a uniform magnetic field $\mathbf{H}_0$, and along the y-axis a uniform electric field $\mathbf{E}_0$, determine the velocity distribution in the fluid undergoing a stationary motion.

Solution. The equation of motion of a viscous, conducting fluid (see (5.4)) is

$$\rho\mathbf{a} = \rho\mathbf{F} - \nabla p + (\xi + \eta)\nabla\theta' + \eta\Delta\mathbf{v} + \mathbf{j} \times \mathbf{B}, \qquad (5.91)$$

where ρ is the mass density of the fluid, $\mathbf{a} = \partial\mathbf{v}/\partial t + (\mathbf{v} \cdot \nabla)\mathbf{v}$ is the acceleration of a fluid particle, $\mathbf{F}$ is the density of the force of non-electromagnetic nature (force per unit mass), and $\theta' = \nabla \cdot \mathbf{v}$.

Since the fluid is homogeneous and incompressible, which means $\rho(\mathbf{r}, t) =$ const., the equation of continuity leads to $\theta' = \nabla \cdot \mathbf{v} = 0$. Neglecting the gravitational force, which is very weak as compared to the electromagnetic terms, and recalling that the motion is stationary, the equation of motion (5.91) becomes

$$\nabla p = \eta \Delta \mathbf{v} + \mathbf{j} \times \mathbf{B}. \tag{5.92}$$

To write the components of (5.92), let us first consider the last term. The components of $\mathbf{j}$, according to Ohm's law $\mathbf{j} = \lambda \mathbf{E}'$, are

$$
\begin{aligned}
j_x &= \lambda E_x + \lambda (v_y B_z - v_z B_y) = 0, \\
j_y &= \lambda E_y + \lambda (v_z B_x - v_x B_z) = \lambda E_0 - \mu_0 \lambda v_x(z) H_0, \\
j_z &= \lambda E_z + \lambda (v_x B_y - v_y B_x) = 0.
\end{aligned}
$$

Here we considered the fact that the fluid is moving along the x-axis, and its velocity, at any point, depends only on the position of that point between the two planes, given by the coordinate z: $\mathbf{v} = v_x(z)\mathbf{u}_x$. We have

$$
\begin{aligned}
(\mathbf{j} \times \mathbf{B})_x &= j_y B_z - j_z B_y = j_y B_0 = \mu_0 \lambda H_0 [E_0 - \mu_0 v_x(z) H_0], \\
(\mathbf{j} \times \mathbf{B})_y &= j_z B_x - j_x B_z = 0, \\
(\mathbf{j} \times \mathbf{B})_z &= j_x B_y - j_y B_x = -\mu_0 \lambda H_x [E_0 - \mu_0 v_x(z) H_0].
\end{aligned}
$$

Projecting (5.92) on axes, one then obtains

$$
\begin{aligned}
\frac{\partial p}{\partial x} &= \eta \frac{\partial^2 v_x}{\partial z^2} + \mu_0 \lambda H_0 (E_0 - \mu_0 v_x H_0), \\
\frac{\partial p}{\partial y} &= 0, \\
\frac{\partial p}{\partial z} &= -\mu_0 \lambda H_x (E_0 - \mu_0 v_x H_0).
\end{aligned} \tag{5.93}
$$

Due to the geometry of the problem, there is no pressure gradient along the x-axis. In addition, since $v_x = v_x(z)$, we have $\partial^2 v_x / \partial z^2 = d^2 v_z / dz^2$, and (5.94)$_1$ yields

$$\frac{d^2 v_x}{dz^2} - \frac{1}{d_0^2} v_x = -\frac{1}{d_0^2} \frac{E_0}{\mu_0 H_0}, \quad \text{where} \quad \frac{1}{d_0^2} = \frac{\lambda \mu_0^2 H_0^2}{\eta}. \tag{5.94}$$

This is an ordinary second-order inhomogeneous differential equation, with constant coefficients. Again we shall write the general solution of (5.94) as the sum of the general solution of the homogeneous equation and a particular solution of the inhomogeneous equation. The general solution of the homogeneous equation is

$$v_x^0(z) = V_1 e^{z/d_0} + V_2 e^{-z/d_0},$$

or, in an alternative form,

$$v_x^0(z) = W_1 \sinh \frac{z}{d_0} + W_2 \cosh \frac{z}{d_0}. \qquad (5.95)$$

Here V_1, V_2, and W_1, W_2 are constants of integration. A particular solution of the inhomogeneous equation is of the form $v_x^p = W_3$, where W_3 is a real, non-zero constant. One easily finds

$$W_3 = \frac{E_0}{\mu_0 H_0}. \qquad (5.96)$$

The general solution of (5.94) therefore is

$$v_x(z) = v_x^0(z) + v_x^p = W_1 \sinh \frac{z}{d_0} + W_2 \cosh \frac{z}{d_0} + \frac{E_0}{\mu_0 H_0}. \qquad (5.97)$$

The arbitrary constants of integration W_1, W_2 are obtained using the boundary conditions: $v_x(z = 0) = 0$, $v_x(z = d) = v_0$. Thus,

$$v_x(0) = W_2 + \frac{E_0}{\mu_0 H_0} = 0,$$

$$v_x(d) = W_1 \sinh \frac{d}{d_0} + W_2 \cosh \frac{d}{d_0} + \frac{E_0}{\mu_0 H_0} = v_0.$$

Solving this system of algebraic equations, one finds

$$W_1 = \frac{v_0 - \frac{E_0}{\mu_0 H_0}\left(1 - \cosh \frac{d}{d_0}\right)}{\sinh \frac{d}{d_0}},$$

$$W_2 = -\frac{E_0}{\mu_0 H_0}. \qquad (5.98)$$

Now, we are able to write the solution of (5.94) as

$$v_x(z) = \frac{v_0 - \frac{E_0}{\mu_0 H_0}\left(1 - \cosh \frac{d}{d_0}\right)}{\sinh \frac{d}{d_0}} \sinh \frac{z}{d_0} - \frac{E_0}{\mu_0 H_0} \cosh \frac{z}{d_0} + \frac{E_0}{\mu_0 H_0}$$

$$= \frac{v_0}{\sinh \frac{d}{d_0}} \sinh \frac{z}{d_0} + \frac{E_0}{\mu_0 H_0}\left[1 - \cosh \frac{z}{d_0} - \frac{\left(1 - \cosh \frac{d}{d_0}\right)}{\sinh \frac{d}{d_0}} \sinh \frac{z}{d_0}\right]$$

$$= \frac{v_0}{\sinh \frac{d}{d_0}} \sinh \frac{z}{d_0} + \frac{E_0}{\mu_0 H_0}$$

$$\times \left[1 - \frac{\left(\sinh \frac{d}{d_0} \cosh \frac{z}{d_0} - \sinh \frac{z}{d_0} \cosh \frac{d}{d_0}\right) + \sinh \frac{z}{d_0}}{\sinh \frac{d}{d_0}}\right].$$

Using the trigonometric formula

$$\sinh(x - y) = \sinh x \, \cosh y - \sinh y \, \cosh x,$$

we finally obtain the velocity distribution as

$$v_x(z) = v_0 \frac{\sinh \frac{z}{d_0}}{\sinh \frac{d}{d_0}} + \frac{E_0}{\mu_0 H_0} \left[1 - \frac{\sinh \frac{d-z}{d_0} + \sinh \frac{z}{d_0}}{\sinh \frac{d}{d_0}} \right]. \tag{5.99}$$

For weak magnetic fields, i.e. $d \ll d_0 = \sqrt{\frac{\eta}{\lambda} \frac{1}{\mu_0 H_0}}$, one can use the series expansion

$$\sinh x = \frac{x}{1!} + \frac{x^3}{3!} + \cdots + \frac{x^{2n+1}}{(2n+1)!} + \cdots,$$

and retain only the first term. The result is

$$v_x(z) = v_0 \frac{z}{d}. \tag{5.100}$$

This means that, if the magnetic field is weak, the velocity increases linearly with the distance between the two planes, being 0 at $z = 0$ and v_0 at $z = d$.

If, on the contrary, the magnetic field is strong, which means $d \gg d_0 = \sqrt{\frac{\eta}{\lambda} \frac{1}{\mu_0 H_0}}$, we may approximate

$$\frac{\sinh \frac{z}{d_0}}{\sinh \frac{d}{d_0}} \simeq 0,$$

$$\frac{\sinh \frac{d-z}{d_0}}{\sinh \frac{d}{d_0}} = \frac{e^{\frac{d-z}{d_0}} - e^{\frac{z-d}{d_0}}}{e^{\frac{d}{d_0}} - e^{-\frac{d}{d_0}}} = \frac{e^{-\frac{z}{d_0}} - e^{-\frac{2d-z}{d_0}}}{1 - e^{-\frac{2d}{d_0}}} \simeq e^{-\frac{z}{d_0}},$$

and in this case (5.99) leads to the exponential variation of the velocity with distance,

$$v_x(z) = \frac{E_0}{\mu_0 H_0} \left(1 - e^{-\frac{z}{d_0}} \right) = \frac{E_0}{\mu_0 H_0} \left(1 - e^{-\frac{\mu_0 H_0 \sqrt{\lambda}}{\sqrt{\eta}} z} \right). \tag{5.101}$$

5.6 Proposed Problems

1. Show that the expression $I_2 = \int_V \mathbf{v} \cdot \mathbf{B} \, d\tau$, where V is a fixed volume occupied by an ideal magnetofluid, is also a first integral of the system of equations (5.8).
2. Show that the equation of motion of a viscous magnetofluid (5.4), assuming that $\mathbf{F} = 0$, can also be written as a momentum conservation equation

$$\frac{\partial}{\partial t}(\rho\, v_i) = -\frac{\partial \Pi_{ik}}{\partial x_k}, \quad i, k = 1, 2, 3,$$

where

$$\Pi_{ik} = \rho\, v_i v_k + p\, \delta_{ik} + \frac{1}{2}(\mathbf{H} \cdot \mathbf{B})\delta_{ik} - H_i B_k - \sigma'_{ik}$$

is the stress tensor of the system composed of fluid and field, and σ'_{ik} is a term representing the viscosity.

3. Given a plasma composed of elastically coupled charged particles, with different elastic constants for mutually orthogonal directions, find the electric permittivity tensor of the medium. The particle density is N.

4. A plane electromagnetic wave given by $\mathbf{E} = \mathbf{E}_0 e^{i(kx-\omega t)}$, $\mathbf{B} = \mathbf{B}_0 e^{i(kx-\omega t)}$ propagates in a neutral plasma (metal, ionosphere). The following forces are considered to act on the electrons (of density N and charge $-e$):
   (i) the electric force: $-e\mathbf{E}$;
   (ii) the friction force: $-(m/\tau)\mathbf{v}$, where τ is the average time between two collisions of electrons and ions;
   (iii) the quasi-elastic force: $-m\omega_0^2\mathbf{r}$.
   Find the dispersion relation $n = n(\omega)$.

5. Consider a magnetofluid moving orthogonal to a magnetic field $\mathbf{B}$. As a result of the fluid motion, the magnetic lines of force will become deformed. Find the variation of magnetic energy in such a process and show that the deformation of the magnetic lines corresponds to strengthening of the magnetic field.

6. Using the representation of the electromagnetic field in terms of the generalized antipotentials $\mathcal{M}$, ψ (see (3.143)), write the Hamiltonian density $\mathcal{H}$ and determine the equation of motion of a compressible, non-viscous, infinitely long conducting fluid, undergoing an isentropic motion in the external electromagnetic field $(\mathbf{E}, \mathbf{B})$. Hamilton's canonical equations for continuous systems (fields) are

$$\dot{q}_i = \frac{\partial \mathcal{H}}{\partial p_i},$$

$$\dot{p}_i = -\frac{\partial \mathcal{H}}{\partial q_i} + \frac{\partial}{\partial x_k}\left(\frac{\partial \mathcal{H}}{\partial q_{i,k}}\right),$$

where q_i stand for the generalized coordinates, and p_i for the associated generalized momenta.

7. Using the Clebsch representation for *all* vector fields $(\mathbf{E}, \mathbf{B}, \mathbf{v}, \mathbf{A})$, and taking the Clebsch potentials as independent variational parameters, find the system of equations governing the behaviour of an ideal magnetofluid.
   (*Hint*: According to Clebsch's theorem, for *any* vector field $\mathbf{A}$ is possible to find three scalar quantities ξ, η, and ζ, functions of coordinates and time, so as to have $\mathbf{A} = -\nabla\xi + \eta\nabla\zeta$.)

8. According to Noether's theorem, to each infinitesimal symmetry transformation corresponds a conservation equation. This equation can be written as

$$\frac{\partial}{\partial t} \int_V \gamma d\tau = - \oint_S \mathbf{G} \cdot d\mathbf{S},$$

where

$$\gamma = \left[\mathcal{L} - \frac{\partial \mathcal{L}}{\partial \varphi^{(s)}_{,t}} \varphi^{(s)}_{,t} \right] \delta t - \frac{\partial \mathcal{L}}{\partial \varphi^{(s)}_{,t}} (\delta \mathbf{x} \cdot \nabla) \varphi^{(s)} + \frac{\partial \mathcal{L}}{\partial \varphi^{(s)}_{,t}} \delta \varphi^{(s)} + \delta \Omega_t,$$

$$\mathbf{G} = \frac{-\partial \mathcal{L}}{\partial (\nabla \varphi^{(s)})} \varphi^{(s)}_{,t} \delta t + \left[\mathcal{L} \delta \mathbf{x} - \frac{\partial \mathcal{L}}{\partial (\nabla \varphi^{(s)})} (\delta \mathbf{x} \cdot \nabla) \varphi^{(s)} \right]$$

$$+ \frac{\partial \mathcal{L}}{\partial (\nabla \varphi^{(s)})} \delta \varphi^{(s)} + \delta \boldsymbol{\Omega}.$$

Find an appropriate Lagrangian density $\mathcal{L} \left(x_\alpha, \varphi^{(s)}, \varphi^{(s)}_{,\alpha} \right)$, $\alpha = 1, 2, 3, 4$ and show that the equation of conservation associated with the infinitesimal displacement of the time origin $t \to t' = t + \delta t$ ($\delta t = $ const.) is the equation of conservation of energy of an ideal magnetofluid. Here $x_1 = x$, $x_2 = y$, $x_3 = z$, $x_4 = t$.

9. Resume the preceding application and show that the equation of conservation associated with the infinitesimal displacement of the origin of axes $\mathbf{x} \to \mathbf{x}' = \mathbf{x} + \delta \mathbf{x}$ ($\delta \mathbf{x} = $ const.) is the momentum conservation equation of the system composed of fluid and field.

10. Consider a fully ionized plasma. If τ is the time corresponding to the mean free path of electrons, m is the electron mass, $\mathbf{v}_e$ – its velocity, and only $\mathbf{E}$ is present, the friction force $m \mathbf{v}_e \tau^{-1}$ is compensated by the electric field force $m \mathbf{v}_e \tau^{-1} = - e\mathbf{E}$, which gives the current density

$$\mathbf{j} = - N e \mathbf{v}_e = - \frac{N e^2 \tau}{m} \mathbf{E} = \lambda \mathbf{E}.$$

If both $\mathbf{E}$ and $\mathbf{B}$ act upon electrons, Ohm's law is written as

$$\mathbf{j} = \lambda \mathbf{E} - \frac{N e^2 \tau}{m} \mathbf{v} \times \mathbf{B} = \lambda \mathbf{E} - \frac{e\tau}{m} \mathbf{j} \times \mathbf{B} = \lambda \mathbf{E} - \alpha \, (\mathbf{j} \times \mathbf{B}).$$

Study the relation between the vector quantities $\mathbf{j}$, $\mathbf{E}$, and $\mathbf{B}$ in two cases: $\mathbf{E} \parallel \mathbf{B}$ and $\mathbf{E} \perp \mathbf{B}$.

Part II
Relativistic Formulation
of Electrodynamics

Theory of relativity is a discovery, not an invention

Chapter 6
Special Theory of Relativity

6.1 Experimental Basis of Special Relativity

At the end of the 19th century and the beginning of the 20th century, in physics appeared some revolutionary discoveries: in 1895 Wilhelm Conrad Röntgen (1845–1923) (Nobel Prize 1901) discovered the X-rays; in 1896 Henri Antoine Becquerel (1853–1908) discovered *natural radioactivity*; in 1897 Joseph John Thomson (1856–1940) (Nobel Prize 1906) discovered the *electron*; in 1900 Max Planck (1858–1947) (Nobel Prize 1918) postulated the idea that the energy emitted by a black body could only take on discrete values; in 1905 Albert Einstein (1879–1955) (Nobel Prize 1921) elaborated the *special theory of relativity*. These discoveries marked genuine turning points that shook classical physics and marked out the fundamental directions of further development of physics: quantum theory, theory of elementary particles, special and general theory of relativity, etc.

The inception of the theory of relativity is closely intertwined with the development of electrodynamics of moving media, as well as the attempts to solve the so-called "æther problem". For centuries, scientists like Newton, Maxwell, Hertz, used as an absolute frame of reference a transparent, perfectly elastic, invisible and imponderable medium, called *æther*. This concept played a crucial role in the appearance and development of the theory of relativity.

Once postulated, the existence of *æther* gave rise to a very natural question: is the *æther* dragged by bodies in motion, or is it not? For example, does the Earth in its orbital motion drag with it the *æther*? In other words, is there a "wind" of *æther*? And, still, if this dragging exists, is it total, or partial? Here are the answers to this question, given by three prominent physicists:

– *Augustin-Jean Fresnel* (1788–1827) proved (theoretically) that the *æther* is partially dragged by the moving bodies;

– *Hendrik Antoon Lorentz* (1853–1928) considered the *æther* as being immobile;

– *George Gabriel Stokes* (1819–1903), on the contrary, conceived the *æther* as completely dragged by media in motion.

© Springer-Verlag Berlin Heidelberg 2016
M. Chaichian et al., *Electrodynamics*, DOI 10.1007/978-3-642-17381-3_6

Fresnel's theory, elaborated in 1818, even if it is the first in chronological order, contains the other two theories as particular cases. In 1817, Fresnel had introduced the transverse wave theory of light which could account for all the known phenomena of optics; consequently he conceived the *æther* as solid-like and rigid, yet allowing the free passage of heavenly bodies. In Fresnel's theory, the *æther* flowed through the interstices of material bodies even on the smallest scale; but he did allow for matter to have a small dragging effect on the aether. Let us follow Fresnel's ideas regarding the *æther*.

Let u be the phase velocity of a transversal (e.g. light) wave, propagating in an infinite, elastic, homogeneous, and isotropic medium. The theory of elasticity shows that the velocity u is related to the density ρ of the medium through the formula $u = \sqrt{\mu/\rho}$, where μ is the coefficient of elasticity. For vacuum we should have $c = \sqrt{\mu/\rho_0}$. Fresnel considered ρ as also being the density of *æther* in the medium, while ρ_0 was the density of *æther* in vacuum. Since the refractive index of the medium is $n = c/u = \sqrt{\rho/\rho_0} > 1$, it follows that $\rho > \rho_0$.

Denote by V the velocity of a body with respect to the *æther*, supposed to be fixed. Then V is also the velocity of the *æther* that penetrates the body. Since no accumulations of the *æther* in the body are allowed (the opposite situation would lead to changes in optical properties), the mass of the *æther* entering the body with velocity V has to be equal to that leaving the body with velocity V', at any moment: $\rho V' = \rho_0 V$. Since $\rho_0 < \rho$, it follows that $V' < V$. We also have $n^2 = \rho/\rho_0 = V/V'$, so that the *æther* is dragged along the direction of motion of the body with the velocity $v_d = V - V' = V\left(1 - \frac{1}{n^2}\right)$. This means that a stationary external observer determines the velocity

$$u' = \frac{c}{n} \pm v_d = \frac{c}{n} \pm V\left(1 - \frac{1}{n^2}\right), \qquad (6.1)$$

where the $+$ sign stands for the case when the light and the body move in the same direction, and the $-$ sign when they move in opposite direction. Fresnel's formula (6.1) shows that in case of water, for example, the *drag coefficient* of the *æther* by water is $k = 1 - 1/n^2 = 0.437$.

One also observes that if the *æther* is undragged, the speed of light in the medium would be $u_1 = \frac{c}{n}$, while if it is completely dragged, the speed of light would be $u_2 = \frac{c}{n} + V$. Consequently, according to (6.1), we have $u_1 < u' < u_2$.

Obviously, the "*æther* problem" could have been solved only by the experiment. But, in its turn, the experiment could not offer a unique solution. Indeed, some famous experiments performed especially during the second half of the 19th century, came to certify various assumptions regarding the *æther* dragging. (It is important to notice the fact that all these theories were concerned with the *æther* drag, but none of them was questioning the existence of *æther* itself).

Let us approach, in chronological order, some of the most prominent effects and experiments, and show how they were explained on the basis of the existence of the hypothetical *æther*.

6.1.1 Aberration of Light

The phenomenon of *aberration of light*, also referred to as *astronomical aberration* or *stellar aberration*, was discovered by James Bradley (1693–1762) in 1727. For an observer connected to the Earth, the celestial objects seem to perform an apparent motion about their real locations. Bradley explained the phenomenon as a result of the finite speed of light and the motion of Earth on its orbit around the Sun. The stellar aberration should not be confused with stellar parallax, or with the refraction of light when passing through terrestrial atmosphere.

Let SA be the real direction of propagation of light coming from the star S, and OP the direction of motion of the Earth (observer). To make the explanation easier, we consider that the two directions are orthogonal (see Fig. 6.1).

The star light propagates on the direction SA. If the Earth were stationary, the telescope should be placed along SA in order to have the star at the centre of the field. However, the Earth is in motion, and while the star light travels to the Earth with the speed c, the Earth itself moves on its orbit with the speed V. As a result, due to the displacement of the Earth, to see the star the observer must incline the telescope in the direction of motion of the Earth, until O coincides with O' (the eye of the observer). The star is observed at the position S'. The angle α between the optical axis of the telescope and the direction of the light ray is called *aberration angle*.

As shown in Fig. 6.1, the time for the light ray to pass from A to O' equals the time for the Earth to cover the distance OO' with velocity V, so that $\frac{AO'}{c} = \frac{OO'}{V}$. Since $AO' = AO \sin \theta$ and $OO' = AO \sin \alpha$, it follows that

$$\sin \alpha = \frac{V}{c} \sin \theta. \tag{6.2}$$

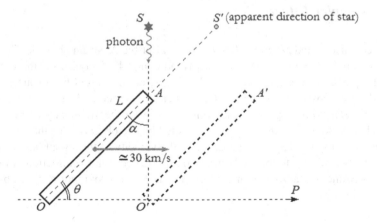

Fig. 6.1 Aberration of light (with exaggerated aberration angle).

The maximum of aberration is produced when the star is situated at its *zenith* (the position of the star on the vertical direction opposite to the net gravitational force at the location of the observer), in which case $\theta \simeq \frac{\pi}{2}$. Since the aberration angle is very small, we may write

$$\sin \alpha \simeq \alpha = \frac{V}{c}. \qquad (6.3)$$

The velocity of the Earth on its elliptical orbit around the Sun is $V = 30\,\mathrm{km \cdot s^{-1}}$, which gives $\alpha \simeq 10^{-4}\,\mathrm{rad} = 20\,\mathrm{arc\ seconds}$. As a result, the star seems to describe, over the course of a year, an elliptic orbit. This phenomenon is encountered for any "fixed" star.

If the instrument (telescope) is filled up with water, according to the theory the aberration angle will be different (say, β). According to the refraction law,

$$n = \frac{\sin \beta}{\sin \alpha'} \simeq \frac{\beta}{\alpha'},$$

or, by means of (6.2) with u instead of c,

$$\beta = n\,\alpha' = n\frac{V}{u}\sin \varphi = n^2\frac{V}{c}\sin \varphi = n^2\alpha. \qquad (6.4)$$

But the experiment invalidates this theoretical result. In 1871, the British astronomer George Biddell Airy (1801–1892) proved that the aberration does not depend on the refracting medium (water), i.e. $\beta = \alpha$. This results led to the conclusion that the medium (water, in this case), being at rest with respect to the Earth, *does not drag the æther*.

6.1.2 Doppler Effect

This effect, also called *Doppler shift*, is named after the Austrian physicist Christian Andreas Doppler (1803–1853), who discovered it in 1842. It consists in the variation of the frequency of a signal (acoustic, luminous, etc.), emitted by a source S and detected by an observer O (see Fig. 6.2), when the source and the observer are in relative motion with respect to each other. Let ν be the frequency of the signal emitted by the source, ν' the frequency detected by the observer, $\mathbf{V}_s$ the velocity of the source, $\mathbf{V}_0$ the velocity of the observer, c the velocity of the signal (all velocities are determined relative to the medium), θ_s the angle between the direction of motion of the source and SO, and θ_0 the angle between the direction of motion of the observer and OS (Fig. 6.2).

Fig. 6.2 Doppler effect: a particular example of observer–source relative motion.

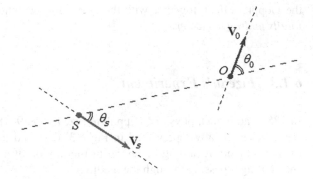

One can distinguish three situations:

1) S – stationary, O – mobile. In this case

$$\nu' = \nu \left(1 \pm \frac{V_0}{c} \cos \theta_0 \right), \qquad (6.5)$$

where the "+" sign is taken when the observer is approaching the source, and "−" when the observer is moving away from the source.

2) O – stationary, S – mobile. In this situation

$$\nu' = \frac{\nu}{1 \mp \frac{V_s}{c} \cos \theta_s}. \qquad (6.6)$$

Here "+" corresponds to the situation when the source is moving towards the observer, and "−" when the source is receding from the observer. If $V_s \ll c$, one can expand in series (6.6) to obtain

$$\nu' = \nu \left(1 \pm \frac{V_s}{c} \cos \theta_s \mp \ldots \right) \simeq \nu \left(1 \pm \frac{V_s}{c} \cos \theta_s \right),$$

and we find the result of the previous case.

3) Both S and O are mobile. The relation between frequencies is

$$\nu' = \nu \frac{1 + \frac{V_0}{c} \cos \theta_0}{1 - \frac{V_s}{c} \cos \theta_s}, \qquad (6.7)$$

where θ_0 and θ_s are the angles, at a certain moment, between the directions of motion of the source and the observer, respectively, with the straight line connecting them.

We note that the frequency detected by the observer does not depend on the relative motion source–observer (as it should, according to the principle of classic relativity), but depends on the absolute motion relative to the *æther* (or the medium, which is the same thing). Since the medium is considered to be fixed, it follows that the Doppler effect, together with the aberration of light, proves the hypothesis of *totally undragged æther*.

6.1.3   Fizeau's Experiment

In 1851, the French physicist Hippolyte Fizeau (1819–1896) carried out the experiment schematically represented in Fig. 6.3, designed to verify formula (6.1) and Fresnel's theory regarding the effect of the motion of a dispersive medium on the speed of light passing through the medium.

The tube C is filled up with water. The light coming from the source S falls on a semitransparent mirror which acts as a beam splitter and is afterwards collimated by the lens L_1. The screen E helps to separate the light into two beams which are reflected on the mirror M and subsequently reunited, so that in Q are produced interference fringes, detectable by an interferometer.

When water is moving in the tube, one beam travels against the flow of water (light is "dragged backwards"), while the other beam travels in the same direction as the water (light is "dragged forward"). The interference fringes suffer a displacement and the *drag coefficient* of the *æther* can be determined. Fizeau found $k = 0.460$, in good agreement with Fresnel's theoretical value. Later, Michelson and Morley in 1886, then Zeeman in 1915, resumed the experiment and showed that there is an almost perfect agreement between the experimental data and Fresnel's hypothesis of partial *æther* drag.

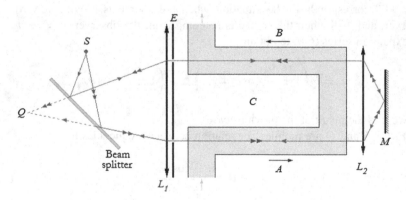

Fig. 6.3   Fizeau's experiment – schematic representation.

6.1.4 Michelson–Morley Experiment

A crucial moment in the evolution of the ideas regarding the *æther* was the experiment imagined and carried out in 1887 by Albert Abraham Michelson (1852–1931) and Edward Williams Morley (1838–1923).

Since the Fizeau experiment, which Michelson and Morley had refined, seemed to prove the existence of the undragged *æther*, the two collaborators decided to conceive an experiment sufficiently accurate to detect the "æther wind". The device they used in the experiment was an *interferometer*, which Michelson had invented earlier and which is schematically represented in Fig. 6.4. Here S is a coherent light source, L – a semi-silvered mirror used as a beam splitter, M_1 and M_2 – two plane mirrors, D – a detector for interference fringes, while the direction SO coincides with the direction of motion of the Earth. The velocity of the Earth through the *æther* is denoted by **V**. Consequently, a wind of *æther* with the velocity $-\mathbf{V}$ should be detected by an observer at rest with respect to the Earth. The arms of the device were of about $l = 1.2$ m (but, as later shown, this detail is not essential).

Let us calculate the time intervals t' and t'' for the light rays to cover the paths OM_2O and OM_1O. According to the theory of *æther* that Michelson and Morley were attempting to prove, the speed of light is c with respect to the *æther* itself. We denote by t_1 the time necessary for the light beam to travel in the longitudinal arm from O to M_2, and by t_2 the time for coming back. The Earth is moving with respect to the *æther* with the velocity **V**. When the light travels from O to M_2, it goes against the wind of *æther*, therefore one expects its speed with respect to Earth to be $c - V$. When the light travels backwards, with the wind of *æther*, its expected speed would be $c + V$. Thus, we may write

$$l = (c - V)t_1, \quad l = (c + V)t_2,$$

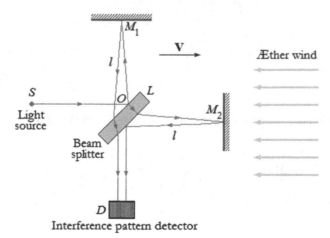

Fig. 6.4 Schematic representation of the Michelson–Morley interferometer.

Fig. 6.5 The scheme of the geometric path LM_1L' of the light beam, with respect to the *æther*.

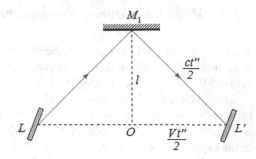

Fig. 6.5 The scheme of the geometric path LM_1L' of the light beam, with respect to the *æther*.

which gives

$$t_1 = \frac{l}{c - V}, \quad t_2 = \frac{l}{c + V},$$

consequently

$$t' = t_1 + t_2 = \frac{2l}{c} \frac{1}{1 - \frac{V^2}{c^2}}. \tag{6.8}$$

For the transverse direction with respect to the *æther* wind, the beam propagates from L, is reflected on M_1 and reaches back to L'. The scheme of the geometric path LM_1L' is given in Fig. 6.5.

The time for the light beam to cover the geometric path LM_1L' equals the time for the Earth to move from L to L', so that

$$LM_1L' = ct'' = 2\,LM_1, \quad LL' = Vt'' = 2\,LO,$$

therefore

$$l^2 = \left(\frac{ct''}{2}\right)^2 - \left(\frac{Vt''}{2}\right)^2,$$

which yields

$$t'' = \frac{2l}{c} \frac{1}{\sqrt{1 - \frac{V^2}{c^2}}}. \tag{6.9}$$

The time difference between the coherent beams interfering in D is then

$$\Delta t = t'' - t' = \frac{2l}{c} \left(\frac{1}{\sqrt{1 - \frac{V^2}{c^2}}} - \frac{1}{1 - \frac{V^2}{c^2}} \right). \tag{6.10}$$

Since $V = 3 \times 10^4$ m · s^{-1}, it follows that $V^2/c^2 \ll 1$, which allows us to expand in power series the square root in (6.10). Keeping only the terms up to V^2/c^2, we obtain

$$\Delta t \simeq -\frac{l}{c}\left(\frac{V}{c}\right)^2.$$

This time difference corresponds to a phase difference between the two beams and, as a result, the telescope D will detect interference fringes.

Rotating the interferometer by 90 degrees, the longitudinal and the transverse beams change places, and OM_1 becomes parallel to V. In this case, the time difference between the two light beams is

$$\Delta t^* \simeq \frac{l}{c}\left(\frac{V}{c}\right)^2.$$

As a result of turning the device, there appears a time difference

$$\Delta T = \Delta t^* - \Delta t = \frac{2l}{c}\left(\frac{V}{c}\right)^2,$$

which should produce a displacement of the fringes by a number of fringes given by

$$N = \frac{\Delta T}{T} = \frac{2l}{\lambda}\left(\frac{V}{c}\right)^2, \qquad (6.11)$$

where T is the period and λ is the wavelength of the monochromatic radiation used in the experiment. Choosing $\lambda = 6 \times 10^{-7}$ m, Michelson should have found $N \simeq 0.1$, i.e. a displacement of at least one tenth of a fringe.

Even if the precision of the measuring system was one-hundredth of an interfringe, Michelson was surprised to detect *no displacement of fringes*. The experiment was resumed, with higher precision and various improvements, by Morley and Miller in 1902–1904, Dayton Clarence Miller (1866–1941) in 1921–1926, and again Michelson et al. in 1929, etc., but the result was always the same. (In fact, measurements showed very small displacements, but these were within the range of the experimental error.)

Under these circumstances, Michelson and Morley were forced to draw one of the two conclusions: either the Earth is immobile, which is absurd, or there is no "æther wind", or, in other words, the *æther* is *completely dragged* by the Earth in its motion. But this last idea came in contradiction with previous experiments and effects of *æther* and its dragging.

Michelson was very disappointed about the "negative result" of his experiment, that became what might be called the "most famous failed experiment to date". But, as the development of science showed, the experiment "offered the most important negative result in the history of science" (John Desmond Bernal).

There were various attempts to explain the unexpected result of the Michelson–Morely experiment. Among these endeavours was the ad-hoc hypothesis suggested in 1889 by the Irish physicist George Francis FitzGerald (1851–1901), and a few years later, in 1892, independently, by H.A. Lorentz. They supposed that the longitudinal arm OM_2 of the interferometer suffers a contraction of ratio $l_{real}/l_{determined} = \sqrt{1 - V^2/c^2}$. In this case, we would have $\Delta T = 0$ in (6.11), and this would explain why the fringes are not displaced. But this hypothesis did not have any theoretical support, and it was not confirmed by experiments specially imagined for this purpose (Rayleigh 1902, Brace 1905, Trouton and Rankine 1908, etc.).

All this experimental work led to a natural conclusion: the existence of *æther*, either immobile or dragged by the moving bodies, cannot be experimentally proved. As we shall see, Einstein's special relativity explains the FitzGerald–Lorentz contraction, being consistent with the null result of the Michelson–Morley experiment.

In 1907, Michelson was awarded the Nobel Prize in Physics for his extremely precise interferometric method, with multiple applications in physics and astronomy. In the presentation speech at the Royal Swedish Academy of Sciences, a great emphasis was put on the fact that Michelson's interferometer "rendered it possible to obtain a non-material standard of length, possessed of a degree of accuracy never hitherto attained. By its means we are enabled to ensure that the prototype of the metre has remained unaltered in length, and to restore it with absolute infallibility, supposing it were to get lost." Indeed, Michelson was the first to measure with an interferometer the standard metre, in 1893. His method led, in 1960, to the definition of the metre in the International System, as equal to 1 650 763.73 wavelengths of the orange-red emission line in the electromagnetic spectrum of the krypton-86 atom in a vacuum. This conceptual and experimental achievement is in no way less important than the merit of the Michelson–Morley experiment in the development of the theory of relativity.

6.2 Principles of Special Relativity

6.2.1 Einstein's Postulates

It has been known for centuries that the classical mechanics laws *are covariant*, i.e. they keep their form when passing from one inertial reference frame to another.

Let $S(Oxyz)$ and $S'(O'x'y'z')$ be two inertial frames. If x, y, z are the Cartesian coordinates of a point P determined at time t by an observer from S, and x', y', z' are the coordinates of the same point measured at time t' by an observer connected to S', then one can write

$$\mathbf{r}' = \mathbf{r} - \mathbf{V}t,$$
$$t' = t. \tag{6.12}$$

Here $\mathbf{r} = (x, y, z)$, $\mathbf{r}' = (x', y', z')$, while $\mathbf{V}$ is the relative velocity of the inertial frames S and S'. The time origin is conveniently chosen.

Relations (6.12) are called the *Galilei–Newton transformation relations*. It can be shown that these transformations fulfill the axioms of an additive Abelian group structure. If the displacement of frames takes place along the common axis $Ox \equiv O'x'$, the transformations (6.12) become, in components,

$$x' = x - Vt,$$
$$y' = y,$$
$$z' = z \tag{6.13}$$
$$t' = t.$$

As an example, consider a group of particles interacting by central-type forces. The equation of motion of a particle i, in S', reads

$$m_i \frac{d\mathbf{v}'_i}{dt'} = -\nabla'_i \sum_k V_{ik}\left(\left|\mathbf{r}'_i - \mathbf{r}'_k\right|\right). \tag{6.14}$$

Using (6.12), we find

$$\mathbf{v}'_i = \mathbf{v}_i - \mathbf{V},$$
$$\nabla'_i = \nabla_i,$$
$$\frac{d\mathbf{v}_i}{dt'} = \frac{d\mathbf{v}_i}{dt},$$
$$\mathbf{r}'_i - \mathbf{r}'_k = \mathbf{r}_i - \mathbf{r}_k,$$

and (6.14) leads to

$$m_i \frac{d\mathbf{v}_i}{dt} = -\nabla_i \sum_k V_{ik}(|\mathbf{r}_i - \mathbf{r}_k|), \tag{6.15}$$

which is the equation of motion of the particle i with respect to S. We conclude that the Galilei–Newton transformation (6.12) leaves unchanged the form of the equation of motion (6.14) (and all the equations of classical mechanics as well). This property is termed *covariance* and represents an expression of the *principle of classical relativity*.

In contrast with classical mechanics laws, the equations governing the wave phenomena *are not covariant* with respect to the transformation (6.12). Take, again, an example. If in S' the well-known wave equation is written as

$$\left(\Delta' - \frac{1}{c^2}\frac{\partial^2}{\partial t'^2}\right)\psi(\mathbf{r}', t') = 0, \tag{6.16}$$

it is straightforward to show that in S this equation becomes

$$\left[\Delta - \frac{1}{c^2}\frac{\partial^2}{\partial t^2} - \frac{2}{c^2}(\mathbf{v} \cdot \nabla)\frac{\partial}{\partial t} - \frac{1}{c^2}(\mathbf{v} \cdot \nabla)^2 \right] \psi(\mathbf{r}, t) = 0. \qquad (6.17)$$

This means that the wave equation (6.16) is not covariant with respect to the Galilei–Newton transformations. It was shown by Hendrik Lorentz that the equations describing wave phenomena are covariant with respect to the transformations (written for $Ox \equiv O'x'$):

$$x' = \frac{1}{\sqrt{1 - \frac{V^2}{c^2}}} (x - Vt),$$

$$y' = y$$

$$z' = z \qquad\qquad\qquad\qquad\qquad\qquad (6.18)$$

$$t' = \frac{1}{\sqrt{1 - \frac{V^2}{c^2}}} \left(t - \frac{V}{c^2} x \right),$$

where c is the velocity of light in vacuum. Lorentz obtained these relations by mathematical considerations in 1904. One year later, Albert Einstein reinterpreted the transformations to be a statement about the nature of both space and time, and he independently re-derived these transformations from his postulates of special relativity.

Under these circumstances, there were only three possibilities:

(*i*)   to maintain unchanged mechanics, and modify electrodynamics so as to be covariant with respect to the Galilei–Newton transformations;
(*ii*) to maintain unchanged electrodynamics, and modify mechanics in order to be covariant with respect to the Lorentz transformations (6.18);
(*iii*) to change both mechanics and electrodynamics.

Some attempts were made to modify electrodynamics. But, at the beginning of the 20th century, the electromagnetic field theory was very well established, so that to change the Maxwellian edifice was an inconceivable idea, and those attempts did not succeed at all. Since the third possibility would have implied a lot of work, scientists chose the second way, and this is one of the auspices under which the theory of relativity came to life.

This theory was elaborated in 1905 by Albert Einstein, in his paper "Zur Elektrodynamik bewegter Körper" ("On the Electrodynamics of Moving Bodies"), published in Annalen der Physik, IV. Folge. 17. Seite 891–921. Juni 1905. Einstein's theory, known as the *special theory of relativity*, or, shorter, *special relativity*, is based on two postulates:

(i) *The laws of physics are the same in all inertial reference frames.* This postulate is an extension from the Newtonian principle of relativity, which states that the laws of mechanics are the same for all observers in uniform motion.

(ii) *The speed of light in empty space is the same in all inertial frames*. This means that the velocity of light in free space appears the same to all observers, regardless of the motion of the source of light and of the observer. This postulate was based on the fact that there were no experiments to contradict the constancy of the speed of light. On the contrary, the astronomical observations performed by Daniel Frost Comstock (1910) and Willem de Sitter (1913) on the double-star orbits, and by Alexey Bonch-Bruyevich (1957) on the light coming from different parts of the Sun, showed that Einstein's second postulate had a strong experimental justification.

As one can see, the *æther* plays no role in Einstein's theory, and Einstein deemed it "superfluous" in his first paper on the subject. Special relativity does not need privileged reference systems to explain physical phenomena. The motion of bodies/waves is not referred to the hypothetical, absolute reference system, the *æther*, but to other bodies.

6.2.2 Lorentz Boosts

In 1904, Lorentz proved that the electromagnetic field equations are covariant with respect to the transformations (6.18). In the same year, the French mathematician, physicist and philosopher of science Henri Poincaré (1854–1912) presented a lecture at the world scientific congress of Saint-Louis (Missouri, September 1904), in which he postulated the "Principle of Relativity": *"The laws of physical phenomena must be the same for a fixed observer and for an observer in rectilinear and uniform motion so that we have no possibility of perceiving whether or not we are dragged in such a motion"*. This principle was essentially based on the negative results of all æther experiments of that time. In 1905, Poincaré showed that Lorentz transformations form a group on a four-dimensional manifold in which the time-coordinate is imaginary, gave a geometric interpretation to the Lorentz transformations, and constructed a tensor representation of the electromagnetic field.

Einstein's merit was, among other things, that he realized the *physical significance* of the Lorentz transformations. Let us show that the transformations (6.18) result *necessarily* from Einstein's postulates.

Taking the reference frames and their relative motion as shown in Fig. 6.6, the transformations we are looking for have to satisfy the following three conditions:

(1) be *symmetric* with respect to the frames S and S', that is the relations expressing x, y, z, t as functions of x', y', z', t' must have the same form as the inverse relations. This condition is required by the *equivalence* of the two frames.
(2) be *linear*, as the Galilei–Newton transformations (6.12), which are the limiting case for velocities much smaller than the speed of light. Such a requirement can be satisfied simultaneously with condition (1), while a non-linear transformation (e.g. quadratic) could not.

Fig. 6.6 Two inertial
reference frames S and S'.

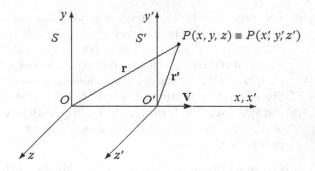

(3) for $V = 0$, one obtains the *identical* transformation ($x' = x$, $y' = y$, $z' = z$, $t' = t$).

Let the transformations we are searching for be generically represented as

$$x_i' = x_i'(x_k, t),$$
$$t' = t'(x_k, t),$$ (6.19)

where $x_1 = x$, $x_2 = y$, $x_3 = z$, and i, $k = 1, 2, 3$. According to condition (2), there also exists the inverse transformation

$$x_i = x_i(x_k', t'),$$
$$t = t(x_k', t').$$ (6.20)

By differentiating (6.19), one obtains

$$dx_i' = \frac{\partial x_i'}{\partial x_k}\, dx_k + \frac{\partial x_i'}{\partial t}\, dt,$$
$$dt' = \frac{\partial t'}{\partial x_k}\, dx_k + \frac{\partial t'}{\partial t}\, dt.$$ (6.21)

Here (dx_i, dt) signifies the distance and the time interval between two infinitely closed points, determined from S, and (dx_i', dt') the same quantities, determined by an observer at rest with respect to S'. Recalling that $Ox \equiv O'x'$, we have

$$dx' = b_{11}\, dx + b_{14}\, dt,$$
$$dy' = dy,$$
$$dz' = dz,$$ (6.22)
$$dt' = b_{41}\, dx + b_{44}\, dt,$$

where

$$b_{ik} \equiv \frac{\partial x_i'}{\partial x_k}.$$ (6.23)

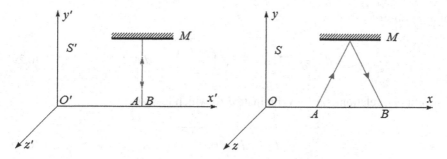

Fig. 6.7 Geometrical representation of the path of a reflected light-beam in two inertial reference systems, S and S'.

Consider a plane mirror M, fixed with respect to S' (Fig. 6.7). A light beam, emitted at a point A situated on $O'x'$ is reflected by M and comes back to the same point. Therefore, an observer from S' determines

$$dx_1' = 0, \qquad dt_1' = 2\frac{dy_1'}{c} = 2\frac{dy_1}{c}. \tag{6.24}$$

The same experiment, recorded by an observer from S, gives

$$dx_1 = V\,dt_1 = 2\sqrt{\left(\frac{c\,dt_1}{2}\right)^2 - dy_1^2}. \tag{6.25}$$

Eliminating dy_2 between the last two relations, one finds

$$dt_1' = dt_1\sqrt{1 - \frac{V^2}{c^2}}. \tag{6.26}$$

With (6.24) and (6.26), the relations (6.22) lead to

$$b_{11}\,V + b_{14} = 0,$$
$$b_{41}\,V + b_{44} = \sqrt{1 - \frac{V^2}{c^2}}. \tag{6.27}$$

Let us now repeat the experiment, but this time with the mirror M fixed with respect to S. The observer from S determines

$$dx_2 = 0,$$
$$dt_2 = 2\frac{dy_2}{c} = 2\frac{dy_2'}{c}, \tag{6.28}$$

and the observer from S'

$$dx_2' = -V\,dt_2' = 2\sqrt{\left(\frac{cdt_2'}{2}\right)^2 - dy_2'^2}. \tag{6.29}$$

The last two relations serve to eliminate dy_2', which leads to

$$dt_2 = dt_2'\sqrt{1 - \frac{V^2}{c^2}}. \tag{6.30}$$

Introducing (6.28)–(6.30) into (6.22), we find

$$b_{44} = \frac{1}{\sqrt{1 - \frac{V^2}{c^2}}},$$

$$b_{14} = -\frac{1}{\sqrt{1 - \frac{V^2}{c^2}}}\,V, \tag{6.31}$$

and (6.27) yields

$$b_{11} = \frac{1}{\sqrt{1 - \frac{V^2}{c^2}}},$$

$$b_{41} = -\frac{1}{\sqrt{1 - \frac{V^2}{c^2}}}\frac{V}{c^2}. \tag{6.32}$$

The transformations (6.22) therefore become

$$dx' = \Gamma(dx - V\,dt),$$
$$dy' = dy,$$
$$dz' = dz, \tag{6.33}$$
$$dt' = \Gamma\left(dt - \frac{V}{c^2}\,dx\right),$$

where

$$\Gamma = \frac{1}{\sqrt{1 - \frac{V^2}{c^2}}}. \tag{6.34}$$

The postulated equivalence of the inertial frames S and S' makes it possible to find the *inverse transformations* by substituting V with $-V$, that is

$$
\begin{aligned}
dx &= \Gamma \left(dx' + V\, dt' \right), \\
dy &= dy', \\
dz &= dz', \\
dt &= \Gamma \left(dt' + \frac{V}{c^2}\, dx' \right).
\end{aligned}
\tag{6.35}
$$

The transformations (6.33) and (6.35) are inverse of each other: if one set (any of them) is called *direct*, the other one is the *inverse* of the first.

Since the coefficients of x and t depend only on V and c, and not on x or t, the transformations (6.33) are *linear*, in agreement with condition (2). We can then integrate and obtain

$$
x' - x_0' = \Gamma \left[x - x_0 - V(t - t_0) \right] \quad \Leftrightarrow \quad \Delta x' = \Gamma(\Delta x - V\, \Delta t), \tag{6.36}
$$
$$
t' - t_0' = \Gamma \left[t - t_0 - \frac{V}{c^2}(x - x_0) \right] \quad \Leftrightarrow \quad \Delta t' = \Gamma \left(\Delta t - \frac{V}{c^2}\Delta x \right).
$$

If, in particular, at the initial moment $t_0 = t_0' = 0$ the two frames have the same origin ($O \equiv O'$) and $x_0 = x_0' = 0$, we can finally write

$$
\begin{aligned}
x' &= \Gamma(x - V t), \\
y' &= y, \\
z' &= z, \\
t' &= \Gamma \left(t - \frac{V}{c^2} x \right),
\end{aligned}
\tag{6.37}
$$

which is nothing else but the Lorentz transformations (6.18). Such transformations between inertial frames in relative motion are customarily called *Lorentz boosts*; together with the set of space rotations, they form the Lorentz group, on which we shall elaborate further in Sect. 7.3.

Observations:

(a) In the limit $\frac{V}{c} \to 0$, the Lorentz transformations (6.37) go to the Galilei–Newton transformations (6.13). Still we draw the reader's attention upon the fact that the relations (6.37) have been deduced assuming the the speed of light c is *invariant*, while non-relativistic physics does not admit a velocity which is the same in all reference frame.

(b) The velocity of light in vacuum, c, is not only an invariant, but also the *speed limit* in the Universe. Indeed, if we had $V > c$, the factor Γ would become imaginary, and all physical laws obtained by these transformations would be meaningless.

Fig. 6.8 Two inertial
frames, one moving with
respect to the other in an
arbitrary direction, with the
velocity **V**.

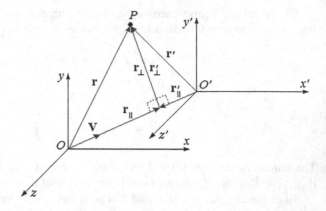

Fig. 6.8 Two inertial frames, one moving with respect to the other in an arbitrary direction, with the velocity **V**.

The Lorentz boost (6.37) is written for the particular case $Ox \equiv O'x'$ ($\mathbf{V} \parallel \mathbf{i}$, where $\mathbf{i}$ is the versor of Ox-axis). Let us now generalize this result for the case when the motion of the inertial frames takes place in an arbitrary direction, defined by the unit vector $\mathbf{v}_0 = \mathbf{V}/V$ ($V = |\mathbf{V}|$).

Take an arbitrary point P, and let $\mathbf{r} = (x, y, z)$ and $\mathbf{r}' = (x', y', z')$ be the radius vectors of P with respect to O and O', respectively. Using Fig. 6.8, we can write

$$\mathbf{r} = \mathbf{r}_\parallel + \mathbf{r}_\perp,$$
$$\mathbf{r}' = \mathbf{r}'_\parallel + \mathbf{r}'_\perp.$$

The vector $\mathbf{r}'_\perp$ is orthogonal to the direction of motion. Thus, according to (6.37), we have

$$\mathbf{r}'_\perp = \mathbf{r}_\perp. \tag{6.38}$$

Since $\mathbf{r}'_\parallel$ lies along $\mathbf{V}$, we also have

$$\mathbf{r}'_\parallel = \Gamma \left(\mathbf{r}_\parallel - \mathbf{V}t \right). \tag{6.39}$$

Therefore

$$\mathbf{r}' = \mathbf{r}_\perp + \Gamma \left(\mathbf{r}_\parallel - \mathbf{V}t \right). \tag{6.40}$$

But $\mathbf{r}_\parallel = |\mathbf{r}_\parallel| \, \mathbf{v}_0 = (\mathbf{r} \cdot \mathbf{v}_0)\mathbf{v}_0$, so that

$$\mathbf{r}_\perp = \mathbf{r} - \mathbf{r}_\parallel = \mathbf{r} - (\mathbf{r} \cdot \mathbf{v}_0)\mathbf{v}_0,$$

and (6.40) yields

$$\mathbf{r}' = \mathbf{r} + (\Gamma - 1)(\mathbf{r} \cdot \mathbf{v}_0)\mathbf{v}_0 - \Gamma \mathbf{V}t. \tag{6.41}$$

In its turn, the time transformation is

$$t' = \Gamma \left(t - \frac{V}{c^2} |\mathbf{r}_\parallel| \right) = \Gamma \left(t - \frac{\mathbf{r} \cdot \mathbf{V}}{c^2} \right). \tag{6.42}$$

The relations (6.41) and (6.42) can also be written as

$$\mathbf{r}' = (\Gamma - 1)(\mathbf{r} \times \mathbf{v}_0) \times \mathbf{v}_0 + \Gamma(\mathbf{r} - \mathbf{V}t),$$
$$t' = \Gamma \left[t - \frac{V}{c^2} (\mathbf{r} \cdot \mathbf{v}_0) \right]. \tag{6.43}$$

These relations are a vectorial generalization of the Lorentz transformations (6.37).

6.3 Some Consequences of the Lorentz Transformations

6.3.1 Relativity of Simultaneity

In classical mechanics, the simultaneity has an absolute character: two events simultaneous in one inertial reference frame S, are simultaneous in any other inertial frame S'. One of the most striking consequences of the Lorentz transformations is that simultaneity, as an absolute concept, has to be abandoned. In contrast, special relativity shows that simultaneity is *relative*, in the sense that it depends on the motion of the reference frames.

In order to discuss any consequences of the Lorentz transformations, we have to define the concept of relativistic reference frame. For this, we shall assume that the postulates of special relativity are valid in every frame; moreover, the speed of light is the maximum speed at which signals are transmitted. Although this assumption is not part of the postulates, it needs to be incorporated in the theory in order for the principle of causality to hold, as we shall see later.

The main difference between a nonrelativistic frame and a relativistic one consists in the procedure for the synchronization of clocks. If we wish to have reliable time measurements in a given frame, any clock at rest in that frame, placed at any point of space, has to show the same time. This synchronization is easily achieved in Galilean mechanics, where instantaneous transmission of signals is conceivable. However, the synchronization of clocks is much more subtle in relativity. The synchronization agent is a light signal, assuming that the speed of light in vacuum is the same in all directions of space. In his first paper on special relativity, Einstein described a synchronization procedure which goes as follows: having placed at the space points A and B two clocks which are stationary with respect to a given reference frame, one sends a light signal from the clock at A, at the time t_A (measured by the clock at A), towards the clock at B. The clock at B receives the signal at the time t_B (measured by the clock at B). This clock is provided with a mirror, which reflects back the light signal to A, where the reflected light is detected at the time t'_A. Light has to travel the same length back and forth, and by assumption its speed is the same, c, whether it goes from A to B, or from B to A. Consequently, the clocks at A and B are synchronized if

$$t_B - t_A = t'_A - t_B,$$

in other words, the time of the clock at B has to be set to

$$t_B = \frac{1}{2}(t_A + t'_A) \tag{6.44}$$

when the light signal is received and reflected.

The synchronization is *reflexive* (each clock is synchronized with itself), *symmetric* (if the clock at A is synchronized with the clock at B, then the clock at B is synchronized with the clock at A), and *transitive* (if the clock at A is synchronized with the clock at B, and the clock at B is synchronized with the clock at C, then the clock at A and the clock at C are synchronized). We emphasize once more that this synchronization procedure is valid for clocks which are at rest with respect to each other, i.e. clocks rigidly attached to a given inertial frame.

With this synchronization procedure, we define a relativistic frame of reference as a coordinate system of rigid axes and a set of synchronized clocks fixed rigidly to this system.

The *relativity of simultaneity* is the concept that simultaneity is not absolute, but dependent on the frame of reference. Let us recall a Gedanken Experiment to illustrate this fact, the famous train-and-platform experiment. Imagine that a train is passing with constant speed the platform of a railway station. The frame of reference attached to the train will be denoted by S and the frame attached to the platform – by S'. A flash of light is shot from the middle of a traincar towards the two ends of it. An observer which is riding with the train measures the light as arriving simultaneously to the front and rear of the traincar (see Fig. 6.9a). For an observer which is stationary on the platform, the rear of the traincar moves towards the point from which the flash of light was shot, while the front moves away. As a result, as measured from the

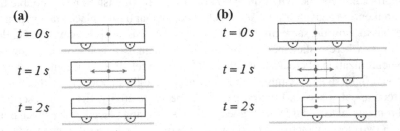

Fig. 6.9 (a) The train-and-platform experiment from the reference frame of an observer on board the train; (b) The same sequence of events in the frame of an observer standing on the platform (length contraction not depicted).

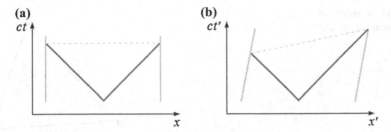

Fig. 6.10 (a) The spacetime diagram of the flashlights propagation in the frame of the observer on the train: the vertical lines represent the wordlines of the ends of the traincar. The light reaches the two ends of the traincar simultaneously; (b) The same diagram in the frame of an observer who sees the train moving to the right. The light reaches the two ends of the traincar at different times.

platform, the light reaches the rear before it reaches the front of the traincar, that is, the same two events are not simultaneous in S' (see Fig. 6.9b).

We can make use of (6.37) to express mathematically the relativity of simultaneity. Let us denote by (x_1, y_1, z_1, t_1) and (x_2, y_2, z_2, t_2) the points in the frame S which represent the arrival of the light signal to the front and to the end of the traincar. As the two events are simultaneous in S, it means that $t_1 = t_2$, i.e. $\Delta t = 0$. An observer fixed with respect to the frame S', according to (6.37), determines the duration $\Delta t' = \Gamma \left(- \frac{V}{c^2} \Delta x\right)$, meaning that the two events *are not* necessarily simultaneous in S'. The events could be simultaneous only if $\Delta x = 0$, which is obvious. In other words, according to the special theory of relativity, it is impossible to say in an absolute sense whether two events occur at the same time, if those events are separated in space. The space-time diagrams corresponding to the two observers are depicted in Fig. 6.10.

Now, to see how this differs from the nonrelativistic approach, let us imagine that two bullets are shot from the middle of the traincar towards the front and the rear, with equal speeds, v. For the observer connected to the traincar, the two bullets arrive simultaneously at the ends of the traincar. For the observer on the platform, the bullet reaching the rear travels a shorter distance than the bullet reaching the front. However, for this observer, the two bullets have *different speeds*: the speed of the bullet shot towards the front is $v + V$, while the speed of the bullet shot towards the rear is $v - V$. As a result, for the observer on the platform the two bullets reach the ends of the traincar simultaneously. The fact that the speed of light does not depend on the speed of the object which emits it is the key to the understanding of the relativity of simultaneity.

It should be emphasized that the *relativity of simultaneity does not contradict the principle of causality*. To prove this statement, we use the method of *reductio ad absurdum*. Suppose that an event occurs at point A, an effect of this event is recorded at point B, while O is an observer which receives light signals from A and

Fig. 6.11 Relativity of
simultaneity and causality.

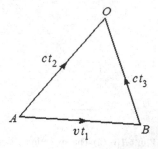

B (Fig. 6.11). If the signal from B reached the observer before the signal from A (the
effect would precede the cause in time), we would have $t_1 + t_3 < t_2$, that is

$$ct_2 - ct_3 > ct_1.$$

On the other hand, in the triangle AOB we have

$$ct_2 - ct_3 < Vt_1.$$

The last two relations yield $V > c$. From the point of view of special relativity, this
is a contradiction.

6.3.2 *Length Contraction*

Let us consider again the two inertial frames S and S' in relative motion along
$Ox \equiv O'x'$, and a rigid bar AB attached to S' and placed along the Ox'-axis, as in
Fig. 6.12.

Fig. 6.12 The length of the
bar AB is contracted when
observed from the reference
system S which is in motion
with respect to the bar.

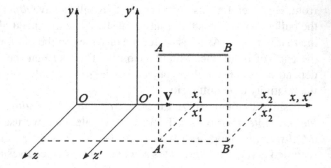

We shall call *proper length* (l_0) of the bar the length as measured by an observer in the reference frame in which the object is at rest, in our case S'. The frame with respect to which the bar is at rest is called *proper frame*. Thus,

$$x'_2 - x'_1 = l_0.$$

The length of the bar, determined by an observer attached to S, is

$$x_2 - x_1 = l.$$

The measurements are performed exclusively by light signals. Since the speed of light is finite, the observer from S must determine the two ends A and B of the bar *at the same time*, $t_1 = t_2$ (otherwise, during the measurements the bar would change its place). Taking $\Delta t = 0$, $\Delta x' = l_0$, $\Delta x = l$, we have

$$l_0 = \Gamma l, \quad \text{or} \quad l = l_0 \sqrt{1 - \frac{V^2}{c^2}} < l_0. \tag{6.45}$$

We arrive at the same result by using the inverse transformation $\Delta x = \Gamma(\Delta x' + V \Delta t')$, where $\Delta t'$ is replaced with its value extracted from $\Delta t = 0 = \Gamma\left(\Delta t' + \frac{V}{c^2}\Delta x'\right)$.

Suppose, now, that the bar is attached to the frame S (proper frame), that is $x_2 - x_1 = l_0$. The observer attached to S' will determine, at the same time ($t'_1 = t'_2$) the length $\Delta x = \Gamma \Delta x'$, or $l = l_0\sqrt{1 - V^2/c^2} < l_0$, which is the same result as (6.45).

The above considerations lead to the following conclusions:

(i) The maximum length of the bar is the proper length. Measured from any other reference frame the bar is shorter, as if being *contracted* along the direction of relative motion. The effect is called *length contraction*, or the *FitzGerald–Lorentz contraction*;

(ii) There is no Lorentz contraction in a direction orthogonal to the relative direction of motion;

(iii) There is no speed greater than the speed of light in vacuum. For $V \geq c$, one of the dimensions of the body would disappear, or even become imaginary.

Observations:

(a) The relativistic contraction is not produced by a force, or by any modification of structure of the body. This effect is exclusively due to the observation by means of light signals. All the experiments imagined to prove a *real* contraction along the direction of motion have failed;

(b) The notion of *reference frame* is an abstraction, used to describe the physical phenomena. There is no material frame apart from moving or steady bodies and media. The inertial frame to which the bar is attached, is the bar itself.

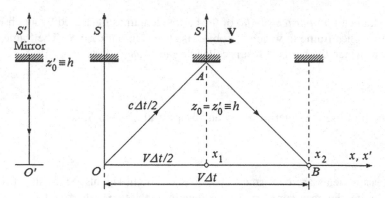

Fig. 6.13 Gedanken Experiment showing that the proper time interval between two events is always less than the time interval recorded from a moving frame.

6.3.3 Time Dilation

Any real phenomenon develops in space and time. Setting the place (*where*) and a time (*when*), we have an *event*.[1] In relativistic theories, space and time are combined into a new mathematical and physical entity, called the *space-time*.[2]

Let us consider again the reference frames S and S' and a Gedanken Experiment to see how durations change when we pass from one frame to another. At the origin O' of the system S' there is a clock and a light source; at a point M on the $O'z'$-axis, at a distance h from the origin, is placed a mirror, at rest in the frame S'. When the origins O and O' of the two frames coincide, a light flash is emitted by the source towards the mirror. As observed from the frame S', the flash is reflected by the mirror and it returns to O' in the time interval

$$\Delta t' = \frac{2h}{c}.$$

This interval is measured by the same clock, which means that the two events (the emission of the flash of light and the detection of the reflected light) took place at the same point in space. This sequence is represented in Fig. 6.13a. A time interval between events registered by the same clock (which implies that the events occurred at the same point in space) is called *proper time*.

---

[1] An *event* is "an occurrence that happens in a small space and lasts a short time" (J.L. Synge).

[2] On the 21st of September 1908, Hermann Minkowski (1864–1909) began his talk at the 80th Assembly of German Natural Scientists and Physicians with the following introduction, which has become by now famous: "The views of space and time which I wish to lay before you have sprung from the soil of experimental physics, and therein lies their strength. They are radical. Henceforth space by itself, and time by itself, are doomed to fade away into mere shadows, and only a kind of union of the two will preserve an independent reality."

However, in the reference frame S the sequence of events will be observed differently, as in Fig. 6.13b: the emission takes place at the point O (which at that moment coincides with O'), but the reflection on the mirror happens at a point of coordinates $(x_1, 0, h)$ and the reception of the reflected light at another point, $(x_2, 0, 0)$. In the frame S, the time of the emission and detection of light are recorded by two synchronized clocks, placed at different points on the Ox-axis. Clearly, the path of the light signal is longer in the frame S and the duration between emission and detection is expressed implicitly by

$$\Delta t = 2\sqrt{h^2 + \left(\frac{V\Delta t}{2}\right)^2},$$

leading to

$$\Delta t = \frac{2h}{c} \frac{1}{\sqrt{1 - \frac{V^2}{c^2}}} = \Gamma \Delta t' > \Delta t'. \tag{6.46}$$

We see from Eq. (6.46) that the time interval between events is the least in the reference frame in which the events take place at the same point in space, in our case S'. The duration between the same events, determined from any other inertial frame, appears as *dilated*. As a plastic image, a clock associated with a non-proper observer moves its hands slower than the clock which belongs to the proper observer. This relativistic effect is called *time dilation*. The proper time is sometimes denoted by the Greek letter τ. We should emphasize that, by Lorentz transformations, we can always find a reference frame in which two given events occur at a given point.

Various interpretations and speculations about time dilation, scientific or philosophical, have been put forward. One of them is the *twin paradox*. Suppose that at the instant when the origins of the inertial frames S and S' coincide, a pair of twins are born. One of them remains on Earth, in the frame S, and the other one, associated with S', begins to travel with a cosmic rocket with the velocity $V \leq c$. For the traveling twin the time will pass faster than for his brother, and after some time – say 1 year for the traveling brother – their eventual meeting would prove a paradoxical fact: the brother remaining on Earth is an old men. One version of the twin paradox appeared in Einstein's first paper on special relativity.

As any other paradox, the *twin paradox* is explicable. The twins can meet *only once*, without any further possibility of comparing their age by direct encounter, because coming back to Earth implies either an acceleration (deceleration), or a change of the shape of trajectory, or simply that the traveling twin switches between two inertial frames on the outbound and inbound journeys. In any case, this problem does not involve only two inertial frames of reference. In fact, in the framework of special relativity one can show that the age of the twin (1), determined by his brother (2) by light signals, is exactly the same with the age of twin (2), determined by his brother (1).

Starting with the French physicist Paul Langevin (1872–1946) in 1911, over the time there have been various explanations of this paradox. These explanations can be grouped into those that focus on the effect of different standards of simultaneity in different frames, and those that designate the acceleration as the main reason. For instance, Max von Laue (1879–1960) argued in 1913 that since the traveling twin must be in two separate inertial frames, one on the way out and another on the way back, this frame switch is the reason for the aging difference. Explanations put forth by Albert Einstein and Max Born (1882–1970) invoked gravitational time dilation to explain the aging as a direct effect of acceleration.

Gravitational time dilation and special relativity were used to explain the Hafele–Keating experiment. This experiment was a test of the theory of relativity. In October 1971, the American physicist Joseph C. Hafele and the American astronomer Richard E. Keating took four cesium-beam atomic clocks aboard commercial airliners. They flew twice around the world, first eastward, then westward, and finally compared the clocks against others that remained at the United States Naval Observatory. It turned out that there were differences between the three sets of clocks, and these differences were consistent with the predictions of Einstein's theory of relativity (special and general relativity). The experiment was made more precise in 1975 by a team of physicists from the University of Maryland, which achieved an accuracy of 1.6 % compared to the theoretical predictions.

The Lifetime of Muons

The lifetime of the μ lepton (muon), measured in its proper frame is $\Delta t_0 \simeq 2.2\,\mu$s. Traveling at relativistic speeds (i.e. speeds close to the speed of light in vacuum), this would allow a survival distance of about 0.66 km at most. In the original cosmic rays experiments by which the muons were first detected, they were known to be produced at higher distances and part of them survived the flight to the Earth's surface. The explanation was given in 1941 by the Italian physicist Bruno Benedetto Rossi (1905–1993): the lifetime of muons as measured from the Earth has to be affected by the time dilation, making it much longer than the proper lifetime. Using the relativistic relation (6.46), with $V/c = 0.99$, he found $\Delta t = \Gamma \Delta t_0 \simeq 2.2 \times 10^{-5}$ s, which corresponds to a survival distance of about 6 km, in very good agreement with the experimental data.

6.3.4   Relativistic Doppler Effect

Consider, again, the two reference frames S and S', with $Ox \equiv O'x'$, and assume that there is a light source at O', which produces monochromatic waves. If the point P is far enough, the waves arriving at P can be considered as being plane waves. Let **s** and **s**$'$ be the unit vectors normal to the wave fronts in P with respect to S and S', respectively. Due to the symmetry of the problem with respect to the axis $Ox \equiv O'x'$, it is sufficient to study the phenomenon in an orthogonal plane, say the plane xOy (Fig. 6.14).

Fig. 6.14 Schematic representation for the relativistic Doppler effect.

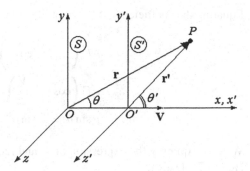

The solution of the wave equation corresponding to a plane wave is

$$\psi \sim e^{i(\omega t - \mathbf{k} \cdot \mathbf{r})} = e^{i\varphi(\mathbf{r}, t)}.$$

Since $\mathbf{k} = \frac{2\pi}{\lambda} \mathbf{s}$, we can write the phase as

$$\varphi(\mathbf{r}, t) = \omega t - \mathbf{k} \cdot \mathbf{r} = 2\pi \nu \left(t - \frac{\mathbf{s} \cdot \mathbf{r}}{c} \right) = 2\pi \nu \left(t - \frac{x \cos \theta + y \sin \theta}{c} \right),$$

because $\mathbf{r} \cdot \mathbf{s} = (x\mathbf{i} + y\mathbf{j}) \cdot (\mathbf{i} \cos \theta + \mathbf{j} \sin \theta)$, where θ is the angle between the x-axis and the direction OP (source–observer) in the frame S. In the frame S', the wave is

$$\psi \sim e^{i(\omega' t' - \mathbf{k}' \cdot \mathbf{r}')} = e^{i\varphi(\mathbf{r}', t')},$$

where

$$\varphi(\mathbf{r}', t') = \omega' t' - \mathbf{k}' \cdot \mathbf{r}' = 2\pi \nu' \left(t' - \frac{x' \cos \theta' + y' \sin \theta'}{c} \right),$$

and θ' is the angle between the x'-axis (along the direction of motion of the source) and $O'P$, which is the direction source–observer in S'. The covariance principle requires that $\varphi(\mathbf{r}, t) = \varphi(\mathbf{r}', t')$, i.e.

$$\nu \left(t - \frac{x \cos \theta + y \sin \theta}{c} \right) = \nu' \left(t' - \frac{x' \cos \theta' + y' \sin \theta'}{c} \right).$$

Using the Lorentz transformation (6.37), this yields

$$\nu \Gamma \left(t' + \frac{V}{c^2} x' \right) - \frac{\nu}{c} \Gamma \left(x' + V t' \right) \cos \theta - \frac{\nu}{c} y' \sin \theta$$

$$= \nu' t' - \frac{\nu'}{c} x' \cos \theta' - \frac{\nu'}{c} y' \sin \theta'.$$

Equating the coefficients of t', x', and y', one obtains

$$\nu \Gamma \left(1 - \frac{V}{c} \cos \theta \right) = \nu',$$

$$\nu \Gamma \left(\cos \theta - \frac{V}{c} \right) = \nu' \cos \theta', \qquad (6.47)$$

$$\nu \sin \theta = \nu' \sin \theta'.$$

Writing explicitly the expression of Γ and recalling that ν' is the proper frequency ν_0, $(6.47)_1$ leads to

$$\nu = \nu_0 \frac{\sqrt{1 - \frac{V^2}{c^2}}}{1 - \frac{V}{c} \cos \theta}. \qquad (6.48)$$

Remark that the relativistic formula (6.48) and the analogous classical one (6.6) differ by the factor $\sqrt{1 - \frac{V^2}{c^2}}$. We distinguish two particular cases:

(i) $\theta = \frac{\pi}{2}$. Relation (6.48) then yields

$$\nu = \nu_0 \sqrt{1 - \frac{V^2}{c^2}} < \nu_0. \qquad (6.49)$$

The effect described by (6.49) is called *transverse Doppler effect*. The classical formula, for $\theta = \frac{\pi}{2}$, gives $\nu = \nu_0$, which means that the transverse Doppler effect is a *purely relativistic effect*. One also observes that (6.49) is the relativistic formula for time dilation. The transverse Doppler effect was experimentally detected in 1938 by Herbert E. Ives (1882–1953) and G.R. Stilwell. The *Ives–Stilwell experiment* tested the contribution of relativistic time dilation to the Doppler shift of light. The result was in good agreement with the formula for the transverse Doppler effect, and was the first direct, quantitative confirmation of the time dilation factor. Together with the Michelson–Morley experiment and the Kennedy–Thorndike experiment, it represents one of the fundamental tests of Einstein's special theory of relativity. Other tests confirming the relativistic Doppler effect are the *Mössbauer rotor experiment* and modern Ives–Stilwell experiments (performed with increased precision). The transverse Doppler effect, predicted by Einstein in 1905, is very small as compared to the longitudinal Doppler effect (the first is expressed in terms of V^2/c^2, while the latter in terms of V/c), and the separation of the two effects required a lot of experimental ingenuity.

(ii) $\theta = 0$. This case is named *longitudinal Doppler effect*. According to (6.48), we have

$$\nu = \nu_0 \frac{\sqrt{1 + \frac{V}{c}}}{\sqrt{1 - \frac{V}{c}}} > \nu_0. \qquad (6.50)$$

If the source is at rest in the frame S, the frequency ν in (6.50) becomes the *proper frequency* ν_0, while ν_0 becomes the frequency determined by the observer attached to S'. Since S moves with the velocity $-\mathbf{V}$ with respect to S', we have

$$\nu_0 = \nu \frac{\sqrt{1 - \frac{V}{c}}}{\sqrt{1 + \frac{V}{c}}} < \nu,$$

i.e. the *same formula*. This result shows that the value of the measured frequency depends only on the *relative velocity between the source and the observer*. For $V/c \ll 1$, formula (6.48) becomes the classical relation (6.6).

From $(6.47)_{1,2}$ we find also

$$\cos \theta' = \frac{\cos \theta - \frac{V}{c}}{1 - \frac{V}{c} \cos \theta},$$

$$\cos \theta = \frac{\cos \theta' + \frac{V}{c}}{1 + \frac{V}{c} \cos \theta'}. \tag{6.51}$$

Each of these relations can be obtained from the other by interchanging $V \leftrightarrow -V$ and $\theta \leftrightarrow \theta'$. To clarify the significance of relations (6.51), consider the particular case $\theta' = \pi/2$, which means

$$\cos \theta = \frac{V}{c}.$$

Since θ is the angle between the direction of motion of the source and the direction source–observer in the frame S, we may write (see Fig. 6.15):

$$\cos \theta = \cos \left(\frac{\pi}{2} - \alpha \right) = \sin \alpha,$$

and, for small angles α,

$$\sin \alpha \approx \alpha = \frac{V}{c},$$

Fig. 6.15 Aberration of light always accompanies the Doppler effect.

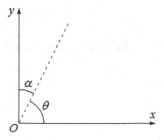

which is nothing else but the *aberration of light* discovered by the English astronomer James Bradley in 1727 (see (6.3)). If $\theta = 0$, then also $\theta' = 0$. In summary:

(i) Relations (6.51) give the relativistic explanation of the aberration of light. The classical theory of the Doppler effect cannot explain this phenomenon;
(ii) The aberration of light always accompanies the Doppler effect. The transverse Doppler effect corresponds to a maximum of aberration, while the longitudinal Doppler effect reveals no aberration.

6.3.5 Composition of Velocities and Accelerations

6.3.5.1 Relativistic Composition of Velocities

Let $v'_x = dx'/dt'$ be the instantaneous velocity of a body (material point), with respect to the reference frame S', along the axis $Ox \equiv O'x'$, and $v_x = dx/dt$ the velocity of the same material point with respect to S. Using the Lorentz transformations (6.33), we have

$$
\begin{aligned}
v'_x &= \frac{dx'}{dt'} = \frac{dx - V\,dt}{dt - \frac{V}{c^2}dx} = \frac{v_x - V}{1 - \frac{V}{c^2}v_x}, \\[2mm]
v'_y &= \frac{v_y}{\Gamma\left(1 - \frac{V}{c^2}v_x\right)}, \\[2mm]
v'_z &= \frac{v_z}{\Gamma\left(1 - \frac{V}{c^2}v_x\right)}.
\end{aligned}
\tag{6.52}
$$

These formulas give the rule of composition of velocities when the motion takes place along the $Ox \equiv O'x'$ axis. A similar procedure using the inverse Lorentz transformations (6.35) leads to

$$
\begin{aligned}
v_x &= \frac{v'_x + V}{1 + \frac{V}{c^2}v'_x}, \\[2mm]
v_y &= \frac{v'_y}{\Gamma\left(1 - \frac{V}{c^2}v'_x\right)}, \\[2mm]
v_z &= \frac{v'_z}{\Gamma\left(1 - \frac{V}{c^2}v'_x\right)}.
\end{aligned}
\tag{6.53}
$$

In the limit of low speeds, $V/c \to 0$, we obtain the expected result

$$
\begin{aligned}
v_x &= v'_x + V, \\
v_y &= v'_y, \\
v_z &= v'_z,
\end{aligned}
$$

that is, the composition law of velocities in classical mechanics.

Suppose that $v'_x = c$. Then, following $(6.53)_1$ we have $v_x = c$, in agreement with Einstein's second postulate.

As an immediate application of the relativistic rule of composition of velocities, let us explain Fresnel's formula (6.1), as well as Fizeau's experiment. If $v'_x = c/n$ is the phase velocity with respect to the water, and the water moves with velocity $V \ll c$ with respect to the Earth, we can write

$$v_x = \frac{\frac{c}{n} + V}{1 + \frac{V}{cn}} = \left(\frac{c}{n} + V\right)\left(1 - \frac{V}{cn} + \dots\right) \simeq \frac{c}{n} + V\left(1 - \frac{1}{n^2}\right),$$

namely relation (6.1).

To write the vector form of the velocity composition rule, we use the Lorentz transformation (6.43), and obtain

$$\mathbf{v}' = \frac{(\Gamma - 1)(\mathbf{v} \times \mathbf{v}_0) \times \mathbf{v}_0 + \Gamma(\mathbf{v} - \mathbf{V})}{\Gamma\left[1 - \frac{V}{c^2}(\mathbf{v} \cdot \mathbf{v}_0)\right]}. \tag{6.54}$$

The inverse Lorentz transformation

$$\mathbf{r} = (\Gamma - 1)(\mathbf{r}' \times \mathbf{v}_0) \times \mathbf{v}_0 + \Gamma\left(\mathbf{r}' + \mathbf{V}t'\right),$$
$$t = \Gamma\left[t' + \frac{V}{c^2}\left(\mathbf{r}' \cdot \mathbf{v}_0\right)\right] \tag{6.55}$$

serves to define the rule of the inverse velocity composition:

$$\mathbf{v} = \frac{(\Gamma - 1)(\mathbf{v}' \times \mathbf{v}_0) \times \mathbf{v}_0 + \Gamma(\mathbf{v}' + \mathbf{V})}{\Gamma\left[1 + \frac{V}{c^2}(\mathbf{v}' \cdot \mathbf{v}_0)\right]}. \tag{6.56}$$

If, in particular, we take $|\mathbf{V}| = V_x = V$, (6.54) and (6.56) go to (6.52) and (6.53).

6.3.5.2 Relativistic Composition of Accelerations

Let $a'_x = dv'_x/dt'$ and $a_x = dv_x/dt$ be the instantaneous accelerations of a particle along the $Ox \equiv O'x'$ axis, determined in the frames S' and S, respectively. Using (6.52), one finds

$$dv'_x = \frac{dv_x}{\Gamma^2\left(1 - \frac{V}{c^2}v_x\right)^2},$$
$$dv'_y = \frac{dv_y + \frac{V}{c^2}(v_y\,dv_x - v_x\,dv_y)}{\Gamma\left(1 - \frac{V}{c^2}v_x\right)^2}, \tag{6.57}$$
$$dv'_z = \frac{dv_z + \frac{V}{c^2}(v_z\,dv_x - v_x\,dv_z)}{\Gamma\left(1 - \frac{V}{c^2}v_x\right)^2}.$$

Therefore,

$$a_x' = \frac{a_x}{\Gamma^3\left(1 - \frac{V}{c^2}v_x\right)^3},$$

$$a_y' = \frac{a_y + \frac{V}{c^2}(v_y a_x - v_x a_y)}{\Gamma^2\left(1 - \frac{V}{c^2}v_x\right)^3},$$

$$a_z' = \frac{a_z + \frac{V}{c^2}(v_z a_x - v_x a_z)}{\Gamma^2\left(1 - \frac{V}{c^2}v_x\right)^3}.$$

(6.58)

The vectorial formula of acceleration composition is found by taking the derivative $\mathbf{a}' = d\mathbf{v}'/dt'$, with $d\mathbf{v}'$ found by means of (6.54). The result is

$$\mathbf{a}' = \frac{\left[1 - \frac{V}{c^2}(\mathbf{v}\cdot\mathbf{v}_0)\right]\left[\mathbf{a} + (\Gamma - 1)(\mathbf{a}\cdot\mathbf{v}_0)\mathbf{v}_0\right]}{\Gamma^2\left(1 - \frac{\mathbf{V}\cdot\mathbf{v}}{c^2}\right)^3}$$

$$+ \frac{\frac{V}{c^2}(\mathbf{a}\cdot\mathbf{v}_0)\left[\mathbf{v} + (\Gamma - 1)(\mathbf{v}\cdot\mathbf{v}_0)\mathbf{v}_0 - \Gamma\mathbf{V}\right]}{\Gamma^2\left(1 - \frac{\mathbf{V}\cdot\mathbf{v}}{c^2}\right)^3},$$

(6.59)

and we leave the proof to the reader.

The acceleration composition formula corresponding to the inverse Lorentz transformations is easily obtained from (6.59) by substituting $\mathbf{V} \leftrightarrow -\mathbf{V}, \mathbf{v} \leftrightarrow \mathbf{v}', \mathbf{a} \leftrightarrow \mathbf{a}'$.

6.4 Solved Problems

Problem 1. Show that the wave equation

$$\frac{\partial^2\psi}{\partial x^2} - \frac{1}{c^2}\frac{\partial^2\psi}{\partial t^2} = 0$$

(6.60)

is covariant with respect to the Lorentz transformations (see (6.37)):

$$x = \Gamma(x' + V t'),$$
$$y = y',$$
$$z = z',$$
$$t = \Gamma\left(t' + \frac{V}{c^2}x'\right).$$

(6.61)

Solution. We have

$$\frac{\partial\psi}{\partial x'} = \frac{\partial\psi}{\partial x}\frac{\partial x}{\partial x'} + \frac{\partial\psi}{\partial t}\frac{\partial t}{\partial x'} = \Gamma\frac{\partial\psi}{\partial x} + \Gamma\frac{V}{c^2}\frac{\partial\psi}{\partial t}.$$

We deduce that the operator correspondence is

$$\frac{\partial}{\partial x'} = \Gamma \frac{\partial}{\partial x} + \Gamma \frac{V}{c^2} \frac{\partial}{\partial t}. \tag{6.62}$$

which leads to

$$\begin{aligned}
\frac{\partial^2 \psi}{\partial x'^2} &= \left(\Gamma \frac{\partial}{\partial x} + \Gamma \frac{V}{c^2} \frac{\partial}{\partial t} \right) \left(\Gamma \frac{\partial \psi}{\partial x} + \Gamma \frac{V}{c^2} \frac{\partial \psi}{\partial t} \right) \\
&= \Gamma^2 \frac{\partial^2 \psi}{\partial x^2} + 2\Gamma^2 \frac{V}{c^2} \frac{\partial^2 \psi}{\partial x \partial t} + \Gamma^2 \frac{V^2}{c^4} \frac{\partial^2 \psi}{\partial t^2}.
\end{aligned} \tag{6.63}$$

In the same way

$$\frac{\partial \psi}{\partial t'} = \frac{\partial \psi}{\partial x} \frac{\partial x}{\partial t'} + \frac{\partial \psi}{\partial t} \frac{\partial t}{\partial t'} = \Gamma V \frac{\partial \psi}{\partial x} + \Gamma \frac{\partial \psi}{\partial t},$$

with the operator correspondence

$$\frac{\partial}{\partial t'} = \Gamma V \frac{\partial}{\partial x} + \Gamma \frac{\partial}{\partial t}. \tag{6.64}$$

Next,

$$\begin{aligned}
\frac{\partial^2 \psi}{\partial t'^2} &= \left(\Gamma V \frac{\partial}{\partial x} + \Gamma \frac{\partial}{\partial t} \right) \left(\Gamma V \frac{\partial \psi}{\partial x} + \Gamma \frac{\partial \psi}{\partial t} \right) \\
&= \Gamma^2 V^2 \frac{\partial^2 \psi}{\partial x^2} + 2\Gamma^2 V \frac{\partial^2 \psi}{\partial x \partial t} + \Gamma^2 \frac{\partial^2 \psi}{\partial t^2}.
\end{aligned} \tag{6.65}$$

It is easily seen that

$$\frac{\partial^2 \psi}{\partial x'^2} - \frac{1}{c^2} \frac{\partial^2 \psi}{\partial t'^2} = \frac{\partial^2 \psi}{\partial x^2} - \frac{1}{c^2} \frac{\partial^2 \psi}{\partial t^2},$$

which completes the proof.

Problem 2. Denoting $\frac{V}{c} = \tanh \theta$ and $u = ct$, show that the Lorentz transformation (6.37) can be written as

$$\begin{aligned}
x' &= x \cosh \theta - u \sinh \theta, \\
u' &= -x \sinh \theta + u \cosh \theta.
\end{aligned} \tag{6.66}$$

Using this result, show that a succession of Lorentz boosts, performed in the same direction (Fig. 6.16), is also a Lorentz boost.

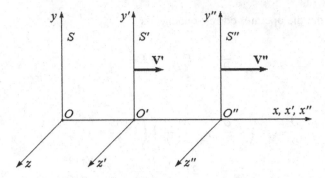

Fig. 6.16 Schematic representation of three inertial reference frames (Problem 2).

Solution. Using the formulas

$$\cosh^2 \theta - \sinh^2 \theta = 1,$$

$$\cosh \theta = \frac{1}{\sqrt{1 - \tanh^2 \theta}} = \frac{1}{\sqrt{1 - \frac{V^2}{c^2}}} = \Gamma,$$

$$\sinh \theta = \sqrt{\cosh^2 \theta - 1} = \frac{V}{c}\Gamma,$$

we can write

$$x' = \Gamma\left(x - \frac{V}{c}ct\right) = x \cosh \theta - u \sinh \theta,$$

$$ct' = u' = \Gamma\left(u - \frac{V}{c}x\right) = -x \sinh \theta + u \cosh \theta,$$

and relations (6.66) are proved.

Next, let us consider three inertial frames S, S', and S'', and suppose that the velocity of S' with respect to S is V, while S'' moves with velocity V' with respect to S', and V'' with respect to S.

One observes that the Lorentz transformation (6.66) can be written in a matrix form as

$$X' = BX,$$

where

$$X = \begin{pmatrix} x \\ u \end{pmatrix}, \quad X' = \begin{pmatrix} x' \\ u' \end{pmatrix}, \quad B = \begin{pmatrix} \cosh \theta & -\sinh \theta \\ -\sinh \theta & \cosh \theta \end{pmatrix}. \qquad (6.67)$$

Using this representation, we also have $X'' = AX'$, so that the total transformation is $X'' = AX' = ABX = CX$, where C is given by

$$C = AB = \begin{pmatrix} \cosh\theta' & -\sinh\theta' \\ -\sinh\theta' & \cosh\theta' \end{pmatrix} \begin{pmatrix} \cosh\theta & -\sinh\theta \\ -\sinh\theta & \cosh\theta \end{pmatrix}.$$

Since the element c_{ij} of the matrix product AB is

$$c_{ij} = \sum_{r=1}^{2} a_{ir}b_{rj}, \quad i, j = 1, 2,$$

and in view of the trigonometric formulas

$$\sinh(\theta + \theta') = \sinh\theta\cosh\theta' + \cosh\theta\sinh\theta',$$
$$\cosh(\theta + \theta') = \cosh\theta\cosh\theta' + \sinh\theta\sinh\theta',$$

we obtain

$$C = \begin{pmatrix} \cosh(\theta + \theta') & -\sinh(\theta + \theta') \\ -\sinh(\theta + \theta') & \cosh(\theta + \theta') \end{pmatrix} = \begin{pmatrix} \cosh\theta'' & -\sinh\theta'' \\ -\sinh\theta'' & \cosh\theta'' \end{pmatrix}, \quad (6.68)$$

where $\theta'' = \theta + \theta'$. The parameter θ is called *rapidity*. This name was given in 1911 by the English physicist Alfred Robb (1873–1936), being an alternative to speed as a method of measuring motion. For low speeds, rapidity and speed are proportional, but for high speeds rapidity becomes very large and tends to infinity for $v = c$.

Returning to our problem, formula (6.68) shows that the successive application of two Lorentz boosts in the same direction is also a Lorentz boost. Note that the rapidity does not obey the rule of relativistic transformation of coordinates and time.

Problem 3. A particle moves with respect to the inertial frame S' with the constant velocity $\mathbf{v}'$ that makes an angle θ' with $Ox \equiv O'x'$. Determine the angle θ between the trajectory of the particle and $Ox \equiv O'x'$ in the inertial frame S.

Solution. Without any loss of generality, we can study the motion of the particle in the xy-plane (Fig. 6.17). The components of the velocity of the particle in the two frames are

$$v_x = v\cos\theta, \qquad v_y = v\sin\theta \quad (S),$$
$$v'_x = v'\cos\theta', \qquad v'_y = v'\sin\theta' \quad (S').$$

But

$$v_x = \frac{v'_x + V}{1 + \frac{V}{c^2}v'_x}, \qquad v_y = \frac{v'_y}{\Gamma\left(1 + \frac{V}{c^2}v'_x\right)},$$

Fig. 6.17 A particle moving
with the constant velocity $\mathbf{v}'$
that makes an angle θ' with
$Ox \equiv O'x'$.

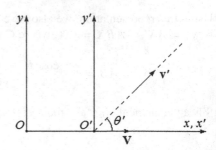

which gives

$$\tan \theta = \frac{v_y}{v_x} = \frac{v'_y}{\Gamma(v'_x + V)} = \frac{1}{\Gamma} \frac{\sin \theta'}{\cos \theta' + \frac{V}{v'}},$$

$$\tan \theta' = \frac{1}{\Gamma} \frac{\sin \theta}{\cos \theta - \frac{V}{v}}. \tag{6.69}$$

Then

$$\cos \theta = \frac{1}{\sqrt{1 + \tan^2 \theta}} = \frac{\cos \theta' + \frac{V}{c}}{1 + \frac{V}{c} \cos \theta'}, \tag{6.70}$$

which is the formula expressing the relativistic Doppler effect (6.51).
   We also have

$$\sin \theta' = \frac{\tan \theta'}{\sqrt{1 + \tan^2 \theta'}} = \frac{\sqrt{1 - \frac{V^2}{c^2}} \sin \theta}{1 - \frac{V}{c} \cos \theta}. \tag{6.71}$$

If $\frac{V}{c} \ll 1$, we may take $\Gamma \approx 1$ and write

$$\sin \theta' \simeq \sin \theta \left(1 - \frac{V}{c} \cos \theta\right)^{-1} \simeq \sin \theta + \frac{V}{c} \sin \theta \cos \theta,$$

or

$$2 \sin \frac{\theta' - \theta}{2} \cos \frac{\theta' + \theta}{2} = \frac{V}{c} \sin \theta \cos \theta. \tag{6.72}$$

Since $\theta' \simeq \theta$, one can approximate $\theta' + \theta = 2\theta$. Denoting $\theta' - \theta = \alpha$, we also realize
that α is very small, so that we may take $\sin \alpha/2 \approx \alpha/2$, and (6.72) becomes

$$\alpha \simeq \frac{V}{c} \sin \theta. \tag{6.73}$$

If the light is coming from an object situated at the zenith of the observer ($\theta = \pi/2$), we finally arrive at the formula for the aberration of light,

$$\alpha \simeq \frac{V}{c}.$$

Problem 4. Maxwell's equations are covariant under the Lorentz transformation of frames (6.37). Find the transformation properties of the electromagnetic field vectors **E** and **B**, in vacuum.

Solution. Following Einstein, let us use Maxwell's source-free equations

$$\frac{\partial \mathbf{B}}{\partial t} = -\nabla \times \mathbf{E}, \qquad \nabla \cdot \mathbf{B} = 0. \tag{6.74}$$

If the motion of the inertial frame takes place along the $Ox \equiv O'x'$ axis, then

$$\frac{\partial}{\partial z} = \frac{\partial}{\partial z'}, \quad \frac{\partial}{\partial y} = \frac{\partial}{\partial y'},$$

and so, in the frame S', we can write

$$\frac{\partial B'_x}{\partial t'} = \frac{\partial E'_y}{\partial z} - \frac{\partial E'_z}{\partial y},$$

$$\frac{\partial B'_y}{\partial t'} = \frac{\partial E'_z}{\partial x'} - \frac{\partial E'_x}{\partial z}, \tag{6.75}$$

$$\frac{\partial B'_z}{\partial t'} = \frac{\partial E'_x}{\partial y} - \frac{\partial E'_y}{\partial x'}.$$

Since $\mathbf{E}'$, $\mathbf{B}'$ are functions of x', t', we can use the operators (6.62) and (6.64) to re-write (6.75), that is

$$\Gamma V \frac{\partial B'_x}{\partial x} + \Gamma \frac{\partial B'_x}{\partial t} = \frac{\partial E'_y}{\partial z} - \frac{\partial E'_z}{\partial y},$$

$$\Gamma V \frac{\partial B'_y}{\partial x} + \Gamma \frac{\partial B'_y}{\partial t} = \Gamma \frac{\partial E'_z}{\partial x} + \frac{V}{c^2} \Gamma \frac{\partial E'_z}{\partial t} - \frac{\partial E'_x}{\partial z}, \tag{6.76}$$

$$\Gamma V \frac{\partial B'_z}{\partial x} + \Gamma \frac{\partial B'_z}{\partial t} = \frac{\partial E'_x}{\partial y} - \Gamma \frac{\partial E'_y}{\partial x} - \frac{V}{c^2} \Gamma \frac{\partial E'_y}{\partial t}.$$

Grouping the terms in the last two equations (6.76), we find

$$\frac{\partial}{\partial t}\left[\Gamma\left(B'_y - \frac{V}{c^2} E'_z \right) \right] = \frac{\partial}{\partial x}[\Gamma(E'_z - V B'_y)] - \frac{\partial E'_x}{\partial z},$$

$$\frac{\partial}{\partial t}\left[\Gamma\left(B'_z + \frac{V}{c^2} E'_y \right) \right] = \frac{\partial E'_x}{\partial y} - \frac{\partial}{\partial x}[\Gamma(E'_y + V B'_z)]. \tag{6.77}$$

Equations (6.77) are covariant with respect to Lorentz transformations, that is they keep their form $(6.75)_{2,3}$, if the field components satisfy the following relations

$$
\begin{aligned}
E_x &= E'_x, \\
E_y &= \Gamma(E'_y + V B'_z), \\
E_z &= \Gamma(E'_z - V B'_y), \\
B_y &= \Gamma\left(B'_y - \frac{V}{c^2} E'_z\right), \\
B_z &= \Gamma\left(B'_z + \frac{V}{c^2} E'_y\right).
\end{aligned}
\tag{6.78}
$$

Replacing V by $-V$, and moving the "primed" index to the "unprimed" components, we also have the inverse transformations:

$$
\begin{aligned}
E'_x &= E_x, \\
E'_y &= \Gamma(E_y - V B_z), \\
E'_z &= \Gamma(E_z + V B_y), \\
B'_y &= \Gamma\left(B_y + \frac{V}{c^2} E_z\right), \\
B'_z &= \Gamma\left(B_z - \frac{V}{c^2} E_y\right).
\end{aligned}
\tag{6.79}
$$

Introducing these results into $(6.76)_1$ and conveniently grouping the terms, we find

$$
\frac{\partial B'_x}{\partial t} = V\left(\frac{\partial B'_x}{\partial x} + \frac{\partial B_y}{\partial y} + \frac{\partial B_z}{\partial z}\right) = \frac{\partial E_y}{\partial z} - \frac{\partial E_z}{\partial y}.
$$

Maxwell's equation $\nabla \cdot \mathbf{B} = 0$ keeps its form if we take

$$
B'_x = B_x.
\tag{6.80}
$$

This relation, together with (6.79), or (6.78), give the whole picture of the transformation relations for the field components. One observes that the field components oriented along the direction of motion of frames *do not change*, while the components orthogonal to this direction *change* according to the above formulas. These formulas can synthetically be written as

$$
\begin{aligned}
\mathbf{E}'_{\parallel} &= \mathbf{E}_{\parallel}, \\
\mathbf{E}'_{\perp} &= \Gamma(\mathbf{E} + \mathbf{V} \times \mathbf{B})_{\perp}, \\
\mathbf{B}'_{\parallel} &= \mathbf{B}_{\parallel}, \\
\mathbf{B}'_{\perp} &= \Gamma\left(\mathbf{B} - \frac{1}{c^2}\mathbf{V} \times \mathbf{E}\right)_{\perp}.
\end{aligned}
\tag{6.81}
$$

To summarize, Maxwell's source-free equations

$$B_{i,t} + \epsilon_{ijk} E_{k,j} = 0, \quad i, j, k = 1, 2, 3 \tag{6.82}$$

are covariant if the space coordinate and time obey the Lorentz transformation (6.37), while the field components **E**, **B** transform according to (6.81). The reader is invited to resume the calculation for Maxwell's equations with sources (taking $\mathbf{j} = 0, \rho = 0$), and show that the result is the same.

In case of small velocities, $V \ll c$, one can approximate

$$\mathbf{E}' \simeq \mathbf{E} + \mathbf{V} \times \mathbf{B},$$
$$\mathbf{B}' \simeq \mathbf{B}. \tag{6.83}$$

Problem 5. Show that the generalized Lorentz transformations (6.43) can be derived by three successive proper space-time transformations.

Solution. Let us consider the inertial frames S and S', $\mathbf{V}$ being their relative velocity. Without loss of generality, we may assume that $\mathbf{V}$ is situated in the xOy-plane (see Fig. 6.18). The transition from S to S' can be performed in three steps:

(i) A counterclockwise rotation of angle θ, in the xOy-plane, until the x-axis becomes parallel to OO'. By this transition one passes from $S(Oxyzt)$ to $S_1(Ox_1y_1z_1t_1) \equiv S_1(Ox_1y_1zt)$. The transformation relations are

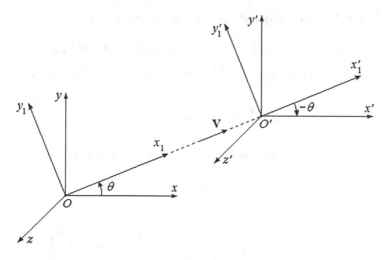

Fig. 6.18 Transition from S to S' in three steps (Problem 5).

$$x_1 = x \cos \theta + y \sin \theta,$$
$$y_1 = -x \sin \theta + y \cos \theta,$$
$$z_1 = z, \tag{6.84}$$
$$t_1 = t.$$

(ii) A Lorentz boost, along the x_1-axis, until O coincides with O', and the frame S_1 becomes $S_1'(O'x_1'y_1'z_1't_1') \equiv S_1'(O'x_1'y_1z_1t_1')$. This transformation is

$$x_1' = \Gamma(x_1 - V t_1),$$
$$y_1' = y_1,$$
$$z_1' = z_1, \tag{6.85}$$
$$t_1' = \Gamma \left(t_1 - \frac{V}{c^2} x_1 \right).$$

(iii) The last transformation is a clockwise rotation of angle $-\theta$ about the x'-axis, until S_1' coincides with S'. This transformation is given by

$$x' = x_1' \cos \theta - y_1' \sin \theta,$$
$$y' = -x_1' \sin \theta + y_1' \cos \theta,$$
$$z' = z_1', \tag{6.86}$$
$$t' = t_1'.$$

Introducing (6.84) into (6.85), and the result into (6.86), we find

$$x' = \left[1 + (\Gamma - 1)\cos^2 \theta\right] x + (\Gamma - 1)y \sin \theta \cos \theta - V\Gamma t \cos \theta,$$
$$y' = (\Gamma - 1)x \sin \theta \cos \theta + \left[1 + (\Gamma - 1)\sin^2 \theta\right] y - V\Gamma t \sin \theta,$$
$$z' = z, \tag{6.87}$$
$$t' = -\frac{V}{c^2}\Gamma x \cos \theta - \frac{V}{c^2}\Gamma y \sin \theta + \Gamma t.$$

These relations can be also obtained by taking the matrix product

$$X' = R_1 B R_2 X = A X, \tag{6.88}$$

where

$$X = \begin{pmatrix} x \\ y \\ z \\ t \end{pmatrix}, \quad X' = \begin{pmatrix} x' \\ y' \\ z' \\ t' \end{pmatrix}, \quad R_1 = \begin{pmatrix} \cos \theta & -\sin \theta & 0 & 0 \\ \sin \theta & \cos \theta & 0 & 0 \\ 0 & 0 & 1 & 0 \\ 0 & 0 & 0 & 1 \end{pmatrix},$$

$$B = \begin{pmatrix} \Gamma & 0 & 0 & -V\Gamma \\ 0 & 1 & 0 & 0 \\ 0 & 0 & 1 & 0 \\ -\frac{V}{c^2}\Gamma & 0 & 0 & \Gamma \end{pmatrix}, \qquad R_2 = \begin{pmatrix} \cos\theta & \sin\theta & 0 & 0 \\ -\sin\theta & \cos\theta & 0 & 0 \\ 0 & 0 & 1 & 0 \\ 0 & 0 & 0 & 1 \end{pmatrix}.$$

Let us now return to (6.87) and observe that

$$\mathbf{r} \cdot \mathbf{V} = xV_x + yV_y = V(x \cos\theta + y \sin\theta),$$

that is

$$x \cos\theta + y \sin\theta = \frac{1}{V}\mathbf{r} \cdot \mathbf{V} = \mathbf{r} \cdot \mathbf{v_0},$$

where $\mathbf{v_0}$ is the unit vector of $\mathbf{V}$. Then we can write (6.87) as follows

$$\begin{aligned} x' &= x + (\Gamma - 1)(\mathbf{r} \cdot \mathbf{v_0}) \cos\theta - V\Gamma t \cos\theta, \\ y' &= y + (\Gamma - 1)(\mathbf{r} \cdot \mathbf{v_0}) \sin\theta - V\Gamma t \sin\theta, \\ z' &= z, \\ t' &= \Gamma \left[t - \frac{V}{c^2}(\mathbf{r} \cdot \mathbf{v_0}) \right]. \end{aligned} \qquad (6.89)$$

If we now multiply relations (6.89) by the unit vectors of the axes x', y', and z', which are the same as those of the axes x, y, and z (the frames are inertial), we arrive at the desired result (see (6.43)):

$$\begin{aligned} \mathbf{r}' &= \mathbf{r} + (\Gamma - 1)(\mathbf{r} \cdot \mathbf{v_0})\mathbf{v_0} - \Gamma \mathbf{V} t, \\ t' &= \Gamma \left[t - \frac{V}{c^2}(\mathbf{r} \cdot \mathbf{v_0}) \right]. \end{aligned}$$

In short, we proved that a boost in an arbitrary direction can be obtained by rotating first the coordinates to align one coordinate axis of each reference system with the relative velocity of the system, followed by a boost in that direction, and then rotating the coordinates back.

6.5 Proposed Problems

1. Show that two Lorentz boosts performed with non-parallel velocities are not commutative, but they are commutative if the velocities are parallel.
2. A bar of length l_0 makes the angle φ_0 with respect to the direction of motion of two inertial reference frames S and S'. Taking one of these frames as a proper frame, determine the length of the bar in the other frame.

3. Show that if two velocities $\mathbf{v}_1$ and $\mathbf{v}_2$ have the same modulus, but are not parallel in the inertial frame S, they do not have the same modulus in the frame S' (except for the case $v_{1x} = v_{2x}$).

4. Using the result of the solved problem (4), show that the expressions $\mathbf{E} \cdot \mathbf{B}$ and $E^2 - c^2 B^2$ are invariant under the Lorentz transformations (6.37). Show that, except for the case when $\mathbf{E}$ and $\mathbf{B}$ are orthogonal, one can find a reference frame in which $\mathbf{E}$ and $\mathbf{B}$ are parallel.

5. Let S and S' be two inertial reference frames. A particle is moving in a straight line and uniformly with respect to S', with the velocity $\mathbf{v}' = (v'_x, v'_y, v'_z)$. Show that the motion of the particle with respect to S is also uniform, and calculate the components of its velocity $\mathbf{v} = (v_x, v_y, v_z)$.

6. Show that the velocity of the particle given in the previous exercise satisfies the following relation:

$$\sqrt{1 - \frac{v^2}{c^2}} = \frac{1}{\Gamma \left(1 + \frac{V}{c^2} v'_x\right)} \sqrt{1 - \frac{v'^2}{c^2}}.$$

7. Using the Lorentz transformation (6.43), find the vector formula of the acceleration composition (6.59).

8. Show that two successive Lorentz boosts in different directions are equivalent to the composition of a Lorentz boost and a three-dimensional space rotation. This fact shows that the Lorentz boosts by themselves cannot form a group structure, unless we include also the space rotations.

9. Two reference frames move uniformly with velocities $\mathbf{v}_1$ and $\mathbf{v}_2$ with respect to a third, arbitrary reference frame. Show that their relative velocity $\mathbf{v}$ satisfies the relation

$$v^2 = \frac{(\mathbf{v}_1 - \mathbf{v}_2)^2 - (\mathbf{v}_1 \times \mathbf{v}_2)^2}{(1 - \mathbf{v}_1 \cdot \mathbf{v}_2)^2}.$$

10. The relative velocity between two inertial frames S and S' is $\mathbf{V}$. In S' a bullet with velocity $\mathbf{v}'$ is shot under the angle θ' with respect to the direction of $\mathbf{V}$. Determine the corresponding angle θ in S. What happens if the bullet is replaced by a photon?

Chapter 7
Minkowski Space

The special theory of relativity was given a remarkable geometric interpretation by Hermann Minkowski (1864–1909). While teaching at the Eidgenössische Polytechnische Schule of Zürich (today the ETH Zürich), one of his students was Albert Einstein. Minkowski presented his ideas on Einstein's theory in some papers and lectures, between 1907 and 1909.

The connection between space and time in Einstein's theory led Minkowski to realize that the special theory of relativity could be represented on a four-dimensional vector space, which is commonly denoted by M_4. In *Minkowski space*, M_4, an *event* is represented by a *point* (*"world point"*), while the motion of a body is represented by a succession of points, called *world line*.

Let (x_1, y_1, z_1, t_1) and (x_2, y_2, z_2, t_2) be the coordinates of two events in Minkowski space, representing the emission and reception of a light signal. The light flash travels the distance $\sqrt{(x_2 - x_1)^2 + (y_2 - y_1)^2 + (z_2 - z_1)^2}$, with the speed c, during the time interval $t_2 - t_1$. Then we can write

$$(x_2 - x_1)^2 + (y_2 - y_1)^2 + (z_2 - z_1)^2 = c^2 (t_2 - t_1)^2.$$

By analogy with the distance between two points in the three-dimensional Euclidean space E_3, one can define the *relativistic interval* between the two events by

$$s_{12}^2 = c^2 t_{12}^2 - l_{12}^2, \tag{7.1}$$

where $t_{12} = t_2 - t_1, l_{12}^2 = (x_2 - x_1)^2 + (y_2 - y_1)^2 + (z_2 - z_1)^2$. Remark that in the case of light signals we have $s_{12} = 0$. If the two events are infinitely close, we may write

$$ds^2 = c^2 dt^2 - dx^2 - dy^2 - dz^2 = c^2 dt^2 - dl^2. \tag{7.2}$$

This symmetric bilinear form represents the Lorentzian inner product of vectors, as well as the *metric* on Minkowski space. There are three possibilities for the interval (7.2): $ds^2 > 0$, if $c\, dt > dl$; $ds^2 < 0$, if $c\, dt < dl$, and $ds^2 = 0$, if $c\, dt = dl$ (the

© Springer-Verlag Berlin Heidelberg 2016
M. Chaichian et al., *Electrodynamics*, DOI 10.1007/978-3-642-17381-3_7

case of a light beam). Such an inner product is called *indefinite* and sometimes we speak about an indefinite metric.

Let us now show that the space-time interval is a *relativistic invariant* with respect to the Lorentz transformations (6.33). Suppose that (7.2) expresses the interval in the reference frame S, while in S' the interval is

$$ds'^2 = c^2 \, dt'^2 - dx'^2 - dy'^2 - dz'^2 = c^2 dt'^2 - dl'^2. \tag{7.3}$$

As the speed of light c is invariant in accordance with Einstein's second postulate, it follows that for light signals ds and ds' vanish, therefore they necessarily satisfy the relation

$$ds^2 = \lambda \, ds'^2. \tag{7.4}$$

where λ is a factor of proportionality. Since the space is homogeneous and isotropic, and the time passes uniformly, λ can depend neither on x, y, z, t, nor on the direction of relative displacement between the frames S and S'. The last possible situation for λ is to depend on the modulus of $\mathbf{V}$. For three inertial frames S, S', S'', we would then have

$$ds^2 = \lambda(V') \, ds'^2 = \lambda(V'') \, ds''^2,$$
$$ds'^2 = \lambda(V_{rel}) \, ds''^2, \tag{7.5}$$

where V_{rel} is the modulus of the relative velocity of frames. Thus,

$$\frac{\lambda(V'')}{\lambda(V')} = \lambda(V_{rel}).$$

But if λ depends on V' and V'', it will also depend on the angle between them, which contradicts our hypothesis. Consequently, λ could only be a constant (equal to 1, according to (7.5)) and (7.4) goes to

$$ds^2 = ds'^2. \tag{7.6}$$

The reader could arrive at the same result by direct calculation, introducing (6.33) into (7.6).

The invariant (7.2) is the *fundamental invariant* of special relativity.

One observes that

$$ds = c \, dt \, \sqrt{1 - \frac{V^2}{c^2}} = cd\tau. \tag{7.7}$$

Comparing (7.7) with (6.46), we conclude that $d\tau$ is the infinitesimal *proper duration*. Since the space-time interval ds is an invariant, it follows that the proper time is also an *invariant*, which is an already known result.

The definition (7.2) of the space-time interval also shows that, if the world line lies in xt-plane, then $ds^2 = c^2 dt^2 - dx^2$, that is *the square of the hypotenuse is equal to the difference of the squares of the other two sides.* Consequently, our space-time is not Euclidean, but *pseudo-Euclidean*. In such a manifold the Euclidean geometry has to be replaced by an adequate geometric representation. The essence of this new geometry will be presented in the following.

7.1   Time-Like and Space-Like Intervals

Let

$$s_{12}^2 = c^2 t_{12}^2 - l_{12}^2,$$
$$s_{12}'^2 = c^2 t_{12}'^2 - l_{12}'^2 \tag{7.8}$$

be the finite intervals between two events determined from the inertial frames S and S', respectively. Since the space-time interval is an invariant, we may write

$$s_{12}^2 = c^2 t_{12}^2 - l_{12}^2 = c^2 t_{12}'^2 - l_{12}'^2. \tag{7.9}$$

Take $l_{12}' = 0$, which means that the two events happen at the same place in S'. Then (7.9) yields

$$s_{12}^2 = c^2 t_{12}^2 - l_{12}^2 = c^2 t_{12}'^2 > 0. \tag{7.10}$$

Such an interval is called *time-like*. Reciprocally, for any time-like interval one can find an inertial reference frame in which the two events occur at the same place.

Take now $t_{12}' = 0$, meaning that the two events occur at the same time in S'. In this case (7.9) leads to

$$s_{12}^2 = c^2 t_{12}^2 - l_{12}^2 = -l_{12}'^2 < 0. \tag{7.11}$$

This type of interval is called *space-like*.

To give an intuitive geometric interpretation to time-like and space-like intervals, let us consider an event placed at the coordinate origin O and analyze the relation between the O–event and all the other possible events in space-time. Since in the xt-plane the light signals are defined by $s^2 = c^2 t^2 - x^2 = 0 \Rightarrow x = \pm c\,t$, the straight lines ac and bd in Fig. 7.1 represent two light signals propagating in opposite directions.

The rectilinear and uniform motion of a particle passing through the space-time point ($x = 0$, $t = 0$) is represented by a straight line which makes an angle α with the ct-axis, such that $\tan \alpha = v/c$, where v is the speed of the particle. Since $v_{max} = c$, it follows that $\alpha < \alpha_{max} = \pi/4$ (see Fig. 7.1). Consequently, the arbitrary motion of a body, represented by the world line AB, can lie *only* inside the domains I and II. For *any* event inside the domain I,

$$\tan \alpha = \frac{x}{ct} = \frac{v}{c} < 1 \quad \Rightarrow \quad c^2 t^2 - x^2 > 0,$$

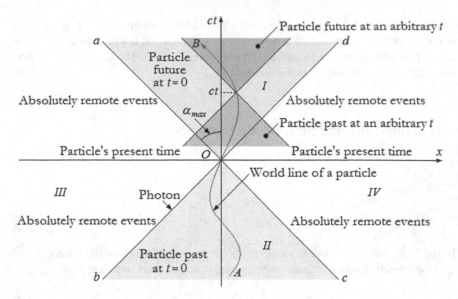

Fig. 7.1 The intersection of the light cone by the plane (x, ct).

which shows that the interval between O and any other event of the domain I is time-like. In other words, the two events cannot be simultaneous in any frame S'. Consequently, any event which belongs to the domain I happens *after* the event of O in any inertial reference frame. For this reason, the domain I is called *absolute future*.

In a similar way it can be shown that all events from the region II happen before the event O, in any reference frame S'. The domain II is named *absolute past*.

Any event situated inside the domains III and IV satisfies the condition

$$\tan \alpha = \frac{x}{ct} = \frac{v}{c} > 1 \quad \Rightarrow \quad c^2 t^2 - x^2 < 0,$$

meaning that the interval between O and any other event from the regions III or IV is *space-like*. We can always find by Lorentz transformations reference frames in which any event of these domains can happen either simultaneously with the event O, or precede O in time, or, finally, succeed in time the event O. These events are called *absolutely remote* with respect to O. They cannot affect and cannot be affected by the event O, i.e. they are not in a causal relation with the event at O.

If we consider all three spatial dimensions, the figure becomes a cone in four dimensions (a hypercone) called *light cone* (see Fig. 7.2, in which two of the three spatial dimensions are depicted). The name comes from the fact that it can be imagined as being generated by the world line $x = c\,t$, representing a light signal. Because signals and other causal influences cannot travel faster than light in relativity, the light cone plays an essential role in defining the *concept of causality*. For a given event O,

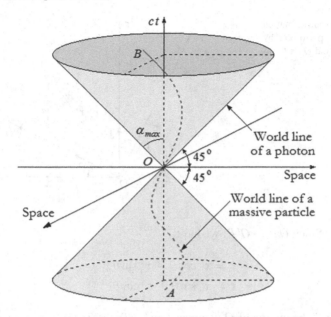

Fig. 7.2 Schematical representation of a hypercone (only two of the three spatial axes are depicted).

the set of events that lie on (or inside) the past light cone of O would also be the set of all events that could send a signal that would have time to reach O and influence it in some way. Events that lie neither in the past nor in the future light cone of O cannot influence or be influenced by O in special relativity. In other words, *two events can be causally related only if the world line which connects them lies inside the light cone.*

7.2 Various Representations of Minkowski Space

7.2.1 Euclidean-Complex Representation

Let us take the time coordinate as imaginary and denote $\tau = ict$. The metric (7.2) then becomes

$$- ds^2 = d\tau^2 + dx^2 + dy^2 + dz^2, \tag{7.12}$$

which is, formally, an Euclidean metric of an Euclidean-complex four dimensional space.

We shall show that in this notation, the transformation from the reference frame S to the frame S' is given by a rotation of angle θ around $O \equiv O'$ in the $x\tau$-plane (see Fig. 7.3). Let us consider an event P in the frames $S(xO\tau)$ and $S'(x'O'\tau')$,

Fig. 7.3 The transformation from S to S' is performed by a rotation around $O \equiv O'$ in the $x\tau$-plane.

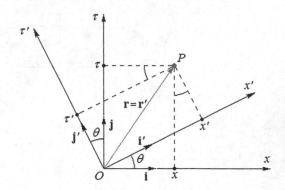

respectively. Since $OP = O'P$, we have

$$x = x' \cos\theta - \tau' \sin\theta,$$
$$\tau = x' \sin\theta + \tau' \cos\theta. \tag{7.13}$$

In the frame S, the motion of O' is given by $x' = 0$, that is

$$x = -\tau' \sin\theta,$$
$$\tau = \tau' \cos\theta,$$

which gives

$$\tan\theta = i\frac{V}{c}. \tag{7.14}$$

The use of the trigonometric formulas

$$\sin\theta = \frac{\tan\theta}{\sqrt{1 + \tan^2\theta}} = i\frac{V}{c}\Gamma, \quad \cos\theta = \frac{1}{\sqrt{1 + \tan^2\theta}} = \Gamma \tag{7.15}$$

and (7.13) yield

$$x = \Gamma(x' + V t'), \qquad t = \Gamma\left(t' + \frac{V}{c^2}x'\right),$$

which is nothing else but the Lorentz transformation (6.37).

Thus, we can say that a Lorentz boost is equivalent to a rotation of imaginary angle in the $x\tau$-plane (or any plane made up by the time coordinate and one spatial coordinate) of Minkowski space in the Euclidean complex representation. In other words, in this representation of Minkowski space, uniform and rectilinear translations along the x-axis in the physical space can be represented as rotations in the $x\tau$-plane, with complex time coordinate τ.

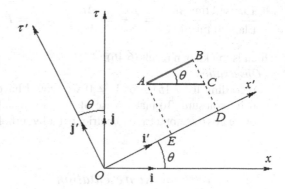

Fig. 7.4 Geometrical interpretation of *length contraction*.

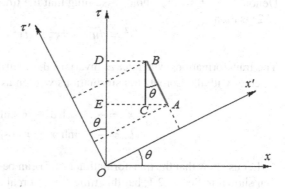

Fig. 7.5 Geometrical interpretation of *time dilation*.

This representation allows one to give a suggestive geometrical interpretation to the *length contraction* and *time dilation*. In Fig. 7.4, consider a bar AB at rest with respect to the frame S' (proper frame). Since the bar is rigidly connected to S', the world lines of the ends A and B are straight lines parallel to $O\tau'$ passing through A and B and intersecting Ox' in E and D. The length of the bar, determined by an observer connected to S, is the segment AC delimited by AE and BD on a straight line parallel to Ox. Then (7.15) and Fig. 7.4 lead to the relation (6.45):

$$\frac{AC}{AB} = \frac{l}{l_0} = \frac{1}{\cos\theta} = \frac{1}{\Gamma} \quad \Rightarrow \quad l = \frac{l_0}{\Gamma} < l_0.$$

Similarly, a graphical representation of the time dilation can be given. Consider a clock at rest in S' (proper frame). In this case, since $x' = \text{const.}$, the world line of the clock is the straight line passing through A and B, where A and B are two arbitrary events recorded by the clock (Fig. 7.5).

The time interval determined from S is the segment BC delimited by BD and AE on a straight line parallel to $O\tau$:

$$\frac{\text{Elapsed time in } S}{\text{Elapsed time in } S'} = \frac{BC}{BA} = \frac{\Delta t}{\Delta t_0} = \cos\theta = \Gamma \quad \Rightarrow \quad \Delta t = \Gamma \Delta t_0 > \Delta t_0,$$

which is the known result (6.46).

Observation:

According to (7.15)$_2$, for $V \neq 0$ we would have $\cos\theta = \Gamma > 1$, which, taken as such, is absurd. To make sense of this, we recall that the angle θ is imaginary, therefore $\cos\theta$ is not a trigonometric, but a hyperbolic function.

7.2.2 Hyperbolic Representation

Denote $u = ct$, by this choice assuming that the time coordinate is *real*. The metric (7.2) is then

$$ds^2 = du^2 - dx^2 - dy^2 - dz^2. \tag{7.16}$$

The transformations which leave invariant the metric (7.16) are all space rotations, together with the Lorentz transformations written as (see (6.66)):

$$x' = +x \cosh\alpha - u \sinh\alpha,$$
$$u' = -x \sinh\alpha + u \cosh\alpha. \tag{7.17}$$

Let us show that the transformation (7.17) can be obtained in a manner similar to that shown in Sect. 7.2.1, but this time using a real angle. Taking $\theta = i\alpha$ in (7.14), we have

$$\tan(i\alpha) = i \tanh\alpha = i\frac{V}{c} \quad \Rightarrow \quad \tanh\alpha = \frac{V}{c},$$
$$\sin(i\alpha) = i \sinh\alpha,$$
$$\cos(i\alpha) = \cosh\alpha,$$
$$\tau = iu,$$

and (7.13) yield

$$x = x' \cosh\alpha + u' \sinh\alpha,$$
$$u = x' \sinh\alpha + u' \cosh\alpha,$$

which is the inverse of the Lorentz transformation (7.17). Therefore, to express the Lorentz transformations by means of *a real angle*, it is necessary to use *hyperbolic functions*.

Let us consider again the plane defined by the space coordinate x and the time coordinate $u = ct$. In this plane, the metric is $s^2 = u^2 - x^2$. This means that the locus of events characterized by the same interval ($s = $ const.) with respect to O is the

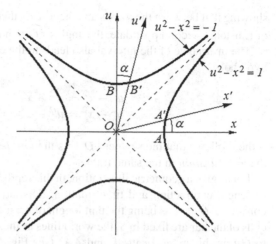

Fig. 7.6 The *hyperbolic representation* of Minkowski space.

equilateral hyperbola $u^2 - x^2 = $ const. Giving different values to the constant, one obtains a family of equilateral hyperbolas, having as asymptotic lines the bisectors of the quadrants of the coordinate system.

Consider the pair of equilateral hyperbolas

$$u^2 - x^2 = 1,$$
$$u^2 - x^2 = -1. \qquad (7.18)$$

Each hyperbola intersects only one coordinate axis. Hyperbola $(7.18)_1$ intersects the u-axis at the points $x = \pm 1$. We conclude that the segment OA in Fig. 7.6 is the *unit length* in the frame S, and OB the *unit duration* in the same frame.

To draw the axes of the frame S', we use the Lorentz transformations (6.37) written in the variables x and u:

$$x' = \Gamma\left(x - \frac{V}{c}u\right),$$
$$u' = \Gamma\left(u - \frac{V}{c}x\right). \qquad (7.19)$$

The equation of the x'-axis is obtained by setting $u' = 0$, which means $u - \frac{V}{c}x = 0$. This is a straight line that makes with the x-axis the angle α, given by

$$\tan\alpha = \frac{u}{x} = \frac{V}{c}.$$

To draw the u'-axis we take $x' = 0$ in (7.19) and have

$$\tan\alpha = \frac{x}{u} = \frac{V}{c},$$

showing that between the u-axis and the u'-axis there is the same angle α. As a result of Einstein's second postulate, the angles $\widehat{xOu}$ and $\widehat{x'Ou'}$ have the same bisector.

The invariance of the interval also leads to the covariance of the equations of the hyperbolas:

$$u^2 - x^2 = u'^2 - x'^2 = +1,$$
$$u^2 - x^2 = u'^2 - x'^2 = -1.$$

It then follows that the segment OA' is the *unit length* in the frame S', and OB' is the *unit duration* in the same frame.

Let us give a geometric illustration, in this representation, of the relativistic effects of length contraction and time dilation. Separating the first quadrant of Fig. 7.6, consider $AO = 1$ as being the unit length of a bar attached to the frame S. Since the ends of the bar are fixed in S, the world lines of these points are straight lines parallel to Ou, in this case Ou itself, and AA'' (see Fig. 7.7). The hyperbola which passes through A intersects the x'-axis in A', therefore $OA' = 1$.

To ensure that an observer of S' simultaneously determines the ends of the bar, the world lines Ou and AA'' must intersect the straight line $u' = $ const., for example the x'-axis (corresponding to $u' = 0$). The length of the same bar, determined by the observer of S', therefore is $OA'' < OA' = 1$. The maximum length of the bar is determined in the proper frame.

To represent the effect of time dilation, consider a clock at rest in S', situated at the origin $O \equiv O'$. Its world line coincides with the u'-axis, therefore OB' is (as previously defined) the unit duration in S'. But in S the points B' and B'' lie on the same straight line $u = $ const., therefore they are simultaneous. Since Ou is the world line for the clock placed at O and attached to S, it follows that OB'' is the duration determined by the observer from S, corresponding to the unit duration determined

Fig. 7.7 Geometrical interpretation of *length contraction* and *time dilation* in the hyperbolic representation of Minkowski space.

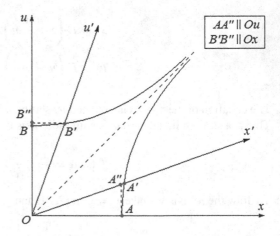

by the observer of S'. Figure 7.7 shows that $OB'' > OB = 1$. In conclusion, the minimum duration is that determined in the proper frame.

7.3 Four-Vectors

7.3.1 Euclidean-Complex Representation

The coordinates x, y, z, $\tau = ict$ of an event in Minkowski space can be considered as the components of a *four-vector*, called *position four-vector*. Denoting these components by x_μ, $\mu = 1, 2, 3, 4$, we choose

$$
\begin{aligned}
x_1 &= x, \\
x_2 &= y, \\
x_3 &= z, \\
x_4 &= ict.
\end{aligned}
\tag{7.20}
$$

With this notation, the metric (7.12) is written as

$$- ds^2 = dx_\mu dx_\mu, \qquad \mu = 1, 2, 3, 4,$$

where we used Einstein's summation convention. The Lorentz transformation (6.37) becomes

$$
\begin{aligned}
x_1' &= \Gamma \left(x_1 + i \frac{V}{c} x_4 \right), \\
x_2' &= x_2, \\
x_3' &= x_3, \\
x_4' &= \Gamma \left(-i \frac{V}{c} x_1 + x_4 \right),
\end{aligned}
\tag{7.21}
$$

or, in condensed form,

$$x_\mu' = a_{\mu\nu} x_\nu, \qquad \mu, \nu = 1, 2, 3, 4,
\tag{7.22}$$

where $a_{\mu\nu}$ are the elements of the matrix of the transformation $x \to x'$,

$$A = \begin{pmatrix} \Gamma & 0 & 0 & i\frac{V}{c} \\ 0 & 1 & 0 & 0 \\ 0 & 0 & 1 & 0 \\ -i\frac{V}{c} & 0 & 0 & \Gamma \end{pmatrix}.
\tag{7.23}$$

A system of quantities V_μ, $\mu = 1, 2, 3, 4$ which, when changing the coordinate system $(S \rightarrow S')$, transform according to (7.22), that is

$$V'_\mu = a_{\mu\nu} V_\nu, \quad \mu, \nu = 1, 2, 3, 4 \tag{7.24}$$

form a *four-vector*. Minkowski space is a vector space.

The inverse of the Lorentz transformation (7.22) is the transformation

$$x_\mu = a'_{\mu\nu} x'_\nu \equiv a_{\nu\mu} x'_\nu, \quad \mu, \nu = 0, 1, 2, 3, \tag{7.25}$$

where $a_{\nu\mu}$ are the elements of the matrix of the inverse transformation

$$A^{-1} = \begin{pmatrix} \Gamma & 0 & 0 & -i\frac{V}{c} \\ 0 & 1 & 0 & 0 \\ 0 & 0 & 1 & 0 \\ i\frac{V}{c} & 0 & 0 & \Gamma \end{pmatrix}. \tag{7.26}$$

From (7.22) and (7.25) we deduce that the transformation matrix A satisfies the orthogonality condition (see Appendix A):

$$a_{\mu\nu} a_{\mu\lambda} = \delta_{\nu\lambda}, \quad \mu, \nu, \lambda = 1, 2, 3, 4. \tag{7.27}$$

This shows that the Lorentz transformation (7.22) is an *orthogonal linear transformation* in Minkowski space, represented in the Euclidean-complex form.

The transformation (7.22) can also be expressed in the matrix form

$$X' = A X, \tag{7.28}$$

where $A = (a_{\mu\nu})$ stands for the matrix (7.23), while X and X' are the one-column matrices

$$X = \begin{pmatrix} x_1 \\ x_2 \\ x_3 \\ x_4 \end{pmatrix}, \quad X' = \begin{pmatrix} x'_1 \\ x'_2 \\ x'_3 \\ x'_4 \end{pmatrix}. \tag{7.29}$$

The inverse of transformation (7.28) is then

$$X = A^{-1} X', \tag{7.30}$$

where A^{-1} is the matrix (7.26) of the inverse transformation (7.25). Obviously,

$$A^{-1} A = A A^{-1} = I, \tag{7.31}$$

where I is the *unit* (or *identity*) 4×4 matrix. This relation can also be written as

$$a_{\mu\nu}a'_{\nu\lambda} = \delta_{\mu\lambda}. \tag{7.32}$$

Multiplying (7.32) by $a_{\mu\rho}$ and using (7.27), we arrive at

$$a'_{\rho\lambda} = a_{\lambda\rho}, \quad \text{i.e.} \quad A^{-1} = A^T, \tag{7.33}$$

hence the elements of the inverse transformation matrix are obtained by transposing the elements of the direct transformation matrix. Clearly, the matrices (7.23) and (7.26) satisfy the conditions (7.31)–(7.33).

7.3.2 Hyperbolic Representation

Starting with the metric in the hyperbolic representation of Minkowski space (7.16), let us choose

$$\begin{aligned} x^0 &= ct, \\ x^1 &= x, \\ x^2 &= y, \\ x^3 &= z, \end{aligned} \tag{7.34}$$

as the components of the position four-vector x^μ.

Unlike the Euclidean-complex representation, which makes no difference between contravariant and covariant vectors (tensors), in hyperbolic representation the choice (7.34) leads to the metric (see Appendix B):

$$ds^2 = g_{\mu\nu}dx^\mu dx^\nu = (dx^0)^2 - (dx^1)^2 - (dx^2)^2 - (dx^3)^2. \tag{7.35}$$

The components of the metric tensor therefore are

$$g_{\mu\nu} = \text{diag}\,(1, -1, -1, -1),$$

that is

$$g_{\mu\nu} = \begin{pmatrix} 1 & 0 & 0 & 0 \\ 0 & -1 & 0 & 0 \\ 0 & 0 & -1 & 0 \\ 0 & 0 & 0 & -1 \end{pmatrix}. \tag{7.36}$$

The Lorentz transformation (7.19) is then written as

$$x'^0 = \Gamma\left(x^0 - \frac{V}{c}x^1\right),$$

$$x'^1 = \Gamma\left(-\frac{V}{c}x^0 + x^1\right),$$

$$x'^2 = x^2, \tag{7.37}$$

$$x'^3 = x^3.$$

The inverse transformation is

$$x^0 = \Gamma\left(x'^0 + \frac{V}{c}x'^1\right),$$

$$x^1 = \Gamma\left(\frac{V}{c}x'^0 + x'^1\right),$$

$$x^2 = x'^2, \tag{7.38}$$

$$x^3 = x'^3,$$

or, in a condensed form,

$$x'^\mu = \Lambda^\mu_{\ \nu} x^\nu,$$

$$x^\mu = \bar{\Lambda}^\mu_{\ \nu} x'^\nu, \tag{7.39}$$

where $\Lambda^\mu_{\ \nu}$ and $\bar{\Lambda}^\mu_{\ \nu}$ are the matrices of the transformations $x \to x'$ and $x' \to x$, respectively (see Appendix B):

$$\Lambda = (\Lambda^\mu_{\ \nu}) = \left(\frac{\partial x'^\mu}{\partial x^\nu}\right) = \begin{pmatrix} \Gamma & -\frac{V}{c}\Gamma & 0 & 0 \\ -\frac{V}{c}\Gamma & \Gamma & 0 & 0 \\ 0 & 0 & 1 & 0 \\ 0 & 0 & 0 & 1 \end{pmatrix},$$

$$\bar{\Lambda} = (\bar{\Lambda}^\mu_{\ \nu}) = \left(\frac{\partial x^\mu}{\partial x'^\nu}\right) = \begin{pmatrix} \Gamma & +\frac{V}{c}\Gamma & 0 & 0 \\ +\frac{V}{c}\Gamma & \Gamma & 0 & 0 \\ 0 & 0 & 1 & 0 \\ 0 & 0 & 0 & 1 \end{pmatrix}.$$

These matrices satisfy the orthogonality condition

$$\Lambda^\mu_{\ \nu} \bar{\Lambda}^\eta_{\ \mu} = \delta^\eta_\nu, \tag{7.40}$$

as well as condition (7.31):

$$\Lambda\bar{\Lambda} = \bar{\Lambda}\Lambda = \Lambda\Lambda^{-1} = \Lambda^{-1}\Lambda = I. \tag{7.41}$$

Also,

$$[\det \Lambda]^2 = 1 \;\Rightarrow\; \det \Lambda = \pm 1. \tag{7.42}$$

In the hyperbolic representation of Minkowski space, a contravariant four-vector transforms according to the rule (7.39):

$$
\begin{aligned}
A'^{\mu} &= \Lambda^{\mu}_{\;\nu}\, A^{\nu}, \\
A^{\mu} &= \bar{\Lambda}^{\mu}_{\;\nu}\, A'^{\nu},
\end{aligned}
\tag{7.43}
$$

while a covariant four-vector obeys the rule

$$
\begin{aligned}
B'_{\mu} &= \bar{\Lambda}^{\nu}_{\;\mu}\, B_{\nu}, \\
B_{\mu} &= \Lambda^{\nu}_{\;\mu}\, B'_{\nu}.
\end{aligned}
\tag{7.44}
$$

In a similar way can be defined contravariant, covariant, and mixed four-tensors of any order (see Appendix B).

Observation:

Sometimes it is convenient to choose $x^0 = t$, in which case the components of the metric tensor are

$$
\begin{aligned}
g^{00} &= \frac{1}{c^2}, \\
g_{00} &= c^2, \\
g^{ii} &= g_{ii} = -1 \text{ (no summation over } i).
\end{aligned}
$$

With this notation, the Lorentz transformation is written as

$$
\begin{aligned}
x'^0 &= \Gamma \left(x^0 - \frac{V}{c^2} x^1 \right), \\
x'^1 &= \Gamma \left(-V x^0 + x^1 \right), \\
x'^2 &= x^2, \\
x'^3 &= x^3.
\end{aligned}
\tag{7.45}
$$

Such a representation is used, as we shall see, in the general theory of relativity.

As far as we are concerned, while working in Minkowski space we shall use the choice $x^0 = ct$, as being the most commonly found in the literature. Here and hereafter, the contravariant space-like components of four-vectors will be considered as components of the usual vectors ($\mathbf{r}$ for x^i, $\mathbf{p}$ for p^i, etc.).

7.3.3 Lorentz Group

We consider the Minkowski space M_4 with $x^0 = ct$, $x^1 = x$, $x^2 = y$, $x^3 = z$ and the metric

$$ds^2 = g_{\mu\nu}dx^\mu dx^\nu,$$

where $g_{\mu\nu} = \text{diag}(+1, -1, -1, -1)$ is the metric tensor. The set of transformations

$$x^\mu \rightarrow x'^\mu = \Lambda^\mu{}_\nu x^\nu, \quad \mu, \nu = 0, 1, 2, 3, \tag{7.46}$$

or, in matrix form,

$$x' = \Lambda x,$$

which leave invariant the interval (or, equivalently, the norm of the four-vectors) in Minkowski space form a group, called the *Lorentz group*. The transformations (7.46) represent "rotations" in the four-dimensional space, i.e. usual three-dimensional rotations and Lorentz boosts (7.37). In the following, by Lorentz transformation we shall understand any element of the Lorentz group.

Besides the Lorentz transformations, the space-time translations

$$x^\mu \rightarrow x'^\mu = x^\mu + a^\mu, \quad \mu, \nu = 0, 1, 2, 3, \tag{7.47}$$

also leave invariant the interval. The translations form an Abelian group.

Thus, an isometry of the interval in Minkowski space is written in general as

$$x^\mu \rightarrow x'^\mu = \Lambda^\mu{}_\nu x^\nu + a^\mu, \quad \mu, \nu = 0, 1, 2, 3, \tag{7.48}$$

or in matrix form

$$x' = \Lambda x + a.$$

where by x, x', and a we denoted the corresponding four-vectors in one-column matrix form. The group of transformations (7.48) is called the *Poincaré group*, or the *inhomogeneous Lorentz group*. The general concept of relativistic invariance means invariance under Poincaré transformations. The Poincaré group plays a fundamental role in the relativistic theory of quantized fields, where particle states are constructed as irreducible unitary representations of this group. This classification of particles was introduced by Eugene Paul Wigner (1902–1995), who received the Noble Prize in Physics in 1963 for this seminal theory.

The invariance of the norm of four-vectors under a Lorentz transformation is written as

$$x'_\mu x'^\mu = x_\nu x^\nu, \quad \mu, \nu = 0, 1, 2, 3,$$

or

$$x'^2 = x^2,$$

where

$$x'^2 = g_{\mu\nu}x'^\mu x'^\nu = (\Lambda x)^T G(\Lambda x) = x^T(\Lambda^T G \Lambda)x$$

and

$$x^2 = g_{\mu\nu}x^\mu x^\nu = x^T G x,$$

with the notation for the metric tensor $G = (g_{\mu\nu})$. We have thus obtained that

$$G = \Lambda^T G \Lambda. \tag{7.49}$$

From (7.49) it follows that $\det G = \det G \, (\det \Lambda)^2$, which, recalling that $\det G \neq 0$, implies

$$\det \Lambda = \pm 1. \tag{7.50}$$

In components, relation (7.49) reads

$$g_{\lambda\rho} = \Lambda^\mu{}_\lambda g_{\mu\nu} \Lambda^\nu{}_\rho.$$

Let $\lambda = \rho = 0$. Then

$$g_{00} = \Lambda^\mu{}_0 g_{\mu\nu} \Lambda^\nu{}_0 = (\Lambda^0{}_0)^2 - (\Lambda^1{}_0)^2 - (\Lambda^2{}_0)^2 - (\Lambda^3{}_0)^2,$$

that is

$$(\Lambda^0{}_0)^2 = 1 + (\Lambda^1{}_0)^2 + (\Lambda^2{}_0)^2 + (\Lambda^3{}_0)^2,$$

leading to

$$(\Lambda^0{}_0)^2 \geq 1.$$

From here we infer that $|\Lambda^0{}_0| \geq 1$, such that

$$\Lambda^0{}_0 \geq +1 \quad \text{or} \quad \Lambda^0{}_0 \leq -1. \tag{7.51}$$

Thus, the Lorentz group can be written as

$$L = \left\{ \Lambda \in GL(4, \mathbb{R}) \mid \Lambda^T G \Lambda = G, \ \det \Lambda = \pm 1, \ |\Lambda^0{}_0| \geq 1 \right\},$$

where $GL(4, \mathbb{R})$ is the *general linear group of degree* 4, which represents the set of the 4×4 real invertible matrices, together with the operation of ordinary matrix multiplication as the group law.

The conditions $\det \Lambda = \pm 1$ and $|\Lambda^0{}_0| \geq 1$ lead to a division of the Lorentz group into four subsets. We introduce the following notions and notations: *proper Lorentz transformations*, denoted by L_+, are the subset of Lorentz transformations with $\det \Lambda = +1$; *improper Lorentz transformations*, denoted by L_-, are the transformations with $\det \Lambda = -1$; *orthochronous Lorentz transformations*, denoted by $L^\uparrow$, is the subset of Lorentz transformations with $\Lambda^0{}_0 \geq 1$; *antichronous Lorentz transformations*, denoted by $L^\downarrow$, are characterized by $\Lambda^0{}_0 \leq -1$. The proper Lorentz transformations preserve the orientation of spatial axes, while the orthochronous transformations preserve the direction of time. Having settled these definitions, the Lorentz group is divided into the following classes:

1) $L_+^\uparrow = \{\Lambda \in L \mid \det \Lambda = +1,\ \Lambda^0{}_0 \geq +1\}$;

2) $L_+^\downarrow = \{\Lambda \in L \mid \det \Lambda = +1,\ \Lambda^0{}_0 \leq -1\}$;

3) $L_-^\uparrow = \{\Lambda \in L \mid \det \Lambda = -1,\ \Lambda^0{}_0 \geq +1\}$;

4) $L_-^\downarrow = \{\Lambda \in L \mid \det \Lambda = -1,\ \Lambda^0{}_0 \leq -1\}$.

The unit matrix, which is the unit of the group, is included among the first class, of proper orthochronous Lorentz transformations. As a result, this subset is the only one which forms a subgroup of the Lorentz group. In other words, the proper orthochronous transformations are those Lorentz transformations which are continuously connected with the unit transformation.

Observations:

(a) For the proper Lorentz transformations the condition $\det \Lambda = +1$ is *necessary*, but not sufficient, as there exist improper Lorentz transformations for which $\det \Lambda = +1$. An example is the space-time inversion described by $\Lambda = -I$, which is an improper transformation, but still satisfies $\det \Lambda = +1$;

(b) For the improper Lorentz transformations, the condition $\det \Lambda = -1$ is *sufficient*, but not necessary. The fact that the condition $\det \Lambda = \pm 1$ does not select unequivocally the proper and improper transformations is due to the indefinite metric on Minkowski space. For example, the space inversion $\Lambda = G = \mathrm{diag}(+1, -1, -1, -1)$ has $\det \Lambda = -1$, while the space-time inversion $\Lambda = -I = \mathrm{diag}(-1, -1, -1, -1)$ has $\det \Lambda = +1$, but both are improper Lorentz transformations.

Infinitesimal Lorentz Transformations
An infinitesimal Lorentz transformation is written as

$$x^\alpha \to x'^\alpha = x^\alpha + \omega^\alpha{}_\beta x^\beta \qquad (7.52)$$

meaning that

$$\Lambda^\alpha{}_\beta = \delta^\alpha{}_\beta + \omega^\alpha{}_\beta, \qquad (7.53)$$

with $\mathcal{O}(\omega^2) = 0$. The quantities $\omega^\alpha{}_\beta$ are the infinitesimal parameters of the transformation. From $g_{\alpha\beta} = \Lambda^\mu{}_\alpha g_{\mu\nu} \Lambda^\nu{}_\beta$ it follows that

$$g_{\alpha\beta} = \left(\delta^\mu{}_\alpha + \omega^\mu{}_\alpha\right) g_{\mu\nu} \left(\delta^\nu{}_\beta + \omega^\nu{}_\beta\right) = g_{\alpha\beta} + \omega_{\beta\alpha} + \omega_{\alpha\beta}.$$

Consequently,

$$\omega_{\beta\alpha} + \omega_{\alpha\beta} = 0, \tag{7.54}$$

which shows that $\omega_{\alpha\beta}$ are antisymmetric quantities. Let us consider the group of proper orthochronous Lorentz transformations, $L_+^\uparrow$. This set forms a Lie group structure, whose elements in exponential parametrization are written as

$$\Lambda = \exp\left(-\frac{i}{2}\omega^{\alpha\beta} L_{\alpha\beta}\right), \tag{7.55}$$

where $L_{\alpha\beta}$ are the *generators* of the group, which are 4×4 matrices.[1]

One can easily see that the generators are antisymmetric, since

$$\omega^{\alpha\beta} L_{\alpha\beta} = \frac{1}{2}\left(\omega^{\alpha\beta} L_{\alpha\beta} + \omega^{\alpha\beta} L_{\alpha\beta}\right) = \frac{1}{2}\omega^{\alpha\beta}\left(L_{\alpha\beta} - L_{\beta\alpha}\right),$$

leading to

$$L_{\alpha\beta} = \frac{1}{2}\left(L_{\alpha\beta} - L_{\beta\alpha}\right) \Rightarrow L_{\alpha\beta} = -L_{\beta\alpha}.$$

Due to this property, which is the only condition on the generators, instead of sixteen generators on the four-dimensional space-time, there are only six. The six distinct components of $\omega_{\alpha\beta}$ represent the *parameters* of the Lorentz group. Physically, three of these parameters are the angles necessary for the unequivocal determination of the relative orientation of the coordinate axes of the inertial frames S and S' (for example, Euler's angles), while the other three parameters are the three components of the relative velocity $\mathbf{V}$ between the two inertial frames.

In the case of infinitesimal transformations, relation (7.55) becomes

$$\Lambda = I - \frac{i}{2}\omega^{\mu\nu} L_{\mu\nu}, \tag{7.57}$$

or, in components,

---

[1] The exponential function with a matrix as an argument is a formal way of writing. The actual meaning of e^M, where M is a $n \times n$ matrix, is the power series expansion

$$e^M = I + M + \frac{1}{2!}M^2 + \frac{1}{3!}M^3 + \ldots = \sum_{k=0}^{\infty} \frac{1}{k!}M^k, \tag{7.56}$$

where I represents the $n \times n$ unit matrix.

$$\Lambda^\alpha{}_\beta = \delta^\alpha{}_\beta - \frac{i}{2}\omega^{\mu\nu}\left(L_{\mu\nu}\right)^\alpha{}_\beta. \tag{7.58}$$

Comparing the latter equation with $\Lambda^\alpha{}_\beta = \delta^\alpha{}_\beta + \omega^\alpha{}_\beta$, we obtain

$$\omega^\alpha{}_\beta = -\frac{i}{2}\omega^{\mu\nu}\left(L_{\mu\nu}\right)^\alpha{}_\beta,$$

leading to

$$\left(L_{\mu\nu}\right)^\alpha{}_\beta = i\left(g_{\nu\beta}\delta^\alpha{}_\mu - g_{\mu\beta}\delta^\alpha{}_\nu\right), \tag{7.59}$$

since

$$\omega^\alpha{}_\beta = g_{\gamma\beta}\omega^{\alpha\gamma} = \frac{1}{2}g_{\gamma\beta}\omega^{\alpha\gamma} + \frac{1}{2}g_{\gamma\beta}\omega^{\alpha\gamma} = \frac{1}{2}\delta^\alpha{}_\mu g_{\nu\beta}\omega^{\mu\nu} + \frac{1}{2}\delta^\alpha{}_\nu g_{\mu\beta}\omega^{\nu\mu}$$
$$= \frac{1}{2}\omega^{\mu\nu}\left(g_{\nu\beta}\delta^\alpha{}_\mu - g_{\mu\beta}\delta^\alpha{}_\nu\right).$$

The Lorentz group has thus six generators: L_{01}, L_{02}, L_{03}, L_{12}, L_{23}, and L_{31}. The group algebra is given by the commutator of the generators, which reads

$$\left[L_{\alpha\beta}, L_{\mu\nu}\right] = -i\left(g_{\alpha\mu}L_{\beta\nu} + g_{\beta\nu}L_{\alpha\mu} - g_{\alpha\nu}L_{\beta\mu} - g_{\beta\mu}L_{\alpha\nu}\right). \tag{7.60}$$

Each of the tensor generators can be put into correspondence with a vector through the relations $L_{ij} = \varepsilon_{ijk}J_k$ and $L_{0i} = K_i, \alpha, \beta = 0, 1, 2, 3; \ i, j, k = 1, 2, 3$. The generators $K_i, \ i = 1, 2, 3$ represent the boosts along the three space directions, while the vectors $J_k, \ k = 1, 2, 3$ generate the space rotations. The generators in the vector form can be shown to satisfy the commutation relations

$$\begin{aligned} \left[J_i, J_j\right] &= i\varepsilon_{ijk}J_k, \\ \left[J_i, K_j\right] &= i\varepsilon_{ijl}K_l, \\ \left[K_i, K_j\right] &= -i\varepsilon_{ijk}J_k. \end{aligned} \tag{7.61}$$

The expressions of the six generators of the Lorentz group are:

$$J_1 = \begin{pmatrix} 0 & 0 & 0 & 0 \\ 0 & 0 & 0 & 0 \\ 0 & 0 & 0 & -i \\ 0 & 0 & +i & 0 \end{pmatrix}, \ J_2 = \begin{pmatrix} 0 & 0 & 0 & 0 \\ 0 & 0 & 0 & +i \\ 0 & 0 & 0 & 0 \\ 0 & -i & 0 & 0 \end{pmatrix}, \ J_3 = \begin{pmatrix} 0 & 0 & 0 & 0 \\ 0 & 0 & -i & 0 \\ 0 & +i & 0 & 0 \\ 0 & 0 & 0 & 0 \end{pmatrix},$$

$$K_1 = \begin{pmatrix} 0 & -i & 0 & 0 \\ -i & 0 & 0 & 0 \\ 0 & 0 & 0 & 0 \\ 0 & 0 & 0 & 0 \end{pmatrix}, \ K_2 = \begin{pmatrix} 0 & 0 & -i & 0 \\ 0 & 0 & 0 & 0 \\ -i & 0 & 0 & 0 \\ 0 & 0 & 0 & 0 \end{pmatrix}, \ K_3 = \begin{pmatrix} 0 & 0 & 0 & -i \\ 0 & 0 & 0 & 0 \\ 0 & 0 & 0 & 0 \\ -i & 0 & 0 & 0 \end{pmatrix}. \tag{7.62}$$

The first of the commutation relations (7.61) represents the commutator of the components of the angular momentum,[2] the second relation (7.61) shows that the set K_i, $i = 1, 2, 3$, transforms like a vector under the action of rotations, while the third commutator (7.61) shows that in general the Lorentz boosts are not commuting.

Introducing the combination of generators

$$S_i = \frac{1}{2} (J_i + i K_i),$$

$$T_i = \frac{1}{2} (J_i - i K_i), \qquad (7.63)$$

the commutation relations (7.61) become

$$\begin{aligned} [S_i, S_j] &= i \, \varepsilon_{ijk} S_k, \\ [T_i, T_j] &= i \, \varepsilon_{ijk} T_k, \\ [S_i, T_j] &= 0. \end{aligned}$$

Clearly, the combinations S_i, $i = 1, 2, 3$ generate one $SU(2)$ group and the combinations T_i, $i = 1, 2, 3$ generate another, such that the Lorentz group can be written as the direct product $SU(2)_S \times SU(2)_T$, where the subscripts refer to the symbols given to the generators. This fact is important in finding the finite dimensional irreducible representations of the Lorentz group, but this issue is beyond the scope of our discussion.

7.4 Relativistic Kinematics

Kinematics is the study of the possible motions of mechanical systems, without considering the cause of motion, i.e. the force. Kinematics is concerned with *how* the body moves on its trajectory; in short, one must know the velocity and the acceleration of the body.

The non-relativistic definitions guide us in the relativistic case as well. However, if we take the derivative of the position vector with respect to time, neither of them have definite transformation properties under the Lorentz transformations. In order to produce a sensible definition, we should take the derivative of the position four-vector with respect to a scalar (i.e. a Lorentz invariant), such that the result is a four-vector. In relativistic kinematics, the *velocity four-vector* and the *acceleration four-vector* are both defined by means of the invariants ds (the interval) or/and $d\tau$ (the proper time). Recalling that

---

[2]The angular momentum conservation is obtained, through Noether's theorem, from the invariance of the action with respect to three-dimensional rotations (see Sect. 8.7.3).

$$ds = c\,dt\sqrt{1 - \frac{v^2}{c^2}} = \frac{c}{\gamma}\,dt = c\,d\tau,\tag{7.64}$$

one defines the *velocity four-vector* of a particle as

$$u^\mu = \frac{dx^\mu}{ds} = \frac{\gamma}{c}\frac{dx^\mu}{dt}, \quad \text{or} \quad \bar{u}^\mu = \frac{dx^\mu}{d\tau} = \gamma\frac{dx^\mu}{dt} = cu^\mu.\tag{7.65}$$

Remark that the components of u^μ are dimensionless, while $\bar{u}^\mu$ have dimension of velocity. We shall mostly use the first definition. The velocity of the particle with respect to the frame S has the components $v^i = dx^i/dt$ and the corresponding Lorentz factor is $\gamma = (1 - v^2/c^2)^{-1/2}$. If the motion is observed from another inertial frame S', moving with respect to the frame S with the velocity $\mathbf{V}$, the transformation between the two frames will involve as usual the factor $\Gamma = (1 - V^2/c^2)^{-1/2}$.

In the hyperbolic representation of the Minkowski space, the components of the velocity four-vector are, respectively

$$u^0 = \frac{\gamma}{c}\frac{dx^0}{dt} = \gamma,$$

$$u^i = \frac{\gamma}{c}\frac{dx^i}{dt} = \frac{\gamma}{c}v^i, \quad i = 1, 2, 3,\tag{7.66}$$

i.e.

$$u^\mu = \gamma\left(1, \frac{\mathbf{v}}{c}\right).$$

The covariant components are

$$u_\mu = g_{\mu\nu}u^\nu = \gamma\left(1, -\frac{\mathbf{v}}{c}\right).\tag{7.67}$$

The last two relations yield

$$u_\mu u^\mu = u_0 u^0 + u_i u^i = \gamma^2 - \frac{v^2}{c^2}\gamma^2 = 1, \quad \mu = 0, 1, 2, 3\,; \; i = 1, 2, 3\,,\tag{7.68}$$

therefore the components of the velocity four-vector are not independent, but must satisfy (7.68). Equation (7.68) also shows that u^μ is a *unit four-vector* tangent to the world line of the moving particle.

The *acceleration four-vector* can be defined either by

$$a^\mu = \frac{du^\mu}{ds} = \frac{\gamma}{c}\frac{du^\mu}{dt}\tag{7.69}$$

with the components

$$a^0 = a_0 = \frac{\gamma}{c} \frac{d\gamma}{dt},$$

$$a^i = -a_i = \frac{\gamma}{c^2} \frac{d}{dt} \left(\gamma v^i \right), \qquad (7.70)$$

or by

$$\overline{a}^\mu = \frac{d\overline{u}^\mu}{d\tau} = \gamma \frac{d}{dt} \left(\gamma \frac{dx^\mu}{dt} \right) = \gamma^2 \frac{d^2 x^\mu}{dt^2} + \frac{1}{2} \frac{d\gamma^2}{dt} \frac{dx^\mu}{dt}.$$

Note that

$$a^\mu u_\mu = a^0 u_0 + a^i u_i = \frac{\gamma^2}{c} \frac{d\gamma}{dt} - \frac{\gamma^2}{c^3} \left(v^2 \frac{d\gamma}{dt} + \gamma v^i \frac{dv^i}{dt} \right) = 0, \qquad (7.71)$$

which means that the velocity and the acceleration four-vectors are *orthogonal*.

7.5 Relativistic Dynamics in Three-Dimensional Approach

7.5.1 Notions, Quantities, and Fundamental Relations

Using the analytical mechanics formalism, let us first define some fundamental quantities of relativistic dynamics: momentum, energy, mass, force, etc.

Consider a free particle of proper mass (also called *rest mass*) m_0. Let us find the expression of the action for a free relativistic particle. As the action integral should not depend on the choice of the inertial reference frame, it means that the action has to be Lorentz invariant, i.e. to transform like a scalar under Lorentz transformations. Moreover, since the equations of motion must be differential equations maximum of second order, it follows that the integrand should be a differential expression of the first order. The only quantity which describes the free particle motion and satisfies the above conditions is the interval ds. Therefore we shall take the action to be the integral over the interval, multiplied by a positive constant α which takes care of the dimensionality and has to be determined. In this case, the principle of Maupertuis is written as

$$\delta S = -\alpha \, \delta \int_a^b ds = 0, \qquad (7.72)$$

where a and b are two events. The minus sign is necessary because the variational problem shows that in the case of the free particle, the straight trajectory is the longest worldline between two events. By taking the negative of the interval under the integral, one insures that the maximum is transformed into a minimum. We can re-write the action in a more familiar manner, as integral over time, using the relation (7.64) and considering that the integration limits a and b correspond to the time moments t_1 and t_2, respectively:

$$S = -\alpha \int_{t_1}^{t_2} \sqrt{1 - \frac{v^2}{c^2}} \, dt = 0, \tag{7.73}$$

For small velocities ($v \ll c$), we may take

$$\sqrt{1 - \frac{v^2}{c^2}} \simeq 1 - \frac{1}{2}\frac{v^2}{c^2},$$

which yields

$$S = -\alpha c \int_{t_1}^{t_2} dt + \frac{\alpha}{2c} \int_{t_1}^{t_2} v^2 dt. \tag{7.74}$$

The first term of (7.74) does not affect the form of the equations of motion and may be omitted. Comparing (7.74) with the action of a free non-relativistic particle,

$$S = \int L dt = \int T dt = \frac{1}{2}m_0 \int v^2 dt, \tag{7.75}$$

it follows that $\alpha = m_0 c$ and the action (7.73) becomes

$$S = -m_0 c^2 \int_{t_1}^{t_2} \sqrt{1 - \frac{v^2}{c^2}} \, dt. \tag{7.76}$$

The Lagrangian of the free particle therefore is

$$L = -m_0 c^2 \sqrt{1 - \frac{v^2}{c^2}} = -\frac{m_0 c^2}{\gamma}. \tag{7.77}$$

The relativistic linear momentum of the particle is obtained by the usual procedure:

$$p^i = \frac{\partial L}{\partial v_i} = m_0 \gamma v^i, \quad i = 1, 2, 3, \quad \text{or} \quad \mathbf{p} = m_0 \gamma \mathbf{v}. \tag{7.78}$$

For $v \ll c$, one retrieves the classical formula $\mathbf{p} = m_0 \mathbf{v}$.

To find the relativistic energy of the particle we shall use the Hamiltonian definition $H = p^i \dot{q}^i - L$, where $\dot{q}^i = v^i$ are the generalized velocities. In case of conservative systems, the Hamiltonian is the *total energy* E of the system, so that

$$E = p^i v^i - L = m_0 \gamma v^i v^i + \frac{m_0 c^2}{\gamma} = m_0 \gamma v^2 + \frac{m_0 c^2}{\gamma},$$

that is,

$$E = m_0 \gamma c^2. \tag{7.79}$$

For small velocities ($v \ll c$), the series expansion of γ leads to

$$E = m_0 c^2 \left(1 - \frac{v^2}{c^2}\right)^{-1/2} \simeq m_0 c^2 \left(1 + \frac{1}{2}\frac{v^2}{c^2}\right),$$

or

$$E = m_0 c^2 + \frac{1}{2}m_0 v^2 = E_0 + T. \tag{7.80}$$

The relativistic kinetic energy, T, is no longer defined as $T = \frac{1}{2}mv^2$, like in the non-relativistic case (even if m is the so-called motion mass, $m = m_0\gamma$), but it is defined as the difference between the total energy, E, and the *rest energy*, or *proper energy* $E_0 = m_0 c^2$, which results from (7.79) for $v = 0$ (i.e. $\gamma = 1$):

$$T = E - E_0 = (\gamma - 1)m_0 c^2. \tag{7.81}$$

Thus, Eq. (7.79) shows that the total energy of the particle is different from zero even if the particle is at rest ($v = 0$). The notion of *rest energy* is valid not only for particles, but also for macroscopic bodies, in which case m_0 is the mass of the whole body.

Unlike classical mechanics, in relativistic mechanics $m_0 c^2$ does not mean the sum of the rest energies of the constitutive particles, since $m_0 c^2$ also contains the interaction energy of particles. Therefore, we cannot write $m_0 = \sum_{i=1}^{N} m_{0i}$, where N is the total number of constitutive particles, meaning that *in relativistic mechanics the mass conservation law is not valid*. The valid law is the law of conservation of *total energy*.

Using (7.78) and (7.79), one finds the following relation between the energy, mass, and velocity of a free particle:

$$\mathbf{p} = \frac{E}{c^2}\mathbf{v}. \tag{7.82}$$

According to (7.78) and (7.79), if $v = c$, the energy and momentum of the particle become infinite, showing that a massive particle ($m_0 \neq 0$) cannot move with the speed of light in vacuum. On the other hand, if $m_0 = 0$ and $v < c$, the momentum and the energy of the particle vanish. To clarify the special case of a particle with $v = c$ and $m_0 = 0$, let us square (7.78) and (7.79), then subtract term by term. The result is

$$\frac{E^2}{c^2} = \mathbf{p}^2 + m_0^2 c^2. \tag{7.83}$$

A particle with velocity $v = c$ has, according to (7.82), the momentum $|\mathbf{p}| = E/c$; by virtue of (7.83), this means $m_0 = 0$. Conversely, a particle with $m_0 = 0$ has, because of (7.83), the momentum $|\mathbf{p}| = E/c$, in which case, according to (7.82), the velocity of the particle equals c. Consequently, massless particles ($m_0 = 0$) move in vacuum with the speed of light c.

Particles with $E \gg m_0 c^2$ are called *ultrarelativistic* (the acronym is UR), while those for which $E \ll m_0 c^2$ are called *non-relativistic* (the acronym being NR).

The reader is invited to show that between the time derivatives of E and $\mathbf{p}$ there exists the following relation:

$$\frac{dE}{dt} = \mathbf{v} \cdot \frac{d\mathbf{p}}{dt}, \tag{7.84}$$

which is very useful in many applications.

7.5.2 Variation of Mass with Velocity

By analogy with Newtonian mechanics, the *force* is defined as the time derivative of the relativistic momentum

$$\mathbf{F}^N = \frac{d\mathbf{p}}{dt} = \frac{d}{dt} (m_0 \gamma \mathbf{v}), \tag{7.85}$$

where the upper index N stands for "Newtonian". In classical mechanics, this relation (with $\gamma = 1$) serves to give an unequivocal definition of *mass*, as being the ratio between the magnitude of the force acting on a body, and the magnitude of the acceleration which the body acquires under the action of the force. In relativistic mechanics, the definition of mass by means of (7.85) is not unequivocal, as we shall see shortly.

Let us calculate the derivative in (7.85):

$$\mathbf{F}^N = m_0 \mathbf{v} \frac{d\gamma}{dt} + m_0 \gamma \frac{d\mathbf{v}}{dt} = m_0 \gamma^3 \frac{\mathbf{v}}{c^2} \left(\mathbf{v} \cdot \frac{d\mathbf{v}}{dt} \right) + m_0 \gamma \frac{d\mathbf{v}}{dt},$$

and take the x_i-component

$$F_i^N = m_0 \gamma^3 \frac{v_i}{c^2} \left(v_k \frac{dv_k}{dt} \right) + m_0 \gamma \frac{dv_i}{dt}.$$

Denoting $\mathbf{a}^N = d\mathbf{v}/dt$, we have

$$F_i^N = m_0 \gamma^3 \frac{v_i}{c^2} \left(v_k a_k^N \right) + m_0 \gamma a_i^N$$

$$= m_0 \gamma^3 a_k^N \left[\frac{1}{c^2} v_i v_k + \left(1 - \frac{v^2}{c^2} \right) \delta_{ik} \right], \tag{7.86}$$

or, in short,

$$F_i^N = m_{ik} a_k^N,$$ (7.87)

where

$$m_{ik} = m_0 \gamma^3 \left[\frac{1}{c^2} v_i v_k + \left(1 - \frac{v^2}{c^2} \right) \delta_{ik} \right]$$ (7.88)

is the *mass tensor*.

Further, let us multiply (7.86) by v_i. We obtain

$$F_i^N v_i = m_0 \gamma^3 a_k^N v_k,$$

that is, if $\mathbf{F}^N \perp \mathbf{v}$, then $\mathbf{a}^N \perp \mathbf{v}$ as well. We can distinguish two limit cases:

(a) $\mathbf{F}^N \perp \mathbf{v}$. Since $\mathbf{a}^N \cdot \mathbf{v} = 0$, (7.86) reduces to

$$F_i^N = m_0 \gamma a_i^N,$$

so that

$$\frac{|\mathbf{F}^N|}{|\mathbf{a}^N|} = m_0 \gamma = \frac{m_0}{\sqrt{1 - \frac{v^2}{c^2}}} = m_t,$$ (7.89)

which is called *transverse mass*.

(b) $\mathbf{F}^N \parallel \mathbf{v}$. Taking $\mathbf{F}^N = \lambda \mathbf{v}$, one obtains

$$F_i^N v_i = \lambda v^2 = m_0 \gamma^3 a_k^N v_k$$

and (7.86) yields

$$F_i^N = \frac{\lambda}{c^2} v^2 v_i + m_0 \gamma a_i^N = \frac{v^2}{c^2} F_i^N + m_0 \gamma a_i^N,$$

or

$$F_i^N = m_0 \gamma^3 a_i^N,$$

hence

$$\frac{|\mathbf{F}^N|}{|\mathbf{a}^N|} = m_0 \gamma^3 = m_l,$$ (7.90)

called *longitudinal mass*.

This result is confusing, since it shows that the mass in not uniquely determined, but depends on the values of the angles between force and velocity. This conclusion is not confirmed by experiments, even if by *relativistic mass* one usually understands the transverse mass. Even Einstein, in 1949, noticed: "*It is not good to introduce the concept of the mass* $M = m/\sqrt{1 - v^2/c^2}$ (Einstein's notation) *of a moving body for which no clear definition can be given. It is better to introduce no other mass concept than the rest mass. Instead of introducing M it is better to mention the expression for the momentum and energy of a body in motion"*. (The concepts of *longitudinal* and *transverse* mass of the electron were introduced by Lorentz, in 1904, in his paper *"Electromagnetic Phenomena in a System Moving with Any Velocity Less than That of Light"*, in *Proceedings of the Royal Academy of Amsterdam* 6 (1904): 809.)

We should mention that one can define a notion of relativistic mass as the ratio between the moduli of momentum and velocity,

$$m = m_0 \gamma = \frac{|\mathbf{p}|}{|\mathbf{v}|}. \tag{7.91}$$

The so-called "transverse mass" is traditionally named *relativistic mass*, and the experiments validate relation (7.91); still, in our further discussions we shall avoid the use of this concept. Unless otherwise specified, by *mass* we shall always mean the *invariant mass* of the body or particle.

7.5.3 Relationship Between Mass and Energy

According to (7.79), each body of mass m_0 possesses $m_0 c^2$ of "rest energy", which potentially is available for conversion into other forms of energy. Such a conversion occurs in ordinary chemical reactions, but much larger conversions occur in nuclear reactions. If in a process appears a variation Δm of the mass, this will be accompanied by a variation of energy given by

$$\Delta E = c^2 \Delta m. \tag{7.92}$$

Einstein's *mass–energy relation* was verified not only by the atomic bomb, but also by more recent developments of particle physics. The nucleons, for example, are formed of subatomic particles known as *quarks*, which are massive, bound together by massless *gluons*. The contribution of the valence quarks to the nucleon mass is however very small, most of the mass coming from the large amount of energy associated with the strong nuclear force.

In 1948, Einstein was explaining the equivalence of mass and energy as follows: *"It followed from the special theory of relativity that mass and energy are both but different manifestations of the same thing – a somewhat unfamiliar conception for the average mind. Furthermore, the equation $E = mc^2$, in which energy is put equal to mass, multiplied by the square of the velocity of light, showed that very small amounts of mass may be converted into a very large amount of energy and vice versa. The* *mass and energy were in fact equivalent, according to the formula mentioned above. This was demonstrated by Cockcroft and Walton in 1932, experimentally."*

7.6 Relativistic Dynamics in Four-Dimensional Approach

Using the variational formalism and tensor calculus, we shall resume some definitions and formulas obtained in the previous section. First, let us re-write the principle of Maupertuis,

$$\delta S = -m_0 c\, \delta \int_a^b ds = 0. \tag{7.93}$$

Suppose that $x_\mu = x_\mu(\lambda)$ are the parametric equations of the integration path (world line) between $\lambda = a$ and $\lambda = b$. Since $ds = (dx_\mu dx^\mu)^{1/2}$ and recalling (7.65), we find

$$\delta S = -m_0 c \int_a^b \frac{dx_\mu\, \delta(dx^\mu)}{\sqrt{dx_\nu\, dx^\nu}} = -m_0 c \int_a^b u_\mu d(\delta x^\mu),$$

or, integrating by parts,

$$\delta S = -m_0 c\, u_\mu\, \delta x^\mu\big|_a^b + m_0 c \int_a^b a_\mu \delta x^\mu\, ds, \tag{7.94}$$

where $a_\mu = du_\mu/ds$ is the acceleration four-vector.

If the initial and final events a and b are fixed, then the variations of the trajectory at the initial and final points vanish, $\delta x^\mu|_a = \delta x^\mu|_b = 0$. Thus, the postulate $\delta S = 0$, for δx^μ arbitrary, gives $a_\mu = 0$, implying $u_\mu = $ const., which is obvious for a free particle. To find δS as a function of coordinates, we suppose that only one end point is fixed, say a, ($\delta x^\mu|_a = 0$), while the point b describes the world line $\delta x^\mu|_b = \delta x^\mu$. Taking into consideration only those trajectories which satisfy the equations of motion ($a_\mu = 0$), we have from (7.94):

$$\delta S = -m_0\, c\, u_\mu \delta x^\mu. \tag{7.95}$$

By definition, the four-vector

$$p_\mu = -\frac{\partial S}{\partial x^\mu} = m_0\, c\, u_\mu. \tag{7.96}$$

is the covariant *momentum four-vector* of the free particle. Its components are

$$p_0 = m_0 c u_0 = m_0 c \gamma = \frac{E}{c},$$

$$p_i = -p^i = -m_0 \gamma \, v^i, \qquad (7.97)$$

while the components of the contravariant vector are

$$p^0 = g^{00} p_0 = p_0 = \frac{E}{c},$$

$$p^i = m_0 \gamma \, v^i. \qquad (7.98)$$

Resuming, the contravariant and covariant components of the momentum four-vector are

$$p^\mu = \left(\frac{E}{c}, \ \mathbf{p} \right),$$

$$p_\mu = \left(\frac{E}{c}, -\mathbf{p} \right). \qquad (7.99)$$

Recall from non-relativistic analytical mechanics that $\frac{\partial S}{\partial x}$, $\frac{\partial S}{\partial y}$, and $\frac{\partial S}{\partial z}$ are the components of the momentum three-vector, while $-\frac{\partial S}{\partial t}$ is the energy E of the particle. Taking into account the minus sign in (7.96), we find that the space components of the contravariant four-momentum are associated with the three-dimensional momentum. Thus, the rule is to associate the contravariant components of a four-vector to the corresponding three-dimensional vectors, with the "correct" positive sign.

As one observes, the time component of the momentum four-vector is connected to the energy, while the spatial components represent the three-dimensional relativistic momentum. For this reason, p^μ is called *energy-momentum four-vector*. Among other things, the significance of its components shows that the conservation of both energy and momentum can be simultaneously expressed by the law of conservation of p^μ.

When passing from one inertial frame to another, in relative motion with velocity $\mathbf{V}$ along the x-axis, the components of the energy-momentum four-vector transform according to (7.37), i.e.

$$p'^0 = \Gamma \left(p^0 - \frac{V}{c} p^1 \right),$$

$$p'^1 = \Gamma \left(-\frac{V}{c} p^0 + p^1 \right),$$

$$p'^2 = p^2,$$

$$p'^3 = p^3,$$

or

$$E' = \Gamma(E - V p_x),$$
$$p'_x = \Gamma\left(-\frac{V}{c^2}E + p_x\right),$$
$$p'_y = p_y, \tag{7.100}$$
$$p'_z = p_z.$$

In view of (7.68), we have on the one hand

$$p^\mu p_\mu = (m_0 c u^\mu)(m_0 c u_\mu) = m_0^2 c^2, \tag{7.101}$$

and on the other

$$p^\mu p_\mu = \frac{E^2}{c^2} - |\mathbf{p}|^2.$$

The last two relations give

$$\frac{E^2}{c^2} = |\mathbf{p}|^2 + m_0^2 c^2,$$

which we have already derived (see (7.83)).

7.6.1 Hamilton–Jacobi Equation

By virtue of (7.96), we can write (7.101) as

$$\frac{\partial S}{\partial x^\mu}\frac{\partial S}{\partial x_\mu} = g^{\mu\nu}\frac{\partial S}{\partial x^\mu}\frac{\partial S}{\partial x^\nu} = m_0^2 c^2, \tag{7.102}$$

or

$$\left(\frac{\partial S}{\partial x}\right)^2 + \left(\frac{\partial S}{\partial y}\right)^2 + \left(\frac{\partial S}{\partial z}\right)^2 - \frac{1}{c^2}\left(\frac{\partial S}{\partial t}\right)^2 + m_0^2 c^2 = 0. \tag{7.103}$$

This is the *Hamilton–Jacobi equation* in relativistic mechanics. To write it in a form leading to the classical equation in the limit $(v/c) \to 0$, we remember that the relativistic energy contains the term $m_0 c^2$, which is missing in classical mechanics. Then, the time components of (7.96) suggests to take $S' = S - m_0 c^2 t$ instead of S, and (7.103) becomes

$$\frac{\partial S'}{\partial t} + \frac{1}{2m_0}\left[\left(\frac{\partial S'}{\partial x}\right)^2 + \left(\frac{\partial S'}{\partial y}\right)^2 + \left(\frac{\partial S'}{\partial z}\right)^2\right] - \frac{1}{2m_0 c^2}\left(\frac{\partial S'}{\partial t}\right)^2 = 0. \tag{7.104}$$

In the limit $(v/c) \rightarrow 0$, the last term vanishes and (7.104) becomes the classical Hamilton–Jacobi equation.

7.6.2 Force Four-Vector

Using the definition of the momentum four-vector (7.96), we define the *force four-vector*, also called the *force-power four-vector*, f^μ, by

$$f^\mu = \frac{dp^\mu}{ds} = m_0 c a^\mu. \tag{7.105}$$

(The definition can also use derivative with respect to proper time, instead: $\overline{f}^\mu = dp^\mu/d\tau = m_0 c \, du^\mu/d\tau = m_0 d\overline{u}^\mu/d\tau = m_0 \overline{a}^\mu$). The time and space components of f^μ are

$$f^0 = f_0 = m_0 c \, a^0 = m_0 \gamma \frac{d\gamma}{dt} = \frac{\gamma}{c^2} \frac{dE}{dt},$$

$$f^i = -f_i = \frac{\gamma}{c} \frac{dp^i}{dt} = \frac{\gamma}{c} \left(F^N \right)^i. \tag{7.106}$$

Multiplying (7.106) by u_μ and using (7.71), we find

$$f^\mu u_\mu = f^0 u_0 + f^i u_i = 0,$$

which yields

$$f^0 = -\frac{f^i u_i}{u_0} = -\frac{\gamma}{c^2} v_i \frac{dp^i}{dt} = \frac{\gamma}{c^2} v^i \frac{dp^i}{dt}. \tag{7.107}$$

Comparing (7.107) with (7.106)$_1$, we regain (7.84). It is worthwhile to mention that in our representation the product $v_i v^i$ is not an invariant. In this respect, one also observes that the components F_i^N of the Newtonian force *do not represent* the spatial components of a four-vector.

7.6.3 Angular Momentum Four-Tensor

In Newtonian mechanics, in the case of an isolated system not only the energy and momentum are conserved, but also the angular momentum $\mathbf{L} = \sum \mathbf{r} \times \mathbf{p}$, where the summation refers to all the particles of the system. The conservation of angular momentum is a consequence of the invariance of the Lagrangian of an isolated system with respect to rotations (when all the particles of the system perform a rotation of angle θ, about a certain axis, i.e. the system rotates as a whole). This invariance is an expression of the isotropy of space.

To transpose this property in Minkowski space, let x^μ be the position four-vector associated with a particle. By an infinitesimal rotation we pass to the coordinates x'^μ, related to the original coordinates x^μ by

$$x'^\mu = x^\mu + x_\nu \delta\omega^{\mu\nu}, \tag{7.108}$$

where $\delta\omega^{\mu\nu}$ are the components of an infinitesimal four-tensor. Since under the rotation (7.108) the norm of the position four-vector x^μ must remain invariant, i.e.

$$x'^\mu x'_\mu = x^\mu x_\mu,$$

it then follows that

$$(x^\mu + x_\nu \delta\omega^{\mu\nu})(x_\mu + x^\lambda \delta\omega_{\mu\lambda}) = x^\mu x_\mu,$$

or, if we neglect the terms of second order in $\delta\omega_{\mu\nu}$,

$$x^\mu x^\nu \delta\omega_{\mu\nu} = 0.$$

Since this relation must be satisfied for arbitrary x^μ, and because the quantity $x^\mu x^\nu$ – from the algebraic point of view – is a symmetric second-order tensor, the infinitesimal four-tensor $\delta\omega_{\mu\nu}$ must be antisymmetric,

$$\delta\omega_{\mu\nu} = -\delta\omega_{\nu\mu}.$$

Indeed, if $S^{ij} = S^{ji}$ is a symmetric second-order tensor, and $A_{ij} = -A_{ji}$ is an antisymmetric second-order tensor (in the same pair of indices, $i\ j$), then

$$S^{ij} A_{ij} = \frac{1}{2}\left(S^{ij} A_{ij} + S^{ij} A_{ij}\right) = \frac{1}{2}\left(S^{ij} A_{ij} - S^{ji} A_{ji}\right)$$
$$= \frac{1}{2}\left(S^{ij} A_{ij} - S^{ij} A_{ij}\right) = 0.$$

According to (7.95), the variation of the action under the infinitesimal transformation (7.108) is

$$\delta S = \delta\omega_{\mu\nu} \sum p^\mu x^\nu,$$

or, if we recall that $\delta\omega_{\mu\nu}$ is antisymmetric and use (B.21),

$$\delta S = \frac{1}{2}\delta\omega_{\mu\nu} \sum (p^\mu x^\nu - p^\nu x^\mu).$$

The summation symbol refers to all the particles of the system. Since the system is isolated, the rotation parameters $\delta\omega_{\mu\nu}$ are cyclic coordinates (do not appear in the Lagrangian), implying that the generalized momenta associated with these parameters, defined as

$$\tilde{p}^{\nu\mu} = \frac{\partial S}{\partial\omega_{\mu\nu}} = \frac{1}{2}\sum(p^{\mu}x^{\nu} - p^{\nu}x^{\mu}),$$

are conserved. Consequently, the components of the antisymmetric second-order tensor

$$L^{\mu\nu} = \sum(x^{\mu}p^{\nu} - x^{\nu}p^{\mu}), \tag{7.109}$$

called *angular momentum four-tensor*, are conserved. Its six independent components are

$$L^{ik} = \sum(x^{i}p^{k} - x^{k}p^{i}) = \epsilon^{ikm}L_{m}, \quad (L^{12} = L_{3}, \text{ etc.}),$$

$$L^{i0} = \sum\left(x^{i}\frac{E}{c} - ctp^{i}\right), \quad i,k,m = 1,2,3, \tag{7.110}$$

where L_{m} are the components of the three-dimensional pseudovector associated with the antisymmetric tensor L^{ik}. Since the system of particles is isolated, the angular momentum and the energy of the system are conserved. Dividing $(7.110)_2$, written in vector form, by $\sum E$ (which is also conserved), one obtains

$$\frac{\sum E\mathbf{r}}{\sum E} - \frac{c^{2}\sum\mathbf{p}}{\sum E}t = \text{const.} \tag{7.111}$$

This shows that the point described by the radius-vector

$$\mathbf{R} = \frac{\sum E\mathbf{r}}{\sum E} \tag{7.112}$$

moves uniformly with velocity

$$\mathbf{v} = \frac{c^{2}\sum\mathbf{p}}{\sum E}. \tag{7.113}$$

Relation (7.111) expresses the *relativistic law of motion* of the centre of mass of a closed system of particles. Formula (7.112) gives the relativistic definition of the radius-vector of the centre of mass, while (7.113) represents its velocity.

7.7 Some Applications of Relativistic Mechanics

7.7.1 Collision Between Two Particles

7.7.1.1 Laboratory Frame

Consider an elastic collision between two particles of masses m_1, m_2, with the energies E_1, E_2 before collision, and E'_1, E'_2 after collision. If one of the particles, say particle 2, is in the proper frame, its momentum before collision is $\mathbf{p}_2 = 0$. When one of the colliding particles is at rest, we say that we study the motion in the *laboratory frame*, denoted by L. The four-momentum conservation law is written as

$$p_1^\mu + p_2^\mu = p_1'^\mu + p_2'^\mu, \tag{7.114}$$

where

$$p_1^\mu = (E_1/c, \mathbf{p}_1), \quad p_2^\mu = (m_2 c, 0),$$
$$p_1'^\mu = (E'_1/c, \mathbf{p}'_1), \quad p_2'^\mu = (E'_2/c, \mathbf{p}'_2), \tag{7.115}$$

Separating the space and time parts, we have

$$\mathbf{p}_1 = \mathbf{p}'_1 + \mathbf{p}'_2,$$
$$E_1 + E_2 = E'_1 + E'_2, \quad \text{with} \quad E_2 = m_2 c^2. \tag{7.116}$$

Let us project now (7.116) on two directions: one along $\mathbf{p}_1$, and the other orthogonal to $\mathbf{p}_1$ (see Fig. 7.8). Denoting $|\mathbf{p}_1| = p_1$, $|\mathbf{p}'_1| = p'_1$, and $|\mathbf{p}'_2| = p'_2$, the result is

$$p'_1 \cos\theta + p'_2 \cos\varphi = p_1,$$
$$p'_1 \sin\theta - p'_2 \sin\varphi = 0. \tag{7.117}$$

Fig. 7.8 Geometry of the relativistic collision between two particles in the laboratory frame.

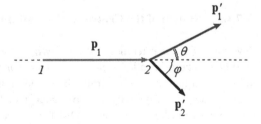

The angles θ and φ, between the direction of the incident particle and the directions of the particles after collision, are called scattering angles. We thus note that there are four unknown quantities, p_1', p_2', θ, φ, which characterize the two particles after collision.

Let us determine, for example, the scattering angles of the two particles. To find θ, we re-write (7.114) as

$$p_1^\mu + p_2^\mu - p_1'^\mu = p_2'^\mu \tag{7.118}$$

and square it, recalling that $p_1^\mu p_{1\mu} = p_1'^\mu p_{1\mu}' = m_1^2 c^2$ and $p_2^\mu p_{2\mu} = p_2'^\mu p_{2\mu}' = m_2^2 c^2$. We thus obtain

$$m_1^2 c^4 + E_1 m_2 c^2 - E_1' m_2 c^2 - E_1 E_1' + c^2 \mathbf{p}_1 \cdot \mathbf{p}_1' = 0,$$

leading to the relation between the scattering angle θ and the energies of the particles between and after collision:

$$\cos\theta = \frac{E_1'(E_1 + m_2 c^2) - E_1 m_2 c^2 - m_1^2 c^4}{\sqrt{E_1^2 - m_1^2 c^4}\sqrt{E_1'^2 - m_1^2 c^4}}. \tag{7.119}$$

Similar calculations give

$$\cos\varphi = \frac{(E_1 + m_2 c^2)(E_2' - m_2 c^2)}{\sqrt{E_1^2 - m_1^2 c^4}\sqrt{E_2'^2 - m_2^2 c^4}}. \tag{7.120}$$

If the incident particle is more massive than the target particle, i.e. $m_1 > m_2$, the angle θ under which the incident particle is scattered is bounded according to the relation

$$\sin\theta_{\text{max}} = \frac{m_2}{m_1}. \tag{7.121}$$

This coincides with the classical result, and we leave its proof to the reader (see Problem 6). Customarily, the laboratory frame is used in the analysis of scatterings in which the target particle is more massive than the incident one.

7.7.1.2 System of the Centre of Momentum (COM)

By definition, the reference system in which the total momentum of the interacting particles is zero is called *centre of momentum frame*,[3] denoted by COM. We shall denote the quantities in this frame by the subindex 0. Thus, for a two-particle collision in the COM-frame, $\mathbf{p}_{01} = -\mathbf{p}_{02} \equiv \mathbf{p}_0$ (see Fig. 7.9). The law of conservation of momentum implies that in the COM-frame the momenta of the two particles after

---

[3]This reference system is also called in literature *system of the centre of inertia* or the *"C-system"*.

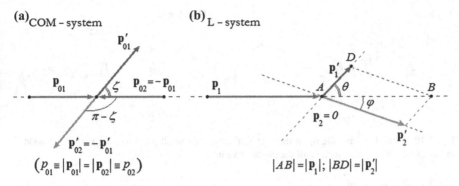

Fig. 7.9 Geometry of the relativistic collision between two particles in the centre of momentum (COM) and laboratory (L) frames.

the collision remain equal and in opposite directions, and the law of conservation of energy ensures that the absolute values of the momenta after collision are the same as before.

The velocity $\mathbf{V}$ of the COM-system with respect to the L-system can be easily obtained using the relativistic relation between the energy, momentum, and velocity of a free particle (7.82). As the particle of mass m_2 is at rest in the L-frame, the total energy of the system of two colliding particles is $E = E_1 + E_2 = E_1 + m_2 c^2$, and the total momentum is $\mathbf{p} = \mathbf{p}_1 + \mathbf{p}_2 = \mathbf{p}_1$, such that relation (7.82) leads to

$$V = \frac{\mathbf{p}}{E} c^2 = \frac{\mathbf{p}_1 c^2}{E_1 + m_2 c^2}. \tag{7.122}$$

Let us denote by ζ the scattering angle in the COM-frame (see Fig. 7.9a), that is the angle by which the momentum vectors $\mathbf{p}_{01}$ and $\mathbf{p}_{02}$ are rotated compared to the direction on which the particles moved before the collision.

When the angle ζ goes through all the possible values, from 0 to 2π, the tip of the vector $\mathbf{p}'_1$ denoted by D in Fig. 7.9b describes an ellipse (see Fig. 7.10), whose major axis is on the direction AB (the direction of the incident particle in the L-frame). The point B is obviously fix due to momentum conservation, and it is located on the ellipse, while the point A (the origin of the vector $\mathbf{p}_1$) can be

i) outside the ellipse, if $m_1 > m_2$ (Fig. 7.10a);
ii) inside the ellipse, if $m_1 < m_2$ (Fig. 7.10b);
iii) on the ellipse, if $m_1 = m_2$ (Fig. 7.11).

To show that the tip of the vector $\mathbf{p}'_1$ describes an ellipse in the L-frame, we make use of the relations (7.100), but for the inverse transformation, in this case from the COM-frame to the L-frame, written for the energies and momenta after collision:

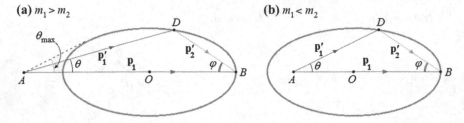

(a) $m_1 > m_2$ **(b)** $m_1 < m_2$

Fig. 7.10 In the L-frame, the tip of the vector $\mathbf{p}_1'$ describes an ellipse when the scattering angle ζ in the COM-frame goes through all the possible values.

Fig. 7.11 The minimum angle between the directions of two particles of equal masses, after the collision, in the L-frame.

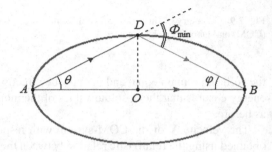

$$E = \Gamma\left(E_0 + V p_{0x}\right),$$

$$p_x = \Gamma\left(p_{0x} + \frac{V}{c^2}E_0\right),$$

$$p_y = p_{0y},$$

$$p_z = p_{0z}, \tag{7.123}$$

where $\Gamma = \left(1 - V^2/c^2\right)^{-1/2}$, with V given by (7.122).
For the particle of mass m_1, the relations $(7.123)_{2,3}$ read:

$$p_{1x}' = \Gamma\left(p_{0x} + \frac{V}{c^2}E_0\right) = \Gamma\left(p_0 \cos\zeta + \frac{V}{c^2}E_0\right),$$

$$p_{1y}' = p_{0y} = p_0 \sin\zeta = p_0 \sin\zeta, \tag{7.124}$$

which, upon elimination of the angle ζ, lead to

$$\left(p_{1x}'\sqrt{1 - \frac{V^2}{c^2}} - \frac{V}{c^2}E_0\right)^2 + p_{1y}'^2 = p_0^2,$$

that is

$$\left(\frac{p'_{1x} - \frac{V}{c^2}\Gamma E_0}{p_0 \Gamma}\right)^2 + \left(\frac{p'_{1y}}{p_0}\right)^2 = 1.$$

This is the equation of an ellipse, with the semi-major axis $p_0\Gamma$ and the semi-minor axis p_0, whose centre O (see Fig. 7.10b) is displaced by $VE_0\Gamma/c^2$ with respect to the point A (which is described by the condition $\mathbf{p}'_1 = 0$).

Since the speed of the particle of mass m_2 with respect to the COM-frame is V, it follows that

$$p_0 \equiv p_{01} = p_{02} = \frac{m_2 V}{\sqrt{1 - \frac{V^2}{c^2}}} = m_2 V\Gamma,$$

such that the semi-major and semi-minor axes can be expressed in terms of the momentum p_1 and energy E_1 of the incident particle with mass m_1 in the L-frame:

$$p_0 = \frac{m_2 c p_1}{\sqrt{m_1^2 c^2 + m_2^2 c^2 + 2m_2 E_1}},$$

$$p_0 \Gamma = m_2 V \Gamma^2 = \frac{m_2 V}{1 - \frac{V^2}{c^2}} = \frac{m_2 p_1 (E_1 + m_2 c^2)}{m_1^2 c^2 + m_2^2 c^2 + 2m_2 E_1}. \tag{7.125}$$

If $\theta = 0$, then $\mathbf{p}'_1$ coincides with $\mathbf{p}_1$, and the segment AB is equal to $p_1 = |\mathbf{p}_1|$. Comparing p_1 with the major axis of the ellipse, it follows that the point A will be located outside the ellipse if $m_1 > m_2$ (see Fig. 7.10a), or inside, if $m_1 < m_2$ (see Fig. 7.10b). As can be seen on Fig. 7.10a, when $m_1 > m_2$, the angle θ cannot be larger than a certain maximum value (corresponding to the vector $\mathbf{p}'_1$ being tangent to the ellipse), which is given by (7.121).

If the masses of the colliding particles are equal, than the point A is on the ellipse, and the angle between the directions of the particles after the collision cannot go below a limit value. As it can be observed on Fig. 7.11, the minimum value of the above-mentioned angle corresponds to the situation when the point D is located at the end of the minor axis and, using (7.125) with $m_1 = m_2 = m$, we find

$$\tan \frac{\Phi_{min}}{2} = \frac{|OD|}{|OA|} = \frac{p_0}{p_0\Gamma} = \sqrt{\frac{2mc^2}{E_1 + mc^2}},$$

leading to the minimum angle between the particles after collision in L-frame:

$$\Phi_{min} = \arccos\left(\frac{E_1 - mc^2}{E_1 + 3mc^2}\right). \tag{7.126}$$

The scattering angle ζ in the COM-frame determines completely the scattering in any frame of reference (in particular, in the COM- and L-frames), being the only undetermined parameter after the application of energy and momentum conservation laws. Let us find the final energies of the particles in the L-frame in terms of the parameter ζ. To this end, we shall use relation (7.118). Squaring (7.118) and taking into account that the scalar product of two four-vectors is an invariant, we shall express the resulting terms conveniently in the two frames, COM and L. Thus, we find

$$m_1^2 c^2 + p_{1\mu} p_2^\mu - p_{2\mu} p_1^{\prime\mu} - p_{1\mu} p_1^{\prime\mu} = 0. \tag{7.127}$$

We express the first two scalar products in the L-frame, and the third in the COM-frame. Using (7.115), we can write

$$p_{1\mu} p_2^\mu = m_2 E_1,$$
$$p_{2\mu} p_1^{\prime\mu} = m_2 E_1'. \tag{7.128}$$

The four-momenta of the particles before and after collision, in the COM-frame, are

$$p_{01}^\mu = (E_{01}/c, \mathbf{p}_{01}), \quad p_{02}^\mu = (E_{02}/c, \mathbf{p}_{02}),$$
$$p_{01}^{\prime\mu} = (E_{01}'/c, \mathbf{p}_{01}'), \quad p_{02}^{\prime\mu} = (E_{02}'/c, \mathbf{p}_{02}'), \tag{7.129}$$

such that

$$p_{1\mu} p_1^{\prime\mu} \left(= p_{01\mu} p_{01}^{\prime\mu} \right) = \frac{E_{01} E_{01}'}{c^2} - \mathbf{p}_{01} \cdot \mathbf{p}_{01}' = E_{01}^2 - p_0^2 \cos\zeta$$
$$= p_0^2 (1 - \cos\zeta) + m_1^2 c^2, \tag{7.130}$$

where we took into account the fact that the energies of the particles before and after collision are the same in the COM-frame, i.e. $E_{01}' = E_{01}$. Plugging (7.128) and (7.130) into (7.127), we obtain

$$E_1' = E_1 - \frac{p_0^2}{m_2} (1 - \cos\zeta). \tag{7.131}$$

Now we have to express p_0^2 in terms of quantities which characterize the collision in the L-frame. To this end, we make use of the invariance of the scalar product of four-vectors and equate the value of $p_{1\mu} p_2^\mu$ in the L-frame (given by the first relation in (7.128)) with its value in the COM-frame, which is (see (7.129))

$$p_{01\mu} p_{02}^\mu = E_{01} E_{02}/c^2 - \mathbf{p}_{01} \cdot \mathbf{p}_{02},$$

obtaining

$$m_2 E_1 = \frac{E_{01} E_{02}}{c^2} - \mathbf{p}_{01} \cdot \mathbf{p}_{02}.$$

Taking into account that $\mathbf{p}_{02} = -\mathbf{p}_{01} (= -\mathbf{p}_0)$ and

$$E_{01} = \sqrt{p_0^2 c^2 + m_1^2 c^4}, \quad E_{02} = \sqrt{p_0^2 c^2 + m_2^2 c^4},$$

it follows that

$$p_0^2 = \frac{m_2^2 (E_1^2 - m_1^2 c^4)}{m_1^2 c^2 + m_2^2 c^2 + 2 m_2 E_1}. \tag{7.132}$$

Introducing (7.132) into (7.131), we finally obtain

$$E_1' = E_1 - \frac{m_2 (E_1^2 - m_1^2 c^4)}{m_1^2 c^2 + m_2^2 c^2 + 2 m_2 E_1} (1 - \cos \zeta). \tag{7.133}$$

The final energy of the second particle is found from the law of conservation of energy written in the L-frame: $E_1 + m_2 c^2 = E_1' + E_2'$, and it reads

$$E_2' = m_2 c^2 + \frac{m_2 (E_1^2 - m_1^2 c^4)}{m_1^2 c^2 + m_2^2 c^2 + 2 m_2 E_1} (1 - \cos \zeta). \tag{7.134}$$

The second term in the relations (7.133) and (7.134) represents the energy lost by the particle of mass m_1 and transferred to the particle of mass m_2. Obviously, the maximum transferred energy is obtained for $\zeta = \pi$ and it is given by the expression

$$E_{2max}' - m_2 c^2 = E_1 - E_{1min}' = \frac{2 m_2 (E_1^2 - m_1^2 c^4)}{m_1^2 c^2 + m_2^2 c^2 + 2 m_2 E_1}. \tag{7.135}$$

From here we can find the ratio between the minimum kinetic energy of the incident particle after the collision and its initial kinetic energy:

$$\frac{E_{1min}' - m_1 c^2}{E_1 - m_1 c^2} = \frac{(m_1 - m_2)^2 c^2}{m_1^2 c^2 + m_2^2 c^2 + 2 m_2 E_1}. \tag{7.136}$$

According to non-relativistic mechanics, if $m_2 \gg m_1$ (the mass of the particle at rest is large compared to the mass of the incident particle), then the lighter particle can transfer only a negligible amount of energy to the very massive one. As one can see from (7.135), this is not valid in the relativistic case; indeed, for sufficiently large energies E_1, the fraction of the transferred energy can approach unity. For this it is not enough that the velocity of the lighter particle be close to the speed of light in vacuum, c, but it is necessary that $E_1 \sim m_2 c^2$, in other words the light particle has to have a total energy comparable with the rest energy of the massive particle.

In the opposite situation, when $m_2 \ll m_1$ (a heavy particle scatters on a light particle at rest), the non-relativistic result is that the energy transfer is again negligible. In this case, the relativistic calculation shows that the fraction of the energy transferred to the other particle is significant only if $E_1 \sim m_1^2 c^2 / m_2$, meaning that the total energy of the incident particle has to be large compared to its rest energy (i.e. ultra-relativistic). Indeed, in the limit of large energies E_1, the ratio (7.136) tends to zero, and E'_{1min} tends to the constant value $E'_{1min} = \frac{(m_1^2+m_2^2)c^2}{2m_2}$ (which can be easily shown using the second equality in the relation (7.135)).

7.7.2 Compton Effect

The *Compton effect* is the increase in the wavelength of an X-ray or gamma-ray radiation, when it interacts with matter. The phenomenon was first observed by Arthur Compton (1892–1962) in 1923. This increase in wavelength is caused by the interaction of the radiation with the weakly bound electrons in matter, where the scattering takes place. The Compton effect illustrates one of the fundamental aspects of the interactions between radiation and matter and displays the quantum nature of electromagnetic radiation. Arthur Compton was awarded the Nobel Prize in Physics in 1927 for this discovery. In the following, we shall present the relativistic kinematics involved in the explanation of the Compton effect, and obtain the formula expressing the *Compton wavelength shift*. The rigorous relativistic quantum study of Compton's effect is performed in the framework of quantum field theory.

Consider a photon colliding with an electron, the latter being at rest, i.e. in the laboratory frame. Denoting by p_e^μ, p_p^μ the momentum four-vectors of the electron and photon before collision, and by $p_e'^\mu$, $p_p'^\mu$ the four-momenta of the particles after collision, we may write the four-momentum conservation as

$$p_e^\mu + p_p^\mu = p_e'^\mu + p_p'^\mu, \quad \mu = 0, 1, 2, 3. \tag{7.137}$$

To express the four-momenta of the particles, we recall that the photon has zero mass, and the energy of the photon is related to its frequency by the famous Planck formula, $E = h\nu$, where h is the Planck constant. The same formula can be written also as $E = \hbar\omega$, where $\hbar = h/2\pi$ is the reduced Planck constant and $\omega = 2\pi\nu$ is the angular frequency of the photon. The absolute value of the photon's momentum is $|\mathbf{p}_p| = \hbar\omega/c$. With these clarifications, we can write the four-momenta appearing in (7.137) as follows:

$$\begin{aligned}
p_e^\mu &= (m_0 c, 0), \\
p_p^\mu &= \left(\frac{1}{c}\hbar\omega, \mathbf{p}_p\right), \\
p_e'^\mu &= \left(m_0\gamma\, c, \mathbf{p}_e'\right), \\
p_p'^\mu &= \left(\frac{1}{c}\hbar\omega', \mathbf{p}_p'\right),
\end{aligned} \tag{7.138}$$

where m_0 is the mass of the electron, ω and ω' are the angular frequencies of the incident and scattered photons, respectively, while $\mathbf{k}$ and $\mathbf{k}'$ are the corresponding wave-vectors. Squaring (7.137) and observing that

$$(p_e)^\mu (p_e)_\mu = (p'_e)^\mu (p'_e)_\mu = m_0^2 c^2,$$
$$(p_p)^\mu (p_p)_\mu = (p'_p)^\mu (p'_p)_\mu = 0, \quad (m_{0p} = 0),$$

we have

$$(p_e)^\mu (p_p)_\mu = (p'_e)^\mu (p'_p)_\mu .$$

On the other hand, the multiplication of (7.137) by $(p'_p)_\mu$ leads to

$$(p'_e)^\mu (p'_p)_\mu = (p_e)^\mu (p'_p)_\mu + (p_p)^\mu (p'_p)_\mu .$$

The last two relations give

$$(p_e)^\mu (p_p)_\mu = (p_e)^\mu (p'_p)_\mu + (p_p)^\mu (p'_p)_\mu , \qquad (7.139)$$

or, using (7.138),

$$\left(\frac{1}{c} m_0 c^2\right)\left(-\frac{1}{c}\hbar\omega\right) = \left(\frac{1}{c} m_0 c^2\right)\left(-\frac{1}{c}\hbar\omega'\right)$$
$$+ p_p p'_p \cos\theta + \left(\frac{1}{c}\hbar\omega\right)\left(-\frac{1}{c}\hbar\omega'\right),$$

where θ is the angle under which the photon is scattered. Grouping the terms in the last relation, we are left with

$$m_0(\omega' - \omega) = \frac{1}{\hbar} p_p p'_p \cos\theta - \frac{\hbar}{c^2}\omega\omega'.$$

From here we obtain the shift in wavelength between the scattered and the incident photons:

$$\Delta\lambda = \lambda' - \lambda = \frac{h}{m_0 c}(1 - \cos\theta) = 2\Lambda \sin^2\frac{\theta}{2}, \qquad (7.140)$$

where

$$\Lambda = \frac{h}{m_0 c}$$

is called *Compton wavelength*. The Compton wavelength of a particle of mass m_0 represents the wavelength of a photon whose energy is equal to the rest energy of that particle. For example, the Compton wavelength for the electron is $\Lambda_e = 2.43 \times 10^{-12}$ m.

Relation (7.140) is the well-known *Compton scattering formula*. It shows that the wavelength shift does not depend on the wavelength of the incident radiation or on the target material. It depends *only* on the scattering angle, and attains its maximum for $\theta = \pi$.

7.7.3 Cherenkov Effect

The Cherenkov effect is the emission of light from a transparent substance like water or glass when a charged particle (e.g. an electron) travels through the material with a speed faster than the phase velocity of light in that material. Denoting by v the velocity of the electrons and by $u = c/n$ the phase velocity of light in the medium, this implies $u < v < c$. Cherenkov radiation had been theoretically predicted by the English scientist Oliver Heaviside (1850–1925) around 1888–1889.

The effect is named after the Russian physicist Pavel Cherenkov (1904–1990), who discovered the phenomenon in 1934 in the laboratory led by Sergey Vavilov (1891–1951) (some authors call it *Vavilov–Cherenkov effect*) at the Lebedev Physical Institute in Moscow. The effect was interpreted theoretically by the Russian physicists Igor Tamm (1895–1971) and Ilya Frank (1908–1990) in 1937. In 1958, Cherenkov, Tamm, and Frank received the Nobel Prize in Physics "for the discovery and interpretation of the effect". The Cherenkov effect has numerous applications, especially in high energy physics.

The effect was discovered by bombarding a transparent medium (water) with gamma-rays, which gives rise to a nice bluish light (see Fig. 7.12). In fact, most Cherenkov radiation is not in the visible, but in the ultraviolet part of the spectrum. In vacuum, a uniformly moving electron does not radiate, but in a dielectric medium the velocity of the electron can exceed the phase velocity of light in that medium $u = c/n$, which means to exceed the propagation of its own field. This field detaches from the electron and spreads into the medium as a specific radiation. The Cherenkov radiation is analogous to the formation of Mach waves in mechanics: all the spherical wavefronts expand at the speed of sound and bunch along the surface of a cone. In the case of a wave source in a fluid like water or air, the cone signifies a shock wave and is referred to as the *Mach cone*, thus named after the Austrian physicist and philosopher

Fig. 7.12 The *bluish* glow of the Cherenkov radiation appearing when gamma-rays pass through water (color figure online).

Fig. 7.13 The "visual" boom accompanying the sonic boom that appears when a supersonic aircraft exceeds the speed of sound. This image clearly exhibits the *Mach cone* in mechanics.

Ernst Mach (1838–1916). The process can be compared to that of a shock wave of sound generated when an airplane exceeds the speed of sound in air. This shock wave has not only an audible effect (the sonic boom), but sometimes a visible one too (the "visual" boom), as can be seen in Fig. 7.13. The visual effect is the result of water condensing and getting trapped between two high-pressure surfaces of air flowing off the aircraft. As one can see in Fig. 7.14, the angle θ between the direction of motion of the electron and the direction of the emission of radiation is given by

$$\cos\theta = \frac{\frac{c}{n}t}{vt} = \frac{u}{v} \leq 1, \tag{7.141}$$

which means $v \geq u$. Therefore in vacuum, where $u = c > v$, the Cherenkov effect is impossible. The electron which produces the Cherenkov radiation is called *superluminal*.

To obtain the radiation condition when taking into account also the recoil electron, we use the four-momentum conservation law in the process of emission of a photon by an electron which moves uniformly in a medium.

If p_e^μ, $p_e'^\mu$ are the four-momenta of the electron before and after emission, and $p_p'^\mu$ is the momentum four-vector of the emitted photon, we have

$$(p_e)^\mu = (p_e')^\mu + (p_p')^\mu. \tag{7.142}$$

We write this relation as $(p_e')^\mu = (p_e)^\mu - (p_p')^\mu$, then square it, and obtain

$$(p_e')^\mu (p_e')_\mu = (p_e)^\mu (p_e)_\mu + (p_p')^\mu (p_p')_\mu - 2(p_e)^\mu (p_p')_\mu.$$

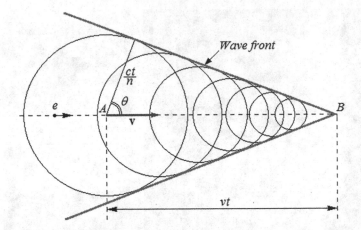

Fig. 7.14 Schematic representation of the Cherenkov effect.

Since $(p'_e)^\mu (p'_e)_\mu = (p_e)^\mu (p_e)_\mu = m_0^2 c^2$ (with m_0 denoting the mass of the electron), and $(p'_p)^\mu (p'_p)_\mu \neq 0$ (the photon moves in a medium, which means that its velocity is smaller than c), we have

$$(p'_p)^\mu (p'_p)_\mu = 2(p_e)^\mu (p'_p)_\mu,$$

or, expanding over $\mu = 0, 1, 2, 3$:

$$(p'_p)^0 (p'_p)_0 + (p'_p)^i (p'_p)_i = 2 \left[(p_e)^0 (p'_p)_0 + (p_e)^i (p'_p)_i \right], \quad i = 1, 2, 3.$$

Recalling that $(p'_p)^\mu = \left(\frac{\hbar \omega}{c}, \hbar \mathbf{k} \right)$, where $\mathbf{k}$ is the wave vector of the photon, the above equation becomes:

$$\frac{\hbar^2 \omega^2}{c^2} - \hbar^2 k^2 = 2 \left(\frac{1}{c^2} E_e \hbar \omega - p_e \hbar k \cos \theta \right),$$

where $p_e = |\mathbf{p}_e|$ and $k = |\mathbf{k}|$. Then

$$\cos \theta = \frac{1}{p_e \hbar k} \left[\frac{E_e}{c^2} \hbar \omega + \frac{1}{2} \hbar^2 k^2 \left(1 - \frac{\omega^2}{c^2 k^2} \right) \right].$$

But, by virtue of (7.83),

$$\frac{1}{p_e} \frac{E_e}{c} = \frac{\sqrt{p_e^2 + m_0^2 c^2}}{\sqrt{p_e^2}} = \sqrt{1 + \frac{m_0^2 c^2}{m_0^2 v^2 \gamma^2}} = \sqrt{1 + \frac{c^2}{v^2} \left(1 - \frac{v^2}{c^2} \right)} = \frac{c}{v},$$

leading to

$$\cos \theta = \frac{\omega}{kv} + \frac{\hbar k}{2p_e} \left(1 - \frac{\omega^2}{k^2 c^2} \right). \tag{7.143}$$

Using the formulas

$$\omega = \frac{ck}{n}, \quad p_e = m_0 \gamma v,$$

we can cast (7.143) into the form

$$\cos \theta = \frac{c}{vn} \left[1 + \frac{\hbar \omega}{2m_0 c^2} (n^2 - 1) \sqrt{1 - \frac{v^2}{c^2}} \right]. \tag{7.144}$$

If the energy of the photon is much smaller than the rest energy of the electron ($\hbar \omega \ll m_0 c^2$), which is true for the visible part of the spectrum, (7.144) becomes $\cos \theta \simeq c/nv$, which is the classical condition (7.141). Formula (7.144) also shows that in vacuum ($n = 1$) we would obtain $\cos \theta = c/v > 1$, and one regains the already known conclusion: the Cherenkov effect does not occur in vacuum.

Let us now show that the Cherenkov effect is intimately connected with the Doppler effect. In this respect, consider the four-vector

$$k^\mu = \frac{1}{\hbar} p_p^\mu = \left(\frac{\omega}{c}, \mathbf{k} \right), \tag{7.145}$$

named the *wave four-vector* associated with the photon. If the photon propagates in a medium with refraction index n, then

$$\mathbf{k} = \frac{2\pi}{\lambda} \mathbf{s} = \frac{2\pi \nu}{u} \mathbf{s} = \frac{\omega}{c} n \mathbf{s}$$

and (7.145) becomes

$$k^\mu = \left(\frac{\omega}{c}, \frac{\omega}{c} n \mathbf{s} \right). \tag{7.146}$$

Suppose that the medium is at rest with respect to the reference frame S', and choose the plane $x'O'y'$ as the plane of propagation of the photon. According to (7.37), we then have

$$k'^0 = \Gamma \left(k^0 - \frac{V}{c} k^1 \right),$$

$$k'^1 = \Gamma \left(-\frac{V}{c} k^0 + k^1 \right),$$

$$k'^2 = k^2,$$

$$k'^3 = k^3.$$

If θ is the angle between $\mathbf{s}$ (the direction of propagation of photon) and the x-axis in S, and θ' is the corresponding angle in S', the first two relations yield

$$\omega' = \Gamma\omega\left(1 - \frac{V}{c}n\cos\theta\right),$$

$$\omega' n \cos\theta' = \Gamma\omega\left(n\cos\theta - \frac{V}{c}\right). \tag{7.147}$$

If the source emitting photons is at rest in S' (proper frame), then $\omega' = \omega_0$ (proper frequency), and $(7.147)_1$ gives

$$\omega = \frac{\omega_0\sqrt{1 - \frac{V^2}{c^2}}}{1 - \frac{V}{c}n\cos\theta}, \tag{7.148}$$

which is the *relativistic Doppler formula for the medium with refraction index n* (see (6.48)).

Inspecting (7.148), we notice that for $n > 1$ (glass, water, etc.), the quantity $\frac{V}{c}n\cos\theta$ can be greater than 1, even if $V < c$, and the denominator can become zero, or negative. Since a change of sign in (7.148) implies, at the most, a change of phase

$$\cos(-\omega t) = \cos\omega t,$$

$$\sin(-\omega t) = -\sin\omega t = \cos\left(\omega t + \frac{\pi}{2}\right),$$

the frequency can always be considered a positive quantity. For this reason, the relativistic Doppler formula (for $n > 1$) reads

$$\omega = \frac{\omega_0\sqrt{1 - \frac{V^2}{c^2}}}{\left|1 - \frac{V}{c}n\cos\theta\right|}. \tag{7.149}$$

On the other hand, dividing side by side the two equations in (7.147), we get

$$n\cos\theta' = \frac{n\cos\theta - \frac{V}{c}}{1 - \frac{V}{c}n\cos\theta}, \tag{7.150}$$

which is the relativistic formula of the *aberration of light* for the considered medium (see (6.51)).

Thus, we have found that the medium does not at all affect the transverse Doppler effect: for $\theta = \pi/2$ we found the same formula as for vacuum (see (6.49)). This proves, once more, that the transverse Doppler effect is due only to the relativistic transformation of time intervals.

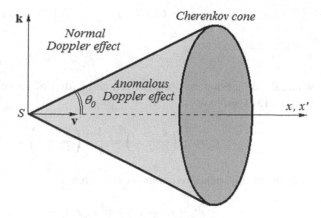

Fig. 7.15 The Cherenkov cone divides the space into the regions of the anomalous and normal Doppler effect.

Suppose, now, that the denominator in (7.149) vanishes, that is

$$1 - \frac{v}{c} n \, \cos\theta_0 = 0,$$

where $V = v$ is the velocity of the electron. Thus, we regained condition (7.141) for the Cherenkov effect. The geometric representation of this condition, for given v and n, is a cone, called *Cherenkov cone*. This cone divides the space into two domains relatively to the Doppler effect. Condition $1 - \frac{v}{c} n \, \cos\theta > 0$ is valid outside the Cherenkov cone, where a *normal Doppler effect* is observed, always found in vacuum, while condition $1 - \frac{v}{c} n \, \cos\theta < 0$ is valid inside the Cherenkov cone and corresponds to an *anomalous Doppler effect* (see Fig. 7.15). One also observes that inside the Cherenkov cone $d\omega/d\theta < 0$, and outside the cone $d\omega/d\theta > 0$.

Among several applications of the Cherenkov effect we mention the well known *Cherenkov counter*. This is a device that identifies particles passing through it by observing a flash of light generated in a manner similar to a sonic boom. Cherenkov counters are superior to some other radiation registrators, like Geiger–Müller counters, due to their *selectivity*: they identify only high energy particles, capable to produce Cherenkov radiation.

7.8 Solved Problems

Problem 1. Deduce the Lorentz boost transformation formula using the invariance property of the space-time interval.

Solution. We assume that the two inertial frames move relative to each other along the x-axis, with the velocity **V**. The transformation we are looking for has to be symmetric and linear, of the form

$$x' = k(x - Vt), \qquad x = k'(x' + Vt'),$$
$$y' = y, \tag{7.151}$$
$$z' = z,$$

where the dimensionless coefficients k and k' depend on velocity only. Extracting t' from (7.151), we have

$$t' = \frac{1}{V}\left(\frac{x}{k'} - x'\right) = \frac{1}{V}\left[\left(\frac{1}{k'} - k\right)x + kVt\right]. \tag{7.152}$$

The interval invariance requirement is written as

$$c^2 t'^2 - x'^2 - y'^2 - z'^2 = c^2 t^2 - x^2 - y^2 - z^2,$$

or, using (7.151) and (7.152),

$$\frac{c^2}{V^2}\left[\left(\frac{1}{k'} - k\right)x + kVt\right]^2 - k^2(x - Vt)^2 = c^2 t^2 - x^2. \tag{7.153}$$

Equating the coefficients of t^2 in (7.153), we obtain

$$k = \pm\frac{1}{\sqrt{1 - \frac{V^2}{c^2}}} = \pm\Gamma. \tag{7.154}$$

The choice of the sign depends on the mutual orientation of the axes x and x'. It is customary to take $+$ if the axes are parallel, and $-$ if they are antiparallel.

The last step is to equate the coefficients of xt (or of x^2) in (7.153). Some very simple algebra leads to

$$k = k' = \Gamma,$$

which completes the proof.

Problem 2. Determine the relativistic, rectilinear motion of a uniformly accelerated particle, knowing that the magnitude w_0 of its acceleration remains constant in the proper frame of reference.

Solution. The acceleration four-vector is defined, according to (7.69) as

$$a^\mu = \frac{du^\mu}{ds} = \frac{\gamma}{c}\frac{du^\mu}{dt}, \quad \text{where} \quad u^\mu = \frac{dx^\mu}{ds} = \frac{\gamma}{c}\frac{dx^\mu}{dt}. \tag{7.155}$$

Note that with this definition, the dimension of the four-acceleration differs from the dimension of the ordinary acceleration w_0 by a factor c^{-2}. If the particle moves along the $Ox \equiv O'x'$-axis, in the proper frame ($v = 0$) we have

$$a^0 = 0,$$
$$a^1 = \frac{w_0}{c^2},$$
$$a^2 = 0,$$
$$a^3 = 0.$$

(7.156)

Since $a^\mu a_\mu = \frac{w_0^2}{c^4} = \text{const.}$, we can write

$$\frac{\gamma^2}{c^2}\left(\frac{d\gamma}{dt}\right)^2 - \frac{\gamma^2}{c^4}\left[\frac{d}{dt}(\gamma v)\right]^2 = \frac{w_0^2}{c^4},$$

where $v = dx^1/dt$. Some simple calculations give

$$\frac{d\gamma}{dt} = \frac{v}{c^2}\gamma^3\frac{dv}{dt},$$
$$\frac{d}{dt}(\gamma v) = \gamma\frac{dv}{dt}\left(1 + \frac{v^2}{c^2}\gamma^2\right) = \gamma^3\frac{dv}{dt},$$
$$\frac{\gamma^6}{c^4}\left(\frac{d\gamma}{dt}\right)^2 = \frac{\gamma^2}{c^4}\left[\frac{d}{dt}(\gamma v)\right]^2 - \frac{\gamma^2}{c^2}\left(\frac{d\gamma}{dt}\right)^2,$$

leading to

$$\frac{d}{dt}(\gamma v) = w_0.$$

(7.157)

Integrating (7.157), we find

$$\frac{v}{\sqrt{1 - \frac{v^2}{c^2}}} = w_0 t + C.$$

Taking as initial conditions $v = 0$ at $t = 0$, one finds $C = 0$. Since $v = dx^1/dt$, we can separate the variables:

$$v = \frac{w_0 t}{\sqrt{1 + \frac{w_0^2 t^2}{c^2}}} = \frac{dx}{dt}.$$

(7.158)

By integrating, with the initial condition $x = 0$ at $t = 0$, we obtain

$$x = \frac{c^2}{w_0}\left(\sqrt{1 + \frac{w_0^2 t^2}{c^2}} - 1\right).$$

(7.159)

If $w_0 t \ll c$, a series expansion of (7.158) and (7.159) gives

$$x = \frac{c^2}{w_0} \left(1 + \frac{1}{2} \frac{w_0^2 t^2}{c^2} + \ldots - 1 \right) \simeq \frac{1}{2} w_0 t^2,$$
$$v = w_0 t, \tag{7.160}$$

which are classical results. If $w_0 t \to \infty$, then $v \to c$.

Problem 3. Determine the world-line of a particle of mass m_0, performing a one-dimensional motion under the action of a constant force F.

Solution. Since the motion is one-dimensional, we may discard the vector symbol in this application, and write

$$F = \frac{dp}{dt} = \frac{d}{dt} \left[m_0 v \left(1 - \frac{v^2}{c^2} \right)^{-1/2} \right]. \tag{7.161}$$

The derivative gives

$$F = m_0 \left(v \frac{d\gamma}{dt} + \gamma \frac{dv}{dt} \right) = m_0 \gamma^3 \frac{dv}{dt},$$

that is

$$\frac{F}{m_0} dt = \frac{dv}{\left(1 - \frac{v^2}{c^2} \right)^{3/2}},$$

and by integrating, with the initial condition $v = 0$ at $t = t_0$, we find

$$\frac{F}{m_0} (t - t_0) = \int_0^v \left(1 - \frac{v^2}{c^2} \right)^{-3/2} dv . \tag{7.162}$$

The integral can be worked out by using the change of variable $v = c \sin \varphi \Rightarrow dv = c \cos \varphi \, d\varphi$, which gives

$$\frac{F}{m_0} (t - t_0) = c \int_0^{\arcsin(v/c)} \frac{d\varphi}{\cos^2 \varphi} = c \tan \varphi \big|_0^{\arcsin(v/c)}$$
$$= c \tan \left[\arcsin \left(\frac{v}{c} \right) \right] = c \frac{\sin \left[\arcsin \left(\frac{v}{c} \right) \right]}{\cos \left[\arcsin \left(\frac{v}{c} \right) \right]} = \frac{v}{\sqrt{1 - \left(\frac{v}{c} \right)^2}},$$

or, by squaring the result,

$$v^2 \left[1 + \frac{(F/m_0)^2}{c^2} (t - t_0)^2 \right] = (F/m_0)^2 (t - t_0)^2,$$

and

$$v = \frac{dx}{dt} = \frac{(F/m_0)\,(t - t_0)}{\sqrt{1 + \frac{(F/m_0)^2}{c^2}(t - t_0)^2}}.$$

This relation can also be written as

$$dx = \frac{c^2}{F/m_0} d\left[\sqrt{1 + \frac{(F/m_0)^2}{c^2}(t - t_0)^2}\right].$$

Taking as initial condition $x = x_0$ at $t = t_0$, one obtains by integration

$$x - x_0 = \frac{c^2}{F/m_0}\sqrt{1 + \frac{(F/m_0)^2}{c^2}(t - t_0)^2},$$

or, finally,

$$\frac{(x - x_0)^2}{\left(\frac{c^2}{F/m_0}\right)^2} - \frac{(t - t_0)^2}{\left(\frac{c}{F/m_0}\right)^2} = 1. \tag{7.163}$$

This formula shows that the world-line of the particle is a *hyperbola*. For this reason, the motion of a particle under the action of a constant force is sometimes called *hyperbolic*. This motion tends asymptotically to a uniform motion with velocity c.

In the non-relativistic limit, using the same initial conditions, relation (7.163) becomes

$$x - x_0 = \frac{1}{2}\frac{F}{m}(t - t_0)^2, \tag{7.164}$$

which is, obviously, an arc of *parabola*.

Problem 4. Using the energy-momentum conservation law, show that a free electron (i.e. an electron in vacuum) cannot emit or absorb a photon.

Solution. We shall prove that the relativistic energy-momentum conservation relation

$$p_e^\mu = p_e'^\mu + p_p'^\mu \tag{7.165}$$

cannot hold in vacuum. Here p_e^μ is the momentum four-vector of the incident particle (electron) before emission (absorption), and $p_e'^\mu$, $p_p'^\mu$ are the four-momenta of the emitted electron and photon, respectively.

Squaring relation (7.165), we have

$$(p_e)^\mu(p_e)_\mu = (p_e')^\mu(p_e')_\mu + (p_p')^\mu(p_p')_\mu + 2(p_e')^\mu(p_p')_\mu. \tag{7.166}$$

Since

$$(p_e)^\mu(p_e)_\mu = (p_e')^\mu(p_e')_\mu = m_0^2 c^2,$$

and

$$(p'_p)^\mu (p'_p)_\mu = 0,$$

it follows that

$$(p'_e)^\mu (p'_p)_\mu = 0. \tag{7.167}$$

We use (7.99) to write:

$$(p'_e)^0 = \frac{E'_e}{c}, \quad |\mathbf{p}'_p| = \frac{E'_p}{c}, \quad (p'_p)^0 = \frac{E'_p}{c},$$

and, with the notation $p_e = |\mathbf{p}_e|$, (7.167) becomes

$$\frac{1}{c} E'_e E'_p - p'_e E'_p \cos\theta = 0.$$

Using now the energy-momentum dispersion relation (7.83) for the recoil electron,

$$\frac{E'_e}{c} = \sqrt{|\mathbf{p}'_e|^2 + m_0^2 c^2},$$

we obtain

$$E'_p \left[p'_e \cos\theta - \sqrt{|\mathbf{p}'_e|^2 + m_0^2 c^2} \right] = 0.$$

Since $E'_p \neq 0$, we must have

$$\cos\theta = \sqrt{1 + \frac{m_0^2 c^2}{|\mathbf{p}'_e|^2}} \geq 1, \tag{7.168}$$

which is absurd.

Problem 5. Study the one-dimensional motion of a particle of mass m_0, under the action of a quasi-elastic force.

Solution. Let $F = -k_0 x$, with $k_0 > 0$, be the quasi-elastic force. According to Newtonian mechanics, the particle would perform a harmonic oscillatory motion of frequency $\omega_0 = \sqrt{k_0/m_0}$, the solution of the equation of motion being

$$x = a \sin\omega_0(t - t_0). \tag{7.169}$$

Since the particle is relativistic, the equation of motion is

$$\frac{d}{dt}(m_0 \gamma v) = -k_0 x, \quad \text{with} \quad \gamma = \left(1 - \frac{v^2}{c^2}\right)^{-1/2}. \tag{7.170}$$

This equation admits the *total energy first integral*

$$m_0 c^2 (\gamma - 1) + \frac{1}{2} m_0 \omega_0^2 x^2 = W_0.$$ (7.171)

To determine W_0, we make use of the initial conditions: at $t = 0$, $x = a$ and $v = 0$. Then $W_0 = \frac{1}{2} m_0 \omega_0^2 a^2$, and the first integral (7.171) leads to

$$v = c \left\{ 1 - \left[1 + \frac{\omega_0^2}{2c^2} (a^2 - x^2) \right]^{-2} \right\}^{1/2} = \frac{dx}{dt}.$$ (7.172)

We note that the velocity of the particle obeys the condition $v \leq c$, while the constant a signifies the amplitude of the periodic motion $(-a \leq x \leq a)$.

To integrate (7.172), it is convenient to make the notations

$$A^2 = a^2 + \frac{2c^2}{\omega_0^2},$$

$$A'^2 = a^2 + \frac{4c^2}{\omega_0^2},$$ (7.173)

and (7.172) becomes

$$\frac{(A^2 - x^2) dx}{\sqrt{(a^2 - x^2)(A'^2 - x^2)}} = c \, dt.$$ (7.174)

Setting

$$k^2 = \frac{a^2}{A'^2} = \left(1 + \frac{4c^2}{\omega_0^2 a^2} \right)^{-1} < 1,$$ (7.175)

the integration of (7.174) gives

$$\left(\frac{A^2}{A'} - A' \right) F(\varphi, k) + A' E(\varphi, k) = c(t - t_0),$$ (7.176)

where $\varphi = \arcsin(x/a)$, while

$$F(\varphi, k) = \int_0^\varphi \frac{d\psi}{\sqrt{1 - k^2 \sin^2 \psi}},$$

$$E(\varphi, k) = \int_0^\varphi \sqrt{1 - k^2 \sin^2 \psi} \, d\psi$$ (7.177)

are the incomplete elliptic integral of the first and second kind, respectively.

If we denote

$$k'^2 = 1 - k^2 = \frac{4c^2}{\omega_0^2 A'^2},$$
(7.178)

and take into account (7.174) and (7.175), relation (7.176) can also be written as

$$\frac{1}{k'}\left[2E\left(\arcsin\frac{x}{a}, k\right) - k'^2 F\left(\arcsin\frac{x}{a}, k\right)\right] = \omega_0(t - t_0).$$
(7.179)

In the limit of small velocities $(v/c \to 0)$, we have $k \to 0, k' \to 1, E(\varphi, 0) = F(\varphi, 0) = \varphi$, and (7.179) reduces to (7.169), as expected.

The period of the classical harmonic motion is $T_0 = 2\pi/\omega_0$. The relativistic motion is also periodic, its period being given by the equation

$$\frac{1}{k'}\left[2E(\varphi + 2\pi, k) - k'^2 F(\varphi + 2\pi, k)\right] = \omega_0(t + T - t_0).$$
(7.180)

But

$$F(\varphi + 2\pi, k) = F(\varphi, k) + F(2\pi, k),$$
$$E(\varphi + 2\pi, k) = E(\varphi, k) + E(2\pi, k),$$

and

$$F(2\pi, k) = 4F(k),$$
$$E(2\pi, k) = 4E(k),$$

where

$$F(k) = F\left(\frac{\pi}{2}, k\right),$$
$$E(k) = E\left(\frac{\pi}{2}, k\right)$$

are complete elliptic integrals of the first and second kind, respectively. Then (7.179) and (7.180) yield

$$\frac{\omega}{\omega_0} = \frac{\pi}{2}\frac{k'}{2\,E(k) - k'^2 F(k)},$$
(7.181)

where $\omega = 2\pi/T$ is the relativistic angular frequency.

Thus, the angular frequency ω is a function of m_0, k_0, and a. If ω_0 and a are chosen in such a way that $\omega_0 a \ll c$, then $k \simeq \frac{\omega_0 a}{2c} \ll 1$, and we can use the series expansions

$$E(k) \simeq \frac{\pi}{2} \left(1 - \frac{1}{4}k^2 \right),$$

$$F(k) \simeq \frac{\pi}{2} \left(1 + \frac{1}{4}k^2 \right),$$

$$k' \simeq 1 - \frac{1}{2}k^2.$$

In this case, formula (7.181) becomes

$$\frac{\omega}{\omega_0} \simeq 1 - \frac{3}{4}k^2 \simeq 1 - \frac{3}{16}\frac{\omega_0^2 a^2}{c^2}. \tag{7.182}$$

Since the particle covers in a period the path $4a$, and its velocity cannot be greater than c, there exists a minimum time for the particle to move on that distance, $T_{min} = 4a/c$, or, equivalently, a *maximum angular frequency*

$$\omega_{max} = \frac{2\pi}{T_{min}} = \frac{\pi c}{2a}. \tag{7.183}$$

7.9 Proposed Problems

1. Consider two Lorentz transformations, Λ_1 and Λ_2, which differ infinitesimally: $\Lambda_1 = e^X$, $\Lambda_2 = e^{X+\delta X}$. Using the formula (7.56) show that, in the first order in δX, the expression

$$I + \delta X + \sum_{n=2}^{\infty} \frac{1}{n!} \underbrace{\left[X, \left[X, \ldots [X, \delta X \,] \, \right] \ldots \right]}_{\substack{n-1 \\ \text{brackets}}},$$

(where $[a, b] = ab - ba$), represents the Lorentz transformation $\Lambda = \Lambda_2 \Lambda_1^{-1}$.
2. In the proper reference frame, S', the angular momentum of a body does not depend on the point with respect to which it is defined. However, in another inertial frame, S, which moves with the velocity V with respect to S', the value of the angular momentum depends on this choice. Find the relation between the angular momenta of the body in S and S', if in both cases the angular momentum is defined with respect to the centre of mass of the body in S'.
3. A particle of mass m_1 and velocity v_1 collides with a particle of mass m_2 which is at rest. The result is a composite particle. Determine the mass m and the velocity v of the resulting composite particle.
4. A particle of mass m_0 moves under the action of the force $\mathbf{F}$. Determine the acceleration $\mathbf{a}$ of the particle in terms of force and velocity.
5. Two particles of masses m_1 and m_2, and energies E_1 and E_2 collide elastically. If before the collision the particle 2 is at rest, find the scattering angles in the

laboratory frame of reference, as functions of the energies E'_1 and E'_2 after collision.

6. Show that for a scattering of two particles, 1 and 2, in the laboratory frame in which the particle 2 is at rest, the scattering angle θ of the incident particle 1 (see Fig. 7.8) is bounded by the relation

$$\sin \theta_{max} = \frac{m_2}{m_1}, \qquad (7.184)$$

if the incident particle is more massive than the target, i.e. $m_1 > m_2$.

7. Consider the decay relation $A \rightarrow B + C$. Calculate the energy W_B of the particle B, if A is at rest in the laboratory reference frame. The masses m_A, m_B, and m_C are supposed to be known.

8. A particle of mass m at rest decays into n fragments of masses m_i, $i = 1, 2, \ldots, n$, such that

$$m - \sum_{i=1}^{n} m_i = \Delta m.$$

Show that the maximum kinetic energy of particle i (with the mass m_i) is given by the relation

$$T_i^{max} = c^2 \Delta m \left(1 - \frac{m_i}{m} - \frac{\Delta m}{2m} \right)$$

and calculate the maximum kinetic energies (in MeV) for each of the products of the following decay processes (we consider that the decaying particles are at rest):

$$\mu^- \rightarrow e^- + \nu_\mu + \bar{\nu}_e,$$
$$n^0 \rightarrow p^+ + e^- + \bar{\nu}_e,$$
$$\Lambda^0 \rightarrow p^+ + \pi^-.$$

The masses of the particles are:
$m_{\mu^-} = 105.658$ MeV/c^2, $m_{e^-} = 0.510$ MeV/c^2, $m_{n^0} = 939.565$ MeV/c^2, $m_{p^+} = 938.272$ MeV/c^2, $m_{\Lambda^0} = 1115.683$ MeV/c^2, $m_{\pi^-} = 139.570$ MeV/c^2.
The neutrinos and antineutrinos are considered massless.

9. A particle A decays in motion into two fragments, B and C. Show that the scattering angle θ of the particle B (see Fig. 7.16) is given by the relation

$$\cos \theta = \frac{2 E_A E_B + c^4 \left(m_C^2 - m_A^2 - m_B^2 \right)}{2 \sqrt{\left(E_A^2 - m_A^2 c^4 \right) \left(E_B^2 - m_B^2 c^4 \right)}},$$

Fig. 7.16 Decay of the
particle A into two
fragments, B and C. The
scattering angle of particle B
is denoted by θ.

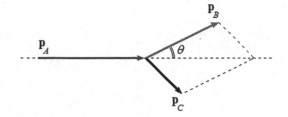

where m_A, m_B, and m_C are the masses of the three particles and E_A and E_B are
the energies of the particles A and B.

10. A plane mirror moves along its normal, in the opposite sense, with a velocity V.
A light ray falls on the mirror at an angle θ. Determine the frequency shift and
the angle of reflection of the reflected ray. (Hint: use the transformation of the
wave four-vector $k^\mu = \left(\frac{\omega}{c}, \mathbf{k}\right)$.)

Chapter 8
Relativistic Formulation of Electrodynamics in Minkowski Space

8.1 Point Charge in Electromagnetic Field

8.1.1 Three-Dimensional Approach

Consider a charged particle of mass m_0 and charge e, moving in the electromagnetic field $\mathbf{E}$, $\mathbf{B}$, defined in terms of the usual electromagnetic potentials V, $\mathbf{A}$. Let us deduce the fundamental quantities characterizing the particle (momentum, energy, etc.), as well as the equation of motion.

In the non-relativistic case, the Lagrangian of the point charge is $L = L_0 + L_{int}$ (see (3.102)), where $L_0 = \frac{1}{2}mv^2$ is the Lagrangian of the free particle, and $L_{int} = -eV + e\mathbf{v} \cdot \mathbf{A}$ is the Lagrangian of interaction between the charge and the field. If one considers the relativistic effect, L_0 will be given by (7.77), and the Lagrangian of the point charge is

$$L = -m_0 c^2 \sqrt{1 - \frac{v^2}{c^2}} - eV + ev_k A_k, \quad k = 1, 2, 3. \tag{8.1}$$

The components p_i of the *generalized momentum*, associated with the generalized coordinates x_i, are

$$p_i = \frac{\partial L}{\partial \dot{q}^i} = \frac{\partial L}{\partial v^i} = m_0 \gamma v_i + e A_i, \quad i = 1, 2, 3, \tag{8.2}$$

or, in vector notation,

$$\mathbf{p} = m_0 \gamma \mathbf{v} + e\mathbf{A} = \mathbf{p}_0 + e\mathbf{A}, \tag{8.3}$$

where $\mathbf{p}_0 = m_0 \gamma \mathbf{v}$ is the momentum of the free particle. The *Hamiltonian* of the particle is

$$H = p_i v_i - L = m_0 \gamma c^2 + eV = E_0 + eV, \tag{8.4}$$

© Springer-Verlag Berlin Heidelberg 2016
M. Chaichian et al., *Electrodynamics*, DOI 10.1007/978-3-642-17381-3_8

where $E_0 = m_0\gamma c^2$ is the mechanical energy of the free particle. Written in terms of generalized coordinates and generalized momenta, using (7.83), the Hamiltonian is

$$H = c\sqrt{(\mathbf{p} - e\mathbf{A})^2 + m_0^2 c^2} + eV \ . \tag{8.5}$$

Replacing here $H = -\partial S/\partial t$ and $\mathbf{p} = \nabla S$, one obtains the Hamilton–Jacobi equation of the point charge:

$$(\nabla S - e\mathbf{A})^2 - \frac{1}{c^2}\left(\frac{\partial S}{\partial t} + eV\right)^2 + m_0^2 c^2 = 0. \tag{8.6}$$

The *equation of motion* is found from the Lagrange equations (see (3.103)):

$$\frac{d}{dt}\left(\frac{\partial L}{\partial v_i}\right) - \frac{\partial L}{\partial x_i} = 0, \tag{8.7}$$

which yield

$$\frac{d}{dt}(m_0\gamma \mathbf{v})_i = -e\frac{\partial V}{\partial x_i} - e\frac{\partial A_i}{\partial t} + ev_k\left(\frac{\partial A_k}{\partial x_i} - \frac{\partial A_i}{\partial x_k}\right),$$

or (see (3.104)):

$$\frac{d}{dt}(m_0\gamma \mathbf{v}) = e\,(\mathbf{E} + \mathbf{v} \times \mathbf{B}), \tag{8.8}$$

which is the equation of motion of the relativistic charged particle.

8.1.2 Covariant Approach

We can also adopt a covariant treatment of the Lagrangian problem. The action obtained by the time integration of the Lagrangian (8.1) reads

$$S = \int_{t_a}^{t_b}\left(-m_0 c^2\sqrt{1 - \frac{v^2}{c^2}} - eV + ev_k A_k\right)dt,$$

$$= -\int_a^b\left(m_0 c^2 d\tau + eA^\mu dx_\mu\right), \tag{8.9}$$

where we defined the *four-vector potential* A^μ of the electromagnetic field as

$$A^\mu = \left(\frac{V}{c}, \mathbf{A}\right). \tag{8.10}$$

Further, we apply the principle of least action:

$$\delta S = -\int_a^b \left[m_0 c^2 \delta \tau + e\delta (A^\mu dx_\mu) \right]$$

$$= -\int_a^b \left(m_0 \frac{dx^\mu \delta dx_\mu}{d\tau} + eA^\mu \delta dx_\mu + e\delta A^\mu dx_\mu \right)$$

$$= -\int_a^b \left(m_0 \bar{u}^\mu d\delta x_\mu + eA^\mu d\delta x_\mu + e\delta A^\mu dx_\mu \right) = 0, \qquad (8.11)$$

where we used the definition (7.65) of the velocity four-vector, $\bar{u}^\mu = \frac{dx^\mu}{d\tau}$. Integrating by parts the first two terms, we find

$$\delta S = \int_a^b \left(m_0 d\bar{u}^\mu \delta x_\mu + e\delta x_\mu dA^\mu - e\delta A^\mu dx_\mu \right)$$

$$- \left(m_0 \bar{u}^\mu + eA^\mu \right) \delta x_\mu |_a^b = 0. \qquad (8.12)$$

The second term vanishes since the ends of the integration domain are fixed, therefore the variations $(\delta x_\mu)_a$ and $(\delta x_\mu)_b$ are zero. Thus, (8.12) becomes

$$\int_a^b \left(m_0 d\bar{u}^\mu \delta x_\mu + e\frac{\partial A^\mu}{\partial x_\nu} \delta x_\mu dx_\nu - e\frac{\partial A^\mu}{\partial x_\nu} \delta x_\nu dx_\mu \right)$$

$$= \int_a^b \left[m_0 \frac{d\bar{u}^\mu}{d\tau} - e\left(\frac{\partial A^\nu}{\partial x_\mu} - \frac{\partial A^\mu}{\partial x_\nu} \right) u_\nu \right] \delta x_\mu d\tau = 0. \qquad (8.13)$$

The arbitrariness of the variations δx_μ implies that the the integrand has to be zero, that is

$$m_0 \frac{d\bar{u}^\mu}{d\tau} = e\left(\frac{\partial A^\nu}{\partial x_\mu} - \frac{\partial A^\mu}{\partial x_\nu} \right) u_\nu. \qquad (8.14)$$

Let us introduce the notation

$$F^{\mu\nu} = \frac{\partial A^\nu}{\partial x_\mu} - \frac{\partial A^\mu}{\partial x_\nu}, \qquad (8.15)$$

to specify the components of a second-order antisymmetric tensor, called the *electromagnetic field tensor*. This tensor plays an essential role in the derivation of the covariant form of the fundamental laws of the electromagnetic field theory. Using (8.15), equation (8.14) becomes

$$\frac{d}{d\tau}(m_0 \bar{u}^\mu) = eF^{\mu\nu} \bar{u}_\nu, \qquad (8.16)$$

or, written in terms of $u^\mu = \frac{dx^\mu}{ds}$,

$$\frac{dp^\mu}{ds} = \frac{d}{ds}(m_0 c u^\mu) = e F^{\mu\nu} u_\nu, \tag{8.17}$$

which is the covariant form of the equation of motion of the point charge, moving in the electromagnetic field $\mathbf{E}$, $\mathbf{B}$.

We shall derive now the generalized momentum four-vector. Customarily, this is done by taking the derivative of the Lagrangian with respect to the time derivative of the generalized coordinate (in our case, x^μ). However, we have not identified the Lagrangian in this formalism and worked only with the action.[1] We can continue in the same manner, by noting the following: we consider the action integral on the generalized trajectory, such that only one of the two end-points is fixed:

$$S(q, \dot{q}, t) = \int_{t_0}^{t} L(q, \dot{q}, t') dt',$$

and take its variation on all *possible* trajectories:

$$\delta S = \sum_{j=1}^{n} \left[\frac{\partial L}{\partial \dot{q}_j} \delta q_j \right]_{t_0}^{t} + \int_{t_0}^{t} \sum_{j=1}^{n} \left[\frac{\partial L}{\partial q_j} - \frac{d}{dt} \left(\frac{\partial L}{\partial \dot{q}_j} \right) \right] \delta q_j \, dt'.$$

The Lagrange equations are satisfied on any of these trajectories and thus the second term vanishes and we arrive at

$$\delta S = \sum_{j=1}^{n} \frac{\partial L}{\partial \dot{q}_j} \delta q_j = \sum_{j=1}^{n} p_j \delta q_j,$$

which yields

$$p_j = \frac{\delta S}{\delta q_j}, \quad j = 1, \ldots, n. \tag{8.18}$$

We apply this method to the action (8.9). Inspecting formula (8.12), we notice that the integral has led to the Euler–Lagrange equations, therefore this term will vanish if we take the variation only on the possible trajectories. The second term survives and we obtain thus the generalized momenta for the relativistic particle in electromagnetic field:

$$P^\mu = \frac{\delta S}{\delta x_\mu} = -m_0 c u^\mu - e A^\mu. \tag{8.19}$$

As $u^\mu u_\mu = 1$, we obtain

---

[1]There exists also the possibility to define a Lorentz-invariant Lagrangian whose integral with respect to an invariant parameter leads to the action of the relativistic particle in the covariant formalism. The procedure is thoroughly and transparently presented in the book of H. Goldstein, *Classical Mechanics* (2nd ed.), Addison-Wesley, 1980.

$$(P^\mu + eA^\mu)(P_\mu + eA_\mu) = m_0^2 c^2$$

and, if we replace above P^μ by $\frac{\partial S}{\partial x_\mu}$, we find the covariant form of the Hamilton–Jacobi equation:

$$\left(\frac{\partial S}{\partial x_\mu} + eA^\mu\right)\left(\frac{\partial S}{\partial x^\mu} + eA_\mu\right) - m_0^2 c^2 = 0. \tag{8.20}$$

8.2 Electromagnetic Field Tensor

The electrodynamic potentials V and $\mathbf{A}$ in vacuum are the solutions of the non-homogeneous d'Alembert equations, in the Lorenz gauge (see (4.198)):

$$\epsilon_0 \mu_0 \frac{\partial^2 V}{\partial t^2} - \Delta V = \frac{\rho}{\epsilon_0},$$

$$\epsilon_0 \mu_0 \frac{\partial^2 \mathbf{A}}{\partial t^2} - \Delta \mathbf{A} = \mu_0 \mathbf{j}, \tag{8.21}$$

which can also be written with the help of the d'Alembertian operator

$$\Box = \frac{1}{c^2} \frac{\partial^2}{\partial t^2} - \Delta, \qquad c^2 = \frac{1}{\epsilon_0 \mu_0},$$

as

$$\Box V = \frac{\rho}{\epsilon_0},$$

$$\Box \mathbf{A} = \mu_0 \mathbf{j}. \tag{8.22}$$

From now on we shall use for derivatives also the notations

$$\partial_\mu = \frac{\partial}{\partial x^\mu}, \qquad \partial^\mu = \frac{\partial}{\partial x_\mu}. \tag{8.23}$$

The symbol of partial derivative $\frac{\partial}{\partial x^\mu}$ is a covariant four-vector, while the the symbol $\frac{\partial}{\partial x_\mu}$ is a contravariant four-vector. Given that the operator nabla is defined as $\nabla = \left(\frac{\partial}{\partial x}, \frac{\partial}{\partial y}, \frac{\partial}{\partial x}\right) = \left(\frac{\partial}{\partial x^1}, \frac{\partial}{\partial x^2}, \frac{\partial}{\partial x^3}\right)$, it follows that in this notation

$$\partial^\mu = \left(\frac{\partial}{\partial x_0}, -\nabla\right). \tag{8.24}$$

Thus, the d'Alembertian is written covariantly as

$$\Box = \partial^\mu \partial_\mu. \tag{8.25}$$

The form (8.22) of the differential equations for the potentials suggests us a covariant formulation, by defining the four-potential as in (8.10) and a four-current as

$$j^\mu = (c\rho, \mathbf{j}). \tag{8.26}$$

Then Eqs. (8.22) become

$$\Box A^\mu = \mu_0 j^\mu. \tag{8.27}$$

In this covariant notation, the Lorenz gauge condition $\nabla \cdot \mathbf{A} + \epsilon\mu \frac{\partial V}{\partial t} = 0$ (see (3.80)) will be written as

$$\partial^\mu A_\mu = 0, \tag{8.28}$$

while the equation of continuity $\frac{\partial \rho}{\partial t} + \nabla \cdot (\rho\mathbf{v}) = 0$ (see (2.13)) becomes

$$\partial^\mu j_\mu = 0. \tag{8.29}$$

All these covariant equations suggest that we should find a way of writing also the electric and magnetic fields, expressed in terms of the scalar and vector potentials,

$$\mathbf{E} = -\nabla V - \frac{\partial \mathbf{A}}{\partial t}, \tag{8.30}$$
$$\mathbf{B} = \nabla \times \mathbf{A},$$

in a covariant manner. We shall express the x-components of the fields $\mathbf{E}$ and $\mathbf{B}$ and show that the covariant expression of the electromagnetic field strength is given through the electromagnetic field tensor $F^{\mu\nu}$ defined in (8.15). As the tensor $F^{\mu\nu}$ is antisymmetric by definition, it has six distinct components, which correspond exactly to the six components of the fields $\mathbf{E}$ and $\mathbf{B}$. Explicitly, we have

$$E_x = E^1 = -\frac{1}{c}\frac{\partial A^1}{\partial t} - \frac{\partial V}{\partial x^1} = -(\partial^0 A^1 - \partial^1 A^0),$$
$$B_x = B^1 = \frac{\partial A_z}{\partial y} - \frac{\partial A_y}{\partial z} = \frac{\partial A^3}{\partial x^2} - \frac{\partial A^2}{\partial x^3} = -(\partial^2 A^3 - \partial^3 A^2). \tag{8.31}$$

Thus, the detailed matrix form of the electromagnetic field strength tensor is

$$F^{\mu\nu} = \partial^\mu A^\nu - \partial^\nu A^\mu = \begin{pmatrix} 0 & -E_x/c & -E_y/c & -E_z/c \\ E_x/c & 0 & -B_z & B_y \\ E_y/c & B_z & 0 & -B_x \\ E_z/c & -B_y & B_x & 0 \end{pmatrix}. \tag{8.32}$$

(The first Lorentz index of the field tensor denotes the line and the second index denotes the column.) The components of the covariant tensor $F_{\mu\nu}$ are then easily found from the formula $F_{\mu\nu} = g_{\mu\rho}g_{\nu\lambda}F^{\rho\lambda}$. If one writes explicitly the components of $F_{\mu\nu}$, one notices that they are obtained from $F^{\mu\nu}$ by putting $\mathbf{E} \to -\mathbf{E}$.

In addition to (8.32), another useful way to write the components of $F^{\mu\nu}$ is the following:

$$F^{0i} = -\frac{1}{c}E^i,$$
$$F^{ij} = \epsilon^{ijk}B_k, \quad i, j, k = 1, 2, 3, \tag{8.33}$$

where ϵ^{ijk} is the usual three-dimensional Levi-Civita symbol (with $\epsilon^{123} = +1$) and

$$E^1 = E_x, \; E^2 = E_y, \; E^3 = E_z,$$
$$B^1 = B_x, \; B^2 = B_y, \; B^3 = B_z. \tag{8.34}$$

8.2.1 Gauge Invariance of $F^{\mu\nu}$

If we make a transformation of A_μ as follows:

$$A'_\mu = A_\mu - \partial_\mu\psi, \tag{8.35}$$

where $\psi(x^\nu)$ is an arbitrary differentiable function of x^ν, we observe that

$$F'_{\mu\nu} = \frac{\partial A'_\nu}{\partial x^\mu} - \frac{\partial A'_\mu}{\partial x^\nu} = \frac{\partial A_\nu}{\partial x^\mu} - \frac{\partial^2\psi}{\partial x^\mu \partial x^\nu} - \frac{\partial A_\mu}{\partial x^\nu} + \frac{\partial^2\psi}{\partial x^\nu \partial x^\mu} = F_{\mu\nu}.$$

Consequently, the transformation (8.35) leaves invariant the tensor $F_{\mu\nu}$. Having in view the significance of the components of $F^{\mu\nu}$, we conclude that this transformation does not modify the fields $\mathbf{E}, \mathbf{B}$. Moreover, according to (8.16) or/and (8.17), the equation of motion of a relativistic charged particle that moves in an external electromagnetic field remains also unchanged under this transformation.

The space and time components of (8.35) are

$$\mathbf{A}' = \mathbf{A} + \nabla\psi,$$
$$V' = V - \frac{\partial\psi}{\partial t}, \tag{8.36}$$

i.e. the gauge transformations of the electrodynamic potentials (3.87) and (3.88), respectively. We thus conclude that $F^{\mu\nu}$ is gauge invariant (as expected, given the significance of its components), and the relativistic equation of motion of the charged particle in electromagnetic field is gauge covariant.

8.2.2   Lorentz Transformations of the Electromagnetic Field

Recall that the proper orthochronous Lorentz transformations are the three-dimensional space rotations and the Lorentz boosts. Under the space rotations, the electric and magnetic field vectors $\mathbf{E}$ and $\mathbf{B}$ transform like any other space vector, for example the radius vector. Now we shall find the transformations of the electromagnetic field under the Lorentz boosts.

Let us derive the *transformation relations* of the components of the tensor $F_{\mu\nu}$. Since it is a second-order tensor, its components transform according to (B.16), that is

$$F'^{\mu\nu} = \Lambda^\mu_{\ \lambda} \Lambda^\nu_{\ \rho} F^{\lambda\rho},\tag{8.37}$$

where the matrix Λ designates a boost in the x-direction, for example. According to (7.40), the only non-zero elements of the transformation matrix Λ are

$$\begin{aligned}\Lambda^0_{\ 0} &= \Lambda^1_{\ 1} = \Gamma,\\ \Lambda^2_{\ 2} &= \Lambda^3_{\ 3} = 1,\\ \Lambda^0_{\ 1} &= \Lambda^1_{\ 0} = -\frac{V}{c}\Gamma\ .\end{aligned}\tag{8.38}$$

Writing only the non-zero terms, we find, for instance,

$$F'^{12} = \Lambda^1_{\ 1}\Lambda^2_{\ 2}F^{12} + \Lambda^1_{\ 0}\Lambda^2_{\ 2}F^{02} = \Gamma\left(F^{12} - \frac{V}{c}F^{02}\right),$$

or, in view of (8.32) (see (6.79)):

$$B'_z = \Gamma\left(B_z - \frac{V}{c^2}E_y\right).$$

Take, now, a time-like component of $F^{\mu\nu}$, for example F^{01}, which yields

$$F'^{01} = \Lambda^0_{\ 1}\Lambda^1_{\ 0}F^{10} + \Lambda^0_{\ 0}\Lambda^1_{\ 1}F^{01},$$

leading to

$$E'_x = E_x.$$

The remaining transformation relations are obtained in the same way. Summarizing, the transformation relations are (see also (6.79) and (6.80)):

$$\begin{aligned} E'_x &= E_x, & B'_x &= B_x,\\ E'_y &= \Gamma(E_y - V B_z), & B'_y &= \Gamma\left(B_y + \frac{V}{c^2}E_z\right), \end{aligned}$$

$$E'_z = \Gamma(E_z + V B_y), \qquad B'_z = \Gamma\left(B_z - \frac{V}{c^2}E_y\right), \tag{8.39}$$

or, in vector form (see also (6.81)),

$$\begin{aligned}
\mathbf{E}'_\| &= \mathbf{E}_\|, \\
\mathbf{E}'_\perp &= \Gamma(\mathbf{E} + \mathbf{V} \times \mathbf{B})_\perp, \\
\mathbf{B}'_\| &= \mathbf{B}_\|, \\
\mathbf{B}'_\perp &= \Gamma\left(\mathbf{B} - \frac{1}{c^2}\mathbf{V} \times \mathbf{E}\right)_\perp.
\end{aligned} \tag{8.40}$$

If the relative displacement of the two reference frames S and S' takes place in an arbitrary direction, of unit vector $\mathbf{v}_0$, the transformation relations are

$$\begin{aligned}
\mathbf{E}' &= (1 - \Gamma)(\mathbf{E} \cdot \mathbf{v}_0)\mathbf{v}_0 + \Gamma(\mathbf{E} + \mathbf{V} \times \mathbf{B}), \\
\mathbf{B}' &= (1 - \Gamma)(\mathbf{B} \cdot \mathbf{v}_0)\mathbf{v}_0 + \Gamma\left(\mathbf{B} - \frac{1}{c^2}\mathbf{V} \times \mathbf{E}\right),
\end{aligned} \tag{8.41}$$

and we leave the proof to the reader.

The analysis of formulas (8.39)–(8.41) leads to the following conclusions:

1. The components of $\mathbf{E}$ and $\mathbf{B}$ orthogonal to the direction of relative motion change when passing from one inertial frame to another, but the parallel components do not change;
2. The fields $\mathbf{E}$ and $\mathbf{B}$ have a *relative character*, which means that they depend on the reference frame. For example, if in the frame S the field is electrostatic ($\mathbf{B} = 0$), in S' we have $\mathbf{B}' \neq 0$; if in S the field is magnetostatic ($\mathbf{E} = 0$), in S' we have $\mathbf{E}' \neq 0$. The "appearance" and "disappearance" of one of the fields depends on the choice of reference frame;
3. In the limit $(V/c) \to 0$, the relations (8.40) go to

$$\mathbf{E}' = \mathbf{E} + \mathbf{V} \times \mathbf{B}, \quad \mathbf{B}' = \mathbf{B},$$

already derived in Chap. 3, while studying the electrodynamics of moving media with *small* velocity compared to the speed of light (see (3.131)).

8.2.3 Invariants of the Electromagnetic Field

An essential role in the study of the electromagnetic field is played by quantities that do not change when changing the reference frame – the so-called *invariants*. The invariants of the electromagnetic field are easily formed with the help of the electromagnetic field tensor $F^{\mu\nu}$. But, first, here are some preliminary considerations.

Let $S_{\mu\nu}$ be a symmetric second-order four-tensor and A_μ an arbitrary four-vector. Then the quantities $B_\mu = S_{\mu\nu}A^\nu$ form a covariant four-vector. If A_μ and B_μ are collinear ($B_\mu = \lambda A_\mu$), then B_μ defines the *principal direction* of $S_{\mu\nu}$, while λ is the *principal value* of $S_{\mu\nu}$. To determine λ, we write

$$S_{\mu\nu}A^\nu = \lambda A_\mu = \lambda g_{\mu\nu}A^\nu, \tag{8.42}$$

that is

$$(S_{\mu\nu} - \lambda g_{\mu\nu})A^\nu = 0,$$

or, using the tensor properties,

$$(S_\mu^\nu - \lambda \delta_\mu^\nu)A_\nu = 0. \tag{8.43}$$

This homogeneous system of algebraic equations admits non-zero solutions for A_ν only if the determinant of the matrix of coefficients is zero, i.e.

$$|S_\mu^\nu - \lambda \delta_\mu^\nu| = 0, \tag{8.44}$$

known as *secular equation*.[2]
Multiplying (8.42) by A^μ, we also have

$$S_{\mu\nu}A^\mu A^\nu = \lambda A^\mu A_\mu = \lambda g_{\mu\nu}A^\mu A^\nu = \text{invariant},$$

meaning that λ is an *invariant*, and so are the roots $\lambda_1, \lambda_2, \lambda_3, \lambda_4$ of the secular equation (8.44), as well as all their possible combinations. There are four distinct combinations of the roots λ_ρ, $\rho = 1, 2, 3, 4$, i.e. *four fundamental invariants* of the symmetric tensor $S_{\mu\nu}$, namely:

$$
\begin{aligned}
J_1 &= \sum \lambda_\rho, \quad \rho = 1, 2, 3, 4, \\
J_2 &= \sum \lambda_\rho \lambda_\sigma, \quad \rho, \sigma = 1, 2, 3, 4, \ \rho < \sigma, \\
J_3 &= \sum \lambda_\rho \lambda_\sigma \lambda_\eta, \quad \rho, \sigma, \eta = 1, 2, 3, 4, \ \rho < \sigma < \eta, \\
J_4 &= \sum \lambda_\rho \lambda_\sigma \lambda_\eta \lambda_\zeta = \lambda_1 \lambda_2 \lambda_3 \lambda_4, \quad \rho < \sigma < \eta < \zeta.
\end{aligned}
\tag{8.45}
$$

In the case of an antisymmetric tensor, such as the electromagnetic field tensor $F_{\mu\nu}$, the roots of the secular equation do not signify principal values of the tensor, but we can use this procedure to find its *invariants*. Putting F_μ^ν instead of S_μ^ν in (8.44) and observing that

---

[2]Since A_ν is non-zero, this means that the matrix $S - \lambda I$ is singular, which in turn means that its determinant is 0 (non-invertible). Thus, the roots of the function $\det(S - \lambda I)$ are the *eigenvalues* of S, so it is clear that this determinant is a polynomial in λ.

$$F_\mu^\nu = g_{\mu\rho} F^{\rho\nu},$$
$$F_\mu{}^\nu = -F^\nu{}_\mu,$$

one obtains the secular equation

$$\begin{vmatrix} -\lambda & -E_x/c & -E_y/c & -E_z/c \\ E_x/c & -\lambda & -B_z & B_y \\ E_y/c & B_z & -\lambda & -B_x \\ E_z/c & -B_y & B_x & -\lambda \end{vmatrix} = 0. \tag{8.46}$$

Expanding the determinant, one easily finds

$$\lambda^4 + \left(B^2 - \frac{1}{c^2} E^2 \right) \lambda^2 - \frac{1}{c^2} (\mathbf{E} \cdot \mathbf{B})^2 = 0, \tag{8.47}$$

which means that the expressions

$$I_1 = B^2 - \frac{1}{c^2} E^2,$$
$$I_2 = \mathbf{E} \cdot \mathbf{B} \tag{8.48}$$

are also invariant.

The above analysis shows that I_1 and I_2 are the only independent invariants. In fact, only $(\mathbf{E} \cdot \mathbf{B})^2$ is a true invariant, since I_2 is not a (Lorentz) scalar, but a *pseudoscalar* (the dot product of a polar and an axial vector). In other words, subject to an improper Lorentz transformation, it changes sign (see Appendix A). I_2 is a Lorentz scalar only for a proper Lorentz transformation.

The invariants (8.48) show that, if $\mathbf{E} \cdot \mathbf{B} = 0$ in some frame S, then this property survives in *any* inertial frame S'. They also show that, if in S the angle between $\mathbf{E}$ and $\mathbf{B}$ is acute (obtuse), then it remains acute (obtuse) in any inertial frame S'.

The invariants (8.48) can be expressed in terms of the electromagnetic field tensor $F^{\mu\nu}$. To this end, we form the Lorentz invariant expression (see Appendix B):

$$F^{\mu\nu} F_{\mu\nu} = F^{ik} F_{ik} + 2 F^{0i} F_{0i},$$

or, in view of (8.32) and (A.10):

$$F^{\mu\nu} F_{\mu\nu} = \left(\epsilon^{ikl} B_l \right) \left(\epsilon_{ikm} B^m \right) + 2 F^{0i} F_{0i} = 2 \left(B^2 - \frac{1}{c^2} E^2 \right) = 2 I_1,$$

that is

$$I_1 = \frac{1}{2} F^{\mu\nu} F_{\mu\nu}. \tag{8.49}$$

Another invariant (in fact, pseudoscalar) is $F^{\mu\nu}\tilde{F}_{\mu\nu}$, where $\tilde{F}_{\mu\nu}$ is the *dual electromagnetic field tensor*,

$$\tilde{F}^{\mu\nu} = \frac{1}{2}\epsilon^{\mu\nu\lambda\rho} F_{\lambda\rho} = \begin{pmatrix} 0 & -B_x & -B_y & -B_z \\ B_x & 0 & E_z/c & -E_y/c \\ B_y & -E_z/c & 0 & E_x/c \\ B_z & E_y/c & -E_x/c & 0 \end{pmatrix}. \tag{8.50}$$

Here $\epsilon^{\mu\nu\lambda\rho}$ is the completely antisymmetric unit pseudotensor of the fourth order, or Levi-Civita symbol, with $\epsilon^{\mu\nu\lambda\rho} = +1$ for an even permutation of $0, 1, 2, 3$. The transition from $F^{\mu\nu}$ to $\tilde{F}^{\mu\nu}$ and vice-versa is schematically shown by

$$\frac{1}{c}\mathbf{E} \leftrightarrow \mathbf{B},$$

$$\mathbf{B} \leftrightarrow -\frac{1}{c}\mathbf{E}.$$

The dual electromagnetic field tensor $\tilde{F}^{\mu\nu}$ can also be expressed as follows:

$$\tilde{F}^{0i} = -B^i,$$
$$\tilde{F}^{ij} = -\frac{1}{c}\epsilon^{ijk} E_k, \quad i, j, k = 1, 2, 3, \tag{8.51}$$

where ϵ^{ijk} is the usual three-dimensional Levi-Civita symbol (with $\epsilon^{123} = +1$).
   We then have

$$F^{\mu\nu}\tilde{F}_{\mu\nu} = F^{ik}\tilde{F}_{ik} + 2F^{0i}\tilde{F}_{0i} = -\frac{4}{c}(\mathbf{E} \cdot \mathbf{B}),$$

therefore

$$I_2 = -\frac{c}{4}F^{\mu\nu}\tilde{F}_{\mu\nu}. \tag{8.52}$$

The sign of the above invariant depends on the convention used for the Levi-Civita symbol. The convention used here is $\epsilon^{0123} = +1$ (some authors use $\epsilon_{0123} = +1$).
   The two invariants I_1 and I_2 can also be derived by using an alternative, three-dimensional method. Consider the complex vector

$$\mathbf{F} = \mathbf{B} + \frac{i}{c}\mathbf{E}. \tag{8.53}$$

Write the field transformations (inverse of (8.39)):

$$E_x = E'_x,$$
$$E_y = \Gamma(E'_y + V B'_z),$$
$$E_z = \Gamma(E'_z - V B'_y),$$

$$B_x = B'_x, \tag{8.54}$$

$$B_y = \Gamma \left(B'_y - \frac{V}{c^2} E'_z \right),$$

$$B_z = \Gamma \left(B'_z + \frac{V}{c^2} E'_y \right),$$

and use them to project **F** on coordinate axes:

$$F_x = B_x + \frac{i}{c} E_x = B'_x + \frac{i}{c} E'_x = F'_x,$$

$$F_y = B_y + \frac{i}{c} E_y = \Gamma \left(B'_y - \frac{V}{c^2} E'_z \right) + \frac{i}{c} \Gamma (E'_y + V B'_z) = \Gamma \left(F'_y + \frac{i}{c} V F'_z \right),$$

$$F_z = B_z + \frac{i}{c} E_z = \Gamma \left(B'_z + \frac{V}{c^2} E'_y \right) + \frac{i}{c} \Gamma (E'_z - V B'_y) = \Gamma \left(F'_z + \frac{i}{c} V F'_y \right).$$

Setting now $\tanh \theta = V/c$, we obtain

$$F_y = F'_y \cosh \theta + i F'_z \sinh \theta,$$
$$F_z = F'_z \cosh \theta - i F'_y \sinh \theta, \tag{8.55}$$

or, if the usual trigonometric functions $\sin \theta$ and $\cos \theta$ are used,

$$F_y = F'_y \cos(i\theta) + F'_z \sin(i\theta),$$
$$F_z = F'_z \cos(i\theta) - F'_y \sin(i\theta). \tag{8.56}$$

It then results that a rotation in the plane (x, t) of the four-dimensional space is, for the vector **F**, equivalent to a rotation of imaginary angle in the plane (x, y) of the three-dimensional space. The totality of possible rotations in the four-dimensional space (including the usual rotations about the axes x, y, z) is equivalent to the totality of rotations of imaginary angles in three-dimensional space (to the six rotation angles in four-space correspond three imaginary angles in three-dimensional space).

The only invariant with respect to rotations that can be formed with the vector **F**, according to (8.55) and (8.56), is its square

$$F^2 = B^2 - \frac{1}{c^2} E^2 + \frac{2i}{c} (\mathbf{E} \cdot \mathbf{B}), \tag{8.57}$$

which shows that the real expressions $B^2 - E^2/c^2$ and $\mathbf{E} \cdot \mathbf{B}$ are the only independent invariants of the four-tensor $F^{\mu\nu}$.

8.3   Covariant Form of the Equation of Continuity

The formalism of tensor calculus allows us to write the laws of conservation of some fundamental quantities (electric charge, electromagnetic field energy, electromagnetic field momentum, etc.) in a covariant form, as the vanishing of the four-divergence of a vector or tensor, depending on the analyzed physical quantity.

The law of conservation of electric charge is expressed (see (2.13)) by the equation of continuity

$$\frac{\partial \rho}{\partial t} + \nabla \cdot (\rho \mathbf{v}) = 0, \tag{8.58}$$

where $\rho = \rho(\mathbf{r}, t)$ is the volume charge density, and $\mathbf{j} = \rho \mathbf{v}$ is the conduction current density. Since charges are considered point-like, the charge density is written in terms of the Dirac delta function (see (1.6)) as

$$\rho = \sum_a q_a \delta(\mathbf{r} - \mathbf{r}_a), \tag{8.59}$$

where summation is taken over all charges, and $\mathbf{r}_a$ is the radius-vector of the charge q_a. The current density is then

$$\mathbf{j} = \rho \mathbf{v} = \sum_a q_a \mathbf{v}_a \delta(\mathbf{r} - \mathbf{r}_a), \tag{8.60}$$

where $\mathbf{v}_a = d\mathbf{r}_a/dt$.

Let us define a four-vector with $j_i = \rho v_i$ as space components. This can be done multiplying $dq = \rho \, d\mathbf{r}$ by dx^μ

$$dq dx^\mu = \rho d\mathbf{r} dx^\mu = \rho \frac{dx^\mu}{dt} dt \, d\mathbf{r} = \rho \frac{dx^\mu}{dt} d\Omega. \tag{8.61}$$

The total charge in a given volum, dq, is of course the same in any reference frame and the element of volume in Minkowski space, $d\Omega = d\mathbf{r} dt = dx \, dy \, dz \, dt$, is also Lorentz invariant. As a result, the left-hand side of (8.61) is clearly a four-vector. Consequently,

$$j^\mu = \rho \frac{dx^\mu}{dt} = (c\rho, \mathbf{j}) \tag{8.62}$$

has to be a four-vector. The time-like and space-like components of j^μ are

$$j^0 = j_0 = c\rho,$$
$$j^i = -j_i = \rho v^i. \tag{8.63}$$

The four-vector defined by (8.62) is the *current density four-vector*, or *four-current*. By means of j^μ, the equation of continuity (8.58) is written as

$$\partial^\mu j_\mu = 0.$$ (8.64)

The *total electric charge* in the space is obviously

$$Q = \int \rho d\mathbf{r} = \frac{1}{c} \int j^0 d\mathbf{r},$$ (8.65)

where the integral is taken over the whole three-dimensional space in a given frame of reference. This expression is not manifestly Lorentz covariant, but it can be cast in a Lorentz covariant form with the help of the current density four-vector j^μ:

$$Q = \frac{1}{c} \int j^\mu dS_\mu,$$ (8.66)

where the integral is over a region on a *space-like three-dimensional hypersurface*[3] for which the elements of surface, in the direction of the surface, are dS_μ. If we took the integral $\frac{1}{c} \int j^\mu dS_\mu$ over an arbitrary hypersurface we would obtain the sum of charges whose wordlines intersect that hypersurface (not necessarily the total charge in space).

The charge Q defined according to (8.66) is clearly a Lorentz invariant. Further, we shall prove that the equivalence of (8.66) and (8.65) takes place, provided that the current density j^μ satisfies the equation of continuity (8.64). To this end, let us consider

$$\oint j^\mu dS_\mu = \int_{S_1+S_2+S_3} j^\mu dS_\mu,$$ (8.67)

where the integral is taken over the closed hypersurface which includes the hyperplanes $S_1(x^0)_1 = $ const., an arbitrary space-like hypersurface S_2, and the "side" time-like surface S_3 (see Fig. 8.1). The integration over the "side" hypersurface is actually an integration over time, for $\mathbf{r}$ fixed. We assume an infinite integration volume, such that S_3 is situated at infinity, i.e. where all the fields and charges are zero. Due to the latter property, the contribution of the "side" hypersurface S_3 to the integral (8.67) vanishes.

In addition, since the normal unit vectors are opposite, the integral (8.67) becomes

$$\oint j^\mu dS_\mu = \int_{(x^0)_1=\text{const.}} j^\mu dS_\mu - \int_{S_2} j^\mu dS_\mu$$

$$= \int_{(x^0)_1=\text{const.}} j^0 dV - \int_{S_2} j^\mu dS_\mu.$$ (8.68)

---

[3]A *hypersurface* is a generalization of an ordinary two-dimensional surface embedded in three-dimensional space, to an $(n-1)$-dimensional surface embedded in n-dimensional space; in our case, $n = 4$.

Fig. 8.1 Intuitive
representation of the
four-volume enveloped by
the hyperplane
$S_1 = (x^0)_1 = $ const., the
arbitrary space-like
hypersurface S_2 and the
"side" time-like hypersurface
S_3, situated at infinity.

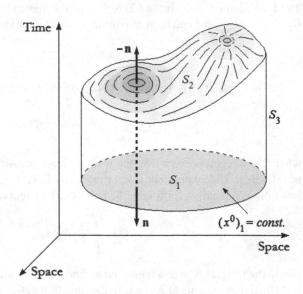

On the other hand, in view of the equation of continuity (8.64) and the divergence
theorem (C.36), we have as well

$$\oint j^\mu \, dS_\mu = \int \frac{\partial j^\mu}{\partial x^\mu} d\Omega = 0. \tag{8.69}$$

Here $d\Omega$ is the element of the four-volume enveloped by the hypersurfaces S_1, S_2,
and S_3. Putting together (8.68) and (8.69), we find the equivalence of the formulas
(8.66) and (8.65), i.e.

$$\int_{(x^0)_1 = \text{const.}} j^0 \, dV = \int_{S_2} j^\mu \, dS_\mu.$$

If the second hypersurface is also a constant-time hyperplane, $S_2 = (x^0)_2 = $ const.,
then

$$\int_{(x^0)_1 = \text{const.}} j^\mu \, dS_\mu - \int_{(x^0)_2 = \text{const.}} j^\mu \, dS_\mu = Q\Big|_{(x^0)_1} - Q\Big|_{(x^0)_2} = 0, \tag{8.70}$$

which expresses the electric charge conservation.

Observation:
This analysis is suitable for an important generalization. If V^μ is a four-vector that
satisfies the equation $\partial V^\mu / \partial x^\mu = 0$, then this property is equivalent to the conservation of the integral $\int V^\mu \, dS_\mu$, extended to a hypersurface which covers the whole

three-dimensional space. This statement remains valid if instead of V^μ one takes a second-order tensor $F^{\mu\nu}$, or a third-order tensor $T^{\mu\nu\lambda}$, etc.

8.4 Covariant Form of Maxwell's Equations

8.4.1 Maxwell's Equations for Vacuum

8.4.1.1 Source Equations

Using the analytical formalism, we shall deduce Maxwell's source equations by means of the Euler–Lagrange formalism for continuous systems.

The action S of the electromagnetic field, in the presence of electric charges, is formed of two parts: one corresponding to the free field S_0 and the other describing the interaction between sources and the field S_{int}. Since

$$S = \int L dt = \frac{1}{c} \int \mathcal{L} d\mathbf{r} \,(c\, dt) = \frac{1}{c} \int \mathcal{L}\, dx^0 dx^1 dx^2 dx^3 = \frac{1}{c} \int \mathcal{L}\, d\Omega, \quad (8.71)$$

the composition of the action S is also valid for the Lagrangian density $\mathcal{L}$, that is

$$\mathcal{L} = \mathcal{L}_0 + \mathcal{L}_{int}, \quad (8.72)$$

where (see (3.108)–(3.109)):

$$\mathcal{L}_0 = \frac{1}{2}\epsilon_0 E^2 - \frac{1}{2\mu_0} B^2 \quad (8.73)$$

and

$$\mathcal{L}_{int} = -\rho V + \mathbf{j} \cdot \mathbf{A}. \quad (8.74)$$

The two Lagrangian densities can be expressed in terms of the four-potential A^μ and the electromagnetic field tensor $F^{\mu\nu}$. Using (8.49):

$$I_1 = \frac{1}{2} F^{\mu\nu} F_{\mu\nu} = B^2 - \frac{1}{c^2} E^2,$$

as well as Maxwell's relation $\epsilon_0 \mu_0 = \frac{1}{c^2}$, we may write

$$\mathcal{L}_0 = -\frac{1}{4\mu_0} F^{\mu\nu} F_{\mu\nu}. \quad (8.75)$$

Similarly, one finds

$$\mathcal{L}_{int} = -j^\mu A_\mu. \quad (8.76)$$

The Lagrangian density of the system composed of the field and sources is then

$$\mathcal{L} = -\frac{1}{4\mu_0} F^{\eta\lambda} F_{\eta\lambda} - j^\lambda A_\lambda. \tag{8.77}$$

The corresponding Euler–Lagrange equations, with the components of A^μ taken as dynamical variables, read

$$\frac{\partial \mathcal{L}}{\partial A_\mu} - \frac{\partial}{\partial x^\nu}\left(\frac{\partial \mathcal{L}}{\partial A_{\mu,\nu}}\right) = 0, \quad \mu, \nu = 0, 1, 2, 3. \tag{8.78}$$

Recalling that $F_{\eta\lambda} = \partial_\eta A_\lambda - \partial_\lambda A_\eta$, we have

$$\frac{\partial \mathcal{L}}{\partial A_{\mu,\nu}} = -\frac{1}{2\mu_0} F^{\eta\lambda} \frac{\partial F_{\eta\lambda}}{\partial A_{\mu,\nu}} = -\frac{1}{2\mu_0} F^{\eta\lambda}(\delta_\lambda^\mu \delta_\eta^\nu - \delta_\eta^\mu \delta_\lambda^\nu) = \frac{1}{\mu_0} F^{\mu\nu}.$$

Besides,

$$\frac{\partial \mathcal{L}}{\partial A_\mu} = -j^\mu,$$

and thus (8.78) yield

$$\partial_\nu F^{\nu\mu} = \mu_0 j^\mu, \quad \mu, \nu = 0, 1, 2, 3, \tag{8.79}$$

which is the *covariant form of Maxwell's source equations*. Indeed, giving values to the Lorentz index μ in (8.79) we have:
1) $\mu = 0$:

$$\frac{\partial F^{k0}}{\partial x^k} = \mu_0 j^0,$$

that is (see (8.33)):

$$\frac{1}{c}\frac{\partial E^k}{\partial x^k} = \mu_0 c \rho,$$

or

$$\nabla \cdot (\epsilon_0 \mathbf{E}) = \rho,$$

which is one of Maxwell's source equations.
2) $\mu = i$:

$$\frac{\partial F^{0i}}{\partial x^0} + \frac{\partial F^{ki}}{\partial x^k} = \mu_0 j^i,$$

or, in view of (8.33),

$$-\frac{1}{c^2}\frac{\partial E^i}{\partial t} - \epsilon^{kil}\partial_k B_l = \mu_0 j^i \quad \Leftrightarrow \quad -\frac{1}{c^2}\frac{\partial E^i}{\partial t} + (\nabla \times \mathbf{B})^i = \mu_0 j^i. \tag{8.80}$$

Relation (8.80) can be cast in vector form:

$$\frac{1}{\mu_0} \nabla \times \mathbf{B} = \mathbf{j} + \epsilon_0 \frac{\partial \mathbf{E}}{\partial t},$$

which is the other Maxwell source equation. Denoting

$$\frac{1}{\mu_0} F^{\mu\nu} = G_0^{\mu\nu},$$ (8.81)

Maxwell's source equations can also be written as

$$\frac{\partial G_0^{\nu\mu}}{\partial x^\nu} = j^\mu, \quad \mu, \nu = 0, 1, 2, 3.$$ (8.82)

8.4.1.2 Source-Free Equations

Taking the partial derivative with respect to x^λ of (8.15):

$$\frac{\partial F_{\mu\nu}}{\partial x^\lambda} = \frac{\partial^2 A_\nu}{\partial x^\lambda \partial x^\mu} - \frac{\partial^2 A_\mu}{\partial x^\lambda \partial x^\nu},$$

then permuting indices, we obtain two more similar relations. Adding all these relations term by term, one obtains

$$\frac{\partial F_{\mu\nu}}{\partial x^\lambda} + \frac{\partial F_{\nu\lambda}}{\partial x^\mu} + \frac{\partial F_{\lambda\mu}}{\partial x^\nu} = 0.$$ (8.83)

It is not difficult to prove that the l.h.s. of (8.83) is a completely antisymmetric third-order tensor. Since the number of distinct components of such a tensor is (see Appendix B) $C_4^3 = 4$, (8.83) contains four distinct equations. If we suitably multiply (8.83) by $\frac{1}{2}\epsilon^{\mu\nu\lambda\sigma}$ and perform summation over three pairs of indices, we have

$$\epsilon^{\mu\nu\lambda\sigma} \frac{\partial F_{\lambda\sigma}}{\partial x^\nu} = 0,$$

or, in view of (8.50):

$$\partial_\nu \tilde{F}^{\mu\nu} = 0,$$ (8.84)

which are *Maxwell's source-free equations*. In order to prove this, let us recall (8.51), that is

$$\tilde{F}^{0i} = -B^i,$$
$$\tilde{F}^{ij} = -\frac{1}{c}\epsilon^{ijk} E_k, \quad i, j, k = 1, 2, 3,$$

where ϵ^{ijk} is the usual three-dimensional Levi-Civita symbol ($\epsilon^{123} = +1$) Giving values to μ in (8.84), we find

1) $\mu = 0$:

$$\frac{\partial \tilde{F}^{0i}}{\partial x^i} = 0 \iff -\frac{\partial B^i}{\partial x^i} = 0 \iff \nabla \cdot \mathbf{B} = 0.$$

2) $\mu = i$:

$$\frac{\partial \tilde{F}^{i0}}{\partial x^0} + \frac{\partial \tilde{F}^{ik}}{\partial x^k} = 0,$$

or

$$\frac{1}{c}\frac{\partial B^i}{\partial t} + \frac{1}{c}\epsilon^{ikj}\frac{\partial E^j}{\partial x^k} = 0,$$

which can be re-written as

$$\frac{1}{c}\frac{\partial B^i}{\partial t} + \frac{1}{c}(\nabla \times \mathbf{E})^i = 0,$$

or, in vector form

$$\frac{\partial \mathbf{B}}{\partial t} + \nabla \times \mathbf{E} = 0.$$

Observation:

The equation of continuity (8.64) can be straightforwardly obtained by means of Maxwell's source equations (8.79). To show this, we take the partial derivative of (8.79) with respect to x^μ:

$$\frac{\partial^2 F^{\mu\nu}}{\partial x^\mu \partial x^\nu} = \mu_0 \frac{\partial j^\mu}{\partial x^\mu}.$$

Because $F^{\mu\nu}$ is an antisymmetric second-order tensor and $\partial^2/\partial x^\mu \partial x^\nu$ is a symmetric second-order tensor, their tensorial product (double contraction) is zero, and the above relation yields

$$\partial_\mu j^\mu = 0.$$

8.4.2   Maxwell's Equations for Media

For convenience we re-write Maxwell's equations for linear, isotropic, polarizable media:

$$\nabla \times \left(\frac{\mathbf{B}}{\mu_0} - \mathbf{M}\right) = \mathbf{j} + \frac{\partial}{\partial t}(\epsilon_0 \mathbf{E} + \mathbf{P}),$$

$$\nabla \cdot (\epsilon_0 \mathbf{E} + \mathbf{P}) = 0, \tag{8.85}$$

with

$$\nabla \times \mathbf{M} + \frac{\partial \mathbf{P}}{\partial t} = \mathbf{j}_p,$$
$$\nabla \cdot \mathbf{P} = -\rho_p, \tag{8.86}$$

where $\mathbf{j}_p$ and ρ_p are the polarization (bound) current density and charge density, respectively.

The form of Eqs. (8.86) is similar to that of Maxwell's equations for vacuum. This observation allows us to write (8.86) in a covariant form. To this end, we define the antisymmetric four-tensor $M^{\mu\nu}$ called *polarization four-tensor*, as being given by

$$M^{0i} = cP^i,$$
$$M^{ik} = \epsilon^{ikl} M_l, \quad i,k,l = 1,2,3, \tag{8.87}$$

where $\epsilon^{123} = +1$, as well as the *polarization four-vector*

$$j_p^\mu = (c\rho_p, \mathbf{j}_p). \tag{8.88}$$

Then Eqs. (8.86) can be written in the compact form

$$\frac{\partial M^{\mu\nu}}{\partial x^\mu} = j_p^\nu. \tag{8.89}$$

Equation (8.85) then reads

$$\frac{\partial F^{\mu\nu}}{\partial x^\mu} = \mu_0 (j^\nu + j_p^\nu) \tag{8.90}$$

or

$$\frac{\partial G^{\mu\nu}}{\partial x^\mu} = j^\nu, \tag{8.91}$$

where the antisymmetric tensor

$$G^{\mu\nu} = \frac{1}{\mu_0} F^{\mu\nu} - M^{\mu\nu} \tag{8.92}$$

is the *excitation four-tensor*. Equations (8.91) are called *Maxwell–Minkowski equations*.

The time-like and space-like components of $G^{\mu\nu}$ are

$$G^{0i} = -\frac{1}{c\mu_0} E^i - cP^i = -cD^i,$$
$$G^{ik} = \epsilon^{ikl} \left(\frac{1}{\mu_0} B_l - M_l \right) = \epsilon^{ikl} H_l,$$

where ϵ^{ijk} is the usual three-dimensional Levi-Civita symbol. As a 4×4 matrix, the excitation four-tensor $G^{\mu\nu}$ is written as follows:

$$G^{\mu\nu} = \begin{pmatrix} 0 & -cD_x & -cD_y & -cD_z \\ cD_x & 0 & -H_z & H_y \\ cD_y & H_z & 0 & -H_x \\ cD_z & -H_y & H_x & 0 \end{pmatrix}. \tag{8.93}$$

In vacuum, $M^{\mu\nu} = 0$ and we fall back on (8.82).

Performing calculations similar to those leading to the field transformation relations (8.39), for the components of $G^{\mu\nu}$ we find

$$\mathbf{H}'_{\|} = \mathbf{H}_{\|}, \qquad \mathbf{H}'_{\perp} = \Gamma(\mathbf{H} - \mathbf{V} \times \mathbf{D})_{\perp},$$

$$\mathbf{D}'_{\|} = \mathbf{D}_{\|}, \qquad \mathbf{D}'_{\perp} = \Gamma\left(\mathbf{D} + \frac{1}{c^2}\mathbf{V} \times \mathbf{H}\right)_{\perp}. \tag{8.94}$$

In this way we justify, among other things, the analysis presented in Sect. 3.9 for the propagation of electromagnetic field through media moving with velocity *smaller* than the velocity of light.

Observation:
The source-free equations are not affected by the presence of the polarizable medium; they keep their form (8.83).

8.5   Covariant Form of Constitutive Relations

Using the tensor formalism, let us now obtain the four-dimensional form of the *constitutive relations*. For a linear, homogeneous, and isotropic medium, these relations are (see Sect. 3.2):

$$\mathbf{D} = \epsilon\mathbf{E},$$

$$\mathbf{B} = \mu\mathbf{H}, \tag{8.95}$$

$$\mathbf{j} = \lambda(\mathbf{E} + \mathbf{E}_{ext}),$$

which in this case are valid in the frame in which the medium is at rest (proper frame). Since the first two formulas imply relations only between the field tensors $F^{\mu\nu}$ and $G^{\mu\nu}$, while in the differential form of the generalized Ohm's law appears the current density $\mathbf{j}$, we shall analyze them separately.

8.5.1 Relation Between $F^{\mu\nu}$ and $G^{\mu\nu}$

Formula $(8.95)_1$, written in the reference frame in which the medium is at rest, is

$$D^i_{(0)} = \epsilon E^i_{(0)},$$

or, taking into account (8.32) and (8.93),

$$\frac{1}{c^2} G^{0i}_{(0)} = \epsilon F^{0i}_{(0)}.$$

We multiply now this equation by $u^{(0)}_\mu = (1, 0, 0, 0)$, i.e. the time-like component of the velocity four-vector in the proper frame, with the result:

$$\frac{1}{c^2} G^{0i}_{(0)} u^{(0)}_0 = \epsilon F^{0i}_{(0)} u^{(0)}_0,$$

or, since $u^{(0)}_k = 0$, $k = 1, 2, 3$,

$$\frac{1}{c^2} G^{\mu i}_{(0)} u^{(0)}_\mu = \epsilon F^{\mu i}_{(0)} u^{(0)}_\mu. \tag{8.96}$$

Obviously, we may write as well

$$\frac{1}{c^2} G^{k0}_{(0)} u^{(0)}_k = \epsilon F^{k0}_{(0)} u^{(0)}_k. \tag{8.97}$$

Putting together (8.96) and (8.97) and taking into consideration the antisymmetry of the two tensors, we find

$$\frac{1}{c^2} G^{\mu\nu}_{(0)} u_\mu = \epsilon F^{\mu\nu}_{(0)} u_\mu.$$

But a covariant tensor relation written in an inertial reference frame is valid in *any* inertial frame, so we may write in general

$$\frac{1}{c^2} G^{\mu\nu} u_\mu = \epsilon F^{\mu\nu} u_\mu. \tag{8.98}$$

We proceed in a similar way with $(8.95)_2$. In the proper reference frame, it reads

$$B^i_{(0)} = \mu H^i_{(0)}.$$

Since $G^{\mu\nu}$ is antisymmetric, we can define its dual

$$\tilde{G}^{\mu\nu} = \frac{1}{2} \epsilon^{\mu\nu\eta\theta} G_{\eta\theta}, \tag{8.99}$$

with

$$\tilde{G}^{\mu\nu} = \begin{pmatrix} 0 & -H_x & -H_y & -H_z \\ H_x & 0 & cD_z & -cD_y \\ H_y & -cD_z & 0 & cD_x \\ H_z & cD_y & -cD_x & 0 \end{pmatrix}. \tag{8.100}$$

Using (8.50) and (8.100), then multiplying by $u_0^{(0)}$ and proceeding as above, we obtain

$$\tilde{F}^{\mu\nu} u_\mu = \mu \tilde{G}^{\mu\nu} u_\mu. \tag{8.101}$$

Equations (8.98) and (8.101) are the covariant form of the constitutive relations (8.95)$_1$ and (8.95)$_2$.

Let us now write the explicit relations between the fields **E**, **D**, **H**, **B**, for media moving with relativistic velocities. These relations can be obtained either by expanding relations (8.98) and (8.101), or expressing (8.95)$_{1,2}$ in the rest frame:

$$\mathbf{D}_{(0)} = \epsilon \mathbf{E}_{(0)},$$
$$\mathbf{B}_{(0)} = \mu \mathbf{H}_{(0)},$$

then using (8.39) and (8.94). Some simple manipulations lead to the following two relations:

$$\mathbf{D} + \frac{1}{c^2} \mathbf{V} \times \mathbf{H} = \epsilon(\mathbf{E} + \mathbf{V} \times \mathbf{B}),$$
$$\mathbf{B} - \frac{1}{c^2} \mathbf{V} \times \mathbf{E} = \mu(\mathbf{H} - \mathbf{V} \times \mathbf{D}), \tag{8.102}$$

called *Minkowski's equations*. These relations are essentially different from the Eqs. (8.95)$_{1,2}$ for stationary media, since they simultaneously imply all vector fields **E**, **D**, **H**, **B**.

Eliminating either **B** or **D** between the relations (8.102), one finds two vector relations for **E**, **D**, **H**:

$$\mathbf{D}_\parallel = \epsilon \mathbf{E}_\parallel,$$
$$\left(1 - \frac{\epsilon\mu}{\epsilon_0\mu_0} \frac{V^2}{c^2}\right) \mathbf{D}_\perp = \epsilon \left(1 - \frac{V^2}{c^2}\right) \mathbf{E}_\perp + (\epsilon\mu - \epsilon_0\mu_0)(\mathbf{V} \times \mathbf{H}),$$

or for **E**, **H**, **B**:

$$\mathbf{B}_\parallel = \mu \mathbf{H}_\parallel,$$
$$\left(1 - \frac{\epsilon\mu}{\epsilon_0\mu_0} \frac{V^2}{c^2}\right) \mathbf{B}_\perp = \mu \left(1 - \frac{V^2}{c^2}\right) \mathbf{H}_\perp + (\epsilon\mu - \epsilon_0\mu_0)(\mathbf{V} \times \mathbf{E}).$$

Remark that, if $\mathbf{D} \parallel \mathbf{E}$ and $\mathbf{B} \parallel \mathbf{H}$ in an isotropic medium at rest with respect to a frame S', in any other inertial frame S this property is not valid anymore.

In the non-relativistic limit, by neglecting the terms in V^2/c^2 and $n^2 V^2/c^2$ (see Maxwell's relation), we have

$$\mathbf{D} = \epsilon\mathbf{E} + \frac{1}{c^2}(n^2 - 1)(\mathbf{V} \times \mathbf{H}),$$

$$\mathbf{B} = \mu\mathbf{H} + \frac{1}{c^2}(n^2 - 1)(\mathbf{V} \times \mathbf{E}). \tag{8.103}$$

In vacuum ($n = 1$), from (8.103) we recover $\mathbf{D} = \epsilon_0\mathbf{E}$ and $\mathbf{B} = \mu_0\mathbf{H}$, as expected.

8.5.2 Covariant Form of Ohm's Law

To write the four-dimensional form of Ohm's law (8.95)$_3$, we assume that $\mathbf{E}_{ext} = 0$. The following considerations are not at all affected by this assumption.

Proceeding like in the previous cases, we write first (8.95)$_3$ in the proper reference frame:

$$j^i_{(0)} = \lambda E^i_{(0)} = \lambda c F^{0i}_{(0)} u_0^{(0)} = \lambda c F^{\mu i}_{(0)} u_\mu^{(0)}. \tag{8.104}$$

Unlike the other constitutive relations considered in Sect. 8.5.1, this relation cannot be directly generalized to four dimensions. To attain our purpose, we shall consider the current density $\mathbf{j}$ as the sum of conduction and convection current densities. Written in the proper frame, this relation is

$$j^i_{(0)} = j^i_{(0)}(cond) + j^i_{(0)}(conv). \tag{8.105}$$

The two current densities can be chosen as

$$j^\mu_{(0)}(cond) = \left(0, \mathbf{j}_{(0)}\right),$$
$$j^\mu_{(0)}(conv) = \left(c\rho_{(0)}, 0\right), \tag{8.106}$$

where $\rho_{(0)}$ is the charge density in the proper reference frame. We then have

$$j^0_{(0)}(conv) = c\rho_{(0)} = c\rho_{(0)}u^0_{(0)}.$$

We multiply the above relation by $u_0^{(0)}$, recalling that $u_\nu^{(0)} u_{(0)}^\nu = 1$ and $u_i^{(0)} = 0$; then

$$c\rho_{(0)} = j^0_{(0)}u_0^{(0)} = j^\nu_{(0)}u_\nu^{(0)}.$$

In this case

$$j^0_{(0)}(conv) = \left(j^\nu_{(0)}u_\nu^{(0)}\right)u^0_{(0)},$$

or, since $j^\mu_{(0)}(conv)$ has only the time-component,

$$j^\mu_{(0)}(conv) = \left(j^\nu_{(0)} u^{(0)}_\nu\right) u^\mu_{(0)}. \tag{8.107}$$

Substituting (8.104) and (8.107) into (8.105), we obtain

$$j^\mu_{(0)} + \left(j^\nu_{(0)} u^{(0)}_\nu\right) u^\mu_{(0)} = \lambda c F^{\mu\nu}_{(0)} u^{(0)}_\nu.$$

This tensor relation, being valid in the proper reference frame, is valid in *any* inertial reference frame, that is

$$j^\mu + (j^\nu u_\nu) u^\mu = \lambda c F^{\mu\nu} u_\nu, \tag{8.108}$$

which is the *covariant form of Ohm's law*.

It is instructive to express the space-like and time-like components of Ohm's law (8.108). Thus, for
1) $\mu = 0$, we have

$$j^0 + (j^\nu u_\nu) u^0 = \lambda c F^{0i} u_i,$$

or (using the relations $j^0 = c\rho$, $u^0 = \gamma$, $F^{0i} = -\frac{1}{c} E^i$, and $u_i = -\frac{\gamma}{c} v_i$):

$$c\rho + (j^\nu u_\nu)\gamma = \lambda c \left(-\frac{1}{c} E^i\right)\left(-\frac{\gamma}{c} v_i\right) = \frac{\lambda\gamma}{c}(\mathbf{v}\cdot\mathbf{E})$$

$$= \frac{\lambda\gamma}{c}(\mathbf{v}\cdot\mathbf{E}^*), \tag{8.109}$$

where $\mathbf{E}^* = \mathbf{E} + \mathbf{v}\times\mathbf{B}$.
2) $\mu = i$, we have

$$j^i + (j^\nu u_\nu) u^i = \lambda c\left(F^{i0}u_0 + F^{ik}u_k\right),$$

or (using the expressions of the components of u^μ, $F^{\mu\nu}$, and u_μ):

$$j^i + (j^\nu u_\nu)\frac{\gamma}{c} v^i = \lambda c\left(\frac{\gamma}{c}E^i + \frac{\gamma}{c}\epsilon^{ikl}v_k B_l\right) = \lambda\gamma(\mathbf{E} + \mathbf{v}\times\mathbf{B})^i$$

$$= \lambda\gamma E^{*i}. \tag{8.110}$$

Relation (8.109) is used to extract the expression $(j^\nu u_\nu)\gamma$, and introduce it into (8.110). In vector form, the result is

$$\mathbf{j} - \rho\mathbf{v} = \lambda\gamma\left[\mathbf{E}^* - \frac{\mathbf{v}}{c^2}(\mathbf{v}\cdot\mathbf{E}^*)\right]. \tag{8.111}$$

This is the relativistic, three-dimensional form of Ohm's law for moving media.

If $v \ll c$, then $\gamma \simeq 1$ and the last term in (8.111) can be neglected. The result is

$$\mathbf{j} - \rho\mathbf{v}(= \mathbf{j}_c) = \lambda(\mathbf{E} + \mathbf{v}\times\mathbf{B}),$$

which is Ohm's law for slowly moving media (see (3.146)). The difference $\mathbf{j} - \rho \mathbf{v} = \mathbf{j}_c$ represents, in this case, the conduction current.

If, finally, $\mathbf{v} = 0$, we arrive at the well-known form of Ohm's law for stationary media, $\mathbf{j} = \lambda \mathbf{E}$.

Observation:
One can easily verify that the space-like components of the force four-vector satisfy the following relation of transformation

$$\mathbf{F}'^N = \gamma \left[\mathbf{F}^N + \frac{\mathbf{v}}{c^2} (\mathbf{F}^N \cdot \mathbf{v}) \right], \tag{8.112}$$

which is similar to (8.111).

8.6 Four-Potential and Its Differential Equations

The differential equations (8.22) of the electrodynamics potential in vacuum, in the Lorenz gauge, can be cast in the manifestly covariant form

$$\Box A^\mu = \mu_0 j^\mu. \tag{8.113}$$

To determine the solution of (8.113), we shall use again the *Green function method*. In this case, we have to find the Green function of the d'Alembertian operator, i.e. the solutions of the equation

$$\Box_x G(x, x') = \delta^{(4)}(x - x'), \tag{8.114}$$

where by x and x' we understand the corresponding four-vectors, and

$$\delta^{(4)}(x - x') = \delta(x^0 - x'^0)\delta(\mathbf{r} - \mathbf{r}') \tag{8.115}$$

is the four-dimensional Dirac δ function. In the absence of discontinuity surfaces, the Green function can only depend on the difference $z^\mu = x^\mu - x'^\mu$, and thus (8.114) becomes

$$\Box_z G(z) = \delta^{(4)}(z). \tag{8.116}$$

To solve this equation we shall Fourier transform it. For the Green function we obtain

$$G(z) = \int g(k) e^{-ikz} d^4k, \tag{8.117}$$

Fig. 8.2 The two simple
poles of the integrand in the
expression (8.120).

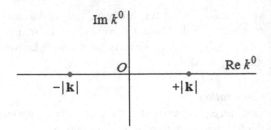

where

$$kz = k^\mu z_\mu = k^0 z_0 + k^i z_i = k^0 z_0 - \mathbf{k} \cdot (\mathbf{r} - \mathbf{r}') = k^0 z_0 - \mathbf{k} \cdot \mathbf{R}.$$

Since the Fourier transform of $\delta^{(4)}(z)$ is

$$\delta^{(4)}(z) = \frac{1}{(2\pi)^4} \int e^{-ikz} d^4k, \tag{8.118}$$

we find

$$g(k) = -\frac{1}{(2\pi)^4} \frac{1}{k^2}, \tag{8.119}$$

where

$$k^2 = k^\mu k_\mu = (k^0)^2 - \mathbf{k}^2.$$

Therefore

$$G(z) = -\frac{1}{(2\pi)^4} \int \frac{e^{-ikz}}{k^2} d^4k, \tag{8.120}$$

which is nothing else but the four-dimensional form of (4.206) (Fig. 8.2).

The derivation of the solution of equation (8.113) is similar to that developed in Sect. 4.9.[4] Various physically significant Green's functions are obtained by choosing suitably the integration contours and including one or the other pole, or both. For example, in order to obtain the retarded Green's function, we choose the contour as in Fig. 8.3 and the integral analogous to (4.209) reads

$$\int_{-\infty}^{+\infty} \frac{e^{-k^0 z_0}}{k^2} dk^0 = \begin{cases} 0, & t < t' \\ \frac{2\pi c}{|\mathbf{k}|} \sin(c|\mathbf{k}|\tau), & t > t', \end{cases}$$

$$= \theta(z^0) \frac{2\pi c}{|\mathbf{k}|} \sin(|\mathbf{k}|z^0),$$

----

[4] In comparing the formulas with those in Sect. 4.9, note that the symbol k there means $|\mathbf{k}|$, while in this section it signifies the four-vector k^μ.

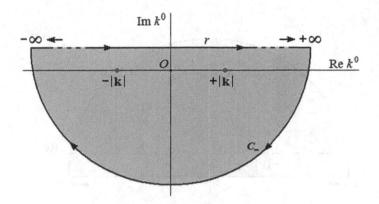

Fig. 8.3 Integration contour for the retarded (causal) Green's function.

where $\theta\left(z^0\right) = \theta\left(x^0 - x'^0\right)$ is the *Heaviside step function*, also called the *unit step function*:

$$\theta\left(z^0\right) = \begin{cases} +1, & z^0 > 0, \\ 0, & z^0 < 0. \end{cases} \qquad (8.121)$$

The *retarded Green's function* then reads

$$G_{ret}(x - x') = \frac{c}{4\pi R} \theta\left(x^0 - x'^0\right) \delta\left(x^0 - x'^0 - R\right). \qquad (8.122)$$

The retarded Green's function is also called *causal Green's function*, because the perturbation is produced at the moment t', which precedes in time the moment t of observation.

In a similar way, choosing the contour depicted in Fig. 8.4, one obtains the *advanced Green function*

$$G_{adv}(x - x') = \frac{c}{4\pi R} \theta\left[-\left(x^0 - x'^0\right)\right] \delta\left(x^0 - x'^0 + R\right). \qquad (8.123)$$

Using (E.27), we can write

$$\delta\left[(x - x')^2\right] = \delta\left[(x^\mu - x'^\mu)(x_\mu - x'_\mu)\right] = \delta\left[|\mathbf{r} - \mathbf{r}'|^2 - \left(x^0 - x'^0\right)^2\right]$$

$$= \delta\left[\left(x^0 - x'^0 - R\right)\left(x^0 - x'^0 + R\right)\right]$$

$$= \frac{1}{2R}\left[\delta\left(x^0 - x'^0 - R\right) + \delta\left(x^0 - x'^0 + R\right)\right]. \qquad (8.124)$$

Since the functions $\theta\left(x^0 - x'^0\right)$ and $\theta\left[-\left(x^0 - x'^0\right)\right] = \theta\left(x'^{(0)} - x^{(0)}\right)$ select one or the other of the two terms in the right-hand side of relation (8.124), we have

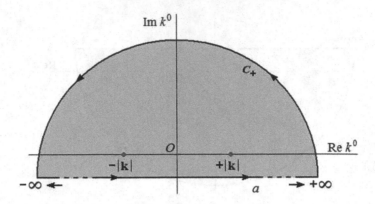

Fig. 8.4 Integration contour for the advanced Green's function.

$$G_{ret}(x - x') = \theta\left(x^0 - x'^0\right)\delta\left[(x - x')^2\right],$$
$$G_{adv}(x - x') = \theta\left(x'^0 - x^0\right)\delta\left[(x - x')^2\right]. \qquad (8.125)$$

The Heaviside functions $\theta\left(x^0 - x'^0\right)$ and $\theta\left(x'^0 - x^0\right)$, which are not covariant, become covariant with respect to the proper orthochronous Lorentz transformations when taken in combination with the corresponding delta functions. Thus, (8.125) provides manifestly covariant expressions for the Green functions. The theta and delta functions in (8.125) show that the retarded and advanced Green functions are nonzero only inside the lightcone ahead and behind the source.

The general solution of the wave equation in vacuum (8.113) obeyed by the four-potential of the electromagnetic field $A^\mu(x)$ can be written now with the help of the causal Green function:

$$A^\mu(x) = A^\mu_{in}(x) + \mu_0 \int G_{ret}(x - x')j^\mu(x')d^4x', \qquad (8.126)$$

or by means of the advanced Green function:

$$A^\mu(x) = A^\mu_{out}(x) + \mu_0 \int G_{adv}(x - x')j^\mu(x')d^4x', \qquad (8.127)$$

where A^μ_{in} and A^μ_{out} are two particular solutions (corresponding to the two situations) of the homogeneous wave equation. In (8.126) appears the retarded Green's function. In this case, in the limit $x^0 \to -\infty$, the integral over the sources vanishes, assuming that the sources are localized in space and time (there are no sources at infinity). We see then that the four-potential A^μ_{in} of the free field has the significance of *incident* or *incoming* four-potential, specified by $x^0 \to -\infty$. Similarly, in (8.127) with the advanced Green function, the solution A^μ_{out} is the asymptotic *emerging* or *outgoing* four-potential, specified for $x^0 \to +\infty$. The radiation fields are defined as the difference between the outgoing and incoming fields, that is

$$A_{rad}^{\mu}(x) = A_{in}^{\mu}(x) - A_{out}^{\mu}(x) = \mu_0 \int G(x - x') j^{\mu}(x') d^4x', \qquad (8.128)$$

where

$$G(x - x') = G_{ret}(x - x') - G_{adv}(x - x'). \qquad (8.129)$$

8.7 Conservation Laws of Electrodynamics in Covariant Formulation

As we have mentioned in Sect. 8.3, tensor calculus gives the possibility of writing in covariant form some fundamental laws of the electromagnetic field: the law of conservation of electromagnetic energy, electromagnetic momentum, and electromagnetic angular momentum.

We shall first present the general theory, applicable to *any* field, and then apply it to the electromagnetic field. In this respect, we presume the field as of unspecified tensor order.

To set the stage for the main discussion, let us start again from the Euler–Lagrange equations,

$$\frac{\partial \mathcal{L}}{\partial \varphi^{(s)}} - \frac{\partial}{\partial x^{\mu}} \left(\frac{\partial \mathcal{L}}{\partial(\partial_{\mu} \varphi^{(s)})} \right) = 0, \quad s = 1, 2, ..., h; \ \mu = 0, 1, 2, 3, \qquad (8.130)$$

where the Lagrangian density $\mathcal{L}$ is, in general, a function of the independent variables $x \equiv x^{\mu}$, the dynamical variables (fields) $\varphi^{(s)}(x)$, and their derivatives with respect to time, $\dot{\varphi}^{(s)}(x)$. The variables x count, in this case, the number of degrees of freedom of the system, which is infinite. We shall assume only local interactions, i.e. the Lagrangian density is a function of one point only, $\mathcal{L}(x)$. Having only time-derivatives of the fields in the Lagrangian is, however, an overly restrictive condition, since such a model cannot cover wave propagation. To relax this constraint, we allow interactions between the field at a space point $\mathbf{x}$ and its infinitesimally close neighbours, since the maximum speed of propagation of signals is finite and equal to the speed of light in vacuum. We thus consider

$$\mathcal{L} = \mathcal{L} \left(x, \varphi^{(s)}(x), \dot{\varphi}^{(s)}(x), \varphi(ct, \mathbf{x} + d\mathbf{x}) \right).$$

For infinitesimal displacements we can write

$$\varphi(ct, \mathbf{x} + d\mathbf{x}) = \varphi(ct, \mathbf{x}) + \frac{\partial \varphi}{\partial x^i}(ct, \mathbf{x} + d\mathbf{x}) dx^i,$$

therefore the Lagrangian can be cast into a manifestly covariant form as a function

$$\mathcal{L}(x) = \mathcal{L} \left(x, \varphi^{(s)}(x), \partial_{\mu} \varphi^{(s)}(x) \right).$$

We cannot have higher order derivatives of $\varphi^{(s)}$ because they would lead to equations of motion of order higher than two; however, the equations of physics (for example, Newton's law, the wave propagation equation, Schrödinger's equations) are differential equations of second order, and not higher, in the time variable.

The requirement of relativistic invariance of the theory is equivalent to requiring the Lorentz invariance of the action. Since the action of a system of fields extending infinitely in space and time is

$$S = \int d^4x \mathcal{L}(x),$$

with the integrals covering the whole space-time, and the volume element d^4x is Lorentz invariant, we deduce that the Lagrangian density has to be also a Lorentz scalar. This gives us an excellent framework for constructing Lagrangian densities from fields of various tensorial orders.

Another property that we shall require is the invariance of action with respect to translations in space and time. This invariance is the mathematical expression of the assumed homogeneity of space-time, just as the Lorentz invariance expresses the postulated isotropy of space-time. The way to account for the translational invariance of the Lagrangian density is by discarding the explicit x-dependence. Under a translation transformation

$$x^\mu \rightarrow x'^\mu = x^\mu + a^\mu, \tag{8.131}$$

where a^μ is an arbitrary constant four-vector, the fields transform trivially as

$$\varphi^{(s)}(x) \rightarrow \varphi'^{(s)}(x+a) = \varphi^{(s)}(x), \tag{8.132}$$

in other words all fields are scalars under translations.[5] It is easy to see that

$$\mathcal{L}'\left(x+a, \varphi'^{(s)}(x+a), \partial_\mu \varphi'^{(s)}(x+a)\right) = \mathcal{L}'\left(x+a, \varphi^{(s)}(x), \partial_\mu \varphi^{(s)}(x)\right)$$
$$\neq \mathcal{L}\left(x, \varphi^{(s)}(x), \partial_\mu \varphi^{(s)}(x)\right),$$

and that this non-invariance is due to the explicit dependence of the Lagrangian density on x.

Putting together the results inferred from the physical requirements of Poincaré invariance, locality, and correspondence with the classical non-relativistic theory, we deduce that the Lagrangian density is a Lorentz scalar, with a functional dependence of the form

$$\mathcal{L}(x) = \mathcal{L}\left(\varphi^{(s)}, \partial_\mu \varphi^{(s)}\right). \tag{8.133}$$

---

[5]The reason is that the translation group is a real continuous Abelian group (it is obvious that any two translations commute among themselves) and as such it admits only scalar, or trivial, irreducible representation. Proving this statement is beyond the scope of this book.

Since the tensor order of the field is not specified, by (s) we mean a group of indices which:

(a) does not exist if the tensor order of the field is zero (scalar field): $\varphi^{(s)} \equiv \varphi$;
(b) is composed of a single index in case of a first-order tensor, i.e. a vector: $\varphi^{(s)} \equiv \varphi^\mu$. Such a field is, for example, the electromagnetic field;
(c) is formed of two indices to characterize a second-order tensor field (such as gravitational field): $\varphi^{(s)} \equiv \varphi^{\mu\nu}$, etc.

8.7.1 Noether's Theorem

The connection between symmetries under continuous transformations and conservation laws is known as *Noether's theorem*. It is due to Emmy Noether (1882–1935), a German-Jewish mathematician who proved it in 1918. The theorem has had long-lasting influence on the development of modern theoretical physics. Starting with the second half of the 20th century, symmetry has been the most dominant concept in formulating the fundamental laws of physics. Actually, after the 1905 papers of Einstein on special relativity, symmetries have been viewed as essential properties of Nature, while the dynamical laws are regarded as mere consequences of the symmetries.

Let us consider a system of local fields $\varphi^{(s)}(x)$, described by the Lagrangian

$$\mathcal{L}(x) = \mathcal{L}\left(\varphi^{(s)}(x), \partial_\mu \varphi^{(s)}(x)\right).$$

Suppose we have an N-parametric group of *continuous transformations* of the space-time:

$$x^\mu \rightarrow x'^\mu = x^\mu + \delta x^\mu, \quad \delta x^\mu = X^\mu_\kappa \omega_\kappa, \tag{8.134}$$

$$\varphi^{(s)}(x) \rightarrow \varphi'^{(s)}(x') = \varphi^{(s)}(x) + \delta \varphi^{(s)}(x), \quad \delta \varphi^{(s)}(x) = \Phi^{(s)}_\kappa \omega_\kappa, \tag{8.135}$$

where ω_κ, $\kappa = 1, 2, \ldots, N$ are the parameters of the infinitesimal transformations and X^μ_κ and $\Phi^{(s)}_\kappa$ are the matrix generators of the transformations for the coordinates and fields, respectively.

The theory defined by $\mathcal{L}(x)$ is *invariant*, or symmetric, under the transformations (8.134) and (8.135) if the action does not change, which can be expressed by the equality

$$\mathcal{L}'(x') \, d^4 x' = \mathcal{L}(x) \, d^4 x. \tag{8.136}$$

Noether's theorem can be stated as follows: If the theory is invariant under any continuous N-parametric group of transformations which satisfy (8.136), than there are correspondingly N conserved quantities, commonly called charges.

We shall prove now the theorem starting from the invariance condition (8.136), which leads to

$$\delta \left(\mathcal{L}(x) \, d^4x \right) = \delta(\mathcal{L}(x)) d^4x + \mathcal{L}(x)\delta(d^4x) = 0. \tag{8.137}$$

We shall consider one by one the two terms in (8.137). Since a finite continuous transformation can be written as an infinite sum of infinitesimal transformations, it is sufficient to analyze only the latter ones. All the formulas will be considered up to the first order in the transformation parameters ω_k.

Under a transformation of coordinates, the functional form of the fields and of the Lagrangian changes, as one can see from Eqs. (8.135) and (8.136), where upon transformation we wrote φ' instead of φ and $\mathcal{L}'$ instead of $\mathcal{L}$. The form variation is denoted by $\bar{\delta}$ and in the case of the Lagrangian, it is written as

$$\bar{\delta}\mathcal{L} = \mathcal{L}'(x) - \mathcal{L}(x). \tag{8.138}$$

Note that this variation is defined at a given space-time point x. The total variation of the Lagrangian $\delta(\mathcal{L}(x))$ can be written as

$$
\begin{aligned}
\delta(\mathcal{L}(x)) &= \mathcal{L}'(x') - \mathcal{L}(x) = \mathcal{L}'(x') - \mathcal{L}'(x) + \mathcal{L}'(x) - \mathcal{L}(x) \\
&= (\partial_\nu \mathcal{L}'(x))\delta x^\nu + \bar{\delta}(\mathcal{L}(x)) \\
&= (\partial_\nu \mathcal{L}(x))\delta x^\nu + \bar{\delta}(\mathcal{L}(x)),
\end{aligned} \tag{8.139}
$$

where the last equality is valid in the considered order in the transformation parameters. Thus, the total variation of the Lagrangian density is due to the form variation $\bar{\delta}(\mathcal{L}(x))$ and to the variation of the coordinates, $(\partial_\nu \mathcal{L}(x))\delta x^\nu$. In its turn, the form variation of the Lagrangian $\bar{\delta}(\mathcal{L}(x))$ is due to the form variation of the fields $\varphi^{(s)}(x)$ and their derivatives:

$$
\begin{aligned}
\bar{\delta}(\mathcal{L}(x)) &= \frac{\partial \mathcal{L}}{\partial \varphi^{(s)}} \bar{\delta}\varphi^{(s)} + \frac{\partial \mathcal{L}}{\partial(\partial_\mu \varphi^{(s)})} \bar{\delta}(\partial_\mu \varphi^{(s)}) \\
&= \left[\frac{\partial \mathcal{L}}{\partial \varphi^{(s)}} - \partial_\nu \left(\frac{\partial \mathcal{L}}{\partial(\partial_\nu \varphi^{(s)})} \right) \right] \bar{\delta}\varphi^{(s)} + \partial_\nu \left(\frac{\partial \mathcal{L}}{\partial(\partial_\nu \varphi^{(s)})} \bar{\delta}\varphi^{(s)} \right).
\end{aligned} \tag{8.140}
$$

The second term in the last line of the above equation was written making use of the fact that the form variation $\bar{\delta}$ is taken at a fixed point, consequently it commutes with the derivative operator ∂_ν. The form variation of the fields is, by the same reasoning as it was shown for the Lagrangian density (see (8.139)),

$$
\begin{aligned}
\bar{\delta}\varphi^{(s)} &= \delta\varphi^{(s)} - (\partial_\nu \varphi^{(s)})\delta x^\nu \\
&= \left[\Phi_\kappa^{(s)} - (\partial_\nu \varphi^{(s)})X_\kappa^\nu \right] \omega_\kappa,
\end{aligned} \tag{8.141}
$$

where we took into account (8.134) and (8.135) in writing the last equality. Introducing (8.140) and (8.141) into (8.139) and observing that the factor in the square brackets vanishes due to the Euler–Lagrange equations (8.130), we obtain:

$$\delta(\mathcal{L}(x)) = \partial_\nu \left(\frac{\partial \mathcal{L}}{\partial(\partial_\nu \varphi^{(s)})} \left[\Phi_\kappa^{(s)} - (\partial_\rho \varphi^{(s)}) X_\kappa^\rho \right] \right) \omega_\kappa + (\partial_\nu \mathcal{L}(x)) X_\kappa^\nu \omega_\kappa. \quad (8.142)$$

Let us now consider the second term in (8.137). The variation of the volume element d^4x under the transformations (8.134) and (8.135) is

$$d^4 x' = |\det J| \, d^4 x, \quad (8.143)$$

where J is the Jacobian matrix of the coordinate transformation (8.134):

$$\det J = \det \left(\frac{\partial x'^\nu}{\partial x^\mu} \right) = \det \left(\delta_\mu^\nu + \partial_\mu (X_\kappa^\nu \omega_\kappa) \right) = 1 + \partial_\nu (X_\kappa^\nu \omega_\kappa) + \mathcal{O}(\omega^2)$$
$$(8.144)$$

(assuming that the parameters of the transformations may depend on the space-time point). Using this expression in (8.143), we find that the second term of (8.137) becomes

$$\mathcal{L}(x) \delta(d^4 x) = \mathcal{L}(x) \partial_\nu (X_\kappa^\nu \omega_\kappa) d^4 x. \quad (8.145)$$

Now we can express the invariance condition (8.137) in terms of fields and their variations, using (8.142) and (8.145), with the result:

$$0 = \partial_\nu \left(\frac{\partial \mathcal{L}}{\partial(\partial_\nu \varphi^{(s)})} \Phi_\kappa^{(s)} - \frac{\partial \mathcal{L}}{\partial(\partial_\nu \varphi^{(s)})} (\partial_\rho \varphi^{(s)}) X_\kappa^\rho \right) \omega_\kappa + \partial_\nu \left(\mathcal{L} X_\kappa^\nu \right) \omega_\kappa \quad (8.146)$$
$$= -\partial_\nu \left\{ \left[\frac{\partial \mathcal{L}}{\partial(\partial_\nu \varphi^{(s)})} (\partial_\rho \varphi^{(s)}) - \mathcal{L}(x) \delta_\rho^\nu \right] X_\kappa^\rho - \frac{\partial \mathcal{L}}{\partial(\partial_\nu \varphi^{(s)})} \Phi_\kappa^{(s)} \right\} \omega_\kappa$$

Since the parameters ω_κ are arbitrary, Eqs. (8.146) can be satisfied only if the coefficients of ω_κ vanish. Thus, we obtain N continuity equations,

$$\partial_\nu J_\kappa^\nu = 0, \quad \kappa = 1, 2, \dots, N, \quad (8.147)$$

where the currents J_κ^ν, called *Noether currents*, have the expressions

$$J_\kappa^\nu(x) = \left[\frac{\partial \mathcal{L}}{\partial(\partial_\nu \varphi^{(s)})} (\partial_\rho \varphi^{(s)}) - \mathcal{L} \delta_\rho^\nu \right] X_\kappa^\rho - \frac{\partial \mathcal{L}}{\partial(\partial_\nu \varphi^{(s)})} \Phi_\kappa^{(s)}. \quad (8.148)$$

Finally, we define the corresponding N charges:

$$Q_\kappa(t) = \int_V d^3 x \, J_\kappa^0, \quad \kappa = 1, 2, \dots, N, \quad (8.149)$$

where V is the space volume occupied by the system of fields. The charges so defined are conserved, i.e. time-independent, as can be easily shown:

$$
\begin{aligned}
\frac{dQ_\kappa}{dt}(t) &= \int_V d^3x \, \partial_0 J_\kappa^0(x) \\
&= \int_V d^3x \, \left(\partial_\mu J_\kappa^\mu - \partial_i J_\kappa^i \right) \\
&= - \int_V d^3x \, \partial_i J_\kappa^i = - \int_{\partial V} dS_i \, J_\kappa^i,
\end{aligned}
$$

where we used the equation of continuity and the divergence theorem (C.36). Assuming that on the boundary surface of V there are no charges, nor currents, that is

$$
J_\kappa^i|_{\partial V} = 0,
$$

we obtain

$$
\frac{dQ_\kappa}{dt} = 0, \quad \kappa = 1, 2, \ldots, N.
$$

The Noether theorem is thus proven. The great usefulness of the formalism is that it not only states the existence of conserved quantities once the system is symmetric under certain transformations, but it also gives a prescription to derive those conserved quantities.

In 1911, the German mathematician Gustav Herglotz (1881–1953) used the variational method to derive the conservation of energy-momentum and angular momentum from the Poincaré invariance. This work was one of the most important precursors to Emmy Noether's celebrated theorem. In the following, we shall use the method of Noether and Herglotz to find the conserved energy-momentum tensor and angular momentum tensor for a general system of fields.

8.7.2 Energy-Momentum Tensor

The conservation of energy and momentum is a consequence of the space-time translational invariance. This is a well-known result in the non-relativistic mechanics of particles. In the following, we shall derive the expression of the energy-momentum tensor using Noether's prescription. Recalling the translation transformations (8.131) and (8.132), which we re-write for convenience, with a^μ replaced by ω^μ:

$$
x^\mu \rightarrow x'^\mu = x^\mu + \omega^\mu,
$$
$$
\varphi^{(s)}(x) \rightarrow \varphi'^{(s)}(x + a) = \varphi^{(s)}(x),
$$

and comparing them with formulas (8.134) and (8.135), we observe that

$$\delta x^\mu = \omega^\mu, \quad \delta\varphi^{(s)}(x) = 0,$$

that is

$$X^\rho_\kappa = \delta^\rho_\nu, \quad \Phi^{(s)}_\kappa = 0.$$

There are four Noether currents in this case, corresponding to the translations in each space-time direction. Their expressions are:

$$J_\mu^{\ \nu} \equiv T_\mu^{\ \nu} = \frac{\partial \mathcal{L}}{\partial(\partial_\nu \varphi^{(s)})}(\partial_\mu \varphi^{(s)}) - \mathcal{L}\delta^\nu_\mu. \tag{8.150}$$

These currents form actually a Lorentz tensor, called the *canonical energy-momentum tensor*. The property of symmetry (or antisymmetry) of a second-order tensor, closely connected with some physical quantities, can be easily analyzed only if the two indices have the same variance. In our case, the contravariant form of the canonical energy-momentum tensor is

$$T^{\mu\nu} = \frac{\partial \mathcal{L}}{\partial_\nu \varphi^{(s)}}\partial^\mu \varphi^{(s)} - \mathcal{L}g^{\mu\nu}. \tag{8.151}$$

We shall adopt the contravariant form and derive the conserved charges, by putting $\nu = 0$ and integrating over the space volume V occupied by the field, in a given Lorentz frame. The first charge is obtained for $\mu = 0$:

$$T^{00} = \frac{\partial \mathcal{L}}{\partial(\partial_0 \varphi^{(s)})}\partial^0 \varphi^{(s)} - \mathcal{L}. \tag{8.152}$$

We observe that the momentum densities, canonically conjugated to the fields $\varphi^{(s)}$ are

$$\pi^{(s)} = \frac{\partial \mathcal{L}}{\partial(\partial_0 \varphi^{(s)})} \tag{8.153}$$

and the Hamiltonian density is

$$\mathcal{H} = \pi^{(s)}\partial_0 \varphi^{(s)} - \mathcal{L}, \tag{8.154}$$

Thus, we deduce that T^{00} signifies the energy density of the system of fields, commonly denoted by w, and

$$E = \int T^{00}d^3x \tag{8.155}$$

represents the total energy of that system, which is conserved.

We expect that the space components, obtained for $\nu = i, i = 1, 2, 3$, i.e.

$$P^i = \text{const.} \int T^{i0}d^3x$$

will give the total momentum of the system of fields. Recalling that for a relativistic particle the time-component of the energy-momentum vector is $p^0 = \frac{1}{c}E$, where E is the energy of the particle, we infer that the constant above is the inverse of the speed of light in vacuum, $1/c$. Thus, the conserved three-momentum of the system of fields is given by

$$P^i = \frac{1}{c} \int d^3x \, T^{i0} = \frac{\partial \mathcal{L}}{\partial(\partial_0 \varphi^{(s)})} \partial^i \varphi^{(s)}. \qquad (8.156)$$

The quantities

$$p^i = \frac{1}{c} T^{i0} \qquad (8.157)$$

form a three-dimensional vector, called *momentum density of the system of fields.*

The energy-momentum tensor (8.151) does not have a definite symmetry property in the Lorentz indices, therefore it contains sixteen independent elements. We can now find the physical significance of the other elements, for which $\nu \neq 0$. Let us take, next, $\mu = 0$ in $\partial_\nu T^{\mu\nu} = 0$:

$$\frac{\partial T^{0i}}{\partial x^i} + \frac{\partial T^{00}}{\partial x^0} = 0,$$

then integrate over the three-dimensional domain of volume V:

$$\frac{\partial}{\partial t} \int_V T^{00} d^3x = -c \int_V \frac{\partial T^{0i}}{\partial x^i} d^3x = -c \oint_{\partial V} T^{0i} dS_i,$$

where divergence theorem has been applied. The left-hand side of this equation gives the time variation of the total energy of the field, therefore the vector

$$\pi^i = c \, T^{0i} \qquad (8.158)$$

stands for the *energy flux density*, which is the energy passing in unit time through the closed surface $\partial V = S$ bounding the integration domain of volume V.

Finally, we consider $\mu = i$ in $\partial_\nu T^{\mu\nu} = 0$, and integrate over the space volume:

$$\frac{\partial}{\partial t} \left(\frac{1}{c} \int_V T^{i0} d^3x \right) = -\int_V \frac{\partial T^{ik}}{\partial x^k} d^3x = -\oint_{\partial V} T^{ik} dS_k.$$

In view of (8.157), this relation shows that

$$\pi^{ik} = T^{ik} \qquad (8.159)$$

represents the *momentum flux density* of the field.

Summarizing, here is the significance of all the sixteen components of the tensor $T^{\mu\nu}$:

$$
\begin{aligned}
T^{00} &= w &\rightarrow& \quad \text{energy density;} \\
T^{i0} &= cp^i &\rightarrow& \quad c \times \text{momentum density;} \\
T^{0i} &= \tfrac{1}{c}\pi^i &\rightarrow& \quad \tfrac{1}{c} \times \text{energy flux density;} \\
T^{ik} &= \pi^{ik} &\rightarrow& \quad \text{momentum flux density.}
\end{aligned}
\tag{8.160}
$$

Thus, we have justified the name of *energy-momentum tensor* given to $T^{\mu\nu}$. Since, in fact, its components have the significance of *densities*, the name of *energy-momentum density tensor* would be more appropriate, but this term is not used in scientific literature.

The formulas (8.155) and (8.156) define the energy-momentum vector in the proper reference frame. The Lorentz covariant definition is

$$
P^\mu = \frac{1}{c} \int T^{\mu\nu} dS_\nu,
\tag{8.161}
$$

where the integration is over a region on a space-like hypersurface for which the elements of surface, in the direction of the surface normal, are dS_ν, as discussed in Sect. 8.3.

It is important to observe that the tensor $T^{\mu\nu}$ is not uniquely determined by $\partial_\nu T^{\mu\nu} = 0$ and (8.161). Indeed, if to $T^{\mu\nu}$ we add a tensor $\partial B^{\mu\lambda\nu}/\partial x^\lambda$, where $B^{\mu\lambda\nu}$ is a third-order tensor, antisymmetric in its last two indices $\left(B^{\mu\lambda\nu} = -B^{\mu\nu\lambda}\right)$, i.e.

$$
T_B^{\mu\nu} = T^{\mu\nu} + \frac{\partial}{\partial x^\lambda}\left(B^{\mu\lambda\nu}\right),
\tag{8.162}
$$

we have

$$
\frac{\partial T_B^{\mu\nu}}{\partial x^\nu} = \frac{\partial T^{\mu\nu}}{\partial x^\nu} + \frac{\partial^2}{\partial x^\nu \partial x^\lambda}\left(B^{\mu\lambda\nu}\right) = \frac{\partial T^{\mu\nu}}{\partial x^\nu} = 0,
$$

because the second-order derivative is symmetric in the index pair (ν, λ), while $B^{\mu\lambda\nu}$ is antisymmetric.

Substituting (8.162) into (8.161), we obtain

$$
P_B^\mu = \int T_B^{\mu\nu} dS_\nu = \int T^{\mu\nu} dS_\nu + \int \partial_\lambda B^{\mu\lambda\nu} dS_\nu.
$$

But

$$
\int \partial_\lambda B^{\mu\lambda\nu} dS_\nu = \frac{1}{2}\int \left(\partial_\lambda B^{\mu\lambda\nu} + \partial_\lambda B^{\mu\lambda\nu}\right) dS_\nu
$$
$$
= \frac{1}{2}\int \left(\partial_\lambda B^{\mu\lambda\nu} dS_\nu + \partial_\nu B^{\mu\nu\lambda} dS_\lambda\right) = \frac{1}{2}\int \left(\partial_\lambda B^{\mu\lambda\nu} dS_\nu - \partial_\nu B^{\mu\lambda\nu} dS_\lambda\right),
$$

or, in view of (C.19),

$$\int \partial_\lambda B^{\mu\lambda\nu} \, dS_\nu = \frac{1}{2} \int B^{\mu\lambda\nu} \, d\tilde{\sigma}_{\nu\lambda}. \tag{8.163}$$

where $d\tilde{\sigma}_{\nu\lambda}$ is an element of the two-dimensional surface bounding the three-dimensional hypersurface of element dS_ν. This two-surface is extended to infinity and, since we assume the field vanishing at infinity, the integral (8.163) vanishes as well and we are left with $P_B^\mu = P^\mu$. Therefore, the four-momentum remains unchanged under the transformation (8.162). Consequently, the transformation (8.162) modifies neither the equation of conservation

$$\partial_\nu T^{\mu\nu} = 0, \tag{8.164}$$

nor the four-momentum (8.161).

8.7.3 Angular Momentum Tensor

The angular momentum of a system of fields is obtained from Noether's theorem as conserved charge when the system is invariant under Lorentz transformations. Under infinitesimal Lorentz transformations, the space-time points and the fields transform as follows:

$$x^\mu \to x'^\mu = x^\mu + \delta x^\mu = x^\mu + X^\mu_{\alpha\beta}\, \omega^{\alpha\beta}, \tag{8.165}$$

$$\varphi^{(s)}(x) \to \varphi'^{(s)}(x') = \varphi^{(s)}(x) + \delta\varphi^{(s)}(x) = \varphi^{(s)}(x) + \Phi^{(s)}_{\alpha\beta}\omega^{\alpha\beta}, \tag{8.166}$$

where $\omega_{\alpha\beta}$ are the antisymmetric parameters of the Lorentz transformations (see (7.54)). Note that the generic index κ in (8.134) and (8.135) is replaced by the pair of Lorentz indices $\alpha\beta$. We have to find the quantities $(X_{\alpha\beta})^\mu_\nu$ and $\Phi^{(s)}_{\alpha\beta}$, in order to utilize them in the formula (8.148) for Noether's conserved current. The form of $X^\mu_{\alpha\beta}$ can be easily read off from Eqs. (7.52)–(7.59). In the language of Sect. 7.3.3,

$$x'^\mu = x^\mu + \omega^\mu_{\ \nu} x^\nu = x^\mu - \frac{i}{2}\omega^{\alpha\beta}(L_{\alpha\beta})^\mu_{\ \nu} x^\nu,$$

with

$$(L_{\alpha\beta})^\mu_{\ \nu} = i\left(g_{\beta\nu}\delta^\mu_{\ \alpha} - g_{\alpha\nu}\delta^\mu_{\ \beta}\right).$$

Comparing these equations with (8.165), we find

$$X^\mu_{\alpha\beta} = \frac{1}{2}\left(x_\beta \delta^\mu_{\ \alpha} - x_\alpha \delta^\mu_{\ \beta}\right). \tag{8.167}$$

The infinitesimal transformations of the fields (8.166) have to be of the form

$$\varphi'^{(s)}(x') = \varphi^{(s)}(x) + \frac{1}{2}\Sigma_{\alpha\beta}^{(s,r)}\omega^{\alpha\beta}\varphi^{(r)}(x), \tag{8.168}$$

where the summation over the repeated indices r and s labeling the fields is implied. The coefficients $\Sigma_{\alpha\beta}^{(s,r)}\omega^{\alpha\beta}$ are antisymmetric in the indices α and β, just like $\omega^{\alpha\beta}$, and they are determined by the concrete transformation properties of the fields under the Lorentz group (in other words, by the representation of the Lorentz group assigned to those fields). They define the intrinsic angular momentum of the fields, i.e. the *spin*. We call $\Sigma^{(s,r)}$ a *spin matrix*. Comparing (8.166) and (8.168), we find

$$\Phi_{\alpha\beta}^{(s)} = \frac{1}{2}\Sigma_{\alpha\beta}^{(s,r)}\varphi^{(r)}(x). \tag{8.169}$$

Having settled the transformations of the space-time points and fields, we can write down the conserved Noether currents, which will be the components of a order three tensor:

$$\begin{aligned} J_{\alpha\beta}{}^{\nu}(x) &= \left[\frac{\partial\mathcal{L}}{\partial(\partial_{\nu}\varphi^{(s)})}(\partial_{\rho}\varphi^{(s)}) - \mathcal{L}\delta_{\rho}^{\nu}\right]X_{\alpha\beta}^{\rho} - \frac{\partial\mathcal{L}}{\partial(\partial_{\nu}\varphi^{(s)})}\Phi_{\alpha\beta}^{(s)} \\ &= -\frac{1}{2}\left[\left(x_{\alpha}T_{\beta}{}^{\nu} - x_{\beta}T_{\alpha}{}^{\nu}\right) + \frac{\partial\mathcal{L}}{\partial(\partial_{\nu}\varphi^{(s)})}\Sigma_{\alpha\beta}^{(s,r)}\varphi^{(r)}\right], \end{aligned} \tag{8.170}$$

which was obtained by plugging (8.167) and (8.169) into (8.148). The second term,

$$S_{\alpha\beta}{}^{\nu} = \frac{\partial\mathcal{L}}{\partial(\partial_{\nu}\varphi^{(s)})}\Sigma_{\alpha\beta}^{(s,r)}\varphi^{(r)}, \tag{8.171}$$

is called *canonical spin tensor*.

The angular momentum tensor is defined as

$$\mathcal{M}^{\alpha\beta\nu} = \left(x^{\alpha}T^{\beta\nu} - x^{\beta}T^{\alpha\nu}\right) + S^{\alpha\beta\nu} \tag{8.172}$$

and it satisfies the continuity equation

$$\partial_{\nu}\mathcal{M}^{\alpha\beta\nu} = 0. \tag{8.173}$$

The conserved charges are obtained by putting $\nu = 0$ and integrating over the whole three-dimensional space:

$$M^{\alpha\beta} = \frac{1}{c}\int d^{3}x\left(x^{\alpha}T^{\beta 0} - x^{\beta}T^{\alpha 0} + S^{\alpha\beta 0}\right), \tag{8.174}$$

or covariantly

$$M^{\alpha\beta} = \frac{1}{c}\int\left(x^{\alpha}T^{\beta\nu} - x^{\beta}T^{\alpha\nu} + S^{\alpha\beta\nu}\right)dS_{\nu}, \tag{8.175}$$

and they represent the components of the total angular momentum, the first two terms signifying the orbital angular momentum, while the last signifies the contribution of the intrinsic angular momentum, or spin.

8.7.4 Belinfante Energy-Momentum Tensor

In 1939, Frederik Belinfante (1913–1991) proposed a method of symmetrization of the energy-momentum tensor, based on a combination of the canonical energy-momentum tensor and canonical spin tensor. Taking advantage of the fact that the continuity equation remains the same if one performs a transformation of the canonical energy-momentum tensor of the form (see (8.162)):

$$T_B^{\mu\nu} = T^{\mu\nu} + \frac{\partial}{\partial x^\lambda}\left(B^{\mu\lambda\nu}\right), \tag{8.176}$$

with $B^{\mu\lambda\nu} = -B^{\mu\nu\lambda}$, one defines the *Belinfante tensor*:

$$B^{\mu\lambda\nu} = \frac{1}{2}\left(S^{\mu\lambda\nu} - S^{\mu\nu\lambda} - S^{\lambda\nu\mu}\right). \tag{8.177}$$

This expression has the required antisymmetry property in the last two indices.

The tensor $T_B^{\mu\nu}$ in (8.176) is called the *Belinfante energy-momentum tensor* and has the remarkable property that it is a symmetric tensor. A symmetric energy-momentum tensor is necessary when matter fields are coupled to gravity, in the framework of the general theory of relativity.

Using the Belinfante energy-momentum tensor, the expressions for the four-momentum and angular momentum become:

$$P^\mu = \frac{1}{c}\int d^3x\, T_B^{\mu 0}, \tag{8.178}$$

$$M^{\mu\nu} = \frac{1}{c}\int d^3x\left(x^\mu T_B^{\nu 0} - x^\nu T_B^{\mu 0}\right). \tag{8.179}$$

8.7.5 Energy-Momentum Tensor of the Electromagnetic Field

Let us now particularize the above general analysis to the case of the electromagnetic field. The Lagrangian density $\mathcal{L}$ of the free electromagnetic field is (see (8.75)):

$$\mathcal{L} = -\frac{1}{4\mu_0}F^{\mu\nu}F_{\mu\nu}.$$

In our case, the field functions $\varphi^{(s)}$ are the components of the four potential A^μ, $\mu = 0, 1, 2, 3$. According to (8.151), the canonical energy-momentum tensor $T_{(c)}^{\mu\nu}$ is then

$$T_{(c)}^{\mu\nu} = \frac{1}{4\mu_0} F^{\rho\sigma} F_{\rho\sigma} g^{\mu\nu} + \frac{\partial \mathcal{L}}{\partial(\partial_\nu A_\lambda)}(\partial_\mu A_\lambda).$$

But

$$\frac{\partial \mathcal{L}}{\partial(\partial_\nu A_\lambda)} = -\frac{1}{\mu_0} F^{\nu\lambda},$$

and we obtain

$$T_{(c)}^{\mu\nu} = \frac{1}{4\mu_0} F^{\rho\sigma} F_{\rho\sigma} g^{\mu\nu} - \frac{1}{\mu_0} F^{\nu\lambda}(\partial_\mu A_\lambda). \tag{8.180}$$

This tensor is not symmetric, but it can be symmetrized by using the Belinfante procedure. To this end, we need the canonical spin tensor for the electromagnetic field, which is easily read off from the Lorentz transformation property of the field, i.e.

$$A^\mu(x) \to A'^\mu(x') = A^\mu + \omega^\mu_{\ \nu} A^\nu(x) = A^\mu(x) - \frac{i}{2}\omega^{\alpha\beta}(L_{\alpha\beta})^\mu_{\ \nu} A^\nu(x),$$

with

$$(L_{\alpha\beta})^\mu_{\ \nu} = i\left(g_{\beta\nu}\delta^\mu_{\ \alpha} - g_{\alpha\nu}\delta^\mu_{\ \beta}\right).$$

in full analogy with the transformation properties of the coordinate four-vector x^μ. Thus, the spin matrix for the electromagnetic field, according to (8.168), is

$$(\Sigma_{\alpha\beta})^\mu_{\ \nu} = g_{\beta\nu}\delta^\mu_{\ \alpha} - g_{\alpha\nu}\delta^\mu_{\ \beta}. \tag{8.181}$$

Introducing this spin matrix into the general formula (8.175), we obtain the canonical spin tensor of the electromagnetic field:

$$S^{\alpha\beta\nu} = \frac{\partial \mathcal{L}}{\partial(\partial_\nu A_\lambda)}(\Sigma^{\alpha\beta})_{\lambda\sigma} A^\sigma = -\frac{1}{\mu_0}\left(A^\alpha F^{\beta\nu} - A^\beta F^{\alpha\nu}\right), \tag{8.182}$$

which leads, by (8.177), to the Belinfante tensor

$$B^{\mu\lambda\nu} = -\frac{1}{\mu_0} A^\mu F^{\lambda\nu}. \tag{8.183}$$

According to the procedure of symmetrizing the energy-momentum tensor, we have to add to the canonical tensor the divergence of the Belinfante tensor, i.e.

$$\partial_\lambda B^{\mu\lambda\nu} = \frac{1}{\mu_0}\left(\frac{\partial A^\mu}{\partial x^\lambda} F^{\lambda\nu} + A^\mu \frac{\partial F^{\lambda\nu}}{\partial x^\lambda}\right). \tag{8.184}$$

From Maxwell's equations (8.79), written for $j^\mu = 0$ (the field is source-free), the last term of (8.184) vanishes. In addition,

$$\frac{\partial A^\lambda}{\partial x_\mu} - \frac{\partial A^\mu}{\partial x_\lambda} = F^{\mu\lambda},$$

and thus, formulas (8.176), (8.180), and (8.184) lead to the symmetric energy-momentum tensor of the electromagnetic field

$$T^{\mu\nu} = \frac{1}{4\mu_0} F^{\lambda\eta} F_{\lambda\eta} g^{\mu\nu} - \frac{1}{\mu_0} F^{\mu\lambda} F^\nu{}_\lambda. \tag{8.185}$$

(We have omitted the subscript B for convenience.) It has also the property of gauge invariance (i.e. invariance under the transformations (8.35)), which can be immediately seen if we recall that the electromagnetic field tensor is gauge invariant.

Since $T^{\mu\nu}$ is symmetric, it should have $(C_4^2) = C_5^2 = \frac{4\cdot5}{1\cdot2} = 10$ distinct components. But, as it can easily be shown, its trace is zero,

$$Tr\{T\} = \sum_{\mu=0}^{3} T^\mu{}_\mu = 0, \tag{8.186}$$

and the number of distinct components reduces to nine. Following the general theory, let us find the expressions and the physical significance of these components.

In view of (8.33) and (8.49), the component T^{00} is

$$T^{00} = \frac{1}{4\mu_0} F^{\mu\nu} F_{\mu\nu} - \frac{1}{\mu_0} F^{0i} F^0{}_i = \frac{1}{4\mu_0} 2 \left(B^2 - \frac{E^2}{c^2} \right) + \frac{1}{\mu_0 c^2} E^2$$

$$= \frac{1}{2} \left(\epsilon_0 E^2 + \frac{1}{\mu_0} B^2 \right) = w_{em}, \tag{8.187}$$

where

$$w_{em} = \frac{1}{2} \epsilon_0 E^2 + \frac{1}{2\mu_0} B^2 \tag{8.188}$$

is the *energy density* of the electromagnetic field.

Also,

$$T^{i0} \left(= T^{0i} \right) = -\frac{1}{\mu_0} F^{i\lambda} F^0{}_\lambda = -\frac{1}{\mu_0} F^{ik} F^0{}_k = -\frac{1}{\mu_0} \left(-\epsilon^{ikl} B_l \right) \left(\frac{1}{c} E_k \right)$$

$$= \epsilon_0 c (\mathbf{E} \times \mathbf{B})^i = c (\mathbf{D} \times \mathbf{B})^i = c\, p^i_{em}, \tag{8.189}$$

where

$$\mathbf{P}_{em} = \mathbf{D} \times \mathbf{B} \tag{8.190}$$

is the *momentum density* of the electromagnetic field (see (3.63)).

The space-like components of $T^{\mu\nu}$ are (see (8.33) and (8.49)):

$$T^{ik} = \frac{1}{4\mu_0} F^{\mu\nu} F_{\mu\nu} g^{ik} - \frac{1}{\mu_0} F^{i\lambda} F_{j\lambda} g^{kj}$$

$$= \frac{1}{2\mu_0} \left(B^2 - \frac{E^2}{c^2} \right) g^{ik} - \frac{1}{\mu_0} F^{i0} F_{j0} g^{kj} - \frac{1}{\mu_0} F^{is} F_{js} g^{kj}$$

$$= \frac{1}{2\mu_0} \left(B^2 - \frac{E^2}{c^2} \right) g^{ik} + \epsilon_0 E^i E_j g^{kj} - \frac{1}{\mu_0} \left(\delta^i_j \delta^l_r - \delta^i_r \delta^l_j \right) B_l B^r g^{kj}.$$

Since $g^{ik} = -\delta^{ik}$, we obtain

$$T^{ik} = -\frac{1}{2\mu_0} \left(B^2 - \frac{E^2}{c^2} \right) \delta^{ik} - \epsilon_0 E^i E_j \delta^{kj} + \frac{1}{\mu_0} B^l B_l \delta^{ik} - \frac{1}{\mu_0} B^i B_j \delta^{kj}$$

$$= -\frac{1}{2} \mathbf{H} \cdot \mathbf{B}\, \delta^{ik} + \frac{1}{2} \mathbf{E} \cdot \mathbf{D}\, \delta^{ik} - E^i D^k + \mathbf{H} \cdot \mathbf{B}\, \delta^{ik} - H^i B^k$$

$$= \frac{1}{2} (\mathbf{E} \cdot \mathbf{D} + \mathbf{H} \cdot \mathbf{B})\, \delta^{ik} - E^i D^k - H^i B^k$$

$$= w_{em} \delta^{ik} - E^i D^k - H^i B^k = T^{ik}_{Maxwell}, \tag{8.191}$$

where $T^{ik}_{Maxwell}$ is *Maxwell's stress tensor* (3.67) and represents, as we know, the momentum density flux of the electromagnetic field.

Finally, recalling that $T^{\mu\nu}$ is symmetric, its components T^{0i} are obtained directly from (8.189), but this time in the form of the components of *Poynting's vector*:

$$T^{0i} = T^{i0} = \frac{1}{c} (\mathbf{E} \times \mathbf{H})^i = \frac{1}{c} \Pi^i. \tag{8.192}$$

Written in compact form, the energy-momentum tensor of the electromagnetic field is

$$T^{\mu\nu}_{em} = \begin{pmatrix} w_{em} & \cdots & c p^i_{em} \\ \vdots & & \vdots \\ \frac{1}{c} \Pi^i_{em} & \cdots & T^{ik}_{Maxwell} \end{pmatrix}. \tag{8.193}$$

8.7.6   Laws of Conservation of Electromagnetic Field
in the Presence of Sources

So far we have assumed that the electromagnetic field is free, that is neither charges, nor currents are present. If sources are considered, the equation of conservation (8.164) will change its form. Supposing that the sources do not interact, let us determine the appropriate equation of conservation.

Denote by $T_\mu^{\nu(f)}$ the components of the energy-momentum tensor of the electromagnetic field itself (the superscript index (f) comes from "field"):

$$T_\mu^{\nu(f)} = \frac{1}{4\mu_0} F^{\lambda\eta} F_{\lambda\eta} \delta_\mu^\nu - \frac{1}{\mu_0} F_{\mu\lambda} F^{\nu\lambda}, \qquad (8.194)$$

and take the partial derivative with respect to x^ν. Using Maxwell's source equations (8.79) and taking into account that $F^{\lambda\eta}\,\delta\left(F_{\lambda\eta}\right) = F_{\lambda\eta}\,\delta\left(F^{\lambda\eta}\right)$, we have

$$\frac{\partial T_\mu^{\nu(f)}}{\partial x^\nu} = \frac{1}{2\mu_0} F^{\lambda\eta} \frac{\partial F_{\lambda\eta}}{\partial x^\mu} - \frac{1}{\mu_0} \frac{\partial F_{\mu\lambda}}{\partial x^\nu} F^{\nu\lambda} - \frac{1}{\mu_0} F_{\mu\lambda} \frac{\partial F^{\nu\lambda}}{\partial x^\nu}$$

$$= \frac{1}{2\mu_0} F^{\nu\lambda} \frac{\partial F_{\nu\lambda}}{\partial x^\mu} - \frac{1}{\mu_0} F^{\nu\lambda} \frac{\partial F_{\mu\lambda}}{\partial x^\nu} - F_{\mu\lambda} j^\lambda,$$

or, if the second term is written as a sum of two halves, and we make a convenient change of indices by taking into account the antisymmetry property of $F_{\mu\lambda}$,

$$\frac{\partial T_\mu^{\nu(f)}}{\partial x^\nu} = \frac{1}{2\mu_0} F^{\nu\lambda} \left(\frac{\partial F_{\nu\lambda}}{\partial x^\mu} + \frac{\partial F_{\lambda\mu}}{\partial x^\nu} + \frac{\partial F_{\mu\nu}}{\partial x^\lambda} \right) - F_{\mu\lambda} j^\lambda.$$

According to the relation (8.83), the sum between parentheses vanishes, and we are left with

$$\frac{\partial T_\mu^{\nu(f)}}{\partial x^\nu} = F_{\lambda\mu} j^\lambda,$$

or, equivalently,

$$\frac{\partial T^{\nu\mu(f)}}{\partial x^\nu} = F^{\lambda\mu} j_\lambda = -F^{\mu\lambda} j_\lambda, \quad \lambda, \mu, \nu = 0, 1, 2, 3. \qquad (8.195)$$

As expected, the energy-momentum tensor of the field alone is not conserved. Recalling the definition of the current density four-vector (see (8.63)):

$$j^0 = j_0 = c\rho,$$
$$j^i = -j_i = \rho v^i,$$

we then have

1) $\mu = 0$:

$$\frac{\partial T^{00(f)}}{\partial x^0} + \frac{\partial T^{i0(f)}}{\partial x^i} = F^{i0} j_i,$$

or

$$\frac{1}{c}\frac{\partial w_{em}}{\partial t} + \frac{1}{c}\frac{\partial \Pi^i_{em}}{\partial x^i} = \left(\frac{1}{c}E^i\right)(-\rho v_i),$$

where we have used (8.33), (8.34), (8.63), (8.187), (8.189), and (8.192). Integrating over the three-dimensional domain D of volume V, we obtain

$$\frac{d}{dt}W_{em} = -\oint_S \mathbf{\Pi} \cdot d\mathbf{S} - \int_V \rho \mathbf{v} \cdot \mathbf{E}\, d\mathbf{r} = -\oint_S \mathbf{\Pi} \cdot d\mathbf{S} - \int_V \mathbf{j} \cdot \mathbf{E}\, d\mathbf{r}, \qquad (8.196)$$

that is *Poynting's theorem* (3.47).

2) $\mu = i$:

$$\frac{\partial T^{0i(f)}}{\partial x^0} + \frac{\partial T^{ki(f)}}{\partial x^k} = F^{0i} j_0 + F^{ki} j_k,$$

or

$$\frac{1}{c^2}\frac{\partial \Pi^i}{\partial t} + \frac{\partial \Pi^{ki}}{\partial x^k} = \left(-\frac{1}{c}E^i\right)(c\rho) + \left(-\epsilon^{kil} B_l\right)(-\rho v_k) = -\rho\left[E^i + (\mathbf{v} \times \mathbf{B})^i\right],$$

where we have used (8.33), (8.34), (8.63), (8.189), (8.191), and (8.192). But

$$\rho\,(\mathbf{E} + \mathbf{v} \times \mathbf{B})^i = (\rho \mathbf{E} + \mathbf{j} \times \mathbf{B})^i = (\mathbf{f}_{em})^i$$

is the x^i-component of the *electromagnetic force density* acting on sources, and the *electromagnetic force* is (see (3.60)):

$$\mathbf{F}_{em} = \int_V (\rho \mathbf{E} + \mathbf{j} \times \mathbf{B})\, d\mathbf{r} = \frac{d\mathbf{P}_{mec}}{dt}.$$

Integrating over a three-dimensional, fixed domain D of volume V, bounded by the surface S, we then have

$$\frac{d}{dt}(\mathbf{P}_{mec} + \mathbf{P}_{em})^i = -\oint_S \Pi^{ki} dS_k = -\oint_S \Pi^{ik} dS_k, \qquad (8.197)$$

or, in dyadic form,

$$\frac{d}{dt}(\mathbf{P}_{mec} + \mathbf{P}_{em}) = -\oint \{\Pi\} \cdot d\mathbf{S}, \qquad (8.198)$$

expressing the equation of conservation of momentum of the system formed of field and sources.

We shall show that the tensor Eq. (8.195) can be written as a conservation equation $\partial T^{\mu\nu}/\partial x^{\mu} = 0$, where $T^{\mu\nu}$ is given by a sum of two tensors, one being $T^{\mu\nu(f)}$, and the other being the energy-momentum tensor of the system of particles. Let us determine the last one.

We consider a system of point-like charges. Their electric charge density is given by (8.269),

$$\rho = \sum_a q_a \delta(\mathbf{r} - \mathbf{r}_a),$$

where summation is taken over all charges, and $\mathbf{r}_a$ is the radius-vector of the particle of charge q_a and mass m_a, while the mass density is defined analogously as

$$\rho_m = \sum_a m_a \delta(\mathbf{r} - \mathbf{r}_a). \tag{8.199}$$

Since for one particle the energy-momentum four-vector is $P^{\mu} = m_0 c u^{\mu}$, the density of the energy-momentum vector for the system of particles will be $p^{\mu} = \rho_m c u^{\mu}$. According to (8.160), for the electromagnetic field the space-like components of the four-momentum density are given by the elements T^{i0}. By analogy, we shall identify for the system of charges p^i by $\frac{1}{c} T^{i0(p)}$, i.e.

$$p^i = \rho_m c u^i \equiv \frac{1}{c} T^{i0(p)},$$

which gives

$$T^{i0(p)} = c^2 \rho_m u^i, \tag{8.200}$$

or

$$T^{i0(p)} = c \rho_m u^i \frac{dx^0}{dt}.$$

We infer then the form of the general element of $T^{\mu\nu(p)}$:

$$T^{\mu\nu(p)} = c \rho_m u^{\mu} \frac{dx^{\nu}}{dt}. \tag{8.201}$$

By analogy with the current density four-vector, we define the *mass density four-vector* as

$$J^{\nu} = \rho_m \frac{dx^{\nu}}{dt} \tag{8.202}$$

such that (8.201) becomes

$$T^{\mu\nu(p)} = c u^{\mu} J^{\nu}. \tag{8.203}$$

Now we take the derivative of (8.203) with respect to x^ν:

$$\frac{\partial T^{\mu\nu(p)}}{\partial x^\nu} = c\frac{\partial u^\mu}{\partial x^\nu}J^\nu + cu^\mu\frac{\partial J^\nu}{\partial x^\nu} = cJ^\nu\frac{\partial u^\mu}{\partial x^\nu} = c\rho_m\frac{dx^\nu}{dt}\frac{\partial u^\mu}{\partial x^\nu} = c\rho_m\frac{du^\mu}{dt}, \quad (8.204)$$

where we have used the mass conservation equation $\partial J^\nu/\partial x^\nu = 0$.

On the other hand, the equation of motion of a point charge in the electromagnetic field $\mathbf{E}$, $\mathbf{B}$ is (see (8.17)):

$$m_0c\frac{du^\mu}{ds} = eF^{\mu\nu}u_\nu.$$

If the mass and charge are continuously distributed, this equation becomes

$$\rho_m\,c\frac{du^\mu}{ds} = \rho F^{\mu\nu}u_\nu$$

or

$$\rho_m c\frac{du^\mu}{dt} = \rho F^{\mu\nu}\frac{dx_\nu}{dt} = F^{\mu\nu}j_\nu.$$

Introducing this result into (8.204), we find

$$\frac{\partial T^{\mu\nu(p)}}{\partial x^\nu} = F^{\mu\nu}j_\nu. \qquad (8.205)$$

Adding (8.195) and (8.205), we finally obtain the desired equation of conservation

$$\partial_\nu\left(T_\mu^{\nu(f)} + T_\mu^{\nu(p)}\right) = \partial_\nu T_\mu^\nu = 0, \qquad (8.206)$$

where $T_\mu^\nu = T_\mu^{\nu(f)} + T_\mu^{\nu(p)}$ is the energy-momentum tensor of the whole system composed of field and particles.

The equation of continuity (8.206) expresses the *laws of conservation of energy, momentum, and angular momentum of the whole system.*

Observation:
Denoting by $\rho_{m0} = \rho_m/\gamma$ the rest mass density, the energy-momentum tensor associated with the sources can be written as

$$T^{\mu\nu(p)} = c\rho_m u^\mu\frac{dx^\nu}{ds}\frac{ds}{dt} = \rho_{m0}c^2u^\mu u^\nu. \qquad (8.207)$$

This form is frequently used in fluid mechanics.

8.8   Elements of Relativistic Magnetofluid Dynamics

8.8.1   Fundamental Equations

Using the above analysis, let us now write the fundamental equations of relativistic magnetofluid dynamics for an ideal magnetofluid (the dissipative processes are negligible).

A part of these equations, i.e. Maxwell's equations, have already been written in a covariant form (see (8.79) and (8.84)):

$$\frac{\partial F^{\nu\mu}}{\partial x^\nu} = \mu_0 j^\mu, \quad \frac{\partial \tilde{F}^{\mu\nu}}{\partial x^\nu} = 0. \tag{8.208}$$

Another fundamental equation is the *equation of motion*. To write it in a covariant form, we must first determine the energy-momentum tensor for a macroscopic fluid. The composition of this tensor has to be more complex than (8.207), since in deriving (8.207) we did not take into account the *mechanical* interactions between particles, that is the *pressure*. Conversely, the tensor we seek has to reduce to (8.207) when mechanical interactions are neglected.

An ideal fluid moves through space-time with constant four-velocity with respect to any inertial frame. To start with, we shall consider the rest frame of the fluid. The mechanics of continuous media shows that in an ideal fluid can exist only normal tensions, such that the stress tensor T_{ik} is proportional to the isotropic pressure, $p > 0$:

$$T_{ik}^{(fluid)} = p \, \delta_{ik}. \tag{8.209}$$

Since the physical dimension of pressure is energy density, while the general theory (see (8.160)) indicates that one component of the energy-momentum tensor signifies the energy density, which will be denoted by ϵ, we come to the conclusion that T^{00} and T^{ik} are components of the energy-momentum tensor of the fluid, $T^{\mu\nu(fluid)}$. Written in the proper reference frame, this tensor is

$$T_0^{\mu\nu\,(fluid)} = \begin{pmatrix} \epsilon & 0 & 0 & 0 \\ 0 & p & 0 & 0 \\ 0 & 0 & p & 0 \\ 0 & 0 & 0 & p \end{pmatrix}. \tag{8.210}$$

By performing a Lorentz transformation to a frame with respect to which the fluid has the the velocity u^μ, we find the general expression

$$T^{\mu\nu(fluid)} = (p + \epsilon)u^\mu u^\nu - p \, g^{\mu\nu}, \tag{8.211}$$

which reduces to (8.210) for $u^\mu = (1, 0, 0, 0)$.

If the velocities of the particles composing the fluid are small as compared to the speed of light, then $T^{\mu\nu(fluid)}$ takes a simpler form. In this case, the energy density ϵ is approximately equal to the rest energy density $\rho_{m0}c^2$, where ρ_{m0} is the proper mass density (more precisely, the sum of the masses of particles per unit proper volume). In its turn, the pressure p determined by the microscopic motion of molecules is also small as compared with the rest energy density. Under these assumptions, (8.211) reduces to

$$T^{\mu\nu(fluid)} = T^{\mu\nu(p)} = \rho_{m0}c^2 u^\mu u^\nu,$$

which coincides with (8.207), as expected.

The energy-momentum tensor of the system composed of fluid and electromagnetic field is then

$$T^{\mu\nu} = T^{\mu\nu(f)} + T^{\mu\nu(fluid)}$$

$$= \frac{1}{4\mu_0} F^{\lambda\eta} F_{\lambda\eta} g^{\mu\nu} - \frac{1}{\mu_0} F^{\mu\lambda} F^\nu_\lambda + (p + \epsilon)u^\mu u^\nu - p\, g^{\mu\nu}, \quad (8.212)$$

and the relation

$$\frac{\partial T^{\mu\nu}}{\partial x^\nu} = 0 \quad (8.213)$$

is the covariant form of the *equation of motion* (the momentum conservation law) of the considered ideal magnetofluid, under the action of the electromagnetic field. As we already know, (8.213) also expresses the conservation of energy and angular momentum of the system.

To these equations we have to add:

- *the equation of continuity*:

$$\frac{\partial j^\mu}{\partial x^\mu} = \frac{\partial}{\partial x^\mu}\left(\rho_m \frac{dx^\mu}{dt}\right) = \frac{\partial}{\partial x^\mu}\left(\rho_m \frac{dx^\mu}{ds}\frac{ds}{dt}\right) = 0,$$

that is

$$\frac{\partial}{\partial x^\mu}(\rho_{m0}u^\mu) = 0; \quad (8.214)$$

- *the entropy conservation law.* Let us denote by σ the entropy density of the fluid. Then σu^μ is the entropy current and its conservation is expressed as usual by a continuity equation:

$$u^\mu \frac{d\sigma}{dx^\mu} = 0; \quad (8.215)$$

- *the equation of state.* In our case, this equation takes the form of the fundamental equation of reversible thermodynamic processes,

$$T dS = dU + p dV,$$

or, written for densities (σ for entropy S, ε for internal energy U, and w for the enthalpy $H = U + pV$):

$$dw = Td\sigma + dp, \tag{8.216}$$

where

$$w = \varepsilon + p. \tag{8.217}$$

Equations (8.208), (8.213)–(8.216) represent the *fundamental system of equations of relativistic magnetofluid dynamics* for an ideal magnetofluid.

8.8.2   Bernoulli's Equation in Relativistic Magnetofluid Dynamics

It is convenient to write Eq. (8.213), using (8.217), as

$$\frac{\partial}{\partial x^\nu}(wu_\mu u^\nu - p\delta_\mu^\nu) = -\frac{\partial T_\mu^{\nu(f)}}{\partial x^\nu}. \tag{8.218}$$

Since $u^\mu \frac{\partial}{\partial x^\mu} = \frac{d}{ds}$, and assuming $\rho_{m0} = $ const. (incompressible fluid), we have

$$\frac{d}{ds}(wu_\mu) - \frac{\partial p}{\partial x^\mu} = \frac{d}{ds}(wu_\mu) - \frac{\partial w}{\partial x^\mu}\bigg|_{\sigma=\text{const.}} = -\frac{\partial T_\mu^{\nu(f)}}{\partial x^\nu}, \tag{8.219}$$

where the equation of continuity (8.214) has been used.

It is not difficult to verify that the homogeneous equation

$$\frac{d}{ds}(wu_\mu) - \frac{\partial w}{\partial x^\mu}\bigg|_{\sigma=\text{const.}} = 0 \tag{8.220}$$

has the solution

$$\frac{\partial \varphi}{\partial x^\mu} = w\,u_\mu, \tag{8.221}$$

where $\varphi = \varphi(x^\nu)$ is an arbitrary function of coordinates. The proof is carried out multiplying (8.221) by u^μ, then introducing the result into (8.220).

For $\mu = i$, Eq. (8.221) corresponds to the *vorticity-free condition* of classical fluid mechanics, $\mathbf{v} = \nabla\varphi$, while $\mu = 0$ leads to the *Bernoulli equation* in relativistic hydrodynamics,

$$\frac{\partial \varphi}{\partial t} - c\gamma w = 0. \tag{8.222}$$

To integrate the inhomogeneous Eq. (8.218) we consider the solution as a linear combination

$$\frac{\partial \varphi}{\partial x^\mu} = w u_\mu + \lambda_\mu,$$ (8.223)

where $\lambda_\mu = \lambda_\mu(x^\nu)$. To determine λ_μ, we take the partial derivative with respect to x^ν of (8.223) and we use the Schwarz integrability condition for $\varphi(x^\nu)$:

$$\frac{\partial^2 \varphi}{\partial x^\nu \partial x^\mu} = \frac{\partial^2 \varphi}{\partial x^\mu \partial x^\nu},$$

i.e.

$$\frac{\partial}{\partial x^\nu}(w u_\mu + \lambda_\mu) = \frac{\partial}{\partial x^\mu}(w u_\nu + \lambda_\nu).$$

Next, we multiply this result by u^ν and we get

$$\frac{d}{ds}(w u_\mu) + \frac{d}{ds}(\lambda_\mu) = \frac{\partial w}{\partial x^\mu} - w u_\nu \frac{\partial u^\nu}{\partial x^\mu} + u^\nu \frac{\partial}{\partial x^\mu}(\lambda_\nu),$$

or, using the relation $u^\mu u_\mu = 1$,

$$\frac{d}{ds}(w u_\mu) - \frac{\partial w}{\partial x^\mu}\bigg|_{\sigma=\text{const.}} = -\frac{d}{ds}(\lambda_\mu) + u^\nu \frac{\partial}{\partial x^\mu}(\lambda_\nu).$$

Suppose that λ_ν does not explicitly depend on time. The time-like component of the last equation is then

$$\frac{d}{ds}(w u_0) - \frac{\partial w}{\partial x^0}\bigg|_{\sigma=\text{const.}} = -\frac{d}{ds}(\lambda_0).$$ (8.224)

Comparing (8.224) with the fourth component of (8.219), we find

$$\lambda_0 = \int \frac{\partial T_0^{\nu(f)}}{\partial x^\nu} ds,$$ (8.225)

where the integral extends on the world line on which we have previously chosen the origin. Replacing (8.225) into (8.223), we finally arrive at

$$\frac{\partial \varphi}{\partial t} + c\gamma w - c \int \frac{\partial T_0^{\nu(f)}}{\partial x^\nu} ds = 0,$$

or, after performing the summation over the index ν,

$$\frac{\partial \varphi}{\partial t} + c\gamma w - \int \left[\frac{\partial w_{em}}{\partial t} + \nabla \cdot (\mathbf{E} \times \mathbf{H}) \right] ds = 0.$$ (8.226)

The integro-differential equation (8.226) represents the *generalization of Bernoulli's equation* in relativistic magnetofluid dynamics. If the electromagnetic field is absent, we get back to equation (8.222), as expected.

8.9   Solved Problems

Problem 1. Using Maxwell's equations and the Lorentz transformation (6.37), determine the transformation relations of the electric charge and current densities when passing from one inertial frame S to another S', moving with respect to each other with the velocity $\mathbf{V}$ in the x-direction. Show that the four-volume Ω and the electric charge q are Lorentz invariants.

Solution. Taking $\epsilon \simeq \epsilon_0$, $\mu \simeq \mu_0$, Maxwell's source equations are

$$\nabla \times \mathbf{H} = \mathbf{j} + \frac{\partial \mathbf{D}}{\partial t} = \rho_e \mathbf{v} + \frac{\partial \mathbf{D}}{\partial t}, \qquad (8.227)$$

$$\nabla \cdot \mathbf{D} = 0. \qquad (8.228)$$

Using (6.62), (6.64), (6.79), and (8.228), we have

$$
\begin{aligned}
\rho_e' &= \frac{\partial D_x'}{\partial x'} + \frac{\partial D_y'}{\partial y'} + \frac{\partial D_z'}{\partial z'} \\
&= \left(\Gamma \frac{\partial}{\partial x} + \Gamma \frac{V}{c^2} \frac{\partial}{\partial t} \right) D_x + \epsilon_0 \frac{\partial}{\partial y} \left[\Gamma \left(E_y - V B_z \right) \right] + \epsilon_0 \frac{\partial}{\partial z} \left[\Gamma \left(E_z + V B_y \right) \right] \\
&= \Gamma \nabla \cdot \mathbf{D} + \Gamma \frac{V}{c^2} \left[\frac{\partial D_x}{\partial t} - \frac{1}{\mu_0} (\nabla \times \mathbf{B})_x \right] = \Gamma \rho_e \left(1 - \frac{V}{c^2} v_x \right).
\end{aligned} \qquad (8.229)
$$

In the same way, we find

$$j_x' = \rho_e' v_x' = \Gamma \rho_e \left(1 - \frac{V}{c^2} v_x \right) \frac{v_x - V}{1 - \frac{V}{c^2} v_x} = \Gamma \left(j_x - \rho_e V \right) = \Gamma \rho_e \left(v_x - V \right). \qquad (8.230)$$

Since $|\mathbf{V}| = V_x = V$, and using (6.45) and (6.46), we obtain the invariance of the element of volume in Minkowski space:

$$
\begin{aligned}
d\Omega' &= dx'^0 dx'^1 dx'^2 dx'^3 = c\,dx'\,dy'\,dz'\,dt' \\
&= c\,dx \sqrt{1 - \frac{V^2}{c^2}}\, dy\,dz\,dt \frac{1}{\sqrt{1 - \frac{V^2}{c^2}}} = c\,dx\,dy\,dz\,dt = d\Omega. \qquad (8.231)
\end{aligned}
$$

Also,

$$\rho_e' = \frac{q'}{\Delta \tau'} = q' \frac{\Delta x'^0}{\Delta \Omega'} = \Gamma \rho_e \left(1 - \frac{V}{c^2} v_x \right) = \Gamma \left(1 - \frac{V}{c^2} v_x \right) q \frac{\Delta x^0}{\Delta \Omega},$$

where $\Delta \tau'$ is the element of three-dimensional space volume in the frame S'. Since $\Delta x'^0 = \Gamma \Delta x^0 \left(1 - \frac{V}{c^2} v_x\right)$, we find

$$\frac{q'}{\Delta \Omega'} = \frac{q}{\Delta \Omega} \Leftrightarrow q' = q. \tag{8.232}$$

Problem 2. Determine the electric and magnetic components of the electromagnetic field created by an electric point charge q, in uniform motion along a straight line with velocity $\mathbf{v}$.

Solution. Suppose that the charge q is situated at the origin O' of the proper reference frame S', being in uniform, straight motion along the $Ox \equiv O'x'$ axis. In S' we obviously have

$$\mathbf{E}' = \frac{q}{4\pi\epsilon_0} \frac{\mathbf{r}'}{r'^3},$$

$$\mathbf{B}' = 0. \tag{8.233}$$

According to the discussion in Sect. 8.2.2, since the x-components of $\mathbf{E}$ and $\mathbf{B}$ are not affected by the relative motion of the frames, we have from (8.39):

$$E_x = E'_x = \frac{q}{4\pi\epsilon_0} \frac{x'}{r'^3} = \gamma \frac{1}{4\pi\epsilon_0} \frac{q}{s^3} (x - vt),$$

$$E_y = \Gamma \left(E'_y + v B'_z\right) = \gamma \frac{1}{4\pi\epsilon_0} \frac{q}{s^3} y,$$

$$E_z = \Gamma \left(E'_z - v B'_y\right) = \gamma \frac{1}{4\pi\epsilon_0} \frac{q}{s^3} z, \tag{8.234}$$

where $s^2 = [\gamma(x - vt)]^2 + y^2 + z^2$, as well as

$$B_x = B'_x = 0,$$

$$B_y = \gamma \left(B'_y - \frac{v}{c^2} E'_z\right) = -\frac{v}{c^2} E_z = -\frac{1}{4\pi\epsilon_0} \gamma \frac{v}{c^2} \frac{q}{s^3} z = -\frac{\mu_0}{4\pi} \gamma \frac{qv}{s^3} z,$$

$$B_z = \gamma \left(B'_z + \frac{v}{c^2} E'_y\right) = \frac{v}{c^2} E_y = \frac{\mu_0}{4\pi} \gamma \frac{qv}{s^3} y. \tag{8.235}$$

Thus, we find that in the frame S the expressions of the electric and magnetic fields produced by the moving charge are

$$\mathbf{E} = \gamma \frac{q}{4\pi\epsilon_0} \frac{\mathbf{r} - \mathbf{v}t}{s^3},$$

$$\mathbf{B} = \gamma \frac{\mu_0 q}{4\pi} \frac{\mathbf{v} \times \mathbf{r}}{s^3} = \frac{1}{c^2} \mathbf{v} \times \mathbf{E}. \tag{8.236}$$

Problem 3. Study the relativistic motion of an electron placed in a uniform electric field **E**. At the time $t_0 = 0$, the particle is situated at the common origin of the proper reference frame and laboratory frame, and it has the momentum **p**(0).

Solution. Taking the proper time τ as parameter in the equation of motion (see (8.16)), and taking into account that the electron has negative charge (we consider $e > 0$), we have

$$\frac{d}{d\tau}(m_0 \bar{u}^\mu) = -e F^{\mu\nu} \bar{u}_\nu, \quad \mu, \nu = 0, 1, 2, 3. \tag{8.237}$$

We consider the x-direction parallel to the electric field **E**. Since in this case the only non-zero components of $F^{\mu\nu}$ are

$$F^{01} = -F^{10} = -\frac{1}{c} E_x = -\frac{1}{c} E,$$

the components of the tensor equation

$$\frac{d}{d\tau}(m_0 u^\mu) = -e F^{\mu\nu} u_\nu, \quad \mu, \nu = 0, 1, 2, 3,$$

(which are obtained from (8.237) taking into account that $\bar{u}^\mu = cu^\mu$ and $\bar{u}_\mu = cu_\mu$) are

$$
\begin{aligned}
m_0 \frac{du^0}{d\tau} &= m_0 \frac{du_0}{d\tau} = (-e) F^{01} u_1 = e \frac{1}{c} E u_1, \\
m_0 \frac{du^1}{d\tau} &= -m_0 \frac{du_1}{d\tau} = (-e) F^{10} u_0 = -e \frac{1}{c} E u_0, \\
m_0 \frac{du^2}{d\tau} &= -m_0 \frac{du_2}{d\tau} = 0, \\
m_0 \frac{du^3}{d\tau} &= -m_0 \frac{du_3}{d\tau} = 0.
\end{aligned}
\tag{8.238}
$$

We have arrived at a system of four equations for the four components of the velocity four-vector $u_\mu = u_\mu(\tau)$. The second equation gives

$$\frac{du_0}{d\tau} = \frac{c}{eE} m_0 \frac{d^2 u_1}{d\tau^2}. \tag{8.239}$$

The first equation (8.238) then leads to

$$\frac{d^2 u_1}{d\tau^2} = \omega_0^2 u_1, \tag{8.240}$$

where

$$\omega_0 = \frac{eE}{m_0 c}. \tag{8.241}$$

The solution of (8.240) is

$$u_1 = C_0 \sinh \omega_0 \tau + C_1 \cosh \omega_0 \tau, \qquad (8.242)$$

where C_0 and C_1 are two constants of integration. Introducing (8.242) into the second equation (8.238), we find

$$u_0 = C_0 \cosh \omega_0 \tau + C_1 \sinh \omega_0 \tau. \qquad (8.243)$$

The rest of the equations (8.238) give

$$\begin{aligned} u_2 &= C_2, \\ u_3 &= C_3. \end{aligned} \qquad (8.244)$$

To determine the constants of integration, we use the initial conditions: $x_1(0) = x_2(0) = x_3(0) = 0$, at $\tau_0 = t_0 = 0$. Recalling (7.96) and (7.99), we write the four-velocity in terms of the components of the four-momentum:

$$\begin{aligned} u_0 &= \frac{p_0}{m_0 c} = \frac{W}{m_0 c^2}, \\ u_i &= \frac{p_i}{m_0 c}, \quad i = 1, 2, 3, \end{aligned} \qquad (8.245)$$

and then

$$\begin{aligned} u_0(0) &= \frac{W_0}{m_0 c^2}, \\ u_i(0) &= -\frac{p_i(0)}{m_0 c}, \quad i = 1, 2, 3, \end{aligned} \qquad (8.246)$$

where W_0 is the total energy of the particle at the initial time. The constants of integration are then

$$\begin{aligned} C_0 &= \frac{W_0}{m_0 c^2}, \\ C_i &= -\frac{p_i(0)}{m_0 c}, \quad i = 1, 2, 3, \end{aligned} \qquad (8.247)$$

and the solution (the four-velocity) becomes

$$\begin{aligned} u_0 &= -\frac{p_x(0)}{m_0 c} \sinh \omega_0 \tau + \frac{W_0}{m_0 c^2} \cosh \omega_0 \tau, \\ u_1 &= -\frac{p_x(0)}{m_0 c} \cosh \omega_0 \tau + \frac{W_0}{m_0 c^2} \sinh \omega_0 \tau, \end{aligned}$$

$$u_2 = -\frac{p_y(0)}{m_0 c}, \tag{8.248}$$

$$u_3 = -\frac{p_z(0)}{m_0 c}.$$

Since

$$u^0 = u_0 = \frac{1}{c}\frac{dx^0}{d\tau} = \frac{dt}{d\tau},$$

$$u^i = -u_i = \frac{1}{c}\frac{dx^i}{d\tau}, \quad i = 1, 2, 3, \tag{8.249}$$

we integrate (8.248) with respect to τ. The result is

$$t = -\frac{p_x(0)}{m_0 \omega_0 c}\cosh\omega_0\tau + \frac{W_0}{m_0\omega_0 c^2}\sinh\omega_0\tau + C_0',$$

$$x = \frac{p_x(0)}{m_0\omega_0}\sinh\omega_0\tau - \frac{W_0}{m_0\omega_0 c}\cosh\omega_0\tau + C_1',$$

$$y = \frac{p_y(0)}{m_0}\tau + C_2', \tag{8.250}$$

$$z = \frac{p_z(0)}{m_0}\tau + C_3'.$$

The constants of integration C_0', C_1', C_2', C_3' are determined from the initial conditions:

$$C_0' = -\frac{p_x(0)}{m_0\omega_0 c},$$

$$C_1' = \frac{W_0}{m_0\omega_0 c},$$

$$C_2' = C_3' = 0. \tag{8.251}$$

Replacing the determined constants of integration into (8.250), we arrive at the parametric solutions of the equations of motion:

$$t = \frac{p_x(0)}{m_0\omega_0 c}(1 - \cosh\omega_0\tau) + \frac{W_0}{m_0\omega_0 c^2}\sinh\omega_0\tau,$$

$$x = \frac{p_x(0)}{m_0\omega_0}\sinh\omega_0\tau + \frac{W_0}{m_0\omega_0 c}(1 - \cosh\omega_0\tau),$$

$$y = \frac{p_y(0)}{m_0}\tau, \tag{8.252}$$

$$z = \frac{p_z(0)}{m_0}\tau.$$

To determine the motion in the laboratory frame of reference S, that is $x = x(t)$, $y = y(t)$, $z = z(t)$, the proper time τ must be expressed in terms of the time t elapsed in S. This can be easily done by means of the last relation (8.252). Using the definition of hyperbolic functions, we have

$$t = \frac{p_x(0)}{m_0\omega_0 c}\left(1 - \frac{\lambda + \lambda^{-1}}{2}\right) + \frac{W_0}{m_0\omega_0 c^2}\left(\frac{\lambda - \lambda^{-1}}{2}\right), \tag{8.253}$$

with $\lambda = e^{\omega_0 \tau}$. Denoting

$$r = \frac{p_x(0)}{m_0\omega_0 c},$$

$$q = \frac{W_0}{m_0\omega_0 c^2},$$

we arrive at the quadratic equation

$$\lambda^2(q - r) - 2(t - r)\lambda - (q + r) = 0, \tag{8.254}$$

whose solutions are

$$\lambda_{1,2} = \frac{t - r \pm \sqrt{(t - r)^2 + (q + r)(q - r)}}{q - r}.$$

The only acceptable solution from the physical point of view is

$$\lambda_1 = \frac{eEt - p_x(0) + \sqrt{[eEt - p_x(0)]^2 + m_0^2 c^2 + p_y^2(0) + p_z^2(0)}}{\frac{W_0}{c} - p_x(0)},$$

which gives

$$\tau = \frac{1}{\omega_0}\ln\frac{eEt - p_x(0) + \sqrt{[eEt - p_x(0)]^2 + m_0^2 c^2 + p_y^2(0) + p_z^2(0)}}{\frac{W_0}{c} - p_x(0)}. \tag{8.255}$$

We are now able to express the coordinates of the particle in terms of the time t. The result is

$$x(t) = \frac{c}{eE}\left[\frac{W_0}{c} - \sqrt{[eEt - p_x(0)]^2 + m_0^2 c^2 + p_y^2(0) + p_z^2(0)}\right], \tag{8.256}$$

$$y(t) = \frac{cp_y(0)}{eE}\ln\frac{eEt - p_x(0) + \sqrt{[eEt - p_x(0)]^2 + m_0^2 c^2 + p_y^2(0) + p_z^2(0)}}{\frac{W_0}{c} - p_x(0)},$$

$$z(t) = \frac{cp_z(0)}{eE}\ln\frac{eEt - p_x(0) + \sqrt{[eEt - p_x(0)]^2 + m_0^2 c^2 + p_y^2(0) + p_z^2(0)}}{\frac{W_0}{c} - p_x(0)}.$$

Let us now show that in the non-relativistic limit, i.e. $p(0) \ll m_0 c$, $t \ll m_0 c/eE$, $W_0 \simeq m_0 c^2$, one must obtain the classical (and expected) result

$$
\begin{aligned}
x(t) &= \frac{p_x(0)}{m_0} t - \frac{eE}{2m_0} t^2, \\
y(t) &= \frac{p_y(0)}{m_0} t, \\
z(t) &= \frac{p_z(0)}{m_0} t.
\end{aligned}
\tag{8.257}
$$

Indeed, in the non-relativistic approximation we have

$$
\begin{aligned}
&\sqrt{[eEt - p_x(0)]^2 + m_0^2 c^2 + p_y^2(0) + p_z^2(0)} \\
&= \sqrt{e^2 E^2 t^2 - 2eEp_x(0)t + m_0^2 c^2 + p^2(0)} \\
&\simeq \sqrt{e^2 E^2 t^2 - 2eEp_x(0)t + m_0^2 c^2} = m_0 c \sqrt{1 - \xi},
\end{aligned}
$$

with

$$
\xi = \frac{2eEp_x(0)t}{m_0^2 c^2} - \frac{e^2 E^2 t^2}{m_0^2 c^2} \ll 1.
$$

Expanding $\sqrt{1 - \xi}$ in Taylor series and retaining only the first two terms, we have

$$
m_0 c \sqrt{1 - \xi} \simeq m_0 c \left(1 - \frac{\xi}{2}\right) = m_0 c \left(1 - \frac{eEp_x(0)t}{m_0^2 c^2} + \frac{1}{2} \frac{e^2 E^2 t^2}{m_0^2 c^2}\right).
$$

With this approximation, $(8.256)_1$ yields

$$
\begin{aligned}
x(t) &= \frac{W_0}{eE} - \frac{c}{eE} \sqrt{[eEt - p_x(0)]^2 + m_0^2 c^2 + p_y^2(0) + p_z^2(0)} \\
&\simeq \frac{W_0}{eE} - \frac{m_0 c^2}{eE} \left(1 - \frac{eEp_x(0)t}{m_0^2 c^2} + \frac{1}{2} \frac{e^2 E^2 t^2}{m_0^2 c^2}\right) \simeq \frac{p_x(0)}{m_0} t - \frac{1}{2} \frac{eE}{m_0} t^2,
\end{aligned}
$$

which proves $(8.257)_1$. To justify $(8.257)_2$, we observe that

$$
\sqrt{[eEt - p_x(0)]^2 + m_0^2 c^2 + p_y^2(0) + p_z^2(0)} \simeq m_0 c,
$$

which allows us to write

$$
\frac{eEt - p_x(0) + \sqrt{[eEt - p_x(0)]^2 + m_0^2 c^2 + p_y^2(0) + p_z^2(0)}}{\frac{W_0}{c} - p_x(0)}
$$

$$\simeq \frac{eEt - p_x(0) + m_0 c}{\frac{W_0}{c} - p_x(0)} \simeq \frac{eEt - p_x(0) + m_0 c}{m_0 c - p_x(0)}$$

$$\simeq \frac{eEt + m_0 c}{m_0 c} = 1 + \frac{eEt}{m_0 c}.$$

Since

$$\ln(1 + \zeta) = \zeta - \frac{\zeta^2}{2} + \frac{\zeta^3}{3} - \frac{\zeta^4}{4} + \cdots = \sum_{n=1}^{\infty} (-1)^{n+1} \frac{\zeta^n}{n},$$

with

$$\zeta = \frac{eEt}{m_0 c} \ll 1,$$

we consider only the first term in the expansion of $\ln(1 + \zeta)$. Then $(8.257)_2$ yields

$$y(t) = \frac{c p_y(0)}{eE} \ln(1 + \zeta) \simeq \frac{c p_y(0)}{eE} \zeta = \frac{p_y(0)}{m_0} t,$$

which completes the proof. Relation $(8.257)_3$ is verified in a similar way.

Thus, we have found that in the non-relativistic case, the motion along the x-axis is uniformly accelerated, with acceleration eE/m_0. The graph of this motion is represented in Fig. 8.5 (the first part of the solid line).

After a sufficiently long time ($t \gg m_0 c/eE$) the particle becomes relativistic ($v \to c$), even if at the beginning of the motion it was not. In the ultrarelativistic limit, the parametric equations of motion are given by

$$x(t) = \frac{m_0 c^2}{eE} - ct,$$

$$y(t) = \frac{c p_y(0)}{eE} \ln \frac{2eEt}{m_0 c}, \tag{8.258}$$

$$z(t) = \frac{c p_z(0)}{eE} \ln \frac{2eEt}{m_0 c}.$$

Indeed, we have

$$\sqrt{[eEt - p_x(0)]^2 + m_0^2 c^2 + p_y(0)^2 + p_z(0)^2} \simeq eEt$$

and (8.256) leads to $(8.258)_1$:

$$x(t) = \frac{c}{eE} \left(\frac{W_0}{c} - eEt \right) = \frac{m_0 c^2}{eE} - ct.$$

Fig. 8.5 Graphical representation of the law of motion $x(t)$ for an electron moving in a uniform electric field **E**.

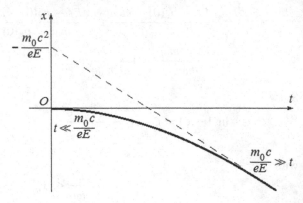

Fig. 8.6 Graphical representation of $y(t)$.

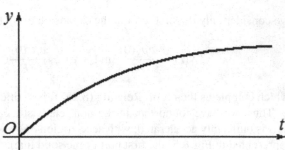

Also,

$$y(t) \simeq \frac{c p_y(0)}{e E} \ln \frac{e E t - p_x(0) + \sqrt{(e E t)^2}}{\frac{m_0 c^2}{c} - p_x(0)} = \frac{c p_y(0)}{e E} \ln \frac{2 e E t}{m_0 c},$$

and similarly for $z(t)$. The graph for the ultrarelativistic limit is represented by the dashed line in Fig. 8.5 for $x(t)$, and in Fig. 8.6 for $y(t)$.

Problem 4. Study the relativistic motion of an electron placed in a uniform magnetic field **B** oriented along the z-axis. At the initial time $t_0 = 0$, the particle is situated at the common origin of the proper reference frame and laboratory frame, and has the momentum $\mathbf{p}(0)$.

Solution. We shall use again the covariant form of the equation of motion for the electron (8.237):

$$\frac{d}{d\tau}(m_0 u^\mu) = - e\, F^{\mu\nu} u_\nu, \quad \mu, \nu = 0, 1, 2, 3. \tag{8.259}$$

This time, the non-zero components of the electromagnetic field tensor $F^{\mu\nu}$ are

$$F^{12} = -F^{21} = -B_z = -B.$$

Then the equations of motion have the form

$$m_0 \frac{du_0}{d\tau} = 0,$$
$$m_0 \frac{du_1}{d\tau} = -eBu_2,$$
$$m_0 \frac{du_2}{d\tau} = +eBu_1, \qquad (8.260)$$
$$m_0 \frac{du_3}{d\tau} = 0.$$

Taking the derivative with respect to τ of the second equation, then introducing the result into the third, we obtain a second-order, linear differential equation for u_1:

$$\frac{d^2 u_1}{d\tau^2} + \omega_0^2 u_1 = 0, \qquad \omega_0 = \frac{eB}{m_0}. \qquad (8.261)$$

The solution of this equation is

$$u_1 = A_1 \sin(\omega_0 \tau + \alpha), \qquad (8.262)$$

α and A_1 being two constants of integration. With (8.262), Eq. (8.260)$_3$ provides the solution u_2:

$$u_2 = -A_1 \cos(\omega_0 \tau + \alpha). \qquad (8.263)$$

The first and the last of the equations (8.260) yield

$$u_0 = A_0,$$
$$u_3 = A_3. \qquad (8.264)$$

According to the initial conditions, at $\tau_0 = t_0 = 0$ we have

$$u_0(0) = \frac{W_0}{m_0 c^2},$$
$$u_1(0) = -\frac{p_x(0)}{m_0 c},$$
$$u_2(0) = -\frac{p_y(0)}{m_0 c}, \qquad (8.265)$$
$$u_3(0) = -\frac{p_z(0)}{m_0 c},$$

that is

$$\frac{W}{m_0 c^2} = A_0,$$

$$-\frac{p_x(0)}{m_0 c} = A_1 \sin \alpha,$$

$$\frac{p_y(0)}{m_0 c} = A_1 \cos \alpha , \qquad (8.266)$$

$$\frac{p_z(0)}{m_0 c} = A_3.$$

The constants α and A_1 are found to be

$$\tan \alpha = -\frac{p_x(0)}{p_y(0)} \quad \Rightarrow \quad \alpha = -\arctan \frac{p_x(0)}{p_y(0)} \qquad (8.267)$$

and

$$A_1 = \frac{1}{m_0 c}\sqrt{p_x^2(0) + p_y^2(0)}. \qquad (8.268)$$

The components of the four-velocity are then

$$u_0 = \frac{W}{m_0 c^2},$$

$$u_1 = \frac{1}{m_0 c}\sqrt{p_x^2(0) + p_y^2(0)} \sin \left(\omega_0 \tau - \arctan \frac{p_x(0)}{p_y(0)} \right),$$

$$u_2 = -\frac{1}{m_0 c}\sqrt{p_x^2(0) + p_y^2(0)} \cos \left(\omega_0 \tau - \arctan \frac{p_x(0)}{p_y(0)} \right),$$

$$u_3 = \frac{p_z(0)}{m_0 c}. \qquad (8.269)$$

Integrating (8.249), we obtain the law of motion:

$$t = \frac{W_0}{m_0 c^2}\tau + b_0,$$

$$x = \frac{\sqrt{p_x(0)^2 + p_y(0)^2}}{eB} \cos \left(\omega_0 \tau - \arctan \frac{p_x(0)}{p_y(0)} \right) + b_1,$$

$$y = \frac{\sqrt{p_x(0)^2 + p_y(0)^2}}{eB} \sin \left(\omega_0 \tau - \arctan \frac{p_x(0)}{p_y(0)} \right) + b_2, \qquad (8.270)$$

$$z = \frac{p_z(0)}{m_0}\tau + b_3.$$

The constants of integration b_ν, $\nu = 0, 1, 2, 3$ are determined, in their turn, from the initial conditions:

$$b_0 = 0,$$

$$b_1 = -\frac{\sqrt{p_x^2(0) + p_y^2(0)}}{eB} \cos\left(-\arctan\frac{p_x(0)}{p_y(0)}\right),$$

$$b_2 = -\frac{\sqrt{p_x^2(0) + p_y^2(0)}}{eB} \sin\left(-\arctan\frac{p_x(0)}{p_y(0)}\right), \qquad (8.271)$$

$$b_3 = 0.$$

Using the trigonometric formulas

$$\sin\xi = \frac{\tan\xi}{\sqrt{1 + \tan^2\xi}},$$

$$\cos\xi = \frac{1}{\sqrt{1 + \tan^2\xi}},$$

one can also write

$$b_1 = -\frac{\sqrt{p_x^2(0) + p_y^2(0)}}{eB} \cos\left(-\arctan\frac{p_x(0)}{p_y(0)}\right)$$

$$= \frac{\sqrt{p_x^2(0) + p_y^2(0)}}{eB} \cos\left(\arctan\frac{p_x(0)}{p_y(0)}\right)$$

$$= -\frac{\sqrt{p_x^2(0) + p_y^2(0)}}{eB} \left[1 + \tan^2\left(\arctan\frac{p_x(0)}{p_y(0)}\right)\right]^{-1/2}$$

$$= -\frac{\sqrt{p_x^2(0) + p_y^2(0)}}{eB} \left[1 + \left(\frac{p_x(0)}{p_y(0)}\right)^2\right]^{-1/2} = -\frac{p_y(0)}{eB}.$$

Also,

$$b_2 = \frac{\sqrt{p_x^2(0) + p_y^2(0)}}{eB} \sin\left(\arctan\frac{p_x(0)}{p_y(0)}\right)$$

$$= \frac{\sqrt{p_x^2(0) + p_y^2(0)}}{eB} \frac{\tan\left(\arctan\frac{p_x(0)}{p_y(0)}\right)}{\sqrt{1 + \tan^2\left(\arctan\frac{p_x(0)}{p_y(0)}\right)}}$$

$$= \frac{\sqrt{p_x^2(0) + p_y^2(0)}}{eB} \frac{\frac{p_x(0)}{p_y(0)}}{\sqrt{1 + \left(\frac{p_x(0)}{p_y(0)}\right)^2}} = \frac{p_x(0)}{eB}.$$

Relations (8.270) then become

$$t(\tau) = \frac{W_0}{m_0 c^2} \tau,$$

$$x(\tau) = \frac{\sqrt{p_x^2(0) + p_y^2(0)}}{eB} \cos\left(\omega_0 \tau - \arctan \frac{p_x(0)}{p_y(0)}\right) - \frac{p_y(0)}{eB},$$

$$y(\tau) = \frac{\sqrt{p_x^2(0) + p_y^2(0)}}{eB} \sin\left(\omega_0 \tau - \arctan \frac{p_x(0)}{p_y(0)}\right) + \frac{p_x(0)}{eB}, \qquad (8.272)$$

$$z(\tau) = \frac{p_z(0)}{m_0} \tau.$$

The last step is to express the space-like coordinates in terms of time t. This is done by means of the last relation of (8.272):

$$\tau = \frac{m_0 c^2}{W_0} t,$$

and we finally obtain

$$x(t) = \frac{\sqrt{p_x^2(0) + p_y^2(0)}}{eB} \cos\left(\frac{\omega_0 m_0 c^2}{W_0} t - \arctan \frac{p_x(0)}{p_y(0)}\right) - \frac{p_y(0)}{eB},$$

$$y(t) = \frac{\sqrt{p_x^2(0) + p_y^2(0)}}{eB} \sin\left(\frac{\omega_0 m_0 c^2}{W_0} t - \arctan \frac{p_x(0)}{p_y(0)}\right) + \frac{p_x(0)}{eB},$$

$$z(t) = \frac{p_z(0) c^2}{W_0} t, \qquad (8.273)$$

or, in a more condensed form,

$$x(t) = -\frac{p_y(0)}{eB} + \frac{p_1(0)}{eB} \cos(\omega_1 t + \alpha),$$

$$y(t) = \frac{p_x(0)}{eB} + \frac{p_1(0)}{eB} \sin(\omega_1 t + \alpha), \qquad (8.274)$$

$$z(t) = \frac{p_z(0) c^2}{W_0} t,$$

with the notations

$$\omega_1 = \omega_0 \frac{m_0 c^2}{W_0},$$

$$p_1(0) = \sqrt{p_x^2(0) + p_y^2(0)}, \tag{8.275}$$

and recalling that $\arctan[p_x(0)/p_y(0)] = -\alpha$.

The trajectory of the electron is, therefore, a helix rolled up on a right cylinder with the axis oriented along Oz (direction of the field), of radius

$$R = \frac{p_1(0)}{eB} = \frac{\sqrt{p_x^2(0) + p_y^2(0)}}{eB} \tag{8.276}$$

and of constant pitch

$$\delta = \frac{2\pi z}{\omega_1 t} = \frac{2\pi p_z(0)}{\omega_0 m_0} = \frac{2\pi p_z(0)}{eB}. \tag{8.277}$$

Problem 5. If **S** is the intrinsic angular momentum of the electron (the spin) and **M** its magnetic moment, then in the proper frame of the electron one can write

$$\left(\frac{d\mathbf{S}}{dt}\right)_{rest} = \mathbf{M} \times \mathbf{B}'. \tag{8.278}$$

Here $\mathbf{B}' \simeq \mathbf{B} - \frac{1}{c^2}\mathbf{v} \times \mathbf{E}$ is the magnetic field in the rest frame of the electron, **v** is the velocity of the electron in an arbitrary inertial frame, and

$$\mathbf{M} = g_s \frac{e}{2m_0} \mathbf{S}, \tag{8.279}$$

where g_s is the *spin gyromagnetic factor* and m_0 is the mass of the electron. Write Eq. (8.278) in the relativistically-covariant form.

Solution. Recall from Sect. 8.7.3 that the intrinsic angular momentum can be written in Lorentz covariant form. In the following we shall denote by prime superscript the quantities in the rest frame. Let S^μ be an axial four-vector with three independent components, which in the rest frame of the electron reduces to the spin vector, i.e. $S'^\mu = (0, \mathbf{S})$. As any four-vector, the time component of the spin four-vector transforms like $x'^0 = \Gamma\left(x^0 - \mathbf{v}\cdot\mathbf{r}/c\right)$, that is

$$S'^0 = \Gamma\left(S^0 - \frac{1}{c}\mathbf{v}\cdot\mathbf{S}\right). \tag{8.280}$$

Take now the contracted product between S^μ and the velocity four-vector $u^\mu = \left(\Gamma, \frac{\Gamma}{c}\mathbf{v}\right)$:

$$u_\mu S^\mu = u_0 S^0 + u_i S^i = \Gamma\left(S^0 - \frac{1}{c}\mathbf{v}\cdot\mathbf{S}\right). \tag{8.281}$$

From the last two relations, (8.280) and (8.281), we infer that

$$S^{\prime 0} = u_\mu S^\mu. \tag{8.282}$$

According to (8.282), the condition $S^{\prime 0} = 0$ can be imposed covariantly as

$$u_\mu S^\mu = 0, \tag{8.283}$$

while (8.280) yields

$$S^0 = \frac{\Gamma}{c} \mathbf{v} \cdot \mathbf{S}. \tag{8.284}$$

Using (8.279), we put (8.278) in the form

$$\left(\frac{d\mathbf{S}}{dt} \right)_{rest} = \frac{g_s e}{2 m_0} \left(\mathbf{S} \times \mathbf{B}' \right). \tag{8.285}$$

To write this equation in a covariant form, we first observe that both sides have to be four-vectors. In the l.h.s., the derivative must be taken with respect to an invariant (either the interval s, or the proper time τ), while the r.h.s. has to contain covariant combinations of only the four-velocity u^μ, four-acceleration a^μ, the spin four-vector S^μ, and the electromagnetic field tensor $F^{\mu\nu}$, and to be linear in S^μ and $F^{\mu\nu}$. Last but not least, in the rest frame the equation has to go to (8.285). The only non-vanishing combinations with four-vector structure, satisfying the above conditions, are

$$F^{\mu\nu} S_\nu, \quad \left(S_\nu F^{\nu\lambda} u_\lambda \right) u^\mu, \quad \left(S_\nu \frac{du^\nu}{ds} \right) u^\mu.$$

Therefore we take as a four-dimensional generalization of (8.285) the equation

$$\frac{dS^\mu}{ds} = C_1 F^{\mu\nu} S_\nu + C_2 \left(S_\nu F^{\nu\lambda} u_\lambda \right) u^\mu + C_3 \left(S_\nu \frac{du^\nu}{ds} \right) u^\mu, \tag{8.286}$$

where the constants C_1, C_2, C_3 are to be determined. Taking the derivative with respect to s of the constraint equation (8.283), and using the relation $u^\mu u_\mu = 1$, we find

$$(C_3 + 1) S_\mu \frac{du^\mu}{ds} + (C_2 - C_1) F^{\nu\mu} S_\nu u_\mu = 0. \tag{8.287}$$

If the electromagnetic field is absent ($F^{\mu\nu} = 0$), while $du^\mu/ds \neq 0$, then $C_3 = -1$. Equation (8.287) therefore becomes

$$(C_2 - C_1) F^{\nu\mu} S_\nu u_\mu = 0, \tag{8.288}$$

which means that in the presence of the electromagnetic field, $C_1 = C_2$. Thus we obtain from (8.286)

$$\frac{dS^\mu}{ds} = C_1 F^{\mu\nu} S_\nu + C_1 S_\nu F^{\nu\lambda} u_\lambda u^\mu - S_\nu \frac{du^\nu}{ds} u^\mu. \tag{8.289}$$

Recalling that in the rest reference frame $S'^\mu = (0, \mathbf{S})$ and $u'^\mu = (1, 0)$, let us write the i-component of (8.289):

$$\frac{dS^i}{ds} = \frac{dS^i}{cdt'} = C_1 (\mathbf{S} \times \mathbf{B}')^i,$$

or

$$\left(\frac{d\mathbf{S}}{dt}\right)_{rest} = cC_1 (\mathbf{S} \times \mathbf{B}'). \tag{8.290}$$

This equation is identical to (8.285) if one takes

$$C_1 = \frac{g_s e}{2m_0 c}, \tag{8.291}$$

and the equation of motion (8.286) becomes

$$\frac{dS^\mu}{ds} = \frac{g_s e}{2m_0 c}(F^{\mu\nu} S_\nu + S_\nu F^{\nu\lambda} u_\lambda u^\mu) - S_\nu \frac{du^\nu}{ds} u^\mu. \tag{8.292}$$

Using the covariant form of the equation of motion of a charged particle in an external electromagnetic field,

$$\frac{dp^\nu}{ds} = m_0 c \frac{du^\nu}{ds} = e F^{\nu\lambda} u_\lambda,$$

we finally cast the equation of spin motion for the electron in the form

$$\frac{dS^\mu}{ds} = \frac{e}{m_0 c} \left[\frac{g_s}{2} F^{\mu\nu} S_\nu + \left(\frac{g_s}{2} - 1\right) S_\nu F^{\nu\lambda} u_\lambda u^\mu \right]. \tag{8.293}$$

This equation is known as *Bargmann–Michel–Telegdi (BMT) equation* and describes relativistically the spin precession of high-velocity particles. The equation is named after Valentine Bargmann (1908–1989), Louis Michel (1923–1999), and Valentine Telegdi (1922–2006). It is the key formula used in high-energy physics experiments to compute the $g_s - 2$ factor, or anomalous magnetic moment of the electron and muon. Its use is based on the fact that, according to the quantum mechanical treatment of the electron by Dirac equation, the gyromagnetic ratio g_s is exactly equal to 2. However, quantum field theory gives higher order corrections, such that $g_s - 2 \neq 0$. Roughly, the idea of the experiment is to have electrons move in an orthogonal magnetic field. The electrons describe a circle with the cyclotron frequency $\omega_c = \frac{e}{m_0} \frac{1}{\Gamma} |\mathbf{B}|$. When the electrons complete one circle, their velocity returns to the initial direction, but the spin has precessed according to the first term in formula (8.293) by an amount proportional to $g_s - 2$. The experiment is done with electrons whose spins are initially polarized in the direction of motion. Due to the precession phenomenon, they develop (very slowly) a component of polarization transverse to the direction of motion. The g_s factor of the electron is measured in very high precision experiments to be 2.00231930436182, with an uncertainty of 5.2×10^{-13}. This is one of the most precisely measured values in physics, and a stringent test of quantum electrodynamics (QED).

8.10   Proposed Problems

1. A particle of mass m and electric charge e is placed in a static combination of electric and magentic fields, $\mathbf{E}, \mathbf{B}$. Find the law of motion, $\mathbf{r} = \mathbf{r}(t)$, and the equation of the trajectory by means of the Hamilton–Jacobi formalism (based on Eq. (8.6)), if at the initial moment the particle is at the point $P_0(x_0, y_0, z_0)$ and it has the velocity $\mathbf{v}_0 = (0, v_0 \sin\alpha, v_0 \cos\alpha)$, while $\mathbf{E} = (0, E, 0)$ and $\mathbf{B} = (B, 0, 0)$.

2. Calculate:

   a) $\tilde{\tilde{F}}^{\mu\nu}$, where $\tilde{F}^{\mu\nu}$ is the dual electromagnetic field tensor;

   b) the contracted product $T^{\mu\nu} T_{\mu\nu}$, where $T^{\mu\nu}$ is the energy-momentum tensor of the electromagnetic field.

3. The electromagnetic field tensor associated to a monochromatic plane wave in vacuum is $F_{\mu\nu} = F^0_{\mu\nu}\, e^{ik^\rho x_\rho}$, where $F^0_{\mu\nu}$ are constant amplitudes and $k^\mu = \left(\frac{\omega}{c}, \mathbf{k}\right)$ is the wave four-vector. Show that $F_{\mu\nu} k^\mu = 0$, and write the space-like and time-like components of this relation.

4. A particle of mass m_0 and charge q performs a relativistic motion in the external electromagnetic field $\mathbf{E}, \mathbf{B}$, in the laboratory frame. Using the covariant formalism presented in Sect. 8.1.2, find the differential equations of motion of the particle. If $\mathbf{E} = (0, 0, E)$, $\mathbf{B} = (0, B, 0)$, $\mathbf{v}_0 = (v_0 \cos\alpha, 0, v_0 \sin\alpha)$ and $\mathbf{r}_0 = (0, 0, 0)$, find the law of motion and the trajectory of the particle.

5. Given the components $\mathbf{E}, \mathbf{B}$ of a uniform electromagnetic field in a fixed inertial frame, with the property $\mathbf{E} \cdot \mathbf{B} > 0$, find the velocity $\mathbf{V}$ of the frame in which $\mathbf{E}$ and $\mathbf{B}$ are parallel.

6. In the inertial reference frame S, the angle between the static fields $\mathbf{E}$ and $\mathbf{B}$ is θ, and $|\mathbf{B}| = k|\mathbf{E}|/c$, while in the inertial frame S' they are parallel and $\mathbf{E} \perp \mathbf{V}$, $\mathbf{B} \perp \mathbf{V}$, where $\mathbf{V}$ is the relative velocity of the frames.

   a) Show that $\alpha = V/c$ satisfies the equation

   $$\alpha^2 - \alpha\frac{1+k^2}{k\sin\theta} + 1 = 0.$$

   Use the following two methods:
   (i) geometrical, starting from the relative position of the vectors $\mathbf{E}, \mathbf{E}', \mathbf{B}, \mathbf{B}',$ and $\mathbf{V}$ and using relations (8.40);

(ii) starting from the general relations of transformation of the vectors **E** and **B**, (8.41);

b) Show that the equation found previously always admits a real solution and calculate $\alpha = \alpha(\theta)$ when $k = 1$. What happens when $\theta = \pi/2$?

7. In the reference frame S the static fields **E** and **B** are orthogonal and have different moduli. Determine

a) the velocity $\mathbf{V}_B$ of an inertial frame S' in which the electric field is zero;

b) the velocity $\mathbf{V}_E$ of an inertial frame S' in which the magnetic field is zero.

8. Show that the four-dimensional Liénard–Wiechert potential can be written as

$$A^{\mu}(x) = \frac{\mu_0 q c}{4\pi} \frac{u^{\mu}(\tau)}{(x - \xi)u}\Bigg|_{\tau = \tau_0},$$

where x, ξ, u are four-vectors, and τ is the proper time of the charged particle. Here ξ is the radius-vector of the particle, whose velocity is $\mathbf{v}(t) = d\xi/dt$, $\xi^{\mu} = (ct_0, \mathbf{r}_0)$, $u^{\mu}(s) = d\xi^{\mu}/ds = \frac{1}{c}\frac{d\xi^{\mu}}{d\tau} = \frac{1}{c}u^{\mu}(\tau)$.

9. A Lagrangian density leading directly to the equations of the electromagnetic field in covariant form, $\Box A^{\mu} = 0$, was proposed by Paul Dirac and Vladimir Fock:

$$\mathcal{L} = -\frac{1}{2\mu_0}\partial_{\mu}A_{\nu}\partial^{\mu}A^{\nu}.$$

To obtain the source equations,

$$\Box A^{\mu} = \mu_0 j^{\mu},$$

one has to consider the Lagrangian density

$$\mathcal{L} = -\frac{1}{2\mu_0}\partial_{\mu}A_{\nu}\partial^{\mu}A^{\nu} - j_{\mu}A^{\mu}. \tag{8.294}$$

a) Write the Euler–Lagrange equations for the Lagrangian density (8.294) and determine in which conditions they coincide with Maxwell's equations;

b) Show that the Lagrangian density (8.294) differs from the one given by Eq. (8.77) by a four-divergence. Does this modify in any way the action or the equations of motion?

10. The system composed of the zero-rest mass vector field (the electromagnetic field) and sources is described by the Lagrangian density (8.77). For the system composed of a massive vector field and sources, an appropriate Lagrangian density was first proposed in 1930 by Alexandru Proca (1897–1955) as

$$\mathcal{L} = -\frac{1}{4\mu_0}F^{\mu\nu}F_{\mu\nu} + \frac{m^2 c^2}{2\mu_0 h^2}A^{\mu}A_{\mu} - j^{\mu}A_{\mu}, \tag{8.295}$$

where h is the Planck constant.

a) Using the Euler–Lagrange equations for the Lagrangian density (8.295), find the *Proca equations* (the equations of motion of the massive vector field in interaction with its sources):

$$\partial_\nu F^{\nu\mu} + \frac{m^2 c^2}{h^2} A^\mu = \mu_0 j^\mu, \qquad \mu, \nu = 0, 1, 2, 3, \tag{8.296}$$

i.e. the analogues of Maxwell's source in covariant form, (8.79);

b) Show that the conservation law for the four-current, expressed as continuity equation in Lorentz covariant form, implies that the massive vector field sat-isfies *necessarily* the condition $\partial_\mu A^\mu = 0$. (Remark that for the massive field this condition is not an arbitrary gauge fixing, but a necessary constraint con-dition. Actually, the action of the massive vector field is not gauge invariant, due to the mass term.) Making use of this constraint, show that A^μ satisfies an inhomogeneous equation of the Klein–Gordon type:

$$\left(\Box + \frac{m^2 c^2}{h^2} \right) A^\mu = \mu_0 j^\mu, \qquad \mu = 0, 1, 2, 3;$$

c) Starting from the static limit[6] of the Proca equations with the constraint $\partial_\mu A^\mu = 0$, and considering that the sources are represented by a single point-like charge q at rest in the origin (in which case only the time component of the four-potential, $A^0 = V/c$ is nonvanishing), show that one obtains the static potential with spherical symmetry, called *Yukawa*[7] *potential*:

$$V(r) = \frac{q}{4\pi\epsilon_0} \frac{e^{-\frac{mc}{h}r}}{r}.$$

---

[6] In the static limit the d'Alembertian operator becomes the Laplacian operator.

[7] This is a potential used in particle and atomic physics and it is also called a *screened Coulomb potential*. The name of this potential comes from the Japanese theoretical physicist (and the first Japanese Nobel laureate) Hideki Yukawa (1907–1981).

Part III
Introduction to General Relativity

Chapter 9
General Theory of Relativity

Einstein developed his ideas about the relativistic approach to gravity over many years, culminating with his 1915 papers presented to the Prussian Academy of Science.

Soon after the elaboration of special relativity, Einstein started pondering about the introduction of accelerated motion, in particular due to gravitational interaction, in his theory. In 1907 a conceptual breakthrough took place, when Einstein realized that Galileo's law of free fall was the key element for expanding the principle of relativity to systems moving non-uniformly relative to each other. The mathematical description of these ideas was still to follow. Its development was partly facilitated by Einstein's personal friendship and collaboration with the Swiss mathematician Marcel Grossmann (1979–1936), who introduced Einstein to fundamental concepts in differential geometry and tensor calculus, developed by Elwin Bruno Christoffel (1829–1900), Gregorio Ricci-Curbastro (1853–1925), and Tullio Levi-Civita (1873–1941), as well as Riemannian geometry, initiated in 1854 by the German mathematician Bernhard Riemann (1826–1866). During the following eight years, Einstein worked out the conceptual and most of the technical details of general relativity, whose epitome are the second-order nonlinear differential equations known as *Einstein's field equations* which connect the curvature of space-time to the energy-momentum tensor of matter.

General relativity is a *geometric theory of gravitation*. In this theory, gravity is not seen as a force, but as an intrinsic distortion of space-time. According to Newtonian gravity, particles' trajectories deviate from the straight line due to the gravitational force. In general relativity, matter and energy deform the space-time into a curved manifold, which encodes all the gravitational effects. Particles still move on the straightest possible lines of the distorted space-time.

Newton's theory of gravitation gave a correct description of the majority of gravitational phenomena, but there were some effects which could not be explained within the Newtonian framework. Einstein's theory was brilliantly confirmed by the astrophysical and experimental observations. The most well-known are:

© Springer-Verlag Berlin Heidelberg 2016
M. Chaichian et al., *Electrodynamics*, DOI 10.1007/978-3-642-17381-3_9

1) Precession of the perihelion of planets;
2) Deviation of light in intense gravitational fields;
3) Redshift of the spectral lines of atoms situated in intense gravitational fields;
4) Retardation of radar signals due to gravity;
5) Direct detection of gravitational waves.

General relativity and the modern theories of gravitation have developed tremendously during the past century, influencing and being influenced by the progress in mathematics, astrophysics, and cosmology. There are many extensive monographs dedicated to this subject. In the following, we shall present an outline of the main concepts and basic techniques, from a historical perspective. We hope to make the reader interested in pursuing this topic at an advanced level.

9.1   Classical Theory of Gravitation

The Newtonian theory of gravitation is based on the law

$$\mathbf{F} = -k\frac{MM'}{r^3}\mathbf{r}, \tag{9.1}$$

which expresses the *force of attraction* between two point masses M and M', situated at a distance r from each other. The value of the constant k depends on the system of units.

The gravitational and electrostatic fields are in many ways alike: they are described by similar laws, and both gravitational and electrostatic forces obey the action and reaction principle. One difference is that the gravitational acceleration does not depend on the mass of the body, while in a given electrostatic field the acceleration of a particle depends on its charge e ($a = eE/m$).

There are two kinds of mass that can be defined for a body:

a)   *inertial mass*, m, encountered in the fundamental equation of Newtonian mechanics, $\mathbf{F} = m\mathbf{a}$;

b)   *gravitational mass*, M, appearing in the formula of the gravitational force, $\mathbf{G} = M\mathbf{g}$.

Since the gravitational acceleration of a body does not depend on its mass, it means that the ratio $C = M/m$ must be the same for all bodies, therefore (9.1) can be written as

$$\mathbf{F} = -k\,C^2\frac{m\,m'}{r^3}\mathbf{r}. \tag{9.2}$$

If one postulates $m = M$, then $C = 1$, and (9.2) becomes

$$\mathbf{F} = -k\,\frac{m\,m'}{r^3}\mathbf{r}. \tag{9.3}$$

The constant k is usually denoted by G. Its value in SI is

$$G = 6.67384(80) \times 10^{-11} \text{N} \cdot \text{m}^2 \cdot \text{kg}^{-2}, \tag{9.4}$$

with relative standard uncertainty 1.2×10^{-4}. Since both relations (9.3) and $F = mg$ express the same gravitational force, we may write

$$\mathbf{F} = -G\frac{m\,m'}{r^3}\mathbf{r} = m'\mathbf{g},$$

which yields

$$\mathbf{g} = -G\frac{m}{r^3}\mathbf{r}. \tag{9.5}$$

Here by $\mathbf{g}$ we mean the *intensity of the gravitational field*, produced at a point determined by the radius vector $\mathbf{r}$, by the point mass m situated at the origin of the coordinate axes.

The equality of the inertial and the gravitational masses of a body is postulated in the *principle of equivalence*, which is one the cornerstones of Einstein's general theory of relativity. One of the famous experiments giving substance to Einstein's ideas was performed by Lórand Eötvös (1848–1919) in 1889 and reported in 1890. Here is the essence of the experiment.

On a body situated on the surface of the Earth act simultaneously the gravity force $\mathbf{F}_g = M\mathbf{g}$, which depends on the gravitational mass, and the centrifugal force of inertia $\mathbf{f} = m\omega^2\mathbf{r}$, which depends on the inertial mass. The ratio $|\mathbf{F}_g|/|\mathbf{f}|$ will then depend on the ratio $C = M/m$. Eötvös placed two bodies, of gravitational masses M_1, M_2 and inertial masses m_1, m_2 on a torsion balance with its rod oriented East-West (to have a maximum moment of the force $\mathbf{f}$). The rod of the balance can rotate in the horizontal plane. The forces $\mathbf{F}_{g1}$, $\mathbf{f}_1$ and $\mathbf{F}_{g2}$, $\mathbf{f}_2$ acting upon the two bodies are in equilibrium.

Denote $C_1 = M_1/m_1$, $C_2 = M_2/m_2$. If C_1 and C_2 were different, the centrifugal forces acting on the two bodies would be different and create a torque capable of producing the rotation of the rod of the balance. Even if the precision of determination was very high (10^{-8}), such an effect was not observed. The experiment was repeated for about 30 years, by Eötvös and other researchers, but the result was always the same. Einstein had postulated the *equivalence between inertial and gravitational masses* in 1907 without knowing about Eötvös's experiment, however he later cited this high precision experiment in support of his ideas.

The similarity between electrostatic and gravitational quantities allows one to establish the following analogy between the fundamental equations which govern the two types of phenomena (G is the *gravitational constant* and $k_e = \frac{1}{4\pi\varepsilon_0}$ is the *electrostatic constant*[1]):

---

[1] All these formulas are written in SI.

$$\mathbf{E} = \frac{\mathbf{F}_e}{q} = k_e \frac{q}{r^3}\mathbf{r} = -\nabla V \qquad\qquad \mathbf{g} = \frac{\mathbf{F}_g}{m} = -G\frac{m}{r^3}\mathbf{r} = -\nabla\varphi$$

$$V = k_e \frac{q}{r} \qquad\qquad \varphi = -G\frac{m}{r}$$

$$V = k_e \sum_{i=1}^{n} \frac{q_i}{r_i} \qquad\qquad \varphi = -G\sum_{i=1}^{n} \frac{m_i}{r_i}$$

$$V = k_e \int \frac{\rho_e(\mathbf{r}')d\mathbf{r}'}{|\mathbf{r} - \mathbf{r}'|} \qquad\qquad \varphi = -G \int \frac{\rho_m(\mathbf{r}')d\mathbf{r}'}{|\mathbf{r} - \mathbf{r}'|}$$

$$\oint \mathbf{E} \cdot d\mathbf{S} = 4\pi k_e \sum_{i=1}^{n} q_i \qquad\qquad \oint \mathbf{g} \cdot d\mathbf{S} = -4\pi G \sum_{i=1}^{n} m_i$$

$$\nabla \cdot \mathbf{E} = 4\pi k_e \rho_e \qquad\qquad \nabla \cdot \mathbf{g} = -4\pi G \rho_m$$

$$\Delta V = -4\pi k_e \rho_e \qquad\qquad \Delta\varphi = 4\pi G \rho_m$$

Consequently, the gravitational potential φ, at some point of the field produced by a point mass (or a continuous/discrete mass distribution) can be determined if one knows the mass density ρ_m (or mass m), by using Poisson's (or Laplace's) equation. This formalism allows one to determine the trajectory of a planet, the geometric elements of the trajectory, etc. The theory is in good agreement with observational data so that, at the beginning of the 19th century, the Newtonian theory of gravitation was providing a good enough frame for the description of gravitational phenomena.

But in 1859 the French mathematician and astronomer Urbain Le Verrier (1811–1877) was the first to realize that the slow precession of Mercury's orbit around the Sun could not be completely explained by Newtonian mechanics. He observed an advance of Mercury's perihelion of approximately 565 arc seconds per century, but when calculating the advance due to the outer planets, he found it to be 527 arc seconds per century, leaving a residual 38 arc seconds that he was not able to explain using Newton's theory. Later, in 1895, Simon Newcomb (1835–1909) gave an improved value for the precession aberration, namely about 43 arc seconds per century and also realized that the phenomenon was happening with other planets, like Mars and Venus.

There were several attempts to explain the perihelion advance *within* the Newtonian theory. Here are two of them:

1) *The phenomenon could be produced by one or a ring of several unidentified small planets*. But a single planet big enough to explain the perihelion advance should have been identified. On the other hand, the existence of a ring of small planets could not have explained, eventually, the perihelion advance of a *single* planet (Mercury).

2) *The phenomenon could be caused by a broad non-sphericity of the Sun or of its crown*, but the astronomical observations did not confirm this hypothesis.

To solve the problem, Newcomb suggested in 1895 to modify Newton's law (9.1), by replacing r^2 by r^n, with $n = 2.0000001612$. Another proposal was to multiply instead by a factor $(1 + \alpha/r^n)$, or by $e^{-\alpha r}(\alpha > 0)$. But these corrections cannot be applied to *all* planets.

All these failed attempts led to the conclusion that the perihelion advance cannot be explained within the frame of Newtonian theory. A new theory was necessary, able to give a consistent and unitary explanation of the phenomenon. The explanation of the perihelion anomaly was one of the first successes of Albert Einstein's relativistic theory of the gravitational field. We shall return to this calculation in Sect. 9.10.1.

9.2 Principles of General Relativity

Einstein's relativistic theory of the gravitational field, known as the *general theory of relativity*, is based on three postulates:

1) **Equivalence principle**: *The gravitational field and the field of inertial forces, in a reference system conveniently accelerated, are equivalent in small enough regions of space-time.*
2) **Covariance principle**: *The form of physical laws under arbitrary differentiable coordinate transformations is invariant.*
3) **Local Lorentz invariance**: *The rules of special relativity apply locally for all inertial observers.*

Let us briefly justify these postulates. First of all, we observe that we can only locally use inertial frames in general relativity, because gravitation is *omnipresent*. For example, a frame attached to a body moving in the gravitational field is not inertial, but in this frame the effect of the gravitational field is eliminated.

In this context, Einstein imagined the following experiment. A person, in an elevator, is able to realize the existence of the gravitational field by feeling his own weight. If the elevator falls freely, the gravitational field is canceled by the field of inertial forces on the elevator (conveniently accelerated). Hence, the observer is not able to realize whether he is situated in an inertial frame in the absence of gravitation, or freely falls down in gravitational field. He ascertains the validity of the principle of inertia, but not the existence of the gravitational field. The acceleration generated by the gravitational field is independent of the mass of the body, just as the acceleration due to inertial forces.

One can then conclude that in a small region of space, the inertial and gravitational forces are *indiscernible*. It is necessary to specify that the phenomenon takes place in a limited place in space, because on the one hand the gravitational field is a source-field, while the field of inertial forces is source-free, and on the other hand the gravitational field goes to zero at infinity, while the field of inertial forces is either zero, or infinite at infinity. This conclusion was given by Einstein the status of a *principle*, known as the *equivalence principle*.

Consider a limited domain of the space-time continuum, where a gravitational field acts. For such a system the metric is (see Appendix C):

$$ds^2 = g_{\mu\nu}dx^\mu dx^\nu, \quad \mu, \nu = 0, 1, 2, 3, \tag{9.6}$$

where $g_{\mu\nu}$ is the metric tensor. In an inertial frame, as we know, the components of $g_{\mu\nu}$ are constant, while the interval (9.6) is Lorentz-invariant.

In a non-inertial frame, the components of the metric tensor become functions of coordinates, and, consequently, ds^2 cannot be written anymore as a sum of the squared coordinate differentials. For example, when passing from the reference frame S to the frame S', which rotates about the $Oz \equiv Oz'$ axis with constant angular velocity ω, the space coordinates transform according to

$$x = x' \cos \omega t - y' \sin \omega t,$$
$$y = x' \sin \omega t + y' \cos \omega t,$$
$$z = z'.$$

To determine the metric in the new coordinates, we replace the differentials of the above coordinates in the Minkowski metric,

$$ds^2 = c^2 dt^2 - dx^2 - dy^2 - dz^2,$$

to obtain

$$ds^2 = \left[c^2 - \omega^2 \left(x'^2 + y'^2\right)\right] dt^2 - dx'^2 - dy'^2 - dz'^2 + 2\omega \left(y'dx' - x'dy'\right) dt.$$

Whatever transformation of time we consider, this expression cannot be brought to a Galilean form, such as $\sum_{\mu=1}^4 (dx'^\mu)^2$, which means that the metric tensor cannot be brought to a diagonal form.

This is the situation with the gravitational field: the metric cannot be brought to diagonal form, because the gravitational field cannot be eliminated by any coordinate transformation. The space-time continuum characterized by such a property is called *curved*, the curvature being represented by the *Riemann tensor* (see Sect. 9.6). In special relativity, the Riemann tensor is identically zero, which is expressed by the statement that *Minkowski space is flat*.

According to Einstein's postulates, the existence of a gravitational field implies a metric describing a curved non-Euclidean space. The local equivalence means that on infinitesimal domains the curved space coincides with its *tangent space*, which is Euclidean (or pseudo-Euclidean, in the case of Minkowski space).

A space which satisfies the condition that the metric

$$ds^2 = g_{\mu\nu}dx^\mu dx^\nu$$

is invariant under the general coordinate transformation

$$x'^{\mu} = x'^{\mu}(x^{\nu}), \quad \mu, \, \nu = 0, 1, 2, 3$$

is a *Riemannian space with indefinite metric*. ("Indefinite" means either a bilinear form, a sesquilinear form, or a non-linear functional of a certain degree of homogeneity, defined on the space under consideration.)

Consequently, the space-time manifold with the metric (9.6) is a *Riemannian manifold*. While in Euclidean space an inertial test particle moves along a straight line, in a curved space the shortest distance between two points of the space, for an inertial particle, is the *geodesic* of the space. The geodesic is defined as a curve whose tangent vectors remain parallel if they are transported along the curve.

In Einstein's general relativity, the dynamical laws of physics are therefore replaced by a set of geometric conditions satisfied by the components of the metric tensor $g_{\mu\nu}$, also called *gravitational potentials*, while the forces due to the presence of bodies are replaced by space curving.

To understand Einstein's fundamental ideas, one needs some preliminary mathematical preparations. Since the following approach makes use of elements of Riemannian geometry, we recommend the reader to go through the Appendices B and C.

9.3 Geodesics

According to the local equivalence principle, a particle in inertial motion (i.e. not accelerated) moves along a line which corresponds to the straight line of Euclidean space. This line is called *geodesic*. Let us establish the differential equations of geodesic lines.

Let x^{μ}, $\mu = 0, 1, 2, 3$ be the general coordinates that define the position of a particle in the Riemannian space R_4, and $x^{\mu} = x^{\mu}(s)$ – the parametric equations of a curve passing through two given world-points P_1 and P_2, where s is the arc length of the curve. By definition, this curve is a geodesic if the distance between the world-points P_1 and P_2, measured on the curve, has a minimum value. The curve satisfying this condition is given by the variational principle

$$\delta \int_{P_1}^{P_2} ds = \delta \int_{P_1}^{P_2} \left(g_{\mu\nu}dx^{\mu}dx^{\nu}\right)^{1/2} = 0,$$

or, if we denote $\dot{x}^{\mu} = dx^{\mu}/ds$,

$$\delta \int_{P_1}^{P_2} \left(g_{\mu\nu}\dot{x}^{\mu}\dot{x}^{\nu}\right)^{1/2} ds = 0.$$

The functional

$$I = \int_1^2 f\left(x^\mu(s), \dot{x}^\mu(s), s\right) ds$$

has an extremum only if f satisfies the Euler–Lagrange equations

$$\frac{d}{ds}\left(\frac{\partial f}{\partial \dot{x}^\lambda}\right) - \frac{\partial f}{\partial x^\lambda} = 0 . \tag{9.7}$$

Replacing $f = \left(g_{\mu\nu}\dot{x}^\mu\dot{x}^\nu\right)^{1/2}$ into (9.7), and using the constraint $f \equiv 1$, we have

$$\frac{\partial f}{\partial \dot{x}^\lambda} = \frac{1}{2}\left(g_{\mu\nu}\dot{x}^\mu\dot{x}^\nu\right)^{-1/2}\left[g_{\mu\nu}\left(\delta^\mu_\lambda\dot{x}^\nu + \delta^\nu_\lambda\dot{x}^\mu\right)\right] = g_{\lambda\nu}\dot{x}^\nu,$$

$$\frac{\partial f}{\partial x^\lambda} = \frac{1}{2}\left(g_{\mu\nu}\dot{x}^\mu\dot{x}^\nu\right)^{-1/2}\frac{\partial g_{\mu\nu}}{\partial x^\lambda}\dot{x}^\mu\dot{x}^\nu = \frac{1}{2}\frac{\partial g_{\mu\nu}}{\partial x^\lambda}\dot{x}^\mu\dot{x}^\nu, \tag{9.8}$$

leading to

$$\frac{d}{ds}\left(g_{\lambda\nu}\dot{x}^\nu\right) - \frac{1}{2}\frac{\partial g_{\mu\nu}}{\partial x^\lambda}\dot{x}^\mu\dot{x}^\nu = 0. \tag{9.9}$$

Denoting

$$\varphi = \frac{1}{2}g_{\mu\nu}\dot{x}^\mu\dot{x}^\nu, \tag{9.10}$$

we have also

$$\frac{d}{ds}\left(\frac{\partial \varphi}{\partial \dot{x}^\lambda}\right) - \frac{\partial \varphi}{\partial x^\lambda} = 0,$$

showing that the variational equations

$$\delta \int_1^2 \left(g_{\mu\nu}\dot{x}^\mu\dot{x}^\nu\right)^{1/2} ds = 0$$

and

$$\delta \int_1^2 g_{\mu\nu}\dot{x}^\mu\dot{x}^\nu ds = 0$$

are equivalent.

Going back to Eq. (9.9), let us carry out the derivative in the first term. Some convenient index manipulation gives

$$\frac{d}{ds}\left(g_{\lambda\nu}\dot{x}^\nu\right) = \frac{\partial g_{\lambda\nu}}{\partial x^\mu}\dot{x}^\mu\dot{x}^\nu + g_{\lambda\nu}\ddot{x}^\nu$$

$$= \frac{1}{2}\left(\frac{\partial g_{\lambda\nu}}{\partial x^\mu} + \frac{\partial g_{\lambda\mu}}{\partial x^\nu}\right)\dot{x}^\mu\dot{x}^\nu + g_{\lambda\nu}\ddot{x}^\nu,$$

and thus (9.9) becomes

$$g_{\lambda\nu}\ddot{x}^{\nu} + \frac{1}{2}\left(\frac{\partial g_{\lambda\nu}}{\partial x^{\mu}} + \frac{\partial g_{\mu\lambda}}{\partial x^{\nu}} - \frac{\partial g_{\mu\nu}}{\partial x^{\lambda}}\right)\dot{x}^{\mu}\dot{x}^{\nu} = 0. \tag{9.11}$$

With the notation

$$\Gamma_{\mu\nu,\lambda} = \frac{1}{2}\left(\frac{\partial g_{\lambda\nu}}{\partial x^{\mu}} + \frac{\partial g_{\mu\lambda}}{\partial x^{\nu}} - \frac{\partial g_{\mu\nu}}{\partial x^{\lambda}}\right), \tag{9.12}$$

equation (9.11) yields

$$g_{\lambda\nu}\ddot{x}^{\nu} + \Gamma_{\mu\nu,\lambda}\dot{x}^{\mu}\dot{x}^{\nu} = 0. \tag{9.13}$$

The quantities $\Gamma_{\mu\nu,\lambda}$, also denoted by $[\mu\nu, \lambda]$ are called *Christoffel symbols of the first kind*. Multiplying (9.13) by $g^{\lambda\rho}$ and recalling (B.30), $g^{\lambda\rho}g_{\lambda\nu} = \delta_{\nu}^{\rho}$, we finally find

$$\ddot{x}^{\rho} + \Gamma_{\mu\nu}^{\rho}\dot{x}^{\mu}\dot{x}^{\nu} = 0, \tag{9.14}$$

where the quantities

$$\Gamma_{\mu\nu}^{\rho} = g^{\lambda\rho}\Gamma_{\mu\nu,\lambda} = \frac{1}{2}g^{\lambda\rho}\left(\partial_{\mu}g_{\lambda\nu} + \partial_{\nu}g_{\mu\lambda} - \partial_{\lambda}g_{\mu\nu}\right), \tag{9.15}$$

also denoted as $\left\{{\rho \atop \mu\nu}\right\}$, are the *Christoffel symbols of the second kind*, also called *connection coefficients*. The definitions (9.12) and (9.15) show that the Christoffel symbols of the second kind are symmetric in the lower indices:

$$\Gamma_{\mu\nu}^{\lambda} = \Gamma_{\nu\mu}^{\lambda}. \tag{9.16}$$

The name is given in honour of the German mathematician Elwin Bruno Christoffel (1829–1900).

Equations (9.14) are the *differential equations of geodesic lines in the Riemannian space R_4*. They also represent the equations of motion of a particle in the gravitational field. The quantities

$$a^{\rho} = \ddot{x}^{\rho} + \Gamma_{\mu\nu}^{\rho}\dot{x}^{\mu}\dot{x}^{\nu} \tag{9.17}$$

are the contravariant components of the acceleration four-vector. In contrast with special relativity, the quantities $\ddot{x}^{\rho}$ are *not four-vectors*. This property can be easily verified by writing their relations of transformation. The same procedure can be used to show that the *Christoffel symbols of the first and second kind are not four-tensors*, as we shall show further. The covariant components of the acceleration four-vector are found in the usual manner:

$$a_{\lambda} = g_{\lambda\rho}a^{\rho} = g_{\lambda\rho}\ddot{x}^{\rho} + \Gamma_{\mu\nu,\lambda}\dot{x}^{\mu}\dot{x}^{\nu}. \tag{9.18}$$

Remark that the acceleration of a particle in the gravitational field depends *only* on the geometric properties of the Riemannian space, that is on the metric tensor and its derivatives with respect to coordinates. As we shall see, the components of $g_{\mu\nu}$ play the role of potentials of the gravitational field, while the quantities $\Gamma^\rho_{\mu\nu}$ determine the *intensity of the field*.

Equations (9.14) and (9.18) emphasize the fact that motion on a geodesic is *unaccelerated motion*, i.e. a particle moving only under the effect of gravity (freely falling on a geodesic) is unaccelerated.

9.4   Covariant Derivatives

9.4.1   Levi-Civita Connection

Let A_μ, $\mu = 0, 1, 2, 3$ be the components of a covariant four-vector, defined on the Riemannian space R_4. Using the transformation rule between two systems of coordinates (see (B.13)):

$$A'_\mu = \frac{\partial x^\nu}{\partial x'^\mu} A_\nu = \underline{x}^\nu_\mu A_\nu,$$ (9.19)

the derivative $\partial A'_\mu / \partial x'^\lambda$ and the total differential dA'_μ transform according to

$$\frac{\partial A'_\mu}{\partial x'^\lambda} = \frac{\partial}{x'^\lambda} \left(\frac{\partial x^\nu}{\partial x'^\mu} \right) A_\nu + \frac{\partial x^\nu}{\partial x'^\mu} \frac{\partial A_\nu}{\partial x^\rho} \frac{\partial x^\rho}{\partial x'^\lambda}$$

$$= \frac{\partial^2 x^\nu}{\partial x'^\lambda \partial x'^\mu} A_\nu + \underline{x}^\nu_\mu \underline{x}^\rho_\lambda \frac{\partial A_\nu}{\partial x^\rho}.$$ (9.20)

Multiplying (9.20) by $dx'^\lambda = \bar{x}^\lambda_\sigma dx^\sigma$, we also have

$$dA'_\mu = \frac{\partial^2 x^\nu}{\partial x'^\lambda \partial x'^\mu} \frac{\partial x'^\lambda}{\partial x^\rho} A_\nu dx^\rho + \underline{x}^\nu_\mu dA_\nu.$$ (9.21)

Relations (9.20) and (9.21) show that in curvilinear coordinates the derivatives of a vector do not transform, in general, like a tensor, while the differential of a vector does not transform, in general, like a vector. This observation is also valid for contravariant vectors. The derivative *is* a tensor, and the differential *is* a vector only in the case of linear transformations, when the second derivative $\frac{\partial^2 x^\nu}{\partial x'^\lambda \partial x'^\mu}$ is zero.

We wish to obtain a tensor which plays, in general (arbitrary) coordinates, the same role as the partial derivative ∂_μ in Galilean coordinates. To achieve this, we have to find the law of transformation of derivatives when passing from Galilean to general coordinates.

To understand the principle of the procedure, let us first consider an Euclidean three-dimensional space. Define in this space an arbitrary vector field **A**, and suppose that $\mathbf{A}(P)$ and $\mathbf{A}(Q)$ are the values of $\mathbf{A}$ at two infinitely close points P and Q. To calculate the differential $d\mathbf{A}$, the origins of the two vectors have to be at the same point, say Q. This is accomplished by displacing $\mathbf{A}(P)$ parallel to itself, until its origin coincides with Q. Then we can write (see Fig. 9.1):

$$d\mathbf{A} = \mathbf{A}(Q) - \mathbf{A}(P). \tag{9.22}$$

The parallel transport of the vector $\mathbf{A}$ is expressed by

$$\mathbf{A} \cdot \boldsymbol{\tau} = \text{const.}, \tag{9.23}$$

where $\boldsymbol{\tau}$ is the unit vector of $\mathbf{PQ}$. Equation (9.23) expresses the fact that during the transport of the vector, the angle between the vector and the direction on which we transport it remains constant, and this is exactly what we mean by parallel transport.

On curved space, the concept of parallel transport is not unequivocal, but it depends on the path on which the vector is parallel transported. An example of parallel transport on a two-sphere is illuminating. Let us consider, as in Fig. 9.2, a vector $\mathbf{A}$ on the

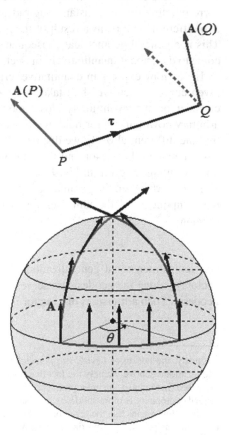

Fig. 9.1 Parallel transport of the vector $\mathbf{A}$ on an Euclidean three-dimensional space (flat space).

Fig. 9.2 Parallel transport on a two-sphere. The result of parallel transport on a curved space depends in general on the path on which it is performed.

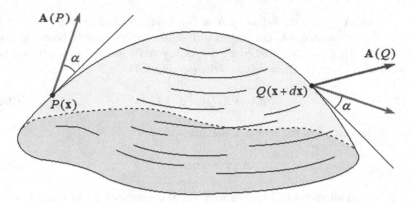

Fig. 9.3 Intuitive representation of the *Levi-Civita parallel transport* on a two-dimensional surface.

equator of the sphere, tangent to a line of constant longitude. Let us parallel transport it to the north pole along that line. Then consider a different path: transport the vector on the equator by an angle θ and then again towards the north pole along the corresponding line of constant longitude. Quite obviously, at the north pole arrived two different vectors, as a result of the parallel transport along two different paths. This is a general geometrical consequence of the curvature of the space. We shall come to describe it quantitatively in Sect. 9.6.

Let us now express in quantitative terms the result of parallel transport along a given curve, which we shall take to be a geodesic of the Riemannian space. We consider on it two infinitely close points $P(x^\mu)$ and $Q(x^\mu + dx^\mu)$. Let A_μ be an arbitrary covariant vector field, with values $A_\mu(P)$ and $A_\mu(Q)$ at the two points. To find the differential of A_μ we displace $A_\mu(P)$ to the point Q in such a way that the angle between the vector and the tangent to the geodesic is always the same (an intuitive image is given in Fig. 9.3).

If $x^\mu = x^\mu(s)$ are the parametric equations of the geodesic, then $dx^\mu/ds = \dot{x}^\mu$ are the components of the unit four-vector of the tangent to the geodesic, while the relation

$$A_\mu \dot{x}^\mu = \text{const.} \qquad (9.24)$$

is a four-dimensional generalization of condition (9.23): the displacement of the vector A_μ along the geodesic preserves the angle between A_μ and the tangent to the geodesic. Such a transport of a vector is called *Levi-Civita parallel transport*.[2]

---

[2]Actually, the notion of *Levi-Civita parallel transport* includes two components: the first is the one mentioned above, namely the fact that under parallel transport the inner product of vectors is preserved (in our case, the inner product between the vector A_μ and the tangent to the geodesic); the second component is more subtle, as it is the requirement that the vectors do not "twist". Roughly speaking, this means that the result of parallel transporting a vector field X along a vector field Y is the 'same' as parallel transporting Y along X.

Let us find the variation δA_μ of A_μ, when transporting it in Levi-Civita manner. Relation (9.24) yields

$$\dot{x}^\mu \delta A_\mu + A_\mu \delta \dot{x}^\mu = 0. \qquad (9.25)$$

Since $\dot{x}^\mu$ is tangent to the geodesic, its components must satisfy Eq. (9.14):

$$\ddot{x}^\sigma + \Gamma^\sigma_{\mu\nu} \dot{x}^\mu \dot{x}^\nu = 0,$$

which gives

$$d\dot{x}^\mu = -\Gamma^\mu_{\nu\lambda} \dot{x}^\nu dx^\lambda (= \delta \dot{x}^\mu),$$

and (9.25) becomes

$$\dot{x}^\mu \delta A_\mu = -A_\mu \delta \dot{x}^\mu = \Gamma^\mu_{\nu\lambda} A_\mu \dot{x}^\nu dx^\lambda.$$

Some index manipulation allows one to identify the coefficients of $\dot{x}^\mu$ of both sides, to get

$$\delta A_\mu = \Gamma^\nu_{\mu\lambda} A_\nu dx^\lambda. \qquad (9.26)$$

Thus, the Levi-Civita variation of the components of A_μ depends on the connection coefficients of the Riemannian space. On flat space, in Cartesian coordinates, the Christoffel symbols of the second kind are all zero, therefore $\delta A_\mu = 0$, and we fall back on (9.22).

As a result of parallel transport, the components of A_μ at the point Q are

$$A_\mu + \delta A_\mu = A_\mu + \Gamma^\nu_{\mu\lambda} A_\nu dx^\lambda.$$

The difference

$$DA_\mu = A_\mu(Q) - [A_\mu(P) + \delta A_\mu] = dA_\mu - \Gamma^\nu_{\mu\lambda} A_\nu dx^\lambda, \qquad (9.27)$$

where $dA_\mu = A_\mu(Q) - A_\mu(P)$ is the usual differential, is called the *absolute differential of the covariant four-vector* A_μ. Since $dA_\mu = (\partial A_\mu / \partial x^\nu) dx^\nu$, we can write also

$$DA_\mu = \left(\partial_\nu A_\mu - \Gamma^\lambda_{\mu\nu} A_\lambda \right) dx^\nu. \qquad (9.28)$$

Being the difference of two four-vectors, the absolute differential is also a four-vector. Since dx^ν, in its turn, is a four-vector, it results that the expression between parentheses is a second-order four-tensor. It is called the *covariant derivative of the covariant four-vector* A_μ:

$$\nabla_\nu A_\mu \equiv A_{\mu;\nu} = \partial_\nu A_\mu - \Gamma^\lambda_{\mu\nu} A_\lambda. \qquad (9.29)$$

Sometimes the covariant derivative along a curve is called *absolute* or *intrinsic* derivative. It is also denoted as $A_{\mu|\nu}$. The covariant derivative ∇_ν expressed by (9.29), with the connection coefficients given by (9.15), is also called *Levi-Civita connection* (but sometimes also *Christoffel connection* or *Riemannian connection*).

Similarly one can define the covariant derivative of a contravariant four-vector. To this end, one uses the fact that a scalar like $A_\mu B^\mu$ is invariant under parallel transport (recall that the inner product of vectors is preserved), that is

$$\delta(A_\mu B^\mu) = A_\mu \delta B^\mu + B^\mu \delta A_\mu = 0,$$

or, by means of (8.37) and a suitable change of summation indices,

$$\delta B^\mu = -\Gamma^\mu_{\nu\lambda} B^\nu dx^\lambda. \tag{9.30}$$

The absolute differential of B^μ therefore is

$$DB^\mu = dB^\mu - \delta B^\mu = \left(\partial_\nu B^\mu + \Gamma^\mu_{\nu\lambda} B^\lambda\right) dx^\nu. \tag{9.31}$$

The mixed tensor

$$\nabla_\nu B^\mu \equiv B^\mu_{;\nu} = \partial_\nu B^\mu + \Gamma^\mu_{\nu\lambda} B^\lambda \tag{9.32}$$

is the *covariant derivative of the contravariant four-vector* B^μ.

In inertial systems, all the 40 distinct quantities $\Gamma^\lambda_{\mu\nu}$ are zero, and the covariant derivative reduces to the common one.

The covariant derivative can be applied to tensors of any order and any variance. Take, for example, the contravariant tensor $T^{\mu\nu}$. Its components transform like the product $A^\mu B^\nu$ (see Appendix B, Sect. B.3). The variation of these quantities under parallel transport is

$$\delta(A^\mu B^\nu) = (\delta A^\mu)B^\nu + A^\mu(\delta B^\nu) = -(\Gamma^\mu_{\sigma\rho} A^\sigma dx^\rho)B^\nu - A^\mu(\Gamma^\nu_{\sigma\rho} B^\sigma dx^\rho),$$

consequently, the variation of $T^{\mu\nu}$ upon parallel transport is given by

$$\delta T^{\mu\nu} = -(\Gamma^\mu_{\sigma\rho} T^{\sigma\nu} + \Gamma^\nu_{\sigma\rho} T^{\mu\sigma})dx^\rho.$$

The absolute differential of $T^{\mu\nu}$ is then

$$DT^{\mu\nu} = d T^{\mu\nu} - \delta T^{\mu\nu} = \left(\partial_\rho T^{\mu\nu} + \Gamma^\mu_{\sigma\rho} T^{\sigma\nu} + \Gamma^\nu_{\sigma\rho} T^{\mu\sigma}\right) dx^\rho, \tag{9.33}$$

leading to the covariant derivative of $T^{\mu\nu}$ as

$$T^{\mu\nu}_{;\rho} = \partial_\rho T^{\mu\nu} + \Gamma^\mu_{\sigma\rho} T^{\sigma\nu} + \Gamma^\nu_{\sigma\rho} T^{\mu\sigma}, \tag{9.34}$$

which is a third-order tensor, once-covariant and twice-contravariant. In a similar way one can define the covariant derivative of a covariant tensor $T_{\mu\nu}$,

$$T_{\mu\nu;\rho} = \partial_\rho T_{\mu\nu} - \Gamma^\lambda_{\rho\nu} T_{\mu\lambda} - \Gamma^\lambda_{\rho\mu} T_{\lambda\nu}, \tag{9.35}$$

and of a mixed tensor,

$$T^\nu_{\mu;\rho} = \partial_\rho T^\nu_\mu + \Gamma^\nu_{\lambda\rho} T^\lambda_\mu - \Gamma^\lambda_{\rho\mu} T^\nu_\lambda. \tag{9.36}$$

We can give a general rule to obtain the covariant derivative of an arbitrary tensor $T^{\cdots}_{\cdots}$ with respect to x^ρ: to the usual derivative one adds a term $(-\Gamma^\lambda_{\mu\rho} T^{\cdots}_{\cdots\lambda\cdots})$ for each covariance index μ $(T^{\cdots}_{\cdots\mu\cdots})$, and a term $(\Gamma^\mu_{\rho\lambda} T^{\cdots\lambda\cdots}_{\cdots})$ for each contravariance index ν $(T^{\cdots\nu\cdots}_{\cdots})$.

Observation:
We can define a *directional covariant derivative* along a curve $x^\mu(s)$ by

$$\frac{D}{ds} = \frac{dx^\rho}{ds} \nabla_\rho. \tag{9.37}$$

If we divide by ds the expression (9.26), we obtain

$$\frac{d}{ds} A_\mu - \Gamma^\nu_{\mu\lambda} A_\nu \frac{dx^\lambda}{ds} = 0. \tag{9.38}$$

This is called the *equation of parallel transport*. It is equivalent to writing

$$\frac{dx^\lambda}{ds} (\partial_\lambda A_\mu - \Gamma^\nu_{\mu\lambda} A_\nu) = 0,$$

or, in view of (9.29) and (9.37),

$$\frac{DA_\mu}{ds} = \frac{dx^\rho}{ds} \nabla_\rho A_\mu = 0. \tag{9.39}$$

The same would be valid for any parallel-transported tensor,

$$\frac{DT^{\mu_1\mu_2\cdots}_{\nu_1\nu_2\cdots}}{ds} = \frac{dx^\rho}{ds} \nabla_\rho T^{\mu_1\mu_2\cdots}_{\nu_1\nu_2\cdots} = 0. \tag{9.40}$$

Thus, the parallel transport of a tensor along the path $x^\mu(s)$ is equivalent to the requirement that the covariant derivative of the tensor along the path vanishes.

9.4.2 Transformation Properties of the Connection Coefficients

Using the concept of absolute differential, let us analyze the tensor character of the connection coefficients $\Gamma^\mu_{\nu\kappa}$. We consider a contravariant four-vector A^ρ and write its absolute differential

$$DA^\rho = dA^\rho + \Gamma^\rho_{\nu\kappa} A^\nu dx^\kappa. \tag{9.41}$$

Being a four-vector, the absolute differential transforms according to (B.8):

$$DA'^{\rho} = \bar{x}^{\rho}_{\lambda} DA^{\lambda},$$

or, in view of (9.41),

$$dA'^{\rho} + \Gamma'^{\rho}_{\nu\kappa} A'^{\nu} dx'^{\kappa} = \frac{\partial x'^{\rho}}{\partial x^{\lambda}} (dA^{\lambda} + \Gamma^{\lambda}_{\nu\kappa} A^{\nu} dx^{\kappa}). \tag{9.42}$$

The differential dA'^{ρ} is obtained in a way similar to that leading to (9.21):

$$dA'^{\rho} = \frac{\partial^2 x'^{\rho}}{\partial x^{\nu} \partial x^{\kappa}} A^{\nu} dx^{\kappa} + \frac{\partial x'^{\rho}}{\partial x^{\nu}} dA^{\nu}. \tag{9.43}$$

Introducing (9.43) into (9.42) and using the transformation relations for A'^{ν} and dx'^{κ}, we have

$$\frac{\partial^2 x'^{\rho}}{\partial x^{\nu} \partial x^{\kappa}} A^{\nu} dx^{\kappa} + \frac{\partial x'^{\rho}}{\partial x^{\nu}} dA^{\nu} + \Gamma'^{\rho}_{\nu\kappa} \frac{\partial x'^{\nu}}{\partial x^{\lambda}} \frac{\partial x'^{\kappa}}{\partial x^{\sigma}} A^{\lambda} dx^{\sigma}$$
$$= \frac{\partial x'^{\rho}}{\partial x^{\lambda}} dA^{\lambda} + \Gamma^{\lambda}_{\nu\kappa} \frac{\partial x'^{\rho}}{\partial x^{\lambda}} A^{\nu} dx^{\kappa}.$$

Now we conveniently interchange the summation indices $\lambda \leftrightarrow \nu$, $\sigma \leftrightarrow \kappa$, then identify the coefficients of $A^{\nu} dx^{\kappa}$ on both sides. The result is

$$\Gamma^{\lambda}_{\nu\kappa} \frac{\partial x'^{\rho}}{\partial x^{\lambda}} = \Gamma'^{\rho}_{\lambda\sigma} \frac{\partial x'^{\lambda}}{\partial x^{\nu}} \frac{\partial x'^{\sigma}}{\partial x^{\kappa}} + \frac{\partial^2 x'^{\rho}}{\partial x^{\nu} \partial x^{\kappa}},$$

or, multiplying by $\partial x^{\mu}/\partial x'^{\rho}$ and performing summation over ρ:

$$\Gamma^{\mu}_{\nu\kappa} = \Gamma'^{\rho}_{\lambda\sigma} \frac{\partial x'^{\lambda}}{\partial x^{\nu}} \frac{\partial x'^{\sigma}}{\partial x^{\kappa}} \frac{\partial x^{\mu}}{\partial x'^{\rho}} + \frac{\partial^2 x'^{\rho}}{\partial x^{\nu} \partial x^{\kappa}} \frac{\partial x^{\mu}}{\partial x'^{\rho}}. \tag{9.44}$$

This relation shows that, in general, the quantities $\Gamma^{\mu}_{\nu\kappa}$ are *not tensors* (except for the case when the coordinate transformation is *linear*, and the last term of the r.h.s vanishes). This was of course expected, since the role of the connection coefficients is to compensate for the fact that the partial derivative of a vector on a curved space does not transform as a tensor, but it has an extra term, as in (9.20). Thus, the addition of two objects which are not tensors, Eq. (9.29), gives in the end a tensor, $A_{\nu;\mu}$.

If a tensor has vanishing components in one coordinate system, it has vanishing components in *any* coordinate system. Since $\Gamma^{\mu}_{\nu\kappa}$ are not tensors, they can be chosen in such a way that all Christoffel symbols are zero along a given curve, or at a given point. Let such a point be the coordinate origin, and let the coordinate transformation be

$$x'^{\rho} = x^{\rho} + \frac{1}{2} (\Gamma^{\rho}_{\nu\kappa})_0 x^{\nu} x^{\kappa}, \tag{9.45}$$

where $(\Gamma^\rho_{\nu\kappa})_0$ are given. We then have

$$\frac{\partial x'^\rho}{\partial x^\nu} = \delta^\rho_\nu + (\Gamma^\rho_{\nu\kappa})_0 x^\kappa \quad \Rightarrow \quad \left(\frac{\partial x'^\rho}{\partial x^\nu}\right)_0 = \delta^\rho_\nu; \quad \frac{\partial^2 x'^\rho}{\partial x^\nu \partial x^\kappa} = (\Gamma^\rho_{\nu\kappa})_0. \qquad (9.46)$$

Introducing these results into (9.44), we find

$$\Gamma^\mu_{\nu\kappa} = \Gamma'^\rho_{\lambda\sigma} \frac{\partial x'^\lambda}{\partial x^\nu} \frac{\partial x'^\sigma}{\partial x^\kappa} \frac{\partial x^\mu}{\partial x'^\rho} + (\Gamma^\rho_{\nu\kappa})_0 \frac{\partial x^\mu}{\partial x'^\rho}.$$

At the origin, this relation becomes

$$(\Gamma^\mu_{\nu\kappa})_0 = (\Gamma'^\rho_{\lambda\sigma})_0 \, \delta^\lambda_\nu \, \delta^\sigma_\kappa \, \delta^\mu_\rho + (\Gamma^\rho_{\nu\kappa})_0 \, \delta^\mu_\rho = (\Gamma'^\mu_{\nu\kappa})_0 + (\Gamma^\mu_{\nu\kappa})_0$$

which yields

$$(\Gamma'^\mu_{\nu\kappa})_0 = 0. \qquad (9.47)$$

Such a coordinate system is called *locally inertial*, or *locally-geodesic*. In fact, at each point, there exist coordinate systems in which the Christoffel symbols vanish at that point.

9.4.3 Other Connections and the Torsion Tensor

The Levi-Civita connection corresponding to the covariant derivative (9.32) is not the only connection that can exist on a curved space-time. It is nevertheless the connection that one obtains by defining geodesics as curves of minimal distance between two points, and in this sense the most intuitive. However, one can define covariant derivatives ∇ more generally, by asking that they generalize the notion of partial derivative on flat space while satisfying a set of natural conditions. These conditions are (below T_1 and T_2 are tensors):

1) linearity: $\nabla(T_1 + T_2) = \nabla T_1 + \nabla T_2$;
2) Leibniz rule: $\nabla(T_1 \otimes T_2) = (\nabla T_1) \otimes T_2 + T_1 \otimes (\nabla T_2)$;
3) commutativity with contractions: $\nabla_\mu(T^\nu_{\nu\rho}) = (\nabla T)_\mu{}^\nu{}_{\nu\rho}$;
4) reduction to partial derivatives on scalars: $\nabla_\mu \varphi = \partial_\mu \varphi$.

With these conditions fulfilled, various connections can be defined by specifying in a given coordinate system the set of connection coefficients $\Gamma^\rho_{\mu\nu}$. In general, the connection coefficients need not be symmetric in the lower indices, hence in four dimensions there will be $64(= 4^3)$ independent components. Thus, any number of connections can be defined on a Riemannian space, each of them with its corresponding covariant derivative. However, once the metric is given, this defines a unique connection, as we shall show below.

To this end, let us first note that the difference of two connections is a tensor. Consider a vector V^λ and the connections ∇_μ and $\tilde{\nabla}_\mu$, specified by the connection coefficients $\Gamma^\rho_{\mu\nu}$ and $\tilde{\Gamma}^\rho_{\mu\nu}$, respectively. Let us now take the difference of the two covariant derivatives:

$$\nabla_\mu V^\rho - \tilde{\nabla}_\mu V^\rho = \partial_\mu V^\rho + \Gamma^\rho_{\mu\nu} V^\nu - \partial_\mu V^\rho - \tilde{\Gamma}^\rho_{\mu\nu} V^\nu = S^\rho_{\mu\nu} V^\nu. \qquad (9.48)$$

Clearly, the left-hand side is a tensor, by definition. The right-hand side has to be a tensor as well, consequently $S^\rho_{\mu\nu}$ is a tensor. This simple calculation leads to the conclusion that by adding a tensor to a set of connection coefficients, we obtain another connection:

$$\Gamma^\rho_{\mu\nu} = \tilde{\Gamma}^\rho_{\mu\nu} + S^\rho_{\mu\nu}. \qquad (9.49)$$

Then, starting from a connection specified by the Christoffel symbols $\Gamma^\rho_{\mu\nu}$, we can define an antisymmetric tensor by the relation

$$T^\rho_{\mu\nu} = \Gamma^\rho_{\mu\nu} - \Gamma^\rho_{\nu\mu} = \Gamma^\rho_{[\mu\nu]}. \qquad (9.50)$$

This is known as the *torsion tensor*. Clearly, if the connection is *symmetric* in the lower indices, i.e.

$$\Gamma^\rho_{\mu\nu} = \Gamma^\rho_{(\mu\nu)}, \qquad (9.51)$$

the torsion tensor vanishes and the connection is called *torsion-free*.

Another important notion in Riemannian geometry is the *metric compatibility* of the connection. By definition, a connection is metric compatible if the covariant derivative of the metric with respect to that connection is everywhere zero, in other words the connection preserves the metric tensor:

$$\nabla_\rho g_{\mu\nu} = 0. \qquad (9.52)$$

A metric-compatible connection is called also *metric connection*.

The *fundamental theorem of Riemannian geometry* states that on a Riemannian manifold endowed with a metric $g_{\mu\nu}$, there exists a unique torsion-free metric connection, which is the Levi-Civita connection of the given metric.

To prove this result, we shall derive in terms of the metric the unique connection $\Gamma^\rho_{\mu\nu}$ which satisfies the requirements of the theorem. We write the metric compatibility condition (9.52) for three different permutations of the indices:

$$\begin{aligned}
\nabla_\rho g_{\mu\nu} &= \partial_\rho g_{\mu\nu} - \Gamma^\lambda_{\rho\mu} g_{\lambda\nu} - \Gamma^\lambda_{\rho\nu} g_{\mu\lambda} = 0, \\
\nabla_\mu g_{\nu\rho} &= \partial_\mu g_{\nu\rho} - \Gamma^\lambda_{\mu\nu} g_{\lambda\rho} - \Gamma^\lambda_{\mu\rho} g_{\nu\lambda} = 0, \\
\nabla_\nu g_{\rho\mu} &= \partial_\nu g_{\rho\mu} - \Gamma^\lambda_{\nu\rho} g_{\lambda\mu} - \Gamma^\lambda_{\nu\mu} g_{\rho\lambda} = 0.
\end{aligned} \qquad (9.53)$$

By adding the last two relations and subtracting the result from the first and subsequently using the symmetry of the connection (9.51), we find:

$$\partial_\rho g_{\mu\nu} - \partial_\mu g_{\nu\rho} - \partial_\nu g_{\rho\mu} + 2\Gamma^\lambda_{\mu\nu} g_{\lambda\rho} = 0. \tag{9.54}$$

Multiplying by $g^{\sigma\rho}$, we finally obtain the connection coefficients

$$\Gamma^\sigma_{\mu\nu} = \frac{1}{2} g^{\sigma\rho} \left(\partial_\mu g_{\nu\rho} + \partial_\nu g_{\rho\mu} - \partial_\rho g_{\mu\nu} \right), \tag{9.55}$$

which is exactly the expression of the Levi-Civita connection coefficients obtained in (9.15).

In general relativity, the Levi-Civita connection is the only one used, as it arises naturally from the metric and any other connection can be obtained from it by the addition of an appropriately chosen tensor.

Coming back to the notion of geodesic, we can think about it also as a curve on which the tangent vector is parallel transported with respect to a given connection ∇_μ, not necessarily symmetric. This is another way of generalizing the straight line, since on a straight line the tangent vector is parallel transported. However, with this definition, there are as many equations of geodesics as there are connection. Using our definition of directional covariant derivative (9.37) and recalling that for a curve $x^\mu(s)$ the tangent vector is $\dot{x}^\mu = dx^\mu/ds$, the condition that the latter be parallel transported is

$$\frac{D}{ds}\frac{dx^\mu}{ds} = \frac{dx^\rho}{ds} \nabla_\rho \left(\frac{dx^\mu}{ds} \right) = 0. \tag{9.56}$$

Using above (9.32) with arbitrary connection coefficients we obtain the equation

$$\ddot{x}^\rho + \Gamma^\rho_{\mu\nu} \dot{x}^\mu \dot{x}^\nu = 0, \tag{9.57}$$

which is indeed the equation of a geodesic (9.14), assuming that the parallel transport is performed with respect to an arbitrary connection ∇_μ specified by the Christoffel symbols $\Gamma^\rho_{\mu\nu}$. The second term on the left-hand side of (9.57) is symmetric in x^μ and x^ν, therefore we may as well write (9.57) as

$$\ddot{x}^\rho + \Gamma^\rho_{(\mu\nu)} \dot{x}^\mu \dot{x}^\nu = 0.$$

We thus conclude that only the symmetric part of an arbitrary connection contributes to the geodesics.

Remark that the two definitions for a geodesic coincide if and only if the connection is the Levi-Civita connection.

Returning briefly to the torsion tensor, we note that in Einstein's general relativity it does not play any significant role. However, it has a prominent role in the Einstein–Cartan theory of gravitation, thus named in honour of Élie Cartan (1869–1951), who proposed it first in 1922. The torsion tensor can be coupled to the spin of matter, just as the curvature tensor is coupled to the energy and momentum of the matter, as we shall see in Sect. 9.7. Although the Einstein–Cartan theory is still a classical theory of gravitation, the coupling between gravity and the spin of matter fields is crucial in developing a quantum theory of gravity.

Further on, as we do not intend to exceed the limits of general relativity, by covariant derivative we shall always understand the one associated with the Levi-Civita connection.

9.5   Equations of Electrodynamics in the Presence of Gravitation

With this preparation, we can write now the fundamental equations of electrodynamics in a covariant form which takes into account the presence of the gravitational field. We shall focus on Maxwell's equations, the equation of continuity, and the equation of motion of a point charge.

9.5.1   Maxwell's Equations

In general relativity, the components x^μ of the position four-vector are usually chosen as

$$x^0 = ct, \quad x^1 = x, \quad x^2 = y, \quad x^3 = z.$$

Since the invariant form of the four-volume element, in general coordinates, is

$$\sqrt{-g}\, dx^0\, dx^1\, dx^2\, dx^3 = \sqrt{-g} d\Omega,$$

(see (C.46)), the action of the electromagnetic field, when both sources and gravitational field are present, is expressed as

$$S = \int \left(-\frac{1}{4\mu_0} F^{\lambda\rho} F_{\lambda\rho} - j^\lambda A_\lambda \right) \sqrt{-g}\, d\Omega. \tag{9.58}$$

As the four-potential A_μ is a vector, clearly $A_{\mu,\nu}$ is no more a vector on curved space-time, so the original definition of the electromagnetic field strength $F_{\mu\nu}$ has to be replaced by a definition with covariant derivatives, i.e.

$$F_{\mu\nu} = A_{\mu;\nu} - A_{\nu;\mu}. \tag{9.59}$$

Using Eq. (9.29) and the symmetry of the connection coefficients in the lower indices, Eq. (9.16), one can easily show (see (C.62)) that

$$F_{\mu\nu} = A_{\mu;\nu} - A_{\nu;\mu} = A_{\mu,\nu} - A_{\nu,\mu}. \tag{9.60}$$

Thus, the expression of the electromagnetic field strength in terms of the four-potential is practically unchanged in the presence of gravity.

Maxwell's source equations, in the presence of gravitation, are then obtained by using the Euler–Lagrange equations in which partial derivatives are replaced by covariant derivatives, and the Lagrangian density is

$$\mathcal{L} = \left(-\frac{1}{4\mu_0}F^{\lambda\rho}F_{\lambda\rho} - j^\lambda A_\lambda\right)\sqrt{-g}. \tag{9.61}$$

Remark that the Lagrangian density defined as above is not a bona fide scalar, but the factor in brackets is. However, when multiplying (9.61) by the four-volume element $d\Omega$, the result is a scalar, consequently the action (9.58) is well defined. Recalling that g depends on coordinates only and not on the electromagnetic field, we have

$$\frac{\partial\mathcal{L}}{\partial A_{\mu;\nu}} = \frac{\partial\mathcal{L}}{\partial A_{\mu,\nu}} = \frac{1}{\mu_0}\sqrt{-g}F^{\mu\nu}, \qquad \frac{\partial\mathcal{L}}{\partial A_\mu} = -\sqrt{-g}j^\mu,$$

therefore Euler–Lagrange equations lead to

$$\frac{1}{\sqrt{-g}}\frac{\partial}{\partial x^\nu}(\sqrt{-g}\,F^{\nu\mu}) = \mu_0 j^\mu, \quad \mu, \nu = 0, 1, 2, 3,$$

or, observing that the l.h.s. is the covariant four-divergence of the antisymmetric tensor $F^{\nu\mu}$, we obtain

$$F^{\nu\mu}{}_{;\nu} = \mu_0 j^\mu, \quad \mu, \nu = 0, 1, 2, 3. \tag{9.62}$$

These are *Maxwell's source equations in the presence of gravitation*. Multiplying (9.62) by $g_{\mu\rho}$, and using the fact that the covariant derivative of the metric tensor is zero, we also have

$$F^\nu{}_{\rho;\nu} = \mu_0 j_\rho. \tag{9.63}$$

Due to (9.60), one observes that *Maxwell's source-free equations, in the presence of gravitation*, have the same form as in the absence of gravitation, Eq. (8.83), which means

$$\frac{\partial F_{\mu\nu}}{\partial x^\lambda} + \frac{\partial F_{\nu\lambda}}{\partial x^\mu} + \frac{\partial F_{\lambda\mu}}{\partial x^\nu} = 0. \tag{9.64}$$

or, using covariant notation,

$$F_{\mu\nu;\lambda} + F_{\nu\lambda;\mu} + F_{\lambda\mu;\nu} = 0, \tag{9.65}$$

as the reader can easily verify.

9.5.2 Equation of Continuity

Using the procedure presented in Sect. 8.3, we multiply by dx^μ the infinitesimal charge $dq = \rho_e d\tau = \rho_e dx^1 dx^2 dx^3$:

$$dq\, dx^\mu = \rho_e\, d\tau\, dx^\mu = \rho_e \frac{d\tau\, cdt}{\sqrt{-g}} \frac{dx^\mu}{cdt} \sqrt{-g} = \frac{1}{\sqrt{-g}} \rho_e \frac{dx^\mu}{cdt} (\sqrt{-g}\, d\Omega).$$

Since $dq\, dx^\mu$ is a four-vector and $(\sqrt{-g}\, d\Omega)$ an invariant, it follows that

$$j^\mu = \frac{1}{\sqrt{-g}} \rho_e \frac{dx^\mu}{dt} = \frac{1}{\sqrt{-g}} \rho_e c \frac{dx^\mu}{dx^0} \tag{9.66}$$

is also a four-vector, called *current density four-vector*. On flat space-time, the expression (9.66) is reduced to (8.62). To write the *equation of continuity in covariant form and in the presence of gravitation*, we use the definition of covariant four-divergence in general coordinates (C.55), and obtain

$$j^\mu_{;\,\mu} = \frac{1}{\sqrt{-g}} \frac{\partial}{\partial x^\mu} (\sqrt{-g}\, j^\mu) = 0, \tag{9.67}$$

where j^μ is defined by (9.66).

9.5.3 Equation of Motion of a Point Charge

If both electromagnetic and gravitational fields act simultaneously on a particle of rest mass m_0 and electric charge e, its equation of motion is found by replacing the usual differential du^μ by the absolute differential $Du^\mu = du^\mu + \Gamma^\mu_{\nu\lambda} u^\nu\, dx^\lambda$:

$$m_0 c \left(\frac{du^\mu}{ds} + \Gamma^\mu_{\nu\lambda} u^\nu \frac{dx^\lambda}{ds} \right) = e F^{\mu\nu} u_\nu, \tag{9.68}$$

or, equivalently,

$$m_0 c \left(\ddot{x}^\mu + \Gamma^\mu_{\nu\lambda} \dot{x}^\nu \dot{x}^\lambda \right) = e F^\mu_{\ \nu} u^\nu. \tag{9.69}$$

If the electromagnetic field is absent, Eq. (9.69) reduces to the Eq. (9.14) of geodesic lines, as expected.

9.6 Riemann Curvature Tensor

If in flat space a vector is parallel transported along a closed curve, at the end of the path the vector coincides with itself. In a curved space, the vector's orientation may not coincide to its original orientation when its origin returns to the initial position.

As an example, consider a two-dimensional curved manifold, which means an arbitrary curved surface, and on this surface take a closed curve, formed by three segments of geodesics, as in Fig. 9.4.

Let **V** be a vector with its origin at A, and parallel transport it along the closed contour $ABCA$, in the sense shown by arrows. The vector **V**‴ obtained as a result of the parallel transport does not coincide with **V**, consequently its variation is not zero.

Consider now a covariant four-vector A_μ, $\mu = 0, 1, 2, 3$ and use a similar procedure (this time on a 4-dimensional curved manifold), letting the vector perform a Levi-Civita parallel transport along a closed contour. If ΔA_μ is the variation of vector components as a result of the parallel transport along the closed curve, then according to (9.26) we have

$$\Delta A_\mu = \oint \delta A_\mu = \oint \Gamma^\lambda_{\mu\nu} A_\lambda dx^\nu. \tag{9.70}$$

In view of the generalized Stokes theorem (C.37), we can write

$$\Delta A_\mu = \frac{1}{2} \int \left[\frac{\partial}{\partial x^\nu}(\Gamma^\lambda_{\mu\rho} A_\lambda) - \frac{\partial}{\partial x^\rho}(\Gamma^\lambda_{\mu\nu} A_\lambda) \right] d\sigma^{\nu\rho}$$

$$= \frac{1}{2} \int \left(\frac{\partial \Gamma^\lambda_{\mu\rho}}{\partial x^\nu} A_\lambda - \frac{\partial \Gamma^\lambda_{\mu\nu}}{\partial x^\rho} A_\lambda + \Gamma^\lambda_{\mu\rho} \frac{\partial A_\lambda}{\partial x^\nu} - \Gamma^\lambda_{\mu\nu} \frac{\partial A_\lambda}{\partial x^\rho} \right) d\sigma^{\nu\rho}.$$

Fig. 9.4 The parallel transport of the vector **V** along a suitably chosen closed path ABCA in a curved space (here, a 2-dimensional curved manifold).

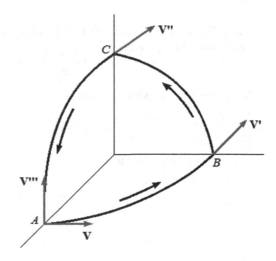

Using (9.26), i.e. $\delta A_\lambda = \Gamma^\kappa_{\lambda\rho} A_\kappa \, dx^\rho$, we find

$$\frac{\partial A_\lambda}{\partial x^\rho} = \Gamma^\kappa_{\lambda\rho} A_\kappa, \quad \frac{\partial A_\lambda}{\partial x^\nu} = \Gamma^\kappa_{\lambda\nu} A_\kappa$$

and then, after a convenient change of summation indices,

$$\Delta A_\mu = \frac{1}{2} \int \left(\frac{\partial \Gamma^\lambda_{\mu\rho}}{\partial x^\nu} - \frac{\partial \Gamma^\lambda_{\mu\nu}}{\partial x^\rho} + \Gamma^\kappa_{\mu\rho} \Gamma^\lambda_{\kappa\nu} - \Gamma^\kappa_{\mu\nu} \Gamma^\lambda_{\kappa\rho} \right) A_\lambda \, d\sigma^{\nu\rho}.$$

Denoting

$$R^\lambda{}_{\mu\nu\rho} = \partial_\nu \Gamma^\lambda_{\mu\rho} - \partial_\rho \Gamma^\lambda_{\mu\nu} + \Gamma^\kappa_{\mu\rho} \Gamma^\lambda_{\kappa\nu} - \Gamma^\kappa_{\mu\nu} \Gamma^\lambda_{\kappa\rho}, \tag{9.71}$$

we can write also

$$\Delta A_\mu = \frac{1}{2} \int R^\lambda{}_{\mu\nu\rho} A_\lambda \, d\sigma^{\nu\rho}. \tag{9.72}$$

The quantities (9.71) are the components of a mixed tensor, once contravariant and three times covariant, called the *Riemann tensor*, or the *Riemann–Christoffel curvature tensor*. If all components of $R^\lambda{}_{\mu\nu\rho}$ are zero, then $\Delta A_\mu = 0$, which corresponds to a flat space. Conversely, *in a flat space the curvature tensor is identically zero*.

As we have seen, in a curved space one can choose a locally-inertial coordinate system in which all the components of the Christoffel symbols are zero at some conveniently chosen point. However, the components of the Riemann tensor *are not zero at that point*, because the derivatives of $\Gamma^\lambda_{\mu\nu}$ do not vanish together with $\Gamma^\lambda_{\mu\nu}$.

Properties of the Curvature Tensor

1) According to the definition (9.71), the curvature tensor is antisymmetric in the last two covariance indices:

$$R^\rho{}_{\mu\nu\lambda} = -R^\rho{}_{\mu\lambda\nu}. \tag{9.73}$$

2) It satisfies the following identity:

$$R^\rho{}_{\mu\nu\lambda} + R^\rho{}_{\lambda\mu\nu} + R^\rho{}_{\nu\lambda\mu} = 0. \tag{9.74}$$

3) The covariant curvature tensor is obtained by the usual procedure:

$$R_{\kappa\mu\nu\lambda} = g_{\kappa\rho} R^\rho{}_{\mu\nu\lambda}. \tag{9.75}$$

By means of (9.71) we have

$$R_{\mu\nu\lambda\rho} = g_{\mu\kappa}\left(\frac{\partial\Gamma_{\nu\rho}^\kappa}{\partial x^\lambda} - \frac{\partial\Gamma_{\nu\lambda}^\kappa}{\partial x^\rho} + \Gamma_{\nu\rho}^\sigma\Gamma_{\sigma\lambda}^\kappa - \Gamma_{\nu\lambda}^\sigma\Gamma_{\sigma\rho}^\kappa\right)$$

$$= \frac{\partial}{\partial x^\lambda}\left(\Gamma_{\nu\rho,\mu}\right) - \frac{\partial}{\partial x^\rho}\left(\Gamma_{\nu\lambda,\mu}\right) - \Gamma_{\nu\rho}^\kappa\left(\Gamma_{\kappa\lambda}^\sigma g_{\mu\sigma} + \Gamma_{\mu\lambda}^\sigma g_{\sigma\kappa}\right)$$

$$+ \Gamma_{\nu\lambda}^\kappa\left(\Gamma_{\rho\kappa}^\sigma g_{\mu\sigma} + \Gamma_{\rho\mu}^\sigma g_{\sigma\kappa}\right) + g_{\mu\kappa}\Gamma_{\nu\rho}^\sigma\Gamma_{\sigma\lambda}^\kappa - g_{\mu\kappa}\Gamma_{\nu\lambda}^\sigma\Gamma_{\sigma\rho}^\kappa.$$

After some index manipulation, we finally obtain

$$R_{\mu\nu\lambda\rho} = \frac{1}{2}\left(\frac{\partial^2 g_{\mu\rho}}{\partial x^\nu \partial x^\lambda} + \frac{\partial^2 g_{\nu\lambda}}{\partial x^\mu \partial x^\rho} - \frac{\partial^2 g_{\mu\lambda}}{\partial x^\nu \partial x^\rho} - \frac{\partial^2 g_{\nu\rho}}{\partial x^\mu \partial x^\lambda}\right)$$

$$+ g_{\sigma\kappa}\left(\Gamma_{\mu\rho}^\sigma\Gamma_{\nu\lambda}^\kappa - \Gamma_{\mu\lambda}^\sigma\Gamma_{\nu\rho}^\kappa\right), \tag{9.76}$$

and the resulting properties

$$R_{\mu\nu\lambda\rho} = -R_{\nu\mu\lambda\rho},$$
$$R_{\mu\nu\lambda\rho} = -R_{\mu\nu\rho\lambda}, \tag{9.77}$$
$$R_{\mu\nu\lambda\rho} = +R_{\lambda\rho\mu\nu}.$$

That is, the curvature tensor is antisymmetric in the first two indices and in the last two indices, and symmetric upon the interchange of the first pair of indices with the second pair. Consequently, those components with $\mu = \nu$, or $\lambda = \rho$, are zero.

4) The covariant curvature tensor satisfies an identity similar to (9.74):

$$R_{\mu\nu\lambda\rho} + R_{\mu\rho\nu\lambda} + R_{\mu\lambda\rho\nu} = 0, \tag{9.78}$$

called the *first Bianchi identity*, or the *algebraic Bianchi identity*.

5) The curvature tensor satisfies also the identity

$$R_{\mu\nu\rho;\kappa}^\lambda + R_{\mu\kappa\nu;\rho}^\lambda + R_{\mu\rho\kappa;\nu}^\lambda = 0, \tag{9.79}$$

known as the *second Bianchi identity* or *differential Bianchi identity*. To prove it, we use a locally-geodesic coordinate system (see Sect. 9.4.2). In such a frame, all connection coefficients $\Gamma_{\mu\nu}^\lambda$ are zero and, according to (9.71), we have

$$(R_{\mu\nu\lambda}^\rho)_0 = \left(\frac{\partial\Gamma_{\mu\lambda}^\rho}{\partial x^\nu}\right)_0 - \left(\frac{\partial\Gamma_{\mu\nu}^\rho}{\partial x^\lambda}\right)_0.$$

The covariant derivative is composed of the usual derivative plus terms containing $\Gamma^{\lambda}_{\mu\nu}$, which are zero, that is

$$\left(R^{\rho}_{\mu\nu\lambda;\kappa}\right)_0 = \left(R^{\rho}_{\mu\nu\lambda,\kappa}\right)_0 = \left(\frac{\partial^2 \Gamma^{\rho}_{\mu\lambda}}{\partial x^{\nu}\partial x^{\kappa}}\right)_0 - \left(\frac{\partial^2 \Gamma^{\rho}_{\mu\nu}}{\partial x^{\lambda}\partial x^{\kappa}}\right)_0 .$$

Writing two more relations obtained by cyclic permutation of the indices ν, λ, κ and adding the results, we arrive at the Bianchi identity (9.79).

6) Contracting the mixed Riemann tensor (9.71), one can construct the second-order covariant tensor

$$R_{\mu\lambda} = R^{\nu}_{\mu\nu\lambda} = \frac{\partial \Gamma^{\nu}_{\mu\lambda}}{\partial x^{\nu}} - \frac{\partial \Gamma^{\nu}_{\mu\nu}}{\partial x^{\lambda}} + \Gamma^{\rho}_{\mu\lambda}\Gamma^{\nu}_{\rho\nu} - \Gamma^{\rho}_{\mu\nu}\Gamma^{\nu}_{\rho\lambda}, \tag{9.80}$$

called the *Ricci curvature tensor*. (For a Levi-Civita connection, one can show that the above contraction is the only independent one.) According to the definition (9.80), the Ricci tensor is symmetric:

$$R_{\mu\lambda} = R_{\lambda\mu}. \tag{9.81}$$

The contracted product

$$R = g^{\mu\lambda} R_{\mu\lambda} \tag{9.82}$$

is an invariant, called *scalar curvature*, or *curvature invariant*, or *Ricci scalar*. We emphasize that a manifold can be *curved*, though all the components of the Ricci tensor are zero. (The Ricci scalar is the simplest curvature invariant of a Riemannian manifold. Some other scalars can also be formed, like the *Kretschmann invariant* $K = R_{\mu\nu\lambda\rho}R^{\mu\nu\lambda\rho}$, etc.).

7) To determine the number of distinct components of the Riemann tensor, it is useful to introduce the following correspondence

$$\begin{array}{cccccc} 0 & 1 & 2 & 3 & 4 & 5 \\ 01 & 02 & 03 & 12 & 13 & 23 \end{array}$$

The Riemann tensor can then be written as $R_{\Psi\Phi} = R_{\Phi\Psi}$ (Ψ, $\Phi = 0, \ldots, 5$). Here we have omitted the components which are zero according to (9.77). $R_{\Psi\Phi}$ is a second-order, symmetric tensor, defined on a six-dimensional manifold. The number of independent components of this tensor should be $\left(C^2_n\right) = n(n+1)/2 = 6(6+1)/2 = 21$ (see (B.38)), but the components with all different indices must satisfy the relation (9.78):

$$R_{0123} + R_{0312} + R_{0231} = 0.$$

This relation diminishes by one the number of distinct components of the curvature tensor, and the final result is 20 components.

9.7 Einstein's Equations

The gravitational field can be expressed in terms of its potentials, which are the components of the metric tensor, just as the electromagnetic field is defined with the help of the potential four-vector. The difference is that the electromagnetic field is a first-order tensor field, while the gravitational field is a second-order tensor field. The components of $g_{\mu\nu}$ are called *gravitational potentials*.

Going further with the analogy between the two fields, we presume that the gravitational potentials satisfy a system of second order, partial differential equations, like the equations $\Box A^\nu = \mu_0 j^\nu$, where A^ν is the electromagnetic potential four-vector.

To establish these equations, we use the analytical formalism. Since the desired partial differential equations have to be of the second order, the Lagrangian density must be an invariant which contains the components $g_{\mu\nu}$ of the metric tensor, and at most their first derivative with respect to coordinates.[3] But an invariant fulfilling all these conditions does not exist, and we have to use a different strategy. The idea is to use the curvature invariant R, the only independent invariant that can be constructed from the Riemann tensor, which contains second derivatives of the metric. Even if it contains, linearly, second partial derivatives of $g_{\mu\nu}$, at the end of the procedure these terms separate into a four-divergence, which can be omitted. In variational calculus, such a case is called *degenerate*.

In 1915, David Hilbert (1862–1943) proposed as Lagrangian of the gravitational field the Ricci scalar R and derived the equations of motion of the metric, known as *Einstein's equations*. Einstein himself had adopted a different approach in deriving the same equations. For this reason, the action of the gravitational field,

$$S_G = \frac{1}{2\kappa} \int R\sqrt{-g}\, d^4x, \qquad (9.83)$$

is called *Einstein–Hilbert action*. Here, $\kappa = 8\pi G/c^4$, where G is the gravitational constant and c is the speed of light in vacuum (this choice of the constant κ is justified by the correspondence principle, which requires that Newton's gravitation is obtained as a limiting case). Besides the action of the gravitational field, we have to include also the action of all the other fields interacting with the gravitational field. They will be called in the following *matter fields* and their action denoted by S_M. Here, *matter* is a generic term, and by matter fields we understand any non-gravitational fields which carry energy, including fields of radiation. The rule for constructing the Lagrangian density $\mathcal{L}_M$ of the matter fields on curved space-time is simple: take the Lagrangian density of the corresponding fields on the flat space, replace the partial derivatives by covariant derivatives and multiply the result by $\sqrt{-g}$. We have seen this rule used in writing the Lagrangian density (9.61) of the electromagnetic

---

[3] However, we have seen that at any point of the Riemannian manifold we can set the metric to the Minkowski one, which means that the derivatives of the metric are set to zero. So we have to involve necessarily in the Lagrangian also second derivatives of the metric.

field. Thus, the total action of the system composed of the gravitational and matter fields is

$$S = \int \left[\frac{1}{2\kappa} R + \mathcal{L}_M \right] \sqrt{-g} \, d^4x. \tag{9.84}$$

When applying the action principle, we may consider going directly to the Euler–Lagrange equations written for the metric $g_{\mu\nu}$ as dynamical variable and its covariant derivative. But recall that the Levi-Civita connection satisfies the condition of metric compatibility, i.e. $\nabla_\rho g_{\mu\nu} = 0$. Hence, the strategy to adopt is to vary directly the action (9.84) with respect to the inverse metric:

$$0 = \delta S \tag{9.85}$$
$$= \int \left[\frac{1}{2\kappa} \frac{\delta(\sqrt{-g}R)}{\delta g^{\mu\nu}} + \frac{\delta(\sqrt{-g}\mathcal{L}_M)}{\delta g^{\mu\nu}} \right] \delta g^{\mu\nu} d^4x$$
$$= \int \left[\frac{1}{2\kappa} \left(\frac{\delta R}{\delta g^{\mu\nu}} + \frac{R}{\sqrt{-g}} \frac{\delta\sqrt{-g}}{\delta g^{\mu\nu}} \right) + \frac{1}{\sqrt{-g}} \frac{\delta(\sqrt{-g}\mathcal{L}_M)}{\delta g^{\mu\nu}} \right] \delta g^{\mu\nu} \sqrt{-g} \, d^4x.$$

The variation with respect to the inverse metric is taken for convenience. As $g^{\mu\rho} g_{\rho\nu} = \delta_\nu^\mu$, and the variation of δ_ν^μ is always zero, it follows that the variations of the metric and its inverse are in the relation

$$g^{\mu\rho} \delta g_{\rho\nu} = -g_{\mu\sigma} \delta g^{\sigma\nu}, \tag{9.86}$$

consequently the stationary points with respect to either of them are the same.

Since the variations $\delta g^{\mu\nu}$ in (9.85) are arbitrary, we find the equation of motion for the metric in the form

$$\frac{\delta R}{\delta g^{\mu\nu}} + \frac{R}{\sqrt{-g}} \frac{\delta\sqrt{-g}}{\delta g^{\mu\nu}} = -2\kappa \frac{1}{\sqrt{-g}} \frac{\delta(\sqrt{-g}\mathcal{L}_M)}{\delta g^{\mu\nu}}. \tag{9.87}$$

The right-hand side of this equation is proportional to the energy-momentum tensor of the matter fields, which is by definition

$$T_{\mu\nu} = \frac{-2}{\sqrt{-g}} \frac{\delta(\sqrt{-g}\mathcal{L}_M)}{\delta g^{\mu\nu}}. \tag{9.88}$$

Now let us focus on the left-hand side and perform the required variations. The first term involves the variation of the Ricci scalar, which can be traced back to the variation of the Riemann tensor, defined in (9.71), i.e.

$$R^\rho{}_{\sigma\mu\nu} = \partial_\mu \Gamma^\rho_{\nu\sigma} - \partial_\nu \Gamma^\rho_{\mu\sigma} + \Gamma^\rho_{\mu\lambda} \Gamma^\lambda_{\nu\sigma} - \Gamma^\rho_{\nu\lambda} \Gamma^\lambda_{\mu\sigma}.$$

Its variation is

$$\delta R^\rho{}_{\sigma\mu\nu} = \partial_\mu \delta \Gamma^\rho_{\nu\sigma} - \partial_\nu \delta \Gamma^\rho_{\mu\sigma}$$
$$+ \delta \Gamma^\rho_{\mu\lambda} \Gamma^\lambda_{\nu\sigma} + \Gamma^\rho_{\mu\lambda} \delta \Gamma^\lambda_{\nu\sigma} - \delta \Gamma^\rho_{\nu\lambda} \Gamma^\lambda_{\mu\sigma} - \Gamma^\rho_{\nu\lambda} \delta \Gamma^\lambda_{\mu\sigma}. \tag{9.89}$$

We note that $\Gamma^\rho_{\nu\mu}$ and $\Gamma'^\rho_{\nu\mu} = \Gamma^\rho_{\nu\mu} + \delta \Gamma^\rho_{\nu\mu}$ represent two sets of connection coefficients. Recall that in (9.49) we showed that two sets of connection coefficients differ by a tensor, as they are the difference of two covariant derivatives. Thus, $\delta \Gamma^\rho_{\nu\mu}$ is a tensor and we can calculate its covariant derivative:

$$\nabla_\lambda(\delta \Gamma^\rho_{\nu\mu}) = \partial_\lambda(\delta \Gamma^\rho_{\nu\mu}) + \Gamma^\rho_{\sigma\lambda} \delta \Gamma^\sigma_{\nu\mu} - \Gamma^\sigma_{\nu\lambda} \delta \Gamma^\rho_{\sigma\mu} - \Gamma^\sigma_{\mu\lambda} \delta \Gamma^\rho_{\nu\sigma}. \tag{9.90}$$

It is straightforward to show that the variation of the Riemann tensor expressed by (9.89) is equal to the difference of two such terms,

$$\delta R^\rho{}_{\sigma\mu\nu} = \nabla_\mu(\delta \Gamma^\rho_{\nu\sigma}) - \nabla_\nu(\delta \Gamma^\rho_{\mu\sigma}). \tag{9.91}$$

Consequently, the variation of the Ricci tensor $R_{\mu\nu} = R^\rho{}_{\mu\rho\nu}$ is

$$\delta R_{\mu\nu} = \delta R^\rho{}_{\mu\rho\nu} = \nabla_\rho(\delta \Gamma^\rho_{\nu\mu}) - \nabla_\nu(\delta \Gamma^\rho_{\rho\mu}), \tag{9.92}$$

which is called the *Palatini identity* in honour of the Italian mathematician Attilio Palatini (1889–1949), who proved it in 1919 in a paper in which he was introducing an alternative variational formalism for obtaining Einstein's equations.

Now we can easily find the variation of the Ricci scalar, defined as the trace of the Ricci tensor:

$$R = g^{\mu\nu} R_{\mu\nu}.$$

Its variation with respect to the inverse metric $g^{\mu\nu}$ is found using the Palatini identity (9.92) and the metric compatibility of the covariant derivative, $\nabla_\sigma g^{\mu\nu} = 0$:

$$\delta R = R_{\mu\nu} \delta g^{\mu\nu} + g^{\mu\nu} \delta R_{\mu\nu} \tag{9.93}$$
$$= R_{\mu\nu} \delta g^{\mu\nu} + \nabla_\sigma \left(g^{\mu\nu} \delta \Gamma^\sigma_{\nu\mu} - g^{\mu\sigma} \delta \Gamma^\rho_{\rho\mu} \right). \tag{9.94}$$

Let us write the last term as $\nabla_\sigma V^\sigma = \nabla_\sigma (g^{\mu\nu} \delta \Gamma^\sigma_{\nu\mu} - g^{\mu\sigma} \delta \Gamma^\rho_{\rho\mu})$. Recall formula (C.55), written here in terms of V^σ:

$$\nabla_\sigma V^\sigma = \frac{1}{\sqrt{-g}} \frac{\partial}{\partial x^\sigma} \left(\sqrt{-g} V^\sigma \right).$$

Introducing this term under the integral in (9.85), we obtain

$$\int \nabla_\sigma V^\sigma \sqrt{-g} \, d^4x = \int \frac{\partial}{\partial x^\sigma} \left(\sqrt{-g} V^\sigma \right) d^4x.$$

Using the Stokes theorem, this integral yields a boundary term, which is set to zero as the variations of the metric $\delta g^{\mu\nu}$ vanish on the boundary. Thus, the first term in (9.87) becomes

$$\frac{\delta R}{\delta g^{\mu\nu}} = R_{\mu\nu}. \tag{9.95}$$

Let us move further to the second term in (9.87). We note that

$$\delta(\sqrt{-g}) = -\frac{1}{2\sqrt{-g}}\delta g. \tag{9.96}$$

Expanding formally the determinant g by minors,

$$g = g_{\mu\nu}G^{\mu\nu},$$

where $G^{\mu\nu}$ is the algebraic complement of the element $g_{\mu\nu}$ of g, we obtain[4]

$$\delta g = G^{\mu\nu}\delta g_{\mu\nu}.$$

The components of the inverse metric, $g^{\mu\nu}$, are given by the minors of the determinant of $g_{\mu\nu}$ divided by g (see (B.29)):

$$g^{\mu\nu} = \frac{1}{g}G^{\mu\nu},$$

which, combined with the previous formula, gives

$$\delta g = g g^{\mu\nu}\delta g_{\mu\nu}.$$

Since $g_{\mu\nu}g^{\mu\nu} = \delta^{\mu}_{\mu} = 4$, we have $g_{\mu\nu}\delta g^{\mu\nu} + g^{\mu\nu}\delta g_{\mu\nu} = 0$, leading to $g^{\mu\nu}\delta g_{\mu\nu} = -g_{\mu\nu}\delta g^{\mu\nu}$. Thus, δg can be written as

$$\delta g = -g g_{\mu\nu}\delta g^{\mu\nu}. \tag{9.97}$$

Putting together (9.96) and (9.97) we find

$$\delta(\sqrt{-g}) = -\frac{1}{2}\sqrt{-g}g_{\mu\nu}\delta g^{\mu\nu}. \tag{9.98}$$

Using (9.98), we can put the energy-momentum tensor given by (9.88) in the form

$$T_{\mu\nu} = \frac{-2}{\sqrt{-g}}\frac{\delta(\sqrt{-g}\mathcal{L}_M)}{\delta g^{\mu\nu}} = -2\frac{\delta\mathcal{L}_M}{\delta g^{\mu\nu}} + g_{\mu\nu}\mathcal{L}_M. \tag{9.99}$$

---

[4]This is a formal writing, because $G^{\mu\nu}$ is not a contravariant tensor, but the product between the minor corresponding to $g_{\mu\nu}$ and the sign $(-1)^{\mu+\nu}$.

We can write now the equations of motion of the metric field (9.87), using (9.95) and (9.98), in the form known as *Einstein's equations*:

$$R_{\mu\nu} - \frac{1}{2}R\,g_{\mu\nu} = \kappa T_{\mu\nu}. \tag{9.100}$$

Einstein's equations are a set of ten coupled *nonlinear* partial differential equations for the metric components. Nonlinearity distinguishes general relativity from other physical theories (Maxwell's equations are linear in the electric and magnetic fields; Schrödinger's equation is linear in the wave function, etc.) and makes it so difficult to quantize, as the superposition principle used customarily in quantum mechanics and the theory of quantized fields is no more valid.

In the case of *weak gravitational fields*, Einstein's equations can be linearized in the first approximation. Since the field is weak, the space is only slightly curved, which means that $g_{\mu\nu}$ differs by a small amount from the metric tensor of Minkowski space, which is denoted by $\eta_{\mu\nu}$. Transposed mathematically, these considerations lead to the linear approximation of the general relativity, which allows the quantization of a weak gravitational field.

By analogy with the equation of motion of the electromagnetic field, $\Box A^{\nu} = \mu_0 j^{\nu}$, we interpret the right-hand side of Einstein's equations (9.100) as the *sources* of the gravitational field. Thus, the physical meaning of the equations is that matter and energy lead to the curvature of space-time.

Let us now show that in the limit of small velocities and implicitly weak gravitational field, Einstein's equations (9.100) reduce to Poisson's equation (see Sect. 9.1),

$$\Delta\varphi = -4\pi G\rho_m, \tag{9.101}$$

where φ is the Newtonian gravitational potential and ρ_m is the density of mass. To do this, we write Einstein's equations in a slightly different form. Contracting (9.100) with $g^{\mu\nu}$, we obtain

$$R = -\kappa T,$$

where $R = R^{\mu}_{\mu}$, $T = T^{\mu}_{\mu}$, and Eq. (9.100) can be written, upon contraction with the inverse metric tensor, as

$$R^{\nu}_{\mu} = \kappa\left(T^{\nu}_{\mu} - \frac{1}{2}T\delta^{\nu}_{\mu}\right). \tag{9.102}$$

In the non-relativistic limit ($\frac{v}{c} \to 0$), the motion of a free particle of mass m is studied by means of the Lagrangian function (7.77):

$$L_0 = -mc\sqrt{1 - \frac{v^2}{c^2}} \simeq -mc^2 + \frac{1}{2}mv^2.$$

If the particle moves under the action of the gravitational field, then the Lagrangian is

$$L = L_0 - m\varphi = \frac{1}{2}mv^2 - mc^2 - m\varphi. \tag{9.103}$$

We shall use (9.103) to find the metric.

The corresponding action is then

$$S = \int L dt = -mc \int \left(c - \frac{v^2}{2c} + \frac{\varphi}{c}\right) dt. \tag{9.104}$$

Comparing this result and (7.76), we realize that in our limit case

$$ds = \left(c - \frac{v^2}{2c} + \frac{\varphi}{c}\right) dt.$$

Squaring this relation and omitting the vanishing terms in the limit $\frac{v}{c} \to 0$, we have

$$ds^2 = (c^2 + 2\varphi)dt^2 - |d\mathbf{r}|^2 = \left(1 + \frac{2\varphi}{c^2}\right)c^2 dt^2 - |d\mathbf{r}|^2, \tag{9.105}$$

where $d\mathbf{r} = \mathbf{v}\, dt$. From here we immediately deduce that

$$g_{00} = 1 + \frac{2\varphi}{c^2}. \tag{9.106}$$

The rest of the metric components are not important for our further considerations. Suppose, next, that T_μ^ν is given by (8.207), where the rest mass density $\tilde{\rho}_0$ will be denoted (here and further) by ρ_m:

$$T_\mu^\nu = \rho_m c^2 u_\mu u^\nu.$$

Since the motion is slow, we may take

$$u^0 = u_0 = 1,$$
$$u^i = \frac{\gamma}{c}v^i \simeq 0.$$

Thus, the only non-zero component of T_μ^ν is

$$T_0^0 = T = \rho_m c^2, \tag{9.107}$$

which we plug into (9.102) to obtain

$$R_0^0 = \kappa \left(T_0^0 - \frac{1}{2}T\delta_0^0\right) = \frac{1}{2}\kappa\rho_m c^2. \tag{9.108}$$

Using now the definition (9.80) of the Riemann tensor, one observes that in our approximation the terms containing products of $\Gamma^\lambda_{\mu\nu}$ are negligible. Also, the terms involving derivatives with respect to x^0 are small as compared to those containing derivatives with respect to x^i, $i = 1, 2, 3$. Consequently, in this approximation,

$$R_{00} \simeq R^0_0 \simeq \partial\Gamma^i_{00}/\partial x^i.$$

According to (9.106),

$$\Gamma^i_{00} \simeq -\frac{1}{2} g^{ij} \frac{\partial g_{00}}{\partial x^j} = \frac{1}{c^2} \frac{\partial\varphi}{\partial x^i},$$

leading to

$$R^0_0 = \frac{1}{c^2} \frac{\partial^2\varphi}{\partial x^i \partial x^i} = \frac{1}{c^2} \Delta\varphi. \tag{9.109}$$

Combining (9.108) and (9.109) we obtain

$$\Delta\varphi = \frac{1}{2} \kappa \rho_m c^4,$$

or, using $\kappa = 8\pi G/c^4$,

$$\Delta\varphi = 4\pi G \rho_m,$$

which is Poisson's equation for the gravitational potential φ. Thus, the proof is complete. The solution of this equation, for a continuous mass distribution is

$$\varphi = -G \int \frac{\rho_m(\mathbf{r}')}{|\mathbf{r} - \mathbf{r}'|} d\mathbf{r}',$$

as we already know.

Observations:

(a) The tensor

$$G_{\mu\nu} = R_{\mu\nu} - \frac{1}{2} R\, g_{\mu\nu} \tag{9.110}$$

is called the *Einstein's tensor*. With this notation, Eqs. (9.100) become

$$G_{\mu\nu} = \kappa\, T_{\mu\nu}. \tag{9.111}$$

(b) In free space $T_{\mu\nu} = 0$, and Einstein's equations (9.100) are

$$R_{\mu\nu} - \frac{1}{2} R\, g_{\mu\nu} = 0. \tag{9.112}$$

Equivalently, if we start from (9.102), we obtain Einstein's equations in vacuum in the form

$$R_{\mu\nu} = 0. \tag{9.113}$$

(c) It can be shown that $\nabla_\nu G^\nu_\mu = 0$. This leads to $\nabla_\nu T^\nu_\mu = 0$, expressing the fact that the energy-momentum tensor is conserved.

(d) We should point out that the symmetry of the left-hand side of Einstein's equations (9.100) implies that the energy-momentum tensor is symmetric. Recall from Sect. 8.7.1 that the *canonical energy-momentum tensor* obtained from Noether's theorem is not symmetric, and in Sect. 8.7.4 we described Belinfante's prescription for its symmetrization. However, using the procedure presented in this section, one obtains automatically the symmetric form of the energy-momentum tensor, (9.88) or (9.99). This procedure can be applied also on flat space-time, by going to general curvilinear coordinates as an intermediate step in the derivation.

(e) Einstein proposed a modification of his original theory by introducing in 1917 the so-called *cosmological constant* Λ, in order to describe the static universe, and wrote his equations in the form

$$R_{\mu\nu} - \frac{1}{2}Rg_{\mu\nu} + \Lambda g_{\mu\nu} = \frac{8\pi G}{c^4}T_{\mu\nu}, \tag{9.114}$$

which can be derived from the action

$$S = \int \left[\frac{1}{2\kappa}(R - 2\Lambda) + \mathcal{L}_M \right] \sqrt{-g}\, d^4x.$$

Naturally, the term containing the constant Λ is the simplest possible Lagrangian that can be constructed on a curved space.

In 1912, Vesto Slipher (1875–1969) discovered that the light received from far away galaxies is redshifted. Later on, it was found in 1927 by Georges Lemaître (1894–1966) and in 1929 by Edwin Hubble (1889–1953) that the redshifts are roughly proportional to the distances to those far-away galaxies. This is known as *Hubble's law* and it marked the introduction of the expanding space paradigm. At that moment, Einstein abandoned the cosmological constant. However, the discovery of the accelerating expansion of the Universe in 1998, simultaneously by two independent projects (Supernova Cosmology Project and High-Z Supernova Search Team) renewed the interest in the cosmological constant.

The simplest explanation for the accelerating expansion is the existence of a hypothetical form of energy, called *dark energy*, which permeates the whole space. The dark energy has to have *negative pressure*, distributed isotropically in space, in order to drive the accelerated expansion in spite of the attractive nature of the gravitational force. According to the latest release of the *Planck mission team* in 2015, out of the total energy of the observable Universe dark energy represents 68.3 %, while 26.8 % is dark matter, and 4.9 % is ordinary (baryonic) matter.

The term containing Λ in the field equations (9.114) can be moved algebraically to the other side,

$$R_{\mu\nu} - \frac{1}{2}Rg_{\mu\nu} = \frac{8\pi G}{c^4}T_{\mu\nu} - \Lambda g_{\mu\nu}.$$

If by $T_{\mu\nu}$ we understand the energy-momentum tensor of matter, by the term $-\Lambda g_{\mu\nu}$ we may understand the energy-momentum tensor of *vacuum*,

$$T_{\mu\nu}^{(vac)} = -\frac{c^4}{8\pi G}\Lambda g_{\mu\nu}, \tag{9.115}$$

because in gravity absolute values of energy are important, and not only energy differences as in the case of the other fundamental forces. In other words, the energy density of the vacuum is

$$\rho^{(vac)} = \frac{c^4}{8\pi G}\Lambda.$$

This is equivalent to adding a vacuum term to the Lagrangian in (9.84),

$$\mathcal{L}^{(vac)} = -\rho^{(vac)}.$$

Comparing (9.115) to the energy-momentum tensor of a perfect fluid,

$$T^{\mu\nu(fluid)} = (p + \rho)u^\mu u^\nu - p\, g^{\mu\nu},$$

we see that

$$p^{(vac)} = -\rho^{(vac)},$$

as expected for the dark energy. A positive vacuum energy density resulting from a positive cosmological constant implies a negative pressure, and vice versa. According to the observational data, the energy density corresponding to the cosmological constant is of the order

$$\rho_{obs}^{(vac)} \leq 10^{-9}\mathrm{J} \cdot \mathrm{m}^{-3}.$$

In contrast, quantum field theoretical considerations which we shall not detail here, lead to an expectation of

$$\rho_{QFT}^{(vac)} \approx 10^{111}\mathrm{J} \cdot \mathrm{m}^{-3}.$$

This huge discrepancy is a topic of active research.

9.8   Central Gravitational Field. Schwarzschild Metric

We shall now solve Einstein's equations in the simplest case of a gravitational field
with central symmetry, produced by a point-like static source.

The geometry of the problem suggests to choose spherical coordinates, with the
origin at the centre of the field. If the source of the field is placed in vacuum, the
components of the energy-momentum tensor $T_{\mu\nu}$ are everywhere zero, except at the
origin of the coordinates. Einstein's equations outside of the source are then

$$R_{\mu\nu} - \frac{1}{2}R\, g_{\mu\nu} = 0. \tag{9.116}$$

To find the metric, i.e. the solution of the above equations, we shall take advantage
of the central symmetry of the field and, also, of the fact that the source is static. The
latter feature indicates that all the components of the metric are time-independent,
and there are no time-space cross terms $(dtdx_i + dx_i dt)$ in the metric. The spherical symmetry suggests that we can consider the Minkowski (flat space) metric in
spherical coordinates,

$$ds^2_{\text{Minkowski}} = c^2 dt^2 - dr^2 - r^2\, d\Omega^2,$$

where $d\Omega^2 = d\theta^2 + \sin^2\theta\, d\varphi^2$, and simply multiply the terms by different coefficients, all of them depending only on the radial coordinate r. Taking into account all
these aspects, we consider the solution of (9.116) of the form

$$\begin{aligned} ds^2 &= A(r)\, dt^2 - B(r)\, dr^2 - r^2\, d\Omega^2 \\ &= A(r)\, dt^2 - B(r)\, dr^2 - r^2\, d\theta^2 - r^2 \sin^2\theta\, d\varphi^2. \end{aligned} \tag{9.117}$$

We could add also a coefficient for the term proportional to $d\Omega^2$, but that can be
absorbed into the definition of r, so we shall continue with this simplified expression.

Now we have to determine the function $A(r)$ and $B(r)$ by solving Einstein's equations. To do this, we must determine the connection coefficients $\Gamma^\lambda_{\mu\nu}$, the components
of the Ricci tensor $R_{\mu\nu}$, and the curvature invariant R.

The first step is to choose the coordinates. As customary in general relativity, we
take $x^0 = ct, x^1 = r, x^2 = \theta, x^3 = \varphi$. Then (9.117) yields

$$g_{00} = A, \quad g_{11} = -B, \quad g_{22} = -r^2, \quad g_{33} = -r^2 \sin^2\theta,$$

$$g^{00} = \frac{1}{A}, \quad g^{11} = -\frac{1}{B}, \quad g^{22} = -\frac{1}{r^2}, \quad g^{33} = -\frac{1}{r^2 \sin^2\theta}. \tag{9.118}$$

$$g_{\mu\nu} = g^{\mu\nu} = 0, \text{ for } \mu \neq \nu.$$

The Christoffel symbols can be determined either directly, by using the definition
(9.15), or indirectly (but with less effort) by the help of the equation of geodesic lines
(9.14). We shall use the second option.

According to (9.117), the variational principle $\delta \int ds = 0$ can be written as

$$\delta \int \frac{ds^2}{ds^2} ds = \delta \int f ds = 0,$$

where

$$f = A\dot{t}^2 - B\dot{r}^2 - r^2\dot{\theta}^2 - r^2 \sin^2 \theta \, \dot{\varphi}^2 \equiv 1. \tag{9.119}$$

Here the "dot" over letters means derivative with respect to s. Applying Euler's equations (9.7),

$$\frac{d}{ds}\left(\frac{\partial f}{\partial \dot{x}^\mu}\right) - \frac{\partial f}{\partial x^\mu} = 0, \quad \mu = 0, 1, 2, 3, \tag{9.120}$$

we find

$$\ddot{t} + \frac{A'}{A}\dot{r}\dot{t} = 0,$$

$$\ddot{r} + \frac{A'}{2B}\dot{t}^2 + \frac{B'}{2B}\dot{r}^2 - \frac{r}{B}\dot{\theta}^2 - \frac{r^2 \sin^2 \theta}{B}\dot{\varphi}^2 = 0,$$

$$\ddot{\theta} + \frac{2}{r}\dot{r}\dot{\theta} - \frac{\dot{\varphi}^2}{2}\sin 2\theta = 0, \tag{9.121}$$

$$\ddot{\varphi} + \frac{2}{r}\dot{r}\dot{\varphi} + \frac{2\dot{\theta}}{\tan \theta}\dot{\varphi} = 0,$$

where $A' = dA/dr$, $B' = dB/dr$. The identification of (9.121) with the equation of geodesic lines (9.14) gives us the Christoffel symbols:

$$\Gamma^0_{10} = \Gamma^0_{01} = \frac{A'}{2A},$$

$$\Gamma^1_{00} = \frac{A'}{2B}, \quad \Gamma^1_{11} = \frac{B'}{2B}, \quad \Gamma^1_{22} = -\frac{r}{B}, \quad \Gamma^1_{33} = -\frac{r}{B}\sin^2 \theta,$$

$$\Gamma^2_{12} = \Gamma^2_{21} = \frac{1}{r}, \quad \Gamma^2_{33} = -\frac{1}{2}\sin 2\theta, \tag{9.122}$$

$$\Gamma^3_{13} = \Gamma^3_{31} = \frac{1}{r}, \quad \Gamma^3_{23} = \Gamma^3_{32} = \cot \theta \,.$$

The components of the Ricci tensor are calculated according to the definition (9.80):

$$R_{\mu\nu} = R^\lambda_{\mu\lambda\nu} = \frac{\partial \Gamma^\lambda_{\mu\nu}}{\partial x^\lambda} - \frac{\partial \Gamma^\lambda_{\mu\lambda}}{\partial x^\nu} + \Gamma^\rho_{\mu\nu}\Gamma^\lambda_{\rho\lambda} - \Gamma^\rho_{\mu\lambda}\Gamma^\lambda_{\rho\nu}.$$

For example, the component R_{00} is

$$R_{00} = \frac{\partial \Gamma_{00}^1}{\partial r} + \Gamma_{00}^1 \left(\Gamma_{10}^0 + \Gamma_{11}^1 + \Gamma_{12}^2 + \Gamma_{13}^3 \right) - 2\Gamma_{00}^1 \Gamma_{10}^0$$

$$= \frac{1}{2} \frac{A''B - A'B'}{G^2} + \frac{A'}{2B} \left(\frac{B'}{2B} + \frac{2}{r} + \frac{A'}{2a} \right) - 2 \frac{A'}{2B} \frac{A'}{2A}$$

$$= \frac{1}{2B} \left(A'' - \frac{A'B'}{2B} + \frac{2A'}{r} - \frac{A'^2}{2A} \right).$$

In the same way, the other three components are found:

$$R_{11} = -\frac{A''}{2A} + \frac{A'B'}{4AB} + \frac{A'^2}{4A^2} + \frac{B'}{rB},$$

$$R_{22} = 1 + \frac{rB'}{2B^2} - \frac{1}{B} \left(1 + \frac{rA'}{2A} \right),$$

$$R_{33} = \sin^2 \theta \left[1 + \frac{rB'}{2B^2} - \frac{1}{B} \left(1 + \frac{rA'}{2A} \right) \right].$$

Using these results, the curvature invariant (9.82) is found to be

$$R = -2 \left(\frac{A''}{2AB} - \frac{A'^2}{4A^2B} - \frac{A'B'}{4AB^2} - \frac{B'}{rB^2} + \frac{1}{r^2B} - \frac{1}{r^2} + \frac{A'}{rBA} \right).$$

Since the tensors $R_{\mu\nu}$ and $g_{\mu\nu}$ are diagonal, Einstein's equations are

$$R_{\mu\mu} - \frac{1}{2} R g_{\mu\mu} = 0 \quad \text{(no summation over } \mu\text{)},$$

that is

$$\frac{1}{r} - \frac{B}{r} + \frac{A'}{A} = 0,$$

$$\frac{A'B'}{2AB} + \frac{A'^2}{2A^2} - \frac{A''}{A} - \frac{A'}{rA} + \frac{B'}{rB} = 0, \qquad (9.123)$$

$$\frac{1}{r} - \frac{1}{rB} + \frac{B'}{B^2} = 0.$$

The last equation, written as

$$\frac{d}{dr} \left(\frac{1}{B} \right) + \frac{1}{r} \left(\frac{1}{B} \right) = \frac{1}{r},$$

can be easily integrated by separation of variables. Its solution is

$$\frac{1}{B} = -\frac{\lambda}{r} + 1,$$

where $-\frac{\lambda}{r}$ is the general solution of the homogeneous equation, with λ a constant of integration, and "1" is a particular solution of the non-homogeneous equation. Therefore

$$B(r) = \left(1 - \frac{\lambda}{r}\right)^{-1}. \tag{9.124}$$

With this solution, Eq. (9.123)$_1$ gives as a result of integration

$$\ln r - \ln(r - \lambda) + \ln A = \ln c^2,$$

where c is an integration constant. One then obtains

$$A(r) = c^2 \left(1 - \frac{\lambda}{r}\right). \tag{9.125}$$

With these results for B and A, the metric (9.117) is brought to the form

$$ds^2 = c^2 \left(1 - \frac{\lambda}{r}\right) dt^2 - \frac{dr^2}{1 - \frac{\lambda}{r}} - r^2 d\theta^2 - r^2 \sin^2 \theta \, d\varphi^2. \tag{9.126}$$

Now, we have to find the integration constants c and λ. For $r \to \infty$, the metric (9.126) approaches the Minkowski metric,

$$ds^2 = c^2 \, dt^2 - dr^2 - r^2 \, d\theta^2 - r^2 \sin^2 \theta \, d\varphi^2,$$

which shows that the integration constant c is the velocity of light in vacuum.

For determining the constant λ we use the weak field approximation. We saw in Sect. 9.7, formula (9.105), that in the limit of non-relativistic velocities and weak gravitational fields, i.e. far from the gravitational source, the component g_{00} of the metric has the expression

$$g_{00} = 1 + \frac{2\varphi}{c^2},$$

where φ is the Newtonian gravitational potential, which in our case is produced by a point-like mass M placed at the origin of coordinates, i.e.

$$\varphi = -G\frac{M}{r}.$$

550 General Theory of Relativity

Consequently, we immediately find that

$$\lambda \equiv r_S = \frac{2GM}{c^2}, \tag{9.127}$$

called *Schwarzschild radius* or *gravitational radius*. Thus, we have obtained the vacuum solution of Einstein's equation for a gravitational field with central symmetry as

$$ds^2 = c^2 \left(1 - \frac{r_S}{r}\right) dt^2 - \frac{dr^2}{1 - \frac{r_S}{r}} - r^2 d\theta^2 - r^2 \sin^2 \theta \, d\varphi^2. \tag{9.128}$$

This is the *Schwarzschild solution (metric)* for Einstein's equations. It can be proven (*Birkhoff's theorem*) that the Schwarzschild solution is the most general spherically symmetric vacuum solution of Einstein's field equations. The solution was discovered in 1915, only one month after the publication of Einstein's theory on general relativity. Its discoverer was the German physicist and astronomer Karl Schwarzschild (1873–1916).

There are two singularities in the Schwarzschild metric, at $r = 0$ and $r = r_S$. The latter singularity can be transformed away with a change of coordinates, but the former remains as a true singularity. Since the Schwarzschild metric is only expected to be valid for radii larger than the radius R of the massive body which acts as source, there is no problem as long as $R > r_S$. For ordinary stars and planets this is always the case. For instance, the radius of the Sun is approximately $700,000 \, \text{km}$, while its Schwarzschild radius is only $3 \, \text{km}$.

The coordinates used to write the Schwarzschild metric (9.128) are called *Schwarzschild coordinates*. The singularity at $r = r_S$ divides the Schwarzschild coordinates into two disconnected patches:

1) the *outer patch* with $r > r_S$ is the one that is related to the gravitational fields of stars and planets;
2) the *inner patch* $0 < r < r_S$, which contains the singularity at $r = 0$, is completely separated from the outer patch by the singularity at $r = r_S$.

In Schwarzschild coordinates there is no physical connection between the two patches, which may be viewed as separate solutions. As already mentioned, the singularity at $r = r_S$ is not a true singularity, but a *coordinate singularity*. As the name implies, the singularity arises from the choice of coordinates or coordinate conditions. For example, the tortoise coordinates are defined by the relation

$$r^* = r + r_S \ln \left| \frac{r}{r_S} - 1 \right|,$$

satisfying

$$\frac{dr^*}{dr} = \left(1 - \frac{r_S}{r}\right)^{-1}.$$

Thus, r^* approaches $-\infty$ as $r \to r_S$. By replacing the time coordinate t with $v = t + r^*/c$ or $u = t - r^*/c$, we obtain the so-called ingoing and outgoing Eddington–Finkelstein coordinates. In these coordinates, the Schwarzschild metric reads:

$$ds^2 = \left(1 - \frac{r_S}{r}\right) c^2 dv^2 - 2c \, dv \, dr - r^2 d\Omega^2,$$

or

$$ds^2 = \left(1 - \frac{r_S}{r}\right) c^2 du^2 + 2c \, du \, dr - r^2 d\Omega^2,$$

for the ingoing and outgoing cases, respectively, where $d\Omega^2 = d\theta^2 + \sin^2 \theta d\varphi^2$ is the standard metric on a unit radius two-sphere.[5] Remark that in this formulation, the metric has no singularity for $r = r_S$.

There are also other systems of coordinates (Lemaître coordinates, Kruskal–Szekeres coordinates, Novikov coordinates, or Gullstrand–Painlevé coordinates) in which the metric becomes regular at $r = r_S$ and the two patches can be related to each other.

For $r = 0$ the singularity cannot be removed and this is called a *gravitational singularity*. The criterion for establishing which singularity is physical is that the curvature becomes infinite. Moreover, the quantities which we analyze must be independent of the choice of coordinates, i.e. scalars. Any scalar derived from the Riemann tensor (which measures the curvature of space-time) would be a reasonable indicator of a singularity. One can construct various such invariants and it is not necessary that they are all simultaneously divergent as we approach a certain point, but it is enough to have one of them going to infinity. One such important quantity is the *Kretschmann invariant*, which is given by

$$K = R_{\mu\nu\lambda\rho} R^{\mu\nu\lambda\rho} = \frac{12 r_S^2}{r^6} = \frac{48 G^2 M^2}{c^4 r^6}. \tag{9.129}$$

Thus, at $r = 0$ the curvature becomes infinite, indicating the presence of a singularity. At this point the metric, and space-time itself, are no longer well-defined. Such singularities are a generic feature of general relativity, as proven in the 1960s by

---

[5]Incidentally, if we allow the mass parameter M to turn into a function of the corresponding null coordinate, $M(u)$ or $M(v)$, we obtain the simplest non-static generalization of the non-radiative Schwarzschild solution, known as the *Vaidya metric* (ingoing and outgoing):

$$ds^2 = \left(1 - \frac{r_S(v)}{r}\right) c^2 \, dv^2 - 2 \, dv \, dr - r^2 \left(d\theta^2 + \sin^2 \theta \, d\varphi^2\right),$$

$$ds^2 = \left(1 - \frac{r_S(u)}{r}\right) c^2 \, du^2 + 2 \, du \, dr - r^2 \left(d\theta^2 + \sin^2 \theta \, d\varphi^2\right).$$

The Vaidya metric describes the non-empty external space-time of a spherically symmetric and non-rotating star which is either emitting or absorbing *null dust* (sometimes called *null fluid*, i.e. a fluid for which the Einstein tensor is null). It is named after the Indian physicist and mathematician Prahalad Chunnilal Vaidya (1918–2010) and it is also called the *radiating/shining Schwarzschild metric*.

Roger Penrose (b. 1931) and Stephen Hawking (b. 1942). Although for some time they were deemed non-physical, they are now known to exist and are called *black holes* – a name coined by John Wheeler (1911–2008).

The Schwarzschild solution, taken to be valid for all $r > 0$, describes the *Schwarzschild black hole*. The surface $r = r_S$ demarcates what is called the *event horizon* of the black hole. It represents the point past which particles and light can no longer escape the gravitational field. Any physical object whose radius R becomes less than or equal to the Schwarzschild radius will undergo *gravitational collapse* and become a black hole.

We shall not expound further on the subject of black holes. This is a fascinating topic to which whole books and an impressive amount of research are devoted. We just note that in 1972, the research on gravity took an entirely new turn with the introduction of the notion of entropy of a black hole by Jacob Bekenstein (1947–2015). This groundbreaking work revolutionized the quest for quantum gravity and it is still the subject of intensive theoretical research.

9.9   Other Solutions of Einstein's Equations

During the past century the mathematical formalism needed for finding and characterizing solutions of Einsteins field equations has been thoroughly developed. After these solutions are found, they are classified by their symmetry group, their algebraic structure (Petrov type) or other invariance properties such as special subspaces or tensor fields and embedding properties.

In the following we present some of the most important metrics obtained as exact solutions, with physical or cosmological significance.

Reissner–Nordström Metric (Reissner 1916; Nordström 1918)

The Reissner–Nordström metric is a static solution of the Einstein–Maxwell field equations, describing the gravitational field created by a spherically symmetrical body, with mass M and total electric charge Q. The solution was found by Hans Reissner (1874–1967) (a German aeronautical engineer who was passionate of mathematical physics), and the Finnish theoretical physicist Gunnar Nordström (1881–1923).

The line element for the Reissner–Nordström metric is given by

$$ds^2 = \left(1 - \frac{r_S}{r} + \frac{r_Q^2}{r^2} \right) c^2 dt^2 - \left(1 - \frac{r_S}{r} + \frac{r_Q^2}{r^2} \right)^{-1} dr^2$$
$$- r^2 \left(d\theta^2 + \sin^2 \theta \, d\varphi^2 \right), \qquad (9.130)$$

where $r_S = \frac{2GM}{c^2}$ is the Schwarzschild radius, while $r_Q^2 = \frac{Q^2 G}{4\pi\varepsilon_0 c^4}$ is a characteristic length scale, determined by the fact that the body which creates the gravitational field is charged with the total electric charge Q. In the limit $Q \to 0$ (or, equivalently, $r_Q \to 0$) one re-obtains the Schwarzschild metric.

Kerr Metric (Kerr 1963)

The Kerr solution describes exterior gravitational fields of stationary rotating axisymmetric isolated and uncharged sources; till now, no satisfactory interior solutions are known.

The Kerr solution was found by the New Zealand mathematician Roy Kerr (b. 1934) by a systematic study of algebraically special vacuum solutions. The Kerr metric characterizing the space-time outside a body of mass M and angular momentum J is written as follows, in Boyer–Lindquist coordinates (r, θ, φ):

$$ds^2 = \left(1 - \frac{r_S r}{\Sigma^2}\right) c^2 dt^2 - \frac{\Sigma^2}{\Delta} dr^2 - \Sigma^2 d\theta^2 \qquad (9.131)$$
$$- \left(r^2 + \alpha^2 + \frac{\alpha^2 r_S r \sin^2 \theta}{\Sigma^2}\right) \sin^2 \theta d\varphi^2 + \frac{2\alpha r_S r \sin^2 \theta}{\Sigma^2} c\, dt\, d\varphi.$$

In the above relation, Σ and Δ are given by

$$\Sigma^2 = r^2 + \alpha^2 \cos^2 \theta,$$
$$\Delta = r^2 - r_S r + \alpha^2.$$

The Boyer–Lindquist coordinates are a generalization of the coordinates used for the Schwarzschild metric. The coordinate transformation from Boyer–Lindquist coordinates r, θ, φ to Cartesian coordinates x, y, z is given by (Boyer and Lindquist, 1967):

$$x = \sqrt{r^2 + \alpha^2} \sin \theta \cos \varphi,$$
$$y = \sqrt{r^2 + \alpha^2} \sin \theta \sin \varphi,$$
$$z = r \cos \theta.$$

The Kerr metric is extremely important in the study of black holes, since it is believed that most of them are spinning, just as the stars from whose gravitational collapse they were born.

Kerr–Newman Metric (Kerr 1963; Newman 1965)

Soon after the discovery of the Kerr solution, the American physicist Ezra T. Newman (b. 1929) generalized it to include also the electric charge. Thus, the Kerr–Newman solution is both the spinning generalization of Reissner–Nordström and the electrically charged version of the Kerr metric. In Boyer–Lindquist coordinates, the line element for this metric is given by the expression

$$ds^2 = (cdt - \alpha \sin^2 \theta \, d\varphi)^2 \frac{\Delta}{\Sigma^2} - \left(\frac{dr^2}{\Delta} + d\theta^2\right) \Sigma^2 \qquad (9.132)$$

$$- \left[(r^2 + \alpha^2) \, d\varphi - \alpha cdt\right]^2 \frac{\sin^2 \theta}{\Sigma^2},$$

where α and Σ have the same expressions as for the Kerr metric, while Δ is given by

$$\Delta = r^2 - r_S r + r_Q^2 + \alpha^2 = \Delta_{\text{Kerr}} + r_Q^2,$$

where

$$r_Q^2 = \frac{Q^2 G}{4\pi\varepsilon_0 c^4}$$

has the same significance as in the case of the Reissner–Nordström metric.

Friedmann–Lemaître–Robertson–Walker (FLRW) metric (Friedmann 1922, 1924; Lemaître 1927; Robertson 1929, 1935, 1936; Walker 1936)

The name of this metric was given after the scientists who found and studied it, namely, the Russian physicist and mathematician Alexander Friedmann (1888–1925), the Belgian physicist Georges Lemaître (1894–1966), the American mathematician and physicist Howard P. Robertson (1903–1961), and the British mathematician Arthur G. Walker (1909–2001).

Alexander Friedmann studied the Einstein equations as applied to the Universe, assuming a homogeneous and isotropic density, and he concluded that there are two possible solutions: the closed and the open models. The latter leads to a perpetual expansion. At the boundary between the open and the closed models, there is the flat solution. Physically, the condition for open, closed, or flat Universe is determined by the density of matter or energy. We shall discuss in more detail this model, as it is at the core of modern *Big Bang cosmology*.

If the distance between two galaxies is taken as $d(t) = R(t)d_0$, their relative speed can be written as $v = [\dot{R}(t)/R(t)]d(t)$, i.e., the speed is proportional to the separation between the two galaxies, with a proportionality factor $H(t) = \dot{R}(t)/R(t)$ which is called the Hubble parameter. Its present value is usually represented by H_0 and called *Hubble's constant*. We call $R(t)$ the *cosmic scale factor*, and here we take it to be dimensionless, while d_0 has the dimension of length. Below we shall consider $R(t)$ frequently as containing implicitly the d_0 factor and having dimensions of length. Concerning $H(t)$, it has the dimension of inverse time.

We shall discuss the problem of the motion of a galaxy by using the Newtonian mechanics, but taking into account Hubble's law. Let us consider the mass of the galaxy as m, under the gravitational attraction of the rest of the Universe, of mass M. As $M \gg m$, one has $M + m \simeq M$ and the total energy is

$$\frac{1}{2}mv^2 - \frac{GMm}{r} = E. \qquad (9.133)$$

Let us write $v = \dot{R}(t) = H(t)R(t)$ and $r = R$, where $H(t)$ is the Hubble parameter and R is the radius of the Universe. For a spherical mass distribution, the total mass is $M = \frac{4}{3}\pi R^3 \rho$, where ρ is the average mass density of the Universe, and we substitute this expression into (9.133). This gives

$$\frac{\dot{R}^2(t)}{2} - \frac{4\pi\rho G R^2(t)}{3} = \frac{E}{m} = \frac{-K}{2}. \tag{9.134}$$

This is a non-relativistic way of obtaining Einstein's equation from the Friedmann model for the expansion of the homogeneous and isotropic Universe. The latter is identical to the one obtained using the relativistic formalism starting from the Robertson–Walker metric, which is a metric compatible with the conditions of homogeneity and isotropy (these conditions are sometimes called *cosmological principle*):

$$ds^2 = c^2 dt^2 - R^2(t)\left[\frac{dr^2}{1 - kr^2} + r^2(d\theta^2 + \sin^2\theta d\varphi^2)\right]. \tag{9.135}$$

Here $k = -1, 0, 1$ correspond to open, flat, and closed cosmologies, respectively. Observe that K in (9.134) has the dimension of the square of a velocity, while k in (9.135) is dimensionless, because $R(t)$ has the dimension of length, and r is dimensionless. Then we have $K \sim kc^2$. According to (9.134), the critical condition to bring the expansion asymptotically to a halt occurs for $k = 0$, that is to say, for the density

$$\rho_c = \frac{3H^2}{8\pi G}. \tag{9.136}$$

With the present-day value of the Hubble parameter, H_0, the value of ρ_c is of the order of 10^{-29} g $\cdot$ cm^{-3}.

But the Robertson–Walker metric does not tell us anything about the time dependence of the scale factor $R(t)$. To obtain this information, one must solve not only the Einstein equations, that is, Eqs. (9.134) and (9.138) below, but also the equation of conservation of energy and the equation of state. Let us discuss the simplest case of a flat Universe. If we expand $R(t)$ in a power series around the reference time t_0, taken as the present time, we get $R(t) = R(t_0)[1 + H_0(t - t_0) - \frac{1}{2}q(t_0)H_0^2(t - t_0)^2 + \cdots]$, where the so-called *deceleration parameter* is given by

$$q(t) = -\frac{\ddot{R}(t)R(t)}{\dot{R}^2(t)}. \tag{9.137}$$

This quantity was estimated to be of the order of -0.5 at present, indicating that the expansion of the Universe is accelerated. The value of the deceleration parameter is a major topic in the present day cosmological research.

Together with Eq. (9.134) we must consider the other Einstein equation,

$$\ddot{R}(t) = -\frac{4\pi G}{3} R(t) \left(\rho + \frac{3p}{c^2} \right).$$ (9.138)

For $\rho > 0$ and $p > 0$, the acceleration $\ddot{R}$ is negative, and consistent with a positive deceleration. But as will be pointed out later, dark energy may provide a negative value for the factor $(\rho + 3p/c^2)$, producing an accelerated expansion of the Universe. We shall omit the discussion of this case and continue with the solutions for standard cosmology. We denote $\Omega = \rho/\rho_c$. Then we can write Eq. (9.134) in terms of the Hubble parameter as follows:

$$H^2(\Omega - 1) = KR^{-2}(t).$$ (9.139)

If one assumes the pressure to be negligible compared with the density, that is to say $p \simeq 0$, simple solutions of the Friedmann model are found. In the flat case ($k = 0$, $q_0 < 0.5$, $\Omega = 1$), one has

$$R(t) = [3GM/\pi]^{1/3} t^{2/3}, \qquad H = 2/3t.$$ (9.140)

In the closed case ($k = +1$, $q_0 > 0$, $\Omega > 1$), the Universe has a finite volume, but it is unbounded (this corresponds to the previously mentioned space which can be regarded as a generalization of the spherical surface to three dimensions). In such a case, one obtains solutions in terms of a parameter η, defined by $d\eta = R(t)dt$:

$$R(\eta) = (2GM/3\pi c^2)(1 - \cos \eta), \qquad t(\eta) = (2GM/3\pi c^3)(\eta - \sin \eta).$$ (9.141)

In both the open ($k = -1$, $\Omega < 1$) and the flat cases, the Universe is infinite and unbounded. In the open case, one has

$$R(\eta) = (2GM/3\pi c^2)(\cosh \eta - 1), \qquad t(\eta) = (2GM/3\pi c^3)(\sinh \eta - \eta).$$
(9.142)

In none of the three cases is the Universe static, and it should be either expanding or contracting. Expansion is interpreted as meaning that the galaxies separate with increasing speed because their mutual separation increases. But if this occurs, there should be a *redshift* in the spectra of light coming from remote galaxies. The effect was observed for the first time in 1912 by Vesto Slipher (1875–1969) at the Lowell Observatory in Flagstaff, Arizona. We discussed the consequences of this observation at the end of Sect. 9.7.

Einstein Metric (Einstein 1917)

This metric, describing *Einstein's static universe*, can be obtained as a special case of the Friedmann–Lemaître–Robertson–Walker solution with cosmological constant, and is classified as *spatially-homogeneous perfect fluid solution*. Einstein introduced the cosmological constant in 1917 in order to compensate for the attractive force of gravity and find a static solution, according to the then dominating paradigm of a static Universe.

The line elements of Einstein's metric is obtained from the FLRW case (9.135) with $k = 1$ and $R(t) = R_0$,

$$ds^2 = c^2 dt^2 - R_0^2 \left[\frac{dr^2}{1 - r^2} + r^2 (d\theta^2 + \sin^2 \theta d\varphi^2) \right], \quad (9.143)$$

where $R_0 = \frac{1}{\sqrt{\Lambda_0}} = \frac{c}{\sqrt{4\pi G \rho_0}}$ and $p_0 = 0$. In the above relations, ρ_0 is the energy density and p_0 is the pressure of the perfect fluid (dust). Note that the cosmological constant has a precise value with respect to the energy density, which renders the solution static. However, the solution is unstable, meaning that any small deviation of either Λ_0, or ρ_0, or R_0 from the above prescribed values would turn the Universe into one which either expands or collapses forever. For this reason, the solution was abandoned rather early as a plausible description of the Universe.

Lemaître–Tolman–Bondi (LTB) Metric (Lemaître 1933; Tolman 1934; Bondi 1947)

Initially, the Lemaître–Tolman metric was obtained as an exact spherically symmetric solution of Einstein's equations for *cosmic dust* (matter with $p = 0$). Later on, due to its inherent inhomogeneities, it gave rise to a whole class of models, known as *LTB models*, some of them being still the subject of active research as an appealing alternative to the prevailing interpretation of the acceleration of the universe in terms of a Λ-CDM model with a dominant dark energy component. Since we observe light rays from the past light cone, not the expansion of the Universe, spatial variation in matter density and Hubble rate can have the same effect on redshift as acceleration in a perfectly homogeneous Universe.

The Lemaître–Tolman–Bondi metric was found by Georges Lemaître, in 1933, then by the American mathematical physicist Richard C. Tolman (1881–1948), in 1934. It was studied later by the Anglo-Austrian mathematician and cosmologist Hermann Bondi (1919–2005), in 1947.

The line element for the *Lemaître–Tolman–Bondi metric*, describing a spherically symmetric cloud of cosmic dust, finite or infinite, which can expand or collapse under the action of its own gravitational field, is given, in comoving coordinates,[6] by the expression:

$$ds^2 = c^2 dt^2 - \frac{A'^2(r, t)}{1 - k(r)} dr^2 - A^2(r, t) \left(d\theta^2 + \sin^2 \theta \, d\varphi^2 \right), \quad (9.144)$$

where $A(r, t)$ is a function with dimension of length in SI, $A' = \frac{\partial A}{\partial r}$, while $k(r)$ is a function associated with the curvature of $t = $ const. hypersurfaces. All the functions depend on r, therefore the metric is inhomogeneous. In the limit $A(r, t) \to R(t)r$ and $k(r) \to kr^2$, one obtains the FLRW metric (9.135).

---

[6] To choose spatial coordinates to comove with the matter means to take $\frac{dx^i}{dt} = 0$.

   The pressure of the cosmic dust is zero, while its mass density ρ_M is given by the relation

$$\frac{8\pi G}{c^2}\rho_M = \frac{M'(r)}{A^2(r,t)A'(r,t)},$$

where $M(r)$ is a non-negative function that is fixed by the boundary condition (it has SI units of length). One can define, by comparison with the homogeneous FLRW model, a *local Hubble rate* by $H(r,t) = \frac{\dot{A}(r,t)}{A(r,t)}$, where $\dot{A} = \frac{dA}{dt}$.

De Sitter and anti de Sitter metrics (de Sitter 1932)

A special set of spaces which can be obtained as solutions of Einstein's field equations are the *maximally symmetric* ones. The concept of maximal symmetry is defined by analogy with the symmetry of the Euclidean n-dimensional space, whose isometries are the translations and rotations in n dimensions. The number of independent translations is n, while the number of independent rotations is $\frac{1}{2}n(n-1)$. In the case of curved spaces, these operations cannot be defined globally, but only in the neighbourhood of an arbitrary fixed point. A maximally symmetric manifold with $\frac{1}{2}n(n+1)$ independent symmetries is a *maximally symmetric space*. One special feature of the maximally symmetric spaces is that the curvature is the same at any point. For any such space, at any point and in any coordinate system, the following relation is valid:

$$R_{\mu\nu\sigma\rho} = K(g_{\mu\sigma}g_{\nu\rho} - g_{\mu\rho}g_{\nu\sigma}),$$

where

$$K = \frac{R}{n(n-1)},$$

with R the constant Ricci scalar.

   Thus, the curvature scalar essentially determines the maximally symmetric spacetimes and classifies them into three categories: *Minkowski space* $(R=0, K=0)$, *de Sitter space* $(R>0, K>0)$, and *anti de Sitter space* $(R<0, K<0)$. The de Sitter space is the simplest solution of Einstein's field equations with $\Lambda > 0$ and it was found by the Dutch mathematician, physicist, and astronomer Willem de Sitter (1872–1934) in 1932.

   Let us note still that, by taking the trace of the above equation, we find that the Ricci tensor is proportional to the metric in maximally symmetric spaces, i.e., in four dimensions,

$$R_{\mu\nu} = 3Kg_{\mu\nu}, \quad R = 12K.$$

Moreover, Einstein's tensor is also proportional to the metric,

$$G_{\mu\nu} = R_{\mu\nu} - \frac{1}{2}Rg_{\mu\nu} = -3Kg_{\mu\nu}$$

and, by Einstein's field equations (9.100), so is the energy-momentum tensor:

$$T_{\mu\nu} = -\frac{3K}{\kappa} g_{\mu\nu}.$$

Recall that (see Eq. (9.115)) an energy-momentum tensor proportional to the metric is specific to the vacuum energy, i.e. to the cosmological constant. Consequently, de Sitter and anti de Sitter space-times are maximally symmetric *vacuum solutions* of Einstein's equations. The cosmological constant, positive for de Sitter and negative for anti de Sitter, is the only source of curvature for the space-time. The metric can be put in the form

$$ds^2 = \left(1 - Kr^2\right) c^2 dt^2 - \frac{dr^2}{1 - Kr^2} - r^2 \left(d\theta^2 + \sin^2\theta \, d\varphi^2\right), \qquad (9.145)$$

in the so-called *static coordinates*. This line element corresponds to a *de Sitter space* (dS) if $K > 0$, or to an *anti de Sitter space* (AdS) if $K < 0$. In this form, it becomes apparent that the (anti) de Sitter space-time is a *static* space, i.e. a space-time whose geometry does not change in time (due to time translation symmetry) and which is irrotational (due to rotation symmetry). However, the physical spatial dimensions expand according to a FLRW model, where the cosmic scale factor is given by $R(t) = e^{Ht}$, with H – the constant Hubble parameter, proportional to the square root of the cosmological constant, $H \sim \sqrt{\Lambda}$. It has a *cosmological horizon* surrounding any observer. Unlike the black hole *event horizon*, which is a *future horizon* (i.e. the particles can enter the black hole crossing the horizon, but they cannot cross it in the opposite direction), the *cosmological horizon* in a Big Bang-type cosmology is a *past horizon* (i.e. the particles can only cross the horizon from inside out).

Schwarzschild–De Sitter Metric (Schwarzschild 1915; de Sitter 1932)

The Schwarzschild–de Sitter metric describes the gravitational field of a spherically symmetric body of mass M in a Universe with cosmological constant Λ.

The line element for the de Sitter–Schwarzschild metric can be written as

$$ds^2 = f(r)c^2 dt^2 - f^{-1}(r)dr^2 - r^2 \left(d\theta^2 + \sin^2\theta \, d\varphi^2\right), \qquad (9.146)$$

where

$$f(r) = 1 - \frac{r_S}{r} - \frac{1}{3}\Lambda r^2.$$

An observer who has not "fallen" into the black hole, but who can still see it in spite of the expansion of the Universe, is caught in between the two horizons, the black hole event horizon corresponding to the Schwarzschild solution and the cosmological one, corresponding to the de Sitter solution.

9.10 Tests of General Relativity

9.10.1 Precession of the Perihelion of Planets

We mentioned in Sect. 9.1 that the "anomalous" precession of the perihelion of Mercury (and of other planets) could not be fathomed within the framework of the Newtonian theory of gravitation. Here we shall show how the discrepancy between the experimental data and the non-relativistic theory of gravitation is explained in general relativity.

Suppose that the central field created by the Sun, of mass M, is described by the Schwarzschild metric (9.128). A planet gravitating around the Sun will describe a geodesic of Riemannian space with the metric (9.128). Let us determine the trajectory of this planet.

We should apply the variational principle $\delta \int f ds = 0$, with f given by

$$f = c^2 \left(1 - \frac{r_S}{r}\right) \dot{t}^2 - \frac{\dot{r}^2}{1 - \frac{r_S}{r}} - r^2 \left(\dot{\theta}^2 + \dot{\varphi}^2 \sin^2 \theta\right), \qquad (9.147)$$

and solve a system of Euler-type equations, together with the constraint equation $f \equiv 1$, which arises from the definition of the function f. Since this procedure is very difficult, we shall use two simplifying elements, based on experimental data:

a) A planet acted upon by a central force describes a plane trajectory ($\varphi = $ const., $\dot{\varphi} = 0$);
b) The trajectory of the planet is, to the first approximation, an ellipse with the Sun at one of the foci. This means that instead of solving Eulerian equations, we shall calculate the deviation of the trajectory from the elliptical shape.

We obtain three differential equations of motion by applying the Euler equations (9.120) with respect to the remaining three independent variables r, θ, t, the function f being given by

$$f = c^2 \left(1 - \frac{r_S}{r}\right) \dot{t}^2 - \frac{\dot{r}^2}{1 - \frac{r_S}{r}} - r^2 \dot{\theta}^2. \qquad (9.148)$$

The differential equations corresponding to the variables t and θ are then

$$\frac{\partial f}{\partial \dot{t}} = 2c^2 \left(1 - \frac{r_S}{r}\right) \dot{t} = 2A,$$

$$\frac{\partial f}{\partial \dot{\theta}} = -2r^2 \dot{\theta} = -2B,$$

where A and B are constants, while the third differential equation is the constraint equation

$$f = c^2 \left(1 - \frac{r_S}{r}\right) \dot{t}^2 - \frac{\dot{r}^2}{1 - \frac{r_S}{r}} - r^2 \dot{\theta}^2 \equiv 1.$$

We then have the following system of differential equations:

$$c^2 \left(1 - \frac{r_S}{r}\right) \dot{t}^2 - \frac{\dot{r}^2}{1 - \frac{r_S}{r}} - r^2 \dot{\theta}^2 = 1,$$

$$c^2 \left(1 - \frac{r_S}{r}\right) \dot{t} = A, \tag{9.149}$$

$$r^2 \dot{\theta} = B.$$

To find the differential equation of the trajectory $r = r(\theta)$, we first observe that

$$\dot{r} = \frac{dr}{ds} = \frac{dr}{d\theta} \frac{d\theta}{ds} = \frac{B}{r^2} \frac{dr}{d\theta} = -B \frac{d}{d\theta} \left(\frac{1}{r}\right), \tag{9.150}$$

where we used $(9.150)_3$. The remaining equations can be now combined to eliminate $\dot{t}$ and $\dot{\theta}$ from $(9.149)_1$, which yields

$$\left(\frac{d\sigma}{d\theta}\right)^2 = r_S \sigma^3 - \sigma^2 + \frac{r_S}{B^2} \sigma + \frac{1}{B^2} \left(\frac{A^2}{c^2} - 1\right),$$

where we made the notation $\sigma = 1/r$. The separation of variables leads to

$$d\theta = \frac{d\sigma}{\sqrt{r_S \sigma^3 - \sigma^2 + \frac{r_S}{B^2} \sigma + \frac{1}{B^2} \left(\frac{A^2}{c^2} - 1\right)}} = \frac{d\sigma}{\sqrt{F(\sigma)}}. \tag{9.151}$$

Denoting by $\sigma_1, \sigma_2, \sigma_3$ the roots of the equation $F(\sigma) = 0$, we can write

$$d\theta = \frac{d\sigma}{\sqrt{r_S (\sigma - \sigma_1)(\sigma - \sigma_2)(\sigma - \sigma_3)}}, \tag{9.152}$$

leading to an elliptic integral. Since there is no analytical solution, we shall exploit it by recalling that, in the first approximation, the trajectory is an ellipse. In polar coordinates, with the origin at one focus and the angular coordinate θ measured from the major axis, the equation of the ellipse is

$$\sigma = \frac{1}{r} = \frac{1 + r \cos \theta}{p}, \tag{9.153}$$

where $p = b^2/a = a(1 - e^2)$ is the focal parameter of the ellipse (with a and b the semi-major and semi-minor axes, $e = c/a$ the eccentricity, and $c^2 = a^2 - b^2$).

Observing that $F(\sigma) = (d\sigma/d\theta)^2$, we realize that the roots $\sigma_1, \sigma_2, \sigma_3$ are, at the same time, extreme values of the function $\sigma = \sigma(\theta)$. Then we may choose

$$\sigma_2 = \sigma_{max} = \frac{1}{r_{min}} = \frac{1+e}{p} \quad \text{(perihelion: } \theta = 0\text{)},$$

$$\sigma_3 = \sigma_{min} = \frac{1}{r_{max}} = \frac{1-e}{p} \quad \text{(aphelion: } \theta = \pi\text{)}.$$

Then, σ is a periodical function, varying between σ_2 and σ_3, which suggests the choice

$$\sigma = \frac{\sigma_2 + \sigma_3}{2} + \frac{\sigma_2 - \sigma_3}{2} \cos \varphi, \tag{9.154}$$

or, if we plug in the expressions for σ_2 and σ_3,

$$\sigma = \frac{1 + e \cos \varphi}{p}, \tag{9.155}$$

which is also the equation of an ellipse. To determine the deviation of the trajectory from the classical Newtonian form, one must find the relation between θ and φ.

We may write

$$\sigma - \sigma_2 = \frac{e}{p}(\cos \varphi - 1),$$

$$\sigma - \sigma_3 = \frac{e}{p}(\cos \varphi + 1),$$

$$(\sigma - \sigma_2)(\sigma - \sigma_3) = -\frac{e^2}{p^2} \sin^2 \varphi,$$

$$d\sigma = -\frac{e}{p} \sin \varphi \, d\varphi,$$

and (9.152) becomes

$$d\theta = \mp \frac{d\varphi}{\sqrt{r_S (\sigma_1 - \sigma)}}. \tag{9.156}$$

Taking into account the fact that the orbits of planets are very close to perfect circles ($\sigma_2 - \sigma_3 \ll \sigma_2 + \sigma_3$), one can approximate

$$\sigma \simeq \frac{\sigma_2 + \sigma_3}{2},$$

and (9.156) becomes

$$d\theta = \mp \frac{d\varphi}{\sqrt{r_S \left(\sigma_1 - \frac{\sigma_2 + \sigma_3}{2}\right)}}. \tag{9.157}$$

We observe that

$$\sigma_1 - \frac{\sigma_2 + \sigma_3}{2} = \sigma_1 + \sigma_2 + \sigma_3 - \frac{3}{2}(\sigma_2 + \sigma_3) = \frac{1}{r_S} - \frac{3}{p},$$

that is

$$r_S\left(\sigma_1 - \frac{\sigma_2 + \sigma_3}{2}\right) = 1 - \frac{3r_S}{p} = 1 - \frac{3r_S}{a\left(1 - e^2\right)}. \qquad (9.158)$$

The variation $\Delta\theta$ of θ, for a 2π variation of φ, is obtained by integrating (9.157) from 0 to 2π. Using (9.158) we then have

$$\Delta\theta = 2\pi\left[1 - \frac{3r_S}{a(1 - e^2)}\right]^{-1/2} \simeq 2\pi\left[1 + \frac{3}{2}\frac{r_S}{a(1 - e^2)}\right],$$

or, by replacing r_S with (9.127):

$$\Delta\theta - 2\pi = \Delta\omega = \frac{6\pi GM}{c^2 a(1 - e^2)}, \qquad (9.159)$$

which gives the *advance of the perihelion* of a planet after a full revolution around the Sun. The trajectory is, therefore, a rosette with elliptical loops, with the same focus on the Sun (Fig. 9.5).

The perihelion advance (9.159) is determined by an observer fixed with respect to the Sun. Since the measurements are made from the Earth, we have to consider the motion of the Earth around the Sun. If T_{Earth} and T_{planet} are the revolution periods of the Earth and the observed planet, then the advance of the perihelion of the planet during one terrestrial year is

$$\Delta\Omega = \frac{T_{\text{Earth}}}{T_{\text{planet}}}\Delta\omega,$$

Fig. 9.5 Trajectory of a planet around the Sun (supposing the Sun does not have a translation motion through the Universe).

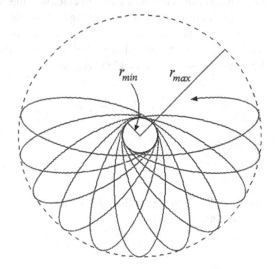

and for one terrestrial century

$$\Delta\Omega = 100\frac{T_{\text{Earth}}}{T_{\text{planet}}}\,\Delta\omega. \tag{9.160}$$

With the mass of the Sun $M_\odot = 1.98 \times 10^{30}$ kg, and $T_{\text{Earth}} = 365.25$ days, we finally obtain

$$\Delta\Omega = \frac{2.094 \times 10^{16}}{a(1-e^2)T_{\text{planet}}} \text{ arc seconds per century.} \tag{9.161}$$

We note that the relativistic effect of the perihelion precession depends on the semi-major axis a of the orbit trajectory: the smaller a, the greater $\Delta\Omega$. The closest planet to the Sun is Mercury ($T_{\text{Mercury}} = 88$ days, $a = 5.8 \times 10^{10}$m, $e = 0.2056$) and, according to (9.161),

$$\Delta\Omega_{\text{Mercury}} = 42.89 \text{ arc seconds per century,}$$

in very good agreement with the observational data ($42.45'' \pm 0.94''$).

9.10.2 Deflection of Light by the Sun

The phenomenon of *deflection of light in strong gravitational fields* was first signaled by Henry Cavendish (1731–1810) in 1784 (in an unpublished manuscript), then by Johann Georg von Soldner (1776–1833) in 1804, and by Siméon Poisson (1781–1840) in 1833. These calculations were made within the framework of Newtonian mechanics. Einstein was the first to calculate the correct value for the light bending.

Let us consider a light ray propagating in the Sun's gravitational field. Its path, according to general relativity, is a geodesic of the Riemannian space, characterized by $ds = 0$ (the so-called *null geodesic*). Admitting that the trajectory is plane, Schwarzschild's metric (9.128) becomes

$$c^2\left(1 - \frac{r_S}{r}\right)dt^2 - \frac{dr^2}{1 - \frac{r_S}{r}} - r^2 d\theta^2 = 0.$$

We also have

$$c^2\left(1 - \frac{r_S}{r}\right)dt = \frac{A}{B}r^2\,d\theta,$$

where $A/B = $ const. (see (9.149)). Eliminating the time from the last two equations, one obtains

$$\left(\frac{d\sigma}{d\theta}\right)^2 = r_S\sigma^3 - \sigma^2 + \frac{A^2}{B^2}\frac{1}{c^2},$$

or, by taking the derivative with respect to θ,

$$\frac{\partial^2 \sigma}{\partial \theta^2} + \sigma = \frac{3r_S}{2}\sigma^2. \tag{9.162}$$

To integrate this equation we shall use the method of successive approximations. As it can be remarked, the effect of gravitation is expressed by the right-hand side, which means that equation $\sigma'' + \sigma = 0$ gives the trajectory of light in the absence of gravitation. The solution of this equation is, obviously, a straight line:

$$\sigma_0 = \frac{1}{R}\cos\theta, \tag{9.163}$$

where σ_0 is the zeroth order approximation, and $1/R$ a constant of integration. To obtain the first-order approximation to the solution, we replace (9.163) into the right-hand side of (9.162) and obtain

$$\frac{\partial^2 \sigma}{\partial \theta^2} + \sigma = \frac{3r_S}{2}\frac{1}{R^2}\cos^2\theta,$$

with the solution

$$\sigma_1 = \frac{r_S}{2R^2}\left(\cos^2\theta + 2\sin^2\theta\right).$$

In the first-order approximation, the solution of Eq. (9.162) therefore is

$$\sigma = \sigma_0 + \sigma_1 = \frac{1}{R}\cos\theta + \frac{r_S}{2R^2}\left(\cos^2\theta + 2\sin^2\theta\right), \tag{9.164}$$

and represents the parametric equation of the trajectory of the light ray in the gravitational field of the Sun.

It is more convenient to use Cartesian coordinates instead of the polar ones. By means of the transformation equations $x = r\cos\theta$, $y = r\sin\theta$, Eq. (9.164) takes the form

$$x = R - \frac{r_S}{2R}\frac{x^2 + 2y^2}{\sqrt{x^2 + y^2}}, \tag{9.165}$$

which is a hyperbola, with the Sun at its focus. The value of the deviation is given by the angle between the asymptote to the trajectory and the straight line $x = R$ (see Fig. 9.6), namely

$$\tan\frac{\alpha}{2} = \lim_{y\to\infty}\frac{R - x}{y} = \frac{r_S}{2R}\lim_{y\to\infty}\frac{x^2 + 2y^2}{\sqrt{x^2 + y^2}} = \frac{r_S}{r}.$$

For small angles,

$$\tan\frac{\alpha}{2} \simeq \frac{\alpha}{2} = \frac{r_S}{R},$$

Fig. 9.6 Deflection of a
light beam by the Sun.

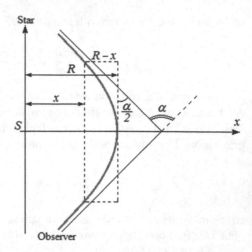

which yields

$$\alpha \simeq 2\frac{r_S}{R} = \frac{4GM}{c^2R}. \tag{9.166}$$

Approximating R with the radius of the Sun, one obtains $\alpha = 1.75$ arc seconds. The
observations performed in 1919 by Arthur Eddington (1822–1944) and his collaborators during a total solar eclipse (in Africa), so that the stars near the Sun could
be observed, produced a spectacular confirmation of Einstein's theory. Astronomers
now refer to this displacement of light as *gravitational lensing*.

The observations were repeated during other total Solar eclipses over the time.
All were in excellent agreement with Einstein's predictions.

9.10.3   Gravitational Redshift

The light (or other form of electromagnetic radiation) originating from a source
placed in a stronger gravitational field is found to be of longer wavelength (shifted
towards red) when detected by an observer placed in a region of weaker gravitational field (e.g. terrestrial conditions). This effect is called *gravitational redshift*
and it must not be confounded with the relativistic Doppler effect (explained within
the Minkowski flat space geometry), or with the cosmological redshift due to the
expansion of the Universe.

The gravitational redshift appears as a direct consequence of Einstein's equivalence principle, as a prediction which was confirmed by observation. Consider a
star of mass M, whose gravitational field is described by the Schwarzschild metric
(9.128). A clock at rest with respect to M (i.e. $dr = 0$, $d\theta = 0$, $d\varphi = 0$) determines
a proper duration

$$\Delta\tau = \int \frac{ds}{c} = \int_{t_1}^{t_2} \sqrt{1 - \frac{r_S}{r}}\, dt = \sqrt{1 - \frac{r_S}{r}}\, \Delta t.$$

At some other point of the same frame ($dr' = 0$, $d\theta = 0$, $d\varphi = 0$) an identical clock, during the same time interval Δt, will register the proper duration

$$\Delta\tau' = \sqrt{1 - \frac{r_S}{r'}}\, \Delta t.$$

The last two relations give

$$\frac{\Delta\tau}{\sqrt{1 - \frac{r_S}{r}}} = \frac{\Delta\tau'}{\sqrt{1 - \frac{r_S}{r'}}}. \tag{9.167}$$

Thus, proper time runs differently in stronger and weaker gravitational field: the stronger the field, the slower the passage of proper time. This is called *gravitational time dilation*.

Suppose that the first clock lies on the star of radius R (i.e. $r = R$ in (9.167)), and the second on the surface of the Earth, placed at the distance $r' \gg R$. In this case we may approximate

$$\Delta\tau' \simeq \frac{\Delta\tau}{\sqrt{1 - \frac{r_S}{R}}}. \tag{9.168}$$

Such identical clocks exist in the Universe as atoms of any element. They are called *atomic clocks*. Let ν be the frequency of the radiation emitted by an atom situated on the considered star. According to (9.168), the frequency determined from the Earth is

$$\nu' = \nu \sqrt{1 - \frac{r_S}{R}} \simeq \nu \left(1 - \frac{GM}{c^2 R}\right) < \nu,$$

which means a redshift of the spectral line emitted by the atom. In terms of wavelengths, the relative shift is

$$\frac{\Delta\lambda}{\lambda} = -\frac{\Delta\nu}{\nu}.$$

The redshift increases with the ratio M/R. In the case of the Sun, for example, the relative redshift is 2×10^{-6}.

This gravitational redshift predicted by Einstein in 1911 was confirmed by some high-precision observations. Preliminary observations were reported as early as 1925, by the American astronomer Walter Adams (1876–1956). In 1959, Robert Pound (1919–2010) and Glen Rebka (1931–2015) succeeded in testing Einstein's prediction by a very precise experiment at Harvard University, based on the then recently discovered *Mössbauer effect* (the resonant and recoil-free emission and absorption of gamma ray photons by atoms bound in a solid form, discovered in 1958 by Rudolf

Mössbauer (1929–2011)). The distance between the emitter and the absorber was of
only 22.5 m. Pound and Rebka reported a relative variation in the frequency of the
photons emitted by a source of ^{57}Fe in terrestrial gravitational field of

$$\frac{\Delta \nu}{\nu} = -(5.13 \pm 0.51) \times 10^{-15},$$

confirming Einstein's prediction of -4.92×10^{-15} to 10% accuracy. In subsequent
refined versions of the experiment, by 1964, the accuracy was increased to 1%.

9.10.4 Gravitational Time Delay

Another effect of the gravitational time dilation was discovered in 1964 by the American physicist Irwin I. Shapiro (b. 1929). Shapiro's idea was to determine the effect
of the Sun's gravitational field on a radar signal traveling from Earth to the inner
planets, Venus or Mercury, and back. This effect was called by Shapiro "the fourth
test of general relativity".

   According to general relativity, as a result of such a travel the radar signals would
be delayed as compared to the time given by Newton's theory.[7] Thus, if we denote
by t_r the round trip time of the radar signal predicted by general relativity, and by t_n
the corresponding flat-space value, then the delay will be

$$\Delta t_r = t_r - t_n.$$

The *radar signal delay* is maximum when the planet (e.g., Mercury) is in superior
conjunction with the Sun (it is located on the opposite side of the Sun with respect
to Earth) (see Fig. 9.7), since the radar signal has to travel in this case in the intense
gravitational field of the Sun. We shall make the following assumptions:

1) the Earth (E) and the reflecting planet (P) are considered pointlike, and move in
   the same plane on circular orbits;
2) the positional change of the Earth and the planet during the emission and reception
   of the radar signal is neglected;
3) the time delay due to the curvature of the trajectory is neglected;
4) the Sun's gravitational field is described by the Schwarzschild metric, with $r_S = \frac{2GM_\odot}{c^2}$, where $M_\odot$ is the Sun's mass.

   Under these assumptions, one may consider the Sun at the centre of a Cartesian
system of coordinates, with $(-x_e, d)$ the coordinates of the Earth, and (x_p, d) the
coordinates of the planet (Fig. 9.8). Then d is the minimum distance from the Sun to
the radar signal trajectory, while r_p and r_e are the distances from the planet and from

---

[7]The method used here was suggested by D. K. Ross and L. I. Schiff, *Phys. Rev.*, 1215, **141**, 1960.

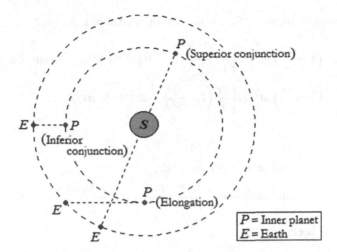

Fig. 9.7 Various positions of the Earth (E) and an inner planet (P), showing that the *gravitational time delay* is maximal when the planet and the Earth are in superior conjunction, since the radar signal has to travel in the intense gravitational field of the Sun.

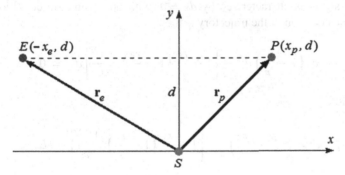

Fig. 9.8 Travel of a radar signal from the Earth (E) to a planet (P) and return. The Sun (S) is at the origin of the reference system.

Earth to the Sun, respectively. A radar signal travels from E to P and returns on the same trajectory (the dashed line). Since

$$r^2 = x^2 + y^2 + z^2,$$
$$dl^2 = dr^2 + r^2 d\theta^2 + r^2 \sin^2\theta \, d\varphi^2 \qquad (9.169)$$
$$= dx^2 + dy^2 + dz^2 = dx_i dx_i = dx_i \, dx_k \, \delta_{ik},$$

the Schwarzschild metric in Cartesian coordinates can be calculated:

$$ds^2 = c^2 \left(1 - \frac{r_S}{r}\right) dt^2 - \left\{ \frac{dr^2}{1 - \frac{r_S}{r}} - dr^2 + \left[dr^2 + r^2 \left(d\theta^2 + \sin^2\theta \, d\varphi^2\right)\right]\right\}$$

$$= c^2 \left(1 - \frac{r_S}{r}\right) dt^2 - \left\{ \frac{r_S}{r} \left(1 - \frac{r_S}{r}\right)^{-1} dr^2 + dx_i \, dx_k \, \delta_{ik}\right\}.$$

But

$$r \, dr = x \, dx + y \, dy + z \, dz,$$

$$dr^2 = \frac{(x \, dx + y \, dy + z \, dz)^2}{r^2} = \frac{x_i x_k \, dx_i \, dx_k}{r^2},$$

and the metric becomes

$$ds^2 = c^2 \left(1 - \frac{r_S}{r}\right) dt^2 - \left[\frac{r_S}{r} \left(1 - \frac{r_S}{r}\right)^{-1} \frac{x_i x_k}{r^2} + \delta_{ik}\right] dx_i dx_k. \qquad (9.170)$$

The radar signal is characterized by $ds = 0$, with our geometric conditions $dy = dz = 0$. The equation of the trajectory is then

$$c^2 \left(1 - \frac{r_S}{r}\right) dt^2 - \left[1 + \frac{r_S}{r} \left(1 - \frac{r_S}{r}\right)^{-1} \frac{x^2}{r^2}\right] dx^2 = 0, \qquad (9.171)$$

which yields

$$dt = \frac{1}{c} \left(1 - \frac{r_S}{r}\right)^{-1/2} \left[1 + \frac{r_S}{r} \left(1 - \frac{r_S}{r}\right)^{-1} \frac{x^2}{r^2}\right]^{1/2} dx. \qquad (9.172)$$

Expanding in series and keeping only the terms of the first order in r_S/r, then integrating between $-x_e$ and x_p and multiplying by 2, we obtain the round trip time t_r of the radar signal, as predicted by general relativity:

$$t_r = 2 \int_{-x_e}^{x_p} dt = t_n + \frac{4GM_\odot}{c^2} \left[\ln \frac{x_p + r_p}{r_e - x_e} - \frac{1}{2} \left(\frac{x_p}{r_p} + \frac{x_e}{r_e}\right)\right] + \dots \qquad (9.173)$$

Here

$$t_n = \frac{2(x_e + x_p)}{c} \qquad (9.174)$$

is the transit time in flat space, according to the Newtonian theory. The transit time delay is then the difference

$$\Delta t_r = t_r - t_n = \frac{4GM_\odot}{c^2} \left[\ln \frac{x_p + r_p}{r_e - x_e} - \frac{1}{2} \left(\frac{x_p}{r_p} + \frac{x_e}{r_e}\right)\right]. \qquad (9.175)$$

Since the determinations are made from the Earth, we have to use instead of t_r the proper time on the Earth, given by

$$t_E = \frac{1}{c} \int_{S_E(0)}^{S_E(t)} dS_E, \tag{9.176}$$

where dS_E is the Schwarzschild metric (9.128) for the terrestrial gravitational field, while $S_E(0)$ and $S_E(t)$ are the positions of the Earth in the four-dimensional manifold for the values $t = 0$ and $t = t$ of the time coordinate. According to (9.176), the proper time for the Earth is

$$t_E = \left(1 - \frac{2GM_E}{c^2 r_e}\right)^{1/2} t_r,$$

where M_E is the Earth mass, and t_r is given by (9.173). A series expansion in which we keep only first order terms in $\frac{GM_E}{c^2 r_e}$ then yields

$$t_E \simeq \left(1 - \frac{GM_E}{c^2 r_e}\right) t_r$$

$$= \left(1 - \frac{GM_E}{c^2 r_e}\right) \left\{ \frac{2(x_s + x_p)}{c} + \frac{4GM_S}{c^2} \left[\ln \frac{x_p + r_p}{r_e - x_e} - \frac{1}{2} \left(\frac{x_p}{r_p} + \frac{x_e}{r_e} \right) \right] \right\}.$$

In this approximation, the transit time delay of the radar signal, determined from the Earth and due to the Sun's gravitational field, is

$$\Delta t = t_E - t_n$$

$$= \frac{4GM_S}{c^3} \left\{ \ln \frac{x_p + r_p}{r_e - x_e} - \frac{1}{2} \left(\frac{x_p}{r_p} + \frac{x_e}{r_e} \right) - \frac{M_E}{M_S} \frac{(x_e + x_p)}{2 r_e} \right\}. \tag{9.177}$$

For Mercury in superior conjunction with the Sun, general relativity gives a delay $\Delta t = 1.6 \times 10^{-4}$ s. The precision of the first experiment of Shapiro's effect was not very high, but in subsequent years the accuracy was increased from over 3 % to less than 1 %. Later on, the experiment was repeated with transponders on space probes. These receive the signal from Earth and after a precisely known delay send it with increased intensity back to Earth. Thus with the Viking Mars probe of 1979 the predictions of the general theory of relativity for this delay in the gravitational field of the Sun could be confirmed to an accuracy of 0.1 %. In 2003, with the space probe Cassini, an accuracy of 0.0012 % was achieved.

9.10.5 Gravitational Waves

One of the most elusive predictions of general relativity was the existence of gravitational waves. Einstein predicted them in 1916 and refined his calculations in 1918.

Their effect is a strain on space-time, which manifests itself through the periodic increase and decrease of distances between free objects.

We shall sketch below the proof that a gravitational field propagates in vacuum as gravitational waves. As Einstein's equations are highly nonlinear, we shall adopt the weak field approximation.

Since, by assumption, the field is weak, we may choose the metric tensor as

$$g_{\mu\nu} = \eta_{\mu\nu} + h_{\mu\nu}, \tag{9.178}$$

where $\eta_{\mu\nu}$ is the Minkowski metric tensor,

$$\eta_{\mu\nu} = \begin{pmatrix} 1 & 0 & 0 & 0 \\ 0 & -1 & 0 & 0 \\ 0 & 0 & -1 & 0 \\ 0 & 0 & 0 & -1 \end{pmatrix}$$

and $h_{\mu\nu}$ is a small perturbation ($|h_{\mu\nu}| \ll 1$), called *normal metric perturbation*. In this approximation, we shall keep in all calculations only terms at most linear in $h_{\mu\nu}$. As a consequence, we can also write

$$g^{\mu\nu} = \eta^{\mu\nu} - h^{\mu\nu},$$

where $h^{\mu\nu} = \eta^{\mu\rho}\eta^{\nu\sigma}h_{\rho\sigma}$. One can see that the indices can be raised and lowered with the Minkowski metric tensor, since the corrections would be higher order in the perturbation. This suggests that linearized general relativity can be viewed as the field theory of a symmetric tensor field $h_{\mu\nu}$ on a flat (Minkowski) space-time background.

Next, we use formula (9.76) for the curvature tensor

$$R_{\mu\nu\lambda\rho} = \frac{1}{2} \left(\frac{\partial^2 g_{\mu\rho}}{\partial x^\nu \partial x^\lambda} + \frac{\partial^2 g_{\nu\lambda}}{\partial x^\mu \partial x^\rho} - \frac{\partial^2 g_{\mu\lambda}}{\partial x^\nu \partial x^\rho} - \frac{\partial^2 g_{\nu\rho}}{\partial x^\mu \partial x^\lambda} \right)$$
$$+ g_{\sigma\kappa} \left(\Gamma^\sigma_{\mu\rho}\Gamma^\kappa_{\nu\lambda} - \Gamma^\sigma_{\mu\lambda}\Gamma^\kappa_{\nu\rho} \right). \tag{9.179}$$

Introducing (9.178) into (9.179) and neglecting the second-order terms in $h_{\mu\nu}$, we are left with

$$R_{\mu\nu\lambda\rho} = \frac{1}{2} \left(\frac{\partial^2 h_{\mu\rho}}{\partial x^\nu \partial x^\lambda} + \frac{\partial^2 h_{\nu\lambda}}{\partial x^\mu \partial x^\rho} - \frac{\partial^2 h_{\mu\lambda}}{\partial x^\nu \partial x^\rho} - \frac{\partial^2 h_{\nu\rho}}{\partial x^\mu \partial x^\lambda} \right). \tag{9.180}$$

We are now able to calculate the components of the Ricci tensor (see (9.80)). Using the Riemann tensor properties, we have

$$R_{\mu\nu} = g^{\lambda\rho}R_{\lambda\mu\rho\nu} \simeq \eta^{\lambda\rho}R_{\lambda\mu\rho\nu}$$

$$= \frac{1}{2}\eta^{\lambda\rho}\left(\frac{\partial^2 h_{\lambda\nu}}{\partial x^\mu \partial x^\rho} + \frac{\partial^2 h_{\mu\rho}}{\partial x^\nu \partial x^\lambda} - \frac{\partial^2 h_{\lambda\rho}}{\partial x^\mu \partial x^\nu} - \frac{\partial^2 h_{\mu\nu}}{\partial x^\rho \partial x^\lambda}\right) \qquad (9.181)$$

$$= \frac{1}{2}\left(-\eta^{\lambda\rho}\frac{\partial^2 h_{\mu\nu}}{\partial x^\rho \partial x^\lambda} + \frac{\partial^2 h_\mu^\lambda}{\partial x^\lambda \partial x^\nu} + \frac{\partial^2 h_\nu^\lambda}{\partial x^\mu \partial x^\lambda} - \frac{\partial^2 h}{\partial x^\mu \partial x^\nu}\right),$$

where $h = \eta^{\lambda\rho}h_{\lambda\rho}$.

It should be mentioned that the decomposition (9.178) does not uniquely define the coordinate system, in the sense that in various coordinate systems the metric can be written as the Minkowski metric plus a small perturbation, but the perturbation would be different from system to system. This means that there is a gauge freedom in the choice of coordinates. Basically, a general coordinate transformation is specified by

$$x^\mu \to x^\mu + \epsilon \xi^\mu(x), \qquad (9.182)$$

where $\xi^\mu(x)$ denotes a general vector field and ϵ is an infinitesimal parameter. Under such a transformation, the metric perturbation $h_{\mu\nu}$ will change to

$$h'_{\mu\nu} = h_{\mu\nu} + 2\epsilon\partial_{(\mu}\xi_{\nu)}, \qquad (9.183)$$

remaining small. The transformation (9.183) is called gauge transformation in the linearized theory. It can be straightforwardly showed that such a transformation leaves invariant the Riemann tensor, consequently the physical space-time. As we are faced with this gauge freedom, we have to fix the coordinate system, partially or completely. We shall start with a partial fixing, by imposing a coordinate condition which selects the so-called *harmonic coordinates*, i.e. the systems of coordinates which satisfy d'Alembert's equation

$$\Box x^\mu = 0, \qquad (9.184)$$

where $\Box$ represents the d'Alembertian operator in terms of covariant derivatives. Writing the condition explicitly, one can easily show that it is equivalent to

$$\Gamma^\alpha_{\mu\nu}g^{\mu\nu} = 0. \qquad (9.185)$$

Replacing the Christoffel symbols and the metric tensor with their expressions in the weak field limit and keeping only terms linear in $h_{\mu\nu}$, we find that the condition (9.184) becomes a Lorenz-type condition,

$$\frac{\partial \bar{h}_\mu^\lambda}{\partial x^\lambda} = 0, \qquad (9.186)$$

where

$$\bar{h}_\mu^\lambda = h_\mu^\lambda - \frac{1}{2}h\,\delta_\mu^\lambda, \tag{9.187}$$

with h the trace of the tensor $h_{\mu\nu}$, is the so-called *trace-reversed metric perturbation*. (The name comes from the fact that $\bar{h} = \eta^{\mu\nu}h_{\mu\nu} = -h$.) With this gauge condition, the number of degrees of freedom of the symmetric tensor field $h_{\mu\nu}$ is reduced from ten to six. Using for the moment the gauge fixing (9.186), we find

$$\frac{\partial h_\mu^\lambda}{\partial x^\lambda} = \frac{1}{2}\frac{\partial h}{\partial x^\mu},$$
$$\frac{\partial^2 h_\mu^\lambda}{\partial x^\lambda \partial x^\nu} = \frac{1}{2}\frac{\partial^2 h}{\partial x^\mu \partial x^\nu},$$
$$2\frac{\partial^2 h_\mu^\lambda}{\partial x^\lambda \partial x^\nu} = \frac{\partial^2 h_\mu^\lambda}{\partial x^\lambda \partial x^\nu} + \frac{\partial^2 h_\nu^\lambda}{\partial x^\lambda \partial x^\mu},$$

and (9.181) reduces to

$$R_{\mu\nu} = -\frac{1}{2}\Box h_{\mu\nu}, \quad \text{where} \quad \Box = \eta^{\lambda\rho}\frac{\partial^2}{\partial x^\lambda \partial x^\rho}. \tag{9.188}$$

With this expression for the Ricci tensor, the Einstein equations (9.100) become:

$$\Box \bar{h}_{\mu\nu} = -2\kappa T_{\mu\nu}. \tag{9.189}$$

In vacuum,

$$\Box \bar{h}_{\mu\nu} = 0. \tag{9.190}$$

Consequently, the *gravitational field propagates as gravitational waves*.

A plane-wave solution is simply written as

$$\bar{h}_{\mu\nu} = A\,\mathbf{e}_{\mu\nu}e^{ik_\alpha x^\alpha},$$

where A is the amplitude of the wave, $\mathbf{e}_{\mu\nu}$ is the polarization tensor, and k_α is the wave vector. The wave equation (9.190) leads to the condition $k_\alpha k^\alpha = 0$, which means that the wave vector is light-like, therefore the gravitational waves in vacuum propagate with the speed of light. The gauge fixing condition (9.186) leads to the orthogonality between the wave vector and the polarization tensor, $\mathbf{e}^{\alpha\beta}k_\alpha = 0$.

Further gauge fixing conditions can be imposed, to fully specify the coordinate system. One of the most transparent gauges is the so-called transverse traceless (TT) gauge, which requires transversality of the waves by imposing $\mathbf{e}^{0\beta} = 0$, implying

$$\mathbf{e}^{ij}k_j = 0, \tag{9.191}$$

as well as traceless amplitudes,

$$e^j_{\ j} = 0. \tag{9.192}$$

In this gauge, $\bar{h}_{\mu\nu} = h_{\mu\nu}$. With the extra conditions, the number of degrees of freedom reduces to two, which is equal to the number of physical polarizations of the graviton.

We can understand the physical meaning of the TT gauge by examining the effect of the passage of a wave on a particle at rest (in flat space, before being affected by the wave). The geodesic equation in the TT gauge gives the acceleration of the particle:

$$\frac{d^2 x^i}{ds^2} = -\Gamma^i_{00} = -\frac{1}{2}(2h_{i0,0} - h_{00,i}) = 0.$$

This shows that the particle does not move, in other words the coordinate system specified by the TT gauge is co-moving with a freely falling particle.

The polarizations of the gravitational waves are easily understood in the TT gauge. Let us consider a wave moving in the z-direction, such that $k^0 = k_z = \omega$ and $k_x = k_y = 0$. From the conditions (9.191) and (9.192) we find then that $e^{0\alpha} = e^{z\alpha} = 0$, while $e^{xx} = -e^{yy}$. These relations show that there are actually only two independent components of the polarization tensor, which can be taken as e^{xx}, denoted by $\oplus$, and e^{xy}, denoted by $\otimes$.

To understand how the gravitational waves distort the space-time, let us consider the simplest case of a purely $\oplus$-polarized wave, for which $e^{xy} = 0$. The associated metric is

$$ds^2 = -dt^2 + (1 + h_+)dx^2 + (1 - h_+)dy^2 + dz^2, \tag{9.193}$$

where $h_+ = \mathcal{A}e^{xx}\exp[-i\omega(t - z)]$. Such a metric produces opposite effects on proper distance on the two transverse axes, contracting one while expanding the other. A purely $\otimes$-polarized wave ($e^{xx} = 0$) is obtained by a 45° rotation of the $\oplus$ polarization. Due to the linearity of the wave equation and of the TT gauge conditions, a general wave will be a linear superposition of these two polarization tensors. For example, circular polarization can be described by

$$e_R = \frac{1}{\sqrt{2}}(e_+ + ie_\times), \qquad e_L = \frac{1}{\sqrt{2}}(e_+ - ie_\times), \tag{9.194}$$

where $e_+ = e^{xx}$ and $e_\times = e^{xy}$ are the two linear polarization tensors and e_R and e_L are polarizations that rotate in the right-handed and left-handed directions respectively.

The detection of gravitational waves is a highly complex experimental undertaking, due to the very weak signals which may reach the Earth even from very powerful cosmic sources. The first indirect evidence for gravitational energy radiation, which is understood as a wave phenomenon, came in 1974 from the so-called Hulse–Taylor binary – a pair of stars, one of which is a pulsar (a radiating neutron star). They each have masses around $1.4M_\odot$ and the distance between them is around 2×10^6 km, of the order of Sun's diameter. They are expected to radiate 10^{22} times the gravitational energy radiated by the Earth–Sun system. This causes the stars to gradually move closer together, in what is known as an *inspiral*, and this has an effect on the observed pulsar's signals.

Russell Hulse (b. 1950) and Joseph Taylor (b. 1941) were awarded the Nobel Prize in 1993 for their measurements which led to the discovery of the first binary pulsar, and allowed them to show that the gravitational radiation predicted by general relativity matched the results of these observations with a precision within 0.2 %.

The search for direct evidence involves mainly detectors based on laser interferometry, like LIGO on Earth ground (Laser Interferometer Gravitational Wave Observatory) with two sites, in Livingston, Louisiana, and Hanford, Washington, or the planned eLISA (Evolved Laser Interferometer Space Antenna) orbiting in space.

The principle of the laser interferometry in gravitational wave detection is conceptually rather simple: as we saw previously, if a $\otimes$ wave propagates in the z-direction, the proper distance will dilate on one of the transverse directions and will increase on the other. If on one of these directions we orient one arm of an interferometer, set initially to show destructive interference, the effect of the gravitational wave propagating in the arm will, for example, increase the travel time of one laser beam, leading to an interference signal. The basic experimental set-up is illustrated in Fig. 9.9.

The technical complexity of the laser interferometers is however staggering, since they have to detect amplitudes (or *strains*, as they are termed in the gravitational literature) of about $\mathcal{A} \approx 10^{-20}$ or less. The strongest source of gravitational waves

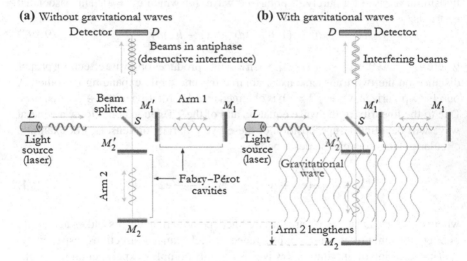

Fig. 9.9 Schematic representation of a laser-interferometer observatory: (**a**) the half-beams produced by the beam splitter S follow optical paths calibrated so that in the absence of gravitational waves, the interference at the detector D is destructive and no signal is observed; (**b**) when a gravitational wave arrives, it disturbs the space-time, changing (in our example, lengthening) the light's path along arm 2; when the beams recombine and arrive at the detector D, an interference signal is registered.

is the coalescence of a black-hole/black-hole binary. A calculation which is beyond the scope of this book shows that the amplitude of the gravitational waves emitted by this system while the black holes circle each other is

$$\mathcal{A} \sim \frac{r_S^2}{rR},$$ (9.195)

where r_S is the Schwarzschild radius (the black holes are assumed to have the same mass M), R is the orbital radius, and r is the distance between the pair of black holes and Earth. Typically, black holes are about 10 times more massive than the Sun and we can consider that the orbital radius is 10 times the Schwarzschild radius, while the distance to the observation point on Earth is of the order of 300 Mpc. This means $r_S \sim 3\,\mathrm{km}$, $R \sim 30\,\mathrm{km}$, and $r \sim 10^{22}\,\mathrm{km}$, leading to the order of magnitude for the amplitude quoted above.

On 14 September 2015, the two detectors of LIGO registered simultaneously a gravitational wave signal – the first direct detection ever achieved[8] (see Fig. 9.10). The source was a pair of black holes merging together. The event happened at a distance of 410 Mpc, which means that it was observed by LIGO 1.3 billion years later. The masses of the two initial black holes were $36 M_\odot$ and $29 M_\odot$, while the mass of the final one was $62 M_\odot$, meaning that an energy of $3 M_\odot c^2$ was radiated in gravitational waves. The observed signal took less then one second: during the first 0.2 s, the signal increased in frequency from 35 to 150 Hz, which is consistent with the theoretical prediction for the inspiral of two orbiting masses. The orbital frequency is half of the gravitational wave frequency, i.e. maximum 75 Hz in this case. This is a very high orbital frequency, that can be achieved without merging only if the orbiting masses are black holes. The signal peaks in amplitude during the merger and then the waveform decays in a manner consistent with the damped oscillations of a (final) black hole to a stationary Kerr configuration.

The LIGO interferometer is basically a Michelson interferometer, but with arms 4 km long. It is the largest interferometer ever built and the most precise, capable of measuring a change in the arm length 10,000 times smaller than a proton. The mirrors placed near the beam splitter cause multiple reflections of the laser beam, increasing the distance traveled in each arm to 1120 km. This system of mirrors forms an optical resonator known as the Fabry–Pérot cavity. Another ingenious stratagem for increasing the sensitivity is the power-boosting mirror, necessary for increasing the power of the laser beams from 200 W to 750 kW. This one-way mirror is placed before the beam splitter. The interferometer is aligned in such a way that almost all the laser light reflected by the arms is directed to the recycling mirror, and through it back to the source (instead of the detector), which greatly increases the power of the beam and consequently the sensitivity of the instrument.

The gravitational wave interferometers are not directional, they survey the whole sky. It is therefore of great importance to have simultaneous observations of gravi-

---

[8]The observation was reported in B.P. Abbott et al. (LIGO Scientific Collaboration and Virgo Collaboration), *Phys. Rev. Lett.* 116, 061102 (2016).

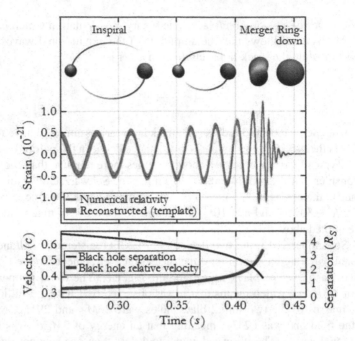

Fig. 9.10 *Top*: Estimated gravitational wave strain amplitude from the event observed by the LIGO Collaboration on 14 September 2015 (Hansford site). The *inset images* show numerical relativity models of the black hole horizons as the black holes coalesce. *Bottom*: The Keplerian effective black hole separation in units of Schwarzschild radii and the effective relative velocity given by the post-Newtonian parameter $v/c = (GM\pi f/c^3)^{1/3}$, where f is the gravitational-wave frequency calculated with numerical relativity and M is the total mass. *Source*: B.P. Abbott et al. (LIGO Scientific Collaboration and Virgo Collaboration), *Phys. Rev. Lett.* 116, 061102 (2016).

tational waves at different locations, in order to confirm the signals and increase the accuracy in determining their origin. At the moment, there are three more observatories: GEO600 near Sarstedt in Germany is already operational, while VIRGO near Pisa in Italy and KAGRA (Kamioka Gravitational Wave Detector) in the Kamioka mines in Japan are under construction. The recent success of LIGO came together with the approval of a third LIGO Observatory in India, which is expected to start operation in 2023.

9.11   Solved Problems

Problem 1. Find the geodesic lines corresponding to the metric

$$ds^2 = \frac{1}{t^2}\left(dt^2 - dx^2\right). \tag{9.196}$$

Solution. Following the usual procedure, we apply the Euler–Lagrange equation for the dynamical variable x:

$$\frac{\partial f}{\partial x} - \frac{d}{ds}\left(\frac{\partial f}{\partial \dot{x}}\right) = 0, \tag{9.197}$$

where

$$f = \frac{1}{t^2}\left(\dot{t}^2 - \dot{x}^2\right), \tag{9.198}$$

fulfilling the condition $f = 1$. Since f does not explicitly depend on x, the integration gives

$$\frac{\dot{x}}{t^2} = A = \text{const.} \tag{9.199}$$

Observing that

$$ds = \frac{1}{t}\left[1 - \left(\frac{dx}{dt}\right)^2\right]^{1/2} dt, \tag{9.200}$$

we then have

$$\dot{x} = \frac{dx}{dt}\frac{dt}{ds} = \frac{dx}{dt}\frac{t}{\left[1 - \left(\frac{dx}{dt}\right)^2\right]^{1/2}}.$$

Using (9.199) and separating variables, we can write also

$$\frac{dx}{dt} = \frac{At}{\sqrt{1 + A^2 t^2}},$$

and, by integrating the equation and a convenient arrangement of terms,

$$\frac{x^2}{a^2} - \frac{t^2}{a^2} = 1, \quad a = \frac{1}{A}. \tag{9.201}$$

Giving values to the constant A, we obtain the geodesics in the shape of *equilateral hyperbolas, tangent to the light cone*.

Problem 2. At the point $\theta = \theta_0$, $\varphi = \varphi_0$ on the surface of the two-dimensional sphere $ds^2 = d\theta^2 + \sin^2\theta\, d\varphi^2$, we have $\mathbf{A} = \mathbf{u}_\theta$. Write the form of vector $\mathbf{A}$ as a result of a parallel transport along the circle $\theta = \theta_0$, as well as its magnitude after transport.

Solution. By definition, a tensor Q (of any order and/or variance) suffers a parallel transport if its absolute differential is zero

$$DQ = dQ + \delta Q = 0.$$

For an arbitrary contravariant vector A^μ, we have

$$DA^\mu = dA^\mu + \Gamma^\mu_{\nu\lambda} A^\nu \, dx^\lambda = 0,$$

or

$$u^\lambda \left(\frac{\partial A^\mu}{\partial x^\lambda} + \Gamma^\mu_{\nu\lambda} A^\nu \right) = 0, \quad u^\lambda = \frac{dx^\lambda}{ds}.$$

If a vector tangent to a curve is parallel transported, then the curve is called *self-parallel*. For $A^\mu = u^\mu$, it follows that

$$\frac{du^\mu}{ds} + \Gamma^\mu_{\nu\lambda} u^\nu u^\lambda = 0,$$

which are the differential equations of the geodesic lines. If the connection coefficients are the Christoffel symbols of the second kind, the manifold is Riemannian. Therefore, *the self-parallel curves of a Riemannian space are geodesics of that space.*

Using this definition, we then have

$$\nabla_\varphi A^i = A^i_{,\varphi} + \Gamma^i_{k\varphi} A^k = 0, \quad i, k = 1, 2. \tag{9.202}$$

With $x^1 = \theta$, $x^2 = \varphi$, and $f = \dot\theta^2 + \dot\varphi^2 \sin^2\theta$, Euler–Lagrange equations give

$$\Gamma^1_{22} \equiv \Gamma^\theta_{\varphi\varphi} = -\sin\theta\cos\theta,$$
$$\Gamma^2_{12} \equiv \Gamma^\varphi_{\theta\varphi} = \cot\theta,$$

and (9.202) yield

$$A^\theta_{,\varphi} - \sin\theta\cos\theta\, A^\varphi = 0,$$
$$A^\varphi_{,\varphi} + \cot\theta\, A^\theta = 0. \tag{9.203}$$

Take now the partial derivative of (9.203)$_1$ with respect to φ:

$$A^\theta_{,\varphi\varphi} = \sin\theta\cos\theta A^\varphi_{,\varphi} = -\cos^2\theta A^\theta. \tag{9.204}$$

The solution of this equation is

$$A^\theta = C_1 \cos(\varphi\cos\theta) + C_2 \sin(\varphi\cos\theta),$$

allowing to determine A^φ:

$$A^\varphi = \frac{1}{\sin\theta} \left[-C_1 \sin(\varphi\cos\theta) + C_2 \cos(\varphi\cos\theta) \right]. \tag{9.205}$$

To fix the constants of integration C_1 and C_2, we use the boundary conditions: at $\varphi = 0$, $\mathbf{A} = \mathbf{u}_\theta$, $A^\theta = 1$, $A^\varphi = 0$. Then $C_1 = 1$, $C_2 = 0$, and thus

$$A^\theta = \cos(\varphi \cos \theta),$$

$$A^\varphi = -\frac{1}{\sin \theta} \sin(\varphi \cos \theta),$$

which leads to

$$\mathbf{A} = \cos(\varphi \cos \theta) \mathbf{u}_\theta - \frac{1}{\sin \theta} \sin(\varphi \cos \theta) \mathbf{u}_\varphi. \tag{9.206}$$

After transport ($\varphi = 2\pi$), $\mathbf{A}$ becomes

$$\mathbf{A}_{\varphi=2\pi} = \cos(2\pi \cos \theta) \mathbf{u}_\theta - \frac{1}{\sin \theta} \sin(2\pi \cos \theta) \mathbf{u}_\varphi \neq \mathbf{u}_\theta. \tag{9.207}$$

But the magnitude of the vector remains the same:

$$
\begin{aligned}
(A_\mu A^\mu)_{\varphi=2\pi} &= g_{\mu\nu} A^\mu A^\nu = g_{11} A^1 A^1 + g_{22} A^2 A^2 \\
&= g_{\theta\theta} A^\theta A^\theta + g_{\varphi\varphi} A^\varphi A^\varphi = \cos^2(2\pi \cos \theta) \\
&+ \sin^2 \theta \left[\frac{1}{\sin^2 \theta} \sin^2(2\pi \cos \theta) \right] = 1 = (A_\mu A^\mu)_0.
\end{aligned}
\tag{9.208}
$$

Problem 3. Show that the covariant four-divergence of Einstein's tensor $G_{\mu\nu} = R_{\mu\nu} - \frac{1}{2} R g_{\mu\nu}$ is zero.

Solution. Using the second Bianchi identity (9.79):

$$\nabla_\sigma R^\rho{}_{\mu\nu\lambda} + \nabla_\nu R^\rho{}_{\mu\lambda\sigma} + \nabla_\lambda R^\rho{}_{\mu\sigma\nu} = 0, \tag{9.209}$$

and recalling the Riemann tensor properties, we interchange the indices ν and λ in the first terms, then multiply by $g^{\mu\nu}$. The result is

$$- g^{\mu\nu} R^\rho{}_{\mu\lambda\nu\,;\,\sigma} + g^{\mu\nu} R^\rho{}_{\mu\lambda\sigma\,;\,\nu} + g^{\mu\nu} R^\rho{}_{\mu\sigma\nu\,;\,\lambda} = 0.$$

Multiply now by δ^λ_ρ and use the fact that the covariant derivative of the metric tensor is zero. We then have

$$- (g^{\mu\nu} R_{\mu\nu})_{;\sigma} + (g^{\mu\nu} R_{\mu\sigma})_{;\nu} + (g^{\mu\nu} R^\lambda{}_{\mu\sigma\nu})_{;\lambda} = 0.$$

The last term is the covariant four-divergence of the mixed Ricci tensor. Indeed,

$$
\begin{aligned}
g^{\mu\nu} R^\lambda{}_{\mu\sigma\nu} &= g^{\mu\nu} g^{\lambda\kappa} R_{\kappa\mu\sigma\nu} = g^{\mu\nu} g^{\lambda\kappa} R_{\mu\kappa\nu\sigma} \\
&= g^{\lambda\kappa} R^\nu{}_{\kappa\nu\sigma} = g^{\lambda\kappa} R_{\kappa\sigma} = R^\lambda{}_\sigma.
\end{aligned}
$$

Thus, we have obtained that

$$- R_{;\sigma} + R^{\nu}_{\sigma;\nu} + R^{\lambda}_{\sigma;\lambda} = 0,$$

or

$$\nabla_{\nu}\left(R^{\nu}_{\sigma} - \frac{1}{2} R\, \delta^{\nu}_{\sigma} \right) = \nabla_{\nu} G^{\nu}_{\sigma} = 0. \tag{9.210}$$

Problem 4. Determine the elementary space-like distance in a uniformly rotating coordinate system.

Solution. Let us start by elucidating the notions of space-like and time-like intervals in general relativity.

Consider two time-like separated close events, that happen at the same point of space. Choosing x^1, x^2, x^3 as space coordinates and $x^0 = ct$ as time coordinate, it follows from the problem statement that $dx^1 = dx^2 = dx^3 = 0$, and thus the metric is

$$ds^2 = c^2 d\tau^2 = g_{\mu\nu} dx^{\mu} dx^{\nu} = g_{00}(dx^0)^2.$$

This means that the proper time separating the two events is

$$d\tau = \frac{1}{c}\sqrt{g_{00}}\, dx^0. \tag{9.211}$$

Recall that in special relativity the elementary space-like distance dl can be determined as the interval between two close events which take place at the same moment, by choosing $dx^0 = 0$. In general relativity this procedure cannot be used, because at different points the proper time τ is differently connected to x^0.

Suppose that a light signal emitted at point $B(x^i + dx^i)$, $i = 1, 2, 3$ is intercepted at the neighbouring point $A(x^i)$, and then transmitted back on the same path. Since the metric is isotropic, we have

$$g_{ik} dx^i dx^k + 2g_{i0} dx^i dx^0 + g_{00}(dx^0)^2 = 0.$$

The roots of this equation in (dx^0) are

$$(dx^0)_1 = -\frac{1}{g_{00}}\left[g_{i0} dx^i + \sqrt{(g_{i0}g_{k0} - g_{ik}g_{00})dx^i dx^k} \right],$$

$$(dx^0)_2 = -\frac{1}{g_{00}}\left[g_{i0} dx^i - \sqrt{(g_{i0}g_{k0} - g_{ik}g_{00})dx^i dx^k} \right],$$

and correspond to the propagation of the light signal in the two directions, between the points A and B of given coordinates. In Fig. 9.11 the solid lines are world lines of the points A and B, while the dashed lines represent the world lines of the light

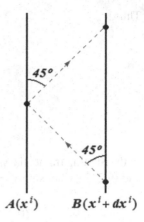

Fig. 9.11 World lines of the points A and B (*solid lines*) and of the light signals (*dashed lines*).

signals. The "time" difference between the emission and reception of the light signal at the same point is

$$(dx^0)_2 - (dx^0)_1 = \frac{2}{g_{00}}\sqrt{(g_{i0}g_{k0} - g_{ik}g_{00})dx^i dx^k}.$$

The "true" elementary proper time is then

$$d\tau = \frac{1}{c}\sqrt{g_{00}}\left[(dx^0)_2 - (dx^0)_1\right] = \frac{2}{c\sqrt{g_{00}}}\sqrt{(g_{i0}g_{k0} - g_{ik}g_{00})\,dx^i\,dx^k},$$

while the elementary proper distance $dl = c\,d\tau/2$ is

$$(dl)^2 = \left(\frac{g_{i0}g_{k0}}{g_{00}} - g_{ik}\right)dx^i dx^k = \gamma_{ik}dx^i dx^k, \tag{9.212}$$

where

$$\gamma_{ik} = \frac{g_{i0}g_{k0}}{g_{00}} - g_{ik} \tag{9.213}$$

is the *three-dimensional metric tensor*.

Now we can give the solution to the problem. Let us perform the transition from the inertial frame S' to another frame S, uniformly rotating around $Oz \equiv Oz'$ axis. The geometry of the problem suggest the use of cylindrical coordinates. Let r', φ', z', t, and r, φ, z, t be the coordinates in the two frames. In S' the metric is

$$ds^2 = c^2 dt^2 - dr'^2 - r'^2 d\varphi'^2 - dz'^2.$$

But $r' = r$, $z' = z$, $\varphi' = \varphi + \omega t$, so that

$$ds^2 = g_{\mu\nu}dx^\mu dx^\nu = (c^2 - \omega^2 t^2)dt^2 - 2\omega r^2\,d\varphi\,dt - dz^2 - r^2 d\varphi^2 - dr^2.$$

Thus,

$$g_{00} = 1 - \frac{\omega^2 r^2}{c^2},$$

$$g_{11} = g_{33} = -1,$$

$$g_{22} = -r^2,$$

$$g_{20} = -\frac{\omega r^2}{c}.$$

In the rotating frame, the space-like distance element is worked out by using (9.212) and (9.213). Choosing $x^0 = ct$, $x^1 = r$, $x^2 = \varphi$, $x^3 = z$, we obtain

$$\gamma_{11} = g_{11} = -1,$$

$$\gamma_{33} = g_{33} = -1,$$

$$\gamma_{22} = \frac{(g_{20})^2}{g_{00}} - g_{22} = r^2 \left(1 - \frac{\omega^2 r^2}{c^2} \right)^{-1}.$$

The solution is then given by the expression

$$(dl)^2 = \gamma_{ik} dx^i dx^k = dr^2 + dz^2 + \frac{r^2 d\varphi^2}{1 - \frac{\omega^2 r^2}{c^2}}. \tag{9.214}$$

Observation:

In the inertial (fixed) frame, the ratio of a circle's circumference (with its centre on the axis of rotation) to its radius in the plane $z = \text{const.}$ is 2π, while in the non-inertial (rotating) frame we have

$$L = \int_0^{2\pi} \frac{R d\varphi}{\sqrt{1 - \frac{\omega^2 R^2}{c^2}}} = \frac{2\pi R}{\sqrt{1 - \frac{\omega^2 R^2}{c^2}}}, \tag{9.215}$$

that is,

$$\frac{L}{R} > 2\pi. \tag{9.216}$$

Problem 5. Show that, irrespective of its mass, a body cannot orbit a Schwarzschild black hole at a distance smaller than $r_{min}^{stable} = 3r_S$, on a stable orbit, or $r_{min}^{unstable} = 3r_S/2$, on an unstable orbit.

Solution. The trajectory of a particle of mass m, moving in the spherically symmetric gravitational field described by the Schwarzschild metric can be found by means of the relativistic Hamilton–Jacobi equation (7.102):

$$g^{\mu\nu} \frac{\partial S}{\partial x^\mu} \frac{\partial S}{\partial x^\nu} - m^2 c^2 = 0. \tag{9.217}$$

The components of the metric tensor are straightforwardly inferred from the form of the Schwarzschild metric (9.128):

$$ds^2 = g_{\mu\nu}dx^\mu dx^\nu = c^2\left(1 - \frac{r_S}{r}\right)dt^2 - \left(1 - \frac{r_S}{r}\right)^{-1}dr^2 - r^2 d\theta^2 - r^2\sin^2\theta\, d\varphi^2,$$

$$(9.218)$$

where $x^0 = ct$, $x^1 = r$, $x^2 = \theta$, $x^3 = \varphi$, and $r_S = 2GM/c^2$ is the Schwarzschild radius of the central body of mass M, which creates the gravitational field. As the field has central symmetry, the motion takes place in a plane which contains the source of the field (this is a general result, valid for any central field). We shall choose this plane as defined by the condition $\theta = \pi/2$. Then, the metric (9.218) becomes

$$ds^2 = c^2\left(1 - \frac{r_S}{r}\right)dt^2 - \left(1 - \frac{r_S}{r}\right)^{-1}dr^2 - r^2 d\varphi^2,$$

from where

$$g_{00} = 1 - \frac{r_S}{r}, \quad g_{11} \equiv g_{rr} = -\frac{1}{1 - \frac{r_S}{r}}, \quad g_{33} \equiv g_{\varphi\varphi} = -r^2.$$

As $g_{\mu\lambda}g^{\nu\lambda} = \delta^\nu_\mu$, we have also

$$g^{00} = \left(1 - \frac{r_S}{r}\right)^{-1}, \quad g^{11} \equiv g^{rr} = -\left(1 - \frac{r_S}{r}\right), \quad g^{33} \equiv g^{\varphi\varphi} = -\frac{1}{r^2},$$

such that the Hamilton–Jacobi equation (9.217) becomes

$$\left(1 - \frac{r_S}{r}\right)^{-1}\frac{1}{c^2}\left(\frac{\partial S}{\partial t}\right)^2 - \left(1 - \frac{r_S}{r}\right)\left(\frac{\partial S}{\partial r}\right)^2 - \frac{1}{r^2}\left(\frac{\partial S}{\partial \varphi}\right)^2 - m^2 c^2 = 0. \quad (9.219)$$

Using the method of the separation of variables and taking into account that the system is conservative (the Hamiltonian does not depend explicitly on time), while the variable φ is cyclic, we shall seek the solution of (9.219) in the form

$$S \equiv S(t, r, \varphi) = -E_0 t + S_r(r) + L\varphi, \quad (9.220)$$

where E_0 is the total (conserved) energy of the particle of mass m, while L is the angular momentum of the particle with respect to the centre of symmetry. The angular momentum is also conserved in the case of a central force field. Introducing (9.220) into (9.219) we find the following expression for the radial part $S_r(r)$ of the action:

$$S_r(r) = \frac{1}{c}\int \frac{dr}{r}\sqrt{r^2 E_0^2\left(1 - \frac{r_S}{r}\right)^{-2} - c^2\left(m^2 c^2 r^2 + L^2\right)\left(1 - \frac{r_S}{r}\right)^{-1}}. \quad (9.221)$$

In Hamilton–Jacobi formalism, the dependence $r = r(t)$ is given by the equation

$$\frac{\partial S}{\partial E_0} = \text{const.}, \tag{9.222}$$

while the trajectory of the particle is determined by the equation

$$\frac{\partial S}{\partial L} = \text{const.} \tag{9.223}$$

Taking into account the relations (9.220) and (9.221), from (9.222) follows immediately that

$$ct = E_0 \int \left(1 - \frac{r_S}{r}\right)^{-1} \left[(rE_0)^2 - c^2 \left(1 - \frac{r_S}{r}\right)\left(m^2 c^2 r^2 + L^2\right)\right]^{-\frac{1}{2}} r\, dr. \tag{9.224}$$

With the notation

$$U_{\text{eff}}(r) \equiv \frac{c}{r}\sqrt{\left(1 - \frac{r_S}{r}\right)\left(m^2 c^2 r^2 + L^2\right)} = mc^2 \sqrt{\left(1 - \frac{r_S}{r}\right)\left(1 + \frac{L^2}{m^2 c^2 r^2}\right)}, \tag{9.225}$$

the dependence $r = r(t)$ – given in integral form by the relation (9.224) – can be written in differential form as follows:

$$\frac{dr}{dt} = c\left(1 - \frac{r_S}{r}\right)\sqrt{1 - \left[\frac{U_{\text{eff}}(r)}{E_0}\right]^2}. \tag{9.226}$$

This equation indicates that the function $U_{\text{eff}} = U_{\text{eff}}(r)$ plays the role of an *effective potential energy*, in the sense that, by analogy with the classical theory, the condition $E_0 \geqslant U_{\text{eff}}(r)$ establishes the intervals of values of the radial coordinate r within which the motion of the particle is allowed. In Fig. 9.12 we represented four curves of variation of the reduced effective potential energy U_{eff}/mc^2 as a function of the ratio r/r_S, corresponding to as many values of the angular momentum L.

The minima of the function $U_{\text{eff}}(r)$ correspond to the stable orbits of the particle, while the maxima correspond to unstable orbits. The values of the radii of the circular orbits, as well as the corresponding values of the constant quantities E_0 and L are determined by the system of equation

$$\begin{cases} U_{\text{eff}}(r) = E_0, \\ \frac{dU_{\text{eff}}}{dr} = 0, \end{cases} \tag{9.227}$$

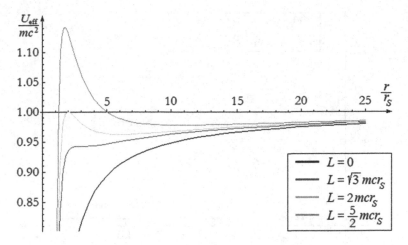

Fig. 9.12 Reduced effective potential energy U_{eff}/mc^2 for various values of the angular momentum L of the particle.

which leads to the following relations that have to be satisfied simultaneously:

$$E_0 = cL \sqrt{\frac{2}{r} \frac{r - r_S}{r r_S}} \tag{9.228}$$

and

$$\frac{r}{r_S} = \frac{L^2}{m^2 c^2 r_S^2} \left(1 \pm \sqrt{1 - \frac{3 m^2 c^2 r_S^2}{L^2}} \right). \tag{9.229}$$

The "+" sign in relation (9.229) corresponds to the stable orbits $\left(\frac{d^2 U_{\text{eff}}}{dr^2} > 0 \right)$, while the sign "−", corresponds to the unstable ones $\left(\frac{d^2 U_{\text{eff}}}{dr^2} < 0 \right)$. The closest stable orbit from the centre of symmetry is characterized by the following parameters:

$$r_{min}^{stable} = 3 r_S = \frac{6 G M}{c^2}, \quad L\big|_{r = r_{min}^{stable}} = \sqrt{3} m c r_S = \frac{2\sqrt{3} G m M}{c},$$

$$E_0\big|_{r = r_{min}^{stable}} = \frac{2}{3}\sqrt{2} m c^2, \tag{9.230}$$

and it corresponds to the point A on Fig. 9.13. The smallest value for the radius of an unstable orbit is $r_{min}^{unstable} = 3 r_S / 2$ and it is obtained in the limit $L \to \infty$, $E_0 \to \infty$ (point B corresponding to the horizontal asymptote figured with dotted line in Fig. 9.13).

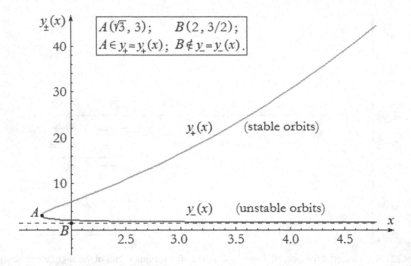

Fig. 9.13 Dependence of $r/r_S \equiv y$ on $L/mcr_S \equiv x$. The upper branch gives the radius for stable, and the lower for unstable orbits. The point B does not belong to the curve $y_- = y_-(x)$, but to the horizontal asymptote.

With the notations

$$\frac{L}{mcr_S} \equiv x, \qquad \frac{r}{r_S} \equiv y, \tag{9.231}$$

the relation (9.229) is written as

$$y_\pm = x\left(x \pm \sqrt{x^2 - 3}\right). \tag{9.232}$$

The graphical representations of $y_\pm = y_\pm(x)$, given by the relations

$$y_+ = x\left(x + \sqrt{x^2 - 3}\right) \text{ and } y_- = x\left(x - \sqrt{x^2 - 3}\right) \tag{9.233}$$

are shown in Fig. 9.13, in which the branch $y_+ = y_+(x)$ corresponds to stable orbits, while the branch $y_- = y_-(x)$ corresponds to the unstable ones. Moreover,

$$\lim_{x \to \infty} y_-(x) = \lim_{x \to \infty} x\left(x - \sqrt{x^2 - 3}\right) = \frac{3}{2}, \tag{9.234}$$

which justifies the value $r_{min}^{unstable} = 3r_S/2$.

We note that the obvious difference compared to the classical case is that in a Newtonian gravitational field there are stable circular orbits at any distance from the centre of force, the radii of these orbits being given by the relation $r = \frac{1}{GM}\left(\frac{L}{m}\right)^2$.

9.12 Proposed Problems

1. Show that the covariant derivative (corresponding to the Levi-Civita connection) of the metric tensor $g_{\mu\nu}$ is zero.
2. Prove the following identities:

   a. $g_{\mu\lambda,\nu} = \Gamma_{\mu\nu,\lambda} + \Gamma_{\lambda\nu,\mu}$;

   b. $g_{\mu\nu}g^{\mu\lambda}_{,\rho} = -g^{\mu\lambda}g_{\mu\nu,\rho}$;

   c. $g^{\mu\lambda}_{,\nu} = -\Gamma^{\mu}_{\rho\nu}g^{\rho\lambda} - \Gamma^{\lambda}_{\rho\nu}g^{\rho\mu}$;

   d. $g_{,\nu} = gg^{\lambda\mu}g_{\lambda\mu,\nu} = -g_{\lambda\mu}g^{\lambda\mu}_{,\nu}$, $g = \det(g_{\mu\nu})$;

   e. $\Gamma^{\mu}_{\nu\mu} = \left(\ln|g|^{1/2}\right)_{,\nu}$.

3. Determine the connection coefficients associated to the metric

$$ds^2 = dr^2 + r^2 d\theta^2 + r^2 \sin^2\theta\, d\varphi^2.$$

4. Calculate the components of the Ricci tensor $R_{\mu\nu}$ in a reference frame in which $g_{00} = 1$ and $g_{i0} = 0$.
5. Show that the two-dimensional manifold with the metric

$$ds^2 = dv^2 - v^2 du^2$$

is flat. Also show that the component P_u of the four-momentum of a unit mass particle is a constant, while the component P_v is not.
6. Prove Bianchi's identity (9.78).
7. If $A = \det(A_{\mu\nu})$, where $A_{\mu\nu}$ is a second-order tensor, show that A is not a scalar. In this case, the relation $A_{;\alpha} = A_{,\alpha}$ is not true. How can the covariant derivative $A_{;\alpha}$ be expressed in terms of $A_{,\alpha}$ and A?
8. Show that the line element

$$ds^2 = R^2 \left[d\alpha^2 + \sin^2\alpha \left(d\theta^2 + \sin^2\theta\, d\varphi^2\right)\right]$$

corresponds to a hypersphere of radius R in a four-dimensional Euclidean space, that is to the locus of points situated at the distance R from some given point.
9. Starting from the standard expression of Fermat's principle,

$$\delta \int k_i dx^i = 0,$$

where the integral is taken along a ray and k_i, $i = 1, 2, 3$ are the components of the wave vector, show that in a constant gravitational field Fermat's principle is written as

$$\delta \int \frac{1}{g_{00}} \left(\sqrt{g_{00}}\, dl - g_{0i}\, dx^i\right) = 0,$$

where dl is given by (9.212) and it represents the element of space-like length along the ray.

10. Find the equations of motion of a particle of mass m, which is moving in the axially symmetric gravitational field described by the Kerr metric, using the method of separation of variables for the Hamilton–Jacobi equation

$$g^{\mu\nu} \frac{\partial S}{\partial x^\mu} \frac{\partial S}{\partial x^\nu} - m^2 c^2 = 0.$$

(Hint: One starts from the expression of the action as $S = -E_0 t + L\varphi + S_r(r) + S_\theta(\theta)$, where E_0 is the total conserved energy of the particle, and L is the component of the angular momentum of the particle along the symmetry axis of the gravitational field.)

Appendix A
Vectors and Vector Analysis

A.1 Vector Algebra

A.1.1. Let a_i, $i = 1, 2, 3$ be the components of a vector $\mathbf{a}$ in the orthonormal basis $\mathbf{u}_i$ of an Euclidean three-dimensional space. Using Einstein's summation convention, the analytical expression of $\mathbf{a}$ is

$$\mathbf{a} = a_i \mathbf{u}_i. \tag{A.1}$$

The analytical expression of the radius-vector is then

$$\mathbf{r} = x_i \mathbf{u}_i. \tag{A.2}$$

    A.1.2. Let $\mathbf{a}$ and $\mathbf{b}$ be two arbitrary vectors. Skipping addition and subtraction, one can define

1. The *scalar product* or *dot product* of the two vectors:

$$\mathbf{a} \cdot \mathbf{b} = (a_i \mathbf{u}_i) \cdot (b_k \mathbf{u}_k) = a_i b_i = ab \cos(\widehat{\mathbf{a}, \mathbf{b}}), \tag{A.3}$$

since

$$\mathbf{u}_i \cdot \mathbf{u}_k = \delta_{ik}. \tag{A.4}$$

2. The *vector product*, or *cross product* of the two vectors:

$$\mathbf{a} \times \mathbf{b} = -\mathbf{b} \times \mathbf{a} = \epsilon_{ijk} a_j b_k \mathbf{u}_i, \qquad |\mathbf{a} \times \mathbf{b}| = ab \sin(\widehat{\mathbf{a}, \mathbf{b}}), \tag{A.5}$$

as well as

$$\mathbf{u}_i \times \mathbf{u}_j = \epsilon_{ijk} \mathbf{u}_k, \qquad \mathbf{u}_s = \frac{1}{2} \epsilon_{sij} \mathbf{u}_i \times \mathbf{u}_j, \tag{A.6}$$

© Springer-Verlag Berlin Heidelberg 2016
M. Chaichian et al., *Electrodynamics*, DOI 10.1007/978-3-642-17381-3

where

$$\epsilon_{ijk} = \begin{cases} +1, \text{ if } i,\, j,\, k \text{ are an even permutation of } 1,\ 2,\ 3, \\ -1, \text{ if } i,\, j,\, k \text{ are an odd permutation of } 1,\ 2,\ 3, \\ 0, \text{ if any two indices are equal,} \end{cases} \qquad \text{(A.7)}$$

is the *Levi-Civita permutation symbol* (see Sect. A.5).

One can easily verify the property

$$\epsilon_{ijk}\epsilon_{lmn} = \begin{vmatrix} \delta_{il} & \delta_{im} & \delta_{in} \\ \delta_{jl} & \delta_{jm} & \delta_{jn} \\ \delta_{kl} & \delta_{km} & \delta_{kn} \end{vmatrix}. \qquad \text{(A.8)}$$

For $i = l$, (A.8) becomes

$$\epsilon_{ijk}\epsilon_{imn} = \delta_{jm}\delta_{kn} - \delta_{jn}\delta_{km}, \qquad \text{(A.9)}$$

while for $i = l$, $j = m$ it becomes

$$\epsilon_{ijk}\epsilon_{ijn} = 2\delta_{kn}. \qquad \text{(A.10)}$$

Obviously, $\epsilon_{ijk}\epsilon_{ijk} = 3! = 6$.

A.1.3. Let $\mathbf{a}$, $\mathbf{b}$, $\mathbf{c}$ be three arbitrary vectors. With these vectors one can define the following two types of product:

1. The *mixed product*

$$\mathbf{a} \cdot (\mathbf{b} \times \mathbf{c}) = \epsilon_{ijk}a_i b_j c_k, \qquad \text{(A.11)}$$

in particular,

$$\mathbf{u}_i \cdot (\mathbf{u}_j \times \mathbf{u}_k) = \epsilon_{ijk}. \qquad \text{(A.12)}$$

2. The *double cross product*

$$\mathbf{a} \times (\mathbf{b} \times \mathbf{c}) = (\mathbf{a} \cdot \mathbf{c})\mathbf{b} - (\mathbf{a} \cdot \mathbf{b})\mathbf{c}. \qquad \text{(A.13)}$$

A.2   Orthogonal Coordinate Transformations

The transformation from a three-dimensional orthogonal Euclidean coordinate system $S(Oxyz)$ to another system $S'(O'x'y'z')$ can be accomplished by three main procedures:

i)   *translation* of axes;
ii) *rotation* of axes;
iii) mirror *reflection*.

The first two types of transformations do not change the orientation of the axes of S' with respect to S (*proper transformations*), while under mirror reflection (i.e. $x' = -x$, $y' = y$, $z' = z$) a right-handed coordinate system transforms into a left-handed coordinate system (*improper transformation*). There are also possible combinations of these transformations, with the observation that a succession of two proper/improper transformations gives a proper transformation, while a proper transformation followed by an improper transformation leads to an improper transformation.

Suppose that S and S' have the same origin, and let $\mathbf{u}_i, \mathbf{u}'_i$, $i = 1, 2, 3$ be the orthonormal associated bases. Since

$$\mathbf{r} = x_i \mathbf{u}_i = x'_i \mathbf{u}'_i, \tag{A.14}$$

we have

$$x'_i = a_{ik} x_k, \quad a_{ik} = \mathbf{u}'_i \cdot \mathbf{u}_k,$$
$$x_i = a_{ki} x'_k, \quad a_{ki} = \mathbf{u}'_k \cdot \mathbf{u}_i, \quad i, k = 1, 2, 3, \tag{A.15}$$

as well as

$$\mathbf{u}'_i = a_{ik} \mathbf{u}_k,$$
$$\mathbf{u}_i = a_{ki} \mathbf{u}'_k. \tag{A.16}$$

The coefficients a_{ik} form the *matrix of the orthogonal transformation* (A.15)$_1$, and a_{ki} – the *matrix of the inverse transformation* (A.15)$_2$.

The invariance of the distance between two points,

$$r^2 = x_i x_i = x'_k x'_k,$$

yields the *orthogonality condition*

$$a_{ik} a_{im} = \delta_{km}, \quad i, k, m = 1, 2, 3. \tag{A.17}$$

Using the rule of the product of determinants, we find

$$\det(a_{ik}) = \pm 1. \tag{A.18}$$

On the other hand,

$$\det(a_{ik}) = \epsilon_{ijk} a_{1i} a_{2j} a_{3k}, \quad i, j, k = 1, 2, 3, \tag{A.19}$$

that is

$$\epsilon_{ijk} a_{1i} a_{2j} a_{3k} = \pm 1, \tag{A.20}$$

or

$$\epsilon_{ijk} a_{li} a_{mj} a_{nk} = \pm \epsilon_{lmn} = \epsilon_{lmn} \det(a_{ik}), \quad l, m, n = 1, 2, 3. \tag{A.21}$$

Here the sign "+" corresponds to the case when the frames have the same orientation, and "−" to the case when the orientations are different.

Under a change of orthonormal basis, we have

$$
\begin{aligned}
\mathbf{p} + \mathbf{q} &= p_i \mathbf{u}_i + q_i \mathbf{u}_i \\
&= p_i a_{mi} \mathbf{u}'_m + q_i a_{mi} \mathbf{u}'_m = p'_m \mathbf{u}'_m + q'_m \mathbf{u}'_m = \mathbf{p}' + \mathbf{q}', \\
\mathbf{p} \cdot \mathbf{q} &= p_i q_i = (a_{mi} p'_m)(a_{si} q'_s) = \delta_{ms} p'_m q'_s = p'_s q'_s = \mathbf{p}' \cdot \mathbf{q}', \\
\mathbf{p} \times \mathbf{q} &= \epsilon_{ijk} p_j q_k \mathbf{u}_i = \epsilon_{ijk} a_{mj} a_{sk} a_{li} p'_m q'_s \mathbf{u}'_l = (\mathbf{p}' \times \mathbf{q}') \det(a_{ik}), \\
\mathbf{p} \cdot (\mathbf{q} \times \mathbf{r}) &= \epsilon_{ijk} p_i q_j r_k = \epsilon_{ijk} a_{li} a_{mj} a_{nk} p'_l q'_m r'_n \\
&= \mathbf{p}' \cdot (\mathbf{q}' \times \mathbf{r}') \det(a_{ik}). \tag{A.22}
\end{aligned}
$$

Consequently, the addition (subtraction) and dot product of two vectors do not change when changing the basis, while the cross product and the mixed product (of polar vectors) change sign when the bases have different orientations.

We call *scalars of the first kind* or *scalar invariants* those quantities whose sign does not depend on the basis orientation (e.g., temperature, mass, mechanical work, electric charge, etc.), and *scalars of the second kind* or *pseudoscalars* those quantities which change sign when the basis changes its orientation (e.g., magnetic flux, $d\Phi = \mathbf{B} \cdot d\mathbf{S}$, the moment of a force $\mathbf{F}$ with respect to an axis Δ, $M_\Delta = \mathbf{u}_\Delta \cdot (\mathbf{r} \times \mathbf{F})$, etc.).

We call *polar vector* or *proper vector* a vector which transforms to its negative under the inversion of its coordinate axes (electric field intensity, velocity of a particle, gradient of a scalar, etc.), and *axial vector* or *pseudovector* a vector which is invariant under inversion of coordinate axes (magnetic induction, angular velocity, etc).

A.3   Elements of Vector Analysis

A.3.1   Scalar and Vector Fields

If to each point P of a domain D of the Euclidean space E_3 one can associate a value of a scalar $\varphi(P)$, in D is defined a *scalar field*. If to each point P one can associate a vector quantity $\mathbf{A}(P)$, in D is defined a *vector field*.

A scalar (or vector) field is called *stationary* if φ (or $\mathbf{A}$) do not explicitly depend on time. In the opposite case, the field is *non-stationary*.

A.3.2 Frequently Occurring Integrals

In classical/phenomenological electrodynamics one encounters three types of integrals:

1. *Line integral*

$$\int_{P_1}^{P_2} \mathbf{a} \cdot d\mathbf{s}$$

along a curve C, taken between the points P_1 and P_2, where $\mathbf{a}$ is a vector with its origin on the curve, and $d\mathbf{s}$ is a vector element of C. This integral is called *circulation* of the vector $\mathbf{a}$ along the curve C, between the points P_1 and P_2. If the curve C is closed, the circulation is denoted by

$$\oint_C \mathbf{a} \cdot d\mathbf{s}.$$

2. *Double integral*

$$\iint \mathbf{a} \cdot d\mathbf{S} = \int_S \mathbf{a} \cdot d\mathbf{S} = \int_S \mathbf{a} \cdot \mathbf{n} \, dS,$$

where $\mathbf{a}$ is a vector with its origin on the surface S, $d\mathbf{S}$ is a vector surface element, and $\mathbf{n}$ is the unit vector of the external normal to dS. This integral is called the *flux* of the vector $\mathbf{a}$ through the surface S. If S is a closed surface, the integral is denoted by

$$\oint_S \mathbf{a} \cdot d\mathbf{S}.$$

3. *Triple integral*

$$\iiint \mathbf{a} \, d\tau = \int_V \mathbf{a} \, d\tau = \int_V \mathbf{a} \, d\mathbf{r} = \int_V \mathbf{a} \, dV,$$

where V is the volume of the three-dimensional domain $D \subset E_3$, and $\mathbf{a}$ has its origin at some point of D.

A.3.3 First-Order Vector Differential Operators

The *nabla* or *del* operator is defined as

$$\nabla = \frac{\partial}{\partial x_1} \mathbf{u}_1 + \frac{\partial}{\partial x_2} \mathbf{u}_2 + \frac{\partial}{\partial x_3} \mathbf{u}_3. \tag{A.23}$$

By applying it in different ways to scalars and vectors, we obtain the gradient, divergence, and curl.

1. **Gradient.**[1] Let $\varphi(\mathbf{r})$ be a scalar field, with $\varphi(\mathbf{r})$ a continuous function with continuous derivative in $D \subset E_3$. The vector field

$$\mathbf{A} = \operatorname{grad} \varphi = \nabla \varphi = \frac{\partial \varphi}{\partial x_i} \mathbf{u}_i, \quad i = 1, 2, 3 \tag{A.24}$$

is the *gradient* of the scalar field $\varphi(\mathbf{r})$. The vector field $\mathbf{A}$ defined by (A.24) is called *conservative*.

Equipotential surfaces. Consider the fixed surface

$$\varphi(x, y, z) = C(= \text{const.}). \tag{A.25}$$

Then $d\varphi = \nabla \varphi \cdot d\mathbf{r} = 0$ shows that at every point of the surface (A.25) the vector $\nabla \varphi$ is oriented along the normal to the surface. Giving values to C, one obtains a family of surfaces called *equipotential surfaces*, or *level surfaces*.

Field lines. Let $\mathbf{A}(\mathbf{r})$ be a stationary vector field, and C – a curve given by its parametric equations $x_i = x_i(s)$, $i = 1, 2, 3$. If at every point the field $\mathbf{A}$ is tangent to the curve C, then the curve is a *line of the vector field* $\mathbf{A}$. The differential equations of the field lines are obtained by projecting on axes the vector relation $\mathbf{A} \times d\mathbf{r} = 0$, where $d\mathbf{r}$ is a vector element of the field line.

Directional derivative. Let us project the vector $\nabla \varphi$ unto the direction defined by the unit vector $\mathbf{u}$. Since $d\mathbf{r} = \mathbf{u}|d\mathbf{r}| = \mathbf{u}\,ds$, where ds is a curve element in the direction $\mathbf{u}$, we have

$$(\nabla \varphi) \cdot \mathbf{u} = (\nabla \varphi)_u = \frac{d\varphi}{ds}. \tag{A.26}$$

If, in particular, $\mathbf{u}$ is the unit vector $\mathbf{n}$ of the external normal to the surface (A.25), then

$$\nabla \varphi_n = \frac{d\varphi}{dn} \geq 0, \tag{A.27}$$

meaning that the gradient is oriented along the normal to the equipotential surfaces, and points in the direction of the greatest rate of increase of the scalar field.

---

[1] As a matter of fact, *gradient, divergence*, and *curl* are not *bona fide* first-order vector differential operators, but the results of the nabla operator (Hamilton's operator) action, in different ways, upon scalar and/or vector fields. Note that the Laplacian $\triangle = \nabla \cdot \nabla = \nabla^2$ and the d'Alembertian $\square = \frac{1}{v^2} \frac{\partial^2}{\partial t^2} - \triangle$ are true differential operators, but of the second order.

2. **Divergence**. Consider a continuously differentiable vector field $\mathbf{A}(\mathbf{r})$. The scalar-valued function

$$\operatorname{div} \mathbf{A} = \nabla \cdot \mathbf{A} = \frac{\partial A_i}{\partial x_i} = \partial_i A_i = A_{i,i}, \quad i = 1, 2, 3 \qquad (\text{A.28})$$

is called the *divergence* of the vector field $\mathbf{A}(\mathbf{r})$. If the vector field $\mathbf{A}(\mathbf{r})$ satisfies the condition

$$\nabla \cdot \mathbf{A} = 0, \qquad (\text{A.29})$$

then it is called *source-free* or *solenoidal*. The lines of such a field are *closed curves* (e.g. the magnetostatic field).

3. **Curl**. The curl (or *rotor*) of a vector field $\mathbf{A}(\mathbf{r})$ is another vector field $\mathbf{B}(\mathbf{r})$ defined as

$$\mathbf{B} = \nabla \times \mathbf{A} = \mathbf{u}_i \, \epsilon_{ijk} \partial_j A_k. \qquad (\text{A.30})$$

If the field $\mathbf{A}(\mathbf{r})$ has the property

$$\nabla \times \mathbf{A} = 0, \qquad (\text{A.31})$$

then it is called *irrotational* or *curl-free* (e.g. the velocity field of a fluid that moves in laminar flow regime).

A.3.4 Fundamental Theorems

A.3.4.1 Divergence theorem

Consider a spatial domain D bounded by the surface S, and a vector field $\mathbf{A}(\mathbf{r})$ of class C^1 in D and C^0 in $\overline{D}$ (the closure of the domain D, consisting of all the points of D and its boundary surface S). It can be shown that

$$\oint_S \mathbf{A} \cdot \mathbf{n} \, dS = \int_V \nabla \cdot \mathbf{A} \, d\tau, \qquad (\text{A.32})$$

which is the mathematical expression of the *divergence theorem*. It is also called the Green–Gauss–Ostrogradsky theorem, after Johann Carl Friedrich Gauss (1777–1855), George Green (1793–1841), and Mikhail Ostrogradsky (1801–1862). Here $\mathbf{n}$ is the unit vector of the external normal to S.

Let us now diminish the surface S, so that the volume V becomes smaller and smaller. In the limit, we have

$$\nabla \cdot \mathbf{A} = \lim_{\Delta\tau \to 0} \frac{1}{\Delta\tau} \oint_S \mathbf{A} \cdot \mathbf{n} \, dS. \qquad (\text{A.33})$$

This formula can be considered as the definition of the divergence at a point. It is useful because it expresses divergence independently of any coordinate system (intrinsic relation). If at some point we have $\nabla \cdot \mathbf{A} > 0$, we say that at that point there is a *source*; if $\nabla \cdot \mathbf{A} < 0$, that point is a *sink* ; if $\nabla \cdot \mathbf{A} = 0$, the point is a *node*.

A.3.4.2   Stokes Theorem

Let C be any closed curve in the three-dimensional Euclidean space, and let S be any surface bounded by C. If $\mathbf{A}$ is a vector field of class C^1 in $C \cup S$, then

$$\oint_C \mathbf{A} \cdot d\mathbf{l} = \iint (\nabla \times \mathbf{A}) \cdot d\mathbf{S} = \int_S (\nabla \times \mathbf{A}) \cdot \mathbf{n} \, dS. \tag{A.34}$$

This is the *Stokes theorem*, named after George Stokes (1819–1903) (sometimes called the *Stokes–Ampère theorem*). If $\mathbf{n}$ is the unit vector of the external normal to S, then the circulation on C is given by the right-hand screw rule. In particular, if $\mathbf{A} = \nabla \varphi$, we have

$$\oint_C \mathbf{A} \cdot d\mathbf{l} = \oint_C d\varphi = 0. \tag{A.35}$$

Contracting the surface S until it becomes an infinitesimal quantity, in the limit we may write

$$(\nabla \times \mathbf{A})_n = \mathbf{n} \cdot (\nabla \times \mathbf{A}) = \lim_{\Delta S \to 0} \frac{1}{\Delta S} \oint_C \mathbf{A} \cdot d\mathbf{l}, \tag{A.36}$$

which is the *intrinsic* definition of curl $\mathbf{A}$.

A.3.5   *Some Consequences of the Divergence Theorem and Stokes Theorem*

Let $\mathbf{A}(\mathbf{r}) = \varphi(\mathbf{r}) \, \mathbf{e}$, where $\varphi(\mathbf{r})$ is a scalar field of class C^1 in $D \subset E_3$, and $\mathbf{e}$ is a constant vector. Formula (A.32) then yields

$$\oint_S \varphi \, d\mathbf{S} = \int_V \nabla \varphi \, d\tau, \tag{A.37}$$

while (A.34) leads to

$$\oint_C \varphi \, d\mathbf{l} = \int_S (\mathbf{n} \times \nabla \varphi) \, dS. \tag{A.38}$$

If in (A.32) we choose $\mathbf{A} = \mathbf{e} \times \mathbf{B}$, where $\mathbf{B}$ is a vector field of class C^1 in D, the result is

$$\oint_S \mathbf{n} \times \mathbf{B} \, dS = \int_V \nabla \times \mathbf{B} \, d\tau. \tag{A.39}$$

Consider, now, the vector field $\mathbf{A}$ of the form $\mathbf{A} = \psi \nabla \varphi$, where $\psi \in C^1(D)$ and $\varphi \in C^2(D)$. Formula (A.32) leads to

$$\int_V (\nabla \psi \cdot \nabla \varphi + \psi \Delta \varphi) \, d\tau = \oint_S \psi \frac{\partial \varphi}{\partial n} \, dS. \tag{A.40}$$

Interchanging φ and ψ in (A.40), then subtracting the result from (A.40), we obtain

$$\int_V (\psi \Delta \varphi - \varphi \Delta \psi) \, d\tau = \oint_S \left(\psi \frac{\partial \varphi}{\partial n} - \varphi \frac{\partial \psi}{\partial n} \right) dS, \tag{A.41}$$

where $(\varphi, \psi) \in C^2(D) \cup C^1(\overline{D})$.

Relations (A.40) and (A.41) are known as *Green's identities*.

A.3.6 Useful Formulas

The nabla operator, ∇, is sometimes applied to products of (scalar or vector) functions, or can be found in successive operations ($\nabla \cdot \nabla \varphi$, etc.). Here are some useful formulas, frequently encountered in electrodynamics:

$$\nabla(\varphi \psi) = \varphi \nabla \psi + \psi \nabla \varphi, \tag{A.42}$$

$$\nabla \cdot (\varphi \mathbf{A}) = \varphi \nabla \cdot \mathbf{A} + \mathbf{A} \cdot \nabla \varphi, \tag{A.43}$$

$$\nabla \times (\varphi \mathbf{A}) = \varphi \nabla \times \mathbf{A} + (\nabla \varphi) \times \mathbf{A}, \tag{A.44}$$

$$\nabla \cdot (\mathbf{A} \times \mathbf{B}) = \mathbf{B} \cdot \nabla \times \mathbf{A} - \mathbf{A} \cdot \nabla \times \mathbf{B}, \tag{A.45}$$

$$\nabla(\mathbf{A} \cdot \mathbf{B}) = \mathbf{A} \times (\nabla \times \mathbf{B}) + \mathbf{B} \times (\nabla \times \mathbf{A})$$
$$+ (\mathbf{A} \cdot \nabla)\mathbf{B} + (\mathbf{B} \cdot \nabla)\mathbf{A}, \tag{A.46}$$

$$\nabla \times (\mathbf{A} \times \mathbf{B}) = \mathbf{A}\nabla \cdot \mathbf{B} - \mathbf{B}\nabla \cdot \mathbf{A}$$
$$+ (\mathbf{B} \cdot \nabla)\mathbf{A} - (\mathbf{A} \cdot \nabla)\mathbf{B}, \tag{A.47}$$

$$\nabla \cdot (\nabla \times \mathbf{A}) = 0, \tag{A.48}$$

$$\nabla \times (\nabla \varphi) = 0, \tag{A.49}$$

$$\nabla \cdot \nabla \varphi = \nabla^2 \varphi = \Delta \varphi; \quad \Delta = \frac{\partial^2}{\partial x_i \partial x_i} = \partial_i \partial_i, \tag{A.50}$$

$$\nabla \times (\nabla \times \mathbf{A}) = \nabla(\nabla \cdot \mathbf{A}) - \Delta \mathbf{A}. \tag{A.51}$$

If $\mathbf{r}$ is the radius-vector of some point $P \in E_3$ with respect to the origin of the Cartesian coordinate system $Oxyz$, then

$$\nabla r = \frac{\mathbf{r}}{r} = \mathbf{u}_r, \quad |\nabla r| = 1,$$

$$\nabla \cdot \mathbf{r} = 3,$$

$$\nabla \times \mathbf{r} = 0,$$

$$\nabla \left(\frac{1}{r} \right) = -\frac{\mathbf{r}}{r^3},$$

$$\Delta \left(\frac{1}{r} \right)_{r \neq 0} = 0, \tag{A.52}$$

$$\Delta \left(\frac{1}{r} \right) = -4\pi \delta(\mathbf{r}),$$

$$\Delta \left(\frac{1}{|\mathbf{r} - \mathbf{r}'|} \right) = -4\pi \delta(\mathbf{r} - \mathbf{r}').$$

If $\mathbf{A}$ is a constant vector, then (A.45)–(A.47) and (A.52) yield

$$\nabla \cdot (\mathbf{A} \times \mathbf{r}) = 0, \tag{A.53}$$

$$\nabla(\mathbf{A} \cdot \mathbf{r}) = \mathbf{A}, \tag{A.54}$$

$$\nabla \times (\mathbf{A} \times \mathbf{r}) = 2\mathbf{A}. \tag{A.55}$$

Given the fields $\varphi(r)$ and $\mathbf{A}(r)$, with $r = |\mathbf{r}|$, one can show that

$$\nabla \varphi = \varphi' \mathbf{u}_r,$$

$$\nabla \cdot \mathbf{A}(r) = \mathbf{u}_r \cdot \mathbf{A}', \tag{A.56}$$

$$\nabla \times \mathbf{A}(r) = \mathbf{u}_r \times \mathbf{A}'.$$

where

$$\varphi' = \frac{d\varphi}{dr}, \quad \mathbf{A}' = \frac{d\mathbf{A}'}{dr}, \quad \mathbf{u}_r = \frac{\mathbf{r}}{r}.$$

A.4 Second-Order Cartesian Tensors

A system of three quantities A_i, $i = 1, 2, 3$ which transform according to (A.15) upon a change of basis, that is

$$A'_i = a_{ik} A_k, \quad i, k = 1, 2, 3, \tag{A.57}$$

where a_{ik} satisfy the orthogonality condition (A.17), form a *first-order orthogonal affine tensor*, or an *orthogonal affine vector*.

A system of $3^2 = 9$ quantities T_{ik}, i, $k = 1$, 2, 3 which transform like the product $A_i B_k$, that is according to the rule

$$T'_{ik} = a_{ij} a_{km} T_{jm}, \quad i, j, k, m = 1, 2, 3 \tag{A.58}$$

is a *second-order orthogonal affine tensor*. If in (A.58) we set $i = k$ and use the orthogonality condition (A.17), we have

$$T'_{ii} = a_{ij} a_{im} T_{jm} = \delta_{jm} T_{jm} = T_{mm}. \tag{A.59}$$

The sum $T_{ii} = T_{11} + T_{22} + T_{33}$ is called *trace* (Tr) or *spur* (Sp) of the tensor T_{ik}. The relation (A.59) shows that the trace is invariant under the coordinate change (A.15).

A tensor T_{ik} is called *symmetric* if $T_{ik} = T_{ki}$, and *antisymmetric* if $T_{ik} = -T_{ki}$. In an n-dimensional Euclidean space, E_n, a symmetric tensor has $(C_n^2) = n(n + 1)/2$ distinct components, and an antisymmetric tensor has $C_n^2 = n(n - 1)/2$ distinct components. Therefore, in E_3 a symmetric tensor has 6 independent components, while an antisymmetric tensor has 3 distinct components.

An antisymmetric tensor is characterized by $T_{ii} = 0$ (no summation), i.e. the elements on the principal diagonal are zero.

A tensor with the property $T_{ik} = 0$ $(i \neq k)$ is called *diagonal*. Such a tensor is, for example, the *Kronecker symbol*, also named the *second-order symmetric unit tensor* δ_{ik}, i, $k = 1$, 2, 3. Its components do not change upon a change of coordinates. Indeed,

$$\delta'_{ik} = a_{il} a_{km} \delta_{lm} = a_{il} a_{kl} = \delta_{ik}.$$

Given the antisymmetric tensor A_{ik}, let us denote

$$A_{12} = A_3,$$
$$A_{23} = A_1,$$
$$A_{31} = A_2,$$

or, in a condensed form,

$$A_{ij} = \epsilon_{ijk} A_k,$$
$$A_i = \frac{1}{2} \epsilon_{ijk} A_{jk}, \quad i, j, k = 1, 2, 3. \tag{A.60}$$

The ordered system of objects (quantities, numbers, etc.) A_i form a pseudovector, called the *pseudovector associated with the antisymmetric tensor* A_{ik}. Such a situation is encountered in the case of the cross product of two polar vectors. Taking into account that any second-order tensor can be written as a sum of a symmetric and an antisymmetric tensor,

$$A_{ik} = \frac{1}{2}(A_{ik} + A_{ki}) + \frac{1}{2}(A_{ik} - A_{ki}), \tag{A.61}$$

we can write

$$c_i = (\mathbf{a} \times \mathbf{b})_i = \epsilon_{ijk} a_j b_k = \frac{1}{2}\epsilon_{ijk} c_{jk}, \tag{A.62}$$

where

$$c_{jk} = a_j b_k - a_k b_j. \tag{A.63}$$

Consequently, the vector associated with a second-order antisymmetric tensor is a pseudovector. The antisymmetric tensor A_{ik} and the pseudovector $A_i = \frac{1}{2}\epsilon_{ijk} A_{jk}$ are said to be *dual* to each other.

A.5 Cartesian Tensors of Higher Order

An *order-p tensor* or tensor of type p in E_3 is a system of 3^p components which, under an orthogonal transformation of coordinates, transform according to

$$T'_{i_1 i_2 \dots i_p} = a_{i_1 j_1} a_{i_2 j_2} \dots a_{i_p j_p} T_{j_1 j_2 \dots j_p}, \quad i_1, \dots, i_p, j_1, \dots, j_p = 1, 2, 3. \tag{A.64}$$

An *order-p pseudotensor* in E_3 is a system of 3^p components which, under an orthogonal transformation of coordinates, transform according to

$$T^{*'}_{i_1 i_2 \dots i_p} = [\det(a_{ik})] \, a_{i_1 j_1} a_{i_2 j_2} \dots a_{i_p j_p} T^*_{j_1 j_2 \dots j_p}, \quad i_1, \dots, i_p, j_1, \dots, j_p = 1, 2, 3. \tag{A.65}$$

In other words, under a proper orthogonal transformation, characterized by $\det(a_{ik}) = +1$, pseudotensors transform like tensors, while under an improper orthogonal transformation, for which $\det(a_{ik}) = -1$, there appears a change of sign. Comparing (A.65) and (A.21), we conclude that the Levi-Civita permutation symbol (A.7) is a pseudotensor. It is called the *third-order totally antisymmetric unit pseudotensor*. It will therefore transform according to the rule

$$\epsilon'_{ijk} = [\det(a_{ik})] \, a_{il} a_{jm} a_{kn} \epsilon_{lmn}. \tag{A.66}$$

In view of (A.21), we then have

$$\epsilon'_{ijk} = [\det(a_{ik})]^2 \, \epsilon_{ijk} = \epsilon_{ijk}, \tag{A.67}$$

which says that the components of ϵ_{ijk} do not depend on the choice of orthonormal basis.

Appendix B
Tensors

B.1 *n*-Dimensional Spaces

An *n-dimensional space* S_n is a set of elements, called *points*, which are in biu-nivocal and bicontinuous correspondence with n real variables $x^1, x^2, \ldots, x^n$. The variables $x^1, x^2, \ldots, x^n$ are called *coordinates*. To a set of values of the coordinates corresponds a single point and vice-versa. The number of coordinates defines the dimension of the space.

Let $x'^1, \ldots, x'^n$ be another set of coordinates in S_n and consider the transformation

$$x'^\nu = h(x^\mu), \qquad \mu, \nu = 1, 2, \ldots, n, \tag{B.1}$$

where h is a function mapping S_n to itself. If the Jacobian of the transformation (B.3) is different from zero,

$$J = \frac{\partial(x'^1, \ldots, x'^n)}{\partial(x^1, \ldots, x^n)} \neq 0, \tag{B.2}$$

then the transformation (B.1) is (at least locally) reversible, or *biunivocal*, and we also have

$$x^\mu = h^{-1}(x'^\nu), \qquad \mu, \nu = 1, 2, \ldots, n, \tag{B.3}$$

meaning that $x'^1, \ldots, x'^n$ are also a system of coordinates in S_n.

A transformation like (B.1) or (B.3) is called *coordinate transformation*. The transformations (B.3) and (B.1) are inverse to each other. If the variables x'^μ are linear with respect to x^ν, then the transformation (B.1) is called *linear* of *affine*. (An affine transformation is any transformation that preserves collinearity and ratios of distances.)

If in S_n one defines the notion of distance between two points (called a *metric*), the space S_n is called *metric space*.

© Springer-Verlag Berlin Heidelberg 2016
M. Chaichian et al., *Electrodynamics*, DOI 10.1007/978-3-642-17381-3

Let $f(x^1, \ldots, x^n)$ be a function of coordinates. Under a coordinate transformation (B.1), the functional form of f is changed:

$$f(x^\mu) = f(h^{-1}(x'^\mu)) = (f \circ h^{-1})(x'^\mu) = f'(x'^\mu), \tag{B.4}$$

with the notation $f' = (f \circ h^{-1})$. A function which satisfies

$$f(x^1, \ldots, x^n) = f'(x'^1, \ldots, x'^m), \tag{B.5}$$

is an *invariant*.

Functions of coordinates describing physical quantities can have more complicated transformation properties under coordinate transformations. If the coordinate transformations form a continuous group, we say that those functions transform under a representation of that group. In that case, the system is covariant under the transformation of coordinates, i.e. the equations of motion characterizing the system keep their form upon a coordinate transformation. In that case, invariants are formed also from combinations of different representations of the group of transformations. In Chap. 7 there are many examples of Lorentz invariants formed by combinations of vectors and tensors.

B.2   Contravariant and Covariant Vectors

Differentiating relation (B.1), we have

$$dx'^\nu = \frac{\partial x'^\nu}{\partial x^\mu} dx^\mu = \overline{x}^\nu_\mu dx^\mu, \quad \mu, \nu = 1, 2, \ldots, n, \tag{B.6}$$

where we denoted

$$\overline{x}^\nu_\mu = \frac{\partial x'^\nu}{\partial x^\mu}, \tag{B.7}$$

and used Einstein's summation convention over repeated indices.

A system of n quantities (objects) $A^1, \ldots, A^n$ which transform according to (B.6), i.e.

$$A'^\nu = \overline{x}^\nu_\mu A^\mu, \tag{B.8}$$

form a *first-order contravariant tensor*, or a tensor of type $(1, 0)$, or an *n-dimensional contravariant vector*. The quantities A^μ, $\mu = 1, 2, \ldots, n$ are the vector components in x^μ coordinates, while A'^μ are the components of the same vector in x'^μ coordinates. Observing that

$$dx^\mu = \frac{\partial x^\mu}{\partial x'^\nu} dx'^\nu = \underline{x}^\mu_\nu dx'^\nu, \tag{B.9}$$

where

$$\underline{x}_\nu^\mu = \frac{\partial x^\mu}{\partial x'^\nu},$$ (B.10)

one can define the inverse (relative to (B.8)) transformation

$$A^\mu = \underline{x}_\nu^\mu A'^\nu.$$ (B.11)

Consider now the scalar function $f(x^\mu)$, $\mu = 1, 2, \ldots, n$ and take its partial derivatives with respect to x'^μ:

$$\frac{\partial f}{\partial x'^\mu} = \frac{\partial f}{\partial x^\nu} \frac{\partial x^\nu}{\partial x'^\mu} = \underline{x}_\mu^\nu \frac{\partial f}{\partial x^\nu}, \quad \mu, \nu = 1, 2, \ldots, n.$$ (B.12)

A set of n quantities B_μ, $\mu = 1, 2, \ldots, n$ which transform according to (B.12), that is

$$B'_\mu = \underline{x}_\mu^\nu B_\nu,$$ (B.13)

define a *first-order covariant tensor*, or a tensor of type $(0, 1)$, or an *n-dimensional covariant vector*. The inverse transformation is

$$B_\nu = \overline{x}_\nu^\mu B'_\mu.$$ (B.14)

In the Euclidean space with Cartesian orthogonal coordinates, $\overline{x}_\nu^\mu = \underline{x}_\mu^\nu$, hence there is no distinction between contravariant and covariant vectors.

It can be easily shown that the contracted product of a contravariant and a covariant vector is an invariant. Indeed,

$$A'^\nu B'_\nu = \overline{x}_\nu^\mu \underline{x}_\nu^\lambda A^\mu B_\lambda = \delta_\mu^\lambda A^\mu B_\lambda = A^\mu B_\mu = A_\mu B^\mu.$$ (B.15)

***Observation*:**
In tensor analysis, covariance and contravariance describe how the quantitative description of certain geometrical or physical entities changes when passing from one coordinate system to another. To be coordinate system invariant, vectors like radius-vector, velocity, acceleration, their derivatives with respect to time, etc., must *contra-vary* with the change of basis to compensate. That is, the components must vary oppositely to the change of basis. For a dual vector (covector), like the gradient, to be coordinate system invariant, its components must *co-vary* with the change of basis, that is its components must vary by the same transformation as the basis. If, for example, vectors have units of distance (radius vector), the covectors have units of inverse of distance. This distinction becomes very important in general relativity.

B.3   Second-Order Tensors

A set of n^2 quantities $T^{\mu\nu}$ form a *second-order contravariant tensor* if, upon a coordinate change (B.1), they transform as the product $A^\mu B^\nu$, that is according to

$$T'^{\mu\nu} = \overline{x}^\mu_\lambda \overline{x}^\nu_\rho T^{\lambda\rho},$$
$$T^{\mu\nu} = \underline{x}^\mu_\lambda \underline{x}^\nu_\rho T'^{\lambda\rho}. \tag{B.16}$$

A set of n^2 quantities $U_{\mu\nu}$ form a *second-order covariant tensor* if, upon a coordinate change (B.1), they transform as the product $A_\mu B_\nu$, that is according to

$$U'_{\mu\nu} = \underline{x}^\lambda_\mu \underline{x}^\rho_\nu U_{\lambda\rho},$$
$$U_{\mu\nu} = \overline{x}^\lambda_\mu \overline{x}^\rho_\nu U'_{\lambda\rho}. \tag{B.17}$$

A set of n^2 quantities V^μ_ν form a *second-order mixed tensor* if, upon a coordinate change (B.1), they transform as the product $A^\mu B_\nu$, that is according to

$$V'^\mu_\nu = \overline{x}^\mu_\lambda \underline{x}^\rho_\nu V^\lambda_\rho,$$
$$V^\mu_\nu = \underline{x}^\mu_\lambda \overline{x}^\rho_\nu V'^\lambda_\rho. \tag{B.18}$$

Such a tensor is the Kronecker symbol:

$$\delta'^\mu_\nu = \overline{x}^\mu_\lambda \underline{x}^\rho_\nu \delta^\lambda_\rho = \overline{x}^\mu_\lambda \underline{x}^\lambda_\nu = \frac{\partial x'^\mu}{\partial x^\lambda} \frac{\partial x^\lambda}{\partial x'^\nu} = \frac{\partial x'^\mu}{\partial x'^\nu} = \delta^\mu_\nu. \tag{B.19}$$

A contravariant index of a tensor can be lowered using the metric tensor $g_{\mu\nu}$, and a covariant index can be raised using the inverse metric tensor $g^{\mu\nu}$. One must take care of the order of the indices. If the mixed tensor is symmetric, indices are written on the same vertical line, one below the other.

One can show that the contracted product $T^{\mu\nu} A_\mu B_\nu$ $\mu, \nu = 1, 2, \ldots, n$ is an invariant. Indeed,

$$T'^{\mu\nu} A'_\mu B'_\nu = \overline{x}^\mu_\lambda \overline{x}^\nu_\rho \underline{x}^\sigma_\mu \underline{x}^\kappa_\nu T^{\lambda\rho} A_\sigma B_\kappa = \delta^\sigma_\lambda \delta^\kappa_\rho T^{\lambda\rho} A_\sigma B_\kappa = T^{\lambda\rho} A_\lambda B_\rho. \tag{B.20}$$

The covariant second-order tensor $T_{\mu\nu}$ is *symmetric* if $T_{\mu\nu} = T_{\nu\mu}$, and *antisymmetric* if $T_{\mu\nu} = -T_{\nu\mu}$. These definitions are analogous for contravariant tensors. The property of symmetry (antisymmetry) is invariant under coordinate transformations.

A second-order (contravariant or covariant) tensor can always be decomposed into a sum of symmetric and antisymmetric parts,

$$T_{\mu\nu} = \frac{1}{2}\left(T_{\mu\nu} + T_{\nu\mu}\right) + \frac{1}{2}\left(T_{\mu\nu} - T_{\nu\mu}\right) = S_{\mu\nu} + A_{\mu\nu}, \tag{B.21}$$

where $S_{\mu\nu} = S_{\nu\mu}$ and $A_{\mu\nu} = -A_{\nu\mu}$.

B.4   The Metric Tensor

Consider an m-dimensional Euclidean space E_m and let $y_1, \ldots, y_m$ be the Cartesian coordinates of a point P. The squared distance between P and an infinitely closed point P' is

$$ds^2 = dy_J dy_J, \quad J = 1, 2, \ldots, m. \tag{B.22}$$

Take now in E_m an embedded submanifold R_n ($n < m$) and let $x^1, \ldots, x^n$ be the coordinates of a point in R_n. Obviously,

$$y_J = y_J(x^1, \ldots, x^n), \quad J = 1, 2, \ldots, m. \tag{B.23}$$

The squared distance between two infinitely closed points in R_n is

$$ds^2 = \frac{\partial y_J}{\partial x^\mu} \frac{\partial y_J}{\partial x^\nu} dx^\mu dx^\nu = g_{\mu\nu} dx^\mu dx^\nu, \tag{B.24}$$

where we denoted

$$g_{\mu\nu}(x^1, \ldots, x^n) = \frac{\partial y_J}{\partial x^\mu} \frac{\partial y_J}{\partial x^\nu}, \quad J = 1, 2, \ldots, m; \quad \mu, \nu = 1, 2, \ldots, n. \tag{B.25}$$

The bilinear form (B.24) is *positive definite* and, according to (B.20), it is invariant under coordinate changes. Since dx^μ and dx^ν are contravariant vectors, it results that $g_{\mu\nu}$ is a second-order covariant symmetric tensor, called the *metric tensor*. (The terms *metric* and *line element* are often used interchangeably.)

Since

$$dx_\mu = g_{\mu\nu} dx^\nu, \tag{B.26}$$

we may write

$$ds^2 = dx_\mu dx^\mu.$$

But in this case relation (B.26) is also true for the components of any vector A^μ:

$$A_\mu = g_{\mu\nu} A^\nu, \quad \mu, \nu = 1, 2, \ldots, n. \tag{B.27}$$

Relations (B.27) can be considered as a system of n algebraic equations with n unknowns $A^1, \ldots, A^n$. Solving the system by Cramer's rule, we obtain

$$A^\nu = g^{\nu\lambda} A_\lambda, \quad \nu, \lambda = 1, 2, \ldots, n, \tag{B.28}$$

where

$$g^{\nu\lambda} = \frac{G^{\nu\lambda}}{g} \tag{B.29}$$

are the components of the *contravariant metric tensor*. Here $G^{\nu\lambda}$ is the algebraic complement of the element $g_{\nu\lambda}$ in the determinant

$$g = \det(g_{\nu\lambda}).$$

Since $G^{\nu\lambda} = G^{\lambda\nu}$, we have $g^{\nu\lambda} = g^{\lambda\nu}$. Relation (B.27) also shows that

$$A_\mu = g_{\mu\nu} A^\nu = g_{\mu\nu} g^{\nu\lambda} A_\lambda,$$

which yields

$$g_{\mu\nu} g^{\nu\lambda} = \delta_\mu^\lambda. \tag{B.30}$$

Using the metric tensor, one can lower or raise the indices of any tensor. In an Euclidean space holds the relation $g_{\mu\nu} = \delta_{\mu\nu}$, so that $A_\mu = \delta_{\mu\nu} A^\nu = A^\mu$. Consequently, on such a manifold there is no distinction between contravariant and covariant indices.

B.5   Higher Order Tensors

In an analogous way can be defined tensors of any variance. For example, the system of $n^{\alpha+\beta}$ real quantities $T_{k_1...k_\alpha}^{j_1...j_\beta}$ form a tensor of order $(\alpha + \beta)$, or a type (α, β) tensor, i.e. α-times covariant and β-times contravariant, if its components transform according to

$$T'^{j_1...j_\beta}_{k_1...k_\alpha} = \underline{x}^{p_1}_{k_1} \cdots \underline{x}^{p_\alpha}_{k_\alpha} \overline{x}^{j_1}_{i_1} \cdots \overline{x}^{j_\beta}_{i_\beta} T^{i_1...i_\beta}_{p_1...p_\alpha}. \tag{B.31}$$

B.6   Tensor Operations

B.6.1   Addition

Two tensors can be added (subtracted) only if they have the same order and the same variance. For example, the tensors $U^\mu_{\nu\lambda}$ and $V^\mu_{\nu\lambda}$ can be added to give

$$T^\mu_{\nu\lambda} = U^\mu_{\nu\lambda} + V^\mu_{\nu\lambda}. \tag{B.32}$$

B.6.2   Multiplication

Let $U^{i_1...i_\alpha}_{j_1...j_\beta}$ and $V^{k_1...k_\gamma}_{m_1...m_\delta}$ be two tensors of arbitrary order and variance. Their product is defined by

$$T^{i_1...i_\alpha k_1...k_\gamma}_{j_1...j_\beta m_1...m_\delta} = U^{i_1...i_\alpha}_{j_1...j_\beta} V^{k_1...k_\gamma}_{m_1...m_\delta}, \tag{B.33}$$

being a tensor of order $(\alpha + \beta + \gamma + \delta)$, $(\alpha + \gamma)$-times contravariant and $(\beta + \delta)$-times covariant.

B.6.3 Contraction

Consider a mixed tensor (T) of order $p \geq 2$. The operation of *contraction* consists in setting equal one contravariant index and one covariant index, and summing over them. By one contraction, the order of the tensor reduces by two units. As an example, consider the tensor $T_{\mu}^{\nu\lambda}$, and take $\nu = \mu$. The result is

$$T_{\mu}^{\prime\mu\lambda} = \overline{x}_{\rho}^{\mu}\overline{x}_{\kappa}^{\lambda}\underline{x}_{\mu}^{\sigma}T_{\sigma}^{\rho\kappa} = \delta_{\rho}^{\sigma}\overline{x}_{\kappa}^{\lambda}T_{\sigma}^{\rho\kappa} = \overline{x}_{\kappa}^{\lambda}T_{\rho}^{\rho\kappa}, \tag{B.34}$$

which is the transformation of a contravariant vector.

In general, by setting equal the first γ covariance indices with the first γ contravariance indices of the tensor $T_{j_1\ldots j_\beta}^{i_1\ldots i_\alpha}$, one obtains the tensor $U_{j_{\gamma+1}\ldots j_\beta}^{i_{\gamma+1}\ldots i_\alpha}$, of type $(\alpha - \gamma, \beta - \gamma)$.

B.6.4 Raising and Lowering Indices

Consider the tensor $T_{\nu\lambda\rho}^{\mu}$ and perform the product

$$g_{\mu\sigma}T_{\nu\lambda\rho}^{\mu} = T_{\sigma\nu\lambda\rho}, \tag{B.35}$$

which is a covariant tensor. We call this operation *index lowering* (μ in our case). In the same manner, an index can be *raised*:

$$g^{\nu\sigma}T_{\nu\lambda\rho}^{\mu} = T_{\lambda\rho}^{\mu\sigma}. \tag{B.36}$$

To lower (raise) n indices, this operation has to be accomplished n times.

The operations of lowering and raising indices do not modify the order of a tensor, but only its variance.

B.6.5 Symmetric and Antisymmetric Tensors

Consider the tensor

$$T_{j_1\ldots j_\beta}^{i_1\ldots i_\alpha} \quad (\alpha, \beta \geq 2) \tag{B.37}$$

on an n-dimensional Euclidean space. If its components do not change under a permutation of a group of indices, either of contravariance or of covariance, we call the tensor *symmetric* in that group of indices.

Let us consider, for the beginning, a totally symmetric tensor of type $(\alpha, 0)$. Due to the symmetry, its number of free parameters is naturally lower than that of an arbitrary type $(\alpha, 0)$ tensor. Namely, the number of distinct components is given by the number of combinations with repetitions, into groups of α indices, formed with $1, 2, \ldots, n$. This number is given by

$$\left(C_n^\alpha\right) = C_{n+\alpha-1}^\alpha = \frac{n(n+1)\ldots(n+\alpha-1)}{\alpha!}. \tag{B.38}$$

An analogous formula is obtained for the number of free parameters of a totally symmetric tensor of type $(0, \alpha)$.

If we consider the general tensor (B.37), symmetric in the first γ contravariance indices $i_1, \ldots i_\gamma$, then the number of its distinct components is

$$\left(C_n^\gamma\right) n^{\alpha+\beta-\gamma} = C_{n+\gamma-1}^\gamma n^{\alpha+\beta-\gamma}. \tag{B.39}$$

As an example, let us take the second-order symmetric tensor $T_{\mu\nu}$. According to (B.38), it has distinct $n(n+1)/2$ components (6 components in three-dimensional space, 10 in a four-dimensional space, etc).

If the tensor (B.37) changes its sign when permuting any two indices out of a group of indices, either of contravariance or of covariance, the tensor is called *antisymmetric* in that group of indices. Suppose that (T) is antisymmetric in the first γ covariance indices, and denote by $[j_1 \ldots j_\gamma]$ the group of these indices. Then we have

$$T^{i_1\ldots i_\alpha}_{[j_1\ldots j_\gamma]j_{\gamma+1}\ldots j_\beta} = (-1)^I T^{i_1\ldots i_\alpha}_{(j_1'\ldots j_\gamma')j_{\gamma+1}\ldots j_\beta}, \tag{B.40}$$

where $\left(j_1' \ldots j_\gamma'\right)$ is an arbitrary permutation of indices $j_1 \ldots j_\gamma$, and I the number of inversions of indices needed in order to achieve the final ordering. In this category falls, for example, the Levi-Civita permutation symbol $\epsilon_{\mu\nu\lambda\rho}$.

The number of distinct components of the tensor (B.40) is $C_n^\gamma n^{\alpha+\beta-\gamma}$, i.e.

$$C_n^\gamma n^{\alpha+\beta-\gamma} = \frac{n(n-1)(n-2)\ldots(n-\gamma+1)}{\gamma!} n^{\alpha+\beta-\gamma}. \tag{B.41}$$

For example, the second-order antisymmetric tensor $A_{\mu\nu}$ has $n(n-1)/2$ distinct components (3 in a three dimensional space, 6 in a four-dimensional space, etc.). Since $A_{\mu\nu} = -A_{\nu\mu}$, it results that $A_{\mu\mu} = 0$ (no summation). In general, if a pair of indices are equal in the group of antisymmetry indices, the corresponding tensor component is zero.

A tensor with the property $T_\nu^\mu = 0$ ($\mu \neq \nu$) is called *diagonal*. Such a tensor is the Kronecker symbol δ_ν^μ.

Observation:

The number of distinct components of a tensor diminishes if among components there are supplementary relations. For example, if the symmetric tensor $T_{\mu\nu}$ in Euclidean three-dimensional space has zero trace, the number of distinct components is $6 - 1 = 5$. In general, each independent relation between the components diminishes the number of independent components by one.

B.7 Tensor Variance: An Intuitive Image

As we have seen, the *variance* of tensors can be of two types, *covariance* and *contravariance*. Strictly speaking, one way of defining a co- or contravariant first-order tensor is the one presented in Sect. B.2; in the following, we wish to provide a more intuitive image about these concepts, and at the same time emphasize the necessity of introducing them.

For an easier presentation, we shall refer in the following to a three-dimensional space. Between the simplest, Cartesian type of coordinate system and the most general coordinate system, there are two types of intermediate systems, namely:

1) orthogonal curvilinear coordinate systems (the coordinate axes are curvilinear, but at each point the tangent vectors to the axes form an orthogonal trihedron – see Fig. B.1c);
2) non-orthogonal rectilinear coordinate systems (the coordinate axes are straight lines, but they form angles different from $\pi/2$ – see Fig. B.1d).

The orthogonal curvilinear coordinate systems are discussed in Appendix D, where it is shown that the principal effect of the curving of axes is the appearance of the *Lamé coefficients*. As we shall see, the non-orthogonality of the coordinate axes brings about the necessity of introducing the *variance of tensors*.

To facilitate the graphical presentation, we shall use mainly two-dimensional spaces (the generalization to three dimensions is trivial). Let us then consider on the Euclidean plane a vector **a**, and express its components with respect to an orthogonal and a non-orthogonal coordinate system (Fig. B.2).

In the first system (Fig. B.2a) we write

$$a_1 = \mathbf{a} \cdot \mathbf{u}_1,$$
$$a_2 = \mathbf{a} \cdot \mathbf{u}_2, \tag{B.42}$$

and if we denote $\mathbf{a}_1 = a_1\mathbf{u}_1$, $\mathbf{a}_2 = a_2\mathbf{u}_2$, then we have

$$\mathbf{a} = \mathbf{a}_1 + \mathbf{a}_2, \tag{B.43}$$

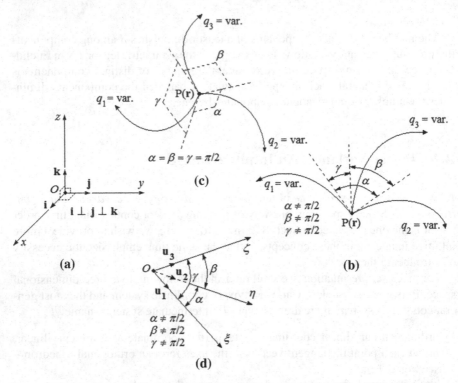

Fig. B.1 Four types of coordinate system: (**a**) Cartesian (orthogonal rectilinear), (**b**) general curvilinear, (**c**) orthogonal curvilinear, and (**d**) non-orthogonal rectilinear.

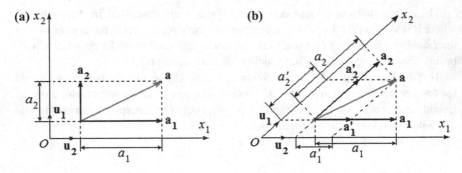

Fig. B.2 Two types of rectilinear/straight coordinate axes: (**a**) orthogonal and (**b**) non-orthogonal.

where $\mathbf{u}_1$ and $\mathbf{u}_2$ are the versors of the two axes, while a_1 and a_2 are the components of the vector $\mathbf{a}$ in this basis (in other words, the *orthogonal projections* of the vector $\mathbf{a}$ on the coordinate axes).

In the second system (Fig. B.2b), we notice that there are two possibilities of defining the components of the vector **a**:

1) by orthogonal projection on the axes, leading to the components denoted by a_1 and a_2;
2) by drawing parallels to the axes through the tip of the vector **a**, leading to the components denoted by a_1' and a_2'.

Thus, in the first case we write

$$a_1 = \mathbf{a} \cdot \mathbf{u}_1,$$
$$a_2 = \mathbf{a} \cdot \mathbf{u}_2, \qquad (B.44)$$

but, with the notation $\mathbf{a}_1 = a_1\mathbf{u}_1$ and $\mathbf{a}_2 = a_2\mathbf{u}_2$, a relation like (B.43) is not valid anymore, i.e.

$$\mathbf{a} \neq \mathbf{a}_1 + \mathbf{a}_2. \qquad (B.45)$$

In the second case, denoting

$$\mathbf{a}_1' = a_1'\mathbf{u}_1,$$
$$\mathbf{a}_2' = a_2'\mathbf{u}_2, \qquad (B.46)$$

we find that a relation of the type (B.43) remains valid, i.e.

$$\mathbf{a} = \mathbf{a}_1' + \mathbf{a}_2', \qquad (B.47)$$

but we cannot write anymore a relation of the type (B.42), since in this case

$$a_1' \neq \mathbf{a} \cdot \mathbf{u}_1,$$
$$a_2' \neq \mathbf{a} \cdot \mathbf{u}_2. \qquad (B.48)$$

In other words, the components a_1' and a_2' are not the scalar products of the vector **a** with the versors of the coordinate axes.

If we wish to have simultaneously valid relations of the type (B.44) and (B.45), we have to introduce a new basis, called the *dual basis*.

Let **u**, **v**, and **w** be three linearly independent vectors in E_3. By definition, they form a basis. Then, any vector **a** in E_3 can be written as

$$\mathbf{a} = \lambda\mathbf{u} + \mu\mathbf{v} + \nu\mathbf{w}, \qquad (B.49)$$

where λ, μ, and ν are three scalars, which are called the *components of the vector* **a** in the basis $\{\mathbf{u}, \mathbf{v}, \mathbf{w}\}$.

Let us introduce three more vectors, denoted by $\mathbf{u}^*$, $\mathbf{v}^*$, and $\mathbf{w}^*$, satisfying the following conditions:

$$\begin{cases} \mathbf{u}^* \cdot \mathbf{u} = 1, \\ \mathbf{u}^* \cdot \mathbf{v} = 0, \\ \mathbf{u}^* \cdot \mathbf{w} = 0, \end{cases} \quad \begin{cases} \mathbf{v}^* \cdot \mathbf{u} = 0, \\ \mathbf{v}^* \cdot \mathbf{v} = 1, \\ \mathbf{v}^* \cdot \mathbf{w} = 0, \end{cases} \quad \begin{cases} \mathbf{w}^* \cdot \mathbf{u} = 0, \\ \mathbf{w}^* \cdot \mathbf{v} = 0, \\ \mathbf{w}^* \cdot \mathbf{w} = 1. \end{cases} \tag{B.50}$$

For instance, the vectors $\mathbf{u}^*$, $\mathbf{v}^*$, and $\mathbf{w}^*$ can be given by the relations

$$\begin{aligned} \mathbf{u}^* &= \frac{\mathbf{v} \times \mathbf{w}}{(\mathbf{u}, \mathbf{v}, \mathbf{w})}, \\ \mathbf{v}^* &= \frac{\mathbf{w} \times \mathbf{u}}{(\mathbf{u}, \mathbf{v}, \mathbf{w})}, \\ \mathbf{w}^* &= \frac{\mathbf{u} \times \mathbf{v}}{(\mathbf{u}, \mathbf{v}, \mathbf{w})}, \end{aligned} \tag{B.51}$$

where $(\mathbf{u}, \mathbf{v}, \mathbf{w}) = \mathbf{u} \cdot (\mathbf{v} \times \mathbf{w})$ is the mixed product of the vectors.

From the definition, it follows that the vectors $\mathbf{u}^*$, $\mathbf{v}^*$, and $\mathbf{w}^*$ are also linearly independent, and thus they form a basis in E_3. Such vectors are called *dual* to the vectors $\mathbf{u}$, $\mathbf{v}$, and $\mathbf{w}$, respectively, while their basis is called *dual basis*. Moreover, one can easily check that the dual of a dual vector is the original vector, i.e.

$$\begin{aligned} \left(\mathbf{u}^*\right)^* &= \mathbf{u}, \\ \left(\mathbf{v}^*\right)^* &= \mathbf{v}, \\ \left(\mathbf{w}^*\right)^* &= \mathbf{w}. \end{aligned} \tag{B.52}$$

Taking the scalar product of the vector (B.49) with the vectors $\mathbf{u}^*$, $\mathbf{v}^*$, and $\mathbf{w}^*$ of the dual basis and using (B.50), one can express the components of the vector in the original basis as

$$\begin{aligned} \lambda &= \mathbf{a} \cdot \mathbf{u}^*, \\ \mu &= \mathbf{a} \cdot \mathbf{v}^*, \\ \nu &= \mathbf{a} \cdot \mathbf{w}^*. \end{aligned} \tag{B.53}$$

Returning to our problems, we notice that, using the vectors of the dual basis, we can write

i) for the components a_1 and a_2 defined by (B.44):

$$\mathbf{a} = \mathbf{a}_1 + \mathbf{a}_2,$$

where

$$\begin{aligned} \mathbf{a}_1 &= a_1 \mathbf{u}_1^*, \quad \text{with} \quad a_1 = \mathbf{a} \cdot \mathbf{u}_1 = \mathbf{a} \cdot \left(\mathbf{u}_1^*\right)^*, \\ \mathbf{a}_2 &= a_2 \mathbf{u}_2^*, \quad \text{with} \quad a_2 = \mathbf{a} \cdot \mathbf{u}_2 = \mathbf{a} \cdot \left(\mathbf{u}_2^*\right)^*. \end{aligned} \tag{B.54}$$

ii) for the components a_1' and a_2' defined by (B.46):

$$a_1' = \mathbf{a} \cdot \mathbf{u}_1^*,$$
$$a_2' = \mathbf{a} \cdot \mathbf{u}_2^*, \qquad (B.55)$$

as well as

$$\mathbf{a} = \mathbf{a}_1' + \mathbf{a}_2', \quad \text{with} \quad \mathbf{a}_1' = a_1' \mathbf{u}_1, \quad \mathbf{a}_2' = a_2' \mathbf{u}_2.$$

The customary notations are different from the above, namely,

$$\begin{cases} a_1' = a^1, \\ a_2' = a^2, \end{cases} \qquad \begin{cases} \mathbf{u}_1^* = \mathbf{u}^1, \\ \mathbf{u}_2^* = \mathbf{u}^2. \end{cases} \qquad (B.56)$$

Thus, the vector $\mathbf{a}$ has two sets of components: one with *lower* indices,

$$a_1 = \mathbf{a} \cdot \mathbf{u}_1,$$
$$a_2 = \mathbf{a} \cdot \mathbf{u}_2, \qquad (B.57)$$

and another with *upper* indices,

$$a^1 = \mathbf{a} \cdot \mathbf{u}^1,$$
$$a^2 = \mathbf{a} \cdot \mathbf{u}^2. \qquad (B.58)$$

The components with *lower* indices are called *covariant*, while the ones with *upper* indices – *contravariant*. The original basis is formed of covariant versors, while the dual basis is formed with contravariant versors. Moreover, in Fig. B.2b one notices that, if the angle between the axes becomes $\pi/2$ (the system becomes orthogonal), then the two types of components coincide.

To conclude, in a non-orthogonal coordinate system, any vector has two sets of components: *covariant components*, by means of which one writes the vector in the dual, or contravariant, basis, and *contravariant components*, used to write the vector in the original, or covariant, basis.

For example, in the case of a three-dimensional non-orthogonal frame, the radius vector of a point can be written as follows:

1) in dual (contravariant) basis, using the covariant components x_1, x_2, and x_3:

$$\mathbf{r} = x_1 \mathbf{u}^1 + x_2 \mathbf{u}^2 + x_3 \mathbf{u}^3;$$

2) in the original (covariant) basis, using the contravariant components x^1, x^2, and x^3:

$$\mathbf{r} = x^1 \mathbf{u}_1 + x^2 \mathbf{u}_2 + x^3 \mathbf{u}_3.$$

Clearly,

$$dr = dx_1 \mathbf{u}^1 + dx_2 \mathbf{u}^2 + dx_3 \mathbf{u}^3,$$
$$dr = dx^1 \mathbf{u}_1 + dx^2 \mathbf{u}_2 + dx^3 \mathbf{u}_3,$$

therefore the metric of the space can be written in the following ways:

$$ds^2 = |dr|^2 = dr \cdot dr$$
$$= \begin{cases} \left(dx_1 \mathbf{u}^1 + dx_2 \mathbf{u}^2 + dx_3 \mathbf{u}^3\right) \cdot \left(dx_1 \mathbf{u}^1 + dx_2 \mathbf{u}^2 + dx_3 \mathbf{u}^3\right), \\ \left(dx^1 \mathbf{u}_1 + dx^2 \mathbf{u}_2 + dx^3 \mathbf{u}_3\right) \cdot \left(dx^1 \mathbf{u}_1 + dx^2 \mathbf{u}_2 + dx^3 \mathbf{u}_3\right), \\ \left(dx_1 \mathbf{u}^1 + dx_2 \mathbf{u}^2 + dx_3 \mathbf{u}^3\right) \cdot \left(dx^1 \mathbf{u}_1 + dx^2 \mathbf{u}_2 + dx^3 \mathbf{u}_3\right). \end{cases}$$

With the new notations, relations (B.50) can be written in a condensed manner as

$$\mathbf{u}^i \cdot \mathbf{u}_j = \delta^i_j, \tag{B.59}$$

hence

$$ds^2 = |dr|^2 = dr \cdot dr = \begin{cases} \left(dx_i \mathbf{u}^i\right) \cdot \left(dx_j \mathbf{u}^j\right) = \left(\mathbf{u}^i \cdot \mathbf{u}^j\right) dx_i dx_j, \\ \left(dx^i \mathbf{u}_i\right) \cdot \left(dx^j \mathbf{u}_j\right) = \left(\mathbf{u}_i \cdot \mathbf{u}_j\right) dx^i dx^j, \\ \left(dx_i \mathbf{u}^i\right) \cdot \left(dx^j \mathbf{u}_j\right) = \left(\mathbf{u}^i \cdot \mathbf{u}_j\right) dx_i dx^j. \end{cases} \tag{B.60}$$

Usually, one denotes the scalar products of the contravariant and covariant versors by

$$\left(\mathbf{u}^i \cdot \mathbf{u}^j\right) = g^{ij},$$
$$\left(\mathbf{u}_i \cdot \mathbf{u}_j\right) = g_{ij}, \tag{B.61}$$

leading to

$$ds^2 = g^{ij} dx_i dx_j = g_{ij} dx^i dx^j = dx_i dx^i. \tag{B.62}$$

Thus, we have arrived at the fundamental metric tensor, with contravariant components, g^{ij}, or covariant components, g_{ij}. Relations (B.61) provide an intuitive interpretation of the elements of the metric tensor.

Finally, from the last equality in (B.62) it follows that

$$dx_i = g_{ij} dx^j, \tag{B.63}$$

which shows that the *lowering of indices* by means of the metric tensor appears naturally. The equality $g^{ij} dx_i dx_j = dx_i dx^i$ from (B.62) leads to

$$dx^i = g^{ij} dx_j, \tag{B.64}$$

i.e. the *raising of indices*.

Appendix C
Representations of Minkowski Space

C.1 Euclidean-Complex Representation

Let $x_1 = x$, $x_2 = y$, $x_3 = z$, $x_4 = ict$ be the coordinates of an event in Minkowski space. The metric then is

$$-ds^2 = dx_\mu dx_\mu = g_{\mu\nu}dx^\mu dx^\nu = dx_1^2 + dx_2^2 + dx_3^2 + dx_4^2$$
$$= dx^2 + dy^2 + dz^2 - c^2 dt^2, \tag{C.1}$$

which means

$$g_{\mu\nu} = \delta_{\mu\nu}, \tag{C.2}$$

corresponding to a pseudo-Euclidean space. Such a representation of Minkowski space is called *Euclidean-complex representation*.

The coordinates x_μ, $\mu = 1, 2, 3, 4$ of an event form a *four-vector*. The *space components* x_1, x_2, x_3 are denoted by x_i, $i = 1, 2, 3$, and the *time component* by x_4. Under a change of coordinates the components of the position (or radius) four-vector transform according to (see (A.57)):

$$x'_\mu = a_{\mu\nu}x_\nu, \quad \mu, \nu = 1, 2, 3, 4. \tag{C.3}$$

If x'_μ are the coordinates of the same event, but determined in the inertial frame S', moving along $Ox \equiv O'x'$ axis, with velocity V with respect to S, then $a_{\mu\nu}$ are the elements of the *Lorentz transformation* matrix

$$A = \begin{pmatrix} \Gamma & 0 & 0 & i\frac{V}{c} \\ 0 & 1 & 0 & 0 \\ 0 & 0 & 1 & 0 \\ -i\frac{V}{c} & 0 & 0 & \Gamma \end{pmatrix}, \tag{C.4}$$

© Springer-Verlag Berlin Heidelberg 2016
M. Chaichian et al., *Electrodynamics*, DOI 10.1007/978-3-642-17381-3

while (C.3) represents, in condensed form, the Lorentz transformation

$$
\begin{aligned}
t' &= \Gamma \left(t - \frac{V}{c^2} x \right), \\
x' &= \Gamma \left(x - V t \right), \\
y' &= y, \\
z' &= z.
\end{aligned}
\tag{C.5}
$$

A system of 4 quantities A_μ, $\mu = 0, 1, 2, 3$ which transform like the coordinates, that is according to

$$
A'_\mu = a_{\mu\nu} A_\nu, \quad \mu, \nu = 1, 2, 3, 4,
\tag{C.6}
$$

form a four-vector. In Euclidean-complex representation the space components of a four-vector are real, and the time component is imaginary.

A second-order four-tensor transforms as the product $A_\mu B_\nu$, that is according to

$$
T'_{\mu\nu} = a_{\mu\lambda} a_{\nu\rho} T_{\lambda\rho}, \quad \lambda, \rho, \mu, \nu = 1, 2, 3, 4.
\tag{C.7}
$$

In the same way, one can define four-tensors of order three, four, etc.

In the theory of relativity, a special role is played by the *totally antisymmetric fourth-order unit pseudotensor* $\epsilon_{\mu\nu\lambda\rho}$. It is defined as being $+1, -1, 0$, as the indices are even, odd, or repeated-index permutation of $1, 2, 3, 4$. The quantities $\epsilon_{\mu\nu\lambda\rho}$ form a *pseudotensor* (sometimes called *axial tensor*) because they exhibit a tensor behaviour under rotations and Lorentz boosts, but are not invariant under parity inversions.

It can be shown that

$$
\epsilon_{\mu\nu\lambda\rho} \epsilon_{\sigma\kappa\xi\zeta} =
\begin{vmatrix}
\delta_{\mu\sigma} & \delta_{\mu\kappa} & \delta_{\mu\gamma} & \delta_{\mu\zeta} \\
\delta_{\nu\sigma} & \delta_{\nu\kappa} & \delta_{\nu\xi} & \delta_{\nu\zeta} \\
\delta_{\lambda\sigma} & \delta_{\lambda\kappa} & \delta_{\lambda\xi} & \delta_{\lambda\zeta} \\
\delta_{\rho\sigma} & \delta_{\rho\kappa} & \delta_{\rho\xi} & \delta_{\rho\zeta}
\end{vmatrix}.
\tag{C.8}
$$

It then results

$$
\begin{aligned}
\epsilon_{\mu\nu\lambda\rho} \epsilon_{\mu\nu\sigma\kappa} &= 2! \left(\delta_{\lambda\sigma} \delta_{\rho\kappa} - \delta_{\lambda\kappa} \delta_{\rho\sigma} \right), \\
\epsilon_{\mu\nu\lambda\rho} \epsilon_{\mu\nu\lambda\sigma} &= 3! \, \delta_{\rho\sigma} = 6 \, \delta_{\rho\sigma}, \\
\epsilon_{\mu\nu\lambda\rho} \epsilon_{\mu\nu\lambda\rho} &= 4! = 24.
\end{aligned}
\tag{C.9}
$$

With the help of $\epsilon_{\mu\nu\lambda\rho}$ one can define the *dual* of an antisymmetric tensor. If $A_{\mu\nu}$ is an antisymmetric tensor, then $A_{\mu\nu}$ and the pseudotensor $\tilde{A}_{\mu\nu} = \frac{1}{2} \epsilon_{\mu\nu\lambda\rho} A_{\lambda\rho}$ are called *dual* to each other. Similarly, the third-order antisymmetric pseudotensor $A_{\mu\nu\lambda} = \epsilon_{\mu\nu\lambda\rho} A_\rho$ and the four-vector A_ρ are dual to each other.

There are four possible types of manifolds that can be embedded in the four-space, which means that there exist four types of integrals:

1) *Line integral*, when the integration is performed along a curve.
2) *Integral over a two-dimensional surface*. In E_3, as surface element, one uses $d\tilde{S}_i$. This is the integration differential and it is the dual of the antisymmetric tensor dS_{ik} (see (A.60)):

$$d\tilde{S}_i = \frac{1}{2!}\epsilon_{ijk}dS_{jk}.$$

From the geometric point of view, $d\tilde{S}_i$ is a vector orthogonal to the surface element and having the same modulus as the elementary area.

In the four-dimensional space,

$$d\tilde{S}_{\mu\nu} = \frac{1}{2}\epsilon_{\mu\nu\lambda\rho}dS_{\lambda\rho}, \quad \mu, \nu, \lambda, \rho = 0, 1, 2, 3, \qquad (C.10)$$

therefore the dual of the tensor $dS_{\mu\nu}$ is also a second-order tensor and, geometrically, represents a surface element equal to $dS_{\mu\nu}$ and orthogonal to it.

3) *Integral over a three-dimensional hypersurface*. In three dimensions, the volume element is constructed as the mixed product of three arc elements corresponding to three coordinate directions which intersect at a point. In four-dimensions, as hypersurface element one defines the antisymmetric tensor $dS_{\mu\nu\lambda}$, together with its dual $d\tilde{S}_\mu$:

$$d\tilde{S}_\mu = \frac{1}{3!}\epsilon_{\mu\nu\lambda\rho}dS_{\nu\lambda\rho},$$
$$dS_{\mu\nu\lambda} = \epsilon_{\mu\nu\lambda\rho}d\tilde{S}_\rho. \qquad (C.11)$$

Geometrically, the four-vector $d\tilde{S}_\mu$ is orthogonal to the hypersurface element, and has the modulus equal to the "area" of this element. In particular, $dS_0 = dx\,dy\,dz$ is the projection of the hypersurface element on the hyperplane $x_0 = $ const.

4) *Integral over the four-dimensional hypervolume*. The volume element in this case is

$$d\Omega = dx_0\,dx_1\,dx_2\,dx_3 = dS_\mu\,dx_\mu, \quad (\text{no summation over } \mu), \qquad (C.12)$$

where the hypersurface element is orthogonal to the arc element dx_μ.

Using these notions, one can generalize the divergence theorem and the Stokes theorem in Minkowski space. In view of (C.12), we have

$$\oint A_\mu dS_\mu = \int \frac{\partial A_\mu}{\partial x_\mu}d\Omega, \qquad (C.13)$$

which generalizes in four dimensions the divergence theorem. Formally, the integral extended over a hypersurface can be transformed into an integral over the four-domain enclosed by the hypersurface by substituting

$$dS_\mu \to d\Omega \, \frac{\partial}{\partial x_\mu}, \quad \mu = 0, 1, 2, 3. \tag{C.14}$$

In a similar way, an integral over a two-dimensional surface, of element $dS_{\mu\nu} = dx_\mu \, dx_\nu$, can be transformed according to

$$\int \frac{\partial A_\mu}{\partial x_\nu} dS_{\mu\nu} = \int \frac{\partial A_\mu}{\partial x_\nu} dx_\mu \, dx_\nu = \oint A_\mu dx_\mu, \tag{C.15}$$

meaning that the circulation along a closed curve in four-dimensional space can be transformed into an integral over the two-dimensional surface bounded by the curve, by substituting

$$dx_\mu \to dS_{\mu\nu} \frac{\partial}{\partial x_\nu}. \tag{C.16}$$

After some index manipulation in (C.15), one obtains

$$\oint A_\mu dx_\mu = \frac{1}{2} \int \left(\frac{\partial A_\nu}{\partial x_\mu} - \frac{\partial A_\mu}{\partial x_\nu} \right) dS_{\mu\nu}. \tag{C.17}$$

One can also establish a formula connecting an integral over a two-dimensional surface, and the boundary three-dimensional surface. As an example, if $A_{\mu\nu}$ is an antisymmetric tensor, we may write

$$\int \frac{\partial A_{\mu\nu}}{\partial x_\nu} dS_\mu = \frac{1}{2} \int \left(\frac{\partial A_{\mu\nu}}{\partial x_\nu} dS_\mu + \frac{\partial A_{\nu\mu}}{\partial x_\mu} dS_\nu \right)$$

$$= \frac{1}{2} \int \left(\frac{\partial A_{\mu\nu}}{\partial x_\nu} dS_\mu - \frac{\partial A_{\mu\nu}}{\partial x_\mu} dS_\nu \right).$$

If one denotes

$$d\tilde{S}_{\mu\nu} \to \frac{1}{2} \left(dS_\mu \frac{\partial}{\partial x_\nu} - dS_\nu \frac{\partial}{\partial x_\mu} \right), \tag{C.18}$$

if follows that

$$\int \frac{\partial A_{\mu\nu}}{\partial x_\nu} dS_\mu = \int A_{\mu\nu} d\tilde{S}_{\mu\nu}. \tag{C.19}$$

C.2 Hyperbolic Representation

An event in Minkowski space can be also defined by the choice $x^0 = ct$, $x^1 = x$, $x^2 = y$, $x^3 = z$, in which case the metric

$$ds^2 = dx_\mu \, dx^\mu = g_{\mu\nu} dx^\mu dx^\nu = c^2 dt^2 - dx^2 - dy^2 - dz^2 \qquad (C.20)$$

gives

$$
\begin{aligned}
g_{00} &= +1, \\
g_{11} &= g_{22} = g_{33} = -1, \qquad\qquad (C.21) \\
g_{\mu\nu}(\mu \ne \nu) &= 0.
\end{aligned}
$$

A system of coordinates in a pseudo-Euclidean (e.g. Minkowski) space in which the line element has the form: $ds^2 = \sum e_\mu^2 dx_\mu^2$, where $e_\mu = \pm 1$, is a *Galilean coordinate system*. We have, therefore, above a Galilean coordinate system, while this representation of Minkowski space is called *hyperbolic*.

Relations (C.21) show that in such a representation one makes distinction between contravariance and covariance indices. For example, in case of a four vector,

$$A_\mu = g_{\mu\nu} A^\nu \Rightarrow \begin{cases} A_0 = g_{00} A^0 = A^0, \\ A_i = g_{ii} A^i = - A^i \quad \text{(no summation)}. \end{cases} \qquad (C.22)$$

The square of a four-vector is

$$A^\mu A_\mu = A^0 A_0 + A^i A_i = A^0 A^0 - A^i A^i = \text{invariant}. \qquad (C.23)$$

In Minkowski space, the components of a contravariant four-vector transform according to (B.8), where the transformation matrix is

$$(\bar{x}^\nu_\mu) \equiv \Lambda^\nu{}_\mu = \begin{pmatrix} \Gamma & -\frac{V}{c}\Gamma & 0 & 0 \\ -\frac{V}{c}\Gamma & \Gamma & 0 & 0 \\ 0 & 0 & 1 & 0 \\ 0 & 0 & 0 & 1 \end{pmatrix}. \qquad (C.24)$$

If, for example, the relative motion of frames S and S' takes place along $Ox \equiv O'x'$-axis, then the contravariant components of A^μ transform according to

$$
\begin{aligned}
A'^0 &= \Gamma\left(A^0 - \frac{V}{c} A^1\right), \\
A'^1 &= \Gamma\left(-\frac{V}{c} A^0 + A^1\right), \\
A'^2 &= A^2, \qquad\qquad\qquad\qquad (C.25) \\
A'^3 &= A^3,
\end{aligned}
$$

622 Appendix C: Representations of Minkowski Space

while the covariant components obey the rule

$$A'_0 = \Gamma \left(A_0 + \frac{V}{c} A_1 \right),$$

$$A'_1 = \Gamma \left(\frac{V}{c} A_0 + A_1 \right),$$

$$A'_2 = A_2, \tag{C.26}$$

$$A'_3 = A_3.$$

In view of (C.21) and (B.30), we can write

$$g_{\mu\nu} = g^{\mu\nu}, \quad \mu, \nu = 0, 1, 2, 3. \tag{C.27}$$

In Galilean coordinates, both the contravariant and covariant components of a four-vector are *real*.

Let us have a look over the relations written in the previous section in Euclidean complex representation. At the beginning, we choose the contravariant Levi-Civita permutation symbol as

$$\epsilon^{0123} = +1. \tag{C.28}$$

It then follows that

$$\epsilon_{\mu\nu\lambda\rho} = g_{\mu\sigma} g_{\nu\kappa} g_{\lambda\upsilon} g_{\rho\zeta} \epsilon^{\sigma\kappa\upsilon\zeta} = -\epsilon^{\mu\nu\lambda\rho}, \tag{C.29}$$

because irrespective of the order of the four different indices, the product of the four metric tensor components is -1. Then we have

$$\epsilon^{\mu\nu\lambda\rho} \epsilon_{\sigma\kappa\upsilon\theta} = - \begin{vmatrix} \delta^\mu_\sigma & \delta^\mu_\kappa & \delta^\mu_\upsilon & \delta^\mu_\theta \\ \delta^\nu_\sigma & \delta^\nu_\kappa & \delta^\nu_\upsilon & \delta^\nu_\theta \\ \delta^\lambda_\sigma & \delta^\lambda_\kappa & \delta^\lambda_\upsilon & \delta^\lambda_\theta \\ \delta^\rho_\sigma & \delta^\rho_\kappa & \delta^\rho_\upsilon & \delta^\rho_\theta \end{vmatrix}. \tag{C.30}$$

Summation over two, three, and four pairs of indices gives

$$\epsilon^{\mu\nu\lambda\rho} \epsilon_{\sigma\kappa\lambda\rho} = -2! \, \delta^{\mu\nu}_{\sigma\kappa} = -2 \left(\delta^\mu_\sigma \delta^\nu_\kappa - \delta^\mu_\kappa \delta^\nu_\sigma \right),$$

$$\epsilon^{\mu\nu\lambda\rho} \epsilon_{\sigma\nu\lambda\rho} = -3! \, \delta^\mu_\sigma = -6 \, \delta^\mu_\sigma, \tag{C.31}$$

$$\epsilon^{\mu\nu\lambda\rho} \epsilon_{\mu\nu\lambda\rho} = -4! = -24.$$

If $A^{\mu\nu}$ is an antisymmetric tensor, its dual is the pseudotensor $\tilde{A}^{\mu\nu} = \frac{1}{2} \epsilon^{\mu\nu\lambda\rho} A_{\lambda\rho}$. The product $\tilde{A}^{\mu\nu} A_{\mu\nu}$ is a pseudoscalar. In the same way, we observe that the antisymmetric pseudotensor $\epsilon^{\mu\nu\lambda\rho} A_\rho$ and the four-vector A^μ are dual to each other.

The symbol of partial derivative $\frac{\partial}{\partial x^\mu}$ is a covariant four-vector, while the symbol $\frac{\partial}{\partial x_\mu}$ is a contravariant four-vector.

Consider now, as we already did in the Euclidean representation of Minkowski space, the four possible types of integrals and the relations between them:

1) *Line integral*, with the arc element dx^μ.
2) *Integral over a two-dimensional surface*, with the surface element

$$d\tilde{S}^{\mu\nu} = \frac{1}{2}\epsilon^{\mu\nu\lambda\rho}dS_{\lambda\rho} \qquad (C.32)$$

which, geometrically, is an area element orthogonal (and quantitatively equal) to $dS_{\mu\nu}$.
3) *Integral over a three-dimensional hypersurface*, of surface element

$$d\tilde{S}^{\mu} = -\frac{1}{3!}\epsilon^{\mu\nu\lambda\rho}dS_{\nu\lambda\rho},$$
$$dS_{\mu\nu\lambda} = \epsilon_{\mu\nu\lambda\rho}dS^{\rho}, \qquad (C.33)$$

such as

$$d\tilde{S}^{0} = -dS_{123} = dS^{123}, \text{ etc.} \qquad (C.34)$$

The four-vector $d\tilde{S}^{\mu}$ has its modulus equal to the area of the hypersurface element, being orthogonal to it.
4) *Integral over a four-dimensional domain*, the elementary hypervolume being

$$d\Omega = dx^{0}dx^{1}dx^{2}dx^{3} = dx^{\mu}dS_{\mu} \quad \text{(no summation)}, \qquad (C.35)$$

where the line element dx^μ and the hypersurface element dS_μ are orthogonal.

The divergence theorem in this representation is

$$\oint A^{\mu}dS_{\mu} = \int \frac{\partial A^{\mu}}{\partial x^{\mu}}d\Omega, \qquad (C.36)$$

while the Stokes theorem takes the form

$$\oint A_{\mu}dx^{\mu} = \frac{1}{2}\int \left(\frac{\partial A_{\nu}}{\partial x^{\mu}} - \frac{\partial A_{\mu}}{\partial x^{\nu}}\right)dS^{\mu\nu}. \qquad (C.37)$$

Finally, the generalization of (C.19) in the hyperbolic representation of Minkowski space is

$$\int A^{\mu\nu}d\tilde{S}_{\mu\nu} = \frac{1}{2}\int \left(\frac{\partial A^{\mu\nu}}{\partial x^{\nu}}dS_{\mu} - \frac{\partial A^{\mu\nu}}{\partial x^{\mu}}dS_{\nu}\right) = \int \frac{\partial A^{\mu\nu}}{\partial x^{\nu}}dS_{\mu}. \qquad (C.38)$$

C.3   Representation in General Curvilinear Coordinates

One can represent the Minkowski space in the most general manner in a system of general curvilinear coordinates. These considerations are especially useful in the study of the gravitational field which, according to general relativity, manifests itself through the curvature of space-time. However, locally, the space-time is flat (of Minkowski type) and the coordinate systems are in their turn locally defined. In general curvilinear coordinates, the metric tensor $g_{\mu\nu}$ depends on the coordinates.

Let us first express the law of transformation of the Levi-Civita symbol when one passes from the Galilean coordinates x^μ to an arbitrary set of general curvilinear coordinates, $x'^\mu = x'^\mu(x^\nu)$, $\mu, \nu = 0, 1, 2, 3$. According to the rule of transformation, we have

$$\epsilon'_{\mu\nu\lambda\rho} = \frac{\partial x^\sigma}{\partial x'^\mu} \frac{\partial x^\kappa}{\partial x'^\nu} \frac{\partial x^\xi}{\partial x'^\lambda} \frac{\partial x^\upsilon}{\partial x'^\rho} \epsilon_{\sigma\kappa\xi\upsilon}, \tag{C.39}$$

where $\epsilon_{\sigma\kappa\xi\upsilon}$ is defined in the Galilean coordinates x^μ, and $\epsilon'_{\mu\nu\lambda\rho}$ in the curvilinear coordinates x'^μ.

If A^ν_μ, $\mu, \nu = 0, 1, 2, 3$ is an arbitrary second-order mixed tensor, it can be shown that

$$A^\sigma_\mu A^\kappa_\nu A^\xi_\lambda A^\upsilon_\rho \epsilon_{\sigma\kappa\xi\upsilon} = A \, \epsilon_{\mu\nu\lambda\rho}, \tag{C.40}$$

where $A = \det(A^\nu_\mu)$. Relation (C.40) is a generalization in four dimensions of (A.21). Then we may write

$$\epsilon'_{\mu\nu\lambda\rho} = \det\left(\frac{\partial x^\mu}{\partial x'^\nu}\right) \epsilon_{\mu\nu\lambda\rho} = \frac{1}{J} \epsilon_{\mu\nu\lambda\rho},$$

where

$$J = \frac{\partial(x'^0, x'^1, x'^2, x'^3)}{\partial(x^0, x^1, x^2, x^3)} \tag{C.41}$$

is the functional determinant of the transformation $x^\mu \to x'^\mu$. Using the transformation rule, we have also

$$g'_{\mu\nu} = \frac{\partial x^\lambda}{\partial x'^\mu} \frac{\partial x^\rho}{\partial x'^\nu} \eta_{\lambda\rho},$$

where $\eta_{\lambda\rho} = \mathrm{diag}(1, -1, -1, -1)$ is the *Minkowski metric tensor*. If we take the determinant of the above relation, we find $g = \frac{1}{J^2} \det(\eta_{\mu\nu})$, where $g = \det(g'_{\mu\nu})$. Since $\det(\eta_{\mu\nu}) = -1$, we have

$$J = \frac{1}{\sqrt{-g}}. \tag{C.42}$$

We then define the antisymmetric unit tensor in curvilinear coordinates by

$$\delta_{\mu\nu\lambda\rho} = \sqrt{-g}\, \epsilon_{\mu\nu\lambda\rho}. \tag{C.43}$$

The transformation rule of the contravariant components $\epsilon^{\mu\nu\lambda\rho}$ is found in a similar way:

$$\epsilon'^{\mu\nu\lambda\rho} = \frac{\partial x'^\mu}{\partial x^\sigma}\frac{\partial x'^\nu}{\partial x^\kappa}\frac{\partial x'^\lambda}{\partial x^\xi}\frac{\partial x'^\rho}{\partial x^\upsilon}\epsilon^{\sigma\kappa\xi\upsilon} = J\,\epsilon^{\mu\nu\lambda\rho},$$

that is

$$\delta^{\mu\nu\lambda\rho} = \frac{1}{\sqrt{-g}}\,\epsilon^{\mu\nu\lambda\rho}, \tag{C.44}$$

with $\epsilon'^{\mu\nu\lambda\rho} = \delta^{\mu\nu\lambda\rho}$. In view of (C.43) and (C.44), relation (C.29) yields

$$\delta_{\mu\nu\lambda\rho} = g\,\delta^{\mu\nu\lambda\rho}. \tag{C.45}$$

If $g = -1$, we find the Galilean formula (C.29).

Let us now write the transformation rule of the four-dimensional elementary volume. In Galilean coordinates, $d\Omega = dx^0 dx^1 dx^2 dx^3$ is an invariant. In the curvilinear coordinates x'^μ the element of volume is $d\Omega' = J d\Omega$. Since the four-volume must be an invariant, in the new coordinates x'^μ not $d\Omega'$, but $\sqrt{-g}\,d\Omega'$ has to be used as integration (hyper)volume element:

$$d\Omega \rightarrow \frac{1}{J}d\Omega' = \sqrt{-g}\,d\Omega'. \tag{C.46}$$

If, as a result of integration over Ω of the quantity $\sqrt{-g}\,\varphi$, with φ a scalar, one obtains an invariant, then $\sqrt{-g}\,\varphi$ is called a *scalar density*. In the same way are defined the notions of *vector density* $\sqrt{-g}\,A^\mu$ and *tensor density* $\sqrt{-g}\,T^{\mu\nu}$, respectively.

The elementary three-dimensional surface is

$$\sqrt{-g}\,dS_\mu = -\frac{1}{3!}\epsilon_{\mu\nu\lambda\rho}\sqrt{-g}\,dS^{\nu\lambda\rho} = -\frac{1}{3!}\delta_{\mu\nu\lambda\rho}\,dS^{\nu\lambda\rho}. \tag{C.47}$$

Similarly, the two-dimensional surface element is

$$\sqrt{-g}\,d\tilde{S}_{\mu\nu} = \frac{1}{2!}\sqrt{-g}\epsilon_{\mu\nu\lambda\rho}\,dS^{\lambda\rho} = \frac{1}{2!}\delta_{\mu\nu\lambda\rho}\,dS^{\lambda\rho}. \tag{C.48}$$

C.4 Differential Operators in General Curvilinear Coordinates

In special relativity the fundamental equations of conservation involve the vector or tensor four-divergence operators, written in terms of the usual derivatives. On curved space-times the usual partial derivatives with respect to coordinates have to be replaced by *covariant derivatives*, as we have explained in Sect. 9.4. Here are the most important differential operators, expressed in curvilinear coordinates.

As an auxiliary step, let us calculate the derivative

$$\frac{\partial}{\partial x^\nu}(\sqrt{-g}) = -\frac{1}{2\sqrt{-g}}\frac{\partial g}{\partial x^\nu} = \frac{\sqrt{-g}}{2g}\frac{\partial g}{\partial x^\nu}. \tag{C.49}$$

But

$$\frac{\partial g}{\partial x^\nu} = \frac{\partial}{\partial x^\nu}\begin{vmatrix} g_{00} & \cdots & g_{03} \\ \vdots & \ddots & \vdots \\ g_{30} & \cdots & g_{33} \end{vmatrix} = \begin{vmatrix} \frac{\partial g_{00}}{\partial x^\nu} & \cdots & \frac{\partial g_{03}}{\partial x^\nu} \\ \vdots & \ddots & \vdots \\ g_{30} & \cdots & g_{33} \end{vmatrix} + \ldots + \begin{vmatrix} g_{00} & \cdots & g_{03} \\ \vdots & \ddots & \vdots \\ \frac{\partial g_{30}}{\partial x^\nu} & \cdots & \frac{\partial g_{33}}{\partial x^\nu} \end{vmatrix}$$

$$= \frac{\partial g_{0\sigma}}{\partial x^\nu}G^{0\sigma} + \ldots + \frac{\partial g_{3\sigma}}{\partial x^\nu}G^{3\sigma} = \frac{\partial g_{\rho\sigma}}{\partial x^\nu}G^{\rho\sigma}, \tag{C.50}$$

where $G^{\rho\sigma}$ is the algebraic complement of the element $g_{\rho\sigma}$ in the determinant g. By means of (B.30), we have also

$$\frac{\partial}{\partial x^\nu}\left(\sqrt{-g}\right) = \frac{\sqrt{-g}}{2}\frac{G^{\rho\sigma}}{g}g_{\rho\sigma,\nu} = \frac{\sqrt{-g}}{2}g^{\rho\sigma}g_{\rho\sigma,\nu},$$

which gives

$$\frac{1}{2}g^{\rho\sigma}g_{\rho\sigma,\nu} = \frac{1}{\sqrt{-g}}\frac{\partial}{\partial x^\nu}\left(\sqrt{-g}\right). \tag{C.51}$$

On the other hand, if in

$$\Gamma^\mu_{\nu\lambda} = \frac{1}{2}g^{\mu\sigma}\left(\frac{\partial g_{\sigma\lambda}}{\partial x^\nu} + \frac{\partial g_{\nu\sigma}}{\partial x^\lambda} - \frac{\partial g_{\lambda\nu}}{\partial x^\sigma}\right)$$

we set $\mu = \lambda$, one obtains

$$\Gamma^\lambda_{\nu\lambda} = \frac{1}{2}g^{\sigma\lambda}g_{\sigma\lambda,\nu} + \frac{1}{2}g^{\sigma\lambda}\left(\frac{\partial g_{\nu\sigma}}{\partial x^\lambda} - \frac{\partial g_{\lambda\nu}}{\partial x^\sigma}\right). \tag{C.52}$$

Since the metric tensor is symmetric, and the expression in parentheses is antisymmetric in the same indices (σ, λ), the last term vanishes. Relations (C.51) and (C.52) then yield

$$\Gamma^\lambda_{\nu\lambda} = \frac{1}{\sqrt{-g}}\frac{\partial}{\partial x^\nu}\left(\sqrt{-g}\right). \tag{C.53}$$

This formula is of help in defining operators such as divergence, gradient, curl, and d'Alembertian in general curvilinear coordinates.

C.4.1 Divergence

Consider the contravariant vector A^ν, and take its covariant derivative

$$\nabla_\mu A^\nu \equiv A^\nu_{;\mu} = A^\nu_{,\mu} + \Gamma^\nu_{\mu\lambda} A^\lambda. \tag{C.54}$$

Setting now $\nu = \mu$ and using (C.53), we find

$$A^\mu_{;\mu} = A^\mu_{,\mu} + \Gamma^\mu_{\mu\lambda} A^\lambda = A^\mu_{,\mu} + \frac{1}{\sqrt{-g}} \frac{\partial}{\partial x^\lambda} \left(\sqrt{-g}\right) A^\lambda,$$

or, if one suitably changes the summation index in the last term,

$$A^\mu_{;\mu} = \frac{1}{\sqrt{-g}} \frac{\partial}{\partial x^\mu} \left(\sqrt{-g} A^\mu\right). \tag{C.55}$$

This is the *covariant four-divergence* of A^μ.

Let us now consider the contravariant tensor $A^{\mu\lambda}$, and take its covariant derivative:

$$A^{\mu\lambda}_{;\nu} = A^{\mu\lambda}_{,\nu} + \Gamma^\lambda_{\sigma\nu} A^{\mu\sigma} + \Gamma^\mu_{\sigma\nu} A^{\sigma\lambda}. \tag{C.56}$$

Setting now $\lambda = \nu$ and using (C.53), we have

$$\nabla_\nu A^{\mu\nu} \equiv A^{\mu\nu}_{;\nu} = A^{\mu\nu}_{,\nu} + \Gamma^\nu_{\sigma\nu} A^{\mu\sigma} + \Gamma^\mu_{\sigma\nu} A^{\sigma\nu}$$

$$= A^{\mu\nu}_{,\nu} + \frac{1}{\sqrt{-g}} \frac{\partial}{\partial x^\nu} \left(\sqrt{-g}\right) A^{\mu\nu} + \Gamma^\mu_{\sigma\nu} A^{\sigma\nu}.$$

In the second term on the r.h.s. we made a convenient change of summation indices. Therefore

$$A^{\mu\nu}_{;\nu} = \frac{1}{\sqrt{-g}} \frac{\partial}{\partial x^\nu} \left(\sqrt{-g} A^{\mu\nu}\right) + \Gamma^\mu_{\sigma\nu} A^{\sigma\nu}, \tag{C.57}$$

which is the *covariant four-divergence* of $A^{\mu\nu}$. If $A^{\mu\nu}$ is antisymmetric, and recalling that $\Gamma^\mu_{\sigma\nu}$ is symmetric in the lower indices, we are left with

$$A^{\mu\nu}_{;\nu} = \frac{1}{\sqrt{-g}} \frac{\partial}{\partial x^\nu} \left(\sqrt{-g} A^{\mu\nu}\right). \tag{C.58}$$

C.4.2 Gradient

Consider the scalar function Φ. Its covariant derivative reduces, obviously, to the usual derivative, which is a covariant vector:

$$\Phi_{;\nu} = \Phi_{,\nu} = A_\nu. \tag{C.59}$$

The contravariant components of A_ν are

$$A^\mu = g^{\mu\nu} A_\nu = g^{\mu\nu} \frac{\partial \Phi}{\partial x^\nu}. \tag{C.60}$$

Introducing (C.50) into (C.55), we obtain the *d'Alembertian*[2] of Φ:

$$A^\mu_{;\mu} = \Box\, \Phi = \frac{1}{\sqrt{-g}} \frac{\partial}{\partial x^\mu} \left(\sqrt{-g}\, g^{\mu\nu} \frac{\partial \Phi}{\partial x^\nu} \right). \tag{C.61}$$

This is the most straightforward method to write the d'Alembertian (or the Laplacian) in any coordinate system (see Appendix D).

C.4.3 Curl

Consider the covariant four-vector A_ν and form the covariant antisymmetric four-tensor

$$F_{\mu\nu} = A_{\nu;\mu} - A_{\mu;\nu}.$$

But

$$A_{\nu;\mu} - A_{\mu;\nu} = \left(A_{\nu,\mu} - \Gamma^\lambda_{\mu\nu} A_\lambda \right) - \left(A_{\mu,\nu} - \Gamma^\lambda_{\nu\mu} A_\lambda \right) = A_{\nu,\mu} - A_{\mu,\nu}.$$

Consequently,

$$F_{\mu\nu} = A_{\nu;\mu} - A_{\mu;\nu} = A_{\nu,\mu} - A_{\mu,\nu}. \tag{C.62}$$

The quantities (C.62) represent a four-dimensional curl, generalizing the notion of three-dimensional curl. One observes that the quantities (C.62) do not depend on the metric.

---

[2]This is also known as the *Laplace–Beltrami operator*. Rigorously speaking, the Laplace–Beltrami operator (C.61) is the generalization of the Laplacian to an elliptic operator defined on a Riemannian manifold, the "usual" d'Alembertian $\Box = \frac{1}{v^2} \frac{\partial^2}{\partial t^2} - \Delta$ being the particular form of the Laplace–Beltrami operator in Minkowski space.

Appendix D
Curvilinear Coordinates

D.1 Curvilinear Coordinates

Let $\mathbf{r}$ be the radius-vector of some point P and x^i, $i = 1, 2, 3$ its Cartesian coordinates. Suppose that there exist three real independent parameters x'^i, so that

$$x^i = x^i(x'^k), \quad i, k = 1, 2, 3.$$ (D.1)

To be (at least locally) reversible, i.e. to have

$$x'^i = x'^i(x^k), \quad i, k = 1, 2, 3,$$ (D.2)

it is necessary that the determinant of the Jacobian matrix be nonvanishing,

$$J = \det\left[\frac{\partial(x^1, x^2, x^3)}{\partial(x'^1, x'^2, x'^3)}\right] \neq 0.$$ (D.3)

If we fix the values of two parameters, say x'^2 and x'^3, we obtain the curve $x'^1 =$ variable. In the same way one can obtain the curves $x'^2 =$ variable and $x'^3 =$ variable. Thus, through each point of space pass three coordinate curves. The parameters x'^i are called *curvilinear coordinates* of the point P.

If at the point P (or any other point) the vectors

$$\mathbf{e}_i = \frac{\partial \mathbf{r}}{\partial x'^i}, \quad i = 1, 2, 3,$$ (D.4)

tangent to the three coordinate curves, form a right orthogonal trihedron, then x'^1, x'^2, x'^3 form an *orthogonal coordinate system*.

© Springer-Verlag Berlin Heidelberg 2016
M. Chaichian et al., *Electrodynamics*, DOI 10.1007/978-3-642-17381-3

D.1.1   Element of Arc Length

An elementary displacement of the point P is written as

$$d\mathbf{r} = \frac{\partial \mathbf{r}}{\partial x'^i} dx'^i = \mathbf{e}_i \, dx'^i. \tag{D.5}$$

Condition (D.3) shows that the three vectors $\mathbf{e}_1$, $\mathbf{e}_2$, $\mathbf{e}_3$ are linearly independent, therefore they form a *basis*. Indeed,

$$(\mathbf{e}_1, \mathbf{e}_2, \mathbf{e}_3) = \left(\frac{\partial \mathbf{r}}{\partial x'^1}, \frac{\partial \mathbf{r}}{\partial x'^2}, \frac{\partial \mathbf{r}}{\partial x'^3} \right) = J$$

and the determinant of this matrix is non-zero. The squared arc element (the metric) is

$$ds^2 = d\mathbf{r} \cdot d\mathbf{r} = \left(\mathbf{e}_i dx'^i \right) \cdot \left(\mathbf{e}_k dx'^k \right) = g_{ik} \, dx'^i \, dx'^k, \tag{D.6}$$

where g_{ik} is the covariant metric tensor.

If we fix x'^2 and x'^3, we obtain the elementary arc length on the coordinate curve $x'^1 = $ variable: $(d_1 s)^2 = g_{11} \left(dx'^1 \right)^2$, that is $(d_1 s) = \sqrt{g_{11}} \left(dx'^1 \right)$. In a similar way, we find two more relations. Therefore

$$\begin{aligned} (d_1 s) &= \sqrt{g_{11}} \left(dx'^1 \right), \\ (d_2 s) &= \sqrt{g_{22}} \left(dx'^2 \right), \\ (d_3 s) &= \sqrt{g_{33}} \left(dx'^3 \right). \end{aligned} \tag{D.7}$$

The elementary arc length is then

$$d\mathbf{s} = \left(\sqrt{g_{11}} \, dx'^1 \right) \mathbf{u}_1 + \left(\sqrt{g_{22}} \, dx'^2 \right) \mathbf{u}_2 + \left(\sqrt{g_{33}} \, dx'^3 \right) \mathbf{u}_3, \tag{D.8}$$

where $\mathbf{u}_1$, $\mathbf{u}_2$, $\mathbf{u}_3$ are the unit vectors of the basis vectors $\mathbf{e}_1$, $\mathbf{e}_2$, $\mathbf{e}_3$.

D.1.2   Area Element

The area element constructed, for example, on the length elements $d_1 \mathbf{s}$ and $d_2 \mathbf{s}$, is

$$d\mathbf{S}_3 = d_1 \mathbf{s} \times d_2 \mathbf{s} = \frac{\partial \mathbf{r}}{\partial x'^1} \times \frac{\partial \mathbf{r}}{\partial x'^2} dx'^1 \, dx'^2,$$
$$|d\mathbf{S}_3| = |\mathbf{e}_1 \times \mathbf{e}_2| \, dx'^1 \, dx'^2. \tag{D.9}$$

Thus, $d\mathbf{S}_3$ is orthogonal to the plane determined by $\mathbf{e}_1$ and $\mathbf{e}_2$ (but not necessarily pointing in the $\mathbf{e}_3$ direction).

D.1.3 Volume Element

This quantity is found by taking the mixed product:

$$d\tau = (d_1\mathbf{s}, d_2\mathbf{s}, d_3\mathbf{s}) = (\mathbf{e}_1, \mathbf{e}_2, \mathbf{e}_3)\, dx'^1\, dx'^2\, dx'^3 = \sqrt{g}\, dx'^1\, dx'^2\, dx'^3 . \quad \text{(D.10)}$$

D.2 First-Order Differential Operators in Curvilinear Coordinates

All formulas obtained in Appendix C for the operators divergence, gradient, and curl are, obviously, valid also in three dimensions. Omitting the "prime" superscript for coordinates, we re-write the specified formulas:

$$\operatorname{div}\mathbf{A} = \frac{1}{\sqrt{g}}\frac{\partial}{\partial x^i}\left(\sqrt{g}A^i\right), \quad i = 1, 2, 3, \quad \text{(D.11)}$$

$$(\operatorname{grad}\Phi)_i = \frac{\partial\Phi}{\partial x^i}, \qquad (\operatorname{grad}\Phi)^i = g^{ik}\frac{\partial\Phi}{\partial x^k}, \quad \text{(D.12)}$$

$$\Delta\Phi = \frac{1}{\sqrt{g}}\frac{\partial}{\partial x^i}\left(\sqrt{g}g^{ik}\frac{\partial\Phi}{\partial x^k}\right). \quad \text{(D.13)}$$

To express the curl-operator, we consider the antisymmetric tensor (see (C.62)):

$$F_{ik} = A_{k,i} - A_{i,k} \quad \text{(D.14)}$$

and let B_i be its associated dual (see (C.43)):

$$B_i = \frac{1}{2}\sqrt{g}\,\epsilon_{ijk}F^{jk} . \quad \text{(D.15)}$$

Therefore, the covariant and contravariant components of curl $\mathbf{A}$ are

$$(\operatorname{curl}\mathbf{A})_i = \frac{1}{2}\sqrt{g}\,\epsilon_{ijk}F^{jk},$$

$$(\operatorname{curl}\mathbf{A})^i = \frac{1}{2}\frac{1}{\sqrt{g}}\epsilon^{ijk}F_{jk}. \quad \text{(D.16)}$$

D.2.1 Differential Operators in Terms of Orthogonal Components

If the basis vectors $\mathbf{e}_1$, $\mathbf{e}_2$, $\mathbf{e}_3$ form an orthogonal trihedron, the curvilinear coordinates are called *orthogonal*. In such a coordinate system the metric tensor g_{ik} is diagonal:

$$g_{ik} = \begin{cases} g_{ii}, & i = k \text{ (no summation)}, \\ 0, & i \neq k. \end{cases} \tag{D.17}$$

In curvilinear orthogonal coordinates one usually utilizes the *orthogonal* (or *physical*) components, instead of contravariant and covariant components of vectors and tensors. To find the transformation relations between the contravariant (covariant) components of a vector $\mathbf{A}$ and its orthogonal components, we represent the vector in the basis $\{\mathbf{e}_k\}_{k=1,2,3}$, then take the dot product by the unit vector of the coordinate curve on which the projection is made. Denoting by $A_{(i)}$ the orthogonal components, we then have

$$A_{(i)} = \mathbf{A} \cdot \mathbf{u}_i = A^k \mathbf{e}_k \cdot \mathbf{u}_i = A^k \sqrt{g_{kk}} \, \mathbf{u}_k \cdot \mathbf{u}_i = A^k \sqrt{g_{kk}} \, \delta_{ik}$$

$$= \sqrt{g_{ii}} \, A^i = \sqrt{g_{ii}} \, g^{ii} A_i = \frac{1}{g_{ii}} \sqrt{g_{ii}} \, A_i = \frac{1}{\sqrt{g_{ii}}} A_i \quad \text{(no summation)},$$

or

$$A^i = \frac{1}{\sqrt{g_{ii}}} A_{(i)},$$

$$A_i = \sqrt{g_{ii}} \, A_{(i)} . \tag{D.18}$$

Using (D.18), we are now able to write the differential operators (D.11)–(D.16) in curvilinear orthogonal coordinates:

$$\operatorname{div} \mathbf{A} = \frac{1}{\sqrt{g}} \left[\frac{\partial}{\partial x^1} \left(\sqrt{\frac{g}{g_{11}}} A_{(1)} \right) + \frac{\partial}{\partial x^2} \left(\sqrt{\frac{g}{g_{22}}} A_{(2)} \right) + \frac{\partial}{\partial x^3} \left(\sqrt{\frac{g}{g_{33}}} A_{(3)} \right) \right],$$

or, since $g = g_{11} \, g_{22} \, g_{33}$,

$$\operatorname{div} \mathbf{A} = \frac{1}{\sqrt{g}} \left[\frac{\partial}{\partial x^1} \left(\sqrt{g_{22} g_{33}} \, A_{(1)} \right) + \frac{\partial}{\partial x^2} \left(\sqrt{g_{33} g_{11}} \, A_{(2)} \right) \right.$$

$$\left. + \frac{\partial}{\partial x^3} \left(\sqrt{g_{11} g_{22}} \, A_{(3)} \right) \right]. \tag{D.19}$$

Also,

$$(\text{grad } \Phi)_i = \frac{\partial \Phi}{\partial x^i} = \sqrt{g_{ii}} \, (\text{grad } \Phi)_{(i)} \, ,$$

$$(\text{grad } \Phi)^i = g^{ii} \frac{\partial \Phi}{\partial x^i} = \frac{1}{\sqrt{g_{ii}}} (\text{grad } \Phi)_{(i)} \, ,$$

which are, in fact, one and the same relation,

$$(\text{grad } \Phi)_{(i)} = \frac{1}{\sqrt{g_{ii}}} \frac{\partial \Phi}{\partial x^i} \, . \tag{D.20}$$

Similarly,

$$\Delta \Phi = \frac{1}{\sqrt{g}} \left[\frac{\partial}{\partial x^1} \left(\sqrt{\frac{g_{22} g_{33}}{g_{11}}} \frac{\partial \Phi}{\partial x^1} \right) + \frac{\partial}{\partial x^2} \left(\sqrt{\frac{g_{33} g_{11}}{g_{22}}} \frac{\partial \Phi}{\partial x^2} \right) \right.$$
$$\left. + \frac{\partial}{\partial x^3} \left(\sqrt{\frac{g_{11} g_{22}}{g_{33}}} \frac{\partial \Phi}{\partial x^3} \right) \right]. \tag{D.21}$$

Finally,

$$(\text{curl } \mathbf{A})_{(i)} = \frac{1}{\sqrt{g_{ii}}} (\text{curl } \mathbf{A})_i = \frac{1}{2} \frac{1}{\sqrt{g_{ii}}} \epsilon_{ijk} F^{jk} \quad \text{(no summation over } i\text{).} \tag{D.22}$$

We shall present the detailed calculation of one component, the other two being obtained by cyclic permutations. For example,

$$(\text{curl } \mathbf{A})_{(1)} = \sqrt{g_{22} g_{33}} \, F^{23} = \sqrt{g_{22} g_{33}} \, g^{22} g^{33} F_{23} = \frac{1}{\sqrt{g_{22} g_{33}}} \left(\frac{\partial A_3}{\partial x^2} - \frac{\partial A_2}{\partial x^3} \right)$$
$$= \frac{1}{\sqrt{g_{22} g_{33}}} \left[\frac{\partial}{\partial x^2} \left(\sqrt{g_{33}} \, A_{(3)} \right) - \frac{\partial}{\partial x^3} \left(\sqrt{g_{22}} \, A_{(2)} \right) \right].$$

The orthogonal components of curl $\mathbf{A}$ therefore are

$$(\text{curl } \mathbf{A})_{(1)} = \frac{1}{\sqrt{g_{22} g_{33}}} \left[\frac{\partial}{\partial x^2} \left(\sqrt{g_{33}} \, A_{(3)} \right) - \frac{\partial}{\partial x^3} \left(\sqrt{g_{22}} \, A_{(2)} \right) \right],$$

$$(\text{curl } \mathbf{A})_{(2)} = \frac{1}{\sqrt{g_{33} g_{11}}} \left[\frac{\partial}{\partial x^3} \left(\sqrt{g_{11}} \, A_{(1)} \right) - \frac{\partial}{\partial x^1} \left(\sqrt{g_{33}} \, A_{(3)} \right) \right], \tag{D.23}$$

$$(\text{curl } \mathbf{A})_{(3)} = \frac{1}{\sqrt{g_{11} g_{22}}} \left[\frac{\partial}{\partial x^1} \left(\sqrt{g_{22}} \, A_{(2)} \right) - \frac{\partial}{\partial x^2} \left(\sqrt{g_{11}} \, A_{(1)} \right) \right].$$

Observation:
Sometimes one uses the notation

$$\sqrt{g_{11}} = h_1,$$
$$\sqrt{g_{22}} = h_2,$$
$$\sqrt{g_{33}} = h_3.$$

The quantities h_i, $i = 1, 2, 3$ are called *Lamé's coefficients*.

D.2.2 Differential Operators in Spherical and Cylindrical Coordinates

D.2.3 Spherical Coordinates

In spherical coordinates, the metric of the Euclidean three-dimensional space reads

$$ds^2 = dr^2 + r^2 \, d\theta^2 + r^2 \sin^2 \theta \, d\varphi^2. \tag{D.24}$$

We choose $x^1 = r$, $x^2 = \theta$, $x^3 = \varphi$ and obtain the components of the metric tensor:

$$g_{11} = \frac{1}{g^{11}} = 1,$$

$$g_{22} = \frac{1}{g^{22}} = r^2, \tag{D.25}$$

$$g_{33} = \frac{1}{g^{33}} = r^2 \sin^2 \theta,$$

which allow us to write the differential operators in spherical coordinates, as follows:

$$\operatorname{div} \mathbf{A} = \frac{1}{r^2 \sin \theta} \left[\frac{\partial}{\partial r} \left(r^2 \sin \theta A_r \right) + \frac{\partial}{\partial \theta} \left(r \sin \theta A_\theta \right) \right.$$
$$\left. + \frac{\partial}{\partial \varphi} \left(r A_\varphi \right) \right], \tag{D.26}$$

$$\operatorname{grad} \Phi = \frac{\partial \Phi}{\partial r} \mathbf{u}_r + \frac{1}{r} \frac{\partial \Phi}{\partial \theta} \mathbf{u}_\theta + \frac{1}{r \sin \theta} \frac{\partial \Phi}{\partial \varphi} \mathbf{u}_\varphi, \tag{D.27}$$

$$\Delta \Phi = \frac{1}{r^2} \left\{ \frac{\partial}{\partial r} \left(r^2 \frac{\partial \Phi}{\partial r} \right) + \frac{1}{\sin \theta} \left[\frac{\partial}{\partial \theta} \left(\sin \theta \frac{\partial \Phi}{\partial \theta} \right) \right. \right.$$
$$\left. \left. + \frac{1}{\sin \theta} \frac{\partial^2 \Phi}{\partial \varphi^2} \right] \right\}, \tag{D.28}$$

$$\text{curl } \mathbf{A} = \frac{1}{r \sin \theta} \left[\frac{\partial}{\partial \theta} \left(A_\varphi \sin \theta \right) - \frac{\partial A_\theta}{\partial \varphi} \right] \mathbf{u}_r + \frac{1}{r \sin \theta} \left[\frac{\partial A_r}{\partial \varphi} \right.$$

$$\left. - \frac{\partial}{\partial r} \left(r A_\varphi \sin \theta \right) \right] \mathbf{u}_\theta + \frac{1}{r} \left[\frac{\partial}{\partial r} \left(r A_\theta \right) - \frac{\partial A_r}{\partial \theta} \right] \mathbf{u}_\varphi , \qquad \text{(D.29)}$$

where $\mathbf{u}_r, \mathbf{u}_\theta, \mathbf{u}_\varphi$ are the unit vectors of the three reciprocally orthogonal directions r, θ, φ.

D.2.4   Cylindrical Coordinates

In cylindrical coordinates, the metric of the Euclidean three-dimensional space reads:

$$ds^2 = d\rho^2 + \rho^2 d\varphi^2 + dz^2. \qquad \text{(D.30)}$$

Choosing $x^1 = \rho, x^2 = \varphi, x^2 = z$, we have

$$g_{11} = \frac{1}{g^{11}} = 1,$$

$$g_{22} = \frac{1}{g^{22}} = \rho^2, \qquad \text{(D.31)}$$

$$g_{33} = \frac{1}{g^{33}} = 1.$$

In view of (D.31), the differential operators in cylindrical coordinates are

$$\text{div } \mathbf{A} = \frac{1}{\rho} \left[\frac{\partial}{\partial \rho} \left(\rho A_\rho \right) + \frac{\partial A_\varphi}{\partial \varphi} + \frac{\partial}{\partial z} \left(\rho A_z \right) \right], \qquad \text{(D.32)}$$

$$\text{grad } \Phi = \frac{\partial \Phi}{\partial \rho} \mathbf{u}_\rho + \frac{1}{\rho} \frac{\partial \Phi}{\partial \varphi} \mathbf{u}_\varphi + \frac{\partial \Phi}{\partial z} \mathbf{k} , \qquad \text{(D.33)}$$

$$\Delta \Phi = \frac{1}{\rho} \left[\frac{\partial}{\partial \rho} \left(\rho \frac{\partial \Phi}{\partial \rho} \right) + \frac{1}{\rho} \frac{\partial^2 \Phi}{\partial \varphi^2} + \rho \frac{\partial^2 \Phi}{\partial z^2} \right], \qquad \text{(D.34)}$$

$$\text{curl } \mathbf{A} = \left(\frac{1}{\rho} \frac{\partial A_z}{\partial \varphi} - \frac{\partial A_\varphi}{\partial z} \right) \mathbf{u}_\rho + \left(\frac{\partial A_\rho}{\partial z} - \frac{\partial A_z}{\partial \rho} \right) \mathbf{u}_\varphi$$

$$+ \left(\frac{\partial A_\varphi}{\partial \rho} + \frac{1}{\rho} A_\varphi - \frac{1}{\rho} \frac{\partial A_\rho}{\partial \varphi} \right) \mathbf{k} , \qquad \text{(D.35)}$$

where $\mathbf{u}_\rho, \mathbf{u}_\varphi, \mathbf{k}$ are the unit vectors of the three reciprocally orthogonal directions ρ, φ, z.

Appendix E
Dirac's δ-Function

E.1 Basic Facts

As we have seen in Sect. 1.2, the spatial density of an electric charge distribution
consisting of a single point charge is zero everywhere, except for the location of the
point charge, where it is infinite. This is an example of a *distribution*. A distribution
is a mathematical expression that is well defined only when integrated with suitable
test functions. While an analytic function $g = g(x)$ is defined as a set of values of
g corresponding to a set of values of x, a distribution is a functional characterized
by the fact that to each function there corresponds a number. In the case of the delta
function, it is a mathematical meaningful object only under an integral, as we shall
soon see.

The *Dirac δ function*[3] can be defined by means of several *bona fide* functions
tending to δ in the limit. Let, for example, $\delta(x, \alpha)$ be a function depending on the
variable x and a parameter $\alpha > 0$, which obeys the conditions

$$\lim_{\alpha \to 0} \delta(x, \alpha) = \begin{cases} 0, & \text{for } x \neq 0, \\ \infty, & \text{for } x = 0, \end{cases}$$

$$\lim_{\alpha \to 0} \int_{-\infty}^{+\infty} \delta(x, \alpha)\, dx = 1. \tag{E.1}$$

---

[3]The delta function appeared for the first time in 1822, in the work of Joseph Fourier (1768–1830)
on the *Analytical theory of heat*. However, Fourier did not identify it specifically as an independent
object. In 1930, Paul Dirac (1902–1984) introduced it as a convenient notation, in his book *The
Principles of Quantum Mechanics*. Its role in Dirac's book was as a continuous analogue of the
Kronecker delta symbol, which gave it also the name of delta function.

© Springer-Verlag Berlin Heidelberg 2016
M. Chaichian et al., *Electrodynamics*, DOI 10.1007/978-3-642-17381-3

Let us denote

$$\delta(x) = \lim_{\alpha \to 0} \delta(x, \alpha),$$

$$\int_{-\infty}^{+\infty} \delta(x)dx = \lim_{\alpha \to 0} \int_{-\infty}^{+\infty} \delta(x, \alpha)\, dx = 1. \tag{E.2}$$

Remark that (E.2)$_2$ is only a notation, because the operations of "limit" and "integration" could be reversed only if $\delta(x, \alpha)$ were uniformly convergent for $x = 0$ and $\alpha \to 0$. The same meaning is understood when writing

$$\int_{-\infty}^{+\infty} f(x)\delta(x)dx = \lim_{\alpha \to 0} \int_{-\infty}^{+\infty} f(x)\delta(x, \alpha)dx, \tag{E.3}$$

where $f(x)$ is a smooth function with compact support. Such a function is called a *test function* of the δ distribution.

Let $[a, b]$ be a closed interval on the x-axis, and x_0 a point on the axis. Then

$$\delta(x - x_0) = \begin{cases} 0, & x \neq x_0, \\ \infty, & x = x_0, \end{cases} \tag{E.4}$$

$$\int_a^b \delta(x - x_0)dx = \begin{cases} 1, & x_0 \in [a, b], \\ 0, & x_0 \notin [a, b]. \end{cases} \tag{E.5}$$

If we extend the domain of integration between $-\infty$ and $+\infty$, then

$$\int_{-\infty}^{+\infty} \delta(x - x_0)dx = 1. \tag{E.6}$$

By definition,

$$\int_{-\infty}^{+\infty} f(x)\, \delta(x - x_0)dx = f(x_0), \tag{E.7}$$

where $f(x)$ is non-zero only for $x \in (a, b)$. For $x_0 = 0$, we have

$$\delta[f] \equiv \int_{-\infty}^{+\infty} f(x)\, \delta(x)dx = f(0). \tag{E.8}$$

Relations (E.7) and (E.8) express the so-called *sifting property* of the δ function. In the above relations, $f(x)$ is a test function and the δ distribution evaluates it at the point 0. The notation $\delta[f]$ emphasizes the fact that δ is a linear functional on the space of test functions, and as such it is *defined* by (E.8).

E.1.1 Representations of the Dirac Delta Function

1) Let $\delta(x, \alpha)$ be given by

$$\delta(x, \alpha) = \frac{1}{\pi} \frac{\alpha}{\alpha^2 + x^2}, \quad \alpha > 0. \tag{E.9}$$

Indeed,

$$\lim_{\alpha \to 0} \delta(x, \alpha) = \lim_{\alpha \to 0} \frac{1}{\pi} \frac{\alpha}{\alpha^2 + x^2} = \begin{cases} 0, & x \neq 0, \\ \infty, & x = 0, \end{cases}$$

$$\int_{-\infty}^{+\infty} \delta(x, \alpha)dx = \frac{1}{\pi} \int_{-\infty}^{+\infty} \frac{\alpha dx}{\alpha^2 + x^2} = \frac{1}{\pi} \arctan \frac{x}{\alpha} \Big|_{\infty}^{\infty} = 1.$$

2) As a second representation, consider

$$\delta(x, \alpha) = \begin{cases} \frac{1}{\alpha}, & x \in \left[-\frac{\alpha}{2}, +\frac{\alpha}{2}\right], \\ 0, & x \notin \left[-\frac{\alpha}{2}, +\frac{\alpha}{2}\right]. \end{cases} \tag{E.10}$$

We have

$$\int_{-\infty}^{+\infty} \delta(x, \alpha) f(x)dx = \frac{1}{\alpha} \int_{-\alpha/2}^{+\alpha/2} f(x)dx.$$

The mean value theorem gives

$$\int_{-\frac{\alpha}{2}}^{+\frac{\alpha}{2}} f(x)dx = f(\xi) \int_{-\alpha/2}^{+\alpha/2} dx = \alpha f(\xi),$$

where $\xi \in \left[-\frac{\alpha}{2}, +\frac{\alpha}{2}\right]$. Therefore

$$\lim_{\alpha \to 0} \int_{-\infty}^{+\infty} \delta(x, \alpha)dx = \int_{-\infty}^{+\infty} \delta(x)dx,$$

because in the limit $\alpha \to 0$ the closed interval $\left[-\frac{\alpha}{2}, +\frac{\alpha}{2}\right]$ reduces to a point, namely the origin. This way, the definition (E.8) is verified.

3) Consider, as a third and last example,

$$\delta(x, \alpha) = \frac{1}{\alpha \sqrt{\pi}} e^{-\frac{x^2}{\alpha^2}}, \tag{E.11}$$

which is a *Gaussian-type distribution*. Since

$$\frac{1}{\alpha \sqrt{\pi}} \int_{-\infty}^{+\infty} e^{-\frac{x^2}{\alpha^2}} dx = \frac{1}{\sqrt{\pi}} \int_{-\infty}^{+\infty} e^{-t^2} dt = 1,$$

it follows that

$$\lim_{\alpha \to 0} \frac{1}{\alpha \sqrt{\pi}} e^{-\frac{x^2}{\alpha^2}} = \delta(x).$$

E.2 Properties of the Dirac Delta Function

1° *Linearity.* Let $f_1(x)$ and $f_2(x)$ be two test function, and a, b two constants. Then

$$\int_{-\infty}^{+\infty} \delta(x)[af_1(x) + bf_2(x)]dx = a \int_{-\infty}^{+\infty} \delta(x) f_1(x)dx + b \int_{-\infty}^{+\infty} \delta(x) f_2(x)dx. \quad \text{(E.12)}$$

This expresses the fact that δ function is a linear functional on the space of test functions.

2° *Distributional derivative(s).* If $g(x)$ is a function with the property $\lim_{x \to \pm\infty} g(x) = 0$, then

$$\int_{-\infty}^{+\infty} f(x) g'(x)dx = [fg]_{-\infty}^{+\infty} - \int_{-\infty}^{+\infty} f'(x) g(x)dx = - \int_{-\infty}^{+\infty} f'(x) g(x)dx. \quad \text{(E.13)}$$

This property of functions is extended to the δ distribution, in order to define the distributional derivative δ'. Thus, the distributional derivative of the δ function is another distribution, δ', acting on smooth test functions with compact support by

$$\delta'[f] = \int_{-\infty}^{+\infty} f(x) \delta'(x)dx = - \int_{-\infty}^{+\infty} f'(x) \delta(x)dx = -f'(0). \quad \text{(E.14)}$$

In general,

$$\int_{-\infty}^{+\infty} f(x) \delta^{(n)}(x)dx = (-1)^n \left[f^{(n)}(x) \right]_{x=0}. \quad \text{(E.15)}$$

3° *Composition with a function.* Consider $y = y[x(t)]$. Then

$$\int_{-\infty}^{+\infty} f(t) \delta[x(t)]dt = \int_{-\infty}^{+\infty} f[t(x)] \delta(x) \frac{dt}{dx}dx. \quad \text{(E.16)}$$

This relation is considered as definition of the distribution $\delta[x(t)]$, where $x(t)$ is a function.

4° *Symmetry.* As an application, let us calculate

$$\int_{-\infty}^{+\infty} f(x)\delta(-x)dx = - \int_{+\infty}^{-\infty} f(-y) \delta(y)dy$$

$$= \int_{-\infty}^{+\infty} f(-y) \delta(y)dy = f(0) = \int_{-\infty}^{+\infty} f(x)\delta(x)dx,$$

where we have used (E.8) in writing the last equality. Since this relation holds for any suitable test function $f(x)$, it follows that

$$\delta(-x) = \delta(x),$$ (E.17)

which means that the *delta function is even*.

5° *Scaling.* Let us make a change of variable $y = ax$, where the constant a is either positive, or negative. If $a > 0$, we have

$$\int_{-\infty}^{+\infty} f(x)\delta(ax)dx = \int_{-\infty}^{+\infty} f\left(\frac{y}{a}\right)\delta(y)\frac{dy}{a} = \frac{1}{a}f(0) = \int_{-\infty}^{+\infty} f(x)\frac{\delta(x)}{a}dx.$$

If $a < 0$, the use of property (E.17) yields $\delta(ax) = \delta(-ax)$. The last relation then gives

$$\delta(ax) = \frac{1}{|a|}\delta(x).$$ (E.18)

6° *Three-dimensional version.* In three dimensions, denoting $\mathbf{r} = (x, y, z)$, the delta function is defined by

$$\delta(\mathbf{r} - \mathbf{r}_0) = \delta(x - x_0)\delta(y - y_0)\delta(z - z_0),$$

$$\delta(\mathbf{r} - \mathbf{r}_0) = \begin{cases} 0, & \mathbf{r} \neq \mathbf{r}_0, \\ \infty, & \mathbf{r} = \mathbf{r}_0, \end{cases}$$ (E.19)

$$\int_D \delta(\mathbf{r} - \mathbf{r}_0)d\mathbf{r} = \begin{cases} 1, & P_0(x_0, y_0, z_0) \in D, \\ 0, & P_0(x_0, y_0, z_0) \notin D, \end{cases}$$ (E.20)

$$\int_D f(\mathbf{r})\delta(\mathbf{r} - \mathbf{r}_0)d\mathbf{r} = \begin{cases} f(\mathbf{r}_0), & \mathbf{r}_0 \in D, \\ 0, & \mathbf{r}_0 \notin D, \end{cases}$$ (E.21)

or, if the domain of integration is extended over the whole space,

$$\int_{-\infty}^{+\infty} f(\mathbf{r})\,\delta(\mathbf{r} - \mathbf{r}_0)d\mathbf{r} = f(\mathbf{r}_0),$$ (E.22)

where $d\mathbf{r} = dx\,dy\,dz$.

7° *Fourier expansion.* Let $f(\mathbf{r}) \equiv f(x, y, z)$ be an arbitrary function of coordinates. Its Fourier transform is

$$F(\mathbf{k}) = \frac{1}{(2\pi)^{3/2}} \int f(\mathbf{r})e^{-i\mathbf{k}\cdot\mathbf{r}}d\mathbf{r},$$ (E.23)

The inverse Fourier transform is

$$f(\mathbf{r}) = \frac{1}{(2\pi)^{3/2}} \int F(\mathbf{k})\,e^{i\mathbf{k}\cdot\mathbf{r}}d\mathbf{k},$$ (E.24)

with $d\mathbf{k} = dk_x\, dk_y\, dk_z$. The last two relations then yield

$$f(\mathbf{r}) = \frac{1}{(2\pi)^3} \int f(\mathbf{r}') e^{i\mathbf{k}\cdot(\mathbf{r}-\mathbf{r}')} d\mathbf{r}'\, d\mathbf{k}$$

$$= \int f(\mathbf{r}') \left[\frac{1}{(2\pi)^3} \int e^{i\mathbf{k}\cdot(\mathbf{r}-\mathbf{r}')}\, d\mathbf{k} \right] d\mathbf{r}'.$$

Using (E.22), one then obtains

$$\delta(\mathbf{r} - \mathbf{r}') = \frac{1}{(2\pi)^3} \int e^{i\mathbf{k}\cdot(\mathbf{r}-\mathbf{r}')} d\mathbf{k}, \qquad (E.25)$$

which is the *Fourier expansion of the three-dimensional delta function*.

8° Suppose that the continuously differentiable function $\varphi(x)$ has n simple roots, i.e.

$$\varphi(x_i) = 0, \quad \varphi'(x_i) \neq 0, \quad i = 1, 2, \ldots, n.$$

Then we have

$$\int_{-\infty}^{+\infty} f(x)\, \delta[\varphi(x)]dx = \sum_{i=1}^{n} \int_{a_{i-1}}^{a_i} f(x)\, \delta[\varphi(x)]dx,$$

where

$$a_0 < x_1 < a_1 < x_2 < a_2 < \ldots < a_{n-1} < x_n < a_n.$$

By means of (E.18), we can write

$$\int_{a_{i-1}}^{a_i} f(x)\, \delta[\varphi(x)]\, dx = \int_{x_i-\epsilon}^{x_i+\epsilon} f(x)\, \delta\left[\varphi'(x_i)(x - x_i)\right] dx = \frac{f(x_i)}{|\varphi'(x_i)|},$$

that is

$$\int_{-\infty}^{+\infty} f(x)\, \delta[\varphi(x)]\, dx = \sum_{i=1}^{n} \frac{f(x_i)}{|\varphi'(x_i)|} = \int_{-\infty}^{+\infty} f(x) \sum_{i=1}^{n} \frac{\delta(x - x_i)}{|\varphi'(x_i)|}dx,$$

which yields

$$\delta[\varphi(x)] = \sum_{i=1}^{n} \frac{\delta(x - x_i)}{|\varphi'(x_i)|}. \qquad (E.26)$$

In particular, if $\varphi(x) = x^2 - a^2$, we have

$$\delta(x^2 - a^2) = \frac{1}{2|a|}\left[\delta(x + a) + \delta(x - a)\right]. \qquad (E.27)$$

In Eq. (1.6) we wrote

$$\rho(\mathbf{r}) = q\, \delta(\mathbf{r} - \mathbf{r}_0).$$

We then can make the following interpretations:

a) $\delta(x)$ is the linear charge density of a charge distribution formed by a single point charge of value $+1$ (Coulomb), situated at the origin of coordinates;

b) $\delta(x - x_0)$ is the linear density of an electric charge of value $+1$, situated at the point of coordinate x_0;

c) $\delta(\mathbf{r} - \mathbf{r}_0)$ is the volume charge density of a distribution formed by a single point charge of value $+1$, situated at the point $P_0(\mathbf{r}_0)$.

Appendix F
Green's Function

Numerous physical phenomena are described by equations of the type

$$L f(\mathbf{r}) = u(\mathbf{r}), \tag{F.1}$$

where L is a linear differential operator, $f(\mathbf{r})$ is an unknown function, and $u(\mathbf{r})$ is a given function. The function $u(\mathbf{r})$ is viewed as a source for the field (output) $f(\mathbf{r})$. Let us write the source as a superposition of delta functions, i.e.

$$u(\mathbf{r}) = \int \delta(\mathbf{r} - \mathbf{r}') u(\mathbf{r}') d\mathbf{r}'. \tag{F.2}$$

The inhomogeneous differential equation (F.1) can be then written as

$$L f(\mathbf{r}) = \int L G(\mathbf{r} - \mathbf{r}') u(\mathbf{r}') d\mathbf{r}',$$

with $G(\mathbf{r} - \mathbf{r}')$ being any solution of the equation

$$L G(\mathbf{r}, \mathbf{r}') = \delta(\mathbf{r} - \mathbf{r}'), \tag{F.3}$$

and being called the *Green function* of the differential operator L. Clearly, the Green function is the impulse response of the inhomogeneous differential equation (F.1). By the principle of superposition, if $G(\mathbf{r}, \mathbf{r}')$ is the solution of the differential equation with δ-function type of source, and the source $u(\mathbf{r})$ is a superposition (F.2) of δ functions, then the solution of Eq. (F.1) is a superposition of Green's functions, i.e.

$$f(\mathbf{r}) = \int G(\mathbf{r}, \mathbf{r}') u(\mathbf{r}') d\mathbf{r}'. \tag{F.4}$$

© Springer-Verlag Berlin Heidelberg 2016
M. Chaichian et al., *Electrodynamics*, DOI 10.1007/978-3-642-17381-3

The electrodynamical phenomena encountered in the text are described by non-homogeneous second-order partial differential equations of the form

$$L f(x) = -\rho(x), \tag{F.5}$$

where the operator L stands for the Laplacian Δ, or the d'Alembertian $\square$, etc., x is a variable (or a group of variables $x_1, x_2, \ldots$), while the scalar function $\rho(x)$ is called the *source density*. If $f(x)$ is a vector function, then $\rho(x)$ is also a vector function. Such equations are, for example, Eq. (4.198) of the electrodynamic potentials in variable regime, where the role of source density is played by the spatial density of electric charge, or the conduction current density, respectively.

The solution of Eq. (F.5) can be formally written as

$$f(x) = -L^{-1}\rho(x). \tag{F.6}$$

The inverse L^{-1} of the differential operator is another differential operator, with the property that

$$L^{-1}L\rho(x) = \rho(x).$$

Using the Dirac delta function, we may write

$$\rho(x) = \int_D \rho(x')\,\delta(x - x')dx'. \tag{F.7}$$

Since the operator L acts on the variable x only, from (F.6) and (F.7) we find

$$f(x) = -\int_D \rho(x')\left[L^{-1}\delta(x - x')\right]dx'. \tag{F.8}$$

According to (F.4), one can introduce Green's function by the operation

$$G(x, x') = -L^{-1}\delta(x - x') + G_0(x, x'), \tag{F.9}$$

where $G_0(x, x')$ verifies the condition $LG_0(x, x') = 0$, that is

$$LG(x, x') = -L\left[L^{-1}\delta(x - x')\right] + LG_0(x, x') = -\delta(x - x'). \tag{F.10}$$

In this case

$$f(x) = \int_D G(x, x')\rho(x')dx'. \tag{F.11}$$

Equation (F.10) gives a method of finding the Green's functions: one usually Fourier transforms the delta function according to (E.25), and (F.10) straightforwardly provides the Fourier transform of the Green function.

Since the delta function is even, it then follows from (F.10) that the function $G(x, x')$ is *symmetric*:

$$G(x, x') = G(x', x).$$

From the physical point of view, this means that a source situated at the point P' produces at the point P the same effect as that produced at P' by a source situated at the point P.

If we use as variables x, y, z, t, then Green's function $G(\mathbf{r}, t; \mathbf{r}', t')$ verifies the following equation (see (4.200)):

$$LG(\mathbf{r}, t; \mathbf{r}', t') = -\delta(x - x')\delta(y - y')\delta(z - z')\delta(t - t'), \qquad (F.12)$$

which means that the equation

$$Lf(\mathbf{r}, t) = -\rho(\mathbf{r}, t) \qquad (F.13)$$

has the following solution

$$f(\mathbf{r}, t) = \int G(\mathbf{r}, t; \mathbf{r}', t')\rho(\mathbf{r}', t')d\mathbf{r}'dt', \qquad (F.14)$$

where $d\mathbf{r}' = dx'dy'dz'$. The problem of finding the solution of (F.13) therefore reduces to the determination of Green's function for the studied case. We encountered such a situation in Sects. 4.9.1 and 8.6.

Bibliography

1. Abraham, M., Becker, R.: The Classical Theory of Electricity and Magnetism. Blackie and Son Ltd., London (1960)
2. Adler, R.B., Chu, L.J., Fano, R.M.: Electromagnetic Fields, Forces and Energy. Wiley, New York (1960)
3. Akhiezer, A.I., Berestetskii, V.B.: Quantum Electrodynamics. Wiley, New York (1965)
4. Alexeev, A.: Recueil de problèmes d'électrodynamique. Mir, Moscou (1980)
5. Batygin, V.V., Toptygin, I.N.: Problems in Electrodynamics, 2nd edn. Academic Press, London (1978)
6. Bergmann, P.G.: Introduction to the Theory of Relativity. Dover Publications, New York (1976)
7. Bleaney, B.I., Bleaney, B.: Electricity and Magnetism, 3rd edn. Oxford University Press, Oxford (1989)
8. Bohm, D.: The Special Theory of Relativity. Routledge, Abingdon (1996). Reprint edition
9. Bondi, H.: Relativity and Common Sense, 2nd edn. Dover, New York (1980)
10. Born, M.: Einstein's Theory of Relativity. Dover, New York (1962)
11. Bradbury, T.C.: Theoretical Mechanics. Wiley, New York (1968)
12. Brau, C.A.: Modern Problems in Classical Electrodynamics. Oxford University Press, Oxford (2003)
13. Calder, N.: Einstein's Universe. Gramercy (1988)
14. Carroll, S.: Spacetime and Geometry: An Introduction to General Relativity. Addison-Wesley, Boston (2003)
15. Chaichian, M., Hagedorn, R.: Symmetries in Quantum Mechanics: From Angular Momentum to Supersymmetry. (Graduate Student Series in Physics). Institute of Physics Publishing, Bristol and Philadelphia (1998)
16. Chaichian, M., Merches, I., Tureanu, A.: Mechanics: An Intensive Course. Springer, Berlin, Heidelberg (2012)
17. Chaichian, M., Perez-Rojas, H., Tureanu, A.: Basic Concepts in Physics. From the Cosmos to Quarks. Springer, Berlin, Heidelberg (2014)
18. Crawford Jr., F.S.: Berkeley Physics Course. Waves, vol. 3. McGraw-Hill, New York (1968)
19. Dirac, P.A.M.: General Theory of Relativity. Princeton University Press, Princeton (1996)
20. Einstein, A.: Relativity: The Special and the General Theory. Broadway Books (1995)
21. Feather, N.: Electricity and Matter. Edinburgh University Press, Edinburgh (1968)
22. Feynman, R.P.: Quantum Electrodynamics. Benjamin, New York (1962)
23. Feynman, R.P.: The Feynman Lectures on Physics. Addison-Wesley, Boston (2005). Originally published as separate volumes in 1964 and 1966

© Springer-Verlag Berlin Heidelberg 2016
M. Chaichian et al., *Electrodynamics*, DOI 10.1007/978-3-642-17381-3

24. Fitzpatrick, R.: Maxwell's Equations and the Principles of Electromagnetism. Jones and Bartlett Publishers, Massachusetts (2008)
25. Fleich, D.: A Students Guide to Maxwell's Equations. Cambridge University Press, Cambridge (2008)
26. Fock, V.A.: The Theory of Space. Time and Gravitation. Macmillan, New York (1964)
27. French, A.P.: Special Relativity. Norton, New York (1968)
28. Fujimoto, M.: Physics of Classical Electromagnetism. Springer, Berlin, Heidelberg (2007)
29. Good Jr., R.H., Nelson, T.J.: Classical Theory of Electric and Magnetic Fields. Academic Press, New York (1971)
30. Good Jr., R.H., Nelson, T.J.: Classical Theory of Electric and Magnetic Fields. Academic Press, New York (1971)
31. Greiner, W.: Classical Electrodynamics. Springer, Berlin, Heidelberg (1998)
32. Griffiths, J.B., Podolský, J.: Exact Space-Times in Einstein's General Relativity. Cambridge University Press, Cambridge (2009)
33. Griffiths, J.B., Podolský, J.: Exact Space-Times in Einstein's General Relativity. Cambridge University Press, Cambridge (2009)
34. Heald, M.A., Marion, J.B.: Classical Electromagnetic Radiation, 3rd edn. Saunders, Philadelphia (1995)
35. Hehl, F.W., Obukhov, Yu.N.: Foundations of Classical Electrodynamics: Charge, Flux, and Metric. Birkhäuser, Boston, MA (2003)
36. Hehl, F.W., Obukhov, YuN: Foundations of Classical Electrodynamics: Charge, Flux, and Metric. Birkhäuser, Boston, MA (2003)
37. Jackson, J.D.: Classical Electrodynamics, 3rd edn. Wiley, New York (1998)
38. Jauch, J.M., Rohrlich, F.: The Theory of Photons and Electrons. Springer, Berlin (1976)
39. Kacser, C.: Introduction to the Special Theory of Relativity. Prentice-Hall, Englewood Cliffs (1967)
40. Kilmister, C.W.: Special Theory of Relativity. Pergamon Press, Oxford (1970)
41. Kinoshita, T.: Quantum Electrodynamics. World Scientific, Singapore (1990)
42. Kompaneyets, A.S.: Theoretical Physics. Mir Publishers, Moscow (1961)
43. Kraus, J.D., Fleisch, D.A.: Electromagnetics with Applications, 5th edn. McGraw Hill, New York (1998)
44. Landau, L.D., Pitaevskii, L.P., Lifshitz, E.M.: Electrodynamics of Continuous Media. Course of Theoretical Physics, 2nd edn. Butterworth-Heinemann, Oxford (1984)
45. Lightman, A.P., Press, W.H., Price, R.H., Teukolski, S.A.: Problem Book in Relativity and Gravitation. Princeton University Press, Princeton (1975)
46. Lightman, A.P., Press, W.H., Price, R.H., Teukolski, S.A.: Problem Book in Relativity and Gravitation. Princeton University Press, Princeton (1975)
47. Lorraine, P., Corson, D.R.: Electromagnetic Fields and Waves. Freeman, New York (1970)
48. Lorentz, H.A., Einstein, A., Minkowski, H., Weyl, H.: The Principle of Relativity: A Collection of Original Memoirs on the Special and General Theory of Relativity. Dover, New York (1968)
49. Low, F.E.: Classical Field Theory: Electromagnetism and Gravitation. Wiley, New York (1997)
50. Maxwell, J.C.: Elementary Treatise on Electricity. Clarendon Press, Oxford (1888)
51. Maxwell, J.C.: Elementary Treatise on Electricity. Clarendon Press, Oxford (1888)
52. Melia, F.: Electrodynamics (Chicago Lectures in Physics). University of Chicago Press, Chicago (2001)
53. Melia, F.: Electrodynamics (Chicago Lectures in Physics). University of Chicago Press, Chicago (2001)
54. Mermin, N.D.: Space and Time in Relativity. McGraw-Hill, New York (1968)
55. Møller, C.: The Theory of Relativity. Oxford University Press, London (1952)
56. Nasar, S.A.: 2000 Solved Problems in Electromagnetics. McGraw-Hill, New York (1992)
57. Oppenheimer, J.R.: Lectures on Electrodynamics. Gordon and Breach Science Publishers, Philadelphia (1970)
58. Panofsky, W.K.H., Phillips, M.: Classical Electricity and Magnetism, 2nd edn. Dover Publications, New York (2005)

59. Pauli, W.: Theory of Relativity. Pergamon, London (1958)
60. Pollack, G., Stump, D.: Electromagnetism. Addison Wesley, Boston (2001)
61. Purcell, E.M.: Berkeley Physics Course. Electricity and Magnetism, 2nd edn. McGraw-Hill, New York (1985)
62. Reitz, J.R., Milford, F.J., Christy, R.W.: Foundations of Electromagnetic Theory, 4th edn. Addison Wesley, Boston (2008)
63. Resnick, R.: Introduction to Special Relativity. Wiley, New York (1968)
64. Rindler, W.: Special Relativity, 2nd edn. Oliver and Boyd, Edinburgh (1966)
65. Schwartz, M.: Principles of Electrodynamics. Dover, New York (1987)
66. Schwinger, J.: Einstein's Legacy. Freeman (1987)
67. Shadowitz, A.: Special Relativity. Saunders, Philadelphia (1968)
68. Skinner, R.: Relativity. Blaisdell Publishing, Waltham, MA (1969)
69. Sommerfeld, A.: Electrodynamics. Academic Press, New York (1964)
70. Stephani, H., Kramer, D., MacCallum, M., Hoenselaers, C., Herlt, E.: Exact Solutions to Einstein's Field Equations, 2nd edn. Cambridge University Press, New York (2003)
71. Synge, J.L.: Relativity: The Special Theory, 2nd edn. Elsevier Science Ltd., Amsterdam (1980)
72. Ugarov, V.A.: Special Theory of Relativity. Mir Publishers, Moscow (1979)
73. Vanderlinde, J.: Classical Electromagnetic Theory. Wiley, New York (1993)
74. Wald, R.M.: General Relativity. The University of Chicago Press, Chicago and London (1984)

Author Index

© Springer-Verlag Berlin Heidelberg 2016
M. Chaichian et al., *Electrodynamics*, DOI 10.1007/978-3-642-17381-3

Subject Index

A

Aberration
  angle, 337
  astronomical, 337
  of light, 337, 364, 371, 424
  precession, 514
  stellar, 337
Absolute
  derivative, 523
  differential, 526
  differential of a covariant four-vector, 523
  future, 380
  past, 380
Absolutely remote events, 380
Advance of the perihelion, 563
Alfvén
  velocity, 306
  wave, 307
Ampère's circuital law, 89
  differential form, 91
  integral form, 91
Analogy between electrostatic and gravitational
    phenomena, 513
Angular momentum
  flux density, 137
  intrinsic, 478
  of the electromagnetic field, 137
  tensor, 477
Anomalous magnetic moment, 505
Antenna
  full-wave, 270, 276, 280
  half-wave, 270, 276, 280
Antipotentials, 143, 145
  generalized, 158
Applicability limit of classical electrodynamics,
    296
Atomic clock, 360, 567
Average power of dipole radiation, 261
Axial vector, 594

B

Bargmann–Michel–Telegdi (BMT) equation,
    505
Belinfante
  energy-momentum tensor, 478
  procedure, 479
  tensor, 478
Beltrami's diffusion equation, 303
Bernoulli equation
  relativistic hydrodynamics, 488
  relativistic magnetofluid dynamics, 490
Bianchi identity
  algebraic, 535
  differential, 535
  first, 535
  second, 535
Big Bang cosmology, 554, 559
Biot–Savart–Laplace law, 84
Birefringence, 202
Birkhoff's theorem, 550
Black hole, 552, 553, 577
  binary, 576
  entropy, 552
  event horizon, 552
  Schwarzschild, 552, 584
  stable/unstable orbit around, 587
Bohr magneton, 99
Boltzmann's equation, 299
Borgnis
  equation, 238
  method, 237
Bound charges, 27
Bowditch curves, 180
Brewster angle, 188
Busch relation, 166

C

Canonical
  energy-momentum tensor, 473, 479

© Springer-Verlag Berlin Heidelberg 2016
M. Chaichian et al., *Electrodynamics*, DOI 10.1007/978-3-642-17381-3

Printed in the United States
By Bookmasters